Introductory
DC/AC Circuits

Introductory
DC/AC Circuits
Third Edition

Nigel P. Cook

PRENTICE HALL
Upper Saddle River, New Jersey 07458

Library of Congress Cataloging-in-Publication Data

Cook, Nigel P.
    Introductory DC/AC circuits / Nigel P. Cook. —3rd ed.
        p.   cm.
    Prev. eds. published under title: Introductory DC/AC electronics.
    Includes index.
    ISBN 0-13-215105-7
    1. Electronics.   I. Cook, Nigel P.–   Introductory DC/AC
electronics.   II. Title.
TK7816.C65   1995
621.381—dc20   95-14463

Cover photo: Phil Matt
Editor: Dave Garza
Production Editor: Sheryl Glicker Langner
Project Management: Elm Street Publishing Services, Inc.
Cover Designer: Tom Mack
Production Manager: Pamela D. Bennett
Marketing Manager: Debbie Yarnell
Supplements Editor: Judith Casillo
Illustrations: Rolin Graphics Inc.

This book was set in Times Ten by The Clarinda Company and
was printed and bound by R. R. Donnelley & Sons Company. The
cover was printed by Phoenix Color Corp.

© 1996 by Prentice-Hall, Inc.
Simon & Schuster / A Viacom Company
Upper Saddle River, New Jersey 07458

Printed in the United States of America
10 9 8 7 6 5 4 3 2

ISBN: 0-13-215105-7

Prentice-Hall International (UK) Limited, *London*
Prentice-Hall of Australia Pty. Limited, *Sydney*
Prentice-Hall of Canada, Inc., *Toronto*
Prentice-Hall Hispanoamericana, S. A., *Mexico*
Prentice-Hall of India Private Limited, *New Delhi*
Prentice-Hall of Japan, Inc., *Tokyo*
Simon & Schuster Asia Pte. Ltd., *Singapore*
Editora Prentice-Hall do Brasil, Ltda., *Rio de Janeiro*

To my loving wife, Dawn,
and beautiful children,
Candice and Jonathan,
whose love inspires me.

This book is dedicated to
all of the innovative, energetic,
and motivational instructors
who go out of their way to
prove that mathematics, science,
and technology are within
everyone's grasp and can be
made easy, interesting, and fun.

Preface

TO THE STUDENT

The early pioneers in electronics were intrigued by the mystery and wonder of a newly discovered science, whereas people today are attracted by its ability to lend its hand to any application and accomplish almost anything imaginable. If you analyze exactly how you feel at this stage, you will probably discover that you have mixed emotions about the journey ahead. On one hand, imagination, curiosity, and excitement are driving you on, while apprehension and reservations may be slowing you down. Your enthusiasm will overcome any indecision you have once you become actively involved in electronics and realize that it is as exciting as you ever expected it to be.

ORGANIZATION OF THE TEXTBOOK

This textbook has been divided into three basic parts. Chapters 1 through 5 introduce you to the world of electronics and the fundamentals of electricity. Chapters 6 through 10 cover direct current, or DC, circuits, and Chapters 11 through 18 cover alternating current, or AC, circuits.

The material covered in this book has been logically divided and sequenced to provide a gradual progression from the known to the unknown, and from the simple to the complex.

DEVELOPMENT, CLASS TESTING, AND REVIEWING

The first phase of development was conducted in the classroom with students and instructors as critics. Each topic was class-tested by videotaping each lesson, and the results were then evaluated and implemented. This feedback was invaluable and enabled me to fine-tune my presentation of topics and instill understanding and confidence in the students.

The second phase of development was to forward a copy of the revised manuscript to several instructors at schools throughout the country. These technical and topical critiques helped to mold the text into a more accurate form.

The third and final phase was to class-test the final revised manuscript and then commission the last technical review in the final stages of production.

ILLUSTRATED TOUR OF TEXTBOOK FEATURES

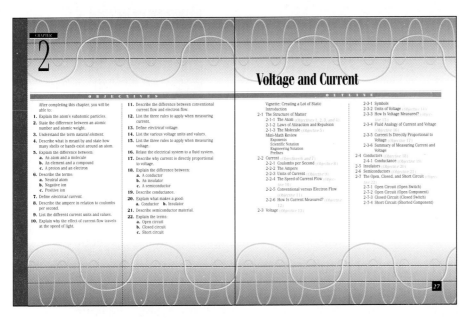

Each chapter opening lists performance-based objectives, with each objective directly correlated to a chapter section on the following page.

Motivational stories detail the people behind the electronics industry, and a conversational introduction reviews what has been previously covered and what is about to be covered.

Chapter 1 includes extensive information about career opportunities for the electronics technician providing interest in and motivation for the material which follows.

Mini-Math Reviews are included in appropriate places to review the necessary mathematics needed for the discussion at hand.

Concept analogies are used throughout the text to take the student from the known to the unknown.

Calculator sequences are included in all examples that introduce new mathematical procedures, and a running glossary defines all new terms in the margin of the text.

Actual circuit examples are included to help the student connect a topic to the real world.

A strong testing, test equipment, and troubleshooting emphasis prepares the student for the working world.

Chapter 5 introduces the student to experimenting in the lab by covering safety, test equipment, protoboards, and soldering tools and techniques.

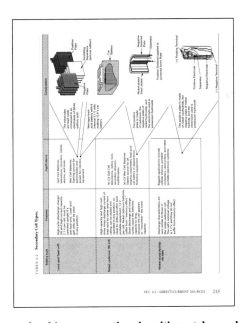

Unique diagrams and tables, coupled with the textbook's conversational writing style, make even traditionally difficult topics easily accessible.

Quick Reference Summary Sheets within the end-of-chapter summary provide a visual reference to all key facts, symbols, circuits, and formulas.

Many new photos have been included to help student recognition.

20. As with series circuits, the total power in a parallel resistive circuit is equal to the sum of all the power losses for each of the resistors in parallel.

21. An open component will cause no current flow within that branch. The total current will decrease, and the voltage across the component will be the same as the source voltage.

22. A shorted component will cause maximum current through that branch. The total current will increase and the voltage source fuse will normally blow.

23. A change in resistance value will cause a corresponding opposite change in branch current and total current.

Branch current	Fixed current divider	Parallel circuit	**NEW TERMS**
Current divider	Kirchhoff's current law		

Multiple-Choice Questions

1. A parallel circuit has _____ path(s) for current to flow.
   a. One **c.** Only three
   b. Two or more **d.** None of the above

2. If a source voltage of 12 V is applied across four resistors of equal value in parallel, the voltage drops across each resistor would be equal to:
   a. 12 V **c.** 4 V
   b. 3 V **d.** 48 V

3. What would be the voltage drop across two 25 Ω resistors in parallel if the source voltage were equal to 9 V?
   a. 50 V **c.** 12 V
   b. 25 V **d.** None of the above

4. If a four-branch parallel circuit has 15 mA flowing through each branch, the total current into the parallel circuit will be equal to:
   a. 15 mA **c.** 30 mA
   b. 60 mA **d.** 45 mA

5. If the total three-branch parallel circuit current is equal to 500 mA, and 207 mA is flowing through one branch and 153 mA through another, what would be the current flow through the third branch?
   a. 707 mA **c.** 140 mA
   b. 653 mA **d.** None of the above

6. A large branch resistance will cause a _____ branch current.
   a. Large **c.** Medium
   b. Small **d.** None of the above are true

7. What would be the conductance of a 1 kΩ resistor?
   a. 10 mS **c.** 2 kΩ
   b. 1 mS **d.** All of the above

---

8. If only two resistors are connected in parallel, the total resistance equals:
   a. The sum of resistance values
   b. Three times the value of one resistor
   c. The product over the sum
   d. All of the above

9. If resistors of equal value are connected in parallel, the total resistance can be calculated by:
   a. One resistor value divided by the number of parallel resistors
   b. The sum of the resistor values
   c. The number of parallel resistors divided by one resistor value
   d. All of the above could be true.

10. The total power in a parallel circuit is equal to the:
   a. Product of total current and total voltage
   b. Reciprocal of the individual power losses
   c. Sum of the individual power losses
   d. Both (a) and (b)
   e. Both (a) and (c)

Essay Questions

11. Describe the difference between a series and a parallel circuit. (8-1)

12. Explain and state mathematically the situation regarding voltage in a parallel circuit. (8-2)

13. State Kirchhoff's current law for parallel circuits. (8-3)

14. What is the current-divider formula? (8-3)

15. List the formulas for calculating the following total resistances: (8-4)
   a. Two resistors of different values
   b. More than two resistors of different values
   c. Equal-value resistors

16. Describe the relationship between branch current and branch resistance. (8-3)

17. Briefly describe total and individual power in a parallel resistive circuit. (8-5)

18. Discuss troubleshooting parallel circuits as applied to: (8-6)
   a. A shorted component **b.** An open component

19. Explain why parallel circuits have a smaller total resistance and larger total current than series circuits. (8-3)

20. Briefly describe why Kirchhoff's voltage law applies to series circuits and why Kirchhoff's current law relates to parallel circuits. (8-3)

Practice Problems

21. Calculate the total resistance of four 30 kΩ resistors in parallel.

22. Find the total resistance for each of the following parallel circuits:
   a. 330 Ω and 560 Ω **c.** 2.2 MΩ, 3 kΩ, and 220 Ω
   b. 47 kΩ, 33 kΩ, and 22 kΩ

An extensive end-of-chapter test bank tests the student's understanding with multiple-choice, essay, and practice problems and troubleshooting questions.

---

23. If 10 V is connected across three 25 Ω resistors in parallel, what will be the total and individual branch currents?

24. If a four-branch parallel circuit has branch currents equal to 25 mA, 37 mA, 220 mA, and 0.2 A, what is the total circuit current?

25. If three resistors of equal value are connected across a 14 V supply and the total resistance is equal to 700 Ω, what is the value of each branch current?

26. If three 75 W light bulbs are connected in parallel across a 110 V supply, what is the value of each branch current? What is the branch current through the other two light bulbs if one burns out?

27. If 33 kΩ and 22 kΩ resistors are connected in parallel across a 20 V source, calculate:
   a. Total resistance **d.** Total power dissipated
   b. Total current **e.** Individual resistor power
   c. Branch currents dissipated

28. If four parallel-connected resistors are each dissipating 75 mW, what is the total power being dissipated?

29. Calculate the branch currents through the following parallel resistor circuits when they are connected across a 10 V supply:
   a. 22 kΩ and 33 kΩ **b.** 220 Ω, 330 Ω, and 470 Ω

30. If 30 Ω and 40 Ω resistors are connected in parallel, which resistor will generate the greatest amount of heat?

31. Calculate the total conductance and resistance of the following parallel circuits:
   a. Three 5 Ω resistors **c.** 1 MΩ, 500 MΩ, 3.3 MΩ
   b. Two 200 Ω resistors **d.** 5 Ω, 3 Ω, 2 Ω

32. Connect the three resistors in Figure 8-29 in parallel across a 12 V battery and then calculate the following:
   a. V_{R1}, V_{R2}, V_{R3} **d.** P_T
   b. I_1, I_2, I_3 **e.** P_1, P_2, P_3
   c. I_T **f.** G_{R1}, G_{R2}, G_{R3}

33. Calculate R_T in Figure 8-30 (a), (b), (c), and (d).

34. Calculate the branch currents through four 60 W bulbs connected in parallel across 110 V. How much is the total current, and what would happen to the total current if one of the bulbs were to burn out? What change would occur in the remaining branch currents?

FIGURE 8-29 Connect in Parallel Across a 12-volt Source.

FIGURE 8-30 Calculate Total Resistance.

35. Calculate the following in Figure 8-31:
   a. I_2 **c.** V_S, I_3, I_2
   b. I_T **d.** R_2, I_1, I_2, P_T

Troubleshooting Questions

36. An open component in a parallel circuit will cause _____ current flow within that branch, which will cause the total current to _____
   a. Maximum, increase **c.** Maximum, decrease
   b. Zero, decrease **d.** Zero, increase

37. A shorted component in a parallel circuit will cause _____ current through a branch, and consequently the total current will _____
   a. Maximum, increase **c.** Maximum, decrease
   b. Zero, decrease **d.** Zero, increase

38. If a 10 kΩ and two 20 kΩ resistors are connected in parallel across a 20 V supply, and the total current measured is 2 mA, determine whether a problem exists in the circuit and, if it does, isolate the problem.

39. What situation would occur and how would we recognize the problem if one of the 20 kΩ resistors in Question 38 were to short?

40. With age, the resistance of a resistor will _____, resulting in a corresponding but opposite change in _____
   a. Increase, branch current **c.** Decrease, source resistance
   b. Change, source voltage **d.** Change, branch current

TO THE INSTRUCTOR

Many people have asked me if there is one single factor involved with student retention. My answer is that whether a student enjoys electronics or hates it, and therefore stays in the program or leaves, is directly dependent on the instructor. I'm sure we have all seen the power we possess over a class of students. If this is used in a positive way, we can retain students despite any obstacles.

The responsibility of being the single most important factor in a student's education can be intimidating, unless we feel that we are fully supported. This support comes in the form of:

- a good textbook and set of teaching aids
- a good lab manual and set of test equipment
- a good course curriculum, and
- some teaching tricks of the trade

TEXTBOOK, LAB MANUAL, AND TEACHING AIDS

In all introductory texts, readability and relativity should be prime concerns to provide a complete understanding of the topics presented. However, in the pursuit of simplicity, boredom is often the price to be paid. Students are more involved in the learning process when they understand how the subject relates to their personal needs, career opportunities, and interests. Consequently, logical, current, and dynamic avenues of approach are needed to generate student enthusiasm, interest, and persistence.

This text has been specifically designed around the above philosophies, with a **complete picture introduction, real-life vignettes** at the beginning of each chapter, the use of **concept analogies** to take the student from the known to the unknown, a constant reference to **actual applications,** and a color insert that describes the many **career opportunities** available in the area of electronics.

The objective of this textbook, and its associated lab manual, is to provide a student with a solid foundation in the theoretical and practical knowledge needed to become an electronics technician. By definition, **an electronics technician must be able to diagnose, isolate, and repair electrical and electronic circuit and system malfunctions.** In order to achieve this goal, he or she must have a thorough knowledge of electronics, test equipment, and troubleshooting techniques. This text has been designed specifically with this objective.

Throughout this text I have also tried to provide to the student the basis for gaining *intelligence* along with an education. All too often education is mistaken for intelligence. Education is simply the acquisition of information. For example, just because I have acquired a first-rate set of woodworking tools does not make me a good carpenter. To be a good carpenter, I must first develop carpentry skills. There are many elements in that nebulous, indefinable quality we call "intelligence"; however, generally it falls into our understanding of **analogies,** which measure our ability to see relationships, and **mathematics, reasoning,** and **logic,** which measure our ability to think logically and make use of the facts we know. Therefore, instead of simply stating a formula and requiring a student to memorize it, which would be simply supplying an education, I have introduced the student to the quantities involved, compared them to something they know, and then studied the relationships to build up a formula in the same way as the pioneers. These analogies will develop a student's ability to see relationships while the study of relationships will develop good reasoning and logic skills, and therefore instill "intelligence." Although this takes a little more time in the early stages, the rewards and ease of understanding in the later stages will compensate for the initial effort.

With the lab manual, Gary Lancaster, Hugh Scriven, and I have tried to create **practical** experiments that inspire and challenge students. Our experiments were purposefully organized to work in conjunction with the textbook and translate all of its theoretical material into practical experience.

Along with the textbook and lab manual you will find a good set of teaching aids that will support your teaching effort.

ELECTRONIC TECHNICIAN TRAINING: INDUSTRY FEEDBACK

To be competitive, "electronic manufacturing companies" and "electronic servicing companies" must make a choice: either they compete with foreign nations on their terms with *lower wages, increased hours,* and *demoralized workers,* or they restructure the work organization and upgrade worker skills for a *high-skill, high-performance organization.* The high school graduate and experienced worker are therefore faced with the same simple choice: *high skills or low wages.* Since "high skills" would be the most obvious choice, the next question is, what skills should the electronics technician have to be competitive in today's and tomorrow's workforce? Posing this question to advisory board members and electronic panels from more than 40 electronic companies, the following feedback was obtained.

1. *Training must provide a core of knowledge for an electronic technician.* Training must produce an electronic technician. By definition, an electronic technician is a person having a complete understanding of the operation, application, and testing of electronic devices, circuits, and systems. Table 1 indicates in priority which training skills industry felt were most useful.

TABLE 1 Areas of Training Most Useful for Job Performance

ELECTRONIC MANUFACTURERS	ELECTRONIC SERVICING
1. Troubleshooting Techniques	1. Customer Relations
2. Use of Test Equipment	2. Troubleshooting Techniques
3. Circuit Recognition and Analysis	3. Use of Test Equipment
4. Adaptability and Teamwork Skills	4. Adaptability and Teamwork Skills
5. Customer Relations	5. Circuit Recognition and Analysis

TABLE 2 Test Equipment Most Useful for Job Performance

1. Digital Multimeter	5. Power Supply
2. Oscilloscope	6. Logic Analyzer
3. Logic Probe and Pulser	7. Signal Generators
4. Automatic Test Equipment	

Table 1 lists test equipment and troubleshooting equipment rate as high priorities. Since test equipment is a technician's tool of the trade, Table 2 lists, in order, the test equipment that industry felt should have the most emphasis.

2. *Technician must be trained to adapt to new technologies.* Training should supply "intelligence along with an education." By definition, education is the acquisition of information and an indication of the ability to make such acquisitions, and intelligence is our ability to think logically and make use of facts we know. This is dependent on our mathematical, logic, and reasoning skills, and our ability to continually learn and see relationships. Concentrating on math skills, logic and reasoning skills, learning skills, and relationship skills develops intelligence, which in turn develops better adaptability and organizational and problem solving skills. Success in these areas will ensure career advancement. To address each of these skills in a little more detail:

   a. *Math Skills:* As a tool, math should be integrated into appropriate position and, once taught, should be instantly applied so as to demonstrate need.

   b. *Logic and Reasoning Skills:* To give only an education, a student would be told to "memorize a formula." To give an education along with intelligence, training should explain why certain quantities are included in a formula and why they are either proportional or inversely proportional. In short, work first on concept comprehension.

   c. *Learning Skills:* The ability to adapt to new technology is directly dependent on an employee's ability to study and learn independently. Reading comprehension is therefore a necessity. To encourage this skill, training must be conversational and interactive, not dry and cold.

   d. *Relationship Skills:* Use analogies to go from the known to the unknown. Learning, retention, and comprehension are enhanced when parallels are explored, and this will in turn develop organizational and relationship skills.

3. *Training must develop better communication, teamwork, and customer service skills.* Industry's drive for high quality and reduced cost has led to the integration of electronic system design, manufacture, and service. There is a need, therefore, for increased teamwork at all levels of personnel, and technicians must have good communication and interpersonal skills. From the standpoint

of customer service skills, training should address how to work under pressure with customers on-site, dress professionally, and explain technical topics to nontechnical people. This can be achieved by having students constantly practice communicating their science (written and spoken).

4. *Training should educate students so that they have a better understanding of business practices, including manufacturing and production processes, and technical and business data.* Training should include tours of electronic companies and details on development procedures, job titles and responsibilities (technical and nontechnical), work ethics, environments, and so on.

5. *The electronic technician's role is evolving due to the complexity of electronic systems, the increase in computer-driven systems, and the economics of unit replacement versus component replacement.* To address these points:

   a. Since computer work stations are becoming an increasingly important tool, training must provide a complete understanding of computer applications in industry.

   b. As equipment becomes more complex, analytical and troubleshooting skills are critical for technicians. However, since yesterday's system is today's component, troubleshooting is moving away from a focus on discrete components to a subsystem and unit (functional) level of maintenance.

   c. Since nearly all design, manufacturing, and service equipment are programmable, the ratio of a technician's involvement in hardware versus software has changed:

   Late 1980s: 80 percent Hardware, 20 percent Software
   Present: 60 percent Hardware, 40 percent Software

   Training should therefore include an introduction to computers and their applications at an early stage in the program. Technicians must be "application software proficient," and, therefore, the following should be covered: DOS, Windows, UNIX, Word Processing, Databases, Spreadsheets, Circuit Simulation Software, Basic Programming (such as C), and Networks.

Most employers agreed that the electronic technician's role is continually evolving to keep pace with the ever-changing electronics industry. As educators, we are also having to continually adapt to the needs of industry so that our graduates have the skills for both career and advancement.

TEACHING TECHNIQUES

Over the past 15 years I have taught thousands of students all aspects of electronics. In this time I have acquired a considerable amount of feedback regarding students' needs and responses to their introduction to the science of electricity and electronics. In this section I would like to share some of my teaching techniques that have enabled me to obtain a high level of student retention.

1. Using concept analogies wherever possible to take the student from the known to the unknown.

2. Keeping the objective "to produce an electrical/electronics technician" in mind at all times, and therefore concentrating on delivering a complete knowledge of electricity, electronics, test equipment, and troubleshooting techniques.

3. Explaining all topics with analogies, reasoning, and logic first, and reinforcing the topic with mathematics second. This provides for giving the students intelligence along with an education.

4. Assuming no prior knowledge of electricity, electronics, or technical mathematics. Starting with the complete picture or world of electronics, and including mini-math reviews where they are needed.

5. Encouraging student enthusiasm, interest, and persistence by including real-life vignettes of entrepreneurs, actual applications, and a career opportunity tour.

In addition, there are a few other teaching methods that have yielded a very good response in my classes, and so I will mention them to you for your consideration.

6. Application Presentations: These 20-minute presentations, generally on the first day of each week, describe and demonstrate systems such as a Jacob's Ladder, laser, computer system, a variety of application software, robots, test equipment, and so on. These presentations seem to generate a lot of student enthusiasm since they demonstrate the exciting end objective of their studies.

I have also turned the tables and had students give presentations on any electrical or electronic equipment. Although the topics seem to be limited to video games, remote controlled toys, and car stereo systems, the presentations have helped develop necessary student communication skills.

7. Schematic Descriptions and Troubleshooting: Once a certain knowledge of electricity, electronics, and test equipment has been mastered, I begin to apply this knowledge to actual system schematics. To cover a circuit fully, we follow four steps.

   a. First we study the purpose of a circuit, its inputs and outputs, and its schematic diagram and circuit description.
   b. Second, we compare the schematic to the actual printed circuit board so that we can identify the components.
   c. Next we study the troubleshooting guides or charts and analyze the process of diagnosing and isolating a problem.
   d. The final step is for me to introduce a problem and have a team of normally two students diagnose, isolate, and repair the problem.

8. First Approximations of a Subject: It was philosopher René Descartes who first stated that to solve any problem you should start with the simple and then proceed to the complex. For Descartes there were three approximations; the **first approximation** was the simplest, the **second approximation** contained more detail, and the **third approximation** was the complex. I have applied this method to my teaching of all topics of electricity and electronics and have noticed a dramatic improvement in student comprehension. To explain this in a little more detail, a first approximation is a general descrip-

tion of a subject in which the key points and purposes are outlined. Following this complete picture description, we step through the details of the chapter, which is a second approximation of the subject. In keeping with the main objective, only the first and second approximations of a subject are needed for a technician since they cover purpose, construction, symbol, operation, characteristics, applications, testing, and troubleshooting. The third approximation of a subject covers design and therefore is only covered by engineering students.

These first approximations give the student a view of the complete picture, instead of having to wait until we finish all of the pieces in the chapter and then trying to connect them all together.

Like me, you may wish to develop first approximation presentations for all topics, especially the more difficult to grasp subjects, such as series–parallel circuits, capacitance, inductance, transistors, and so on.

If you should have any other ideas that you have found in your experience would assist other instructors, please send them to me in care of Prentice Hall.

INSTRUCTOR MOTIVATION: "THE NUTTY PROFESSOR"

I have been teaching electronics for nearly 15 years. In that time, teaching has given me many memorable and rewarding experiences. In contrast, it has been the source of many frustrating moments which have nearly brought me to the brink of madness.

As you are probably already aware, our chosen profession will most often work us day and night and most likely not make us a fortune. The key question is, therefore, why do we do it? The answer is simple. It comes from the personal interactions between the instructor and his or her students. It comes in the form of many intangibles, such as the delight of bringing a drink of knowledge to those who thirst for it, the challenge of the unknown and searching for answers, the pride in turning confusion and bewilderment into understanding and comprehension, and from turning an unskilled person into a working professional. There are mutual benefits to be reaped just from being involved with the learning process. The exchange of knowledge develops for both of us a better understanding of the subject and strong communicative skills, producing a student who is effective at work and in society as a whole.

Teaching is extremely demanding and often very frustrating, but the rewards well compensate the effort. It is a profession seldom understood except by those who really enjoy being involved with life. This vocation may be a "Mad, Mad, Mad World," but it's *my* world.

ACKNOWLEDGMENTS

A special thank-you goes to the professional, creative, and friendly people at Prentice Hall, namely: my editor, David Garza; my marketing manager, Debbie Yarnell; my sales manager, Todd Rossell; my production editors, Sheryl

Langner and Martha Beyerlein; and Judy Casillo, Maria Klimek, and Sylvia Huning.

A special thank-you goes to the husband and wife team of Steve and Marilyn Howe, who worked tirelessly and in a timely fashion to ensure a high level of quality and accuracy.

My appreciation and thanks are extended also to the following instructors who have reviewed and contributed greatly to the development of this textbook: Bob L. Bixler, Austin Community College; James C. Graves, Jr., Indian River Community College; John Hamilton, Spartan School of Aeronautics; William Jenkins, Electronics Institute, Kansas City, Missouri; Gary Lancaster, National Education Center; Leo R. Majewski, Dunwoody Institute; and Marcus S. Rasco, DeVry, Dallas.

Nigel P. Cook

Contents

Chapter 5

PART II

Chapter 6

Chapter 7

Chapter 8

Chapter 9

CHAPTER

1

O B J E C T I V E S

After completing this chapter, you will be able to:

1. State the four basic phenomena on which electronics is based.

2. List some of the components and circuits that exist within electronic equipment.

3. List the seven branches of electronics.

4. Describe the types of equipment that can be found in each of the different branches or fields of electronics.

5. List the persons and companies largely responsible for the progression of electronics.

6. Explain how the unit names of many electrical quantities are derived.

7. Explain the process and people involved in the development of a product.

The World of Electronics

Competitors Beware

During his years as a student at the University of Utah, Nolan Bushnell enjoyed playing the ancient Chinese game called GO. In this game, a word is used frequently throughout as a warning to your opponents that they are in jeopardy of losing with your next move.

With an initial investment of $500, Bushnell launched the video game industry in 1972, developing the first coin-operated Ping-Pong game, Pong. After almost an overnight success, Bushnell sold his company four years later for $15 million; the company was named after the polite Chinese term he learned in college, and as a warning to the competition, ATARI.

Following ATARI, Bushnell stepped up his ambition, taking high-tech to the masses through his Chuck E. Cheese's Pizza Time Theatres.

In most cases, it is easier to build a jigsaw puzzle when you can see the completed picture on the front of the box. The same is true whenever anyone is trying to learn anything new, especially a science that contains many small pieces. In this chapter, you will be introduced to the "world of electricity and electronics." This complete picture introduction will help you see how all of the pieces in the electronics industry fit together.

1-1 THE PRESENT

At present you have probably heard mention of current, voltage, resistance, power, and some of the components, circuits, and systems used in electronics. Figure 1-1 shows how all of these items connect by comparing their relationship to the different parts of a tree. Working from the bottom up, you can see that the foundation of electronics rests on four basic roots or phenomena, known as current, voltage, resistance, and power. These are all related to one another, as can be seen by the formula circle in Figure 1-2.

Moving up the tree, the next block is electronic components, which were developed to manipulate or control voltage and current. Referring to Figure 1-3, you will see a listing of many electronic components, some names of which you may recognize.

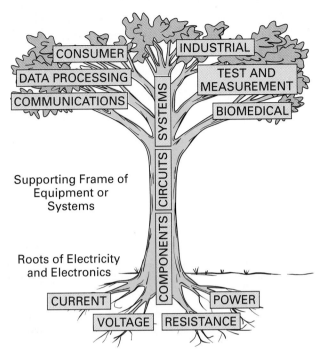

FIGURE 1-1   The Tree of Electronics.

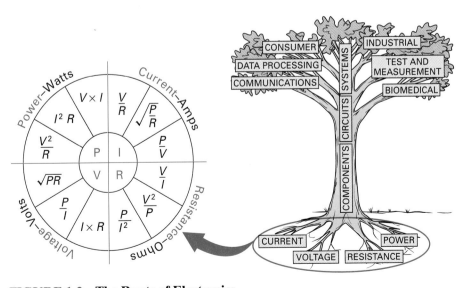

FIGURE 1-2   The Roots of Electronics.

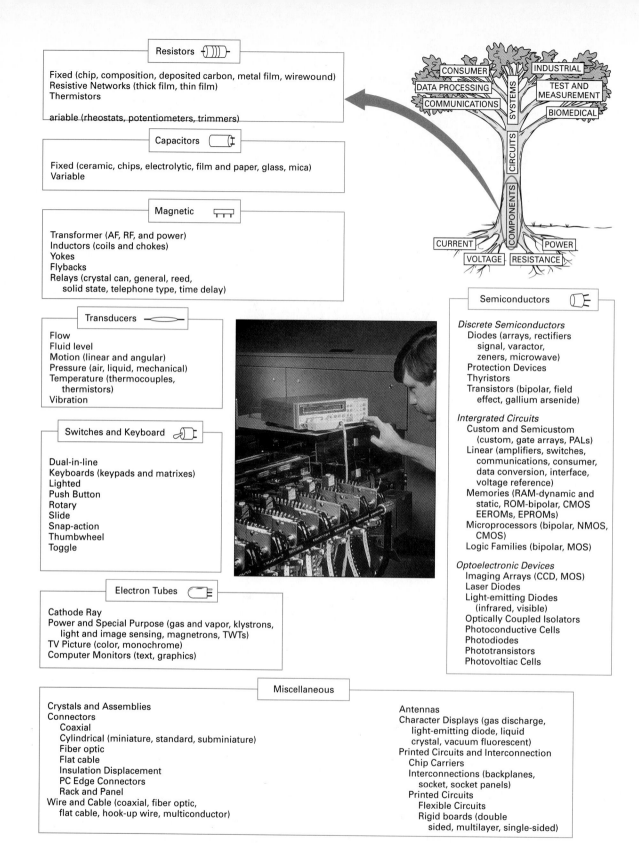

Resistors

Fixed (chip, composition, deposited carbon, metal film, wirewound)
Resistive Networks (thick film, thin film)
Thermistors
Variable (rheostats, potentiometers, trimmers)

Capacitors

Fixed (ceramic, chips, electrolytic, film and paper, glass, mica)
Variable

Magnetic

Transformer (AF, RF, and power)
Inductors (coils and chokes)
Yokes
Flybacks
Relays (crystal can, general, reed,
    solid state, telephone type, time delay)

Transducers

Flow
Fluid level
Motion (linear and angular)
Pressure (air, liquid, mechanical)
Temperature (thermocouples,
    thermistors)
Vibration

Switches and Keyboard

Dual-in-line
Keyboards (keypads and matrixes)
Lighted
Push Button
Rotary
Slide
Snap-action
Thumbwheel
Toggle

Electron Tubes

Cathode Ray
Power and Special Purpose (gas and vapor, klystrons,
    light and image sensing, magnetrons, TWTs)
TV Picture (color, monochrome)
Computer Monitors (text, graphics)

Semiconductors

Discrete Semiconductors
Diodes (arrays, rectifiers
    signal, varactor,
    zeners, microwave)
Protection Devices
Thyristors
Transistors (bipolar, field
    effect, gallium arsenide)

Intergrated Circuits
Custom and Semicustom
    (custom, gate arrays, PALs)
Linear (amplifiers, switches,
    communications, consumer,
    data conversion, interface,
    voltage reference)
Memories (RAM-dynamic and
    static, ROM-bipolar, CMOS
    EEROMs, EPROMs)
Microprocessors (bipolar, NMOS,
    CMOS)
Logic Families (bipolar, MOS)

Optoelectronic Devices
Imaging Arrays (CCD, MOS)
Laser Diodes
Light-emitting Diodes
    (infrared, visible)
Optically Coupled Isolators
Photoconductive Cells
Photodiodes
Phototransistors
Photovoltiac Cells

Miscellaneous

Crystals and Assemblies
Connectors
    Coaxial
    Cylindrical (miniature, standard, subminiature)
    Fiber optic
    Flat cable
    Insulation Displacement
    PC Edge Connectors
    Rack and Panel
Wire and Cable (coaxial, fiber optic,
    flat cable, hook-up wire, multiconductor)

Antennas
Character Displays (gas discharge,
    light-emitting diode, liquid
    crystal, vacuum fluorescent)
Printed Circuits and Interconnection
    Chip Carriers
    Interconnections (backplanes,
        socket, socket panels)
    Printed Circuits
        Flexible Circuits
        Rigid boards (double
            sided, multilayer, single-sided)

FIGURE 1-3   Components. (Photo courtesy of Hewlett-Packard Company.)

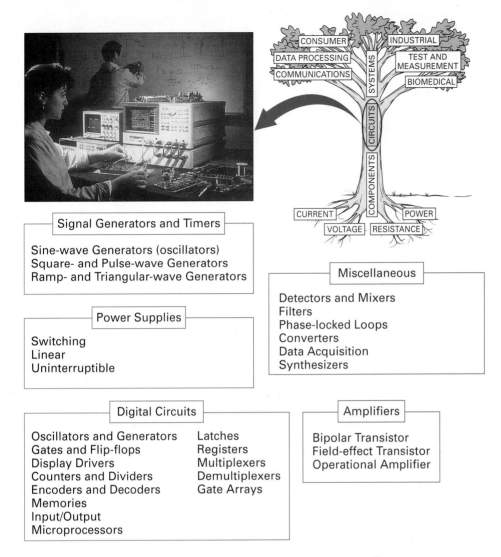

FIGURE 1-4   Circuits. (Photo courtesy of Hewlett-Packard Company.)

When a group of components is used in conjunction with one another, it forms a circuit, examples of which can be seen in Figure 1-4.

Just as components are the building blocks for circuits, circuits are in turn the building blocks for electronic systems or equipment, which can be categorized into one of six basic groups. Let us now examine these six different branches in more detail.

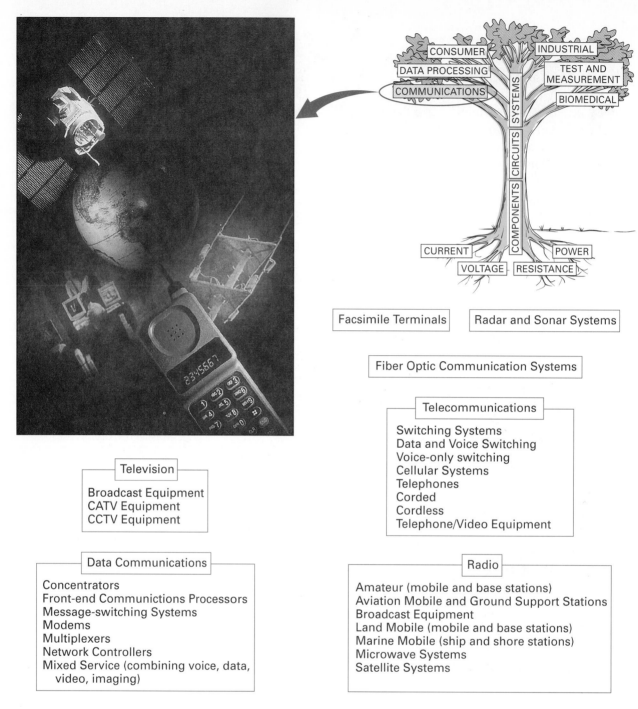

FIGURE 1-5   **Communications Equipment. (Photo courtesy of Hewlett-Packard Company.)**

1-1-1   *Communications Equipment*

Electronic communications allows the transmission and reception of information between two points. Radio and television are obvious communication devices, broadcasting data or entertainment between two points. Figure 1-5 illustrates and lists the different types of communication equipment that exist in what is probably the largest branch of electronics.

1-1-2 Data Processing Equipment

The computer is proving to be one of the most useful of all systems. Its ability to process, store, and manipulate large groups of information at an extremely fast rate makes it ideal for almost any and every application. Systems vary in complexity and capability, ranging from the Cray supercomputer to the home personal computer. The applications of word processing, record keeping, inventory, analysis, and accounting are but a few reasons why data processing systems, illustrated and listed in Figure 1-6, are used extensively.

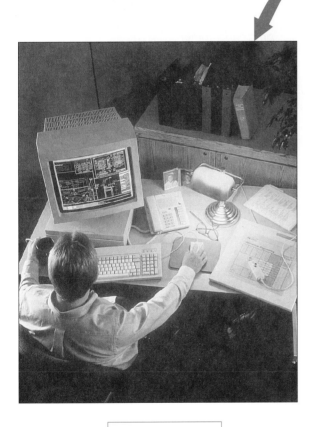

```
                    CONSUMER        INDUSTRIAL
                 DATA PROCESSING    TEST AND
                                    MEASUREMENT
                 COMMUNICATIONS     BIOMEDICAL
                        SYSTEMS
                        CIRCUITS
                        COMPONENTS
         CURRENT                    POWER
                 VOLTAGE  RESISTANCE
```

Data Terminals

CRT Terminals
   ASCII Terminals
   Graphics Terminals (color, monochrome)
Remote Batch Job Entry Terminals

Computer Systems

Microcomputers and Supermicrocomputers
Minicomputers (personal computers) and
   Superminicomputers (technical
   workstations, multiuser)
Mainframe Computers
Supercomputers

I/O Peripherals

Computer Microfilm
Digitizers
Graphics Tablets
Light Pens
Trackball and Mice
Optical Scanning Devices
Plotters
Printers
   Impact
   Nonimpact (laser, thermal, electrostatic, inkjet)

Data Storage Devices

Fixed Disk (14, 8, $5\frac{1}{4}$, and $3\frac{1}{2}$ in.)
Flexible Disk (8, $5\frac{1}{4}$, and $3\frac{1}{2}$ in.)
Optical Disk Drives (read-only, write once, erasable)
Cassette
Cartridge Magnetic Tape ($\frac{1}{4}$ in.)
Cartridge Tape Drives ($\frac{1}{2}$ in.)
Reel-type Magnetic Tape Drives

FIGURE 1-6   Data Processing Equipment. (Photo courtesy of Hewlett-Packard Company.)

1-1-3 Consumer Equipment

Figure 1-7 illustrates and lists some of the current consumer electronic equipment available. From the smart computer-controlled automobiles, which provide navigational information and monitor engine functions and braking, to the compact disk players, video camcorders, satellite TV receivers, and wide-screen stereo TVs, this branch of electronics provides us with entertainment, information, safety, and, in the case of the pacemaker, life.

Audio Equipment

Car
Stereo Equipment
   Compact Systems (miniature components)
   Components (speakers, amps, turntables,
      tuners, tape decks)
Phonographs and Radio Phonographs
Radios (table, clock, portable)
Tape Players/Recorders
Compact Disk Players
Digital Tape Players

Video Equipment

TV Receiver (color, monochrome)
Projection TV Receivers
Video Cassette Recorders (VCRs)
Video Disk Players
Camcorders (8 mm, $\frac{1}{2}$ in.)
Home Satellite Receivers

Personal

Calculators, Cameras, Watches
Telephone Answering Equipment
Personal Computers
Microwave Ovens
Musical Equipment and Instruments
Pacemakers and Hearing Aids
Alarms and Smoke Detectors

Automobile Electronics

Dashboard
Engine Monitoring and Analysis
Computer Navigation Systems
Alarms
Telephones

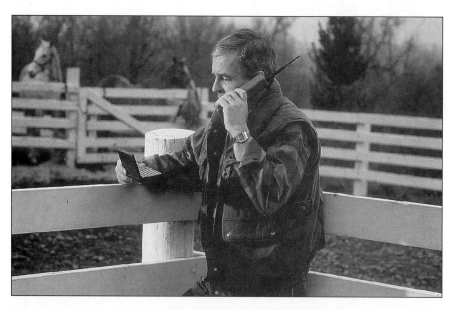

FIGURE 1-7 **Consumer Electronics Equipment. (Photo courtesy of Hewlett-Packard Company.)**

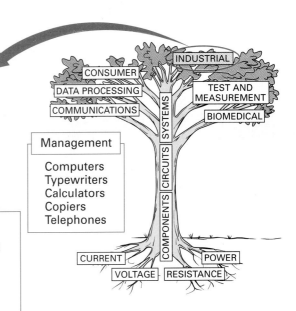

Manufacturing Equipment

Energy Management Equipment

Inspection Systems

Motor Controllers (speed, torque)

Numerical-control Systems

Process Control Equipment (data-aquisition systems, process instrumentation, programmable controllers)

Robot Systems

Vision Systems

Computer-aided Design and Engineering CAD/CAE

Hardware Equipment
    Design Work Stations (PC based, 32 microprocessor based platform, host based)
    Application Specific Hardware

Design Software
    Design Capture (schematic capture, logic fault and timing simulators, model libraries)
    IC Design (design rule checkers, logic synthesizers, floor planners–place and route, layout editors)
    Printed Circuit Board Design Software
    Project Management Software
    Test Equipment

Management

Computers
Typewriters
Calculators
Copiers
Telephones

FIGURE 1-8 Industrial Equipment. (Photo courtesy of Hewlett-Packard Company.)

1-1-4 Industrial Equipment

Almost any industrial company can be divided into three basic sections, all of which utilize electronic equipment to perform their functions. Figure 1-8 illustrates and lists the various types of equipment used with these three sections. The

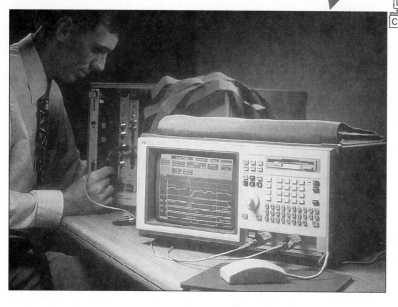

Automated Test Systems

Active and Discrete Component Test Systems
Automated Field Service Testers
IC Testers (benchtop, general purpose, specialized)
Interconnect and Bare pc-Board Testers
Loaded pc-Board Testers (in-circuit, functional, combined)

General Test and Measurement Equipment

Amplifiers (lab)
Arbitrary Waveform Generators
Analog Voltmeters, Ammeters, and Multimeters
Audio Oscillators
Audio Waveform Analyzers and Distortion Meters
Calibrators and Standards
Dedicated IEEE–488 Bus Controllers
Digital Multimeters
Electronic Counters (RF, Microwave, Universal)
Frequency Sythesizers
Function Generators
Pulse/Timing Generators
Signal Generators (RF, Microwave)

Logic Analyzers
Microprocessor Development Systems
Modulation Analyzers
Noise-measuring Equipment
Oscilloscopes (Analog, Digital)
Panel Meters
Personal Computer (PC) Based Instruments
Recorders and Plotters
RF/Microwave Network Analyzers
RF/Microwave Power-measuring Equipment
Spectrum Analyzers
Stand-alone In-circuit Emulators
Temperature-measuring Instruments

FIGURE 1-9   Test and Measurement Equipment. (Photo courtesy of Hewlett-Packard Company.)

manufacturing section will typically use power, motor, and process control equipment, along with automatic insertion, inspection, and vision systems, for the fabrication of a product. The engineering section uses computers and test equipment for the design and testing of a product, while the management section uses electronic equipment such as computers, copiers, telephones, and so on.

1-1-5 Test and Measurement Equipment

The rising complexity of electronic components, circuits, and equipment is causing a demand for sophisticated automatic test equipment for both the manufacturer and customer to test their products. Figure 1-9 illustrates and lists some of the test and measurement equipment which can be basically classified as either stand-alone and/or computer-controlled test instruments.

1-1-6 Biomedical Electronic Equipment

Electronic equipment is used more and more within the biological and medical fields. Figure 1-10 illustrates and lists some of the medical equipment, which can be categorized simply as being either patient care or diagnostic equipment. In the operating room, the endoscope, which is an instrument used to examine the interior of a canal or hollow organ, and the laser, which is used to coagulate, cut, or vaporize tissue with extremely intense light, both reduce the use of invasive surgery. A large amount of monitoring equipment is used both in and out of operating rooms, and this equipment consists of generally large computer-controlled systems that can have a variety of modules inserted (based on the application) to monitor, on a continuous basis, body temperature, blood pressure, pulse rate, and so on. In the diagnostic group of equipment, the clinical laboratory test results are used as diagnostic tools. With the advances in automation and computerized information systems, multiple tests can be carried out at increased speeds. Diagnostic imaging, in which a computer constructs an image of a cross-sectional plane of the body, is probably one of the most interesting equipment areas.

1-1-7 Electrical and Electronic Components, Circuits, and Systems

By definition, **electronics** is the branch of technology or science that deals with the use of **components** to control the flow of **electricity** in a vacuum, gas, liquid, semiconductor, conductor, or superconductor. Both electrical and electronic components, circuits, and systems control electron flow. However, their applications are distinctly different.

To properly manage power, electrical devices must perform such functions as *generating, distributing,* and *converting* electrical power.

To properly manage information, electronic devices must perform such functions as *generating, sensing, storing, retrieving, amplifying, transmitting, receiving,* and *displaying* information.

Electronics
Science related to the behavior of electrons in devices.

Electrical Components
Components, circuits, and systems that manage the flow of *power.*

Electronic Components
Components, circuits, and systems that manage the flow of *information.*

Electricity
Science that states certain particles possess a force field, which with electrons is negative and with protons is positive. Electricity can be divided into two groups: static and dynamic. Static electricity deals with charges at rest, while dynamic electricity deals with charges in motion.

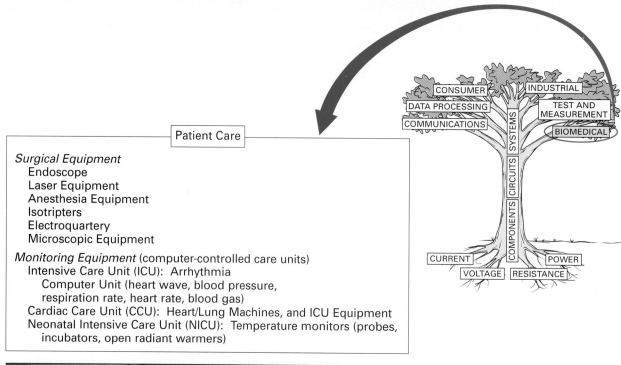

Patient Care

Surgical Equipment
  Endoscope
  Laser Equipment
  Anesthesia Equipment
  Isotripters
  Electroquartery
  Microscopic Equipment

Monitoring Equipment (computer-controlled care units)
  Intensive Care Unit (ICU): Arrhythmia
    Computer Unit (heart wave, blood pressure,
    respiration rate, heart rate, blood gas)
  Cardiac Care Unit (CCU): Heart/Lung Machines, and ICU Equipment
  Neonatal Intensive Care Unit (NICU): Temperature monitors (probes,
    incubators, open radiant warmers)

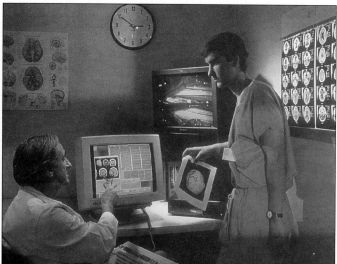

Diagnostic

Diagnosis Equipment
  X-Ray (computed tomography)
  Magnetic Resonance Imaging (MRI)
  Diagnostic Sounder
  Electrocardiograph (EKG)
  Electromyograph
  Electroencephalograph (EEG)
  Coagulograph
  Ultrasound (computed sonography)
  Nuclear Medicine (isotopes, spectroscopy)

Clinical Laboratory
  Automated Clinical Analyzers
  Centrifuge Incubators
  Cell Counters

FIGURE 1-10   Biomedical Equipment. (Photo courtesy of Hewlett-Packard Company.)

Some systems are designed specifically to manage the flow of power and therefore are only electrical, while other systems are designed to manage both power and information. For example, a television contains electrical components and circuits that manage the flow of electrical power from the wall outlet and electronic components and circuits that manage the flow of information or TV signals from the antenna or cable. The electrical circuits are needed since they supply power to the electronic circuits which in turn manage the flow of audio (sound) and video (picture) information signals.

SELF-TEST REVIEW QUESTIONS FOR SECTION 1-1

---

1. Name the four basic roots or phenomena on which the foundation of electronics rests.
2. List three examples of electronic components.
3. Name two circuits.
4. In which branch of electronics would radar be categorized?

1-2 THE PAST

Our way of life has undergone enormous change since the unveiling of electricity and electronics. Each of us, with just a little thought, can easily envisage how difficult it would be to manage in a world without electronics. The people responsible for the discovery of this science are discussed in this section. Figure 1-11 illustrates how a section of the tree of electronics has been extracted so that the rings of time can be viewed. These rings represent the age of the tree, with one ring being equal to approximately one year, and in this illustration the rings are used to summarize the progression of electricity and electronics.

Our history begins in 1600 with an English physician, William Gilbert (1544–1603), who documented many years of research and experiments on magnets and magnetic bodies such as amber and loadstones. Probably his most important discovery was that when rubbed with a cloth, amber would attract lightweight objects. In 1601 he was appointed physician to Queen Elizabeth I at a salary of $150 a year. He was the first to believe that the earth was nothing but a large magnet and that its magnetic field causes a needle to align itself between north and south.

Stephen Gray (1693–1736), also an Englishman, discovered that certain substances would conduct electricity. This lead was picked up in 1730 by Charles du Fay, a French experimenter, who believed that there were two types of electricity, which he called *vitreous* and *resinous* electricity.

One of the best known and most admired men in the latter half of the eighteenth century was the American Benjamin Franklin (1706–1790). He is reknowned for his kite-in-a-storm experiment, which proved that lightning is electricity. Franklin also discovered that there was only one type of electricity— that the two previously believed types were simply two characteristics of electricity. Vitreous was renamed *positive charge* and resinous was called *negative charge,* terms that were invented by Franklin, along with the terms *battery* and *conductor,* which are still used today.

The unit of electrical charge is named the coulomb, in honor of the French physicist Charles A. de Coulomb (1736–1806), who developed the laws of attraction and repulsion between charged bodies.

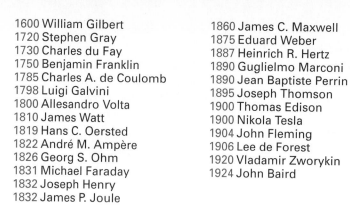

1600 William Gilbert
1720 Stephen Gray
1730 Charles du Fay
1750 Benjamin Franklin
1785 Charles A. de Coulomb
1798 Luigi Galvini
1800 Allesandro Volta
1810 James Watt
1819 Hans C. Oersted
1822 André M. Ampère
1826 Georg S. Ohm
1831 Michael Faraday
1832 Joseph Henry
1832 James P. Joule

1860 James C. Maxwell
1875 Eduard Weber
1887 Heinrich R. Hertz
1890 Guglielmo Marconi
1890 Jean Baptiste Perrin
1895 Joseph Thomson
1900 Thomas Edison
1900 Nikola Tesla
1904 John Fleming
1906 Lee de Forest
1920 Vladamir Zworykin
1924 John Baird

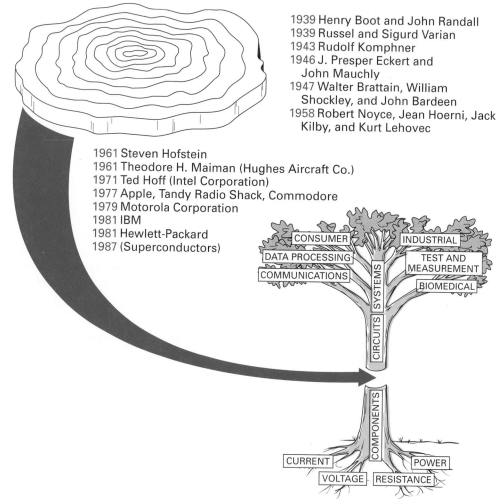

1939 Henry Boot and John Randall
1939 Russel and Sigurd Varian
1943 Rudolf Komphner
1946 J. Presper Eckert and
     John Mauchly
1947 Walter Brattain, William
     Shockley, and John Bardeen
1958 Robert Noyce, Jean Hoerni, Jack
     Kilby, and Kurt Lehovec

1961 Steven Hofstein
1961 Theodore H. Maiman (Hughes Aircraft Co.)
1971 Ted Hoff (Intel Corporation)
1977 Apple, Tandy Radio Shack, Commodore
1979 Motorola Corporation
1981 IBM
1981 Hewlett-Packard
1987 (Superconductors)

FIGURE 1-11   History of Electricity and Electronics.

The galvanometer, which is used to measure electrical current, was named after Luigi Galvani (1737–1798), who conducted many experiments with electrical current, or, as it was known at that time, "galvanism."

The unit of voltage, the volt, was named in honor of Alessandro Volta (1745–1827), an Italian physicist who is famous for his invention of the electric battery. In 1801, he was called to Paris by Napoleon to show his experiment on the generation of electric current.

The unit of power is the watt in honor of a Scottish engineer and inventor, James Watt (1736–1819), for his advances in the field of science.

In 1819, a Danish physicist, Hans C. Oersted (1777–1851), accidentally discovered an interesting phenomenon. Placing a compass near a current-carrying conductor, he noticed that the needle of the compass pointed to the conductor rather than north. He was quick to realize that electricity and magnetism were related, and in honor of his work, the unit oersted was adopted for the unit of magnetic field strength.

The unit of electrical current is the ampere, named in honor of André M. Ampère (1775–1836), a French physicist who pioneered in the study of electromagnetism. After hearing of Oersted's discoveries, he conducted further experiments and discovered that two current-carrying conductors would attract and repel one another, just like two magnets.

Ohm's law, the best known law in electrical circuits, was formulated by Georg S. Ohm (1787–1854), a German physicist. His law was so coldly received that his feelings were hurt and he resigned his teaching post. When his law was finally recognized, he was reinstated. In honor of his accomplishments, the unit of resistance is called the ohm.

In 1831, Michael Faraday (1791–1867), an English physicist, explored further Oersted's discovery of electromagnetism and discovered that a magnetic field could be used to produce electric current. These findings are today referred to as Faraday's laws of electromagnetic induction. A German-born scientist working in Russia extended Faraday's findings and found that the current induced in a conductor is such that it opposes the change in the magnetic field producing it. This is known today as Lenz's law, in honor of Heinrich F. E. Lenz. Michael Faraday also investigated static electricity and the lines of electric force, and it is in acknowledgement of his work in this area that the unit of capacitance is named the farad.

Joseph Henry (1797–1878), an American physicist, also conducted extensive studies into electromagnetism. Henry was the first to insulate the magnetic coil of wire and developed coils for telegraphy and motors. In recognition of his discovery of self-induction in 1832, the unit of inductance is called the henry.

James P. Joule (1818–1889), an English physicist and self-taught scientist, conducted extensive research into the relationships between electrical, chemical, and mechanical effects, which led him to the discovery that one energy form can be converted into another. For his achievements, his name was given to the unit of energy, the joule.

As a small boy, James C. Maxwell (1831–1879) was persistently inquisitive. He built many scientific toys before he was 8. At the age of 14 he wrote a paper on how to construct oval curves, and at 18 two of his papers were published. The supreme achievement of this Scottish physicist however was to translate Faraday's experiments into mathematical notation. This set of mathematical equations, known as Maxwell's equations, shows the relationship between electricity and magnetism.

Eduard W. Weber (1804–1891), a German physicist, made enduring contributions to the modern system of electrical units, and magnetic flux is measured in webers in honor of his work.

Heinrich R. Hertz (1857–1894), a German physicist, was the first to demonstrate the production and reception of electromagnetic (radio) waves. In honor of his work in this field, the unit of frequency is called the hertz.

Studying the experiments of Maxwell and Hertz, Guglielmo Marconi (1874–1937) invented a practical system of telegraphy communication. In an evolutionary process, Marconi extended his distance of communication from 1½ miles in 1896 to 6000 miles in 1902. In September 1899, Marconi equipped two U.S. ships with equipment and used them in the Atlantic Ocean to transmit to America the progress of the America's Cup yacht race.

The electron was first discovered by Jean B. Perrin (1870–1942), a French physicist who was awarded the Nobel prize for physics. Perrin discovered that cathode rays consisted of negatively charged particles, and these particles, which later became known as electrons, were measured by an English physicist, Joseph Thomson (1846–1914).

Thomas Edison (1847–1931) was a self-educated inventor best known for his development of the phonograph and the incandescent lamp. The first version of the phonograph cost $18 and was turned with a hand crank. A decade later it was motor driven, with cylindrical wax and then disk-type records. In 1879, after $40,000 worth of fruitless experiments, he succeeded in developing an incandescent lamp that consisted of a loop of carbonized cotton thread that glowed in a vacuum for 40 hours. In 50 years, he took out 1033 patents. Nikola Tesla (1856–1943) was the inventor of the induction motor and worked to improve power transmission. Working for Edison for a short time, they developed a hatred for one another that prompted Tesla to begin his own business. Edison promoted dc power distribution, while Tesla believed in ac, and eventually Tesla's reasoning was adopted worldwide. In 1912, they were nominated together for the Nobel prize in physics, but Tesla would have nothing to do with Edison, so the prize went to a third party. A theory of Tesla's, which up to this time has not proven possible, is the wireless transmission of electrical power by high-energy electromagnetic or radiant beams.

In 1904 John A. Fleming, a British scientist, saw the value of an effect that was discovered by Edison but for which he saw no practical purpose. The "Edison effect" permitted Fleming to develop the "Fleming valve," which passes current in only one direction. Its operation made it the first device able to convert alternating current into direct current and to detect radio waves.

Although the Fleming valve was an advance, it could not amplify or boost a signal. The "audion," developed by U.S. inventor Lee de Forest (1873–1961), sparked an era known as "vacuum-tube electronics" that brought about transcontinental telephony in 1915, radio broadcasting in 1920, radar in 1936, and television between 1927 and 1946, because of this triode vacuum tube's ability to amplify small signals.

The father of television, Vladamir C. Zworykin, developed the first television picture tube, called the kinescope, in 1920.

John L. Baird (1888–1946) was a British inventor and television pioneer.

He was the first to transmit television over a distance. He reproduced objects in 1924, transmitted recognizable human faces in 1925, demonstrated the first true television in 1926, and in 1939 he developed television in natural color.

During World War II, there was a need for microwave-frequency vacuum tubes. British inventor Henry Boot developed the magnetron in 1939, and in the same year an American brother duo, Russel and Sigurd Varian, invented the klystron. In 1943, the traveling wave tube amplifier was invented by Rudolf Komphner, and up to this day these three microwave vacuum tubes are still used extensively.

In 1946, J. Presper Eckert and John Mauchly unveiled ENIAC, which used over 300,000 vacuum tubes. ENIAC, which is an acronym for "electronic numerical integrator and computer," was the first large-scale electronic digital computer.

In 1947 Walter Brattain, William Shockley, and John Bardeen sparked an even greater era, known as "solid-state electronics," with their invention of the transistor at Bell Laboratories. Transistorized equipment is smaller, cheaper, more reliable, more robust, and consumes less power than its vacuum-tube counterparts.

In 1958, Robert Noyce, Jean Hoerni, Jack Kilby, and Kurt Lehovec all took part in the development of the integrated circuit, which incorporated many transistors and other components on a small chip of semiconductor material.

In 1961, Steven Hofstein devised the field-effect transistor used in MOS (metal oxide semiconductor) integrated circuits, and in the same year Theodore H. Maiman, a scientist working at Hughes Aircraft Company, built the first operational laser using a synthetic ruby crystal.

In 1971, Ted Hoff of Intel Corporation designed a microprocessor, the 4004, that had all the basic parts of a central processor. Intel improved on the 4-bit 4004 microprocessor and unveiled an 8-bit microprocessor in 1974 that could add two numbers in 2.5 millionths of a second.

Three mass-market personal computers emerged in 1977: the Apple II, Radio Shack's TRS-80, and the Commodore PET.

In 1979, Motorola Corporation continued to advance computers by creating a powerful and versatile 16-bit microprocessor that could multiply two numbers in 3.2 billionths of a second.

IBM, who had up to this time dominated the big computer market, entered the personal computer market with the IBM PC in 1981. Also in 1981, Hewlett-Packard unveiled its 32-bit microprocessor to further advance the speed and power of computers.

In scientific laboratories around the world, thousands of scientists began working furiously to develop a new technology that at the beginning of 1987 seemed little more than science fiction. Practical applications of superconductors cannot be fully realized at this time, just as no one could foresee all of the uses for the transistor when it was invented 40 years earlier. In the future, however, it is believed that all power will be distributed over *superconductor cables* because of their low power losses, while all information signals will be distributed over *fiber optic cables* because of their small size and large capacity.

1-3 THE FUTURE

So many things have happened in the past, so many things are happening at present, and so many things are under development to unfold in the future that it has become very difficult to keep tabs on the progression in all fields of electronics.

Electronics is an exciting and rewarding field to be involved in. The main purpose of this chapter is not only to give you an insight into the past and present, but also to give you the reasons why we begin our studies of electronics by discussing current, voltage, resistance, and power. These roots are necessary to understand components, as components need to be comprehended to understand circuits, and in turn circuits are necessary to understand the exciting, wide range of systems that exists in the different branches of electronics.

Your future in the electronics industry begins with this course in DC/AC electronics. To give you an idea of where you are going and what you will be covering, Figure 1-12 breaks up your study of electronics into four basic steps, which correlate to the basic blocks shown in the tree.

1-4 PEOPLE IN ELECTRONICS

In any industrial company, people are required to keep the industry moving, and an electronics company is not any different. The photo essay and associated storyboard demonstrate how a typical electronics company would develop a product; but more important, they show the people involved in this product development process, their job titles and responsibilities, and how each person plays a very important role. Use this photo essay as a career opportunities tour.

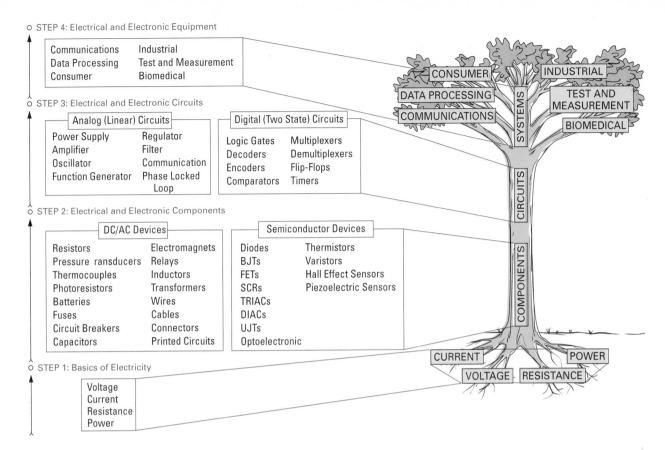

FIGURE 1-12   The Steps Involved in Studying Electricity and Electronics.

1. Electronics is built on four basic roots or phenomena: current, voltage, resistance, and power.

2. Electronic components control the four basic roots or phenomena of electronics.

3. Groups of components are interconnected to perform certain functions, and these arrangements are known as circuits.

4. Just as components are the building blocks for circuits, circuits are in turn the building blocks for electronic equipment.

5. All electronic equipment can be categorized into one of six different branches or groups: communications, data processing, consumer, industrial, test and measurement, and biomedical.

6. The names for almost all the units of electricity and electronics are derived from early experimenters.

7. The four main quantities and units are:

   a. Current, which is measured in amperes or amps
   b. Voltage, which is measured in the unit of volts
   c. Resistance, which is measured in ohms
   d. Power, which is measured in watts

SUMMARY

8. The engineer forms the concepts, designs, and modifications of all electronic circuits or systems.

9. Engineering technicians aid the engineer in the research and development of a product.

10. The technician is an expert in the diagnosis and repair of problems or malfunctions within electronic equipment.

11. The assembler is also an expert in his or her field, which is the fabrication of or manufacturing techniques used in the assembly of the final product.

REVIEW QUESTIONS

Multiple-Choice Questions

1. The unit of charge was named after:
   a. Alessandro Volta   c. Georg Ohm
   b. Heinrich Hertz   d. Charles Coulomb

2. _____ developed the basic laws relating to current flow.
   a. André Ampère   c. Michael Faraday
   b. Heinrich Hertz   d. Intel Corporation

3. The first 16-bit microprocessor was developed by:
   a. Intel Corporation   c. Texas Instruments
   b. Hewlett-Packard   d. Motorola Corporation

4. The first operational laser was built by:
   a. Walter Brattain   c. Theodore Maiman
   b. Guglielmo Marconi   d. James Maxwell

5. The unit of frequency was named in honor of:
   a. Heinrich Hertz   c. Georg Ohm
   b. Jean Perrin   d. Greek experimenters

6. The person largely responsible for the era termed "vacuum-tube electronics" was:
   a. John Fleming   c. Lee de Forest
   b. William Gilbert   d. Benjamin Franklin

7. The carbon filament light bulb was invented by:
   a. Joseph Henry   c. Thomas Edison
   b. James Joule   d. Vladamir Zworykin

8. The persons responsible for the transistor and the subsequent "solid-state electronics" era were:
   a. Hertz and Marconi   c. Noyce, Hoerni, Kilby, and Lehovec
   b. Eckert and Mauchly   d. Brattain, Schockley, and Bardeen

9. The resistor, capacitor, and inductor are all examples of electronic circuits.
   a. True   b. False

10. Amplifiers and oscillators are both examples of electronic _____.
    a. Circuits   c. Both (a) and (b)
    b. Components   d. None of the above

Electronic Technicians in the Product Development Process

In the following 16 pages of this color insert you will be introduced to the many different types of "Electronic Technician." By definition, "An electronics technician must be able to diagnose, isolate, and repair electronic circuit and system malfunctions." In order to achieve this goal an electronic technician must have a thorough knowledge of electronics, test equipment and troubleshooting techniques.

In the basic block diagram below you can see the steps followed by a typical electronics company in the development of an electronic system. As you can see, these companies employ a variety of electronic technicians, all of which play a very important role in the development of an electronic system. In the following pages of this color insert we will examine these electronic technicians in more detail.

Electronic Technicians in the System Development Process

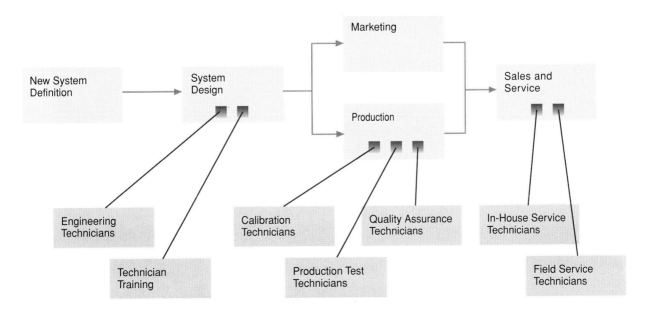

Development of an Electronic Product

The flow chart below expands on the basic block diagram on the previous page, showing the order from top to bottom, in which an electronic product is developed from conception to shipping. This photographic essay also shows the responsibilities of each person and how he or she fits into this process, giving you a clearer picture of the many career opportunities available.

Product Conception

Product Definition

Design Breakdown

Hiring of Personnel

Hardware Design → Software Design → Mechanical Design

ENGINEERS

The engineer is responsible for designing the electronic system and supervising its development. During this process, the engineer relies heavily on the engineering technician to assist in the prototyping of the design and the diagnosis and repair of problems.

Parts Purchasing

Documentation

TECHNICIANS

The technician is the expert in troubleshooting circuit and system malfunctions. Along with a thorough knowledge of all test equipment and how to use it to diagnose problems, the technician is also familiar with how to repair or replace faulty components. Referring to the flow chart, as you will see in the following photographs, technicians are involved in almost every step of product development.

Breadboard Design

Prototype Layout, Construction, & Test

Build Pre-production Units

Quality Engineering Test   Stock Components   Design Test Systems   Technician Training

ASSEMBLERS

The assembler is responsible for putting the electronic product together and is trained in the correct component handling and assembly techniques.

Chassis Fabrication

Component Insertion

Flow Soldering

Wire/Cable Hook-up

Final Assembly

Product Assembly

Product Test & Calibration

Quality Control Test

Product Release

Sales

Shipping

Customer   ←→   Customer Service

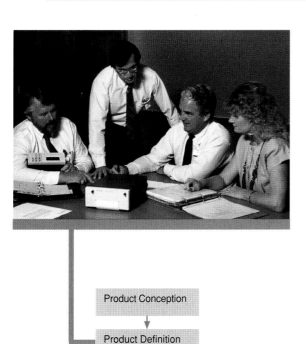

The first step in the process can be seen in this photograph. The key company managers meet to define the new product. This new product will have to meet the needs of the ever-expanding electronics industry. During this meeting, the new product's budget, target dates, and key features will be outlined.

Product Conception

Product Definition

Design Breakdown

Hiring of Personnel

Once the product has been defined, it is the job of engineering to specify and design all major components of the new equipment. In this photograph, the design of the new product is being broken up into smaller tasks by an engineering manager and these task projects are being assigned to different engineers and engineering technicians.

New personnel may be needed by the company to develop and produce this new product. Here, a prospective employee is receiving a technical and personnel interview for a technician position.

```
                    ┌──────────────┬──────────────┬──────────────┐
              ┌─────┴─────┐  ┌──────┴──────┐  ┌──────┴──────┐
              │  Hardware │  │  Software   │  │  Mechanical │
              │   Design  │  │   Design    │  │   Design    │
              └─────┬─────┘  └──────┬──────┘  └──────┬──────┘
                    ▼               ▼                ▼
```

The hardware design engineer shown is entering the
newly designed circuit, which consists of many
interconnected electronic components, into a computer.
The equipments hardware is the physical cables,
components, and circuits you can see inside every piece
of electronic equipment.

At the heart of many electronic products produced
today is a computer controlling operation. The
software design engineer is entering a list of
instructions into the equipments computer memory,
and it is this program of instructions that will control
how the equipment operates.

```
                    │
        ┌───────────┼───────────┐
        ▼           ▼           ▼
┌───────────────┬───────────────┬───────────────┐
│ Hardware Design│ Software Design│ Mechanical Design│
└───────────────┴───────────────┴───────────────┘
        └───────────┬───────────┘
                    ▼
            ┌───────────────┐
            │ Parts Purchasing│
            └───────────────┘
                    │
                    ▼
```

The software program resides within the equipment hardware, and both need to be encased inside a chassis or enclosure. In this photograph, a mechanical design engineer is designing the equipment housing and front panel.

Once the design is complete, the equipment parts have to be ordered so the product can be constructed. In this photograph, the engineering technician on the left is examining the variety of components available from the supplier, while the purchasing agent on the right is comparing costs.

Documentation

From the moment of conception to the final product, the unit is documented with mechanical drawings, schematic diagrams, and written instructions.

All documentation generated on a product is used to produce an operator's and maintenance manual that will accompany the finished unit when shipped. In this photograph, you can see a technical writer overseeing the drafting of some line art by an illustrator.

From sketches supplied by the engineers, a breadboard model of the design is constructed. The breadboard model is an experimental arrangement of a circuit in which the components are temporarily attached to a flat board. In this arrangement, the components can be tested to prove the feasibility of the circuit. A breadboard facilitates making easy changes when they are necessary. Here you can see an engineering technician breadboarding the design.

Breadboard Design

Prototype Layout, Construction, & Test

The engineer does not consider the final location of the components in constructing the breadboard model. At the next stage, however, a prototype working model completely representative of the final, mass-produced product is hand-assembled. The breadboard is replaced by a printed circuit board (PCB). In this scene, an engineering technician is producing a PCB layout from the design schematic diagrams.

The newly constructed prototype seen here is undergoing a complete evaluation of its mechanical and electrical form, design and performance.

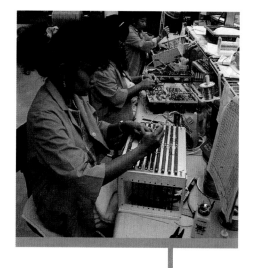

An approved prototype will initiate the production of several pre-production units, which can be seen in this photograph. The pathways these units take through construction will be mapped and used for mass assembly of the new product.

Build Pre-production Units

Quality Engineering Test

Stock Components

Design Test Systems

Technician Training

One of the pre-production units is taken through an extensive series of tests to determine whether it meets the standards listed. In this photograph, a quality assurance (QA) technician is evaluating the new product as it is put through an extensive series of tests.

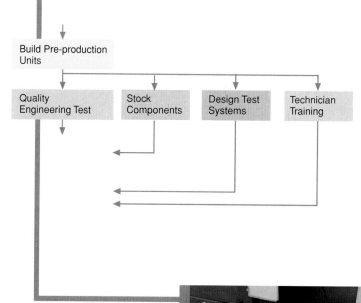

The stockroom holds all of the raw materials needed to build products. In the foreground of this photograph, incoming components are being inspected and values verified, while in the background the parts kits needed for the many phases of assembly are being requisitioned.

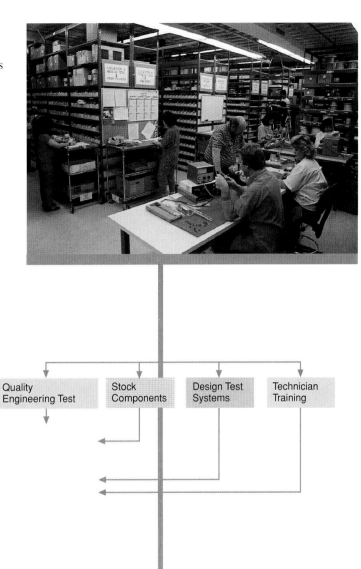

| Quality Engineering Test | Stock Components | Design Test Systems | Technician Training |

In this photograph, prepping of the many electronic components is taking place. In this process, wires, cables, and components are pre-cut to save time during the product assembly.

Once the assembled units come out of production, they will need to be tested. In this phase of the process, a test engineer is designing an automatic test system for the new product.

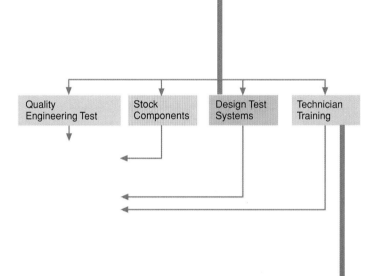

| Quality Engineering Test | Stock Components | Design Test Systems | Technician Training |

Training is a very important function. In this photograph, engineering technicians, production test technicians, quality control technicians, and customer service technicians are being taught the operation and component level maintenance of the new equipment.

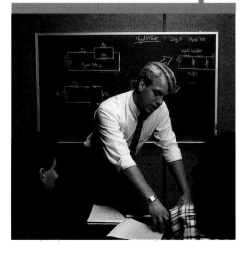

The components which make up the unit are grouped into several areas where complex precision assembly takes place. This photograph of the machine shop shows the chassis or housing for the equipment being fabricated.

In this photograph, all of the electronic components are being inserted into their respective positions within the printed circuit boards.

| Chassis Fabrication |
| Component Insertion |
| Flow Soldering |
| Wire/Cable Hook-up |
| Final Assembly |

Product Assembly

Once the boards have been filled with components, they are run through a flow- or wave-soldering machine. This machine solders the components to the board by moving the printed circuit board over a flowing wave of molten solder in a solder bath.

The wires and cables that interconnect all of the separate boards and units within the equipment are added at this stage.

Chassis Fabrication	Product Assembly
Component Insertion	
Flow Soldering	
Wire/Cable Hook-up	
Final Assembly	

Final assembly of the product is taking place in this photograph. The equipment will have the remaining units inserted, and its front and rear panels will be mounted and connected.

```
┌──────────────┐  ┌──────────────┐  ┌──────────────┐  ┌──────────────┐
│ Quality      │  │ Stock        │  │ Design Test  │  │ Technician   │
│ Engineering  │  │ Components   │  │ Systems      │  │ Training     │
│ Test         │  │              │  │              │  │              │
└──────┬───────┘  └──────┬───────┘  └──────┬───────┘  └──────┬───────┘
       │                 │                 │                 │
┌──────▼───────┐         │                 │                 │
│ Product      │◄────────┘                 │                 │
│ Assembly     │                           │                 │
└──────┬───────┘                           │                 │
       │                                   │                 │
┌──────▼───────┐◄──────────────────────────┘                 │
│ Product Test │◄────────────────────────────────────────────┘
│ & Calibration│
└──────┬───────┘
       │
```

The more complex problems are handled by the production test technicians seen in this photograph. Once the system is fully operational, it is calibrated by a calibration technician.

After leaving assembly, the unit is hooked up to a test system and subjected to various testing procedures. The automatic test equipment (ATE) found at these stations perform many tests that would be too time-consuming for a technician to do manually.

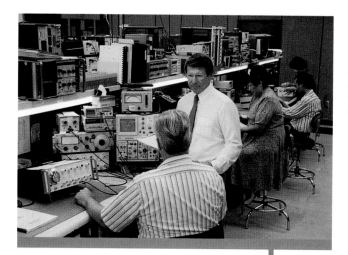

The quality control technicians seen in this photograph are performing the extensive series of electrical and mechanical final inspection tests. These will ensure that the performance standards listed in the unit's specifications are being met.

Quality Control
Test

Product Release

From the time of product definition, the marketing personnel have been planning the advertising brochures and sales approach in preparation for the release of the product.

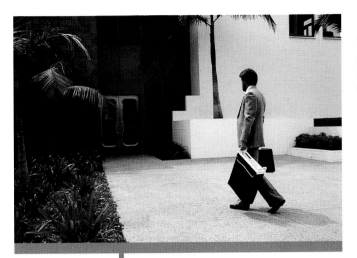

In this photograph, an applications engineer is making a sales call and will demonstrate to the customer the new product with all of its features and possible applications.

Sales

Shipping

Customer

In the background of this photograph, you can see all of the equipment ready to be shipped. In the foreground, a delicate electronic unit is being carefully packaged to prevent damage during transit.

In this photograph, a customer is receiving instruction on the operation of the purchased unit.

Once the customer has received the electronic equipment, customer service provides assistance in maintenance and repair of the unit through direct in-house service or at service centers throughout the world. This photograph shows some in-house service technicians troubleshooting problems on returned units.

In this photograph, another in-house service technician is performing a test on a returned faulty unit containing the latest surface mount devices (SMD).

The field service technician seen here has been requested by the customer to make a service call on a malfunctioning unit that currently is under test.

11. What new technology has just recently been unveiled?

    a. Microelectronics   **c.** Superconductivity
    b. Solid-state electronics   **d.** Two of the above could be true

12. Radio equipment would be classified in what branch of electronics?

    a. Data processing   **d.** Test and measurement
    b. Industrial   **e.** None of the above
    c. Biomedical

13. In which branch of electronics would the oscilloscope be found?

    a. Data processing   **d.** Biomedical
    b. Test and measurement   **e.** None of the above
    c. Military and government

14. Fiber optic cables are used to carry _____.

    a. Power   **c.** Both (a) and (b)
    b. Information   **d.** None of the above

15. Superconductor cables are highly desirable due to:

    a. Their large size   **c.** Their low power losses
    b. Their low efficiency   **d.** All of the above

Essay Questions

16. What are the three sections into which an industrial company can be divided? (1-1)

17. List the seven branches of electronic equipment. (1-1)

18. Give examples of five electronic components and three electronic circuits. (1-1)

19. Give examples of four pieces of equipment in each branch of electronics. (1-1)

20. List three attributes of the human body that medical electronic equipment can measure. (1-1-7)

21. List the four roots of electronics. (1-1)

22. Name the units for the following quantities: (1-1)

    a. Current   **c.** Resistance
    b. Voltage   **d.** Power

23. Who promoted the ac power distribution that is now used worldwide? (1-2)

24. What was the ENIAC? (1-2)

25. List two things for which Thomas Edison is best known. (1-2)

26. State the responsibilities of the following people: (1-4)

    a. Engineer   **b.** Technician   **c.** Assembler

27. What is the difference between a marketing person and a salesperson? (1-4)

28. What tasks do the following people perform? (1-4)

    a. Test engineer   **b.** Calibration technician   **c.** Sustaining engineer

29. At which stage in the product development process would an assembler get involved? (1-4)

30. What responsibilities do the following people have? (1-4)

    a. Engineering/technical writer
    b. Quality assurance technician
    c. Customer service technician (in-house and field service)
    d. Engineering technician

After completing this chapter, you will be able to:

1. Explain the atom's subatomic particles.
2. State the difference between an atomic number and atomic weight.
3. Understand the term *natural element.*
4. Describe what is meant by and state how many shells or bands exist around an atom.
5. Explain the difference between:
   a. An atom and a molecule
   b. An element and a compound
   c. A proton and an electron
6. Describe the terms:
   a. Neutral atom
   b. Negative ion
   c. Positive ion
7. Define *electrical current.*
8. Describe the ampere in relation to coulombs per second.
9. List the different current units and values.
10. Explain why the effect of current flow travels at the speed of light.
11. Describe the difference between conventional current flow and electron flow.
12. List the three rules to apply when measuring current.
13. Define *electrical voltage.*
14. List the various voltage units and values.
15. List the three rules to apply when measuring voltage.
16. Relate the electrical system to a fluid system.
17. Describe why current is directly proportional to voltage.
18. Explain the difference between:
    a. A conductor
    b. An insulator
    c. A semiconductor
19. Describe conductance.
20. Explain what makes a good:
    a. Conductor   b. Insulator
21. Describe semiconductor material.
22. Explain the terms:
    a. Open circuit
    b. Closed circuit
    c. Short circuit

Voltage and Current

Creating a Lot of Static

Coulomb's law states that the force exerted between two electric charges is equal to the product of the charges, and is inversely proportional to the distance between them: $f = Q_1 \times Q_2/d^2$.

Ironically, Coulomb's law was not first discovered by Coulomb, but by Henry Cavendish, a wealthy scientist and philosopher. Cavendish did not publish his discovery, which he made several years before Coulomb discovered the law independently. James Clerk Maxwell published the scientific notebooks of Cavendish in 1879, describing his experiments and conclusions. However, about 100 years had passed, and Coulomb's name was firmly associated with the law. Many scientists demanded that the law be called Cavendish's law, while other scientists refused to change, stating that Coulomb was the discoverer because he made the law known promptly to the scientific community.

I N T R O D U C T I O N

If Chapter 1 were described as the big picture, Chapter 2 would be called the small picture. In this chapter we will examine the smallest and most significant part of electricity and electronics—the **electron.** By having a good understanding of the electron and the atom, we will be able to obtain a clearer understanding of our four basic electrical quantities: **voltage, current, resistance,** and **power.** In this chapter we will be discussing the basic building blocks of matter, the electrical quantities of voltage and current, and the difference between a conductor and insulator.

2-1   THE STRUCTURE OF MATTER

All of the matter on the earth and in the air surrounding the earth can be classified as being either a solid, liquid, or gas. A total of approximately 107 different natural elements exist in, on, and around the earth. An **element,** by definition, is a substance consisting of only one type of atom; in other words, every element has its own distinctive atom, which makes it different from all the other elements. This **atom** is the smallest particle into which an element can be divided without losing its identity, and a group of identical atoms is called an element, shown in Figure 2-1.

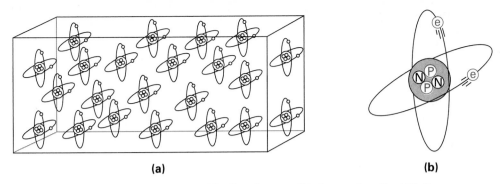

(a)   **(b)**

FIGURE 2-1   **(a) Element: Many Similar Atoms. (b) Atom: Smallest Unit.**

For the sake of discussion, let us take a small amount of either a solid, a liquid, or a gas and divide it into two pieces. Then we divide a resulting piece into two pieces, and keep repeating the process until we finally end up with a tiny remaining part. Viewing the part under the microscope, as shown in Figure 2-2, the substance can still be identified as the original element as it is still made up of many of the original solid, liquid or gas atoms. A small amount of gold, for example, the size of a pinpoint, will still contain several billion atoms. If the element subdivision is continued, however, a point will be reached at which a single atom will remain. Let us now analyze the atom in more detail.

2-1-1   The Atom

The word atom is a Greek word meaning a particle that is too small to be subdivided. At present, we cannot clearly see the atom; however, physicists and researchers do have the ability to record a picture as small as 12 billionths of an

Small part of element under microscope —
original element still made up of many atoms.

Element

Microscope

FIGURE 2-2   **Elements under the Microscope.**

inch (about the diameter of one atom), and this image displays the atom as a white fuzzy ball.

In 1913, a Danish physicist, Neils Bohr, put forward a theory about the atom, and his basic model outlining the **subatomic** particles that make up the atom is still in use today and is illustrated in Figure 2-3. Bohr actually combined the ideas of Lord Rutherford's (1871–1937) nuclear atom with Max Planck's (1858–1947) and Albert Einstein's (1879–1955) quantum theory of radiation.

The three important particles of the atom are the *proton,* which has a positive charge, the *neutron,* which is neutral or has no charge, and the *electron,* which has a negative charge. Referring to Figure 2-3, you can see that the atom consists of a positively charged central mass called the *nucleus,* which is made up of protons and neutrons surrounded by a quantity of negatively charged orbiting electrons.

Table 2-1 lists the periodic table of the elements, in order of their atomic number. The **atomic number** of an atom describes the number of protons that exist within the nucleus.

The proton and the neutron are almost 2000 times heavier than the very small electron, so if we ignore the weight of the electron, we can use the fourth column in Table 2-1 (weight of an atom) to give us a clearer picture of the protons and neutrons within the atom's nucleus. For example, a hydrogen atom, shown in Figure 2-4(a), is the smallest of all atoms and has an atomic number of 1, which means that hydrogen has a one-proton nucleus. Helium, however [Figure 2-4(b)], is second on the table and has an atomic number of 2, indicating that two protons are within the nucleus. The **atomic weight** of helium, however, is 4, meaning that two protons and two neutrons make up the atom's nucleus.

The number of neutrons within an atom's nucleus can therefore be calculated by subtracting the atomic number (protons) from the atomic weight (protons and neutrons). For example, Figure 2-5(a) illustrates a beryllium atom which has the following atomic number and weight.

Subatomic
Particles such as electrons, protons, and neutrons that are smaller than atoms.

Atomic Number
Number of positive charges or protons in the nucleus of an atom.

Atomic Weight
The relative weight of a neutral atom of an element, based on a neutral oxygen atom having an atomic weight of 16.

Beryllium
Atomic number: 4 (protons)
Atomic weight: 9 (protons and neutrons)

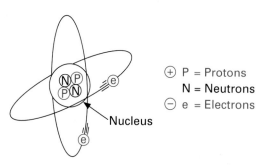

⊕ P = Protons
N = Neutrons
⊖ e = Electrons

FIGURE 2-3   The Atom.

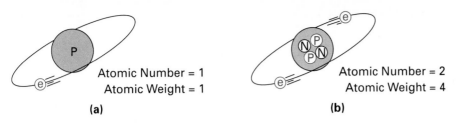

Atomic Number = 1
Atomic Weight = 1

(a)

Atomic Number = 2
Atomic Weight = 4

(b)

FIGURE 2-4   (a) Hydrogen Atom. (b) Helium Atom.

Subtracting the beryllium atom's weight from the beryllium atom number, we can determine the number of neutrons in the beryllium atom's nucleus, as shown in Figure 2-5(a).

In most instances, atoms like the beryllium atom will not be drawn in the three-dimensional way shown in Figure 2-5(a). Figure 2-5(b) shows how a beryllium atom could be more easily drawn as a two-dimensional figure.

A **neutral atom** or *balanced atom* is one that has an equal number of protons and orbiting electrons, so the net positive proton charge is equal but opposite to the net negative electron charge, resulting in a balanced or neutral state.

Neutral Atom
An atom in which the number of positive charges in the nucleus (protons) is equal to the number of negative charges (electrons) that surround the nucleus.

Beryllium	
Atomic Number	**Atomic Weight**
4 Protons	4 Protons + 5 Neutrons

Number of Protons = 4

Number of Neutrons = 9 − 4 = 5

(a)

4 proton nuclei.

2 electrons in the first orbital path.

2 electrons in the second orbital path.

(b)

FIGURE 2-5   Beryllium Atom.

TABLE 2-1 Periodic Table of the Elements

ATOMIC NUMBER	ELEMENT NAME	SYMBOL	ATOMIC WEIGHT	ELECTRONS/SHELL K L M N O P Q							DISCOVERED	COMMENT
				K	L	M	N	O	P	Q		
1	Hydrogen	H	1.007	1							1766	Active gas
2	Helium	He	4.002	2							1895	Inert gas
3	Lithium	Li	6.941	2	1						1817	Solid
4	Beryllium	Be	9.01218	2	2						1798	Solid
5	Boron	B	10.81	2	3						1808	Solid
6	Carbon	C	12.011	2	4						Ancient	Semiconductor
7	Nitrogen	N	14.0067	2	5						1772	Gas
8	Oxygen	O	15.9994	2	6						1774	Gas
9	Fluorine	F	18.998403	2	7						1771	Active gas
10	Neon	Ne	20.179	2	8						1898	Inert gas
11	Sodium	Na	22.98977	2	8	1					1807	Solid
12	Magnesium	Mg	24.305	2	8	2					1755	Solid
13	Aluminum	Al	26.98154	2	8	3					1825	Metal conductor
14	Silicon	Si	28.0855	2	8	4					1823	Semiconductor
15	Phosphorus	P	30.97376	2	8	5					1669	Solid
16	Sulfur	S	32.06	2	8	6					Ancient	Solid
17	Chlorine	Cl	35.453	2	8	7					1774	Active gas
18	Argon	Ar	39.948	2	8	8					1894	Inert gas
19	Potassium	K	39.0983	2	8	8	1				1807	Solid
20	Calcium	Ca	40.08	2	8	8	2				1808	Solid
21	Scandium	Sc	44.9559	2	8	9	2				1879	Solid
22	Titanium	Ti	47.90	2	8	10	2				1791	Solid
23	Vanadium	V	50.9415	2	8	11	2				1831	Solid
24	Chromium	Cr	51.996	2	8	13	1				1798	Solid
25	Manganese	Mn	54.9380	2	8	13	2				1774	Solid
26	Iron	Fe	55.847	2	8	14	2				Ancient	Solid (magnetic)
27	Cobalt	Co	58.9332	2	8	15	2				1735	Solid
28	Nickel	Ni	58.70	2	8	16	2				1751	Solid
29	Copper	Cu	63.546	2	8	18	1				Ancient	Metal conductor
30	Zinc	Zn	65.38	2	8	18	3				1746	Solid
31	Gallium	Ga	69.72	2	8	18	4				1875	Liquid
32	Germanium	Ge	72.59	2	8	18	4				1886	Semiconductor
33	Arsenic	As	74.9216	2	8	18	5				1649	Solid
34	Selenium	Se	78.96	2	8	18	6				1818	Photosensitive
35	Bromine	Br	79.904	2	8	18	7				1898	Liquid
36	Krypton	Kr	83.80	2	8	18	8				1898	Inert gas
37	Rubidium	Rb	85.4678	2	8	18	8	1			1861	Solid
38	Strontium	Sr	87.62	2	8	18	8	2			1790	Solid

Shells or Bands
An orbital path containing a group of electrons that have a common energy level.

For example, Figure 2-6 illustrates a copper atom, which is the most commonly used metal in the field of electronics. It has an atomic number of 29, meaning that 29 protons and 29 electrons exist within the atom when it is in its neutral state.

Orbiting electrons travel around the nucleus at varying distances from the nucleus, and these orbital paths are known as **shells or bands.** The orbital shell nearest the nucleus is referred to as the first or K shell. The second is known as

TABLE 2-1 (*continued*)

ATOMIC NUMBER[a]	ELEMENT NAME	SYMBOL	ATOMIC WEIGHT	ELECTRONS/SHELL							DISCOVERED	COMMENT
				K	L	M	N	O	P	Q		
39	Yttrium	Y	88.9059	2	8	18	9	2			1843	Solid
40	Zirconium	Zr	91.22	2	8	18	10	2			1789	Solid
41	Niobium	Nb	92.9064	2	8	18	12	1			1801	Solid
42	Molybdenum	Mo	95.94	2	8	18	13	1			1781	Solid
43	Technetium	Tc	98.0	2	8	18	14	1			1937	Solid
44	Ruthenium	Ru	101.07	2	8	18	15	1			1844	Solid
45	Rhodium	Rh	102.9055	2	8	18	16	1			1803	Solid
46	Palladium	Pd	106.4	2	8	18	18				1803	Solid
47	Silver	Ag	107.868	2	8	18	18	1			Ancient	Metal conductor
48	Cadmium	Cd	112.41	2	8	18	18	2			1803	Solid
49	Indium	In	114.82	2	8	18	18	3			1863	Solid
50	Tin	Sn	118.69	2	8	18	18	4			Ancient	Solid
51	Antimony	Sb	121.75	2	8	18	18	5			Ancient	Solid
52	Tellurium	Te	127.60	2	8	18	18	6			1783	Solid
53	Iodine	I	126.9045	2	8	18	18	7			1811	Solid
54	Xenon	Xe	131.30	2	8	18	18	8			1898	Inert gas
55	Cesium	Cs	132.9054	2	8	18	18	8	1		1803	Liquid
56	Barium	Ba	137.33	2	8	18	18	8	2		1808	Solid
57	Lanthanum	La	138.9055	2	8	18	18	9	2		1839	Solid
72	Hafnium	Hf	178.49	2	8	18	32	10	2		1923	Solid
73	Tantalum	Ta	180.9479	2	8	18	32	11	2		1802	Solid
74	Tungsten	W	183.85	2	8	18	32	12	2		1783	Solid
75	Rhenium	Re	186.207	2	8	18	32	13	2		1925	Solid
76	Osmium	Os	190.2	2	8	18	32	14	2		1804	Solid
77	Iridium	Ir	192.22	2	8	18	32	15	2		1804	Solid
78	Platinum	Pt	195.09	2	8	18	32	16	2		1735	Solid
79	Gold	Au	196.9665	2	8	18	32	18	1		Ancient	Solid
80	Mercury	Hg	200.59	2	8	18	32	18	2		Ancient	Liquid
81	Thallium	Tl	204.37	2	8	18	32	18	3		1861	Solid
82	Lead	Pb	207.2	2	8	18	32	18	4		Ancient	Solid
83	Bismuth	Bi	208.9804	2	8	18	32	18	5		1753	Solid
84	Polonium	Po	209.0	2	8	18	32	18	6		1898	Solid
85	Astatine	At	210.0	2	8	18	32	18	7		1945	Solid
86	Radon	Rn	222.0	2	8	18	32	18	8		1900	Inert gas
87	Francium	Fr	223.0	2	8	18	32	18	8	1	1945	Liquid
88	Radium	Ra	226.0254	2	8	18	32	18	8	2	1898	Solid
89	Actinium	Ac	227.0278	2	8	18	32	18	9	2	1899	Solid

[a]Rare earth series 58–71 and 90–107 have been omitted

the L, the third is M, the fourth is N, the fifth is O, the sixth is P, and the seventh is referred to as the Q shell. There are seven shells available for electrons (K, L, M, N, O, P, and Q) around the nucleus, and each of these seven shells can only hold a certain number of electrons, as shown in Figure 2-7. The outermost electron-occupied shell is referred to as the **valence shell or ring,** and these electrons are termed *valence electrons*. In the case of the copper atom, a single valence electron exists in the valence N shell.

Valence Shell or Ring
Outermost shell formed by electrons.

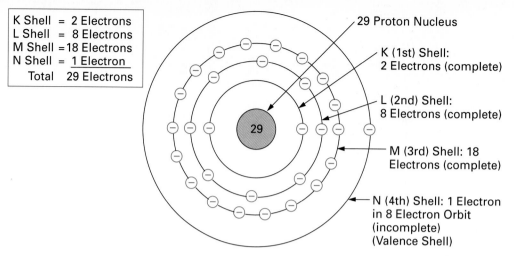

FIGURE 2-6   **Copper Atom.**

All matter exists in one of three states: solids, liquids, and gases. The atoms of a solid are fixed in relation to one another but vibrate in a back-and-forth motion, unlike liquid atoms, which can flow over each other. The atoms of a gas move rapidly in all directions and collide with one another. The far-right column of Table 2-1 indicates whether the element is a gas, a solid, or a liquid.

2-1-2   *Laws of Attraction and Repulsion*

For the sake of discussion and understanding, let us theoretically imagine that we are able to separate some positive and negative subatomic particles. Using these separated protons and electrons, let us carry out a few experiments, the results of which are illustrated in Figure 2-8. Studying Figure 2-8, you will notice that:

1. *Like charges* (positive and positive or negative and negative) repel one another.

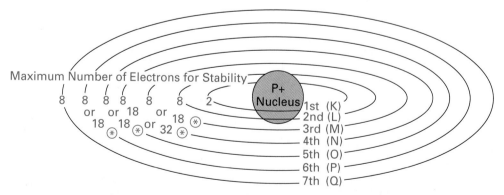

*The maximum number of electrons in these shells is dependent on the element's place ir the periodic table.

FIGURE 2-7   **Electrons and Shells.**

P = Proton (positive)
e = Electron (negative)

FIGURE 2-8 Attraction and Repulsion. (a) Positive Repels Positive. (b) Negative Repels Negative. (c) Unlike Charges Attract.

2. *Unlike charges* (positive and negative or negative and positive) attract one another.

Orbiting negative electrons are therefore attracted toward the positive nucleus, which leads us to the question of why the electrons do not fly into the atom's nucleus. The answer is that the orbiting electrons remain in their stable orbit due to two equal but opposite forces. The centrifugal outward force exerted on the electrons due to the orbit counteracts the attractive inward force trying to pull the electrons toward the nucleus due to the unlike charges.

Due to their distance from the nucleus, valence electrons are described as being loosely bound to the atom. The electrons can, therefore, easily be dislodged from their outer orbital shell by any external force, to become a **free electron.**

Free Electron
An electron that is not in any orbit around a nucleus.

2-1-3 The Molecule

An atom is the smallest unit of a natural element, or an element is a substance consisting of a large number of the same atom. Combinations of elements are known as **compounds,** and the smallest unit of a compound is called a **molecule,** just as the smallest unit of an element is an atom. Figure 2-9 summarizes how elements are made up of atoms and compounds are made up of molecules.

Water is an example of a liquid compound in which the molecule (H_2O) is a combination of an explosive gas (hydrogen) and a very vital gas (oxygen). Table salt is another example of a compound; here the molecule is made up of a highly poisonous gas atom (chlorine) and a potentially explosive solid atom (sodium). These examples of compounds each contain atoms that, when alone, are both poisonous and explosive, yet when combined the resulting substance is as ordinary and basic as water and salt.

Compound
A material composed of united separate elements.

Molecule
Smallest particle of a compound that still retains its chemical characteristics.

SELF-TEST REVIEW QUESTIONS FOR SECTION 2-1

1. Define the difference between an element and a compound.
2. Name the three subatomic particles that make up an atom.
3. What is the most commonly used metal in the field of electronics?
4. State the laws of attraction and repulsion.

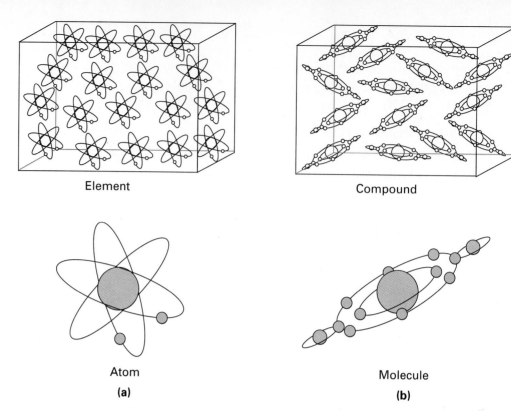

Element

Compound

Atom

Molecule

(a)

(b)

FIGURE 2-9 **(a) An Element Is Made Up of Many Atoms. (b) A Compound Is Made Up of Many Molecules.**

MINI-MATH REVIEW

This mini-math review is designed to remind you of the mathematical details relating to exponents, scientific notation, engineering notation, and prefixes, which will be used in the following section.

Exponents

Many of the values inserted into formulas contain numbers that have *exponents.* An exponent is a smaller number that appears to the right and slightly higher than another number; for example:

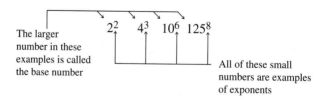

The larger number in these examples is called the base number

2^2 4^3 10^6 125^8

All of these small numbers are examples of exponents

Exponents allow us to save space, as you can see in this example: $5 \times 5 \times 5 \times 5 = 5^4$. Here the exponent (^{4}) is used to indicate how many times its base (5)

is used as a *factor*. You can also see from this example that the other advantage is that *exponents are a sort of math shorthand*. It is much easier to write 5^4 than to write $5 \times 5 \times 5 \times 5$. Using your calculator, you will also find that *exponents allow you to work out problems more quickly*. For example, inputting "⑤⊗⑤⊗⑤⊗⑤" is a lot more time consuming than is simply pressing "⑤Ⓨ④◻". The "y^x" key on your calculator allows you to enter in any base number (y) and any exponent (^{x}). No matter what method is used, however, the result in this example will always equal 625.

$$5 \times 5 \times 5 \times 5 = 5^4 = 625$$

As another example, what does 10^3 mean? It means that the base (10) is to be used as a factor three (^{3}) times; therefore,

$$10^3 = 10 \times 10 \times 10 = 1000$$

Scientific Notation

Many of the sciences, such as electronics, deal with numbers that contain a large number of zeros; for example: 14,000 and 0.000032. By using exponents we can eliminate the large number of zeros to obtain a shorthand version of the same number. This method is called *scientific notation* or *powers of 10*.

As an example, let us remove all of the zeros from the number 14,000 until we are left with simply 14. However, this number (14) is not equal to the original number (14,000) and therefore simply removing the zeros is not an accurate shorthand. Another number needs to be written along with the 14 to indicate what has been taken away—this is called a *multiplier*. The multiplier must indicate what you have to multiply 14 by to get back to 14,000; therefore,

$$14,000 = 14 \times 1000$$

As you know from our discussion on exponents, we can replace the 1000 with 10^3, since $1000 = 10 \times 10 \times 10$. Therefore, the scientific notation for 14,000 is 14×10^3. To convert 14×10^3 back to its original form (14,000), simply remember that each time a number is multiplied by 10, the decimal place is moved one position to the right. In this example 14 is multiplied by 10 three times (10^3 or $10 \times 10 \times 10$) and therefore the decimal point will have to be moved three positions to the right.

$$14 \times 10^3 = 14 \times 10 \times 10 \times 10 = 14.\widehat{0}.\widehat{0}.\widehat{0}. = 14,000$$

As another example, what is the scientific notation for the number 0.000032? If we once again remove all the zeros to obtain the number 32, we will again have to include a multiplier with 32 to indicate what 32 has to be multiplied by to return it to its original form. In this case 32 will have to be multiplied by $\frac{1}{1,000,000}$ (one millionth) to return it to 0.000032.

$$32 \times \frac{1}{1,000,000} = 0.000032$$

This can be verified because when you divide any number by 10, you move the decimal point one position to the left, and therefore to divide any number by 1,000,000, you simply move the decimal point six positions to the left.

$$32 \times \frac{1}{1,000,000} = \frac{32}{1,000,000} = \frac{32}{10 \times 10 \times 10 \times 10 \times 10 \times 10}$$

$$= 0.\overset{6\ 5\ 4\ 3\ 2\ 1}{0\,0\,0\,0\,3\,2}$$

Again, an exponent can be used in place of the $\frac{1}{1,000,000}$ multiplier. It will be

$$\frac{1}{1,000,000} = \frac{1}{10^6} = 0.000001 = 10^{-6}$$

Whenever you divide a number into 1, you get the *reciprocal* of that number. In this example, when you divide 1,000,000 into 1 you get 0.000001, which is equal to a negative exponent of 10^{-6}. This means that the multiplier 10^{-6} is indicating to move the decimal point back (to the left) by six places, and therefore

$$32 \times \frac{1}{1,000,000} = 32 \times 0.000001 = 32 \times 10^{-6} = 0.\overset{6\ 5\ 4\ 3\ 2\ 1}{0\,0\,0\,0\,3\,2}$$

Now that you know exactly what a multiplier means, you only have to remember these simple rules;

- A *negative exponent* tells you how many places *to the left* to move the decimal point.

- A *positive exponent* tells you how many places *to the right* to move the decimal point.

Remember one important point: a negative exponent does not indicate a negative number; it simply indicates a fraction. For example, "4×10^{-3} volts" means that 1 volt has been broken up into 1000 parts and we have 4 of those pieces, or $\frac{4}{1000}$.

To reinforce our understanding, let us try converting a few numbers to scientific notation.

$$230,000,000 = 230,000,\overset{7\ 6\ 5\ 4\ \ 3\ 2\ 1}{000} = 23 \times 10^7$$

$$760,000 = 76\overset{4\ \ 3\ 2\ 1}{0,000} = 76 \times 10^4$$

$$0.0019 = 0.\overset{1\ 2\ 3\ 4}{0019}. = 19 \times 10^{-4}$$

$$0.00085 = 0.\overset{1\ 2\ 3\ 4\ 5}{00085}. = 85 \times 10^{-5}$$

Scientific Notation
A floating-point system in which numbers are expressed as products consisting of a number between 1 and 10 multiplied by an appropriate power of 10.

Engineering Notation
A floating-point system in which numbers are expressed as products consisting of a number that is greater than 1 multiplied by an appropriate power of 10 that is some multiple of three.

Engineering Notation

To be a little more specific, **scientific notation** is expressed as a base number between 1 and 10 with a multiplier that is a power of 10, for example, 3.2×10^{-5}. **Engineering notation** however is expressed as a base number that is greater than 1 with a multiplier that is a power of ten with an exponent that is some multiple of three, for example, 32×10^{-6}.

Prefixes

Although any power of 10 can be used (10^1, 10^2, 10^{-5}, 10^{-2}), most electrical or electronic values use exponents that are multiples of three (10^3, 10^6, 10^9, 10^{12} or 10^{-3}, 10^{-6}, 10^{-9}, 10^{-12}). These multiples have been given names and symbols, as shown below.

VALUE	NAME	SYMBOL		
$10^{12} = 1,000,000,000,000$	tera	T	↑	Multipliers that are greater than 1
$10^{9} = 1,000,000,000$	giga	G		
$10^{6} = 1,000,000$	mega	M		
$10^{3} = 1,000$	kilo	k		
$10^{-3} = 1/1,000$	milli	m		Multipliers that are less than 1 (fractions)
$10^{-6} = 1/1,000,000$	micro	μ		
$10^{-9} = 1/1,000,000,000$	nano	n		
$10^{-12} = 1/1,000,000,000,000$	pico	p	↓	

These multipliers are often called *prefixes*, because they precede units. For example, 25 kilovolts or 25 kV means 25×10^{3} volts or 25,000 volts, or 50 milliamps (50 mA) means 50×10^{-3} amps or $^{50}/_{1000}$ths of one amp.

2-2 CURRENT

The movement of electrons from one point to another is known as *electrical current*. Energy in the form of heat or light can cause an outer shell electron to be released from the valence shell of an atom. Once an electron is released, the atom is no longer electrically neutral and is called a **positive ion,** as it now has a net positive charge (more protons than electrons). The released electron tends to jump into a nearby atom, which will then have more electrons than protons and is referred to as a **negative ion.**

Let us now take an example and see how electrons move from one point to another. Figure 2-10 illustrates a metal conductor connecting two charged objects. The metal conductor could be either gold, silver, or copper, but whichever it is, one common trait can be noted: The valence electrons in the outermost shell are very loosely bound and can easily be pulled from their parent atom.

The negative ions on the right in Figure 2-11 have more electrons than protons, while the positive ions on the left in Figure 2-11 have fewer electrons

Positive Ion
Atom that has lost one or more of its electrons and therefore has more protons than electrons, resulting in a net positive charge.

Negative Ion
Atom that has more than the normal neutral amount of electrons.

Positive Charge or Ions—
More Protons than Electrons
(absence of free electrons).

Negative Charge or Ions—
More Electrons than Protons
(abundance of free electrons).

FIGURE 2-10   Positive and Negative Charges.

than protons and so display a **positive charge.** The strip of metal joining the two charges has its own atoms, which begin in the neutral condition. Let us now concentrate on one of the negative ions. In Figure 2-11(a), the extra electrons in the outer shells of the negative ions on the right side will feel the attraction of the positive ions on the left side and the repulsion of the outer negative ions, or **negative charge.** This will cause an electron in a negative ion to jump away from its parent atom's orbit and land in an adjacent atom to the left within the metal wire conductor, as shown in Figure 2-11(b). This adjacent atom now has an extra electron and is called a negative ion, while the initial parent negative ion becomes a neutral atom, which will now receive an elec-

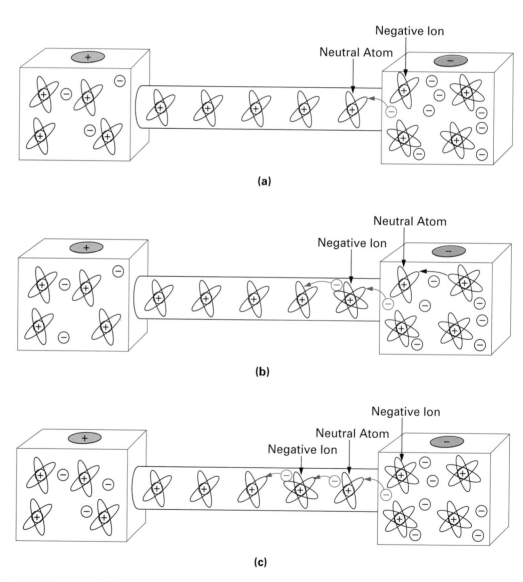

(a)

(b)

(c)

FIGURE 2-11   Electron Migration Due to Forces of Positive Attraction and Negative Repulsion on Electrons.

tron from one of the other negative ions, because their electrons are also feeling the attraction of the positive ions on the left side and the repulsion of the surrounding negative ions.

The electrons of the negative ion within the metal conductor feel the attraction of the positive ions, and eventually one of its electrons jumps to the left and into the adjacent atom, as shown in Figure 2-11(c). This continual movement to the left will produce a stream of electrons flowing from right to left. Millions upon millions of atoms within the conductor pass a continuous movement of billions upon billions of electrons from right to left. This electron flow is known as **electric current.**

To summarize, we could say that as long as a force or pressure, produced by the positive charge and negative charge, exists it will cause electrons to flow from the negative to the positive terminal. The positive side has a deficiency of electrons and the negative side has an abundance, and so a continuous flow or migration of electrons takes place between the negative and positive terminal through our metal conducting wire. This electric current or electron flow is a measurable quantity, as will now be explained.

2-2-1 *Coulombs per Second*

There are 6.24×10^{18} electrons in 1 **coulomb of charge,** as illustrated in Figure 2-12. To calculate coulombs of charge (designated Q), we can use the formula

$$\text{charge, } Q = \frac{\text{total number of electrons } (n)}{6.24 \times 10^{18}}$$

where Q is the electric charge in coulombs.

(a) 6.24×10^{18} Electrons
= 1 Coulomb of Charge

(b) 12.48×10^{18} Electrons
= 2 Coulombs of Charge

FIGURE 2-12   **(a) 1 C of Charge. (b) 2 C of Charge.**

If a total of 3.75×10^{19} free electrons exist within a piece of metal conductor, how many coulombs (C) of charge would be within this conductor?

Solution:

By using the charge formula (Q) we can calculate the number of coulombs (C) in the conductor.

$$Q = \frac{n}{6.24 \times 10^{18}}$$

$$= \frac{3.75 \times 10^{19}}{6.24 \times 10^{18}}$$

$$= 6\ C$$

CALCULATOR SEQUENCE

Step	Keypad Entry	Display Response
1.	3 . 7 5 E (Exponent) 1 9	3.75E19
2.	÷	
3.	6 . 2 4 E 1 8	6.24E18
4.	=	6.0096

A total of 6 C of charge exists within the conductor.

In the opposite calculator sequence you will see how the exponent key (E, EE, or EXP) on your calculator can be used.

EXAMPLE:

A metal wire conductor has 14 C of electrons passing through it. Calculate the number of electrons moving through the conductor.

Solution:

To calculate the number of electrons (n), we have to transpose our original formula. If

$$Q = \frac{n}{6.24 \times 10^{18}}$$

then

$$n = Q \times 6.24 \times 10^{18}$$

The total number of electrons (n) is

$$n = Q \times 6.24 \times 10^{18}$$

$$= 14\ C \times 6.24 \times 10^{18}$$

$$= 8.7 \times 10^{19} \text{ electrons will pass through the conductor}$$

Static
Crackling noise heard on radio receivers caused by electric storms or electric devices in the vicinity.

Ampere (A)
Unit of electric current.

2-2-2 The Ampere

A coulomb is a **static** amount of electric charge. In electronics, we are more interested in electrons in motion, and current is the flow of electrons, which involves time. Coulombs and time are therefore combined to describe the number of electrons and the rate at which they flow. This relationship is called *current* (I) flow and has the unit of **amperes (A).** If 6.24×10^{18} electrons (1 C) were to drift past a specific point on a conductor in 1 second of time, 1 ampere of current is said to be flowing:

$$\boxed{\text{current } (I) = \frac{\text{coulombs } (Q)}{\text{time } (t)}}$$

1 ampere = 1 coulomb per 1 second

$$1 \text{ A} = \frac{1 \text{ C}}{1 \text{ s}}$$

To summarize, we could say that 1 ampere equals a flow rate of 1 coulomb per second, and current is measured in amperes.

■ EXAMPLE:

If 5×10^{19} electrons pass a point in a conductor in 4 s, what is the amount of current flow in amperes?

■ *Solution:*

Current (I) is equal to Q/t. We must first convert electrons to coulombs.

$$Q = \frac{n}{6.24 \times 10^{18}}$$

$$= \frac{5 \times 10^{19}}{6.24 \times 10^{18}}$$

$$= 8 \text{ C}$$

Now, to calculate the amount of current, we use the formula

$$I = \frac{Q}{t}$$

$$= \frac{8 \text{ C}}{4 \text{ s}}$$

$$= 2 \text{ A}$$

This means that 2 A or 1.248×10^{19} electrons (2C) are passing a specific point in the conductor every second.

CALCULATOR SEQUENCE

Step	Keypad Entry	Display Response
1.	[5] [E] (exponent) [1] [9]	5E19
2.	[÷]	
3.	[6] [.] [2] [4] [E] [1] [8]	6.24E18
4.	[=]	8.012
5.	[÷]	
6.	[4]	2.003
7.	[=]	2.003

2-2-3 *Units of Current*

Current within electronic equipment is normally a value in milliamperes or microamperes and very rarely exceeds 1 ampere. Table 2-2 lists all the prefixes related to current. For example, 1 milliampere is one-thousandth of an ampere, which means that if 1 ampere were divided into 1000 parts, 1 part of the 1000 parts would be flowing through the circuit.

TABLE 2-2   Current Units

NAME	SYMBOL	VALUE
Picoampere	pA	$10^{-12} = \dfrac{1}{1,000,000,000,000}$
Nanoampere	nA	$10^{-9} = \dfrac{1}{1,000,000,000}$
Microampere	µA	$10^{-6} = \dfrac{1}{1,000,000}$
Milliampere	mA	$10^{-3} = \dfrac{1}{1000}$
Ampere	A	$10^{0} = 1$
Kiloampere	kA	$10^{3} = 1000$
Megaampere	MA	$10^{6} = 1,000,000$
Gigaampere	GA	$10^{9} = 1,000,000,000$
Teraampere	TA	$10^{12} = 1,000,000,000,000$

■ **EXAMPLE:**

Convert the following:

    a. 0.003 A = _____ mA (milliamperes)

    b. 0.07 mA = _____ µA (microamperes)

    c. 7333 mA = _____ A (amperes)

    d. 1275 µA = _____ mA (milliamperes)

■ *Solution:*

    a. 0.003A = _____ mA. In this example, 0.003 A has to be converted so that it is represented in milliamperes (10^{-3} or ¹⁄₁₀₀₀ of an ampere). The basic algebraic rule to be remembered is that both expressions on either side of the equals must be the same.

LEFT		RIGHT	
Base	Exponent	Base	Exponent
0.003×10^{0} =		_____ $\times 10^{-3}$	

The exponent on the right in this example is going to be decreased 1000 times (10^{0} to 10^{-3}), so for the statement to balance the number on the right will have to be increased 1000 times; that is, the decimal point will have to be moved to the right three places (0.003 or 3). Therefore,

$$0.003 \times 10^{0} = 3 \times 10^{-3}$$

or

$$0.003 \text{ A} = 3 \times 10^{-3} \text{ A or 3 mA}$$

b. 0.07 mA = _____ µA. In this example the exponent is going from milliamperes to microamperes (10^{-3} to 10^{-6}) or 1000 times smaller, so the number must be made 1000 times greater.

$$0.070 \text{ or } 70.0$$

Therefore, 0.07 mA = 70 µA.

c. 7333 mA = _____ A. The exponent is going from milliamperes to amperes, increasing 1000 times, so the number must decrease 1000 times.

$$7333. \text{ or } 7.333$$

Therefore, 7333 mA = 7.333 A.

d. 1275 µA = _____ mA. The exponent is changing from microamperes to milliamperes, an increase of 1000 times, so the number must decrease by the same factor.

$$1275.0 \text{ or } 1.275$$

Therefore, 1275 µA = 1.275 mA.

2-2-4 The Speed of Current Flow

Electrons will in fact move very slowly as they jump from parent to adjacent atom; however, the chain reaction occurs at the speed of light, which is 186,000 miles per second or 300,000,000 meters (m) per second (3×10^8 m/s). This chain reaction is best understood by using an analogy where we relate free electrons within a conductor to a string of Ping-Pong balls within a tube.

If one extra ball is inserted in one end, a ball will appear out of the other end almost instantly. Although each ball within the tube has only moved a small distance, the effect has traveled toward the end of the tube at the speed of light, as seen in Figure 2-13(a).

As far as electrons go, the first electron jump from parent to adjacent atom causes the next to jump, and the next, and next, and so on, as shown in Figure 2-13(b). Just as the first ball causes a chain reaction with the others, and even though the actual speed of the electrons is only a fraction of an inch per second, the effective velocity is at the speed of light.

2-2-5 Conventional versus Electron Flow

Electrons drift from a negative to a positive charge, as illustrated in Figure 2-14. As already discussed, this current is known as **electron flow.**

In the eighteenth and nineteenth centuries, however, when very little was known about the atom, researchers believed wrongly that current was a flow of positive charges. Although this has now been proved incorrect, many texts still use **conventional current flow,** shown in Figure 2–15.

Whether conventional current flow or electron flow is used, the same answers to problems, measurements, or designs are obtained. The key point to remember is that direction is not important, but the amount of current flow is.

Electron Flow
A current produced by the movement of free electrons toward a positive terminal.

Conventional Current Flow
A current produced by the movement of positive charges toward a negative terminal.

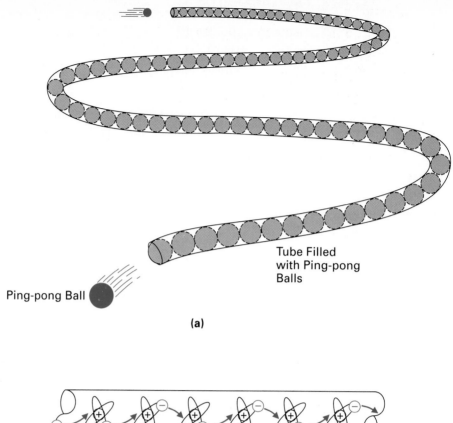

Ping-pong Ball

Tube Filled
with Ping-pong
Balls

(a)

(b)

FIGURE 2-13   Chain Reaction. (a) Ping-Pong Ball Analogy. (b) Electrons.

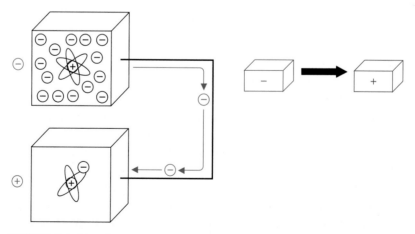

FIGURE 2-14   Electron Current Flow.

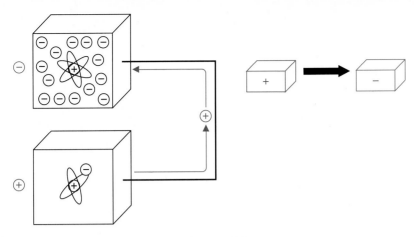

FIGURE 2-15   Conventional Current Flow.

Throughout this book we will be using electron flow so that we can relate back to the atom when necessary; if you wish to use conventional flow, just reverse the direction of the arrows. But to avoid confusion, be consistent with your choice of flow.

2-2-6   How Is Current Measured?

Ammeters (ampere meters) are used to measure the current flow within a circuit, as shown in Figure 2-16. In electronics, where current is generally small, the milliampere (mA) or microampere (µA) scale will generally be used, whereas in electrical high-current circuits, the meter will generally be used on an ampere scale.

Ammeter
Meter placed in the path of current flow to measure the amount.

Three rules must be remembered and applied when measuring current:

1. Always set the selector to the higher scale first (amperes) and then reduce as needed to milli- or microamperes, just in case a larger current than anticipated is within the circuit, in which case the meter could be damaged.

To explain this, let us take an example. If you were to weigh some potatoes on a set of scales and you were not aware of how much they weigh, the largest scale should initially be selected. For example, if you selected the 0 to 10 pound scale and the potatoes weighed 15 pounds, the meter needle would be forced violently to the right side and could damage the scale (this is known as *pegging the meter*). By always selecting the 0 to 50 pound scale and then stepping down to 0 to 40, 0 to 30, and 0 to 20 pounds, as needed, you would almost guarantee never pegging the meter. With the ammeter, we must apply the same philosophy and select a higher scale, then work down if needed.

2. The current meter must be connected so that the positive lead of the ammeter (red) is connected to the positive charge or polarity and the negative lead of the ammeter (black) is connected to the negative charge or

FIGURE 2-16   Measuring Current with the Ammeter.

polarity. Ammeters are sensitive to the polarity (positive or negative) of the charge.

3. If you wish to measure current flowing within a wire, the wire must be opened so that the ammeter measuring the current will be placed in the path of current flow.

SELF-TEST REVIEW QUESTIONS FOR SECTION 2-2

1. What is the unit of current?

2. Define current in relation to coulombs and time.

3. What is the difference between conventional and electron current flow?

4. List the three rules that should be applied when using the ammeter to measure current.

2-3 VOLTAGE

Voltage is the force or pressure exerted on electrons. Referring to Figure 2-17(a) and (b), you will notice two situations. Figure 2-17(a) shows highly concentrated positive and negative charges or potentials connected to one another by a copper wire. In this situation, a large potential difference or voltage is being applied across the copper atom's electrons. This force or voltage causes a large amount of copper atom electrons to move from right to left. Figure 2-17(b) illustrates a low concentration of positive and negative potentials, so a small voltage or pres-

(a)

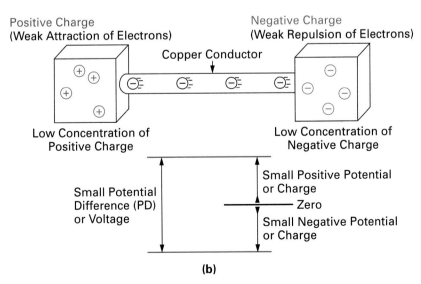

(b)

FIGURE 2-17   (a) Large Potential Difference or Voltage. (b) Small Potential Difference or Voltage.

sure is being applied across the conductor, causing a small amount of force, and therefore current, to move from right to left.

To summarize, then, we could say that a highly concentrated charge produces a high voltage, whereas a low concentrated charge produces a low voltage. Voltage is also appropriately known as the electron moving force or **electromotive force (emf),** and since two opposite potentials exist, one negative and one positive, the strength of the voltage can also be referred to as the amount of **potential difference (PD)** applied across the circuit. Referring back to Figure 2-17(a), we see that a large voltage, electromotive force, or potential difference exists across the copper conductor, while in Figure 2-17(b) a small voltage, potential difference, or electromotive force is exerted across the conductor.

Voltage is the force, pressure, potential difference (PD), or electromotive force (emf) that causes electron flow or current and is symbolized by italic uppercase *V*. The unit for voltage is the volt, symbolized by roman uppercase V. This can become a bit confusing. For example, when the voltage applied to a circuit equals 5 volts, the circuit notation would appear as

$$V = 5 \text{ V}$$

You know the first *V* represents "voltage," not "volt," because 1 volt cannot equal 5 volts. To avoid confusion, some texts and circuits use *E,* symbolizing electromotive force, to represent voltage; for example,

$$E = 5 \text{ V}$$

In this book, however, we maintain the original designation for voltage (*V*).

2-3-1 *Symbols*

A **battery,** illustrated in Figure 2-18, converts chemical energy into electrical energy. At the positive terminal of the battery, positive charges or ions (atoms with more protons than electrons) are present, and at the negative terminal, negative charges or ions (atoms with more electrons than protons) are available to supply electrons for current flow within a circuit. A battery, therefore, chemi-

Electromotive Force (emf)
Force that causes the motion of electrons due to a potential difference between two points.

Potential Difference (PD)
Voltage difference between two points, which will cause current to flow in a closed circuit.

Battery
Dc voltage source containing two or more cells that converts chemical energy into electrical energy.

FIGURE 2-18   Battery.

(a) Device   (b) Symbol

FIGURE 2-19   Batteries. (a) Pictorial. (b) Schematic Symbol.

cally generates negative and positive ions at its respective terminals. The symbol for the battery is illustrated in Figure 2-19.

In Figure 2-20(a) we have illustrated two other devices as they appear physically and in their electronic schematic symbols. The first new device is the light bulb, and the other is a piece of copper conductor wire with an alligator or crocodile clip on either end.

The 9-V battery shown in Figure 2-20(b) chemically generates positive and negative ions. The negative ions at the negative terminal force away the negative electrons, which are attracted by the positive charge or absence of electrons on the other terminal. Proceeding through the copper conductor wire, jumping

FIGURE 2-20   (a) Components. (b) Example Circuit.

from one atom to the next, they eventually reach the bulb. The electrons pass through the bulb, which glows due to this current passing through it. When emerging from the light bulb, the electrons travel through another connector cable and finally reach the positive terminal of the battery.

Studying Figure 2-20(b), you will notice two reasons why the circuit is drawn using symbols rather than illustrating the physical appearance:

1. A circuit with symbols may be drawn faster and more easily.

2. A circuit with symbols has less detail and clutter and is more easily comprehended because of having fewer distracting elements.

2-3-2 Units of Voltage

The unit for voltage is the volt (V). Voltage within electronic equipment is normally measured in volts, whereas heavy-duty industrial equipment normally requires high voltages that are generally measured in kilovolts (kV). Table 2-3 lists all the prefixes and values related to volts.

■ **EXAMPLE:**

Convert the following:

    a. 3000 V = _____ kV (kilovolts)

    b. 0.14 V = _____ mV (millivolts)

    c. 1500 kV = _____ MV (megavolts)

■ *Solution:*

    a. 3000 V = 3 kV or 3×10^3 volts (exponent ↑ 1000, number ↓ 1000)
    b. 0.14 V = 140 mV or 140×10-3 volt (exponent ↓ 1000, number ↑ 1000)
    c. 1500 kV = 1.5 MV or $1.5 \times 10$6 volts (exponent ↑ 1000, number ↓ 1000)

TABLE 2-3 Voltage Units

NAME	SYMBOL	VALUE
Picovolts	pV	$10^{-12} = \dfrac{1}{1,000,000,000,000}$
Nanovolts	nV	$10^{-9} = \dfrac{1}{1,000,000,000}$
Microvolts	μV	$10^{-6} = \dfrac{1}{1,000,000}$
Millivolts	mV	$10^{-3} = \dfrac{1}{1000}$
Volts	V	$10^0 = 1$
Kilovolts	kV	$10^3 = 1000$
Megavolts	MV	$10^6 = 1,000,000$
Gigavolts	GV	$10^9 = 1,000,000,000$
Teravolts	TV	$10^{12} = 1,000,000,000,000$

2-3-3   How Is Voltage Measured?

Voltmeters (voltage meters) are used to measure electrical pressure (voltage) anywhere in an electronic circuit. Figure 2-21 illustrates a voltmeter being used to measure a battery's voltage. A voltmeter is connected across a device, as illustrated in Figure 2-22(a) and (b), with the two leads of the voltmeter touching either side of the component, in this case a light bulb. Figure 2-22(a) indicates how to measure the voltage across light bulb 1 (L1), whereas Figure 2-22(b) indicates how to measure the voltage across light bulb 2 (L2).

Voltmeter
Instrument designed to measure the voltage or potential difference. Its scale can be graduated in kilovolts, volts, or millivolts.

Multirange or Scale Voltmeter

10 mV: This scale can be used to measure any voltage from 0 to 10 mV.

10 V: This scale can be used to measure any voltage from 0 to 10 V.

10 kV: This scale can be used to measure any voltage from 0 to 10 kV.

AVM = Analog Voltmeter
DVM = Digital Voltmeter

FIGURE 2-21   Using the Voltmeter to Measure Voltage. (a) Pictorial. (b) Schematic.

(a)

(b)

FIGURE 2-22   Measuring Components across (a) Lamp 1 and (b) Lamp 2.

Three techniques must be remembered and applied when measuring voltage:

1. To prevent meter damage, always set to the highest scale first (kV) and then reduce as needed to the volt or millivolt range, just in case a larger voltage than anticipated is within the circuit.

2. Always connect the positive lead (red) of the meter to the component lead that is closer to the positive side of the battery, and the negative lead (black) of the meter to the component lead that is closer to the negative side of the battery.

3. If you are measuring the voltage across a component, the voltmeter must be connected across the component and outside of the circuit.

FIGURE 2-23   Fixed-Range Voltmeter.

In some applications, fixed-range voltmeters are used: for example, a voltage indicator within a car showing the condition of the battery (Figure 2-23). In this application, a fixed voltmeter of 0 to 15 V would be sufficient.

2-3-4 *Fluid Analogy of Current and Voltage*

In Figure 2-24(a), a system using a pump, pipes, and a waterwheel is being used to convert electrical energy into mechanical energy. The electrical energy is in the form of voltage, which is applied to the pump to send it into operation and cause water to flow. The pump generates:

1. A high pressure at the outlet port, which forces the water molecules out and into the system
2. A low pressure in the inlet port, which attracts the water molecules into the pump

(a) Fluid System

(b) Electrical System

(a) Fluid System		(b) Electrical System
Pump generates pressure, which is the water moving force.	Equivalent to (≅)	Battery generates voltage, which is the electron moving force.
Water current flow.		Electron current flow.
High pressure or potential.		High voltage or potential.
Low pressure or potential.		Low voltage or potential.

FIGURE 2-24   Comparison between a Fluid System (a) and an Electrical System (b).

The water current flow is in the direction indicated, and the high pressure or potential within the piping will be used to drive the waterwheel around, producing mechanical energy. The remaining water is attracted into the pump due to the suction or low pressure existing on the inlet port. In fact, the amount of water entering the inlet port is the same as the amount of water leaving the outlet port. It can therefore be said that the water flow rate is the same throughout the circuit. The only changing element is the pressure felt at different points throughout the system.

In Figure 2-24(b), an electric circuit containing a battery, conductors, and a bulb is being used to convert electrical energy into light energy. The battery generates a voltage just as the pump generates pressure. This voltage causes electrons to move through conductors, just as pressure causes water molecules to move through the piping. The amount of water flow is dependent on the pump's pressure, and the amount of current or electron flow is dependent on the battery's voltage. Water flow through the wheel can be compared to current flow through the bulb. The high pressure is lost in turning the wheel and producing mechanical energy, just as voltage is lost in producing light energy out of the bulb. We cannot say that pressure or voltage flows: Pressure and voltage exist and cause water and current to flow, and it is this flow that is converted to mechanical energy in our fluid system and light energy in our electrical system. Voltage is the force of repulsion and attraction needed to cause current to flow through a circuit, and without this potential difference or pressure there cannot be current.

2-3-5   *Current Is Directly Proportional to Voltage*

Referring to the fluid system and then the electrical circuit in Figure 2-24, you can easily see that the flow is proportional to the pressure, which means: If the pump were to generate a greater pressure, a larger amount of water would flow through the system.

pressure ↑   water flow ↑

Similarly, if a larger voltage were applied to the electrical circuit, this larger electron moving force (emf) would cause more electrons or current to flow through the circuit, as shown in Figure 2-25.

Current (I) is therefore said to be *directly proportional (∝) to voltage (V),* as a voltage decrease ($V \downarrow$) results in a subsequent current decrease ($I \downarrow$), and similarly, a voltage increase ($V \uparrow$) causes a current increase ($I \uparrow$).

Current is directly proportional to voltage ($I \propto V$)

$V \downarrow$ causes a $I \downarrow$
$V \uparrow$ causes a $I \uparrow$

FIGURE 2-25 **(a) Small Voltage Produces a Small Current. (b) Large Voltage Produces a Large Current.**

As you can see from this example, directly **proportional** (∝) is a phrase which means that one term will change in proportion, or in size, relative to another term.

Proportional
Having the same or constant ratio.

2-3-6 *Summary of Measuring Current and Voltage*

CURRENT (AMMETER)   *VOLTAGE (VOLTMETER)*

1. Always set the meter to the highest scale first.

2. Ensure that the positive lead of the meter is connected closer to the positive side of the voltage supply (for example, battery), and the negative lead of the meter is connected closer to the negative side of the supply voltage.

3. Always connect the ammeter in the path of current flow so that the electrons have to pass through the meter.

3. Always connect the voltmeter across the component so that the pressure or potential difference change across the component can be measured.

Figure 2-26 summarizes these measurements.

FIGURE 2-26 **Measurement of (a) Current and (b) Voltage.**

1. What is the unit of voltage?
2. Convert 3 MV to kilovolts.
3. List the three rules that should be applied when using the voltmeter to measure voltage.
4. What is the relationship between current and voltage?

2-4 CONDUCTORS

A lightning bolt that splits or sets fire to a tree and the operation of your calculator are both electrical results achieved by the flow of electrons. The only difference is that your calculator's circuits control the flow of electrons, unlike the lightning bolt, which is the uncontrolled flow of electrons.

Materials that pass current easily, that is, offer little opposition (low resistance) to current, are called **conductors.** As mentioned previously, the atom has a maximum of seven orbital paths known as shells, which are named K, L, M, N, O, P, and Q, stepping out toward the outermost or valence shell.

Conductors
Length of wire whose properties are such that it will carry an electric current.

Conductors are materials or natural elements whose valence electrons can easily be removed from their parent atoms. They are therefore said to be sources of free electrons, and these free electrons provide us with circuit current. The precious metals of silver and gold are the best conductors. More specifically, a better conductor has:

1. Electrons in shells farthest away from the nucleus; these electrons feel very little nucleus attraction and can be broken away from their atom quite easily.

2. More electrons per atom.

3. An incomplete valence shell. This means that the valence shell does not have in it the maximum possible number of electrons. If the atom had its valence ring complete (full), there would be no holes (absence of an electron) in that shell, so no encouragement for adjacent atom electrons to jump from their parent atom into the next atom would exist, preventing the chain reaction known as current.

Economy must be considered when choosing a conductor. Large quantities of conductors using precious metals are obviously going to send the cost of equipment beyond reach. The conductor must also satisfy some physical requirements, in that we must be able to shape it into wires of different sizes and easily bend it to allow us to connect one circuit to the next.

Copper is the most commonly used conductor, as it meets the following three requirements:

1. It is a good source of electrons.

2. It is inexpensive.

3. It is physically pliable.

Aluminum is also a very popular conductor, and although it does not possess as many free electrons as copper, it has the two advantages of being less expensive and lighter than copper.

2-4-1 Conductance

Conductance is the measure of how good a conductor is at carrying current. Conductance (symbolized G) is equal to the reciprocal of resistance and is measured in the unit siemens (S):

$$\text{conductance } (G) = \frac{1}{\text{resistance } (R)}$$

Conductance (G)
Measure of how well a circuit or path conducts or passes current. It is equal to the reciprocal of resistance.

This means that conductance is inversely proportional to resistance. For example, if the opposition to current flow (resistance) is low, the conductance is high and the material is said to have a good conductance.

$$\text{high conductance } G \uparrow = \frac{1}{R \downarrow \text{ (low resistance)}}$$

Conductance (G) values are given in siemens (S) and resistance (R) in ohms (Ω).

On the other hand, if the resistance of a conducting wire is high ($R \uparrow$), its conductance value is low ($G \downarrow$) and it is called a poor conductor. To summarize, we say that a good conductor has a high conductance value and a very small opposition or resistance to current flow.

EXAMPLE:

A household electric blanket offers 25 ohms of opposition or resistance against current flow. Calculate the conductance of the electric blanket's heating element.

Solution:

$$\text{conductance} = \frac{1}{\text{resistance}}$$

$$G = \frac{1}{R}$$

$$= \frac{1}{25 \text{ ohms}}$$

$$= 40 \text{ mS (millisiemens)}$$

EXAMPLE:

To further reinforce our understanding of a conductor, let us compare a good conductor (copper) to a poor conductor (carbon) as an example. Refer also to carbon and copper in the atomic periodic table (Table 2-1).

TABLE 2-4 Relative Conductivity of Conductors

MATERIAL	CONDUCTIVITY (RELATIVE)
Silver	1.000
Copper	0.945
Gold	0.652
Aluminum	0.575
Tungsten	0.297
Nichrome	0.015

■ *Solution:*

Carbon

1. The valence ring is the L or second shell, which feels a strong nucleus attractive force, discouraging the release of free electrons and thus of circuit current.
2. Only six electrons orbit the carbon atom, making it a poor source of electrons.
3. The valence shell (L, maximum eight electrons) is already half complete (four electrons), so few holes (only four) exist, discouraging the jumping of electrons from parent atoms into adjacent atom holes.

Copper

1. The valence ring is the N or fourth shell, which feels only a weak nucleus attractive force, encouraging the release of free electrons and thus of circuit current.
2. A total of 29 electrons orbit the nucleus, making it a good source of electrons.
3. The valence shell (N, maximum 32 electrons) is incomplete, as only one electron occupies it when neutral, so 31 holes exist, encouraging the jumping of electrons from parent atoms into adjacent atom holes.

To express how good or poor a conductor is, we must specify a reference point. The reference point we use is the best conductor, silver, which has a conductivity value of 1.0. Table 2-4 lists other conductors and their relative conductivity values with respect to the best, silver. The formula for calculating relative conductivity is

$$\text{Relative Conductivity} = \frac{\text{Conductor's Relative Conductivity}}{\text{Reference Conductor's Relative Conductivity}}$$

■ **EXAMPLE:**

What is the relative conductivity of tungsten if copper is used as the reference conductor?

■ *Solution:*

$$tungsten = 0.297$$
$$copper = 0.945$$

$$relative\ conductivity = \frac{conductivity\ of\ conductor}{conductivity\ of\ reference}$$

$$= \frac{0.297}{0.945}$$

$$= 0.314$$

CALCULATOR SEQUENCE

Step	Keypad Entry	Display Response
1.	⓪ . ② ⑨ ⑦	0.297
2.	÷	
3.	⓪ . ⑨ ④ ⑤	0.945
4.	=	0.314

■ **EXAMPLE:**

What is the relative conductivity of silver if copper is used as the reference?

■ *Solution:*

$$relative\ conductivity = \frac{silver}{copper}$$

$$= \frac{1.000}{0.945} = 1.058$$

SELF-TEST REVIEW QUESTIONS FOR SECTION 2-4

1. True or false: A conductor is a material used to block the flow of current.
2. List the three atomic properties that make a better conductor.
3. What is the most commonly used conductor in the field of electronics?
4. Calculate the conductance of a 35 ohm heater element.

2-5   INSULATORS

Materials that are used to block current, that is, offer high resistance or opposition to current flow, are called **insulators.** Just as certain materials permit the easy flow of current and so have good conductivity, certain materials allow small to almost no amount of free electrons to flow. These materials are known as insulators. Insulators can, with sufficient pressure or voltage applied across them, "break down" and conduct current; that is, the voltage, known as the **breakdown voltage,** must be great enough to dislodge the electrons from their close orbital shells (K, L shells) and send them off as free electrons.

Insulators
Materials that have few electrons per atom and those electrons are close to the nucleus and cannot be easily removed.

Breakdown Voltage
The voltage at which breakdown of an insulator occurs.

A good insulator or dielectric should have the maximum possible resistance and conduct no current at all. To express how good or poor an insulator is, we list a voltage that when applied across one-thousandth of an inch of this insulator material, will cause it to break down and conduct a large current. This measure of an insulator is known as its **dielectric strength.** Table 2-5 lists some of the more popular insulators and the value of kilovolts that will cause a centimeter of insulator to break down. From Table 2-5 we say that if 1 centimeter of paper is connected to a variable voltage source, a voltage of 500 kilovolts is needed to break down the paper and cause current to flow.

Dielectric Strength
The maximum potential a material can withstand without rupture.

To calculate the dielectric thickness needed therefore to withstand a certain voltage, we can use the following formula.

$$\text{dielectric thickness} = \frac{\text{voltage to insulate}}{\text{insulator's breakdown voltage}}$$

■ **EXAMPLE:**

What thickness of mica would be needed to withstand 16,000 V?

■ *Solution:*

CALCULATOR SEQUENCE

Step	Keypad Entry	Display Response
1.	①⑥Ⓔ(exponent)③	16E3
2.	÷	
3.	②Ⓔ⑥ (1000 ÷ K = E6)	2E6
4.	=	0.008

mica strength = 2000 kV/cm

$$\text{dielectric thickness} = \frac{16{,}000 \text{ V}}{2000 \text{ kV}} = 0.008 \text{ cm}$$

TABLE 2-5   Breakdown Voltages of Certain Insulators

MATERIAL	BREAKDOWN STRENGTH (kV/cm)
Mica	2000
Glass	900
Teflon	600
Paper	500
Rubber	275
Bakelite	151
Oil	145
Porcelain	70
Air	30

■ **EXAMPLE:**

What maximum voltage could 1 mm of air withstand?

■ *Solution:*

There are 10 mm in 1 cm. If air can withstand 30,000 V/cm, it can withstand 3000 V/mm.

SELF-TEST REVIEW QUESTIONS FOR SECTION 2-5

1. True or false: An insulator is a material used to block the flow of current.

2. What is considered to be the best insulator material?

3. Define *breakdown voltage.*

4. Would the conductance figure of a good insulator be large or small?

2-6 SEMICONDUCTORS

Certain materials are neither insulators (high resistance to current) nor conductors (low resistance to current), but in fact fall between the two, as seen in Figure 2-27. These materials are known as **semiconductors** because they conduct less than a metal conductor, but more than an insulator. Silicon is the most commonly used semiconductor for components such as transistors and integrated

Semiconductors
Conductor with a resistivity figure between that of an insulator and a metal.

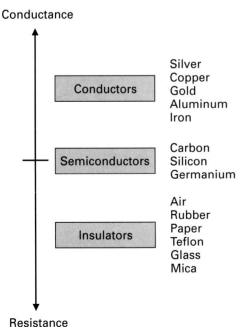

FIGURE 2-27 Elements and Compounds Used in Electronics.

FIGURE 2-28   An Open Switch Causing an Open Circuit. (a) Pictorial. (b) Schematic.

circuits. Semiconductor materials, devices, and circuits are covered in more detail in chapters 19 through 29.

2-7   THE OPEN, CLOSED, AND SHORT CIRCUIT

A simple circuit using a battery, conductors, light bulb, and a switch is shown in Figure 2-28. A new component, the switch, is illustrated in Figure 2-28(a), and the schematic symbol is shown in Figure 2-28(b).

2-7-1   *Open Circuit (Open Switch)*

An opened switch (Figure 2-28) produces an "open" in the circuit, which prevents current flow as the extra electrons in the negative terminal of the battery cannot feel the attraction of the positive terminal due to the break in the path. The opened switch has produced an **open circuit**.

Open Circuit
Break in the path of
current flow.

FIGURE 2-29   An Open Lamp Filament Causing an Open Circuit. (a) Pictorial. (b) Schematic.

2-7-2 Open Circuit (Open Component)

If the switch is now closed so as to make a complete path in the circuit, an open could still exist due to the failure of one of the components. For example, in Figure 2-29, the light bulb filament has burned out; in so doing, it creates an open in the circuit and again there will be no current flow. An open component therefore can also produce an open circuit.

2-7-3 Closed Circuit (Closed Switch)

A closed switch produces a **closed circuit** since current now has a complete path from the negative to positive terminal, as seen in Figure 2-30. The closed switch produces a closed circuit.

Closed Circuit
Circuit having a complete path for current to flow.

2-7-4 Short Circuit (Shorted Component)

A **short circuit** normally occurs when one point is accidentally connected to another. Figure 2-31(a) illustrates the physical appearance of the circuit and how the accident occurred. A set of pliers was accidentally laid across the two contacts connecting the light bulb. Figure 2-31(b) shows the schematic illustration with the effect of the short across the light bulb drawn in. All the current will flow through the metal of the pliers, which offers no resistance or opposition to the current flow; since the light bulb does have some resistance or opposition, almost no current will flow through it, and therefore no light will be produced.

Short Circuit
Also called a short; it is a low-resistance connection between two points in a circuit, typically causing a large amount of current flow.

SELF-TEST REVIEW QUESTIONS FOR SECTIONS 2-6 AND 2-7

1. What is a semiconductor?
2. What is the name of the most commonly used semiconductor material?
3. Does current flow through an open circuit?
4. Describe the difference between a closed circuit and a short circuit.

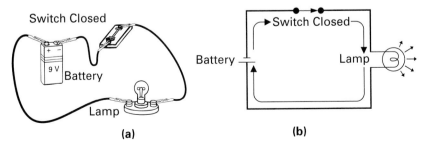

(a) **(b)**

FIGURE 2-30   **Closed Switch Causing a Closed Circuit. (a) Pictorial. (b) Schematic.**

Switch Closed

9 V

Battery

Short Caused by Pliers

(a)

(b)

FIGURE 2-31   Pliers Causing a Short Circuit across a Lamp. (a) Pictorial. (b) Schematic.

SUMMARY

The Structure of Matter

1. All of the matter on the earth and in the air surrounding the earth can be classified as being either a solid, liquid, or gas. A total of approximately 107 different natural elements exist in, on, and around the earth.

2. An *element*, by definition, is a substance consisting of only one type of atom; in other words, every element has its own distinctive atom, which makes it different from all the other elements.

3. The *atom* is the smallest particle into which an element can be divided without losing its identity, and a group of identical atoms is called an element.

4. The word atom is a Greek word meaning a particle that is too small to be subdivided.

5. In 1913, a Danish physicist, Neils Bohr, put forward a theory about the atom, and his basic model outlining the *subatomic* particles that make up the atom is still in use today.

6. The three important particles of the atom are the *proton*, which has a positive charge, the *neutron*, which is neutral or has no charge, and the *electron*, which has a negative charge.

7. The atom consists of a positively charged central mass called the *nucleus*, which is made up of protons and neutrons surrounded by a quantity of negatively charged orbiting electrons.

8. The *atomic number* of an atom describes the number of protons that exist within the nucleus.

9. The proton and the neutron are almost 2000 times heavier than the very small electron.

10. A hydrogen atom is the smallest of all atoms and has an atomic number of 1, which means that hydrogen has a one-proton nucleus.

11. Helium is second on the table and has an atomic number of 2, indicating that two protons are within the nucleus. The *atomic weight* of helium, however, is 4, meaning that two protons and two neutrons make up the atom's nucleus.

12. The number of neutrons within an atom's nucleus can subsequently be calculated by subtracting the atomic number (protons) from the atomic weight (protons and neutrons).

13. A *neutral atom* or *balanced atom* is one that has an equal number of protons and orbiting electrons, so the net positive proton charge is equal but opposite to the net negative electron charge, resulting in a balanced or neutral state.

14. Orbiting electrons travel around the nucleus at varying distances from the nucleus, and these orbital paths are known as *shells or bands.* The orbital shell nearest the nucleus is referred to as the first or K shell. The second is known as the L, the third is M, the fourth is N, the fifth is O, the sixth is P, and the seventh is referred to as the Q shell. There are seven shells available for electrons (K, L, M, N, O, P, and Q) around the nucleus, and each of these seven shells can only hold a certain number of electrons. The outermost electron-occupied shell is referred to as the *valence shell or ring,* and these electrons are termed *valence electrons.*

15. All matter exists in one of three states: solids, liquids, and gases. The atoms of a solid are fixed in relation to one another but vibrate in a back-and-forth motion, unlike liquid atoms, which can flow over each other. The atoms of a gas move rapidly in all directions and collide with one another.

Laws of Attraction and Repulsion

16. *Like charges* (positive and positive or negative and negative) repel one another. *Unlike charges* (positive and negative or negative and positive) attract one another. Orbiting negative electrons are therefore attracted toward the positive nucleus, which leads us to the question of why the electrons do not fly into the atom's nucleus. The answer is that the orbiting electrons remain in their stable orbit due to equal but opposite forces. The centrifugal outward force exerted on the electrons due to the orbit counteracts the attractive inward force trying to pull the electrons toward the nucleus due to the unlike charges.

17. Due to their distance from the nucleus, valence electrons are described as being loosely bound to the atom. The electrons can, therefore, easily be dislodged from their outer orbital shell by any external force, to become a free electron.

The Molecule

18. An atom is the smallest unit of a natural element, or an element is a substance consisting of a large number of the same atom.

19. Combinations of elements are known as *compounds,* and the smallest unit of a compound is called a *molecule,* just as the smallest unit of an element is an atom.

20. Water is an example of a liquid compound in which the molecule (H_2O) is a combination of an explosive gas (hydrogen) and a very vital gas (oxygen). Table salt is another example of a compound; here the molecule is made up of a highly poisonous gas atom (chlorine) and a potentially explosive solid atom (sodium).

Exponents

21. Many of the values inserted into formulas contain numbers that have *exponents*. An exponent is a smaller number that appears to the right and slightly higher than another number.

22. *Exponents allow us to save space,* as you can see in this example:

$$5 \times 5 \times 5 \times 5 = 5^4$$

Here the exponent (^{4}) is used to indicate how many times its base (5) is used as a *factor*. You can also see from this example that the other advantage is that *exponents are a sort of math shorthand*. It is much easier to write 5^4 than to write $5 \times 5 \times 5 \times 5$. The "$y^x$" key on your calculator allows you to enter in any base number (y) and any exponent (^{x}).

Scientific Notation

23. Many of the sciences, such as electronics, deal with numbers that contain a large number of zeros. By using exponents we can eliminate the large number of zeros to obtain a shorthand version of the same number. This method is called *scientific notation* or *powers of 10*.

24. A *negative exponent* tells you how many places *to the left* to move the decimal point. A *positive exponent* tells you how many places *to the right* to move the decimal point. Remember one important point: that a negative exponent does not indicate a negative number; it simply indicates a fraction.

Engineering Notation

25. *Scientific notation* is expressed as a base number between 1 and 10 with an exponent that is a power of 10. *Engineering notation,* however, is expressed as a base number that is greater than 1 with an exponent that is some multiple of three.

Prefixes

26. Although any power of 10 can be used (10^1, 10^2, 10^{-5}, 10^{-2}), most electrical or electronic values use exponents that are multiples of three (10^3, 10^6, 10^9, 10^{12}, or 10^{-3}, 10^{-6}, 10^{-9}, 10^{-12}). These multiples have been given names and symbols, as shown below.

VALUE	NAME	SYMBOL	
$10^{12} = 1,000,000,000,000$	tera	T	Multipliers that are greater than 1
$10^{9} = 1,000,000,000$	giga	G	
$10^{6} = 1,000,000$	mega	M	
$10^{3} = 1,000$	kilo	k	
$10^{-3} = 1/1,000$	milli	m	Multipliers that are less than 1 (fractions)
$10^{-6} = 1/1,000,000$	micro	μ	
$10^{-9} = 1/1,000,000,000$	nano	n	
$10^{-12} = 1/1,000,000,000,000$	pico	p	

These multipliers are often called *prefixes,* because they precede units.

Current (Figure 2-32)

27. The movement of electrons from one point to another is known as *electrical current.*

28. Energy in the form of heat or light can cause an outer shell electron to be released from the valence shell of an atom. Once an electron is released, the atom is no longer electrically neutral and is called a *positive ion,* as it now has a net positive charge (more protons than electrons). The released electron tends to jump into a nearby atom, which will then have more electrons than protons and is referred to as a *negative ion.*

29. Millions upon millions of atoms within the conductor pass a continuous movement of billions upon billions of electrons from right to left. This electron flow is known as *electric current.*

30. As long as a force or pressure, produced by the positive charge and negative charge, exists it will cause electrons to flow from the negative to the positive terminal. The positive side has a deficiency of electrons and the negative side has an abundance, and so a continuous flow or migration of electrons takes place between the negative and positive terminal through our metal conducting wire. This electric current or electron flow is a measurable quantity.

31. There are 6.24×10^{18} electrons in 1 *coulomb of charge.*

32. A coulomb is a *static* amount of electric charge. In electronics, we are more interested in electrons in motion, and current is the flow of electrons, which involves time. Coulombs and time are therefore combined to describe the number of electrons and the rate at which they flow. This relationship is called *current (I)* flow and has the unit of *amperes* (A). If 6.24×10^{18} electrons (1 C) were to drift past a specific point on a conductor in 1 second of time, 1 ampere of current is said to be flowing.

33. Current within electronic equipment is normally a value in milliamperes or microamperes and very rarely exceeds 1 ampere.

FIGURE 2-32   Current.

Electric Charge (Q) in coulombs (C) = $\dfrac{\text{total number of electrons } (n)}{6.24 \times 10^{18}}$

Electric Current (I) in amperes (A) = $\dfrac{\text{coulombs } (Q)}{\text{time } (t)}$

TABLE 2-2   Current Units

NAME	SYMBOL	VALUE
Picoampere	pA	$10^{-12} = \dfrac{1}{1,000,000,000,000}$
Nanoampere	nA	$10^{-9} = \dfrac{1}{1,000,000,000}$
Microampere	μA	$10^{-6} = \dfrac{1}{1,000,000}$
Milliampere	mA	$10^{-3} = \dfrac{1}{1000}$
Ampere	A	$10^{0} = 1$
Kiloampere	kA	$10^{3} = 1000$
Megaampere	MA	$10^{6} = 1,000,000$
Gigaampere	GA	$10^{9} = 1,000,000,000$
Teraampere	TA	$10^{12} = 1,000,000,000,000$

Multirange or Scale Ammeters

10 μA: On this scale the meter will indicate current values between 0 and 10 μA.

10 mA: On this scale the meter will indicate current values between 0 and 10 mA.

10 A: On this scale the meter will indicate current values between 0 and 10 A.

34. Electrons will in fact move very slowly as they jump from parent to adjacent atom; however, the chain reaction occurs at the speed of light, which is 186,000 miles per second or 300,000,000 meters (m) per second (3×10^8 m/s).

35. Electrons drift from a negative to a positive charge. This current is known as *electron flow*. In the eighteenth and nineteenth centuries, however, when very little was known about the atom, researchers believed wrongly that current was a flow of positive charges. Although this has now been proved incorrect many texts still use *conventional current flow*. Whether conventional current flow or electron flow is used, the same answers to problems, measurements, or designs are obtained. The key point to remember is that direction is not important, but the amount of current flow is.

36. *Ammeters* (ampere meters) are used to measure the current flow within a circuit. In electronics, where current is generally small, the milliampere (mA) or microampere (μA) scale will generally be used, whereas in electrical high-current circuits, the meter will generally be used on an ampere scale.

37. Three rules must be remembered and applied when measuring current:
  a. Always set the selector to the higher scale first (amperes) and then reduce as needed to milli- or microamperes, just in case a larger current than anticipated is within the circuit, in which case the meter could be damaged.
  b. The current meter must be connected so that the positive lead of the ammeter (red) is connected to the positive charge or polarity and the negative lead of the ammeter (black) is connected to the negative charge or polarity. Ammeters are sensitive to the polarity (positive or negative) of the charge.
  c. If you wish to measure current flowing within a wire, the wire must be opened so that the ammeter measuring the current will be placed in the path of current flow.

Voltage (Figure 2-33)

38. *Voltage* is the force or pressure exerted on electrons.

39. A highly concentrated charge produces a high voltage, whereas a low concentrated charge produces a low voltage. Voltage is also appropriately known as the electron moving force or *electromotive force (emf),* and since two opposite potentials exist, one negative and one positive, the strength of the voltage can also be referred to as the amount of potential difference (PD) applied across the circuit.

40. Voltage is the force, pressure, *potential difference (PD),* or electromotive force (emf) that causes electron flow or current and is symbolized by italic uppercase *V.* The unit for voltage is the volt, symbolized by roman uppercase V.

41. A *battery* converts chemical energy into electrical energy. At the positive terminal of the battery, positive charges or ions (atoms with more protons than electrons) are present, and at the negative terminal, negative charges

FIGURE 2-33   Voltage.

TABLE 2-3   Voltage Units

NAME	SYMBOL	VALUE
Picovolts	pV	$10^{-12} = \dfrac{1}{1,000,000,000,000}$
Nanovolts	nV	$10^{-9} = \dfrac{1}{1,000,000,000}$
Microvolts	μV	$10^{-6} = \dfrac{1}{1,000,000}$
Millivolts	mV	$10^{-3} = \dfrac{1}{1000}$
Volts	V	$10^{0} = 1$
Kilovolts	kV	$10^{3} = 1000$
Megavolts	MV	$10^{6} = 1,000,000$
Gigavolts	GV	$10^{9} = 1,000,000,000$
Teravolts	TV	$10^{12} = 1,000,000,000,000$

(a)

(b)

Current (I) is directly proportional ($\propto$) to voltage (V).

$$I \propto V$$

or ions (atoms with more electrons than protons) are available to supply electrons for current flow within a circuit.

42. Voltage within electronic equipment is normally measured in volts, whereas heavy-duty industrial equipment normally requires high voltages that are generally measured in kilovolts (kV).

43. *Voltmeters* (voltage meters) are used to measure electrical pressure (voltage) anywhere in an electronic circuit. A voltmeter is connected across a device with the two leads of the voltmeter touching either side of the component.

44. Three techniques must be remembered and applied when measuring voltage:

   a. To prevent meter damage, always set to the highest scale first (kV) and then reduce as needed to the volt or millivolt range, just in case a larger voltage than anticipated is within the circuit.

   b. Always connect the positive lead (red) of the meter to the component lead that is closer to the positive side of the battery, and the negative lead (black) of the meter to the component lead that is closer to the negative side of the battery.

   c. If you are measuring the voltage across a component, the voltmeter must be connected across the component and outside of the circuit.

45. If a larger voltage were applied to the electrical circuit, this larger electron moving force (emf) would cause more electrons or current to flow through the circuit. *Current (I)* is therefore said to be *directly proportional (∝)* to *voltage (V)*, as a voltage decrease $(V\downarrow)$ results in a subsequent current decrease $(I\downarrow)$, and similarly, a voltage increase $(V\uparrow)$ causes a current increase $(I\uparrow)$.

46. Directly *proportional (∝)* is a phrase which means that one term will change in proportion, or in size, relative to another term.

Conductors and Insulators (Figure 2-34)

47. Materials that pass current easily, that is, offer little opposition (low resistance) to current, are called *conductors*. Conductors are materials or natural elements whose valence electrons can easily be removed from their parent atoms. The precious metals of silver and gold are the best conductors.

48. Copper is the most commonly used conductor, as it meets the following three requirements:

   a. It is a good source of electrons.
   b. It is inexpensive.
   c. It is physically pliable.

49. *Conductance* is the measure of how good a conductor is at carrying current. Conductance (symbolized G) is equal to the reciprocal of resistance and is measured in the unit siemens (S): If the opposition to current flow (resistance) is low, the conductance is high and the material is said to have a good conductance. On the other hand, if the resistance of a conducting wire is high $(R\uparrow)$, its conductance value is low $(G\downarrow)$ and it is called a poor con-

FIGURE 2-34   Conductors and Insulators.

Conductors

$$\text{Conductance } (G) = \frac{1}{\text{Resistance } (R)}$$

$$\text{Relative Conductivity} = \frac{\text{Conductor's Relative Conductivity}}{\text{Reference Conductor's Relative Conductivity}}$$

TABLE 2-4   Relative Conductivity of Conductors

MATERIAL	CONDUCTIVITY (RELATIVE)
Silver	1.000
Copper	0.945
Gold	0.652
Aluminum	0.575
Tungsten	0.297
Nichrome	0.015

Insulators

$$\text{Dielectric Thickness} = \frac{\text{Voltage to Insulate}}{\text{Insulator's Breakdown Voltage}}$$

TABLE 2-5   Breakdown Voltages of Certain Insulators

MATERIAL	BREAKDOWN STRENGTH (kV/cm)
Mica	2000
Glass	900
Teflon	600
Paper	500
Rubber	275
Bakelite	151
Oil	145
Porcelain	70
Air	30

ductor. To summarize, we say that a good conductor has a high conductance value and a very small opposition or resistance to current flow.

50. To express how good or poor a conductor is, we must specify a reference point. The reference point we use is the best conductor, silver, which has a conductivity value of 1.0.

51. Materials that are used to block current, that is, offer high resistance or opposition to current flow, are called *insulators*. Just as certain materials permit the easy flow of current and so have good conductivity, certain materials allow small to almost no amount of free electrons to flow. Insulators can, with sufficient pressure or voltage applied across them, "break down" and conduct current. A good insulator or dielectric should have the maximum possible resistance and conduct no current at all. To express how good or poor an insulator is, we list a voltage that when applied across one-thousandth of an inch of this insulator material, will cause it to break down and conduct a large current. This measure of an insulator is known as its *dielectric strength*.

Semiconductors

52. Certain materials are neither insulators (high resistance to current) nor conductors (low resistance to current), but in fact fall between the two. These materials are known as *semiconductors* because they conduct less than a metal conductor, but more than an insulator. Silicon is the most commonly used semiconductor for components such as transistors and integrated circuits.

The Open, Closed, and Short Circuit

53. An opened switch produces an "open" in the circuit, which prevents current flow as the extra electrons in the negative terminal of the battery cannot feel the attraction of the positive terminal due to the break in the path. The opened switch has produced an *open circuit*. If the switch is now closed so as to make a complete path in the circuit, an open could still exist due to the failure of one of the components. An open component can also produce an open circuit. A closed switch produces a *closed circuit* since current now has a complete path from the negative to positive terminal. The closed switch produces a closed circuit.

54. A *short circuit* normally occurs when one point is accidentally connected to another.

NEW TERMS

Ammeter	Conductance	Electron
Ampere	Conductor	Electron current flow
Atom	Conventional current flow	Element
Battery		Insulator
Breakdown voltage	Coulomb	Ion
Charge	Current	Molecule
Closed circuit	Dielectric strength	Negative
Compound	Electromotive force	Neutron

Nucleus   Proton   Siemens
Open circuit   Relative conductivity   Subatomic
Polarity   Resistance   Switch
Positive   Semiconductor   Voltage
Potential difference Short circuit   Voltmeter

REVIEW QUESTIONS

Multiple-Choice Questions

1. Hydrogen is a:
   a. Gas **b.** Solid **c.** Liquid **d.** Other
2. Which of the following is a liquid?
   a. Calcium **b.** Magnesium **c.** Mercury **d.** Helium
3. The atomic number of an atom describes:
   a. The number of neutrons **c.** The number of nuclei
   b. The number of electrons **d.** The number of protons
4. The most commonly used metal in the field of electronics is:
   a. Silver **b.** Copper **c.** Mica **d.** Gold
5. The smallest unit of an element is:
   a. A compound **b.** An atom **c.** A molecule **d.** A proton
6. The smallest unit of a compound is:
   a. An element **b.** A neutron **c.** An electron **d.** A molecule
7. A water molecule is made up of:
   a. 2 parts hydrogen and 1 part oxygen
   b. Chlorine and sodium
   c. 1 part oxygen and 2 parts sodium
   d. 3 parts chlorine and 1 part hydrogen
8. A negative ion has:
   a. More protons than electrons **c.** More neutrons than protons
   b. More electrons than protons **d.** More neutrons than electrons
9. A positive ion has:
   a. Lost some of its electrons **c.** Lost neutrons
   b. Gained extra protons **d.** Gained more electrons
10. One coulomb of charge is equal to:
    a. 6.24×10^{18} electrons **c.** 6.24×10^{8} electrons
    b. 1018×10^{12} electrons **d.** 6.24×10^{81} electrons
11. If 14 C of charge passes by a point in 7 seconds, the current flow is said to be:
    a. 2 A **b.** 98 A **c.** 21 A **d.** 7 A
12. How many electrons are there within 16 C of charge?
    a. 9.98×10^{19} **b.** 14 **c.** 16 **d.** 10.73×10^{19}

13. Current is measured in:

    a. Volts **b.** Coulombs/second **c.** Ohms **d.** Siemens

14. Voltage is measured in units of:

    a. Amperes **b.** Ohms **c.** Siemens **d.** Volts

15. Another word used to describe voltage is:

    a. Potential difference   **c.** Electromotive force (emf)

    b. Pressure   **d.** All of the above

16. Conductors offer a _____ resistance to current flow.

    a. High **b.** Low **c.** Medium **d.** Maximum

17. Conductors have:

    a. Electrons in shells farthest away from the nucleus
    b. Relatively more electrons per atom
    c. An incomplete valence shell
    d. All of the above
    e. None of the above

18. Conductance is the measure of how good a conductor is at passing current, and is measured in:

    a. Siemens **b.** Volts **c.** Current **d.** Ohms

19. Insulators have:

    a. Electrons close to the nucleus   **d.** All of the above
    b. Relatively few electrons per atom **e.** None of the above
    c. An almost complete valence shell

20. An open circuit will cause:

    a. No current flow   **c.** A break in the circuit
    b. Maximum current flow **d.** Both (a) and (c)

Essay Questions

21. Describe the three factors that make a good conductor. (2-4)

22. State the three rules that should be remembered and applied when measuring: (2-3-6)

    a. Current **b.** Voltage

23. Why is current directly proportional to voltage? (2-3-5)

24. List the four most commonly used current units and their values in terms of the basic unit. (2-2-3)

25. What is the speed of light in (a) miles and (b) meters per second? (2-2-4)

26. List the four most commonly used voltage units and their values in terms of the basic unit. (2-3-2)

27. Describe what is meant by: (2-7)

    a. An open circuit **b.** A closed circuit **c.** A short circuit

28. What is the difference between conventional and electron current flow? (2-2-5)

29. Give the unit and symbol for the following:
  a. Voltage (V) is measured in _____ (__). (2-3).
  b. Current (I) is measured in _____ (__). (2-2-2).
  c. Conductance (G) is measured in _____ (__). (2-4-1)
  d. Resistance (R) is measured in _____ (__). (2-4-1)

30. In relation to the structure of matter and the atom, describe: (2-1)

  a. The atom's subatomic particles
  b. An element (2-1)
  c. A compound (2-1-3)
  d. A molecule (2-1-3)

  e. A neutral atom (2-1-1)
  f. A positive ion (2-2)
  g. A negative ion (2-2)

Practice Problems

31. What is the value of conductance in siemens for a 100 ohm resistor?

32. Calculate the total number of electrons in 6.5 C of charge.

33. Calculate the amount of current in amperes passing through a conductor if 3 C of charge passes by a point in 4 s.

34. Convert the following:

  a. 0.014 A = _____ mA   **c.** 0.776 μA = _____ nA

  b. 1374 A = _____ kA   **d.** 0.91 mA = _____ μA

35. Convert the following:

  a. 1473 mV = _____ V   **c.** 0.139 kV = _____ V

  b. 7143 V = _____ kV   **d.** 0.390 MV = _____ kV

36. What is the relative conductivity of copper if silver is used as the reference conductor?

37. What minimum thickness of porcelain will withstand 24,000 V?

38. To insulate a circuit from 10 V, what insulator thickness would be needed if the insulator is rated at 750 kV/cm?

39. Convert the following:

  a. 2000 kV/cm to _____ meters
  b. 250 kV/cm to _____ mm

40. What maximum voltage may be placed across 35 mm of mica without it breaking down?

After completing this chapter, you will be able to:

1. Define *resistance* and *ohm*.
2. Explain Ohm's law and its application.
3. Describe why:
   a. Current is proportional to voltage.
   b. Current is inversely proportional to resistance.
4. Explain what factors affect the resistance of conductors.

5. Explain superconductivity and its advantages.
6. List some of the different types of conductors and their connectors.
7. Describe and define the terms *energy, work,* and *power.*
8. Describe how the wattmeter can be used to measure power.
9. Explain what is meant by the term *kilowatt-hour.*

Resistance and Power

OUTLINE

Genius of Chippewa Falls

In 1960, Seymour R. Cray, a young vice-president of engineering for Control Data Corporation, informed president William Norris that in order to build the world's most powerful computer he would need a small research lab built near his home. Norris would have shown any other employee the door, but Cray was his greatest asset, so in 1962 Cray moved into his lab, staffed by 34 people and nestled in the woods near his home overlooking the Chippewa River in Wisconsin. Eighteen months later the press was invited to view the 14- by 6-foot 6600 supercomputer that could execute 3 million instructions per second and contained 80 miles of circuitry and 350,000 transistors, which were so densely packed that a refrigeration cooling unit was needed due to the lack of airflow.

Cray left Control Data in 1972 and founded his own company, Cray Research. Four years later the $8.8 million Cray-1 scientific supercomputer outstripped the competition. It included some revolutionary design features, one of which is that since electronic signals cannot travel faster than the speed of light (1 foot per billionth of a second) the wire length should be kept as short as possible, because the longer the wire the longer it takes for a message to travel from one end to the other. With this in mind, Cray made sure that none of the supercomputer's conducting wires exceeded 4 feet in length.

In the summer of 1985, the Cray-2, Seymour Cray's latest design, was installed at Lawrence Livermore Laboratory. The Cray-2 was 12 times faster than the Cray-1, and its densely packed circuits are encased in clear Plexiglas and submerged in a bath of liquid coolant. The 60-year-old genius has moved on from his latest triumph, nicknamed "Bubbles," and is working on another revolution in the supercomputer field, because for Seymour Cray a triumph is merely a point of departure.

Voltage, current, resistance, and power are the four basic concepts of prime importance in our study of electronics. In Chapter 2, voltage and current were introduced, and in this chapter we analyze resistance and power.

3-1 WHAT IS RESISTANCE?

Resistance is the opposition to current flow accompanied by the dissipation of heat. In Figures 3-1 and 3-2, we have again used the fluid analogy to explain the concept of resistance. In Figure 3-1(a), a valve has been opened almost completely, so a very small opposition to the water flow exists within the pipe. This small or low resistance within the pipe will not offer much opposition to water flow, so a large amount of water will flow through the pipe and gush from the outlet.

In Figure 3-1(b), a small **resistance** (which is symbolized by the zigzag line) has been placed in the circuit, providing very little resistance to the passage of current flow. This low resistance or small opposition will therefore allow a large amount of current to flow through the conductor, as illustrated by the heavy line. In Figure 3-2(a), the valve is almost completely closed, resulting in a high resistance or opposition to water flow, so only a trickle of water passes through the pipe and out from the outlet. In Figure 3-2(b), a large value of resistance has been placed in the circuit, causing a large opposition to the passage of current. This large resistance allows only a small amount of current to flow through the conductor, as illustrated by the light line.

Resistance
Symbolized *R* and measured in ohms (Ω), it is the opposition to current flow with the dissipation of energy in the form of heat.

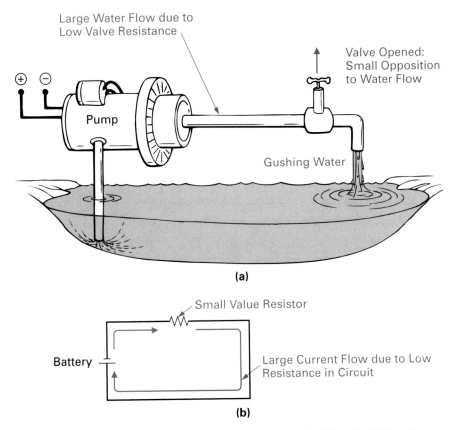

(a)

(b)

FIGURE 3-1   **Low Resistance. (a) Fluid System. (b) Electrical Circuit.**

Small Water Flow due to
High Valve Resistance

Valve Almost Closed:
Large Opposition
to Water Flow

Pump

Trickling Water

(a)

Large Value Resistor

Battery

Small Current Flow due to Large
Resistance in Circuit

(b)

FIGURE 3-2   **High Resistance. (a) Fluid System. (b) Electrical Circuit.**

In both examples of low and high resistance, you may have noticed that resistance and current are inversely proportional (1/∝) to one another; thus, if resistance is high, the current is low, and if resistance is low, the current is high.

Resistance (*R*) is Inversely proportional $\left(\dfrac{1}{\infty}\right)$ **to Current (*I*)**

$$R \quad \frac{1}{\infty} \quad I$$

If resistance is increased by some value, current will be decreased by the same value. For example, if resistance is doubled, current is halved (assuming a constant voltage).

$$R{\uparrow}\frac{1}{\infty}I{\downarrow}$$

In Figures 3-1 and 3-2, the fluid analogy has been used alongside the electrical circuit to help you understand the idea of low and high resistance. In between low and high resistance exist many different values of resistance, and we now need to analyze resistance further to be able to define clearly exactly how much resistance exists within a circuit, and not just say that resistance is generally low or high.

SELF-TEST REVIEW QUESTIONS FOR SECTION 3-1

1. What is resistance?
2. What is the difference between low and high resistance?
3. What is the relationship between current flow and resistance?
4. Would a very high resistance have a small or a large conductance figure?

3-2 THE OHM

Current is measured in amperes, voltage is measured in volts, and resistance is measured in **ohms,** in honor of Georg Simon Ohm and his work with current, voltage, and resistance. The larger the resistance, the larger the value of ohms and the more the resistor will oppose current flow. The ohm is given the symbol Ω, which is the Greek capital letter omega. By definition, 1 ohm is the value of resistance that will allow 1 ampere of current to flow through a circuit when a voltage of 1 volt is applied, as shown in Figure 3-3(b), where the resistor is

Ohm
Unit of resistance, symbolized by the Greek capital letter omega (Ω).

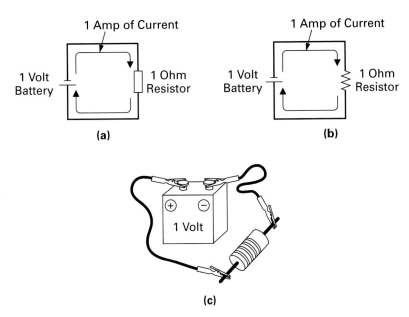

FIGURE 3-3   One Ohm. (a) New and (b) Old Resistor Symbols. (c) Pictorial.

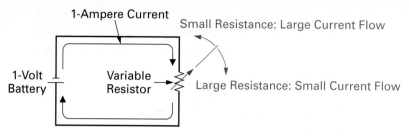

FIGURE 3-4   **1 Ω Allows 1 A to Flow with 1 V Applied.**

drawn as a zigzag; however, in some schematics (circuit diagrams) it can be drawn as a rectangular block, as shown in Figure 3-3(a). The pictorial of this circuit is shown in Figure 3-3(c).

Figure 3-4 reinforces our understanding of the ohm by illustrating a 1-V battery connected across a resistor whose resistance can be either increased or decreased. As the resistance in the circuit is increased, the current will decrease, and as the resistance of the resistor is decreased, the circuit current will increase. If the resistor is adjusted so that exactly 1 A of current is flowing around the circuit, the value of resistance obtained is referred to as 1 ohm (Ω).

3-2-1   Ohm's Law

Ohm's Law
Relationship between the three electrical phenomena of voltage, current, and resistance, which states that the current flow within a circuit is directly proportional to the voltage applied across the circuit and inversely proportional to its resistance.

Current flows in a circuit due to the force or voltage applied; however, the amount of current flow is limited by the amount of resistance in the circuit. In actual fact, the amount of current flow around a circuit is dependent on both voltage and resistance. This relationship between the three electrical properties of current, voltage, and resistance was discovered by Georg Simon Ohm, a German physicist, in 1827. Published originally in 1826, **Ohm's law** states that the current flow in a circuit is directly proportional ($\propto$) to the source voltage applied and inversely proportional ($1/\propto$) to the resistance of the circuit.

Stated in mathematical form, Ohm arrived at this formula:

$$\text{current } (I) = \frac{\text{voltage } (V)}{\text{resistance } (R)}$$

$$\text{current } (I) \propto \text{voltage } (V)$$
$$\text{current } (I) \; \frac{1}{\propto} \; \text{resistance } (R)$$

This mini-math review is designed to remind you of the mathematical details relating to transposition, which will be used in the following section.

Transposition

In the preceding section it was stated that

$$\text{current } (I) = \frac{\text{voltage } (V)}{\text{resistance } (R)} \text{ or } I = \frac{V}{R}$$

Like all formulas, this formula provides a relationship between quantities, or values. In this case we can determine the current flow in a circuit simply by dividing the voltage applied to the circuit, by the circuit's resistance. If we ever need to calculate only current, this formula would be fine on its own. But what if we want to calculate voltage and we know current and resistance, or what if we want to calculate resistance and we know current and voltage. *It is important to know how to rearrange or transpose a formula so that you can solve for any of the formula's quantities—this process is known as transposition.*

As an example, calculate what voltage would cause 2 amperes of current to flow in a circuit having a resistance of 3 ohms.

To solve this problem, you first begin by placing the values in their appropriate position within the formula. In this example,

$$\text{current } (I) = \frac{\text{voltage } (V)}{\text{resistance } (R)}$$

Therefore,

$$2 \text{ amperes} = \frac{V \, ?}{3 \text{ ohms}}$$

or

$$2 = \frac{V}{3}$$

Before we proceed further in our attempt to determine the value of voltage (V), we must first discuss a few ground rules.

1. The equals sign divides the equation or formula into two halves, and in all cases the left half always equals the right half. Therefore,

$$2 = \frac{V}{3}$$

$$\text{left half} \leftarrow | \rightarrow \text{right half}$$

2. Although not always visible, you should always assume that there are lines under every number or letter in mathematics. For example,

$$\frac{2}{-} = \frac{V}{3}$$

Any number beneath the line needs to be divided into the number on the top of the line, as with any fraction.

Now that the basics have been covered, we can transpose the formula. To achieve this operation we use basic algebra and follow two steps:

First step: Move the unknown quantity so that it is above the line.

Second step: Move the unknown quantity so that it stands by itself on either side of the equals sign.

Looking at the first step and our example, you can see that "V" does sit above the line:

$$\frac{2}{} = \frac{V}{3}$$

therefore, we can move onto the second step.

Looking at the second step and our example, you can see that "V" does not stand by itself on either side of the equals sign. To satisfy this step, we must somehow move the "3" under the "V" away from the right side of the equation. *If you do exactly the same thing to both sides of the equation or formula, nothing is changed.* This means that you can add to, subtract from, multiply, or divide each side of the equals by the same number and nothing is changed. To move the "3" in our example away from the right side of the equation, we must do something to both sides of the equation to make it cancel. The solution is to multiply both sides by "3," as follows:

$$\frac{3 \times 2}{} = \frac{V \times 3}{3}$$

Since any number divided by the same number equals 1 ($3 \div 3 = 1$) and $V \times 1 = V$, the second step in our transposition process is achieved with "V" isolated on the right side of the equals sign. To summarize, the process was as follows:

$$2 = \frac{V}{3} \qquad \text{(solve for } V = ?\text{)}$$

$$3 \times 2 = \frac{V \times 3}{3} \qquad \text{(multiply both sides by 3)}$$

$$3 \times 2 = \frac{V \times 3}{3} \qquad \text{(cancel the 3's)}$$

$$3 \times 2 = V \times 1 \qquad (V \times 1 = V)$$

$$3 \times 2 = V$$

or
$$V = 3 \times 2$$

In this example we can see that $V = 3 \times 2 = 6$ V. This means that when a voltage of 6 volts is applied to a circuit containing 3 ohms of resistance, a current of 2 amperes will flow.

Let's return to our original problem, which was: How can you calculate voltage ($V = ?$) if I and R are known, and how do you calculate resistance ($R = ?$) if V and I are known? You can now see that all you have to do is transpose the original formula, as follows:

$$\text{current } (I) = \frac{\text{voltage } (V)}{\text{resistance } (R)}$$

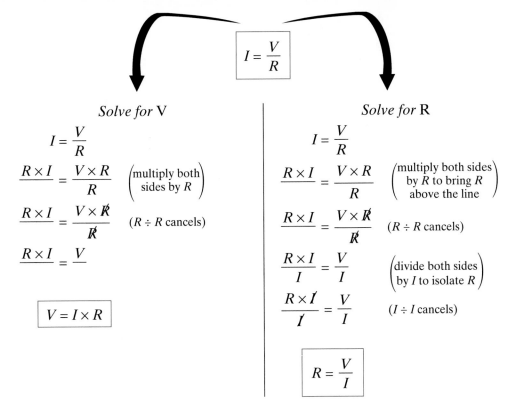

$$I = \frac{V}{R}$$

Solve for V

$$I = \frac{V}{R}$$

$$\frac{R \times I}{} = \frac{V \times R}{R} \quad \left(\begin{array}{c}\text{multiply both} \\ \text{sides by } R\end{array}\right)$$

$$\frac{R \times I}{} = \frac{V \times \cancel{R}}{\cancel{R}} \quad (R \div R \text{ cancels})$$

$$\frac{R \times I}{} = V$$

$$\boxed{V = I \times R}$$

Solve for R

$$I = \frac{V}{R}$$

$$\frac{R \times I}{} = \frac{V \times R}{R} \quad \left(\begin{array}{c}\text{multiply both sides} \\ \text{by } R \text{ to bring } R \\ \text{above the line}\end{array}\right)$$

$$\frac{R \times I}{} = \frac{V \times \cancel{R}}{\cancel{R}} \quad (R \div R \text{ cancels})$$

$$\frac{R \times I}{I} = \frac{V}{I} \quad \left(\begin{array}{c}\text{divide both sides} \\ \text{by } I \text{ to isolate } R\end{array}\right)$$

$$\frac{R \times \cancel{I}}{\cancel{I}} = \frac{V}{I} \quad (I \div I \text{ cancels})$$

$$\boxed{R = \frac{V}{I}}$$

To test if all these three formulas are correct, we can plug into each the values of the preceding example, as follows:

$$I = \frac{V}{R} \qquad V = I \times R \qquad R = \frac{V}{I}$$

$$2\,\text{A} = \frac{6\,\text{V}}{3\,\Omega} \qquad 6\,\text{V} = 2\,\text{A} \times 3\,\Omega \qquad 3\,\Omega = \frac{6\,\text{V}}{2\,\text{A}}$$

As you can see from this example, the transposition has been successful.

3-2-2 The Ohm's Law Triangle

The three forms of Ohm's law can be summarized by placing the three properties within a triangle for easy memory recall, as illustrated in Figure 3-5.

FIGURE 3-5   Ohm's Law Triangle.

(a)

(b)

FIGURE 3-6   Current Flow Is Proportional to Voltage. (a) Pressure or Voltage Increase Causes a Water or Current Flow Increase (V↑, I↑). (b) Pressure or Voltage Decrease Results in a Water or Current Flow Decrease (V↓, I↓).

$$I = \frac{V}{R} \qquad V = I \times R \qquad R = \frac{V}{I}$$

3-2-3 Current Is Proportional to Voltage (I ∝ V)

Referring to Figure 3-6(a), you can see that an increase in the pump pressure will result in an increase in water flow. Similarly, an increase in voltage (electron moving force) will exert a greater pressure on the circuit electrons and cause an increase in current flow. This point is reinforced with Ohm's law:

$$I{\uparrow} = \frac{V{\uparrow}}{R}$$

▧ EXAMPLE:

Referring to Figure 3-7, if resistance remains constant at 1 Ω and the applied voltage equals 2 V, what is the value of current flowing through the circuit?

FIGURE 3-7

▧ Solution:

$$\begin{aligned}
\text{current } (I) &= \frac{\text{voltage } (V)}{\text{resistance } (R)} \\
&= \frac{2\ V}{1\ \Omega} \\
&= 2\ \text{A}
\end{aligned}$$

CALCULATOR SEQUENCE

Step	Keypad Entry	Display Response
1.	[2]	2
2.	[÷]	
3.	[1]	1
4.	[=]	2

▧ EXAMPLE:

Referring to Figure 3-8, if the voltage from Example 3-1 is now doubled to 4 V, what would be the change in current?

FIGURE 3-8

$$\text{current } (I) = \frac{\text{voltage } (V)}{\text{resistance } (R)}$$

$$= \frac{4 \text{ V}}{1 \text{ }\Omega}$$

$$= 4 \text{ A}$$

On the other hand, a decrease in pump pressure or battery voltage will result in a small water or electron force, causing a decrease in current flow, as seen in Figure 3-6(b). This is again reinforced by Ohm's law.

$$I\downarrow = \frac{V\downarrow}{R}$$

Let's once again take an example to prove the point mathematically.

■ **EXAMPLE:**

If the circuit resistance equals 2 Ω and the applied voltage equals 8 V, what would be the circuit current?

■ *Solution:*

$$\text{current } (I) = \frac{\text{voltage } (V)}{\text{resistance } (R)}$$

$$= \frac{8 \text{ V}}{2 \text{ }\Omega}$$

$$= 4 \text{ A}$$

■ **EXAMPLE:**

If the voltage from the example above is now halved to 4 V, what would be the change in circuit current?

■ *Solution:*

$$\text{current } (I) = \frac{\text{voltage } (V)}{\text{resistance } (R)}$$

$$= \frac{4 \text{ V}}{2 \text{ }\Omega}$$

$$= 2 \text{ A}$$

In conclusion, we can say that if resistance were to remain constant and the voltage were to double, the current within the circuit would also double. Similarly, if the voltage were halved, the current would also halve, proving that current and voltage increase or decrease by the same percentage for the same value of resistance, which makes them directly proportional to one another.

3-2-4 *Current Is Inversely Proportional to Resistance* $\left(I \, \dfrac{1}{\propto} \, R \right)$

With the valve opened, as shown in Figure 3-9(a), a small opposition to the water flow will result, so a large water flow exists in a system with low resistance. Similarly, with the electronic circuit a small resistance ($R\downarrow$) allows a large current ($I\uparrow$) flow around the circuit.

Consequently, in both the fluid system and electronic circuit, a decrease in resistance causes an increase in current flow. Ohm's law reinforces this point.

$$(\text{resistance}) \; R\downarrow = \frac{V}{I\uparrow} \; (\text{current})$$

To prove this mathematically, let's take an example.

(a)

(b)

FIGURE 3-9   **Current Is Inversely Proportional to Resistance. (a) Resistance Decrease Causes a Water or Current Flow Increase (R↓, I↑). (b) Resistance Increase Causes a Water or Current Flow Decrease (R↑, I↓).**

I = 4A, 2A

8 V

R = 2Ω, 4Ω

FIGURE 3-10

■ **EXAMPLE:**

If the applied circuit voltage is 8 V and the circuit resistance equals 2 Ω, what is the total amount of current flow?

■ *Solution:*

$$\text{current } (I) = \frac{\text{voltage } (V)}{\text{resistance } (R)}$$

$$= \frac{8 \text{ V}}{2 \text{ Ω}}$$

$$= 4 \text{ A}$$

On the other hand, in Figure 3-9(b) the valve has been almost completely closed, offering a large opposition to the water flow; consequently, a small water flow results in a system with high resistance. With the electronic circuit in Figure 3-10, a large circuit resistance ($R\uparrow$) permits only a small current ($I\downarrow$) to flow around the circuit.

Ohm's law reinforces the illustration in Figure 3-9(b) by showing that a large resistance or opposition causes a small water or current flow.

$$\text{(resistance) } R\uparrow = \frac{V}{I\downarrow} \text{ (current)}$$

To prove this mathematically, let's take an example.

■ **EXAMPLE:**

If the applied voltage is still equal to 8 V and the circuit resistance is now 4 Ω, calculate the value of current flow through the circuit.

■ *Solution:*

$$\text{current } (I) = \frac{\text{voltage } (V)}{\text{resistance } (R)}$$

$$= \frac{8 \text{ V}}{4 \text{ Ω}}$$

$$= 2 \text{ A}$$

In conclusion, we can say that if voltage were to remain constant and the resistance were to double, the current within the circuit would be halved. On the other hand, if the circuit resistance were halved, the circuit current would double, confirming that current is inversely proportional to resistance.

3-3 CONDUCTORS AND THEIR RESISTANCE

Resistors are normally made out of materials that cause an opposition to current flow. Conductors, on the other hand, are not meant to offer any resistance or opposition to current flow. This, however, is not always the case since some are good conductors having a low resistance, while others are poor conductors having a high resistance. Conductance is the measure of how good a conductor is, and even the best conductors have some value of resistance. Up until this time, we have determined a conductor's resistance based on the circuit's *electrical characteristics,*

$$R = \frac{V}{I}$$

With this formula we could determine the conductor's resistance based on the voltage applied and circuit current. Another way to determine resistance is to examine the *physical factors* of a conductor.

The total resistance of a typical conductor, as shown in Figure 3-11, is determined by four main physical factors:

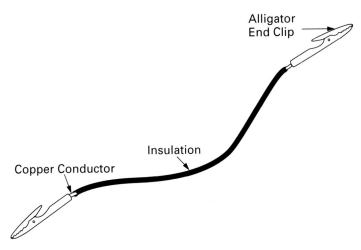

FIGURE 3-11   Copper Conductor.

1. The type of conducting material used for the flow of current
2. The conductor's cross-sectional area
3. The total length of the conductor
4. The temperature of the conductor

Let's now address each of these topics separately and discuss the reason why each will vary a conductor's resistance.

3-3-1 Conducting Material

In our discussion on conductors, we said that copper, for instance, is a better conducting material than carbon as copper had more free electrons ready for current flow; and as we have stated, a good conductor has a large conductance figure and a small resistance value. Carbon, however, has a smaller amount of free electrons than copper, resulting in a small conductance value and a high resistance figure.

3-3-2 Conductor's Cross-Sectional Area

The resistance of a conductor is inversely proportional to the conductor's cross-sectional area, which means: The greater the cross-sectional area, the lower the resistance [Figure 3-12(a)], and similarly, the smaller the cross-sectional area, the higher the resistance [Figure 3-12(b)].

We can relate this to our fluid analogy, shown in Figure 3-13. A larger pipe or conductor results in a larger water flow or current due to the lower resistance. On the other hand, the smaller the pipe or conductor, the less the amount of water or electron flow, due to the larger resistance (Figure 3-14).

We need a unit of measure to compare one thickness of wire conductor to another. Figure 3-15 illustrates a conductor with a diameter of 0.001 inch ($\frac{1}{1000}$ of an inch). This is termed 1 **mil,** and the diameters of all conductors are measured in mils.

As nearly all conductors are circular in cross section, the area of a conductor is measured in **circular mils** (cmil), which can be calculated by squaring the diameter.

Mil
One thousandth of an inch (0.001 in.).

Circular Mil (cmil)
A unit of area equal to the area of a circle whose diameter is 1 mil.

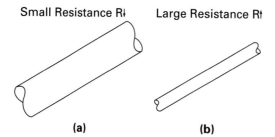

Small Resistance R↓   Large Resistance R↑

(a)   (b)

FIGURE 3-12   Conduction. (a) Large Cross-Sectional Area. (b) Small Cross-Sectional Area.

$$\text{circular mils (cmil)} = \text{diameter}^2$$

■ **EXAMPLE:**

If a conductor has a diameter of 71.96 mils, what is the circular mil area?

■ *Solution:*

CALCULATOR SEQUENCE

$$cmil = \text{diameter}^2$$
$$= 71.96^2$$
$$= 5178 \text{ cmil}$$

Step	Keypad Entry	Display Response
1.	$\boxed{7}\,\boxed{1}\,\boxed{.}\,\boxed{9}\,\boxed{6}$	71.96
2.	$\boxed{X^2}$	5178.24

3-3-3 *Conductor Length*

Increasing the length of the conductor used increases the amount of resistance within the circuit. For example, if a conductor has a resistance of 1 ohm for every 10 feet of conductor, then 30 feet of conductor are going to triple the conductor's resistance within the circuit to 3 ohms.

Large Pipe Has Small Resistance, Yields Large Water Flow.

Large Conductor Has Small Resistance, Yields Large Current Flow.

(a)   **(b)**

FIGURE 3-13 Large Cross-Sectional Area Causes Low Resistance. (a) Large Pipe (Small Resistance) Yields Large Water Flow. (b) Large Conductor (Small Resistance) Yields Large Current Flow.

Small Pipe Has Large Resistance,
Yields Small Water Flow.

(a)

Small Conductor Has
Large Resistance, Yields
Small Current Flow.

(b)

**FIGURE 3-14   Small Cross-Sectional Area Causes High Resistance. (a) Small Pipe
(Large Resistance) Yields Small Water Flow. (b) Small Conductor (Large Resistance)
Yields Small Current Flow.**

Insulation

Metal Conductor

Diameter in Mils (1 Mil = 0.001 inch)

▨ = Cross-Sectional Area in Circular Mils (cmil)

Cross-Sectional Area = Diameter²

FIGURE 3-15   Measuring a Conductor's Cross-Sectional Area.

3-3-4   Physical Resistance Formula

By combining all of these physical factors, we can arrive at a formula for resistance. The resistance of any conductor is calculated by use of the following formula:

$$R = \frac{\rho \times l}{a}$$

where R = resistance of conductor, in ohms
ρ = resistivity of conducting material
l = length of conductor, in feet
a = area of conductor, in circular mils

TABLE 3-1 Material Resistivity

MATERIAL	RESISTIVITY (cmil/ft) IN OHMS
Silver	9.9
Copper	10.7
Gold	16.7
Aluminum	17.0
Tungsten	33.2
Zinc	37.4
Brass	42.0
Nickel	47.0
Platinum	60.2
Iron	70.0

Resistivity, by definition, is the resistance (in ohms) that a certain length of material (in centimeters) will offer to the flow of current. Table 3-1 lists the resistivity of the more commonly used conductors.

Resistivity
Measure of a material's resistance to current flow.

In summary, therefore, resistance can be calculated based either on its physical factors or on electrical performance.

$$\frac{\rho \times l}{a} = R = \frac{V}{I}$$

Physical $\Longleftrightarrow$ Electrical

EXAMPLE:

Calculate the resistance of 333 ft of copper conductor with a conductor area of 3257 cmil.

Solution:

$$R = \frac{\rho \times l}{a}$$

$$= \frac{10.7 \times 333}{3257}$$

$$= 1.09 \ \Omega$$

CALCULATOR SEQUENCE

Step	Keypad Entry	Display Response
1.	[1] [0] [.] [7]	10.7
2.	[×]	
3.	[3] [3] [3]	333
4.	[÷]	3563.1
5.	[3] [2] [5] [7]	3257
6.	[=]	1.09

EXAMPLE:

Calculate the resistance of 1274 ft of aluminum conductor with a diameter of 86.3 mils.

Solution:

$$R = \frac{\rho \times l}{a}$$

If the diameter is equal to 86.3 mils, the circular mil area equals d^2:

$$86.3^2 = 7447.7 \text{ cmil}$$

Referring to Table 3-1 you can see that the resistivity of aluminum is 17.0. Therefore,

$$R = \frac{17 \times 1274}{7447.7}$$

$$= 2.9 \ \Omega$$

3-3-5   Temperature Effects on Conductors

When heat is applied to a conductor, the atoms within the conductor convert this thermal energy into another form of energy, in this case mechanical energy or movement. These random moving atoms cause collisions between the directed electrons (current flow) and the adjacent atoms, resulting in an opposition to the current flow (resistance).

Positive Temperature Coefficient of Resistance
The rate of increase in resistance relative to an increase in temperature.

Metallic conductors are said to have a **positive temperature coefficient of resistance** (+ Temp. Coe. of R), because the greater the heat applied to the conductor, the greater the atom movement, causing more collisions of atoms to occur, and consequently the greater the conductor's resistance.

heat ↑   resistance ↑

3-3-6   Maximum Conductor Current

Any time that current flows through any conductor, a certain resistance or opposition is inherent in that conductor. This resistance will convert current to heat, and the heat further increases the conductor's resistance, causing more heat to

FIGURE 3-16   **Conductor with Insulator.**

be generated due to the opposition. Consequently, a conductor must be chosen carefully for each application so that it can carry the current without developing excessive heat. This is achieved by selecting a conductor with a greater cross-sectional area to decrease its resistance.

Conducting wires are normally covered with a plastic or rubber type of material, known as *insulation,* as shown in Figure 3-16. This insulation is used to protect the users and technicians from electrical shock and also to keep the conductor from physically contacting other conductors within the equipment. If the current through the conductor is too high, this insulation will burn due to the heat and may cause a fire hazard. The National Fire Protection Association has developed a set of standards known as the **American Wire Gauge** for all copper conductors, which lists their diameter, area, resistance, and maximum safe current in amperes. This table is given in Table 3-2. A rough guide for measuring wire size is shown in Figure 3-17.

American Wire Gauge
American wire gauge (AWG) is a system of numerical designations of wire sizes, with the first being 0000 (the largest size) and then going to 000, 00, 0, 1, 2, 3, and so on up to the smallest sizes of 40 and above.

TABLE 3-2 The American Wire Gauge (AWG) for Copper Conductor

AWG NUMBER[a]	DIAMETER (mils)	MAXIMUM CURRENT (A)	Ω/1000 ft
0000	460.0	230	0.0490
000	409.6	200	0.0618
00	364.8	175	0.0780
0	324.9	150	0.0983
1	289.3	130	0.1240
2	257.6	115	0.1563
3	229.4	100	0.1970
4	204.3	85	0.2485
5	181.9	75	0.3133
6	162.0	65	0.3951
7	144.3	55	0.4982
8	128.5	45	0.6282
9	114.4	40	0.7921
10	101.9	30	0.9981
11	90.74	25	1.260
12	80.81	20	1.588
13	71.96	17	2.003
14	64.08	15	2.525
15	57.07		3.184
16	50.82	6	4.016
17	45.26	Wires	5.064
18	40.30	of this 3	6.385
19	35.89	size	8.051
20	31.96	have	10.15
22	25.35	current	16.14
26	15.94	measured	40.81
30	10.03	in mA	103.21
40	3.145		1049.0

[a]The larger the AWG number, the smaller the size of the conductor.

FIGURE 3-17 Wire Gauge Size.

3-3-7 *Superconductivity*

Superconductor
Metal such as lead or niobium that, when cooled to within a few degrees of absolute zero, can conduct current with no electrical resistance.

Conductors have a positive temperature coefficient, which means that if temperature increases, so does resistance; but what happens if the temperature is decreased? In 1911, a Dutch physicist, Heike Onnes, discovered that mercury (a liquid conductor) lost its resistance to electrical current when the temperature was decreased to −459.7 Fahrenheit (0 on the Kelvin temperature scale). Mercury actually became a **superconductor,** allowing a supercurrent to flow and not encounter any resistance and therefore not generate any heat. (Heat dissipated by any resistance can be calculated by the power formula $P = I^2 \times R$. If $R = 0$, the power dissipated by the conductor is zero watts.)

In the spring of 1986, two IBM scientists discovered that a conductor compound made up of barium, lanthanum, copper, and oxygen would superconduct (have no resistance to current flow) at −406°F. Since then other scientists have increased the temperature to a point that now conductor compounds can be made to superconduct at −300°F. Some scientific projects that need to make use of the advantages of superconductivity immerse these conductor compounds in a bath of liquid nitrogen so that they can achieve superconductivity. Liquid nitrogen, however, is difficult to handle, so the real scientific achievement will be to find a conductor compound that will superconduct at room temperature.

When discussing the AWG table (Table 3-2), a maximum value of current was stated for a given thickness of wire. This is because a large current will generate a large amount of heat if the resistance is too large, so to decrease the resistance and therefore heat generated ($P{\uparrow} = I^2R{\uparrow}$), a thicker conductor with a lower resistance is used. Superconductivity allows a standard 1000-A conductor, which would be approximately 2 in. thick, to be replaced by a superconductor about the thickness of a human hair. Not only will conductors be reduced dramatically in size, but the increased efficiency (as no power is being wasted in the form of heat) will result in high energy savings.

FIGURE 3-18 **Wires and Cables. (a) Solid Wire. (b) Stranded Wire. (c) Braided Wire. (d) Printed Wire. (e) Coaxial Cable. (f) Twin Lead.**

3-3-8 *Conductor and Connector Types*

A cable is made up of two or more wires. Figure 3-18 illustrates the different types of **wires** and **cables.** In Figure 3-18(a), (b), and (c), only one conductor exists within the insulation, so they are classified as wires, whereas the cables seen in Figure 3-18(e) and (f) have two conductors. The coaxial and twin-lead cables are most commonly used to connect TV signals into television sets. Figure 3-18(d) illustrates how printed conducting copper, silver, or gold paths exist on a plastic insulating board and are used to connect components such as resistors that are mounted on the other side.

Wires and cables have to connect from one point to another. Some are soldered directly, whereas others are attached to plugs that plug into sockets. A sample of connectors is shown in Figure 3-19.

Wire
Single solid or stranded group of conductors having a low resistance to current flow.

Cable
Group of two or more insulated wires.

FIGURE 3-19 **Connectors. (a) Temporary Connectors. (b) Plugs. (c) Lug and Binding Post. (d) Sockets.**

3-4 ENERGY, WORK, AND POWER

Energy
Capacity to do work.

Work
Work is done anytime energy is transformed from one type to another, and the amount of work done is dependent on the amount of energy transformed.

Joule
The unit of work and energy.

Potential Energy
Energy that has the potential to do work because of its position relative to others.

Kinetic Energy
Energy associated with motion.

The sun provides us with a consistent supply of energy in the form of light. Coal and oil are fossilized vegetation that grew, among other things, due to the sun, and are examples of **energy** that the earth has stored for millions of years. It can be said, then, that all energy begins from the sun. On the earth, energy is not created or destroyed; it is merely transformed from one form to another. The transforming of energy from one form to another is called **work.** The greater the energy transformed, the more work that is done.

The six basic forms of energy are light, heat, magnetic, chemical, electrical, and mechanical energy. The unit for energy is the **joule** (J). "Potential" (position) and "kinetic" (motion) are two terms used when describing energy. A cart on top of a hill has **potential** (position) **energy,** while a cart rolling down a hill has **kinetic** (motion) **energy.** Potential and kinetic energy are best described by looking at the example of a swinging pendulum, as seen in Figure 3-20. When the pendulum is in its upmost position [Figure 3-20(a)], it has potential energy, due to its position relative to the resting position, yet it has no motion and so no kinetic energy. When the pendulum is in the position shown in Figure 3-20(b), it has no potential energy, yet it has motion or kinetic energy. In between these two points, the pendulum possesses a combination of both potential and kinetic energy, as shown in Figure 3-20(c).

Looking at Figure 3-21, let us try to summarize our discussion so far on energy and work. Chemical energy within the battery is converted to electrical energy when the electrons are attracted and repelled and set in motion. To use

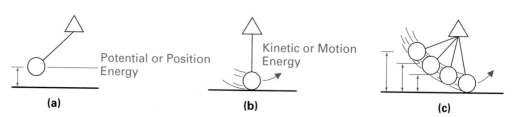

FIGURE 3-20 **Pendulum. (a) Potential Energy. (b) Kinetic Energy. (c) Potential and Kinetic Energy.**

FIGURE 3-21   Energy Transfer.

the two terms, we could say that the battery has the potential energy to set electrons in motion, and these moving electrons are said to possess kinetic energy. This electrical energy drives two devices:

1. The light bulb, which converts electrical energy into light and heat energy
2. The pump, which uses the electrical energy to produce mechanical energy within the pump

The pump has the potential energy to cause water flow or kinetic energy, just as the battery has the potential energy to cause kinetic energy in the form of electron flow. The kinetic energy within the water flow is finally used to turn the waterwheel.

Work is being done everytime one form of energy is transformed to another. A device that converts one form of energy to another is called a **transducer;** in Figure 3-21, the battery, light bulb, and pump were all examples of transducers doing work at each stage of conversion.

The amount of work done is equal to the amount of energy transformed, and in both cases an equal amount of energy was transformed, so an equal amount of work was done. Energy and work have the same symbol (W), the same formula, and the same unit (the joule). Energy is merely the capacity, potential, or ability to do work, and work is done when a transformation of the potential, capacity, or ability takes place.

Transducer
Any device that converts energy from one form to another.

■ **EXAMPLE:**

One person walks around a track and takes 50 minutes, while another person runs around the track and takes 50 seconds. Both were full of energy before they walked or ran around the track, and during their travels around the track they

SEC. 3-4 / ENERGY, WORK, AND POWER   **105**

converted the chemical energy within their bodies into the mechanical energy of movement.

   a. Who exerted the most energy?

   b. Who did the most work?

■ *Solution:*

Both exerted the same amount of energy. The runner exerted all his energy (for example, 100 J) in the short time of 50 seconds, while the walker spaced his energy (100 J) over 50 minutes. They both did the same amount of work. So the only difference between the runner and the walker is time.

3-4-1　Power

Power
Amount of energy converted by a component or circuit in a unit of time, normally seconds. It is measured in units of watts (joules/second).

Power (P) is the rate at which work is performed and is given the unit of **watt** (W), which is joules per second. Thus power involves a time factor.

    Returning to our two persons walking and running around the track, we could say that the number of joules of energy exerted in 1 second by the runner was far greater than the number of joules of energy exerted in 1 second by the walker, although the total energy exerted by both persons around the entire track was equal and therefore the same amount of work was done. Even though the same amount of energy was used, and therefore the same amount of work was done by the runner and the walker, the output power of each was different. The runner exerted a large value of joules/second or watts (high power output) in a short space of time, while the walker exerted only a small value of joules/second or watts (low power output) over a longer period of time.

Watt (W)
Unit of electric power required to do work at a rate of 1 joule/second. One watt of power is expended when 1 ampere of direct current flows through a resistance of 1 ohm.

    Whether discussing a runner, walker, electric motor, heater, refrigerator, light bulb, or compact disk player—power is power. The output power, or power ratings, of electrical, electronic, or mechanical devices can be expressed in watts and describes the number of joules of energy converted every second. The output power of rotating machines is given in the unit *horsepower* (hp), the output power of heaters is given in the unit *British thermal units per hour* (Btu/h), and the output power of cooling units is given in the unit *ton of refrigeration*. Despite the different names, they can all be expressed quite simply in the unit of watts. The conversions are as follows:

$$1 \text{ horsepower (hp)} = 746 \text{ W}$$

$$1 \text{ British thermal unit per hour (Btu/h)} = 0.293 \text{ W}$$

$$1 \text{ ton of refrigeration} = 3.52 \text{ kW (3520 W)}$$

    Now we have an understanding of power, work, and energy. Let's reinforce our knowledge by introducing the energy formula with some mathematical problems related to electronics.

3-4-2　Calculating Energy

The amount of energy stored (W) is dependent on the coulombs of charge stored (Q) and the voltage (V).

$$W = Q \times V$$

where W = energy stored, in joules (J)

Q = coulombs of charge (1 coulomb = 6.24×10^{18} electrons)

V = voltage, in volts (V)

If you consider a battery as an example, you can probably better understand this formula. The battery's energy stored is dependent on how many coulombs of electrons it holds (current) and how much electrical pressure it is able to apply to these electrons (voltage).

■ EXAMPLE:

If a 1-V battery can store 6.24×10^{18} electrons, how much energy is the battery said to have?

■ Solution:

$$W = Q \times V$$
$$= 1\,C \times 1\,V$$
$$= 1\,J \text{ of energy}$$

■ EXAMPLE:

How many coulombs of electrons would a 9-V battery have to store to have 63 J of energy?

■ Solution:

If $W = Q \times V$, then

$$Q = \frac{W}{V}$$

$$\text{coulombs of electrons } (Q) = \frac{\text{energy in joules } (W)}{\text{battery voltage } (V)}$$

$$= \frac{63\,J}{9\,V}$$

$$= 7\,C \text{ of electrons}$$

or

$$7 \times 6.24 \times 10^{18} = 4.36 \times 10^{19} \text{ electrons}$$

CALCULATOR SEQUENCE

Step	Keypad Entry	Display Response
1.	6 3	63
2.	÷	
3.	9	9
4.	=	7
5.	×	7
6.	6 . 2 4 E 1 8	6.24E18
7.	=	4.36E19

3-4-3 Calculating Power

Power, in relation to electricity and electronics, is the rate (t) at which electric energy (W) is converted into some other form. A power formula can therefore be derived as follows:

$$P = \frac{W}{t}$$

where P = power, in watts
W = energy, in joules
t = time, in seconds

Since $W = Q \times V$, we can substitute W in the formula above, to obtain

$$P = \frac{W}{t} = \frac{Q \times V}{t}$$

where Q = coulombs of charge
V = voltage, in volts

Since coulombs of charge $Q = I \times t$, we can substitute Q in the formula above, to obtain

$$P = \frac{Q \times V}{t} = \frac{(I \times t) \times V}{t}$$

By canceling t, we arrive at a final formula for power:

$$P = I \times V$$

where P = power, in watts (W)
I = current, in amperes (A)
V = voltage, in volts (V)

This formula states that the amount of power delivered to a device is dependent on the electrical pressure or voltage applied across the device and the current flowing through the device.

■ **EXAMPLE:**

3 Amperes

9 Volts

FIGURE 3-22   Calculating Power.

Power, in relation to electronics, is the rate at which electric energy is converted into some other form. In our example (Figure 3-22) it will be transformed from electric energy into light and heat energy by the light bulb. Power has the unit of watts, which is the number of joules of energy transformed per second (J/s). If 27 J of electric energy is being transformed into light and heat per second, how many watts of power does the light bulb convert?

$$\text{power} = \frac{\text{joules}}{\text{second}}$$

$$= \frac{27 \text{ J}}{1 \text{ s}}$$

$$= 27 \text{ W}$$

The power output in the previous example could have easily been calculated by merely multiplying current by voltage to arrive at the same result.

$$\text{power} = I \times V = 3 \text{ A} \times 9 \text{ V}$$

$$= 27 \text{ W}$$

Therefore, we could say that the light bulb dissipates 27 W, or 27 J per second.

Looking at the power formula $P = I \times V$, we can say that 1 watt of power is expended when 1 ampere of current flows through a circuit that has 1 volt applied or when 1 ampere flows through 1 ohm.

Like Ohm's law, we can transpose the power formula as follows:

$$P = V \times I$$

Solve for V

$$P = V \times I$$

$$\frac{P}{I} = \frac{V \times I}{I} \quad \left(\begin{array}{c} \text{divide both} \\ \text{sides by } I \end{array} \right)$$

$$\frac{P}{I} = \frac{V \times \cancel{I}}{\cancel{I}} \quad \left(\begin{array}{c} I \div I = 1 \text{, and} \\ V \times 1 = V \end{array} \right)$$

$$\frac{P}{I} = V$$

$$\boxed{V = \frac{P}{I}}$$

This formula can be used to calculate V when P and I are known

Solve for I

$$P = V \times I$$

$$\frac{P}{V} = \frac{V \times I}{V} \quad \left(\begin{array}{c} \text{divide both} \\ \text{sides by } V \end{array} \right)$$

$$\frac{P}{V} = \frac{\cancel{V} \times I}{\cancel{V}} \quad \left(\begin{array}{c} V \div V = 1 \text{, and} \\ I \times 1 = I \end{array} \right)$$

$$\frac{P}{V} = I$$

$$\boxed{I = \frac{P}{V}}$$

This formula can be used to calculate I when P and V are known

MINI-MATH REVIEW

This mini-math review is designed to remind you of the mathematical details relating to substitution, which will be used in following sections.

Substitution

In the previous section we stated that $P = V \times I$, and by transposition we arrived at $V = P/I$ and $I = P/V$. When trying to calculate power we may not always have the values of V and I available. For example, we may know only I and R, or V and R. **Substitution** enables us to obtain alternative power formulas by replacing or substituting one mathematical term with an equivalent mathematical term. For example, since $I = V/R$, we could substitute I in any formula with V/R. Similarly, since $V = I \times R$, we could substitute V in any formula with $I \times R$. The following shows how we can substitute terms in the $P = V \times I$ formula to arrive at alternative power formulas for wattage calculations.

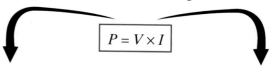

$$P = V \times I$$

$P = V \times I$ (substitute I with V/R)

$P = V \times \dfrac{V}{R}$ ($V \times V = V^2$)

$$P = \dfrac{V^2}{R}$$

This formula can be used to calculate P when V and R are known

$P = V \times I$ (substitute V with $I \times R$)

$P = (I \times R) \times I$ ($I \times I = I^2$)

$$P = I^2 \times R$$

This formula can be used to calculate P when I and R are known

■ **EXAMPLE:**

FIGURE 3-23

In Figure 3-23 a 12-V battery is connected across a 36-Ω resistor. How much power does the resistor dissipate?

■ *Solution:*

Step	Keypad Entry	Display Response
1.	①②	12
2.	X² (square key)	144
3.	÷	
4.	③⑥	36
5.	=	4

$$\text{power} = \dfrac{\text{voltage}^2}{\text{resistance}}$$

$$= \dfrac{12\ V^2}{36\ \Omega} = \dfrac{144}{36}$$

$$= 4\ W$$

Four joules of energy per second is being dissipated.

3-4-4 Measuring Power

Electrical power is measured with a wattmeter. The wattmeter shown in Figure 3-24 has four terminals, which must be connected correctly, as shown in Figure 3-25; in this example we are measuring the amount of power being supplied to a radio.

FIGURE 3-24   **Wattmeter.**

The current terminals of the wattmeter are connected so that the current passing from the battery to the radio has to pass through the wattmeter's current terminals so that the wattmeter can sense current, while the voltage terminals of the wattmeter are connected across the radio so as to sense the radio's voltage supply. The **wattmeter** is therefore like an ammeter and voltmeter in one package, and through a multiplication process ($P = I \times V$), we can obtain a direct reading of the amount of power being supplied to the radio.

Wattmeter
A meter used to measure electric power in watts.

FIGURE 3-25   **Wattmeter Connections or Hookup. (a) Schematic Symbol. (b) Pictorial.**

3-4-5 *The Kilowatt-Hour*

Kilowatt-hour
1000 watts for 1
hour.

Kilowatt-hour meter
A meter used by
electric companies to
measure a customer's
electric power use in
kilowatt-hours.

You and I pay for our electric energy in a unit called the **kilowatt-hour** (kWh). The **kilowatt-hour meter,** shown in Figure 3-26, measures how many kilowatt-hours are consumed, and the electric company then charges accordingly. Since power is the rate at which energy is used, if we multiply power and time, we can calculate how much energy has been consumed.

$$
\text{energy consumed } (W) = \text{power } (P) \times \text{time } (t)
$$

Energy is normally expressed in joules; however, since this formula uses the product of power (in watts) and time (in seconds or hours) we can use one of three other units of energy: the watt-second (Ws), watt-hour (Wh), or kilowatt-hour (kWh). The kilowatt-hour is most commonly used by electric companies. A *kilowatt-hour* of energy is consumed when you use 1000 watts (1 kW) of power in 1 hour.

$$
\text{energy consumed } (\text{kWh}) = \text{power } (\text{kW}) \times \text{time } (\text{h})
$$

FIGURE 3-26 **Kilowatt-hour Meter Being Tested for Accuracy by a Journeyman Meter Tester. (Courtesy of Fred Vaughn, San Diego Gas and Electric.)**

FIGURE 3-26 *(cont.)* **Digital kilowatt-hour meter. (© Tim Stahl 1992, Courtesy of Fred Vaughn, San Diego Gas and Electric.)**

■ **EXAMPLE:**

If a 100-W light bulb is left on for 10 hours, how many kilowatt-hours will we be charged for?

■ *Solution:*

power consumed (kWh) = power (kW) × time
$$\qquad\qquad\text{(hours)}$$
$$= 0.1 \text{ kW} \times 10 \text{ hours}$$
$$(100 \text{ W} = 0.1 \text{ kW})$$
$$= 1 \text{ kWh}$$

CALCULATOR SEQUENCE

Step	Keypad Entry	Display Response
1.	⓪ . ① E ③	0.1E3
2.	✕	
3.	① ⓪	10
4.	=	1E3

■ **EXAMPLE:**

Figure 3-27 illustrates a typical household heater and an equivalent electrical circuit. The heater has a resistance of 7 Ω and the electric company is charging 6 cents/kWh. Calculate:

a. The energy consumed by the heater

b. The cost of running the heater for 7 hours

FIGURE 3-27

■ *Solution:*

a. Power $(P) = \dfrac{V^2}{R} = \dfrac{120^2}{7}$

$$= \dfrac{14400\text{ V}}{7} = \dfrac{14.4\text{ kV}}{7}$$

$$= 2057\text{ W (2 kW)}$$

b. Energy consumed = power (kW) × time (hours)

$$= 2.057 \times 7\text{ h}$$

$$= 14.399\text{ kWh}$$

Cost = kWh × rate = 14.399 × 6 cents

$$= 86\text{ cents}$$

SELF-TEST REVIEW QUESTIONS FOR SECTION 3-4

1. List the six basic forms of energy.
2. What is the difference between energy, work, and power?
3. List the formulas for calculating energy and power.
4. What is 1 kilowatt-hour of energy?

SUMMARY   **Resistance (Figure 3-28)**

1. Resistance is the opposition to current flow accompanied by the dissipation of heat.

2. Resistance and current are inversely proportional ($1/\infty$) to one another; thus, if resistance is high, the current is low, and if resistance is low, the current is high. If resistance is increased by some value, current will be decreased by the same value. For example, if resistance is doubled, current is halved (assuming a constant voltage).

FIGURE 3-28   Resistance.

Resistance (*R*) is inversely proportional $\left(\frac{1}{\infty}\right)$ to Current (*I*)

Ohm's Law

$$\text{Current } (I) = \frac{\text{Voltage } (V)}{\text{Resistance}}$$

$$\text{Resistance } (R) = \frac{\text{Voltage } (V)}{\text{Current } (I)}$$

$$\text{Voltage } (V) = \text{Current } (I) \times \text{Resistance } (R)$$

R = Resistance in Ohms (Ω), *I* = Current in Amps (*A*), *V* = Voltage in Volts (*V*)

One Ohm

(a)   (b)

(c)

3. Current is measured in amperes, voltage is measured in volts, and resistance is measured in *ohms,* in honor of Georg Simon Ohm and his work with current, voltage, and resistance.

4. The larger the resistance, the larger the value of ohms and the more the resistor will oppose current flow.

5. The ohm is given the symbol Ω, which is the Greek capital letter omega. By definition, 1 ohm is the value of resistance that will allow 1 ampere of current to flow through a circuit when a voltage of 1 volt is applied. If a resistor is adjusted so that exactly 1 A of current is flowing around the circuit, the value of resistance obtained is referred to as 1 ohm (Ω).

6. Current flows in a circuit due to the force or voltage applied; however, the amount of current flow is limited by the amount of resistance in the circuit. In actual fact, the amount of current flow around a circuit is dependent on both voltage and resistance. This relationship between the three electrical

properties of current, voltage, and resistance was discovered by Georg Simon Ohm, a German physicist, in 1827.

Published originally in 1826, *Ohm's law* states that the current flow in a circuit is directly proportional ($\propto$) to the source voltage applied and inversely proportional ($1/\propto$) to the resistance of the circuit.

7. Like all formulas, the ohm's law formula provides a relationship between quantities, or values. It is important to know how to rearrange or transpose a formula so that you can solve for any of the formula's quantities—this process is known as transposition.

8. To transpose the ohm's law formula, we use basic algebra and follow two steps: *First step:* Move the unknown quantity so that it is above the line. *Second step:* Move the unknown quantity so that it stands by itself on either side of the equals sign. *If you do exactly the same thing to both sides of the equation or formula, nothing is changed.*

9. If resistance were to remain constant and the voltage were to double, the current within the circuit would also double. Similarly, if the voltage were halved, the current would also halve, proving that current and voltage increase or decrease by the same percentage for the same value of resistance, which makes them directly proportional to one another.

10. If voltage were to remain constant and the resistance were to double, the current within the circuit would be halved. On the other hand, if the circuit resistance were halved, the circuit current would double, confirming that current is inversely proportional to resistance.

Conductors and Their Resistance (Figure 3-29)

11. Resistors are normally made out of materials that cause an opposition to current flow. Conductance is the measure of how good a conductor is, and even the best conductors have some value of resistance.

12. Ohm's law can be used to determine the conductor's resistance based on the voltage applied and circuit current. Another way to determine resistance is to examine the *physical factors* of a conductor. The total resistance of a typical conductor is determined by four main physical factors:

    a. The type of conducting material used for the flow of current
    b. The conductor's cross-sectional area
    c. The total length of the conductor
    d. The temperature of the conductor

13. The resistance of a conductor is inversely proportional to the conductor's cross-sectional area. The greater the cross-sectional area, the lower the resistance, and similarly, the smaller the cross-sectional area, the higher the resistance.

14. As nearly all conductors are circular in cross section, the area of a conductor is measured in *circular mils* (cmil).

15. Increasing the length of the conductor used increases the amount of resistance within the circuit.

16. By combining all of a conductor's physical factors, we can arrive at a formula for resistance.

17. *Resistivity,* by definition, is the resistance (in ohms) that a certain length of material (in centimeters) will offer to the flow of current. In summary, therefore, resistance can be calculated based on either its physical factors or on electrical performance.

18. When heat is applied to a conductor, the atoms within the conductor convert this thermal energy into another form of energy, in this case mechanical energy or movement. These random moving atoms cause collisions between the directed electrons (current flow) and the adjacent atoms, resulting in an opposition to the current flow (resistance).

19. Metallic conductors are said to have a *positive temperature coefficient of resistance* (+ Temp. Coe. of R), because the greater the heat applied to the conductor, the greater the atom movement, causing more collisions of atoms to occur, and consequently the greater the conductor's resistance.

20. Any time that current flows through any conductor, a certain resistance or opposition is inherent in that conductor. This resistance will convert current to heat, and the heat further increases the conductor's resistance, causing more heat to be generated due to the opposition. Consequently, a conductor must be chosen carefully for each application so that it can carry the current without developing excessive heat. This is achieved by selecting a conductor with a greater cross-sectional area to decrease its resistance.

21. Conducting wires are normally covered with a plastic or rubber type of material, known as *insulation.* This insulation is used to protect the users and technicians from electrical shock and also to keep the conductor from physically contacting other conductors within the equipment. If the current through the conductor is too high, this insulation will burn due to the heat and may cause a fire hazard. The National Fire Protection Association has developed a set of standards known as the *American Wire Gauge* for all copper conductors, which lists their diameter, area, resistance, and maximum safe current in amperes.

22. Conductors have a positive temperature coefficient, which means that if temperature increases, so does resistance; but what happens if the temperature is decreased? In 1911, a Dutch physicist, Heike Onnes, discovered that mercury (a liquid conductor) lost its resistance to electrical current when the temperature was decreased to –459.7 Fahrenheit (0 on the Kelvin temperature scale). Mercury actually became a *superconductor,* allowing a supercurrent to flow and not encounter any resistance and therefore not generate any heat.

23. In the spring of 1986, two IBM scientists discovered that a conductor compound made up of barium, lanthanum, copper, and oxygen would superconduct (have no resistance to current flow) at –406° F.

    Some scientific projects that need to make use of the advantages of superconductivity immerse these conductor compounds in a bath of liquid

FIGURE 3-29   Conductors and Their Resistance.

Insulation

Metal Conductor

Diameter in Mils (1 Mil = 0.001 inch)

▨ = Cross-Sectional Area in Circular Mils (cmil)

Cross-Sectional Area = Diameter²

$$R = \frac{\rho \times l}{a}$$

TABLE 3-1   Material Resistivity

R = **Resistance of Conductor in Ohms**
ρ = **Resistivity of Conducting Material**
l = **Length of Conductor in Feet**
a = **Area of Conductor in Circular Mils**

MATERIAL	RESISTIVITY (cmil/ft) IN OHMS
Silver	9.9
Copper	10.7
Gold	16.7
Aluminum	17.0
Tungsten	33.2
Zinc	37.4
Brass	42.0
Nickel	47.0
Platinum	60.2
Iron	70.0

Conductors have a positive temperature coefficient of resistance (*t* ↑ cause *R* ↑).

TABLE 3-2   The American Wire Gauge (AWG) for Copper Conductor

AWG NUMBER[a]	DIAMETER (mils)	MAXIMUM CURRENT (A)	Ω/1000 ft
0000	460.0	230	0.0490
000	409.6	200	0.0618
00	364.8	175	0.0780
0	324.9	150	0.0983
1	289.3	130	0.1240
2	257.6	115	0.1563
3	229.4	100	0.1970
4	204.3	85	0.2485
5	181.9	75	0.3133
6	162.0	65	0.3951
7	144.3	55	0.4982

FIGURE 3-29 *(continued)* **Conductors and Their Resistance.**

TABLE 3-2 The American Wire Gauge (AWG) for Copper Conductor

AWG NUMBER[a]	DIAMETER (mils)	MAXIMUM CURRENT (A)	Ω/1000 ft
8	128.5	45	0.6282
9	114.4	40	0.7921
10	101.9	30	0.9981
11	90.74	25	1.260
12	80.81	20	1.588
13	71.96	17	2.003
14	64.08	15	2.525
15	57.07		3.184
16	50.82	6	4.016
17	45.26	Wires	5.064
18	40.30	of this 3	6.385
19	35.89	size	8.051
20	31.96	have	10.15
22	25.35	current	16.14
26	15.94	measured	40.81
30	10.03	in mA	103.21
40	3.145		1049.0

[a]The larger the AWG number, the smaller the size of the conductor.

nitrogen so that they can achieve superconductivity. Liquid nitrogen, however, is difficult to handle, so the real scientific achievement will be to find a conductor compound that will superconduct at room temperature.

24. A cable is made up of two or more wires.

25. Wires and cables have to connect from one point to another. Some are soldered directly, whereas others are attached to plugs that plug into sockets.

Power (Figure 3-30)

26. The sun provides us with a consistent supply of energy in the form of light. Coal and oil are fossilized vegetation that grew, among other things, due to the sun, and are examples of *energy* that the earth has stored for millions of years. It can be said, then, that all energy begins from the sun.

On the earth, energy is not created or destroyed; it is merely transformed from one form to another. The transforming of energy from one form to another is called *work*. The greater the energy transformed, the more work that is done.

27. The six basic forms of energy are light, heat, magnetic, chemical, electrical, and mechanical energy. The unit for energy is the *joule* (J).

FIGURE 3-30   Power.

$$W = Q \times V$$

$$P = \frac{W}{t}$$

W = Energy Stored in Joules (J)

Q = Coulombs of Charge

V = Voltage in Volts (V)

P = Power in Watts

W = Energy in Joules

t = Time in Seconds

$$P = I \times V$$

P = Power in Watts (W), I = Current in Amps (A), V = Voltage in Volts (V)

$$V = \frac{P}{I}$$

$$I = \frac{P}{V}$$

$P = V \times I$ (substitute I for $\frac{V}{R}$)

$P = V \times (\frac{V}{R})$ ($V \times V = V^2$)

$P = \frac{V^2}{R}$

$V = I \times R$ (substitute R for $\frac{V}{I}$)

$V = I \times \frac{V}{I}$ ($I \times V = P$)

$V = \frac{P}{I}$

$P = V \times I$ (substitute V for $I \times R$)

$P = (I \times R) \times I$ ($I \times I = I^2$)

$P = I^2 \times R$

$P = V \times I$ to $P = \frac{V^2}{R}$

$P \times R = \frac{V^2 \times R}{R}$ (R)

$P \times R = V^2$

$P = V \times I$ to $P = I^2 \times R$

$\frac{P}{R} = \frac{I^2 \times R}{R}$ ($\div R$)

$\frac{P}{R} = I^2$

$\sqrt{\frac{P}{R}} = \sqrt{I^2}$ ($\sqrt{\ }$)

$\sqrt{\frac{P}{R}} = I$ or $I = \sqrt{\frac{P}{R}}$

$\sqrt{P \times R} = \sqrt{V^2}$ ($\sqrt{\ }$)

$\sqrt{P \times R} = V$ or $V = \sqrt{P \times R}$

$P = V \times I$ to $P = \frac{V^2}{R}$

$P \times R = \frac{V^2}{R} \times R$ ($\times R$)

$P \times R = V^2$

$P = V \times I$

$\frac{P}{V} = \frac{V \times I}{V}$ ($\div V$)

$\frac{P}{V} = I$ or $I = \frac{P}{V}$

$P = I^2 \times R$

$\frac{P}{I^2} = \frac{I^2 \times R}{I^2}$ (divide both sides by I^2)

$R = \frac{P}{I^2}$

$\frac{P \times R}{P} = \frac{V^2}{P}$ ($\div P$)

$R = \frac{V^2}{P}$

Energy Consumed (kWh) = Power (kW) $\times$ time (hours)

"Potential" (position) and "kinetic" (motion) are two terms used when describing energy. A cart on top of a hill has *potential* (position) *energy,* while a cart rolling down a hill has *kinetic* (motion) *energy.*

28. Chemical energy within the battery is converted to electrical energy when the electrons are attracted and repelled and set in motion. To use the two terms, we could say that the battery has the potential energy to set electrons in motion, and these moving electrons are said to possess kinetic energy.

29. Work is being done everytime one form of energy is transformed to another. A device that converts one form of energy to another is called a *transducer.*

30. Energy and work have the same symbol (W), the same formula, and the same unit (the joule). Energy is merely the capacity, potential, or ability to do work, and work is done when a transformation of the potential, capacity, or ability takes place.

31. *Power* (P) is the rate at which work is performed and is given the unit of *watt* (W), which is joules per second. Thus power involves a time factor.

32. Whether discussing a runner, walker, electric motor, heater, refrigerator, light bulb, or compact disk player—power is power. The output power, or power ratings, of electrical, electronic, or mechanical devices can be expressed in watts and describes the number of joules of energy converted every second.

    The output power of rotating machines is given in the unit *horsepower* (hp), the output power of heaters is given in the unit *British thermal units per hour* (Btu/h), and the output power of cooling units is given in the unit *ton of refrigeration.* Despite the different names, they can all be expressed quite simply in the unit of watts.

33. The amount of energy stored (W) is dependent on the coulombs of charge stored (Q) and the voltage (V). The battery's energy stored is dependent on how many coulombs of electrons it holds (current) and how much electrical pressure it is able to apply to these electrons (voltage).

34. Power, in relation to electricity and electronics, is the rate (t) at which electric energy (W) is converted into some other form.

35. The $P = I \times V$ formula states that the amount of power delivered to a device is dependent on the electrical pressure or voltage applied across the device and the current flowing through the device. Looking at the power formula $P = I \times V$, we can say that 1 watt of power is expended when 1 ampere of current flows through a circuit that has 1 volt applied or when 1 ampere flows through 1 ohm.

36. When trying to calculate power we may not always have the values of V and I available. For example, we may know only I and R, or V and R. *Substitution* enables us to obtain alternative power formulas by replacing or substituting one mathematical term with an equivalent mathematical term. For example, since $I = V/R$, we could substitute I in any formula with V/R. Similarly, since $V = I \times R$, we could substitute V in any formula with $I \times R$.

37. Electrical power is measured with a wattmeter. The wattmeter has four terminals, which must be connected correctly. The current terminals of the wattmeter are connected so that the wattmeter's current terminals can sense current, while the voltage terminals of the wattmeter are connected to sense the voltage supply. The *wattmeter* is therefore like an ammeter and voltmeter in one package, and through a multiplication process ($P = I \times V$), we can obtain a direct reading of the amount of power being supplied.

38. You and I pay for our electric energy in a unit called the *kilowatt-hour* (kWh). The *kilowatt-hour meter* measures how many kilowatt-hours are consumed, and the electric company then charges accordingly. Since power is the rate at which energy is used, if we multiply power and time, we can calculate how much energy has been consumed.

39. The kilowatt-hour is most commonly used by electric companies. A *kilowatt-hour* of energy is consumed when you use 1000 watts (1 kW) of power in 1 hour.

NEW TERMS

Alligator clip	Kinetic Energy	Resistance
American Wire Gauge (AWG)	Lug	Resistivity
Binding post	Mil	Socket
Cable	Ohm	Superconductor
Circular mil	Phone	Temperature Coefficient of Resistance
Connector	Plug	Transducer
Energy	Positive Temperature Coefficient of Resistance	Watt
Joule	Potential Energy	Wattmeter
Kelvin	Power	Wire
Kilowatt hour	RCA plug and jack	Work
Kilowatt-hour meter		

REVIEW QUESTIONS

Multiple-Choice Questions

1. Resistance is measured in:
   a. Ohms   c. Amperes
   b. Volts   d. Siemens

2. Current is proportional to:
   a. Resistance   c. Both (a) and (b)
   b. Voltage   d. None of the above

3. Current is inversely proportional to:
   a. Resistance   c. Both (a) and (b)
   b. Voltage   d. None of the above

4. If the applied voltage is equal to 15 V and the circuit resistance equals 5 Ω, the total circuit current would be equal to:

    a. 4 A **b.** 5 A **c.** 3 A **d.** 75 A

5. Calculate the applied voltage if 3 mA flows through a circuit resistance of 25 kΩ.

    a. 63 mV **b.** 25 V **c.** 77 µV **d.** 75 V

6. Energy is measured in:

    a. Volts **b.** Joules **c.** Amperes **d.** Watts

7. Chemical energy within a battery is converted into:

    a. Electrical energy   **c.** Magnetic energy
    b. Mechanical energy **d.** Heat energy

8. The water pump has the potential to cause water flow, just as the battery has the potential to cause:

    a. Voltage   **c.** Current
    b. Electron flow **d.** Both (b) and (c)

9. Work is measured in:

    a. Joules **c.** Amperes
    b. Volts   **d.** Watts

10. The device that converts one energy form to another is called a:

    a. Transformer **c.** Transit
    b. Transducer   **d.** Transistor

11. Power is the rate at which energy is transformed and is measured in:

    a. Joules **c.** Volts
    b. Watts **d.** Amperes

12. Power is measured by using a (an):

    a. Ammeter **c.** Ohmmeter
    b. Voltmeter **d.** Wattmeter

13. A good conductor has a:

    a. Large conductance figure **c.** Both (a) and (b)
    b. Small resistance figure   **d.** None of the above

14. The resistance of a conductor is:

    a. Proportional to the length of the conductor
    b. Inversely proportional to the area of the conductor
    c. Both (a) and (b)
    d. None of the above

15. AWG is an abbreviation for:

    a. Alternate Wire Gauge   **c.** American Wave Guide
    b. Alternating Wire Gauge **d.** American Wire Gauge

Essay Questions

16. What is resistance? (3-1)

17. Briefly describe why:
   a. Current is proportional to voltages. (3-2-3)
   b. Current is inversely proportional to resistance. (3-2-4)

18. State Ohm's law. (3-2-1)

19. List the three forms of Ohm's law. (3-2-2)

20. List the six basic forms of energy. (3-4)

21. Briefly describe the following terms:
   a. Potential energy (3-4)
   b. Kinetic energy (3-4)

22. What is a transducer? (3-4)

23. Define power. (3-4-1)

24. Give three formulas for electric power. (3-4-3)

25. Give the units for each of the following:
   a. Energy   **e.** Work
   b. Power   **f.** Current
   c. Voltage   **g.** Charge
   d. Resistance   **h.** Conductance

26. What is the difference between work and power? (3-4)

27. What instrument is used to measure electrical power? (3-4-4)

28. State the formula used by electric companies to determine the amount of power consumed? (3-4-5)

29. What is 1 kilowatt-hour? (3-4-5)

30. List the four factors that determine a conductor's resistance. (3-3)

31. What is a circular mil? (3-3-2)

32. Define the resistivity of a conducting material. (3-3-4)

33. Describe why conductors have a positive temperature coefficient of resistance. (3-3-5)

34. What is the purpose(s) of placing an insulating sheath over conducting wires? (3-3-6)

35. What is the American Wire Gauge? (3-3-6)

36. What is a superconductor? (3-3-7)

37. List some of the advantages of superconductivity. (3-3-7)

38. What is the difference between a wire and a cable? (3-3-8)

39. Give some examples of different wires and cables. (3-3-8)

40. List examples of different conductor connectors. (3-3-8)

Practice Problems

41. An electric heater with a resistance of 6 Ω is connected across a 120-V wall outlet.

    a. Calculate the current flow. (3-2)
    b. Draw the schematic diagram.

42. What source voltage would be needed to produce a current flow of 8 mA through a 16-kΩ resistor? (3-2)

43. If an electric toaster draws 10 A when connected to a power outlet of 120 V, what is its resistance? (3-2)

44. Calculate the power used in Problems 41, 42, and 43. (3-4-3)

45. Calculate the current flowing through the following light bulbs when they are connected across 120 V: (3-4-3)

    a. 300 W   b. 100 W   c. 60 W   d. 25 W

46. If an electric company charges 9 cents/kWh, calculate the cost for each light bulb in Problem 45 if on for 10 hours. (3-4-5)

47. Indicate which of the following unit pairs is larger: (3-4)

    a. Millivolts or volts
    b. Microamperes or milliamperes
    c. Kilowatts or watts
    d. Kilohms or megohms

48. Calculate the resistance of 200 ft of copper having a diameter of 80 mils. (3-3-4)

49. What AWG size wire should be used to safely carry just over 15 A? (3-3-6)

50. Calculate the voltage dropped across 1000 ft of No. 4 copper conductor when a current of 7.5 A is flowing through it. (3-3-6)

51. Calculate the unknown resistance in a circuit when an ammeter indicates that a current of 12 mA is flowing and a voltmeter indicates 12 V. (3-2-4)

52. What battery voltage would use 1000 J of energy to move 40 C of charge through a circuit? (3-4-2)

53. Calculate the resistance of a light bulb that passes 500 mA of current when 120 V is applied. What is the bulb's wattage? (3-4)

54. Which of the following circuits has the largest resistance and which has the smallest? (3-4)

    a. $V = 120$ V, $I = 20$ mA   c. $V = 9$ V, $I = 100$ μA
    b. $V = 12$ V, $I = 2$ A   d. $V = 1.5$ V, $I = 4$ mA

55. Calculate the power dissipated in each circuit in Problem 54. (3-4-3)

56. How many watts are dissipated if 5000 J of energy is consumed in 25 s? (3-4-3)

57. Convert the following:

    a. 1000 W = _____ kW
    b. 0.345 W = _____ mW
    c. 1250×10^3 W = _____ MW
    d. 0.00125 W = _____ μW

58. What is the value of the resistor when a current of 4 A is causing 100 W to be dissipated? (3-4-3)

59. How many kilowatt-hours of energy are consumed in each of the following: (3-4-5)

    a. 7500 W in 1 hour   **c.** 127,000 W for half an hour
    b. 25 W for 6 hours

60. What is the maximum output power of a 12-V, 300-mA power supply? (3-4-3)

After completing this chapter, you will be able to:

1. Describe the difference between a fixed- and a variable-value resistor.

2. Explain the differences between the six basic types of fixed-value resistors: carbon composition, carbon film, metal film, wirewound, metal oxide, and thick film.

3. Identify the different resistor wattage ratings, and their value and tolerance labelling methods.

4. Describe the SIP, DIP, and chip thick-film resistor packages.

5. Explain the following types of variable resistors:
   a. Mechanically adjustable: rheostat and potentiometer.
   b. Thermally adjustable: RTD, TFD, and thermistor.
   c. Optically adjustable: photoresistor.

6. List the three rules to remember when measuring resistance with an ohmmeter.

7. Explain how a resistor's value and tolerance are printed on the body of the component either:
   a. By color coding, or
   b. Typographically

8. Describe the difference between a general-purpose and a precision resistor.

9. Explain why current through any resistance is always accompanied by heat.

10. State the purpose of the filament and ballast resistor.

11. Describe some of the more common resistor problems.

Resistors

Catch the Wave

In the true tradition of the electronics industry, Joel Naive founded his company, Wavetek, in his garage in 1962. As the first of two employees, Joel concentrated on technology while his wife Kathy provided the business counsel. He worked on refining the design of a function generator and also on expanding its capabilities. Thus the Model 101 was born. The prototype was hand-carried to Los Angeles and was exhibited at the West Coast's Electronic Show (WESCON) in September 1962. Joel's total out-of-pocket expenses up to this point—including costs for the trip to LA and a bottle of Scotch for a friend who let Joel sleep on his couch—was $435. The 101 was a complete success. The product line was sold worldwide, and the company grew into an international organization, with annual sales of approximately $100 million.

In fact, Joel's approach to engineering is "simple." He starts each project by writing $E = IR$ on the chalkboard. Then he says, "Let's start with the basics." His approach to design is through practical experimentation in addition to theoretical mathematics.

Besides test and measurement, Joel has a passion for classic automobiles of the 1920s era, which he restores and enters in international rallies. This business he appropriately named Garage Number 2.

Resistance would seem to be an undesirable effect, as it reduces current flow and wastes energy as it dissipates heat. Resistors, however, are probably used more than any other component in electronics, and in this chapter we discuss all the different types.

4-1   RESISTOR TYPES

Resistor
Component made of a material that opposes the flow of current and therefore has some value of resistance.

Conductors offer a certain small amount of resistance; however, in electronics this resistance is not normally enough, so additional resistance is needed to control the amount of current flow. The component used to supply this additional resistance is called a **resistor.**

There are two basic types of resistors: fixed value and variable value. The fixed-value resistor, examples of which can be seen in Figure 4-1, has a value of

FIGURE 4-1 **Fixed-Value Resistors.**

resistance that cannot be changed and is the more common of the two. The variable-value resistor, on the other hand, has a range of values that can be selected generally by mechanically adjusting a control, as illustrated in Figure 4-2.

Let's discuss the different types of fixed- and variable-value resistors.

4-1-1 *Fixed-Value Resistors*

Fixed-value resistors can be divided into six basic categories:

1. Carbon composition
2. Carbon film
3. Metal film
4. Wirewound
5. Metal oxide
6. Thick film

Fixed-Value Resistor
A resistor whose value cannot be changed.

FIGURE 4-2 **Variable-Value Resistors.**

Carbon Composition Resistors

This is the most common and least expensive type of fixed-value resistor, the appearance of which can be seen in Figure 4-3(a). It is constructed by placing a piece of resistive material, with embedded conductors at each end, within an insulating cylindrical molded case, as illustrated in Figure 4-3(b).

This resistive material is called carbon composition because powdered carbon and powdered insulator are bonded into a compound and used as the resistive material. By changing the ratio of powdered insulator to carbon, the value of resistance can be changed within the same area. For example, Figure 4-4 illustrates four fixed-value resistors, ranging from 2 Ω to 10 MΩ.

The 2-Ω resistor is exactly the same size as the 10-MΩ (10 million Ω) resistor. This is achieved by having more powdered carbon and less powdered insulator in the 2 Ω and less powdered carbon and more powdered insulator in the 10 MΩ. The color-coded rings or bands on the resistors in Figure 4-3 are a means of determining the value of resistance. This and other coding systems will be discussed later.

Dissipation
Release of electrical energy in the form of heat.

The physical size of the resistors lets the user know how much power in the form of heat can be **dissipated,** as shown in Figure 4-5. As you already know, resistance is the opposition to current flow or electrons, and this opposition causes heat to be generated when current flows. The amount of heat dissipated per unit of time is measured in watts and each resistor has its own **wattage rating.** The bigger the resistor, the more heat that can be dissipated, so a large resistor could dissipate heat at a rate of 2 W, while a small resistor could dissipate heat at a rate of only ⅛ W. The key point is that heat should be dissipated faster than it is generated; if not, a larger resistor with a larger surface area to dissipate away the additional heat should be used so that the resistor will not burn up.

Wattage Rating
Maximum power a device can safely handle continuously.

Tolerance
Permissible deviation from a specified value, normally expressed as a percentage.

Another factor to consider when discussing resistors is their **tolerance.** Tolerance is the amount of deviation or error from the specified value. For example,

(a)

(b)

FIGURE 4-3 **Carbon Composition Resistors. (a) Appearance. (b) Construction. (Courtesy of Stackpole Electronics, Inc.)**

FIGURE 4-4 **Carbon Composition Resistor Ratios.**

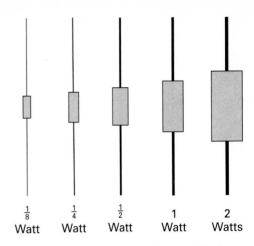

FIGURE 4-5 **Resistor Wattage Rating Guide. All resistor silhouettes are drawn to scale.**

a 1000-Ω (1-kΩ) resistor with a ±10% (plus and minus 10%) tolerance when manufactured could have a resistance anywhere between 900 and 1100 Ω.

$$\pm 10\% \text{ of } 1000 = 100$$

$$10\% - \boxed{1000} + 10\%$$
$$\downarrow \qquad\qquad \downarrow$$
$$900 \qquad\qquad 1100$$

This means that two identically marked resistors when measured could be from 900 to 1100 Ω, a difference of 200 Ω. In some applications, this may be acceptable, although in others, where high precision is required, this deviation could be too large, so more expensive, smaller tolerance resistors are used.

EXAMPLE:

Calculate the amount of deviation of the following resistors:

a. 2.2 kΩ ± 10%

b. 5 MΩ ± 2%

c. 3 Ω ± 1%

■ *Solution:*

a. 10% of 2.2 kΩ = 220 Ω. For + 10%, the value is

$$2200 + 220 \ \Omega = 2420 \ \Omega$$
$$= 2.42 \ \text{k}\Omega$$

For −10%, the value is

$$2200 - 220 \ \Omega = 1980 \ \Omega$$
$$= 1.98 \ \text{k}\Omega$$

The resistor will measure anywhere from 1.98 kΩ to 2.42 kΩ.

CALCULATOR SEQUENCE

Step	Keypad Entry	Display Response
1.	1 0	10
2.	%	10
3.	×	0.10
4.	2 . 2 E 3	2.2E3
5.	=	220

b. 2% of 5 MΩ = 100 kΩ

$$5 \text{ MΩ} + 100 \text{ kΩ} = 5.1 \text{ MΩ}$$
$$5 \text{ MΩ} - 100 \text{ kΩ} = 4.9 \text{ MΩ}$$
Deviation = 4.9 MΩ to 5.1 MΩ

c. 1% of 3 Ω = 0.03 Ω or 30 milliohms (mΩ).

$$3 \text{ Ω} + 0.03 \text{ Ω} = 3.03 \text{ Ω}$$
$$3 \text{ Ω} - 0.03 \text{ Ω} = 2.97 \text{ Ω}$$
Deviation = 2.97 Ω to 3.03 Ω

Carbon Film Resistors

Carbon Film Resistor
Thin carbon film deposited on a ceramic form to create resistance.

Substrate
The mechanical insulating support on which a device is fabricated.

Figure 4-6(a) illustrates the physical appearance of **carbon film resistors.** They are constructed, as shown in Figure 4-6(b), by first depositing a thin layer or film of resistive material (a blend of carbon and insulator) on a ceramic (insulating) **substrate.** The film is then cut to form a helix or spiral. A greater ratio of carbon to insulator will achieve a low-resistance helix; on the other hand, a greater ratio of insulator to carbon will create a higher-resistance helix. Carbon film resistors have smaller tolerance figures (±5% to ±2%), are more stable (maintain same resistance value over a wide range of temperatures), and have less internally generated noise (random small bursts of voltage) than do carbon composition resistors.

Metal Film Resistors

Metal Film Resistor
A resistor in which a film of a metal, metal oxide, or alloy is deposited on an insulating substrate.

Figure 4-7(a) illustrates the physical appearance of some typical **metal film resistors.** They are constructed by spraying a thin film of metal on a ceramic cylinder (or substrate) and then cutting the film to form a substrate, as shown in

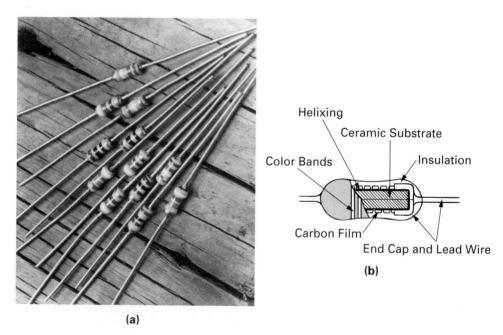

FIGURE 4-6   Carbon Film Resistors. (a) Appearance. (b) Construction. (Courtesy of Stackpole Electronics, Inc.)

(a)	(b)

FIGURE 4-7   Metal Film Resistors. (a) Appearance. (b) Construction. (Courtesy of Stackpole Electronics, Inc.)

Figure 4-7(b). Metal film resistors have possibly the best tolerances commercially available of ±1% to ±0.1%. They also maintain a very stable resistance over a wide range of temperatures (good stability) and generate very little internal noise compared to any carbon resistor.

Wirewound Resistors

Figure 4-8(a) and (c) illustrate the appearance of different types of **wirewound resistors.** They are constructed, as can be seen in Figure 4-8(b), by wrapping a length of wire uniformly around a ceramic insulating core, with terminals making the connections at each end. The length and thickness of the wire are varied to change the resistance, which ranges from 1 Ω to 150 kΩ.

Since current flow is opposed by resistors and this opposition generates heat, the larger the physical size of the resistor, the greater the amount of heat that can be dissipated away and so the greater the current that can be passed through the resistor. Wirewound resistors are generally used in applications requiring low resistance values, which means that the current and therefore power dissipated are high ($R\downarrow = V/I\uparrow$, $P\uparrow = I^2\uparrow R$). Consequently, these resistors are designed to have large surface areas so that they can safely dissipate away the heat. The amount of heat that can be dissipated is measured in watts and is indicated on the resistor. A 10-W resistor can be used to dissipate 20 W if air is blown across its surface by a cooling fan or if it is immersed in a coolant. Wirewound resistors typically have good tolerances of ±1%; however, their large physical size and difficult manufacturing process make them very expensive.

Metal Oxide Resistors

Figure 4-9(a) illustrates the physical appearance of some **metal oxide resistors.** They are constructed, as can be seen in Figure 4-9(b), by depositing an oxide of a metal such as tin onto an insulating substrate. The ratio of oxide (insulator) to tin (conductor) will determine the resistor's resistance.

Wirewound Resistor
Resistor in which the resistive element is a length of high-resistance wire or ribbon, usually nichrome, wound onto an insulating form.

Metal Oxide Resistor
A metal film resistor in which an oxide of a metal (such as tin) is deposited as a film onto the substrate.

(a)

(b)   (c)

FIGURE 4-8   Wirewound Resistors. (a) Appearance. (b) Construction. (c) Other Types. (Courtesy of Stackpole Electronics, Inc.)

(a)   (b)

FIGURE 4-9   Metal Oxide Resistors. (a) Appearance. (b) Construction. (Courtesy of Stackpole Electronics, Inc.)

Metal oxide resistors have excellent temperature stability, which is the ability of the resistor to maintain its value of resistance without change even when temperature is changed. There is a tendency when heat is increased for atoms to move, and in so doing they cause collisions, resulting in an opposition to current; so a temperature increase will cause a resistor's resistance to increase. This increase from the desired resistance is very small with metal oxide resistors.

Thick-Film Resistors

Figure 4-10 illustrates examples of **thick-film resistors.** Figure 4-10(a) and (b) shows the two different types of resistor networks, called SIPs and DIPs. The **single in-line package** is so called because all its lead connections are in a single line, whereas the **dual in-line package** has two lines of connecting pins. The chip resistors shown in Figure 4-10(c) are small thick-film resistors that are approximately the size of a pencil lead.

The SIP and DIP resistor networks are constructed by first screening on the internal conducting strip (silver) that connects the external pins to the resistive material, and then screening on the thick film of resistive paste (bismuth/ruthenate), the blend of which will determine resistance. The chip resistor uses the same resistive film paste material, which is deposited onto an insulating substrate with two end terminations and protected by a glass coat.

The SIP and DIP resistor networks, once constructed, are trimmed by lasers to obtain close tolerances of typically ±2%. Resistance values ranging from 22 Ω to 2.2 MΩ are available, with a power rating of ½ W.

The chip resistor shown in Figure 4-10(c) is commercially available with resistance values from 10 Ω to 3.3 MΩ, a ±2% tolerance, and a ⅛-W heat dissipation capability. They are ideally suited for applications requiring physically small sized resistors, as explained in the inset in Figure 4-10(c). The chip resistor is called a **surface mount technology** (SMT) device. The key advantage of SMT devices over "through-hole" devices is that a through-hole device needs both a hole in the printed circuit board (PCB) and a connecting pad around the hole. With the SMT device, no holes are needed since the package is soldered directly onto the surface of the PCB. Without the need for holes, pads can be placed closer together. This results in a considerable space saving, as you can see in the inset in Figure 4-10(c) by comparing the through-hole resistor to the surface mount chip resistor.

4-1-2 Variable-Value Resistors

In certain applications we may require a variation in resistance while in circuit: for example, in the volume adjustment on a radio or television. The component that achieves this is the variable resistor.

The resistance of **variable resistors** can be varied in one of three ways:

1. Mechanically (user) adjustable:
   a. Rheostat
   b. Potentiometer

Thick-Film Resistors
Fixed-value resistor consisting of a thick-film resistive element made from metal particles and glass powder.

Single In-Line Package (SIP)
Package containing several electronic components (generally resistors) with a single row of external connecting pins.

Dual In-Line Package (DIP)
Package that has two (dual) sets or lines of connecting pins.

Surface Mount Technology
A method of installing tiny electronic components on the same side of a circuit board as the printed wiring pattern that interconnects them.

Variable Resistor
A resistor whose value can be changed.

(a)

Configuration

(b)

Configuration

Through-Hole
Resistor

Surface Mount
Resistor

Printed Circuit
Board

Conductive Strips
Printed on Insulating
Board

Actual Size
Comparison

Glass Coat
Substrate
Resistive Termination
Film

(c)

FIGURE 4-10   Thick-Film Resistors. (a) SIPs: Single In-Line Packages. (b) DIPs: Dual
In-Line Packages. (c) Chips. (Courtesy of Stackpole Electronics, Inc.)

2. Thermally (heat) adjustable:
   a. Thermistor
   b. Resistive temperature detector (RTD)
   c. Thin-film detector (TFD)
3. Optically (light) adjustable:
   a. Photoresistor

Mechanically (User) Adjustable Variable Resistors

This category includes two types, both of which will cause a change in resistance when a shaft is rotated.

Rheostat (two terminals: A and B). Figure 4-11(a) shows the physical appearance of different rheostats, while Figure 4-11(b) illustrates schematic symbols. As can be seen in the construction of a circular rheostat in Figure 4-11(c), one terminal is connected to one side of the track and the other terminal of this two-terminal device is connected to a movable wiper. As the wiper is moved away from the end of the track with the terminal, the resistance between the stationary end terminal and the mobile wiper terminal increases. This is summarized in Figure 4-11(d), where the wiper has been moved down by a clockwise rotation of the shaft. Current would have to flow through a large resistance as it travels from one terminal to the other. On the other hand, as the wiper is moved closer to the end of the track connected to the terminal, the resistance decreases. This is summarized in Figure 4-11(e), which shows that as the wiper is moved up, as a result of turning the shaft counterclockwise, current will see only a small resistance between the two terminals.

Rheostats come in many shapes and sizes, as can be seen in Figure 4-11(a). Some employ a straight-line motion to vary resistance, while others are classified as circular-motion rheostats. The resistive elements also vary; wirewound and carbon tracks are very popular. Cermet, which is a ceramic (insulator)–metal (conductor) mix the ratio of which can be used to produce different values of resistive tracks, is also used. A trimming rheostat is a miniature device used to change resistance by a small amount. Other circular-motion rheostats are available that require between two to ten turns to cover the full resistance range.

Potentiometer (three terminals: A, B, and C). Figure 4-12(a) illustrates the physical appearance of a variety of potentiometers, also called pots (slang), while Figure 4-12(b) shows its schematic symbol. You will probably notice that the difference between a rheostat and potentiometer is the number of terminals; the rheostat has two terminals and the potentiometer has three. With the rheostat, there were only two terminals and the resistance between the wiper and terminal varied as the wiper was adjusted. For the potentiometer illustrated in Figure 4-12(c), you can see that resistance can actually be measured across three separate combinations: between A and B (X), between B and C (Y), and between C and A (Z).

The only difference between the rheostat and the potentiometer in construction is the connection of a third terminal to the other end of the resistive track, as can be seen in Figure 4-12(d), which shows the single-turn potentiometer. Also illustrated in this section of the figure is the construction of a multiturn

Rheostat
Two-terminal variable resistor that, through mechanical turning of a shaft, can be used to vary its resistance and therefore its value of terminal to terminal current.

Potentiometer
Three-lead variable resistor that through mechanical turning of a shaft can be used to produce a variable voltage or potential.

(a) Physical Appearance

(b) Schematic Symbols

(c) Construction

(d) Increasing a Rheostat's Resistance

(e) Decreasing a Rheostat's Resistance

FIGURE 4-11 **Rheostat. (a) Physical Appearance. (b) Schematic Symbols. (c) Construction. (d) Increasing a Rheostat's Resistance. (e) Decreasing a Rheostat's Resistance.**

potentiometer in which a contact arm slides along a shaft and the resistive track is formed into a helix of 2 to 10 coils.

Figure 4-13(a) illustrates a 10-kΩ potentiometer. The resistance measured between terminals *A* and *C* will always be the same (10 kΩ) no matter where we put the wiper, because current still has to travel through the complete resistance between *A* and *C,* as illustrated in Figure 4-13(b) with a 10-kΩ potentiometer.

FIGURE 4-12   Potentiometer. (a) Physical Appearance. (b) Schematic Symbol. (c) Operation. (d) Construction.

The resistance between A and B (X) and B and C (Y) will, however, vary as the wiper's position is moved, as illustrated in Figure 4-14.

If the user mechanically turns the shaft in the clockwise direction, the resistance between A and B increases, while the resistance between B and C decreases. Similarly, if the user mechanically turns the shaft counterclockwise, a decrease occurs between A and B and there is a resulting increase in resistance between B and C. This point is summarized in Figure 4-15.

In some applications, you may see the symbol illustrated in Figure 4-16, where B is connected to C and only two terminals are hooked up by the user. In this situation, the potentiometer is being used as a rheostat.

(a)

Wiper in Mid Position   Wiper in Lower Position   Wiper in Upper Position

(b)

**FIGURE 4-13   A 10-kΩ Potentiometer. (a) Physical Appearance.
(b) End-to-End Resistance of a Potentiometer Remains Constant.**

Linear
Relationship between input and output in which the output varies in direct proportion to the input.

Tapered
Nonuniform distribution of resistance per unit length throughout the element.

Like the rheostat, the potentiometer comes in many different shapes and sizes. Wirewound, carbon, and cermet resistive tracks, circular or straight-line motion, 2 to 10 multiturn, and other variations are available for different applications.

Whether rheostat or potentiometer, the resistive track can be classified as having either a **linear** or a **tapered** (nonlinear) resistance. In Figure 4-17(a) we have taken a 1-kΩ rheostat and illustrated the resistance value changes between A and B for a linear and a tapered one-turn rheostat.

The definition of linear is having an output that varies in direct proportion to the input. The input, in this case, is the user turning the shaft, and the output, as can be seen, is the linearly increasing resistance between A and B.

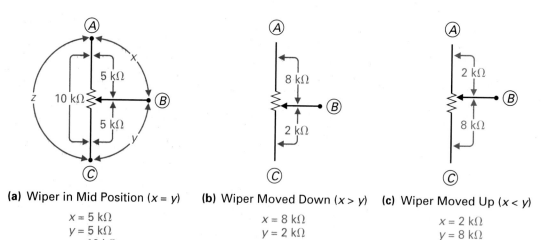

(a) Wiper in Mid Position ($x = y$)
$x = 5\ \text{k}\Omega$
$y = 5\ \text{k}\Omega$
$z = 10\ \text{k}\Omega$

(b) Wiper Moved Down ($x > y$)
$x = 8\ \text{k}\Omega$
$y = 2\ \text{k}\Omega$

(c) Wiper Moved Up ($x < y$)
$x = 2\ \text{k}\Omega$
$y = 8\ \text{k}\Omega$

FIGURE 4-14   Varying a Potentiometer's Resistance.

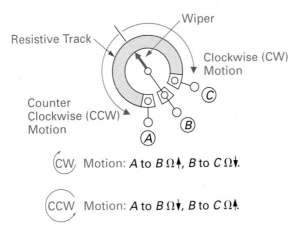

FIGURE 4-15   Potentiometer's Operation.

With a tapered rheostat or potentiometer, the resistance varies nonuniformly along its resistor element, sometimes being greater or less for equal shaft movement at various points along the resistance element, as shown in Figure 4-17(a). Figure 4-17(b) plots the position of the variable resistor's wiper against the resistance between the two output terminals. This graph effectively shows the difference between a linear increase and a nonlinear or tapered increase.

In Figure 4-18 you can see an application of a potentiometer as a volume control in a television set. By increasing or decreasing the potentiometer's resistance, the amount of current passing to the loudspeaker is varied and so is the volume.

Thermally (Heat) Adjustable Variable Resistors

When first discussing variable-value resistors, we talked about the rheostat and potentiometer, both of which have a mechanical input (the user turning the shaft) to produce a change in resistance. A **bolometer** is a device that, instead of changing its resistance when mechanical energy is applied, changes its resistance when heat energy is applied.

Bolometer
Device whose resistance changes when heated.

(a)

(b)

Permanent Connection
Between Ⓑ and Ⓒ

FIGURE 4-16   The Potentiometer as a Rheostat. (a) Schematic Symbol. (b) Physical Appearance.

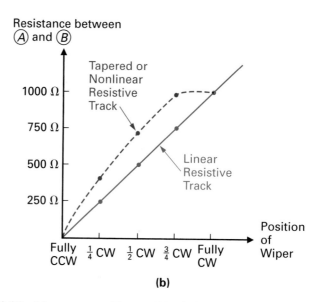

	Fully CCW	$\frac{1}{4}$ CW	$\frac{1}{2}$ CW	$\frac{3}{4}$ CW	Fully CW
1 kΩ Rheostat					
Linear	0 Ω	250 Ω $\left(\begin{array}{c}\frac{1}{4}\text{ of}\\1000\ \Omega\end{array}\right)$	500 Ω $\left(\begin{array}{c}\frac{1}{2}\text{ of}\\1000\ \Omega\end{array}\right)$	750 Ω $\left(\begin{array}{c}\frac{3}{4}\text{ of}\\1000\ \Omega\end{array}\right)$	1000 Ω
Tapered	0 Ω	350 Ω	625 Ω	900 Ω	1000 Ω

These values have been arbitrarily chosen to illustrate a nonlinear change.

(a)

(b)

FIGURE 4-17   Linear versus Tapered Resistive Track.

Thermometry
Relating to the measurement of temperature.

The measurement of temperature **(thermometry)** is probably the most common type of measurement used in industry today. Before discussing our three temperature sensors, let's first consider the four units of temperature measurement detailed in Table 4-1.

Commercially, temperature is normally expressed in degrees Celsius (°C) or degrees Fahrenheit (°F); however, although not commonly known, kelvin (K) and degrees Rankine (°R) are often used in industry, the kelvin being the international unit of temperature.

Potentiometer

FIGURE 4-18   Potentiometer as a Volume Control.

Resistive Temperature Detector (RTD)
A temperature detector consisting of a fine coil of conducting wire (such as platinum) which will produce a relatively linear increase in resistance as temperature increases.

Thin-film Detector (TFD)
A temperature detector containing a thin layer of platinum, and used for very precise temperature readings.

There are basically three different types of temperature detectors, all of which are illustrated in Figure 4-19. The **resistive temperature detector** (RTD) and **thin-film detector** (TFD) shown in Figure 4-19(a) and (b) are both temperature sensors that contain a conducting material such as copper, nickel, or platinum and consequently have a positive temperature coefficient: resistance increases as temperature increases.

EXAMPLE:

Convert the following:

a. $74°F = $ _____ $°C$

b. $45°C = $ _____ $°F$

c. $25°C = $ _____ K

d. $10°F = $ _____ $°R$

TABLE 4-1   Four Temperature Scales and Conversion Formulas

	FAHRENHEIT	CELSIUS	KELVIN	RANKINE	
Absolute zero	−459.69°F	−273.16°C	0 K	0°R	
Melting point of ice (x)	32°F	0°C	273.16 K	491.69°R	$F = (^9/_5 \times C) + 32$
	↓	↓	↓	↓	$C = {}^5/_9 \times (F − 32)$
(Division between x and y)	(180°F)	(100°C)	(100 K)	(180°R)	$R = F + 460$
	↓	↓	↓	↓	$F = R − 460$
Boiling point of water (y)	212°F	100°C	373.16 K	671.69°R	$K = C + 273$
					$C = K − 273$

Solution:

CALCULATOR SEQUENCE

Step	Keypad Entry	Display Response
1.	$\boxed{7}$ $\boxed{4}$	74
2.	$\boxed{-}$	
3.	$\boxed{3}$ $\boxed{2}$	32
4.	$\boxed{=}$	42
5.	$\boxed{STO}$ (store result in memory)	
6.	$\boxed{C/CE}$ (cancel display)	0
7.	$\boxed{5}$	5
8.	$\boxed{÷}$	
9.	$\boxed{9}$	9
10.	$\boxed{=}$	0.55555
11.	$\boxed{×}$	
12.	$\boxed{RCL}$ (Recall value from memory)	42
13.	$\boxed{=}$	23.3

a. $C = \frac{5}{9} \times (F - 32)$

   $= \frac{5}{9} \times (74 - 32)$

   $= 0.555 \times 42 = 23.3°C$

b. $F = (\frac{9}{5} \times C) + 32$

   $= (\frac{9}{5} \times 45) + 32$

   $= (1.8 \times 45) + 32 = 113°F$

c. $K = C + 273$

   $= 25 + 273 = 298 \text{ K}$

d. $R = F + 460$

   $= 10 + 460 = 470°R$

Referring to the construction of the RTD in Figure 4-19(a), you can see that the sensing element consists of a coil of fine wire generally made of platinum, which gives a relatively linear increase in resistance as temperature increases, as indicated in the table in Figure 4-19(a).

The thin-film detector (TFD) shown in Figure 4-19(b) is constructed by placing a thin layer of platinum, for very precise temperature readings, on a ceramic substrate. Because of its small size, the TFD responds rapidly to temperature change and is ideally suited for surface temperature sensing.

Thermistor
Temperature-sensitive semiconductor that has a negative temperature coefficient of resistance (as temperature increases, resistance decreases).

The **thermistor,** on the other hand, contains a semiconductor material that has a negative temperature coefficient: resistance decreases as temperature increases. A variety of thermistors can be seen in Figure 4-19(c). The thermistor is the most common type of temperature sensor and produces rapid and extremely large changes in resistance for very small changes in temperature, as shown in the associated table. The thermistor can be used as a temperature probe inside an oven. As the oven heats up, the thermistor's resistance decreases, and at a certain decreased resistance the thermistor turns off the oven. Once the temperature starts to drop, the thermistor's resistance increases, and this increase of resistance turns on the oven.

temperature ↑   thermistor's R ↓   oven off
temperature ↓   thermistor's R ↑   oven on

Optically (Light) Adjustable Variable Resistors

Photoresistor
Also known as a photoconductive cell or light dependant resistor, it is a device whose resistance varies with the illumination of the cell.

Photo means illumination, and the **photoresistor** is a resistor that is photoconductive. This means that as the material is exposed to light, it will become more conductive and less resistive. Figure 4-20 illustrates the photoresistor, also called a light-dependent resistor (LDR), and its schematic symbol.

(a)

Platinum Winding			
°C	Ohms	°C	Ohms
−200	18.53	+200	175.84
−150	39.65	+250	194.08
−100	60.20	+300	212.03
−50	80.25	+350	229.69
±0	100.0	+400	247.06
+50	119.40	+450	264.14
+100	138.50	+500	280.93
+150	157.32	+550	297.16

(Conductors have a positive temperature coefficient of resistance—temp.↑, R↑)

(b)

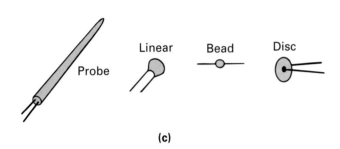

(c)

°C	Ohms
−50	100,000
0	7,500
+50	7,400
+100	100
+150	50
+200	27
+250	10
+300	7.5

(Semiconductors have a negative temperature coefficient of resistance—T↑, R↓)

FIGURE 4-19 Temperature Sensors. (a) Resistive Temperature Detector (RTD). (b) Thin-Film Detector (TFD). (c) Thermistor.

The photoresistor is a thin slice of photoconductive material whose resistance decreases as light is applied. The light energy is absorbed by the atoms within the photoconductive material, causing these atoms to release their valence electrons. This results in an increase in electrons, and therefore current passing through the photoresistor, so its resistance will have decreased. To summarize, we can say:

$$\text{light energy} \uparrow \quad \text{conduction} \uparrow \quad \text{resistance} \downarrow$$

Photoresistors can be used to turn on and off outdoor home security lights. During the day, the natural sunlight decreases the resistance of the photoresistor, and this low resistance is used to keep the lights off. At dusk, the sunlight is almost gone, and the photoresistor's resistance increases. This increased resistance is used to turn on the security lights.

$$\text{sun up} \uparrow \quad \text{photoresistance} \downarrow \quad \text{security lights OFF}$$
$$\text{sun down} \downarrow \quad \text{photoresistance} \uparrow \quad \text{security lights ON}$$

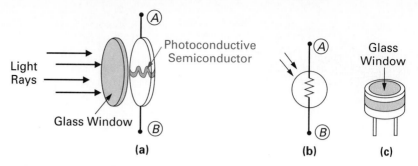

FIGURE 4-20 **Photoresistor. (a) Construction. (b) Schematic Symbol. (c) Physical Appearance.**

SELF-TEST REVIEW QUESTIONS FOR SECTION 4-1

1. List the six types of fixed-value resistors.
2. What is the difference between SIPs and DIPs?
3. Name the two types of mechanically adjustable variable resistors and state the difference between the two.
4. Describe a linear and a tapered potentiometer.
5. True or false: A thermistor has a negative temperature coefficient of resistance.
6. What name is given to the optically adjustable resistor?

4-2 HOW IS RESISTANCE MEASURED?

Ohmmeter

Measurement device used to measure electric resistance.

Resistance is measured with an **ohmmeter.** Figure 4-21(a) illustrates how a multirange analog ohmmeter and a digital ohmmeter can be used to measure the resistance of a resistor. The analog ohmmeter has four ranges. The simplest range has been selected ($R \times 1$), and on this scale the resistance value indicated can be read directly from the meter; in this example the value is 36 Ω. If we had selected the $R \times 10$ range and a value of 72 was obtained, then 72 would have to be multiplied by 10 ($72 \times 10 = 720$) to obtain the value of 720 Ω. If we were on the $R \times 100$ range and a reading of 15 was obtained, then $R \times 100 = 15 \times 100$, which equals 1.5 kΩ. The $R \times 1000$ scale, if selected, means that you would interpret 470 for example as $R \times 1000 = 470 \times 1000$, which equals 470 kΩ.

Three rules must be remembered and applied when measuring resistance.

1. Short the meter leads together and adjust the zero-ohms control so that the pointer is at 0 Ω. This calibrates the meter scale.
2. Turn off the circuit power and remove the component (if practicable) to be measured from circuit to protect the ohmmeter.

(a)

(b)

FIGURE 4-21 Measuring Resistance.

3. Connect the leads across the removed component, adjust the range scale until the pointer is approximately in the middle of the scale (for accuracy), and then multiply this value by the range scale selected.

With the digital auto-ranging ohmmeter shown in Figure 4-21(b), the measurement of resistance is a lot easier since you simply turn the selector to ohms and read off the reading from the display.

SELF-TEST REVIEW QUESTIONS FOR SECTION 4-2

1. Name the instrument used to measure resistance.
2. On the $R \times 1000$ range, how would a reading of 22 be interpreted?
3. On the $R \times 1$ range, how would a reading of 470 be interpreted?
4. List the three rules that should be applied when using the analog ohmmeter.

4-3 RESISTOR CODING

Using the ohmmeter, you can find the value of any resistor. However, manufacturers indicate the value and tolerance of resistors on the body of the component in one of two ways:

1. *Color code:* colored rings or bands
2. *Typographically:* printed alphanumerics (alphabet and numerals)

Examples of both of these can be seen in Figure 4-22.

Alphanumerics
(Alphabet and Numerals)

Color Coded
Rings or Bands

(a)   **(b)**

FIGURE 4-22   **Resistor Value and Tolerance Indication.**
(a) Typographical. (b) Color Code.

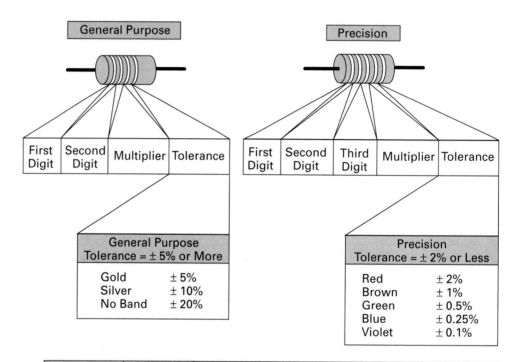

FIGURE 4-23   **General-Purpose and Precision Resistor Color Code.**

4-3-1 General-Purpose and Precision Color Code

There are basically two different types of fixed resistors: general purpose and precision. Resistors with tolerances of ±2% or less are *precision resistors* and have five bands; resistors with tolerances of ±5% or greater have four bands and are referred to as *general-purpose resistors*. The color code and differences between precision and general-purpose resistors are illustrated in Figure 4-23.

When you pick up a resistor, notice that the bands are nearer one end; this end should be held in your left hand. If there are four bands on the resistor, follow the general-purpose resistor code; if five bands are present, adopt the precision resistor code.

General-Purpose Resistor Code

1. The first band on either a general-purpose or precision resistor can never be black, and it is the first digit of our number.
2. The second band indicates the second digit of the number.
3. The third band specifies the multiplier to be applied to the number, which ranges from × 1/100 to 10,000,000.
4. The fourth band describes the tolerance or deviation from the specified resistance, which is ±5% or greater.

Precision Resistor Code

1. The first band, like the general-purpose resistor, is never black and is the first digit of the three-digit number.
2. The second band provides the second digit.
3. The third band indicates the third and final digit of the number.
4. The fourth band specifies the multiplier to be applied to the number.
5. The fifth and final band indicates the tolerance figure of the precision resistor, which is always less than ±2%, which is why precision resistors are more expensive than general-purpose resistors.

EXAMPLE:

Figure 4-24 illustrates a 1/2-W resistor.

   a. Is this a general or a precision resistor?

   b. What is the resistor's value of resistance?

   c. What tolerance does this resistor have, and what deviation plus and minus could occur to this value?

■ *Solution:*

   a. General purpose (four bands)

   b. green   blue × brown

      5   6 × 10 = 560 Ω

Band 1 = Green
Band 2 = Blue
Band 3 = Brown
Band 4 = Gold

FIGURE 4-24

c. Tolerance band is gold, which is ±5%.

$$\text{deviation} = 5\% \text{ of } 560 = 28$$
$$560 + 28 = 588 \ \Omega$$
$$560 - 28 = 532 \ \Omega$$

The resistor could, when measured, be anywhere from 532 to 588 Ω.

EXAMPLE:

State the resistor's value and tolerance and whether it is general purpose or precision for the examples shown in Figure 4-25.

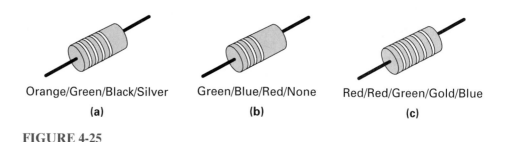

Orange/Green/Black/Silver   Green/Blue/Red/None   Red/Red/Green/Gold/Blue

(a)   (b)   (c)

FIGURE 4-25

Solution:

a. orange   green   black   silver   (four bands = general purpose)
      3   5   ×   1   10%   = 35 Ω ± 10%

b. green   blue   red   none   (four bands = general purpose)
      5   6   × 100   20%   = 5.6 kΩ ± 20%

c. red   red   green   gold   blue   (five bands = precision)
     2   2   5   × 0.1   0.25%   = 22.5 Ω ± 0.25%

4-3-2   *Zero-Ohms Resistor*

Figure 4-26(a) illustrates the only resistor color code that you will probably have trouble with, the zero-ohms resistor, which has one black band. Zero ohms is equivalent to a straight piece of wire. You may ask yourself: Why do they manufacture a resistor of zero ohms since it really is not a resistor at all?

Printed circuit boards like the one shown in Figure 4-26(b) no longer have all their resistor and other components inserted by hand. *Automatic insertion equipment* replaces the human assembler and places the correct component in the correct position in a matter of seconds rather than hours. In some applications, a direct contact between two points is needed, in which case a piece of wire needs to be inserted, as shown in Figure 4-26(b). However, the resistor automatic insertion equipment will only handle resistors, not wire (called jumpers). In the past, manufacturers had to install jumpers manually after all the components were inserted. This caused enormous delays; but now the zero-ohms

(a)

Direct Contact between Two Pads

Printed Circuit Board

Equivalent to

(b)   (c)

FIGURE 4-26   Zero-Ohms Resistor. (a) Appearance. (b) Printed Circuit Board. (c) Zero-Ohms Resistor as Jumper.

ohms resistor can be installed automatically during the resistor insertion process and have exactly the same effect, as shown in Figure 4-26(c).

4-3-3   Other Resistor Identification Methods

If a resistor color code is not found, some other form of marking will always be made on the resistor. The larger fixed, wirewound resistors and nearly all variable resistors normally have the resistance value, tolerance and wattage printed on the resistor, as shown in Figure 4-27(a). SIP and DIP packages are normally coded with both letters and numerals, all of which have particular meaning, as shown in Figure 4-27(b). Chip resistors tend to be too small to have any intelligible form of marking and so will have to be measured with an ohmmeter. They are packaged, as previously shown in Figure 4-10(c), in either polythene bags, paper tape, or plastic magazines, and in all these cases the packaging will be used to specify values, tolerance, and wattage.

Value = 5000 Ω
Tolerance = ± 10%
Power Rating = 5 watt

(a)

Number of Pins
Number of Resistors
Configuration of Resistors
Dot Marks Pin 1 → 6–5–2–330 — Value of Resistors

FIGURE 4-27   Typographical Value Indication.
(a) Standard Variable and Fixed. (b) SIP and DIP.

SELF-TEST REVIEW QUESTIONS FOR SECTION 4-3

1. What tolerance differences occur between a general-purpose and a precision resistor?
2. A red/red/red/gold/blue resistor has what value and tolerance figure?
3. What color code would appear on a 4.7-kΩ, ±10% tolerance resistor?
4. Would a three-band resistor be considered a general-purpose type?

4-4   POWER DISSIPATION DUE TO RESISTANCE

The atoms of a resistor obstruct the flow of moving electrons (current), and this friction between the two causes heat to be generated. Current flow through any resistance therefore will always be accompanied by heat. In some instances, such as a heater or oven, the device has been designed specifically to generate heat. In other applications it may be necessary to include a resistor to reduce current flow, and the heat generated will be an unwanted side effect.

Any one of the three power formulas can be used to calculate the power dissipated by a resistance.

$$P = V \times I \qquad P = I^2 \times R \qquad P = \frac{V^2}{R}$$

However, since power is determined by the friction between current (I) and the circuit's resistance (R), the $P = I^2 \times R$ formula is more commonly used to calculate the heat generated.

EXAMPLE:

What wattage rating should a 33-Ω resistor have if the value of current is 100 mA?

Solution:

$$P = I^2 \times R$$
$$= (100 \text{ mA}^2) \times 33 \text{ Ω}$$
$$= 0.33 \text{ W}$$

It would be safe therefore to use a ½-W (0.5-W) resistor.

EXAMPLE:

A coffee cup warming plate has a power rating of 120 V, 23 W. How much current is flowing through the heating element, and what is the heating element's resistance?

Solution:

$$P = V \times I \qquad I = \frac{P}{V}$$

$$I = \frac{23 \text{ W}}{120 \text{ V}} = 191.7 \text{ mA}$$

$$R = \frac{V}{I} = \frac{120 \text{ V}}{191.7 \text{ mA}} = 626 \text{ Ω}$$

EXAMPLE:

Calculate the current drawn by a 100-W light bulb if it is first used in the home (and therefore connected to a 120-V source) and then used in the car (and therefore connected across a 12-V source).

Solution:

$$P = V \times I \qquad \text{therefore,} \qquad I = \frac{P}{V}$$

In the home,

$$I = \frac{P}{V} = \frac{100 \text{ W}}{120 \text{ V}} = 0.83 \text{ A}$$

In the car,

$$I = \frac{P}{V} = \frac{100 \text{ W}}{12 \text{ V}} = 8.33 \text{ A}$$

This example brings out a very important point in relation to voltage, current, and power. In the home, only a small current needs to be supplied by the source when the applied voltage is large ($P = V \uparrow \times I \downarrow$); however, when the source voltage was small, as in the case of the car, a large current must be supplied by the source in order to deliver the same amount of power ($P = V \downarrow \times I \uparrow$).

4-5 FILAMENT RESISTOR

Filament Resistor
The resistor in a light bulb or electron tube.

Incandescent Lamp
An electric lamp that generates light when an electric current is passed through its filament of resistance, causing it to heat to incandescence.

Figure 4-28(a) illustrates the **filament resistor** within a glass bulb. This component is more commonly known as the household light bulb. The filament resistor is just a coil of wire that glows white hot when current is passed through it and, in so doing, dissipates both heat and light energy.

This **incandescent lamp** is an electric lamp in which electric current flowing through a filament of resistive material heats the filament until it glows and emits light. Figure 4-28(b) shows a test circuit for varying current through a small incandescent lamp. Potentiometer R_1 is used to vary the current flow through the lamp. The lamp is rated at 6 V, 60 mA. Using Ohm's law we can calculate the lamp's filament resistance at this rated voltage and current:

$$\text{filament resistance, } R = \frac{V}{I} = \frac{6 \text{ V}}{60 \text{ mA}} = 100 \text{ }\Omega$$

Cold Resistance
The resistance of a device when cold.

The filament material (tungsten, for example) is like all other conductors in that it has a positive temperature coefficient of resistance. Therefore, as the current through the filament increases, so does the temperature and so does the filament's resistance ($I \uparrow$, temperature $\uparrow$, $R\uparrow$). Consequently, when R_1's wiper is moved to the right so that it produces a high resistance ($R_1\uparrow$), the circuit current will be small ($I \downarrow$) and the lamp will glow dimly. Since the circuit current is small, the filament temperature will be small and so will the lamp's resistance ($I\downarrow$, temperature $\downarrow$, lamp resistance $\downarrow$). This small value of resistance is called the lamp's **cold resistance.** On the other hand, when R_1's wiper is moved to the left so that it produces a small resistance ($R\downarrow$), the circuit current will be large ($I\uparrow$), and the lamp will glow brightly. With the circuit current high, the filament temperature will be high and so will the lamp's resistance ($I\uparrow$, temperature $\uparrow$, lamp resistance $\uparrow$). This large value of resistance is called the lamp's **hot resistance.**

Hot Resistance
The resistance of a device when hot due to the generation of heat by electric current.

Figure 4-28(c) plots the filament voltage, which is being measured by the voltmeter, against the filament current, which is being measured by the ammeter. As you can see, an increase in current will cause a corresponding increase in filament resistance. Studying this graph, you may have also noticed that the lamp has been operated beyond its rated value of 6 V, 60 mA. Although the lamp can be operated beyond its rated value, at, for example 10 V, 80 mA, its life expectancy will be decreased dramatically—from several hundred hours to only a few hours.

Ballast Resistor
A resistor that increases in resistance when current increases. It can therefore maintain a constant current despite variations in line voltage.

A coil of wire similar to the one found in a light bulb, called a **ballast resistor,** is used to maintain a constant current despite variation in voltage. Since the coil of wire is a conductor and conductors have a positive temperature coefficient, an increase in voltage will cause a corresponding increase in current and therefore in the heat generated, which will result in an increase in the wire's resistance. This increase in resistance will decrease the initial current rise. Similarly, a decrease in voltage and therefore current ($I \propto V$) will result in a decrease in heat and wire resistance. This decrease in resistance will permit an increase in current to counteract the original decrease. Current is therefore regulated or maintained constant by the ballast resistor, despite variations in voltage.

(a)

By changing the resistance of the potentiometer, the current flow through the lamp is varied.

R_1

Decrease Resistance | Increase Resistance

This ammeter measures the milliamperes of current flowing through the lamp.

AM

The voltmeter measures the voltage drop across the lamp.

12 V

VM

(b)

$R = V/I = 10\ \text{V}/80\ \text{mA} = 125\ \Omega$

$R = V/I = 6\ \text{V}/60\ \text{mA} = 100\ \Omega$ (Hot Resistance)

$R = V/I = 4\ \text{V}/50\ \text{mA} = 80\ \Omega$

$R = V/I = 1\ \text{V}/20\ \text{mA} = 50\ \Omega$ (Cold Resistance)

(c)

FIGURE 4-28 **Filament Resistor.**

1. Which formula is most commonly used to calculate the heat generated by a resistor?

2. An incandescent lamp has a _____ (large/small) cold resistance and a _____ (large/small) hot resistance.

4-6 TESTING RESISTORS

It is virtually impossible for resistors to short-circuit (no resistance or zero ohms) internally; however, they may be shorted out by some other device, providing a low-resistance path around the resistor. Generally, the resistor's internal resistive elements will begin to develop a higher resistance than its specified value (due to a partial internal open) or in some cases go completely open circuit (maximum resistance or infinite ohms).

Fixed-value resistors tend to go either partially or fully open as the resistive track begins to break down. Variable-value resistors have problems with the wiper making contact with the resistive track at all times. Faulty variable-value resistors in audio systems can normally be detected because they generate a scratchy noise whenever, for example, you adjust the volume or tone control.

The ohmmeter is the ideal instrument for verifying whether or not a resistor is functioning correctly. When checking a resistor's resistance, here are some points that should be remembered.

1. Always calibrate the meter first to zero. If the meter indicates 100 Ω when the two meter leads are shorted together (0 Ω), every reading you take will be in error by 100 ohms.

2. The ohmmeter has its own internal power source (a battery), so always turn off the circuit power and remove from the circuit the resistor to be measured. If this is not done, you will not only obtain inaccurate readings, but you can damage the ohmmeter.

3. The range scales on a digital multimeter often confuse people; for example, if a range is chosen so that the highest reading is 100 kΩ and a 1-MΩ resistor is measured, the meter will indicate an infinite-ohms reading and you may be misled into believing that you have found the problem (an open resistor). In the case of an analog ohmmeter, if the range scale multiplier is not applied, a misleading result is obtained. For example, if the $R \times 100$ range is selected, the resistance reading (R) must be multiplied by 100 to obtain the resistance.

4. Another point to keep in mind is tolerance. For example, if a suspected 1-kΩ (1000-Ω) resistor is measured with an ohmmeter and reads 1.2 kΩ (1200 Ω), it could be within tolerance if no tolerance band is present on the resistor's body (±20%). A 1-kΩ resistor with ±20% tolerance could measure anywhere between 800 and 1200 Ω

5. The ohmmeter's internal battery voltage is really too small to deliver an electrical shock; however, you should avoid touching the bare metal parts of the probes or resistor leads, as your body resistance of approximately 50 kΩ will affect your meter reading.

SELF-TEST REVIEW QUESTIONS FOR SECTION 4-6

1. What problem normally occurs with resistors? (open/shorts)
2. What instrument is used to verify a resistor's resistance?

SUMMARY

1. Conductors offer a certain small amount of resistance; however, in electronics this resistance is not normally enough, so additional resistance is needed to control the amount of current flow. The component used to supply this additional resistance is called a *resistor.*

There are two basic types of resistors: fixed value and variable value. The fixed-value resistor has a value of resistance that cannot be changed and is the more common of the two. The variable-value resistor, on the other hand, has a range of values that can be selected generally by mechanically adjusting a control.

Fixed-Value Resistors (Figure 4-29)

2. *Fixed-value resistors* can be divided into six basic categories:

   a. Carbon composition   **d.** Wirewound
   b. Carbon film   **e.** Metal oxide
   c. Metal film   **f.** Thick film

3. Carbon composition resistors are the most common and least expensive type of fixed-value resistor. It is constructed by placing a piece of resistive material, with embedded conductors at each end, within an insulating cylindrical molded case.

This resistive material is called carbon composition because powdered carbon and powdered insulator are bonded into a compound and used as the resistive material. By changing the ratio of powdered insulator to carbon, the value of resistance can be changed within the same area.

4. The physical size of the resistors lets the user know how much power in the form of heat can be *dissipated.* As you already know, resistance is the opposition to current flow or electrons, and this opposition causes heat to be generated when current flows. The amount of heat dissipated per unit of time is measured in watts and each resistor has its own *wattage rating.* The bigger the resistor, the more heat that can be dissipated. The key point is that heat should be dissipated faster than it is generated; if not, a larger resistor with a larger surface area to dissipate away the additional heat should be used so that the resistor will not burn up.

FIGURE 4-29 Fixed Value Resistors.

Type	Cost	Resistive Material	Advantages/Disadvantage
Carbon Composition	$	Powdered carbon and carbon insulator.	Low Cost but inherently noisy.
Carbon Film	$$	A thin film of carbon and insulator.	Smaller tolerances (±5% to ±2%) and better temperature stability than carbon composition.
Metal Film	$$$	Thin metal film spiral on substrate.	Best tolerances (±1% to ±0.1%) and temperature stability; however, high cost.
Wirewound	$$$$	Length of wire wrapped around ceramic core.	High power rating and good tolerance (±1%); however, very high cost.
Metal Oxide	$$$	Oxide of a metal on an insulating substrate.	Has the best temperature stability; however, high cost.
Thick Film Networks Chips	$$$$ $$	Thick film of resistive paste on insulating substrate.	Small size makes them ideal for resistor networks (SIPs and DIPs) and chip resistors for SMT.

5. Another factor to consider when discussing resistors is their *tolerance*. Tolerance is the amount of deviation or error from the specified value. For example, a 1000-Ω (1-kΩ) resistor with a ± 10% (plus and minus 10%) tolerance when manufactured could have a resistance anywhere between 900 and 1100 Ω. In some applications where high precision is required, smaller tolerance resistors are used.

6. Carbon film resistors have a thin layer or film of resistive material (a blend of carbon and insulator) on a ceramic (insulating) substrate. The film is then cut to form a helix or spiral. Carbon film resistors have smaller tolerance figures (±5% to ±2%), are more stable (maintain same resistance value over a wide

range of temperatures), and have less internally generated noise (random small bursts of voltage) than do carbon composition resistors.

7. *Metal film resistors* are constructed by spraying a thin film of metal on a ceramic cylinder (or *substrate*) and then cutting the film to form a substrate. Metal film resistors have possibly the best tolerances commercially available of ±1% to ±0.1%. They also maintain a very stable resistance over a wide range of temperatures (good stability) and generate very little internal noise compared to any carbon resistor.

8. *Wirewound resistors* are constructed by wrapping a length of wire uniformly around a ceramic insulating core, with terminals making the connections at each end. The length and thickness of the wire are varied to change the resistance, which ranges from 1 Ω to 150 kΩ.

    Since current flow is opposed by resistors and this opposition generates heat, the larger the physical size of the resistor, the greater the amount of heat that can be dissipated away and so the greater the current that can be passed through the resistor.

    Wirewound resistors are generally used in applications requiring low resistance values, which means that the current and therefore power dissipated are high ($R\downarrow = V/I\uparrow$, $P\uparrow = I^2\uparrow R$). The amount of heat that can be dissipated is measured in watts and is indicated on the wirewound resistor.

9. The *metal oxide resistors* are constructed by depositing an oxide of a metal such as tin onto an insulating substrate. The ratio of oxide (insulator) to tin (conductor) will determine the resistor's resistance.

    Metal oxide resistors have excellent temperature stability, which is the ability of the resistor to maintain its value of resistance without change even when temperature is changed.

    There is a tendency when heat is increased for atoms to move, and in so doing they cause collisions, resulting in an opposition to current; so a temperature increase will cause a resistor's resistance to increase. This increase from the desired resistance is very small with metal oxide resistors.

10. The two different types of *thick-film resistor* networks are called SIPs and DIPs. The *single in-line package* is so called because all its lead connections are in a single line, whereas the *dual in-line package* has two lines of connecting pins. The chip resistors are small thick-film resistors that are approximately the size of a pencil lead.

    The SIP and DIP resistor networks are constructed by first screening on the internal conducting strip (silver) that connects the external pins to the resistive material, and then screening on the thick film of resistive paste (bismuth/ruthenate), the blend of which will determine resistance. The chip resistor uses the same resistive film paste material, which is deposited onto an insulating substrate with two end terminations and protected by a glass coat.

    The SIP and DIP resistor networks, once constructed, are trimmed by lasers to obtain close tolerances of typically ±2%. Resistance values ranging from 22 Ω to 2.2 MΩ are available, with a power rating of ½ W.

    The chip resistor is commercially available with resistance values from 10 Ω to 3.3 MΩ, a ±2% tolerance, and a ⅛-W heat dissipation capability.

Variable-Value Resistors

11. The resistance of *variable resistors* can be varied in one of three ways:
    a. Mechanically (user) adjustable:
       1) Rheostat
       2) Potentiometer
    b. Thermally (heat) adjustable:
       1) Thermistor
       2) Resistive temperature detector (RTD)
       3) Thin-film detector (TFD)
    c. Optically (light) adjustable:
       1) Photoresistor

12. Mechanically (user) adjustable variable resistors (Figure 4-30) include two types, both of which will cause a change in resistance when a shaft is rotated. The *rheostat* is a two-terminal variable resistor that, through mechanical turning of a shaft, can be used to vary its resistance and therefore its value of terminal to terminal current.

13. Rheostats come in many shapes and sizes. Some employ a straight-line motion to vary resistance, while others are classified as circular-motion rheostats. The resistive elements also vary; wirewound and carbon tracks are very popular. Cermet, which is a ceramic (insulator)–metal (conductor) mix the ratio of which can be used to produce different values of resistive tracks, is also used. A trimming rheostat is a miniature device used to change resistance by a small amount. Other circular-motion rheostats are available that require between two to ten turns to cover the full resistance range.

14. The *potentiometer* is a three-lead variable resistor that through mechanical turning of a shaft can be used to produce a variable voltage or potential.

15. The difference between a rheostat and potentiometer is the number of terminals; the rheostat has two terminals and the potentiometer has three. With the rheostat, there were only two terminals and the resistance between the wiper and terminal varied as the wiper was adjusted. For the potentiometer, the resistance can actually be measured across three separate combinations: between *A* and *B* (*X*), between *B* and *C* (*Y*), and between *C* and *A* (*Z*).

    The only difference between the rheostat and the potentiometer in construction is the connection of a third terminal to the other end of the resistive track.

16. Like the rheostat, the potentiometer comes in many different shapes and sizes. Wirewound, carbon, and cermet resistive tracks, circular or straight-line motion, 2 to 10 multiturn, and other variations are available for different applications.

17. Whether rheostat or potentiometer, the resistive track can be classified as having either a *linear* or a *tapered* (nonlinear) resistance.

    The definition of linear is having an output that varies in direct proportion to the input. The input, in this case, is the user turning the shaft,

FIGURE 4-30   Mechanically (User) Adjustable Variable Resistor.

Rheostat

(b) Schematic Symbols

Terminal Ⓐ
Large Resistance
Small Current Flow
Terminal Ⓑ

Rheostat Control Turned Clockwise for Large Resistance

(d) Increasing a Rheostat's Resistance

Ⓐ
Small Resistance
Large Current Flow
Ⓑ

Rheostat Control Turned Counterclockwise for Small Resistance

(e) Decreasing a Rheostat's Resistance

Potentiometer

Wiper
Resistive Track

Single-turn Pot   (d)

Terminal Ⓐ
z
Terminal Ⓑ
Terminal Ⓒ

(c)

Resistive Track
Multiturn Pot
Contact Arm

Ⓐ
5 kΩ
10 kΩ   Ⓑ
5 kΩ
Ⓒ

(a) Wiper in Mid Position (x = y)

x = 5 kΩ
y = 5 kΩ
z = 10 kΩ

Ⓐ
8 kΩ
Ⓑ
2 kΩ
Ⓒ

(b) Wiper Moved Down (x > y)

x = 8 kΩ
y = 2 kΩ

Ⓐ
2 kΩ
Ⓑ
8 kΩ
Ⓒ

(c) Wiper Moved Up (x < y)

x = 2 kΩ
y = 8 kΩ

Linear versus Tapered Resistive Tracks

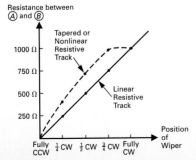

Resistance between Ⓐ and Ⓑ

Tapered or Nonlinear Resistive Track

1000 Ω
750 Ω
500 Ω
250 Ω

Linear Resistive Track

Position of Wiper

Fully CCW   ¼ CW   ½ CW   ¾ CW   Fully CW

163

and the output, as can be seen, is the linearly increasing resistance between *A* and *B*.

With a tapered rheostat or potentiometer, the resistance varies nonuniformly along its resistor element, sometimes being greater or less for equal shaft movement at various points along the resistance element.

18. Thermally (heat) adjustable variable resistor (Figure 4-31). A *bolometer* is a device that, instead of changing its resistance when mechanical energy is applied, changes its resistance when heat energy is applied.

19. The measurement of temperature (*thermometry*) is probably the most common type of measurement used in industry today. Commercially, temperature is normally expressed in degrees Celsius (°C) or degrees Fahrenheit (°F); however, although not commonly known, kelvin (K) and degrees Rankine (°R) are often used in industry, the kelvin being the international unit of temperature.

There are basically three different types of temperature detectors. The *resistive temperature detector* (RTD) and *thin-film detector* (TFD) are both temperature sensors that contain a conducting material such as copper, nickel, or platinum and consequently have a positive temperature coefficient: resistance increases as temperature increases.

The *thermistor,* on the other hand, contains a semiconductor material that has a negative temperature coefficient: resistance decreases as temperature increases. The thermistor is the most common type of temperature sensor and produces rapid and extremely large changes in resistance for very small changes in temperature.

20. Optically (light) adjustable variable resistor (Figure 4-31). *Photo* means illumination, and the *photoresistor* is a resistor that is photoconductive. This means that as the material is exposed to light, it will become more conductive and less resistive.

Photoresistors can be used to turn on and off outdoor home security lights. During the day, the natural sunlight decreases the resistance of the photoresistor, and this low resistance is used to keep the lights off. At dusk, the sunlight is almost gone, and the photoresistor's resistance increases. This increased resistance is used to turn on the security lights.

Measuring Resistance and Resistor Coding (Figure 4-32)

21. Resistance is measured with an *ohmmeter.* The analog ohmmeter has four ranges. The simplest range has been selected ($R \times 1$), and on this scale the resistance value indicated can be read directly from the meter. If we had selected the $R \times 10$ range and a value of 72 was obtained, then 72 would have to be multiplied by 10 ($72 \times 10 = 720$) to obtain the value of 720 Ω. If we were on the $R \times 100$ range and a reading of 15 was obtained, then $R \times 100 = 15 \times 100$, which equals 1.5 k$\Omega$. The $R \times 1000$ scale, if selected, means that you would interpret 470 for example as $R \times 1000 = 470 \times 1000$, which equals 470 k$\Omega$.

FIGURE 4-31 Thermally and Optically Adjustable Variable Resistors.

Thermally (Heat) Adjustable Variable Resistors

	FAHRENHEIT	CELSIUS	KELVIN	RANKINE
Absolute zero	(180°F)	−273.16°C	0 K	0°R
Melting point	↓	0°C	273.16 K	491.69°R
of ice (x)	212°F	↓	↓	↓
(Division between		(100°C)	(100 K)	(180°R)
x and y)		↓	↓	↓
Boiling point		100°C	373.16 K	671.69°R
of water (y)				
−459.69°F				
32°F				

$$F = (^9/_5 \times C) + 32$$
$$C = ^5/_9 \times (F - 32)$$
$$R = F + 460$$
$$F = R - 460$$
$$K = C + 273$$
$$C = K - 273$$

→ Probe Placed in Area Where Temperature Is to Be Sensed

Leads

Wire Coil Element

Actual Size

Platinum Winding			
°C	Ohms	°C	Ohms
−200	18.53	+200	175.84
−150	39.65	+250	194.08
−100	60.20	+300	212.03
−50	80.25	+350	229.69
±0	100.0	+400	247.06
+50	119.40	+450	264.14
+100	138.50	+500	280.93
+150	157.32	+550	297.16

(Conductors have a positive temperature coefficient of resistance—temp.↑, R↑)

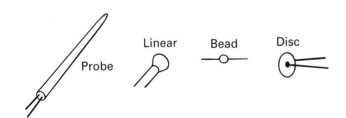

Probe   Linear   Bead   Disc

°C	Ohms
−50	100,000
0	7,500
+50	7,400
+100	100
+150	50
+200	27
+250	10
+300	7.5

(Semiconductors have a negative temperature coefficient of resistance—T↑, R↓)

Optically (Light) Adjustable Variable Resistor

Ⓐ

Light Rays

Photoconductive Semiconductor

Glass Window

Ⓑ

Ⓐ

Glass Window

Ⓑ

FIGURE 4-32   Measuring Resistance and Resistor Coding.

Measuring Resistance

36 Ω Resistors

(a)   (b)

Resistor Color Code

General Purpose

Precision

| | First Digit | Second Digit | Multiplier | Tolerance |

| | First Digit | Second Digit | Third Digit | Multiplier | Tolerance |

General Purpose
Tolerance = ± 5% or More

Gold	± 5%
Silver	± 10%
No Band	± 20%

Precision
Tolerance = ± 2% or Less

Red	± 2%
Brown	± 1%
Green	± 0.5%
Blue	± 0.25%
Violet	± 0.1%

	Color	Digit Value	Multiplier	
Big	Black	0	1 One	1
Beautiful	Brown	1	10 One Zero	10
Roses	Red	2	100 Two Zeros	100
Occupy	Orange	3	1000 Three Zeros	1 k
Your	Yellow	4	10000 Four Zeros	10 k
Garden	Green	5	100000 Five Zeros	100 k
But	Blue	6	1000000 Six Zeros	1 M
Violets	Violet	7	10000000 Seven Zeros	10 M
Grow	Gray	8	–	
Wild	White	9		
So	Silver	–	10^{-2} or 0.01	1/100
Get some	Gold	–	10^{-1} or 0.1	1/10
Now	None	–		

Typographical Valve Indication

Value = 5000 Ω
Tolerance = ± 10%
Power Rating = 5 watt

22. Three rules must be remembered and applied when measuring resistance:

   a. Short the meter leads together and adjust the zero-ohms control so that the pointer is at 0 Ω. This calibrates the meter scale.

   b. Turn off the circuit power and remove the component (if practicable) to be measured from circuit to protect the ohmmeter.

   c. Connect the leads across the removed component, adjust the range scale until the pointer is approximately in the middle of the scale (for accuracy), and then multiply this value by the range scale selected.

23. With the digital auto-ranging ohmmeter the measurement of resistance is a lot easier since you simply turn the selector to ohms and read off the reading from the display.

24. Using the ohmmeter, you can find the value of any resistor. However, manufacturers indicate the value and tolerance of resistors on the body of the component in one of two ways:

   a. *Color code:* colored rings or bands

   b. *Typographically:* printed alphanumerics (alphabet and numerals)

25. There are basically two different types of fixed resistors: general purpose and precision. Resistors with tolerances of ±2% or less are *precision resistors* and have five bands; resistors with tolerances of ±5% or greater have four bands and are referred to as *general-purpose resistors.*

26. For the general-purpose resistor code:

   a. The first band on either a general-purpose or precision resistor can never be black, and it is the first digit of our number.

   b. The second band indicates the second digit of the number.

   c. The third band specifies the multiplier to be applied to the number, which ranges from $\times \frac{1}{100}$ to 10,000,000.

   d. The fourth band describes the tolerance or deviation from the specified resistance, which is ±5% or greater.

27. For the precision resistor code:

   a. The first band, like the general-purpose resistor, is never black and is the first digit of the three-digit number.

   b. The second band provides the second digit.

   c. The third band indicates the third and final digit of the number.

   d. The fourth band specifies the multiplier to be applied to the number.

   e. The fifth and final band indicates the tolerance figure of the precision resistor, which is always less than ±2%, which is why precision resistors are more expensive than general-purpose resistors.

28. Printed circuit boards no longer have all their resistor and other components inserted by hand. *Automatic insertion equipment* replaces the human assembler and places the correct component in the correct position in a matter of seconds rather than hours. The zero-ohms resistor can be installed automatically during the resistor insertion process and have exactly the same effect as a wire jumper.

29. If a resistor color code is not found, some other form of marking will always be made on the resistor. The larger fixed, wirewound resistors and nearly all variable resistors normally have the resistance value, tolerance, and wattage printed on the resistor. SIP and DIP packages are normally coded with both letters and numerals, all of which have particular meaning. Chip resistors tend to be too small to have any intelligible form of marking and so will have to be measured with an ohmmeter. They are packaged in either polythene bags, paper tape, or plastic magazines, and in all these cases the packaging will be used to specify values, tolerance, and wattage.

Power Dissipation Due to Resistance

30. The atoms of a resistor obstruct the flow of moving electrons (current), and this friction between the two causes heat to be generated. Current flow through any resistance therefore will always be accompanied by heat. In some instances, such as a heater or oven, the device has been designed specifically to generate heat. In other applications it may be necessary to include a resistor to reduce current flow, and the heat generated will be an unwanted side effect.

31. Any one of the three power formulas can be used to calculate the power dissipated by a resistance. However, since power is determined by the friction between current (I) and the circuit's resistance (R), and $P = I^2 \times R$ formula is more commonly used to calculate the heat generated.

Filament Resistor

32. The *filament resistor* is more commonly known as the household light bulb. The filament resistor is just a coil of wire that glows white hot when current is passed through it and, in so doing, dissipates both heat and light energy.

   This *incandescent lamp* is an electric lamp in which electric current flowing through a filament of resistive material heats the filament until it glows and emits light.

   The filament material (tungsten, for example) is like all other conductors in that it has a positive temperature coefficient of resistance. Therefore, as the current through the filament increases, so does the temperature and so does the filament's resistance ($I\uparrow$, temperature $\uparrow$, $R\uparrow$).

33. A coil of wire similar to the one found in a light bulb, called a *ballast resistor,* is used to maintain a constant current despite variation in voltage. Since the coil of wire is a conductor and conductors have a positive temperature coefficient, an increase in voltage will cause a corresponding increase in current and therefore in the heat generated, which will result in an increase in the wire's resistance. This increase in resistance will decrease the initial current rise. Similarly, a decrease in voltage and therefore current ($I \propto V$) will result in a decrease in heat and wire resistance. This decrease in resistance will permit an increase in current to counteract the original decrease. Current is therefore regulated or maintained constant by the ballast resistor, despite variations in voltage.

Testing Resistors

34. It is virtually impossible for resistors to short-circuit internally; however, they may be shorted out by some other device, providing a low-resistance path around the resistor. Generally, the resistor's internal resistive elements will begin to develop a higher resistance than its specified value or in some cases go completely open circuit.

Fixed-value resistors tend to go either partially or fully open as the resistive track begins to break down. Variable-value resistors have problems with the wiper making contact with the resistive track at all times.

35. The ohmmeter is the ideal instrument for verifying whether or not a resistor is functioning correctly.

When checking a resistor's resistance, here are some points that should be remembered.

a. Always calibrate the meter first to zero.

b. The ohmmeter has its own internal power source (a battery), so always turn off the circuit power and remove from the circuit the resistor to be measured.

c. The range scales on a digital multimeter often confuse people; for example, if a range is chosen so that the highest reading is 100 kΩ and a 1-MΩ resistor is measured, the meter will indicate an infinite-ohms reading and you may be misled into believing that you have found the problem (an open resistor). In the case of an analog ohmmeter, if the range scale multiplier is not applied, a misleading result is obtained.

d. Another point to keep in mind is tolerance.

e. The ohmmeter's internal battery voltage is really too small to deliver an electrical shock; however, you should avoid touching the bare metal parts of the probes or resistor leads, as your body resistance of approximately 50 kΩ will affect your meter reading.

NEW TERMS

Alphanumeric

Ballast resistor

Carbon composition resistor

Carbon film resistor

Celsius (°C)

Cermet resistor

Chip resistor

Cold resistance

Color code

Dissipation

Dual in-line package (DIP)

Fahrenheit (°F)

Filament resistor

Fixed-value resistor

General-purpose resistor

Helix

Hot resistance

Incandescent lamp

Infinite (∞)

Jumper

Kelvin (K)

Linear

Metal film resistor

Metal oxide resistor

Ohmmeter

Photoresistor

Potentiometer

Precision resistor

Rankine (°R)

Resistive temperature detector (RTD)

Resistor

Rheostat

Single in-line package (SIP)

Substrate

Tapered

Temperature stability

Thermistor

Thermometry

Thick-film resistor

Thin-film detector (TFD)

Tolerance

Variable resistor

Wattage rating

Wirewound resistor

REVIEW QUESTIONS

Multiple-Choice Questions

1. The most common type of fixed-value resistor is the:
   - **a.** Wirewound
   - **b.** Carbon composition
   - **c.** Film
   - **d.** Metal oxide

2. A SIP package has:
   - **a.** A single line of connectors
   - **b.** A double line of connectors
   - **c.** No connectors
   - **d.** None of the above

3. The most commonly used mechanically adjustable variable-value resistor is the:
   - **a.** Rheostat
   - **b.** Thermistor
   - **c.** Potentiometer
   - **d.** Photoresistor

4. A thermistor has a negative temperature coefficient, which means that:
   - **a.** As temperature increases, resistance increases.
   - **b.** As temperature increases, conductance decreases.
   - **c.** As temperature increases, resistance decreases.
   - **d.** All of the above are true.

5. A _____ -band resistor is called a general-purpose resistor, while a _____ -band resistor is known as a precision resistor.
   - **a.** Four, five
   - **b.** Three, four
   - **c.** One, four
   - **d.** Two, seven

6. What would be the power dissipated by a 2-kΩ carbon composition resistor when a current of 20 mA is flowing through it?
   - **a.** 40 W
   - **b.** 0.8 W
   - **c.** 1.25 W
   - **d.** None of the above

7. What of the following is the most common type of temperature detector?
   - **a.** RTD
   - **b.** TFD
   - **c.** Thermistor
   - **d.** Barretter

8. A rheostat is a _____ terminal device, while the potentiometer is a _____ terminal device.
   - **a.** 2, 3
   - **b.** 1, 2
   - **c.** 2, 4
   - **d.** 3, 2

9. Which of the following is the unit of temperature in the International System of Units?
   a. Celsius
   b. Fahrenheit
   c. Kelvin
   d. Rankine

10. Which of the following temperature detectors has a positive temperature coefficient?
    a. RTD
    b. Thermistor
    c. TFD
    d. Both (a) and (b)
    e. Both (a) and (c)

11. Photoresistors are also called:
    a. TFDs
    b. LDRs
    c. RTDs
    d. TGFs

12. Photoresistors have:
    a. A positive temperature coefficient
    b. A negative temperature coefficient
    c. Neither (a) nor (b)

13. Resistance is measured with a (an):
    a. Wattmeter
    b. Milliammeter
    c. Ohmmeter
    d. A 100 meter

14. Which of the following is *not* true?
    a. Precision resistors have a tolerance of > 5%.
    b. General-purpose resistors can be recognized because they have either three or four bands.
    c. The fifth band indicates the tolerance of a precision resistor.
    d. The third band of a general-purpose resistor specifies the multiplier.

15. The first band on either a general-purpose or a precision resistor can never be:
    a. Brown
    b. Red or black stripe
    c. Black
    d. Red

16. What is the name of the resistor used to maintain a constant current despite variations in voltage by changing its resistance?
    a. RTD
    b. Thermistor
    c. Ballast resistor
    d. LDR

17. The term *infinite ohms* describes:
    a. A small finite resistance
    b. A resistance so large that a value cannot be placed on it
    c. A resistance between maximum and minimum
    d. None of the above

18. To calibrate an ohmmeter means:
    a. To make sure that it indicates 100 Ω when 0 Ω is being measured
    b. To zero the meter when the meter leads are touching
    c. To make sure that it indicates 0 Ω when the meter leads are apart
    d. Both (a) and (b)
    e. Both (a) and (c)

19. The typical problem with resistors is that they will:

    a. Develop a partial internal open

    b. Completely open circuit

    c. Develop a higher resistance than specified

    d. All of the above

    e. Both (a) and (c)

20. A 10% tolerance, 2.7 MΩ carbon composition resistor measures 2.99 MΩ when checked with an ohmmeter. It is:

    a. Within tolerance   **d.** Both (a) and (c)

    b. Outside tolerance   **e.** Both (a) and (b)

    c. Faulty

Essay Questions

21. Describe the difference between a fixed- and a variable-value resistor. (4-1)

22. List the six basic categories of fixed-value resistors. (4-1-1)

23. Explain why a resistor's size determines its wattage rating. (4-1-1)

24. Why are resistors given a tolerance figure, and which is best, a small or a large tolerance? (4-1-1)

25. Briefly describe the construction of:

    a. Carbon composition fixed-value resistors (4-1-1)

    b. Rheostats (4-1-2)

    c. Potentiometers (4-1-2)

    d. Multiturn precision potentiometers (4-1-2)

26. Draw the schematic symbol for a rheostat and potentiometer and describe the difference. Also show how a potentiometer can be used as a rheostat. (4-1-2)

27. Define linear and tapered resistive tracks. (4-1-2)

28. List the three rules that should be applied when using an ohmmeter to measure resistance. (4-2)

29. Describe the construction, operation, and temperature coefficient of the following: (4-1-2)

    a. RTD   **c.** Thermistor

    b. Photoresistor   **d.** TFD

30. List the four temperature scales in use today. (4-1-2)

31. Give the color code for the following resistor values: (4-3)

    a. 1.2 MΩ, ± 10%   **c.** 27 kΩ, ± 20%

    b. 10 Ω, ± 5%   **d.** 273 kΩ, ±0.5%

32. Give the resistor values for the following color codes: (4-3)

    a. Orange, orange, black

    b. Red, red, green, red, red

    c. Brown, black, orange, gold

    d. White, brown, brown, silver

33. For what application would the zero-ohms resistor be used? (4-3-2)

34. Describe the difference between a SIP and a DIP resistor package. (4-1-1)

35. What are the common problems encountered with fixed- and variable-value resistors? (4-6)

Practice Problems

36. If a 5.6-kΩ resistor has a tolerance of ±10%, what would be the allowable deviation in resistance above and below 5.6 kΩ? (4-3)

37. (a) If a current of 50 mA is flowing through a 10-kΩ, 25-W resistor, how much power is the resistor dissipating? (b) Can the current be increased, and if so by how much? (4-3)

38. What minimum wattage size could be used if a 12-Ω wirewound resistor were connected across a 12-V supply? (4-3)

39. Calculate the resistance deviation for the tolerances of the resistors listed in Problems 31 and 32. (4-1-1)

40. If a 1-kΩ rheostat has 50 V across it, calculate the different values of current if the rheostat is varied in 100-Ω steps. Referring to Figure 4-29, first calculate and insert the values into the table in Figure 4-33(a), and then plot these values in the graph in Figure 4-33(b).

Troubleshooting Questions

41. Briefly describe some of the problems that can occur with fixed and variable resistors. (4-6)

(a)

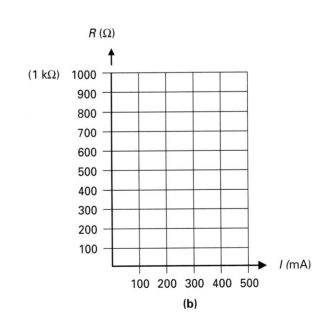

(b)

FIGURE 4-33   Plotting a Rheostat's Current against Resistance.

42. Describe why the tolerance of a resistor can make you think that it has a problem when in fact it does not. (4-6)

43. If a 20-kΩ, 1-W, ±10% tolerance resistor measures 20.39 kΩ, is it in or out of tolerance? (4-1-1)

44. If 33 μA is flowing through a carbon film 4.7-kΩ, ¼-W, ±5% resistor, will it burn up? (4-1-1)

45. What points should be remembered when using the ohmmeter to verify a resistor's value? (4-6)

After completing this chapter, you will be able to:

1. List the precautions that should be observed while troubleshooting electronic circuits or systems.

2. Define the term *body resistance*.

3. Briefly describe first aid, treatment, and resuscitation.

4. Explain the basic operation of the following test instruments:

a. The dc power supply
b. The analog multimeter
c. The digital multimeter

5. Describe the purpose and construction of the protoboard or breadboard.

6. Explain the purpose of most of the commonly used electronic tools, along with soldering principles and techniques.

Experimenting in the Lab

At the Core of Apple

Stephen Wozniak and Steven Jobs met at their Los Altos, California, high school, and due to their common interest, electronics, became friends.

Wozniak was a conservative youth whose obsession with technology left little room for social relationships or studying; in fact, after one year at the University of Colorado, he dropped out with his academic record littered with F's. In contrast to this serious nature, Wozniak was renowned for his high-tech pranks, for which, on one occasion, he spent a night in juvenile hall for wiring up a fake bomb in a friend's locker. In another incident, he devised a way to place a free telephone call to the Pope at the Vatican, identifying himself as then Secretary of State Henry Kissinger.

Jobs, on the other hand, had interests outside electronics. He searched for intellectual, emotional, and spiritual stimulation. After one semester at Reed College, he dropped out to pursue an interest in Eastern religions, which led him to temples in India, searching for the meaning of life.

The first enterprise built by Wozniak and sold by Jobs was an illegal device called a blue box that could crash the telephone system. It generated a set of tones that fooled the computerized telephone switching systems and opened up free, long-distance circuits; these systems allowed "phone phreaks" to take long and illegal joy rides throughout the world's telephone networks.

By selling a Volkswagen van and a programmable calculator, Wozniak and Jobs raised the initial capital of $1300 to start a business in April 1976 called Apple Computer, since Jobs had a passion for the Beatles, who recorded under the Apple record label.

In just five years, Apple Computer grew faster than any other company in history, from a two-man assetless partnership building computers in a family home to a publicly traded corporation earning its youthful entrepreneurs, Jobs (27) and Wozniak (24), fortunes of nearly $200 million each.

This chapter has been designed to introduce you to the laboratory, or lab, where you will be conducting many practical experiments designed to reinforce the topics discussed in theory. To introduce you properly to the lab, we will need to discuss safety, the test equipment you will be using at your lab station, the protoboard or breadboard, and soldering tools and techniques.

In the first section we discuss some of the different safety precautions and procedures that should be observed when working on electrical and electronic equipment. In the second section the operation of the dc power supply, the analog multimeter and the digital multimeter are covered. These pieces of test equipment will be used extensively to power and sense your experimental circuits. In the third section you will be introduced to the protoboard or breadboard, which is designed to accommodate the many experiments described in this book's associated lab manual. The protoboard will hold and interconnect electronic components such as resistors, capacitors, inductors, and the many other components that together form experimental circuits. In the final section you will be introduced to many of the electronic tools that will be needed for experimentation and soldering. In this section we will also introduce you to soldering—the joining together of two metal parts by applying both heat and solder—along with good soldering techniques.

5-1 SAFETY

Safety precautions should always be your first priority when working on electronic equipment, as there is always the possibility of receiving an electric shock. A **shock** is a sudden, uncontrollable reaction as current passes through your body and causes your muscles to contract and deliver a certain amount of pain. Figure 5-1 lists the physiological effects of different amounts of current; as you can see, even a current as small as 10 mA can be fatal.

Any shock is dangerous, since even the mildest could surprise you and cause an involuntary action that could injure you or someone else. For example,

Shock
The sudden pain, convulsion, unconsciousness, or death produced by the passage of electric current through the body.

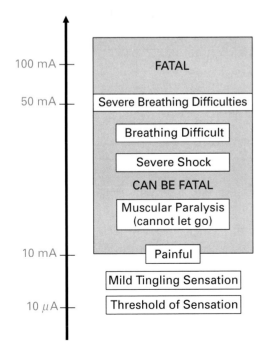

FIGURE 5-1 **Physiological Effects of Electric Current.**

a muscular spasm could throw you against a sharp object or move your arm to a point of higher voltage.

Body Resistance
The resistance of the human body.

5-1-1 Body Resistance

Voltage is the force that causes current to flow through a circuit. The amount of current is dependent on the circuit's resistance and the value of voltage ($I = V/R$). The greatest danger, therefore, is from high voltage points (more than 25 V). The resistance of your body also determines current, since a low resistance will cause a more dangerous high current. The human body has a resistance of about 10,000 to 50,000 Ω, depending on how good a contact you make with the **"live"** (power present) conductor. Skin resistance is generally quite high and will subsequently oppose current flow; however, your resistance is lowered if your skin is wet due to perspiration or if your skin has a cut or an abrasion.

Live
Term used to describe a circuit or piece of equipment that is ON and has current flow within it.

■ **EXAMPLE:**

Calculate the current through a body resistance of 10 kΩ if the body came in contact with 100 V. How much power is the body dissipating?

■ *Solution:*

$$I = \frac{V}{R} = \frac{100\,\text{V}}{10\,\text{k}\Omega} = 10\,\text{mA}$$

$$P = V \times I = 100\,\text{V} \times 10\,\text{mA} = 1\,\text{watt}$$

This value of current will be painful and possibly fatal.

Power is a point to consider. In the example, the body resistance was dissipating 1 watt. A power source may not be able to supply a fatal high power, and its output voltage will be pulled down due to the excessive load current through the body resistance. The current is then reduced to an amount that the power source can deliver.

5-1-2 Precautions to Be Used When Troubleshooting

The following procedures and precautions have been acquired from experienced technicians and should be applied whenever possible.

1. Except when absolutely necessary, do not work on electrical or electronic circuits or equipment when power is on.

2. When troubleshooting inside equipment, people tend to lean one hand on the chassis (the metal framework of the equipment) and hold the test lead or probe in the other hand. If the probing hand comes in contact with a high voltage, current will flow from one hand, through your chest (heart and lungs), and finally through the other hand to the chassis ground. To avoid this dangerous situation, always place the free hand in a pocket or behind your back while testing a piece of equipment with power on.

3. When troubleshooting equipment with power on, try to insulate yourself by not standing on a damp floor or leaning against any metal object and by wearing rubber-soled electrical safety shoes or standing on a rubber mat.

4. Electrolytic and other large-value capacitors (which will be covered in Chapters 13 and 14) can hold a dangerous voltage charge even after the equipment has been turned off. Be sure always to check if these capacitors are fully discharged by shorting the two terminals with a pair of *insulated* pliers or *insulated* test lead.

5. All tools should be well insulated. If not, it is not only dangerous to you, but probes that are not well insulated right down to the tip can cause short circuits between two points, resulting in additional circuit and equipment problems.

6. Always switch off the equipment and disconnect the power (since some equipment has power present even when off) before replacing any components. When removing components that get hot, such as resistors, allow enough time for cooling after the equipment has been turned off.

7. Necklaces, rings, and wristwatches have a low resistance and should be removed when working on equipment.

8. Inspect the equipment before working on it, and if it is in poor condition (frayed, cracked, or burned power cords, chipped plugs), turn off the equipment and replace these hazards.

9. Make sure that someone is present who can render assistance in the event of an emergency.

10. Make it a point to know the location of the power-off switch.

11. Cathode-ray tubes, which are the picture tubes within televisions and computer monitors, are highly evacuated and should therefore be handled with extreme care. If broken, the relatively high external pressure will cause an implosion (burst inward), which will result in the inward metal parts and glass fragments being expelled violently outward.

5-1-3 *First Aid, Treatment, Resuscitation*

Injury to Persons

Injury may be caused in a number of ways:

Shock. Electric shock is the effect produced on the body and in particular on the nerves by an electric current passing through it. Its magnitude depends on the strength of the current, which, in turn, depends on the voltage. Its effect varies according to the ohmic resistance of the body, which varies in different persons, and also according to the parts of the body between which the current flows (contact in the cardiac region can be particularly dangerous). It also depends on the current flow and on the surface resistance of the skin, which is much reduced when the skin is wet and is reduced to zero if the skin is penetrated.

Shock can be felt from voltages as low as 15 V, and at 20 to 25 V most people would experience pain. At about this voltage or a little higher, the victim may find himself unable to let go of the conductor and may suffer burning. It is believed that under certain conditions, death can be caused by voltages as low as 70, but generally the danger below 120 V ac is believed to be small (although not entirely negligible). Most serious and fatal accidents occur at the industrial 200 to 240 V ac and from 25 to 30 mA and over.

Injury can also be caused by a minor shock, not serious in itself but which has the effect of contracting the muscles sufficiently to result in a fall or other reaction.

Burns. Burns can be caused by the passage through the body of a heavy current if the body is in contact with a conductor, or by direct contact with an electrically heated surface. Burns can also be caused by the intense heat generated by the arcing from a short circuit. All cases of burns require immediate medical attention.

Explosion. An explosion can be caused by the ignition of flammable gases by a spark from an electric contact. In all cases where a flammable or ignitable atmosphere or vapor is present, special care is necessary.

Eye injuries. These can be caused by exposure to the strong ultraviolet rays of an electric arc. In these cases, the eyes may become inflamed and painful after a lapse of several hours, and there may be temporary loss of sight. Although very painful, the condition usually passes off within 24 hours. Lasers are also dangerous to the eyes due to their intensely concentrated beam, and therefore specially filtered protective glasses should be worn, as seen in Figure 5-2.

Precautions to protect the eyes must always be taken by wearing protective goggles when clipping leads or soldering. Permanent injury to the eyes can arise from the energy propagated by microwave equipment. No one should look

FIGURE 5-2   Protective Glasses for Laser Experimentation. (Photo courtesy of Hewlett-Packard Company.)

along a waveguide when it is on or examine a highly directional radiator at close distances.

Body injuries from microwave and radio-frequency equipment. The energy in microwave and radio-frequency equipment can damage the body, especially those parts with a low blood supply. The eyes are particularly vulnerable. The highest energy level to which operators should be subject is 1.0 mW/cm^2, and intensities exceeding 10 mW/cm^2 should always be avoided.

Resuscitation

You should familiarize yourself with the various methods of artificial respiration by contacting your local Red Cross for complete instruction. The mouth-to-mouth method of artificial respiration is the most effective of the resuscitation techniques. It is comparatively simple and produces the best and quickest results when done correctly.

Mouth-to-mouth resuscitation method. It is essential to begin artificial respiration without delay. *Do not touch the victim with your bare hands until the circuit is broken.* If this is not possible, *protect yourself* with dry insulating material and pull the victim clear of the conductor.

Step 1: Lay the patient on his or her back and, if on a slope, have the stomach slightly lower than the chest.

Step 2: Make a brief inspection of the mouth and throat to ensure that they are clear of obvious obstructions.

Step 3: Give the patient's head the maximum backward tilt so that the chin is prominent and the neck stretched to give a clear airway, as shown in Figure 5-3(a).

Step 4: Sealing off the patient's nose with your thumb and finger, open your mouth wide and make an airtight seal over the open patient's mouth and then blow, as shown in Figure 5-3(b). (New steps indicate using a plastic bag airway).

Step 5: After exhaling, turn your head to watch for chest movement, while inhaling deeply in readiness for blowing again, as shown in Figure 5-3(c).

Step 6: If the chest does not rise, check that the patient's mouth and throat are free of obstruction and that the head is tilted back as far as possible, and then blow again.

(a)   (b)   (c)

FIGURE 5-3 Mouth-to-Mouth Resuscitation.

1. What is the approximate resistance of the human body?

2. List the precautions that should be taken when troubleshooting.

5-2 TEST EQUIPMENT

As a technician, your tools of the trade will be test equipment. In this section you will be introduced to the basic operation of the *dc power supply,* the *analog multimeter,* and the *digital multimeter* which you will be using in your experiments in the lab.

5-2-1 The DC Power Supply

DC Power Supply
A powerline, generator, battery, power pack, or other dc source of power for electronic equipment.

In the laboratory or repair shop, power can be obtained from a battery, but since ac power is so accessible at the wall outlet, a **dc power supply** is generally used to supply dc voltages. The dc power supply has a large advantage over the battery as a source of dc voltage in that it can quickly provide an accurate output voltage that can easily be varied by a control on the front panel, and it never runs down.

During your electronic studies you will use a dc power supply in the lab to provide different dc output voltages for various experiments. In Figure 5-4(a) you can see how a dc power supply can be used to supply 12 volts (12 V) to a 12-kilohm (12 kΩ) resistor. To explain the basic operation of this piece of equipment, the ON/OFF switch is used to turn on the power supply, while the VOLTS/AMPS switch is used to switch the meter between a voltmeter or ammeter. In the "volts" position, the meter indicates the voltage that is appearing across the output terminals, which in this example is 12 V. If the VOLTS/AMPS switch is placed in the "amps" position, the meter will now monitor the amount of current flowing out of the power supply, which in this example is 1 milliampere (1 mA). The VOLTS ADJUST control is used to either increase or decrease the output voltage. Figure 5-4(b) shows a photograph of a typical dc power supply.

5-2-2 The Analog Multimeter (AMM)

Analog Multimeter
Piece of electronic test equipment that can perform multiple tasks in that it can be used to measure voltage, current, or resistance.

The multimeter is a device for the measurement of current, voltage, and resistance. It is called a multimeter because it is like having three meters all rolled into one, in that it has

a. An ammeter for measuring current

b. A voltmeter for measuring voltage

c. An ohmmeter for measuring resistance

(a)

(b)

FIGURE 5-4   DC Power Supply.

For many years the analog readout multimeter, an example of which is shown in Figure 5-5(a), was the only type available. At present these analog readout multimeters are still being used but are slowly being replaced by digital readout multimeters similar to the type shown in Figure 5-5(b). In the following section, we will cover the digital multimeter.

Analog is a word that means "similar to." For example, a ball is an analog of the world in that the shapes are similar. With an analog multimeter or AMM,

(a)   (b)

FIGURE 5-5   Multimeter. (a) Analog Readout Multimeter. (b) Digital Readout Multimeter.

the amount of pointer deflection across the scale is an analog (or similar to) the magnitude of the electrical property being measured. This type of indicator is therefore called an *analog display* as opposed to a *digital display,* which indicates the magnitude of the electrical property using decimal digits.

Let us now review the steps involved in measuring current, voltage, and resistance.

Measuring Current with the Analog Ammeter

Ammeter

Meter placed in the circuit path to measure the amount of current flow.

Ammeters (ampere meters) are used to measure the current flow within a circuit, as shown in Figure 5-6. In electronics, where current is generally small, the milliampere (mA) or microampere (μA) scale will generally be used, whereas in electrical high-current circuits, the meter will generally be used on an ampere scale.

Three rules must be remembered and applied when measuring current:

1. Always set the selector to the higher scale first (amperes), and then reduce as needed to milli- or microamperes, just in case a larger current than anticipated is within the circuit, in which case the meter could be damaged.

2. The current meter must be inserted in the circuit so that the positive lead of the ammeter (red) is connected to the positive charge or polarity, and the negative lead of the ammeter (black) is connected to the negative charge or polarity. Ammeters are polarity sensitive instruments.

3. If you wish to measure current flowing within a wire, the ammeter measuring the current must be inserted in the path of current flow.

FIGURE 5-6 **Measuring Current with the Analog Ammeter.**

Measuring Voltage with the Analog Voltmeter

Voltmeters (voltage meters) are used to measure electrical pressure (voltage) anywhere in an electronic circuit. Figure 5-7 illustrates a voltmeter being used to measure the voltage drop across a resistor.

A voltmeter is connected across a device, as illustrated in Figure 5-7, with the two leads of the voltmeter touching either side of the component.

Three steps must be remembered and applied when measuring voltage:

1. Always set to the highest scale first (kV) and then reduce as needed to the volt and millivolt range, just in case a larger voltage than anticipated is within the circuit, in which case the meter could be damaged.

Voltmeter
Instrument designed to measure the voltage or potential difference. Its scale can be graduated in kilovolts, volts, or millivolts.

FIGURE 5-7 **Measuring Voltage with the Analog Voltmeter.**

2. Always connect the positive lead (red) of the meter to the component lead that is closer to the positive potential, and the negative lead (black) of the meter to the component lead that is closer to the negative potential. Voltmeters are also polarity sensitive instruments.

3. If you are measuring the voltage across a component, the voltmeter must be connected across the component and outside the circuit.

Measuring Resistance with the Analog Ohmmeter

Ohmmeter
Measurement device used to measure electric resistance.

Resistance is measured with an **ohmmeter.** Figure 5-8 illustrates a multirange ohmmeter with a resistor's resistance being measured. The meter has four ranges. The simplest range has been selected ($R \times 1$), and on this scale the resis-

FIGURE 5-8   Measuring Resistance with the Analog Ohmmeter.

tance value indicated can be read directly from the meter; in this example the value is 36 Ω ($R \times 1 = 36 \times 1 = 36\ \Omega$). If we had selected the $R \times 10$ range, a meter reading of, for example, 71 would have to be multiplied by 10 ($71 \times 10 = 710$) to obtain the value of 710 Ω. If we were on the $R \times 100$ range, then a meter reading of 215 would be interpreted as $R \times 100 = 215 \times 100 = 21500$ or 21.5 kΩ. A meter reading of 14 on the $R \times 1000$ scale, if selected, means that you would interpret 14 as $R \times 1000 = 14 \times 1000$, which equals 14 k$\Omega$.

Three rules must be remembered and applied when measuring resistance.

1. Short the meter leads together and adjust the zero-ohms control so the pointer is at 0 Ω. This calibrates the meter scale.

2. Turn off the power and remove the component (if practicable) to be measured from circuit to protect the ohmmeter.

3. Connect the leads across the component, adjust the range scale until the pointer is approximately in the middle of the scale (for accuracy), and then multiply this value by the range scale selected.

5-2-3   *The Digital Multimeter (DMM)*

The **digital multimeters** (DMM) illustrated in Figure 5-9 are gradually replacing the analog-type multimeter discussed previously because they are easier to read and have greater accuracy.

Analog versus Digital

To compare these two types of multimeters, let us examine their pros and cons in more detail.

Ease of reading.   One of the greatest problems with an analog meter is the errors that occur due to the human factor when reading off the value from the many different scales. The thin analog needle against a calibrated scale is similar to the hands of a clock against the number scale indicating

Digital Multimeter
Multimeter used to measure amperes, volts, and ohms and indicate the results on a digital readout display.

(a)

(b)

FIGURE 5-9   Digital Multimeters. (a) Hand-held. (b) Benchtop.

the hours. When you look at an analog clock, you have to determine where the hands are, which number the hand is nearest to, and so on. With a digital clock, however, the time is read directly from a display, as shown in Figure 5-10(a).

With the analog meter, the decoding of the scale is necessary, while the digital multimeter displays the magnitude, polarity (+ or −), and the units (V, A, or Ω) on typically a four- or five-digit readout.

Accuracy.   DMMs are typically accurate to ±0.01% and have no need for a zero-ohms adjust. Returning to the example of the analog and digital clock, a person reading the time from the traditional analog clock would say that the time is almost 12:30. The wearer of the digital watch, however, will be unquestionably specific and give the time as 12:27. Similarly, an analog

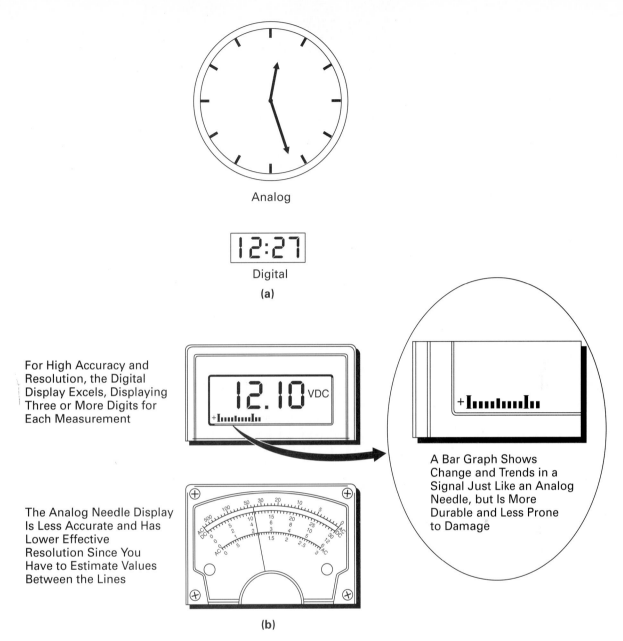

For High Accuracy and Resolution, the Digital Display Excels, Displaying Three or More Digits for Each Measurement

The Analog Needle Display Is Less Accurate and Has Lower Effective Resolution Since You Have to Estimate Values Between the Lines

A Bar Graph Shows Change and Trends in a Signal Just Like an Analog Needle, but Is More Durable and Less Prone to Damage

Analog

Digital

(a)

(b)

FIGURE 5-10   Analog versus Digital. (Courtesy of Fluke Corporation. Reproduced with permission.)

reading on a meter of about 7 V becomes 7.15 V with the far more accurate digital multimeter.

Digital meters have complicated internal circuitry which is why the digital readout meters have a more expensive price tag than the analog readout multimeters. Another disadvantage of the digital multimeter is its slow response to display the amount on the readout once it has been connected in circuit. To compensate for this disadvantage, most DMMs have a **bar graph display** below the digital readout, as shown in Figure 5-10(b), showing the magnitude of the measured quantity using more or less bars.

Bar Graph Display
A left-to-right or low-to-high set of bars that are turned on in successive order based on the magnitude they are meant to represent.

Using the Digital Multimeter

When troubleshooting electronic equipment, the multimeter becomes the technician's eyes, allowing him or her to see the situation and the problem within the circuit or system. Figure 5-11 illustrates a typical autoranging digital multimeter.

For the meter to be of any use, it must first be connected to the circuit or device to be tested. The two test leads for the meter, one of which is red and the other black, must be inserted into the correct meter lead jacks on the meter. The black lead is connected to the meter jack marked "c" or *common*. The red lead is connected to either the meter jack marked "300 mA or 10 A" (for amperes), or V–Ω (for volts or ohms).

Measuring current. Figure 5-12 shows the steps that should be followed when measuring current. The measurement of current is rarely performed when troubleshooting, as the circuit path has to be opened to insert the DMM in series with the current flow. However, if current is to be measured, the red lead is inserted into one of the ampere jacks.

FIGURE 5-11   Typical Digital Multimeter. (Courtesy of Fluke Corporation. Reproduced with permission.)

How to Make Current Measurements

1.
Turn Off the Power to the Circuit ⚠

2.
Cut or Unsolder the Circuit, Creating a Place Where the Meter Probes Can Be Inserted

3.
Select Amps AC (A∼), or Amps DC (A⎓) as Desired

4.
Plug the Black Test Probe into the COM Input Jack. Plug the Red Test Probe into the 10 Amp (10 A) or 300 Milliamp (300 mA) Input Jack Depending on the Expected Value of the Reading

5.
Connect the Probe Tips to the Circuit Across the Break as Shown so That All Current Will Flow Through the Meter (A Series Connection)

6.
Turn the Circuit Power Back On

7.
View the Reading, Being Sure to Note the Unit of Measurement

Note: If Test Leads Are Reversed, a (−) Sign Will Show in the Display

FIGURE 5-12   Measuring Current with a Digital Autoranging Multimeter (Courtesy of Fluke Corporation. Reproduced with permission.)

Measuring voltage.   Figure 5-13 shows the steps that should be followed when measuring voltage. The measurement of voltage and resistance is where the DMM finds its greatest utilization. For voltage or resistance measurement, the red meter lead is inserted into the V–Ω (volt or ohms) meter jack.

Measuring resistance.   Figure 5-14 shows the steps that should be followed when measuring resistance.

Remember that resistance measurements are carried out without power being applied to the component under test, and resistance values can vary by as

How to Make Voltage Measurements

Note: $\frac{1}{1000}$ V = 1 mV

1000 V = 1 kV

1.
Select Volts AC (V∼),
Volts DC (V⎓) or
300 mV⎓ as
Desired

3.
Touch the Probe
Tips to the Circuit
Across a Load or
Power Source as
Shown (In Parallel
to the Circuit) ⚠

2.
Plug the Black
Test Probe into
the COM Input
Jack—Plug the
Red Test Probe
into the V Input
Jack

4.
View the Reading,
Being Sure to
Note the Unit of
Measurement

Note: For DC Readings of the Correct Polarity (+/–), Touch the Red Test Probe to
the Positive Side of the Circuit, and the Black Probe to the Negative Side
or Circuit Ground. If You Reverse the Connections, a DMM with Auto
Polarity Will Merely Display a Minus Sign Indicating Negative Polarity.
With an Analog Meter You Risk Damaging the Meter.

**FIGURE 5-13   Measuring Voltage with a Digital Autoranging Multimeter
(Courtesy of Fluke Corporation. Reproduced with permission.)**

much as 20%, due to the tolerance of certain resistors; so do not be misled if
your meter reading is slightly different from the color band value printed on the
resistor. If a resistor's value is off and exceeds the tolerance, the resistor should
be replaced.

A resistor will rarely short, but typically will open. If a resistor does open,
the DMM display will flash on and off or display OL because the resistor has an
infinite resistance (open circuit).

How to Make Resistance Measurements

Note: 1,000Ω = 1 kΩ
       1,000,000Ω = 1 MΩ

1.
Turn Off Power
to the Circuit ⚠

2.
Select
Resistance (Ω)

3.
Plug the Black
Test Probe into
the COM Input
Jack. Plug
the Red Test
Probe into the
Ω Input Jack

4.
Connect the Probe
Tips Across the
Component or
Portion of the
Circuit for Which
You Want to
Determine
Resistance

5.
View the Reading,
Being Sure to Note
the Unit of
Measurement —
Ohms (Ω),
Kilohms (kΩ), or
Megohms (MΩ)

**FIGURE 5-14   Measuring Resistance with a Digital Autoranging
Multimeter (Courtesy of Fluke Corporation. Reproduced with
permission.)**

SELF-TEST REVIEW QUESTIONS FOR SECTION 5-2

1. What is the purpose of the dc power supply?
2. An ammeter is used to measure _____, a voltmeter is used
   to measure _____, and an ohmmeter is used to measure
   _____.
3. What two key advantages does the DMM have over the AMM?
4. What is meant by the term "autoranging"?

5-3 PROTOBOARDS

Protoboard
An experimental arrangement of a circuit on a board. Also called breadboard.

The solderless prototyping board **(protoboard)** or breadboard is designed to accommodate the many experiments described in this textbook's associated lab manual. This protoboard will hold and interconnect resistors, capacitors, inductors, and many other components, as well as provide electrical power. Figure 5-15(a) shows an experimental circuit wired up on a protoboard.

Figure 5-15(b) shows the top view of a basic protoboard. As you can see by the cross section on the right side, electrical connector strips are within the protoboard. These conductive strips make a connection between the five hole groups. The bus strips have an electrical connector strip running from end to end, as shown in the cross section. They are usually connected to a power supply as seen in the example circuit in Figure 5-16(a). In this circuit, you can see that

(a)

Top View of Protoboard

Conductive Strips within Protoboard

This Bus Line Is Electrically Connected from this End of the Board to the Other by a Metal Strip within the Protoboard

Bus Strips

These Five Holes Are All Electrically Connected by a Metal Strip within the Protoboard

(b)

FIGURE 5-15   Experimenting with the Protoboard.

(a)

(b)

FIGURE 5-16 **Constructing Circuits on the Protoboard.**

the positive supply voltage is connected to the upper bus strip, while the negative supply (ground) is connected to the lower bus strip. These power supply "rails" can then be connected to a circuit formed on the protoboard with hookup wire, as shown in Figure 5-16(a). In this example, three resistors are connected end-to-end, as shown in the schematic diagram in Figure 5-16(b).

Figure 5-17 illustrates how the multimeter could be used to make voltage, current, and resistance measurements of a circuit constructed on the protoboard.

Other protoboards may vary slightly as far as layout, but you should be able to determine the pattern of conductive strips by making a few resistance checks with an ohmmeter.

SELF-TEST REVIEW QUESTIONS FOR SECTION 5-3

1. What is the purpose of the protoboard?
2. What is generally connected to the bus strips?

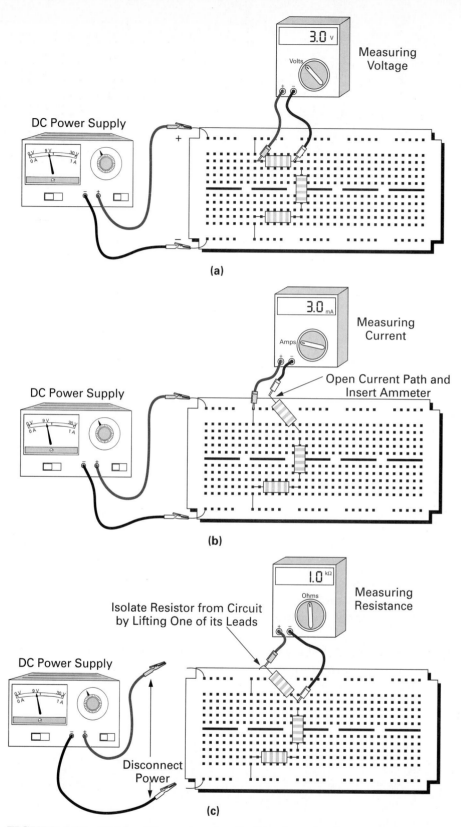

FIGURE 5-17 **Making Measurements of a Circuit on the Protoboard.**

5-4 SOLDERING TOOLS AND TECHNIQUES

Electronic components such as resistors, capacitors, diodes, transistors, and integrated circuits are combined to form circuits, which in turn are combined to form electronic systems. These components have their leads physically interconnected and then bonded with solder. Like the expression the "straw that broke the camel's back," one sloppy solder connection can cause an entire electronic system to fail.

5-4-1 *Soldering Tools*

Soldering is a skill that can be developed with practice and knowledge. A solder connection is the joining together of two metal parts by applying both heat and solder. The heat provided by the soldering iron is at a high enough temperature to melt the solder, making it a liquid that flows onto and slightly penetrates the two metal surfaces that need to be connected. Once the soldering iron and therefore the heat are removed, the solder cools. In so doing, it solidifies and bonds the two metal surfaces, producing a good *electrical and mechanical connection*. This is the purpose of soldering: to bond two metal surfaces together so that they are both electrically and mechanically connected.

Soldering
Process of joining two metallic surfaces to make an electrical contact by melting solder (usually tin and lead) across them.

You should at this time have a set of tools and toolbox with all the components needed for the experiments listed in the lab manual. Figure 5-18 illustrates the set of basic electronic tools that will be needed for soldering and experimentation. These tools include:

1. Solder wick: Used to remove solder from terminals.

FIGURE 5-18 **Basic Electronic Tools.**

2. Heat sink: During the soldering process, this device is clamped onto the lead of especially heat-sensitive components to conduct the heat generated by the soldering iron away from the component.

3. Soldering brush: Used to clean off flux after soldering.

4. Pliers

5. Long-nose pliers: Used for gripping and bending; they are also known as needle-nose pliers.

6. Cutters: Also called dikes; these are available in many different sizes and are used to cut wires or cables.

7. Wire strippers: The strippers are designed to be adjustable or they have a variety of holes for stripping the insulating sleeve off a wire.

8. Solder: 60/40 rosin-core solder is most commonly used in electronics.

9. Nut driver

10. Blade screwdriver

11. Phillips screwdriver

12. Soldering iron

5-4-2 *Wetting*

Wetting
The coating of a contact surface.

Everytime you make a good electrical and mechanical connection, the solder will flow, when heated to its melting temperature, over the lead and terminal to be connected, as shown in Figure 5-19. The solder, in fact, actually penetrates the metals, and this embedding of the solder into the metal is called **wetting.** If the solder feathers out to a thin edge, good wetting is said to have occurred, and it is this that gives the connection its physical strength and electrical connection.

5-4-3 *Solder and Flux*

Solder
Metallic alloy that is used to join two metal surfaces.

Solder is a mixture of tin and lead, both of which have a low melting temperature with respect to other metals. This is necessary so that the soldering iron melts the solder and not the terminals or leads.

FIGURE 5-19 Wetting.

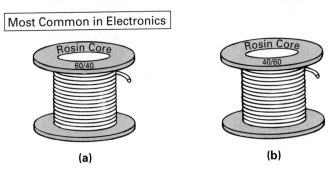

FIGURE 5-20 **Tin/Lead Solder Ratios. (a) 60/40. (b) 40/60.**

Different proportions of tin and lead are available to produce solder with different characteristics. For example, Figure 5-20(a) shows a 60/40 solder, which is a mixture of 60% tin and 40% lead, whereas Figure 5-20(b) shows a 40/60 solder that consists of 40% tin and 60% lead. The proportions of tin to lead determine the melting temperature of the solder; for example, 60/40 solder has a lower melting temperature than 40/60 solder. In electronics, 60/40 is most commonly used because of its low melting temperature, which means that a component lead or terminal will not have to be heated to a high temperature in order to make a connection. A 63/37 solder has an even lower melting temperature and is therefore even safer than 60/40 solder; however, it is more expensive.

Any lead or terminal is always exposed to the air, which forms an invisible insulating layer on the surface of leads, pins, terminals, and any other surface. A chemical substance is needed to remove this layer; otherwise, the solder would not be able to flow and stick to the metal contacts. **Flux** removes this invisible insulating oxide layer, and nearly all solder used in electronics contains the flux inside a type of solder tube, as shown in Figure 5-21. As the solder is applied to a heated connection, the flux will automatically remove any oxide. The two most common types of flux are rosin and acid. Acid-core solder is used only in sheet metal work and should never be used in electronics since it is highly corrosive. In electronics, you should only use rosin-core solder, and this is normally indicated as shown in Figure 5-20.

A variety of solder diameters are available. The larger-diameter solder is used for terminals and large component leads and connections, whereas the

Flux
A material used to remove oxide films from the surfaces of metals in preparation for soldering.

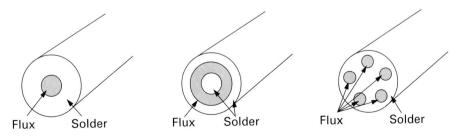

FIGURE 5-21 **Rosin-Core Solder.**

smaller-diameter solder is used for soldering terminals that are very close to one another on a printed circuit board, so the amount of solder being applied to the connection needs to be carefully controlled.

5-4-4 Soldering Irons

Soldering Iron
Tool having an internal heating element that is used for heating a connection to melt solder.

Figure 5-22 illustrates the two basic types of **soldering irons.** They are rated in terms of wattage, which indicates the amount of power consumed. More important, the wattage rating of a soldering iron indicates the amount of heat it pro-

FIGURE 5-22   Types of Soldering Irons. (a) Pencil. (b) Gun.

duces. When an iron is applied to a connection, heat transfer takes place and heat is drained away from the iron to the metals to be connected; so a larger connection will need a larger-wattage soldering iron. A 25- to 60-W pencil iron is ideal for most electronic work.

For your safety and to protect voltage- and current-sensitive components, you should always use a soldering iron with a three-pin plug and three-wire ac power cord. The third ground wire will ground the iron's exposed metal areas and the tip to prevent electrical shock. This grounded tip will protect delicate MOS integrated circuits from leakage electricity and static charges. On the subject of safety, always turn off the equipment before soldering, as the grounded tip may cause a short circuit in the equipment and possible damage.

The tip of a soldering iron can sometimes be changed, and a different temperature tip will allow your iron to be used for different applications. The temperature of the tip selected should be governed by the size of the connection and the temperature sensitivity of the components. In general, a 700°F tip is ideal for most electronic applications; however, for delicate printed circuit boards, use a 600°F tip.

The tip of a soldering iron comes in a variety of shapes and sizes, as shown in Figure 5-22(a). The heat produced by the soldering iron heats the connection and melts the solder so that it flows over the connection. A tip shape should be chosen that will conduct the heat to the connection as quickly as possible, so as not to damage the component or PCB terminal. The tip shape that makes the best contact with the connection, and will therefore conduct the most heat, should be used.

As you regularly use your soldering iron, the tip will naturally accumulate dirt and oxide, and this contamination will reduce its effectiveness. The tip should be regularly cleaned by wiping it across a damp sponge, as shown in Figure 5-23(a). Soldering iron tips are generally made of copper, plated with either iron or nickel, so you should never clean a tip by filing the end, as this will probably remove the plating.

The tip of the soldering iron should always be **tinned** after it has been cleaned. Tinning the tip can be seen in Figure 5-23(b) and is achieved by applying a small layer of rosin-core solder to the tip. This will protect the tip from oxidation and increase the amount of heat transfer to the connection.

Tinned
Coated with a layer of tin or solder to prevent corrosion and simplify the soldering of connections.

FIGURE 5-23 **Cleaning a Soldering Iron. (a) Cleaning the Tip. (b) Tinning the Tip.**

In summary, it is important to remember that *a soldering iron should not be used to melt the solder; its purpose is to heat the connection so that the solder will melt when it makes contact with the connection.*

5-4-5 Soldering Techniques

Before carrying out the six basic soldering steps, wires and components should be prepared. Wires should be wrapped around a terminal to give it a more solid physical support, as shown in Figure 5-24(a). Their insulation should be stripped back so as to leave a small gap. Too large a gap will expose the bare wire and possibly cause a short to another terminal, whereas too small a gap will cause the insulation to burn during soldering. Stranded wire should be tinned, as shown in Figure 5-24(b), and components should be mounted flat on the board, as shown in Figure 5-24(c).

Six-Step Soldering Procedure

To begin, always remember to wear safety goggles and a protective apron, and then proceed.

Step 1: Clean the tip on a damp sponge.

Step 2: Tin the tip if necessary.

Step 3: Heat the connection.

Step 4: Apply the solder to the opposite side of the connection.

Step 5: Leave the tip of the soldering iron on the connection only long enough to melt the solder.

Step 6: Remove the solder and then the iron; then let the melted solder cool and solidify undisturbed.

Once this procedure is complete, you should remove the excess leads with the cutters and remove the flux residue, because it collects dust and dirt that could produce electrical leakage paths. When inspecting your work, you should notice the following:

FIGURE 5-24 Wire and Component Preparation.

FIGURE 5-25 Poor Solder Connections.

1. The connection is *smooth and shiny.*
2. The solder should *feather out to a thin edge.*

Poor Solder Connections

1. *Cold solder joint:* This connection has a dull gray appearance; it makes a poor mechanical and electrical connection and is normally caused due to connector lead movement while the solder was cooling, or to insufficient heat [Figure 5-25(a)]. The cure is to resolder the connection.

2. *Excessive solder joint:* This is caused by too much solder being applied to the connection or by using solder of too large diameter [Figure 5-25(b)]. The cure is to desolder the excess with a wick.

3. *Insufficient solder joint:* Not enough solder was applied, so a poor bond exists between the lead and the terminal [Figure 5-25(c)]. The cure is to resolder as if it were a new joint.

4. *Solder bridge:* This occurs when adjacent terminals or traces on printed circuit boards are connected accidentally as a result of using too much solder [Figure 5-25(d)]. Most can be removed by desoldering the excess with a wick.

5. *Excessive heat:* On printed circuit boards, too much heat can lift the traces or track and ruin the entire board [Figure 5-25(e)].

5-4-6 Desoldering

If a faulty component has to be replaced or if a component or wire was soldered in an incorrect place, it will have to be desoldered. Figure 5-26 illustrates some of the tools used to desolder. The easiest way to remove a component is to remove all the solder and then disconnect the leads. All three tools shown in Figure 5-26 are designed to remove solder.

Use of the braided wick to remove solder can be seen in Figure 5-27(a). The braid contains many small wire strands, and when it is placed between the solder and the iron, the melted solder flows into the braid by capillary action. The solder-filled braid is then cut off and discarded.

Desoldering
Braid

Desoldering
Bulb

Spring-loaded Plunger

FIGURE 5-26   Desoldering Tools.

Desoldering
The process of removing solder from a connection.

The **desoldering** bulb or spring-loaded plunger relies on suction rather than capillary action to remove the solder, as shown in Figure 5-27(b). The steps to follow are:

Step 1:   Squeeze the desoldering bulb or cock the spring-loaded plunger.

Step 2:   Melt the solder with the soldering iron.

Step 3:   Remove the iron's tip.

Step 4:   Quickly insert the tip of the bulb or plunger into the molten solder and then activate the suction.

When as much of the solder as possible has been removed, use the long-nose pliers to hold the lead and remove the component. It may be necessary to use the iron to heat the lead slightly to loosen up the residual solder so that the component lead can be removed with the pliers. Always be careful not to apply an excessive amount of heat or stress to either the component or the board.

5-4-7   *Safety*

Molten solder, like any other liquid, can splash or spill, and soldering irons are even hotter than the solder. Consequently, you should always wear protective safety glasses and an apron to protect yourself from burns. NEVER fling hot solder off an iron; always use the sponge.

(a)   (b)

FIGURE 5-27

SELF-TEST REVIEW QUESTIONS FOR SECTION 5-4

1. A solder connection is the joining together of two metal parts by applying both _____ and _____.
2. In electronics a 60/40 solder is most commonly used and is composed of 60% _____ and 40%_____.
3. Describe the six-step soldering procedure.
4. What devices are available to remove solder?

SUMMARY

Safety

1. Safety precautions should always be your first priority when working on electronic equipment, as there is always the possibility of receiving an electric shock.
2. A shock is a sudden, uncontrollable reaction as current passes through your body and causes your muscles to contract and deliver a certain amount of pain.
3. Any shock is dangerous, since even the mildest could surprise you and cause an involuntary action that could injure yourself or someone else.
4. The human body has a resistance of about 10,000 to 50,000 Ω, depending on how good a contact you make with the "live" (power present) conductor.
5. Skin resistance is generally quite high and will subsequently oppose current flow; however, your resistance is lowered if your skin is wet due to perspiration or if your skin has a cut or an abrasion.

Precautions to Be Used When Troubleshooting

6. Except when absolutely necessary, do not work on electrical or electronic circuits or equipment when power is on.
7. To avoid dangerous situations, always place the free hand in a pocket or behind your back while testing a piece of equipment with power on.
8. When troubleshooting equipment with power on, try to insulate yourself.
9. Electrolytic and other large-value capacitors can hold a dangerous voltage charge even after the equipment has been turned off.
10. All tools should be well insulated.
11. Always switch off the equipment and disconnect the power before replacing any components.
12. Necklaces, rings, and wristwatches have a low resistance and should be removed when working on equipment.
13. Inspect the equipment before working on it, and if it is in poor condition turn off the equipment and replace these hazards.
14. Make sure that someone is present who can render assistance in the event of an emergency.

15. Make it a point to know the location of the power-off switch.

16. Cathode-ray tubes are highly evacuated and should therefore be handled with extreme care.

First Aid, Treatment, Resuscitation

17. Shock can be felt from voltages as low as 15 V, and at 20 to 25 V most people would experience pain. At about this voltage or a little higher, the victim may find himself unable to let go of the conductor and may suffer burning.

18. Injury can also be caused by a minor shock, not serious in itself but which has the effect of contracting the muscles sufficiently to result in a fall or other reaction.

19. Burns can be caused by the passage through the body of a heavy current.

20. An explosion can be caused by the ignition of flammable gases by a spark from an electric contact.

21. Eye injuries can be caused by exposure to the strong ultraviolet rays of an electric arc. Lasers are also dangerous to the eyes due to their intensely concentrated beam. Precautions to protect the eyes must always be taken by wearing protective goggles when clipping leads or soldering.

22. The energy in microwave and radio-frequency equipment can damage the body, especially those parts with a low blood supply.

23. You should familiarize yourself with the various methods of artificial respiration by contacting your local Red Cross for complete instruction.

The DC Power Supply

24. In the laboratory or repair shop, power can be obtained from a battery, but since ac power is so accessible, a dc power supply is generally used to supply dc voltages.

25. The power supply has a large advantage over the battery as a source of dc voltage in that it can quickly provide an accurate output voltage that can easily be varied by a control on the front panel, and it never runs down.

26. During your electronic studies you will use a dc power supply in the lab to provide different dc output voltages for various experiments.

27. The ON/OFF switch is used to turn on the power supply, while the VOLTS/AMPS switch is used to switch the meter between a voltmeter or ammeter. In the volts position, the meter indicates the voltage that is appearing across the output terminals. If the VOLT/AMPS switch is placed in the amps position, the meter will now monitor the amount of current flowing out of the power supply. The VOLTS ADJUST control is used to either increase or decrease the output voltage.

The Analog Multimeter (AMM) (Figure 5-28)

28. The multimeter is a device for the measurement of current, voltage, and resistance.

FIGURE 5-28   Making Measurements with the Analog Multimeter (AMM).

Measuring Current

Multirange or Scale Ammeter

10 μA: On this scale the meter will indicate current values between 0 and 10 μA.

10 mA: On this scale the meter will indicate current values between 0 and 10 mA.

10 A: On this scale the meter will indicate current values between 0 and 10 A.

Measuring Voltage

Multirange or Scale Voltmeter

10 mV: This scale can be used to measure any voltage from 0 to 10 mV.

10 V: This scale can be used to measure any voltage from 0 to 10 V.

10 kV: This scale can be used to measure any voltage from 0 to 10 kV.

Measuring Resistance

36 Ω Resistor

29. It is called a multimeter because it is like having three meters all rolled into one, in that it has

   a. An ammeter for measuring current

   b. A voltmeter for measuring voltage

   c. An ohmmeter for measuring resistance

30. For many years the analog readout multimeter was the only type available. At present these analog readout multimeters are still being used but are slowly being replaced by digital readout multimeters.

31. *Analog* is a word that means "similar to." For example, a ball is an analog of the world in that the shapes are similar. With an analog multimeter or AMM, the amount of pointer deflection across the scale is an analog (or similar to) the magnitude of the electrical property being measured. This type of indicator is therefore called an *analog display* as opposed to a *digital display,* which indicates the magnitude of the electrical property using decimal digits.

32. Ammeters (ampere meters) are used to measure the current flow within a circuit. In electronics, where current is generally small, the milliampere (mA) or microampere (μA) scale will generally be used, whereas in electrical high-current circuits, the meter will generally be used on an ampere scale.

33. Three rules must be remembered and applied when measuring current:

   a. Always set the selector to the higher scale first, and then reduce as needed.

   b. The current meter must be inserted in the circuit so that the positive lead of the ammeter (red) is connected to the positive polarity, and the negative lead of the ammeter (black) is connected to the negative polarity.

   c. The ammeter measuring the current must be connected in the path of current flow.

34. Voltmeters (voltage meters) are used to measure electrical pressure (voltage) anywhere in an electronic circuit.

35. A voltmeter is connected across a device with the two leads of the voltmeter touching either side of the component.

36. Three steps must be remembered and applied when measuring voltage:

   a. Always set to the highest scale first and then reduce as needed.

   b. Always connect the positive lead (red) of the meter to the component lead that is closer to the positive potential, and the negative lead (black) of the meter to the component lead that is closer to the negative potential.

   c. If you are measuring the voltage across a component, the voltmeter must be connected across the component and outside the circuit.

37. Resistance is measured with an ohmmeter.

38. Three rules must be remembered and applied when measuring resistance.

   a. Short the meter leads together and adjust the zero-ohms control so the pointer is at 0 Ω. This calibrates the meter scale.

b. Turn off the power and remove the component (if practicable) to be measured from circuit to protect the ohmmeter.

c. Connect the leads across the component, adjust the range scale until the pointer is approximately in the middle of the scale (for accuracy), and then multiply this value by the range scale selected.

The Digital Multimeter (DMM) (Figure 5-29)

39. The digital multimeters (DMM) are gradually replacing the analog-type multimeter because they are easier to read and have greater accuracy.

40. One of the greatest problems with an analog meter is the errors that occur due to the human factor when reading off the value from the many different scales.

41. When troubleshooting electronic equipment, the multimeter becomes the technician's eyes, allowing him or her to see the situation and the problem within the circuit or system.

42. The two test leads for the meter, one of which is red and the other black, must be inserted into the correct meter lead jacks on the meter. The black lead is connected to the meter jack marked "c" or *common*. The red lead is connected to the meter jack marked A for amperes or V-Ω for volts or ohms.

43. The solderless prototyping board (protoboard) or breadboard is designed to accommodate experiments. This protoboard will hold and interconnect resistors, capacitors, inductors, and many other components, as well as provide electrical power.

Soldering Tools and Techniques

44. Electronic components such as resistors, capacitors, diodes, transistors, and integrated circuits are combined to form circuits, which in turn are combined to form electronic systems. These components have their leads physically interconnected and then bonded with solder.

45. A solder connection is the joining together of two metal parts by applying both heat and solder.

46. The heat provided by the soldering iron is at a high enough temperature to melt the solder, making it a liquid that flows onto and slightly penetrates the two metal surfaces that need to be connected.

47. Once the soldering iron and therefore the heat are removed, the solder cools. In so doing, it solidifies and bonds the two metal surfaces, producing a good *electrical and mechanical connection*.

48. The solder actually penetrates the metals, and this embedding of the solder into the metal is called wetting.

49. Solder is a mixture of tin and lead, both of which have a low melting temperature with respect to other metals. This is necessary so that the soldering iron melts the solder and not the terminals or leads.

50. Different proportions of tin and lead are available to produce solder with different characteristics. A 60/40 solder is a mixture of 60% tin and 40% lead.

FIGURE 5–29 Making Measurements with the Digital Multimeter (DMM).

Measuring Current

How to Make Current Measurements

1.
Turn Off the Power
to the Circuit ⚠

2.
Cut or Unsolder the
Circuit, Creating a Place
Where the Meter
Probes Can Be
Inserted

3.
Select Amps
AC (A∼), or
Amps DC (A⎓)
as Desired

4.
Plug the Black Test
Probe into the COM
Input Jack. Plug the
Red Test Probe into the
10 Amp (10 A) or
300 Milliamp (300 mA)
Input Jack Depending on the
Expected Value of the Reading

5.
Connect the Probe
Tips to the Circuit
Across the Break
as Shown so That
All Current Will
Flow Through the
Meter (A Series
Connection)

6.
Turn the Circuit
Power Back On

7.
View the Reading,
Being Sure to Note
the Unit of
Measurement

Note: If Test Leads
Are Reversed, a
(–) Sign Will Show
in the Display

Measuring Voltage

Note: $\frac{1}{1000}$ V = 1 mV
1000 V = 1 kV

1.
Select Volts AC (V∼),
Volts DC (V⎓) or
300 mV⎓as
Desired

2.
Plug the Black
Test Probe into
the COM Input
Jack. Plug the
Red Test Probe
into the V Input
Jack

3.
Touch the Probe
Tips to the Circuit
Across a Load or
Power Source as
Shown (In Parallel
to the Circuit) ⚠

4.
View the Reading,
Being Sure to
Note the Unit of
Measurement

Note: For DC Readings of the Correct Polarity (+/–), Touch the
Red Test Probe to the Positive Side of the Circuit, and
the Black Probe to the Negative Side or Circuit Ground.
If You Reverse the Connections, A DMM with Auto
Polarity Will Merely Display a Minus Sign Indicating
Negative Polarity. With an Analog Meter You Risk
Damaging the Meter.

Measuring Resistance

Note: 1,000Ω = 1 kΩ
1,000,000Ω = 1 MΩ

1.
Turn Off Power
to the Circuit ⚠

2.
Select
Resistance (Ω)

3.
Plug the Black
Test Probe into
the COM Input
Jack. Plug
the Red Test
Probe into the
Ω Input Jack

4.
Connect the Probe
Tips Across the
Component or
Portion of the
Circuit for Which
You Want to
Determine
Resistance

5.
View the Reading,
Being Sure to Note
the Unit of
Measurement —
Ohms (Ω),
Kilohms (kΩ), or
Megohms (MΩ)

51. The proportions of tin to lead determine the melting temperature of the solder.

52. In electronics, 60/40 is most commonly used because of its low melting temperature, which means that a component lead or terminal will not have to be heated to a high temperature in order to make a connection.

53. A 63/37 solder has an even lower melting temperature is therefore even safer than 60/40 solder; however, it is more expensive.

54. Any lead or terminal is always exposed to the air, which forms an invisible insulating layer on the surface of leads, pins, terminals, and any other surface. A chemical substance is needed to remove this layer; otherwise, the solder would not be able to flow and stick to the metal contacts. Flux removes this invisible insulating oxide layer, and nearly all solder used in electronics contains the flux inside a type of solder tube.

55. The two most common types of flux are rosin and acid. Acid-core solder is used only in sheet metal work and should never be used in electronics since it is highly corrosive. In electronics, you should only use rosin-core solder.

56. A variety of solder diameters is available. The larger-diameter solder is used for terminals and large component leads and connections, whereas the smaller-diameter solder is used for soldering terminals that are very close to one another on a printed circuit board, so the amount of solder being applied to the connection needs to be carefully controlled.

57. Soldering irons are rated in terms of wattage, which indicates the amount of power consumed. More important, the wattage rating of a soldering iron indicates the amount of heat it produces. When an iron is applied to a connection, heat transfer takes place and heat is drained away from the iron to the metals to be connected; so a larger connection will need a larger-wattage soldering iron. A 25- to 60-W pencil iron is ideal for most electronic work.

58. For your safety and to protect voltage- and current-sensitive components, you should always use a soldering iron with a three-pin plug and three-wire ac power cord.

59. Always turn off the equipment before soldering, as the grounded tip may cause a short circuit in the equipment and possible damage.

60. The tip of a soldering iron can sometimes be changed, and a different temperature tip will allow your iron to be used for different applications.

61. In general, a 700°F tip is ideal for most electronic applications; however, for delicate printed circuit boards, use a 600°F tip.

62. The tip of a soldering iron comes in a variety of shapes and sizes. The tip shape that makes the best contact with the connection, and will therefore conduct the most heat, should be used.

63. As you regularly use your soldering iron, the tip will naturally accumulate dirt and oxide, and this contamination will reduce its effectiveness. The tip should be regularly cleaned by wiping it across a damp sponge.

64. The tip of the soldering iron should always be tinned after it has been cleaned. Tinning the tip will protect the tip from oxidation and increase the amount of heat transfer to the connection.

65. It is important to remember that *a soldering iron should not be used to melt the solder; its purpose is to heat the connection so that the solder will melt when it makes contact with the connection.*

66. Wires should be wrapped around a terminal to give it a more solid physical support. Their insulation should be stripped back so as to leave a small gap. Too large a gap will expose the bare wire and possibly cause a short to another terminal, whereas too small a gap will cause the insulation to burn during soldering. Stranded wire should be tinned, and components should be mounted flat on the board.

67. Always remember to wear safety goggles and a protective apron, and then proceed with the six step soldering procedure, which is:

Step 1   Clean the tip on a damp sponge.

Step 2   Tin the tip if necessary.

Step 3   Heat the connection.

Step 4   Apply the solder to the opposite side of the connection.

Step 5   Leave the tip of the soldering iron on the connection only long enough to melt the solder.

Step 6   Remove the solder and then the iron; then let the melted solder cool and solidify undisturbed.

Once this procedure is complete, you should remove the excess leads with the cutters and remove the flux residue, because it collects dust and dirt that could produce electrical leakage paths.

68. When inspecting your work, you should notice the following:

    a. The connection is *smooth and shiny.*

    b. The solder should *feather out to a thin edge.*

69. A *cold solder joint* has a dull gray appearance; it makes a poor mechanical and electrical connection and is normally caused due to connector lead movement while the solder was cooling, or to insufficient heat. The cure is to resolder the connection.

70. An *excessive solder joint* is caused by too much solder being applied to the connection or by using solder of too large diameter. The cure is to desolder the excess with a wick.

71. An *insufficient solder joint* happens because not enough solder was applied so a poor bond exists between the lead and the terminal. The cure is to resolder as if it were a new joint.

72. A *solder bridge* occurs when adjacent terminals or traces on printed circuit boards are connected accidentally as a result of using too much solder. Most can be removed by desoldering the excess with a wick.

73. On printed circuit boards, too much heat can lift the traces or track and ruin the entire board.

74. If a faulty component has to be replaced or if a component or wire was soldered in an incorrect place, it will have to be desoldered.

75. The easiest way to remove a component is to remove all the solder and then disconnect the leads.

76. The braided wick contains many small wire strands, and when it is placed between the solder and the iron, the melted solder flows into the braid by capillary action. The solder-filled braid is then cut off and discarded.

77. The desoldering bulb or spring-loaded plunger relies on suction rather than capillary action to remove the solder.

78. The steps to follow to desolder are:

Step 1 Squeeze the desoldering bulb or cock the spring-loaded plunger.

Step 2 Melt the solder with the soldering iron.

Step 3 Remove the iron's tip.

Step 4 Quickly insert the tip of the bulb or plunger into the molten solder and then activate the suction.

79. When as much of the solder as possible has been removed, use the long-nose pliers to hold the lead and remove the component.

80. It may be necessary to use the iron to heat the lead slightly to loosen up the residual solder so that the component lead can be removed with the pliers.

81. Always be careful not to apply an excessive amount of heat or stress to either the component or the board.

82. Molten solder, like any other liquid, can splash or spill, and soldering irons are even hotter than the solder.

NEW TERMS

Ammeter	Digital multimeter	Shock
Analog multimeter	Flux	Solder
Autoranging	Function generator	Soldering
Body resistance	Live	Soldering iron
Breadboard	Ohmmeter	Triggering
Calibration	Oscilloscope	Voltmeter
Dc power supply	Overload	Wetting
Desoldering	Protoboard	

REVIEW QUESTIONS

Multiple-Choice Questions

1. What would be the current through a body resistance of 20 kΩ if it were to come in contact with 24 V?

   a. 12 mA   **c.** 480 kA
   b. 1.2 mA   **d.** 833.3 A

2. Could the current flow in Question 1 be considered fatal?

   a. Yes   **b.** No

3. Which test instrument is used to measure current, voltage, and resistance?

   a. A dc power supply   **c.** A function generator
   b. A multimeter   **d.** An oscilloscope

4. Which instrument is designed to deliver a dc voltage for experimentation?

   a. A dc power supply   **c.** A function generator
   b. A multimeter   **d.** An oscilloscope

5. In which modes will a multimeter use its internal battery?

   a. Voltage   **c.** Resistance
   b. Current   **d.** None of the above

6. _____ is needed within solder to remove an invisible insulating layer so that the solder can flow and stick to the metal contacts.

   a. Wetting   **c.** Tin
   b. Flux   **d.** Lead

7. The embedding of solder into the metal contacts is called:

   a. Wetting   **c.** Tinning
   b. Flux   **d.** None of the above

8. _____ the tip of a soldering iron after it has been cleaned will protect it from oxidation and increase the amount of heat transfer to the connection.

   a. Fluxing   **c.** Tinning
   b. Wetting   **d.** None of the above

9. What tool can be used to desolder a connection?

   a. A bulb   **c.** A spring-loaded plunger
   b. A braid   **d.** All of the above

10. What is 60/40 rosin?

    a. A rosin-core 60% lead, 40% tin solder
    b. A 60% rosin, 40% lead solder
    c. A 60% tin, 40% lead solder with a rosin core
    d. All of the above

Essay Questions

11. What is the resistance of a human body? (5-1-1)

12. Why is it important for electronic technicians to understand resuscitation? (5-1-3)

13. What advantages does the dc power supply have over a battery as a voltage source? (5-2-1)

14. Choosing either an analog multimeter or digital multimeter, briefly describe the steps involved in measuring: (5-2-2)

    a. Current   **b.** Voltage   **c.** Resistance

15. What advantages does the DMM have over the AMM? (5-2-3)

16. Briefly describe the construction and function of a protoboard. (5-3)

17. Define the following soldering names and terms: (5-4)

    a. Wetting   **c.** Tinning

    b. Rosin 60/40   **d.** Desoldering

18. List the six-step soldering procedure. (5-4-5)

19. How should a good solder connection appear? (5-4-5)

20. List the four-step desoldering procedure. (5-4-6)

Practice Problems

21. If an analog ammeter indicates 15 on the 300 mA range, what value of current is present?

22. If an analog ohmmeter displays 47 on the $R \times 1000$ range, what is the resistance being measured?

23. If a digital voltmeter displays 200 on the 500 mV range, what is the value of voltage being measured?

24. If a DMM displays "OL," what does it indicate?

25. What current is flowing through the heating coil of a 60 W, 120 V pencil soldering iron?

CHAPTER

6

OBJECTIVES

After completing this chapter, you will be able to:

1. Describe how direct current can be generated:
   a. Mechanically with friction and pressure
   b. Thermally with a thermocouple
   c. Optically with a photocell
   d. Magnetically with a generator
   e. Chemically with a battery
   f. Electrically with a power supply

2. Understand the nature of magnetism and why certain materials will magnetize and certain materials will not magnetize.

3. Describe the magnetic field and the rules behind magnetic attraction and repulsion.

4. Explain how magnetic energy can be used to generate direct current.

5. State the differences between the three artificial magnets.

6. Describe the operation of both a primary and a secondary battery.

7. State the difference between a primary and a secondary cell.

8. Describe the features, applications, and construction of all the popular primary and secondary cells.

9. Explain the advantages of connecting batteries in series and parallel.

10. Define what is meant by a battery's:
    a. Internal resistance
    b. Maximum power transfer
    c. Optimum power transfer

11. Describe the operation and use of various types of fuses, circuit breakers, and switches.

Direct Current (DC)

The First Computer Bug

Mathematician Grace Murray Hopper, an extremely feisty and independent U.S. naval officer, was assigned to the Bureau of Ordnance Computation Project at Harvard during World War II. As Hopper recalled, "We were not programmers in those days, the word had not yet come over from England. We were 'coders,'" and with her colleagues she was assigned to compute ballistic firing tables on the Harvard Mark 1 computer. In carrying out this task, Hopper developed programming method fundamentals that are still in use.

Hopper is also credited, on a less important note, with creating a term frequently used today falling under the category of computer jargon. During the hot summer of 1945, the computer developed a mysterious problem. Upon investigation, Hopper discovered that a moth had somehow strayed into the computer and prevented the operation of one of the thousands of electromechanical relay switches. In her usual meticulous manner, Hopper removed the remains and taped and entered it into the logbook. In her own words, "From then on, when an officer came in to ask if we were accomplishing anything, we told him we were 'debugging' the computer," a term that is still used to describe the process of finding problems in a computer program.

Direct current, abbreviated dc, is a flow of continuous current in only one direction and it is what we have been dealing with up to this point. A dc current is produced when a dc voltage source, such as a battery, is connected across a closed circuit. The requirement for a dc voltage source is that its output voltage polarity remain constant, and the fixed polarity of the voltage source will produce a unidirectional (one direction) current through a dc circuit. In this chapter, we will see how a dc voltage can be generated using several different methods, and we will examine fuses, circuit breakers, and switches.

6-1 SOURCE AND LOAD

Figure 6-1(a) illustrates a dc circuit with a battery connected across a light bulb. In this circuit, the potential energy of the battery produces kinetic energy or current that is used to produce light energy from the bulb. The bat-

FIGURE 6-1 Load on a Voltage Source.
(a) Pictorial. (b) Schematic.

tery is therefore the **source** in this circuit and the bulb is called a **load,** which by definition is a device that absorbs the energy being supplied and converts it into the desired form. The **load resistance** will determine how hard the voltage source has to work. For example, the bulb filament has a resistance of 300 Ω. The connecting wires have a combined resistance of 0.02 Ω (20 mΩ), and since this value is so small compared to the filament's resistance, it will be ignored, as it will have no noticeable effect. Consequently, a 300-Ω load resistance will permit 10 mA of **load current** to flow when a 3 V source is connected across the dc circuit ($I = V/R = 10$ mA). A change in filament resistance will result in a change in current; so if the load resistance is increased, the load current decreases ($I \downarrow = V/R \uparrow$), and conversely, if the load resistance is decreased, the load current increases ($I \uparrow = V/R \downarrow$) for a constant voltage source.

Figure 6-2 illustrates a heater, light bulb, motor, computer, robot, television, and microwave oven. All these and all other devices or pieces of equipment are connected to some form of supply (for example, a battery). The supply does not see all the circuitry and internal workings of the equipment; it simply sees the whole device or piece of equipment as some value of resistance. The resistance of the equipment and the supply voltage determine the value of current flow from the supply to the device or equipment. The battery or any other current source sees these devices or pieces of equipment as having some value of resistance, which is referred to as the device's *load resistance.*

Source
Device that supplies the signal power or electric energy to a load.

Load
A component, circuit, or piece of equipment connected to the source. The load resistance will determine the load current.

Load Resistance
The resistance of the load.

Load Current
The current that is present in the load.

◾ **EXAMPLE:**

A component or piece of equipment offers a 390-Ω load resistance in a 9-V circuit. What current will flow in this circuit?

(a)

Load Current (I_L)

Heavy Load ($R_L\downarrow$, $I_L\uparrow$)
Light Load ($R_L\uparrow$, $I_L\downarrow$)

DC
Voltage
Source

Load
Resistance (R_L)

(b)

FIGURE 6-2 **(a) Load Resistance of Equipment. (b) Schematic Symbols and Terms.**

▨ *Solution:*

$$\text{current } (I) = \frac{\text{voltage } (V)}{\text{resistance } (R)}$$

$$= \frac{9\text{ V}}{390\ \Omega}$$

$$= 23\text{ mA}$$

In summary, the phrase *load resistance* describes the device or equipment's circuit resistance, whereas the phrase *load current* describes the amount of current drawn by the device or equipment. A device that causes a large load current to flow, due to its small load resistance, is called a large or heavy load because it is heavily loading down or working the supply or battery. On the other hand, a device that causes a small load current to flow, due to its large load resistance, is referred to as a small or light load because it is only lightly loading down or working the supply or battery.

Load current and load resistance are therefore inversely proportional to one another in that a large load resistance ($R_L\uparrow$) causes a small load current ($I_L\downarrow$), referred to as a light load, and vice versa.

SELF-TEST REVIEW QUESTIONS FOR SECTION 6-1

1. Define *load resistance, load current, dc voltage source,* and *dc current.*
2. True or false: A large load current will flow when a large load resistance is connected across the supply.
3. What is the relationship between load resistance and load current?
4. Describe a heavy load resistance.

6-2   DIRECT-CURRENT SOURCES

The six basic forms of energy are mechanical, heat, light, magnetic, chemical, and electrical. To produce direct current (electrical energy), one of the other forms of energy must be used as a source and then converted or transformed into electrical energy in the form of a dc voltage that can be used to produce direct current.

6-2-1   *Mechanically Generated DC: Mechanical* —————→ *Electrical*

When initially discussing current, we said that a positive charge was a material that has a deficiency or lack of electrons, while a negative charge has an excess of electrons. These positive and negative charges are known as *static* (stationary) *charges* or *static electricity.* Attraction and repulsion occur with positive and negative charges, as reviewed in Figure 6-3.

FIGURE 6-3   Attraction and Repulsion. (a) Positive Repels Positive. (b) Negative Repels Negative. (c) Unlike Charges Attract.

Friction

Friction is a form of mechanical energy, and Figure 6-4 illustrates how the electrons in a glass rod (insulator) can be forced out of their orbits and captured by another material, such as wool, by the work of rubbing. A static negative charge now exists in the wool, while a static positive charge exists in the glass. These mechanically generated static charges can be discharged in one of three ways, producing electrical energy, as illustrated in Figure 6-5.

Figure 6-5(a) illustrates the discharge of a static charge by placing a conductor between the two charges and allowing electrons to flow from one material to the other. Once a state of equilibrium is achieved, the two static charges will have neutralized and no further current will flow. If the connection has a small resistance, the discharge current will be brief and intense; on the other hand, if the connection has a large resistance, the discharge current will be prolonged and less intense.

Figure 6-5(b) illustrates the discharge of a static charge by contact. You may have noticed that your body can, through friction, build up a static charge that is discharged by contact once you touch another oppositely charged or neutralized material, giving you a shock.

Figure 6-5(c) illustrates how large static charges will discharge through the air, causing an arc between the two static charges known as lightning. Moving clouds in a storm cause friction between the molecules in the air. The buildup of a natural static charge in a cloud, if large enough, can cause a discharge or release of the electrical energy to the ground.

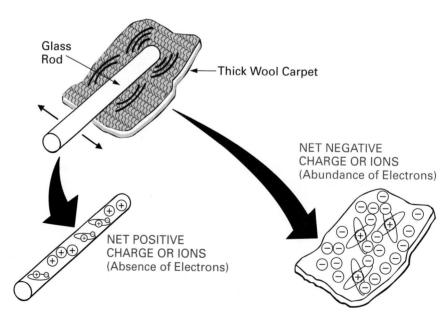

FIGURE 6-4   Using Friction to Generate Static Electricity: Mechanical Energy to Electrical Energy.

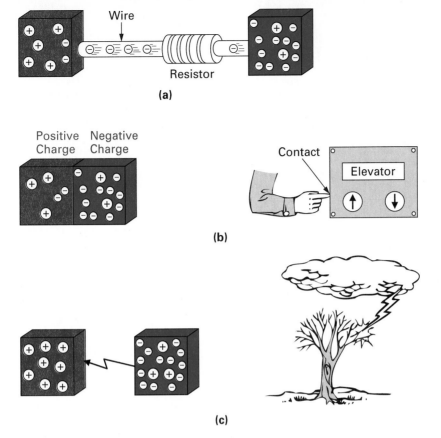

**FIGURE 6-5   Static Discharge. (a) Connection Static Discharge.
(b) Contact Static Discharge. (c) Arc Static Discharge.**

Pressure

Mechanical energy in the form of **pressure** can also be used to generate electrical energy. Quartz and Rochelle salts are solid compounds that will produce electricity when a pressure is applied, as shown in Figure 6-6. Quartz is a natural or artificially grown crystal composed of silicon dioxide, from which thin slabs or plates are carefully cut and ground. Rochelle salt is a crystal of sodium potassium tartrate. They are both piezoelectric crystals. **Piezoelectric effect** is the generation of a voltage between the opposite faces of a crystal as a result of pressure being applied. This effect is made use of in most modern cigarette lighters. A voltage is produced between the opposite sides of a crystal (usually a crystalline lead compound) when it is either struck, pressured, or twisted. The resulting voltage is discharged across a gap to ignite the gas, which is released simultaneously by the operating button. The flame then burns for as long as the button is depressed. Piezoelectric effect also describes the reverse of what has just been discussed, as it was discovered that crystals would not only generate a voltage when mechanical force was applied, but would generate mechanical force when a voltage was applied.

Figure 6-7(a) illustrates the physical appearance of a piezoresistive

Pressure
The application of force to something by something else in direct contact with it.

Piezoelectric Effect
The generation of a voltage between the opposite sides of a piezoelectric crystal as a result of pressure or twisting. Also, the reverse effect in which the application of voltage to opposite sides causes deformation to occur at the frequency of the applied voltage.

FIGURE 6-6 **Direct Current from Pressure Using Crystals. (a) Quartz Crystal. (b) Physical Appearance. (c) Construction. (d) Operation.**

diaphragm pressure sensor. Piezoresistance of a semiconductor is described as a change in resistance due to a change in the applied strain, and therefore semiconductor resistors can be used as pressure sensors. Four piezoresistors [Figure 6-7(b)] are buried in the surface of a thin, circular (silicon) diaphragm that has two pads connected to it making the connection. When pressure is applied, a stress or strain is applied to the diaphragm and therefore to the buried piezoresistors. The resistance between the pads will change depending on the amount of pressure applied to the diaphragm, and consequently a change in pressure (mechanical input) will cause a corresponding change in resistance (electrical output). Unlike quartz, which generates a voltage, the piezoresistive sensor needs an excitation voltage applied, and its value will determine the value of output voltage. Figure 6-7(c) illustrates the operation of this pressure transducer, which could be used, for example, to measure either the cooling system, hydraulic transmission, or fuel injection pressure in an automobile, which all vary in the 0 to 100 psi (pounds per square inch) range. The piezoresistive transducer would receive its excitation voltage from the car's battery (12 V) and produce a 0- to 100-mV output voltage, which would be sent to the car's computer, where the data would be analyzed and, if necessary, acted upon, as shown in Figure 6-7(d).

(a)

(b)

(c)

(d)

FIGURE 6-7 Direct Current from Pressure Using Piezoresistive Sensors. (a) Physical Appearance. (b) Construction. (c) Operation. (d) Application.

(a)

Two Metals	Temperature Range (°F)	Voltage Generated for Temperature Range (mV)
Iron vs. Copper–Nickel	32 to 1382	0 to 42.3
Nickel–Chromium vs. Nickel–Aluminum	–328 to 2282	–5.9 to 50.6
Platinum–6% Rhodium vs. Platinum–30% Rhodium	32 to 3092	0 to 12.4

(b)

(c)

(d)

FIGURE 6-8   Direct Current from Heat Using the Thermocouple. (a) Operation. (b) Typical Thermocouples. (c) Indicating Temperature. (d) Water Heater Application.

6-2-2 Thermally Generated DC:
Heat ⎯⎯⎯⎯⎯⎯⟶ Electrical

Heat can be used to generate an electric charge, as illustrated in Figure 6-8(a). When two dissimilar metals, such as copper and iron, are welded together and heated, an electrical charge is produced. This junction is called a **thermocouple,** and the heat causes the release of electrons from their parent atoms, resulting in a meter deflection indicating the generation of a charge. The size of the charge is proportional to the temperature difference between the two metals. Figure 6-8(b) lists some typical thermocouples and their voltages generated when heat is applied, and shows the physical appearance of two types.

    Thermocouples like the one in Figure 6-8(c) are normally used to indicate temperature on a display calibrated in degrees. The thermocouple seen in Figure 6-8(d) is being heated by the gas pilot light, and the electrical energy generated will allow a valve to open and let gas through to the water heater. If, by accident, the pilot is extinguished by a gust of wind, the thermocouple will not receive any more heat and therefore will not generate electrical energy, which will consequently close the electrically operated valve and prevent a large explosion.

Thermocouple
Temperature transducer consisting of two dissimilar metals welded together at one end to form a junction that when heated will generate a voltage.

SELF-TEST REVIEW QUESTIONS FOR SECTION 6-2-2

1. What is the name given to the device used to convert heat to dc?
2. Name an application in which the transducer in Question 1 could be used.

6-2-3 Optically Generated DC:
Light ⎯⎯⎯⎯⎯⎯⟶ Electrical

A **photovoltaic cell,** shown in Figure 6-9(a), also makes use of two dissimilar metals or semiconductor materials in its transformation of light to electrical energy. Figure 6-9(b) illustrates the construction and operation of a photoelectric cell, also called a **solar cell.**

    A light-sensitive metal or semiconductor is placed behind a transparent piece of dissimilar metal. When the light illuminates the light-sensitive material, a charge is generated, causing current to flow and the meter to indicate current flow. This phenomenon, known as **photovoltaic action,** finds application as a light meter for photographic purposes and as an electrical power source for use by the satellite shown in Figure 6-9(c).

    Electric power companies make use of solar cells to convert sunlight into electrical energy for the consumer. Out in space the solar cell may reach 75% efficiency, whereas on the earth they are only about 20% efficient, as the sun's rays are inhibited by the earth's atmosphere. The power obtained is on average about 100 milliwatts per square centimeter. Solar panels can also be found on calculators and emergency freeway telephones. Figure 6-10 shows how a satellite is generally covered with solar panels so that it can generate power to drive its electronic communication circuits.

Photovoltaic Cell
A solid-state device that generates electricity when light is applied. Also called a solar cell.

Solar Cell
Photovoltaic cell that converts light into electric energy. They are especially useful as a power source for space vehicles.

Photovoltaic Action
The development of a voltage across the junction of two dissimilar materials when light is applied.

FIGURE 6-9 **Direct Current from Light Using the Solar Cell. (a) Physical Appearance. (b) Construction and Operation. (c) Applications.**

SELF-TEST REVIEW QUESTIONS FOR SECTION 6-2-3

1. What is the name of the component used to convert light energy to dc?
2. Name an application in which the transducer in Question 1 could be used.

6-2-4 Magnetically Generated DC: Magnetic ⟶ Electrical

Before discussing how magnetism can be used to generate electrical energy, let us first discuss the principles of magnetism.

FIGURE 6-10 Solar Panels. (Photo courtesy of Hewlett-Packard Company.)

The Nature of Magnetism

In the ancient city of Magnesia in Asia Minor, stones composed of iron oxide were found to have the strange ability to attract and repel one another. It was also noted that when suspended from a string, these stones would always point in only one direction. Historians are not sure whether the Chinese or Europeans first began using these "leading stones," or **lodestones** as they are now known, for long-distance land and sea travel, but the original compass was probably a lodestone suspended from a string. Later it was found that a metal needle could be magnetized by stroking it across a lodestone, and so the next form of compass consisted of a magnetized needle attached to a piece of wood floating in a bowl of water; hence the origin of the term *needle of a compass*.

The lodestone shown in Figure 6-11(a) is an example of a natural magnet, unlike the needle, which is called an artificial magnet because it must be artificially magnetized in order to possess a magnetic field, but they both have the ability to point in only one direction by interacting with the earth's natural magnetic field, as shown in Figure 6-11(b).

Magnetism was recorded by the Greeks as early as 800 B.C., but it was not until 1269 that Petrus Peregrinus de Maricout found that the invisible magnetic lines produced by lodestones intersected at two points, which he called *poles*. In 1600, William Gilbert discovered that the magnetic field of a lodestone was lost when it was heated, but reappeared when the lodestone cooled. In 1819, Hans Christian Oersted discovered that any current-carrying conductor generated a

Lodestone
A magnetite stone possessing magnetic polarity.

(a)

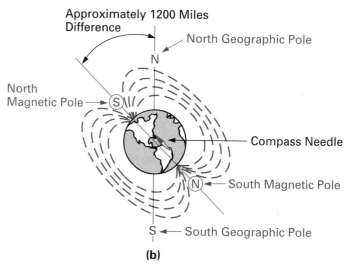

Approximately 1200 Miles
Difference

North Geographic Pole

N

North
Magnetic Pole → ⓈⅢ

Compass Needle

Ⓝ ← South Magnetic Pole

S ← South Geographic Pole

(b)

FIGURE 6-11   Natural Magnets. (a) Lodestone. (b) Earth.

Electromagnetism
Relates to the
magnetic field
generated around a
conductor when
current is passed
through it.

magnetic field, and this interrelationship between electricity and magnetism became known as **electromagnetism.**

In 1838, a German mathematician and physicist, Johann Gauss, published a paper on the earth's natural magnetic field, and in honor of his contribution the flux density of a magnetic field is measured in *gauss*. A magnetic flux density of 1 gauss (G) is produced 1 centimeter away from a wire carrying a current of 5 A. To give some examples, the field strength of a magnetized needle may only be a fraction of a gauss, whereas a lodestone 1 inch long may have a strength of 100 G. The earth's magnetic field has a magnetic field strength of 0.3 G at the equator and 0.7 G at either pole. Although this sounds small, a simple experiment can be conducted to prove how influential this invisible magnetic field is. If a screwdriver is positioned so that it points north and the screwdriver is tapped with a hammer several times, the screwdriver will actually become magnetized by the earth's magnetic field, as illustrated in Figure 6-12(a). The screwdriver is in fact an artificial magnet that was magnetized by a natural magnet (the earth). The atoms within the steel were originally misaligned. However, after being tapped and influenced by the earth's magnetic field, the atoms became aligned to reinforce one another, and will serve as a simple magnet as shown in Figure 6-12(b).

FIGURE 6-12   Magnetizing a Screwdriver. (a) Method. (b) Result.

Magnetism and the Atom

As you are already aware, an atom consists of electrons spinning around a nucleus; some atoms have few, while others have many electrons orbiting the atom's nucleus. A question arises when we find out that we can magnetize a screwdriver made of iron or nickel but cannot magnetize a screwdriver made of copper or tungsten. Why will certain materials magnetize whereas others will not? The answer is found in our smallest subatomic particle, the electron. Each electron is in fact a miniature magnet spinning around the nucleus. An electron at rest will not generate a magnetic field; however, orbiting electrons produce a magnetic field because of their high orbital velocity. Referring to your atomic periodic table of the elements (Table 2-1), you will notice that tungsten, for example, has 74 electrons (will not magnetize) and iron has 26 electrons (will magnetize). With tungsten, 37 electrons rotate around the nucleus in the clockwise direction, while 37 electrons orbit in the counterclockwise direction. The clockwise magnetic field will cancel with the counterclockwise magnetic field, so the net magnetic field is zero, which is why tungsten cannot be magnetized.

Iron has fewer electrons (26); however, 10 of the iron electrons spin clockwise and 16 spin counterclockwise, producing a net counterclockwise magnetic field of 6 electrons, which is why iron can be magnetized. If approximately 100 trillion atoms of the same magnetic polarity are all arranged in one direction, they are called a **domain.** Domains within unmagnetized materials are all misaligned, as shown in Figure 6-13(a), while aligned domains produce magnetized materials, shown in Figure 6-13(b).

The Magnetic Field

The magnetic field produced by a magnet is an invisible force concentrated at two opposite points called the **magnetic poles,** shown in Figure 6-14. The north pole of the magnet, is the pole that would point south toward the earth's south magnetic pole if the magnet were free to rotate. The south pole of the magnet would point in a northerly direction if free to rotate. As you can see by again referring to Figure 6-14, the magnetic field is made up of invisible **flux lines** that leave the north pole and return at the south.

Domain
Also known as a magnetic domain, it is a movable magnetized area in a magnetic material.

Magnetic Poles
Points of a magnet from which magnetic lines of force leave (north pole) and arrive (south pole).

Magnetic Flux
The magnetic lines of force produced by a magnet.

FIGURE 6-13　(a) Misaligned Domains. (b) Aligned Domains.

Attraction and Repulsion

When analyzing magnetic lines of flux within the magnetic field, it is noted that the law of attraction and repulsion could be applied to magnetic poles, just as it can be to positive and negative charges. This is summarized in Figure 6-15. A north against a north pole and south against a south pole will cause a force of repulsion between the two like poles [Figure 6-15(a)], while a south pole against

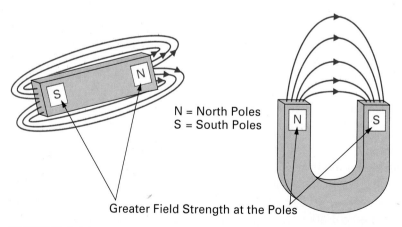

FIGURE 6-14　Magnetic Field (Made Visible for Demonstration).

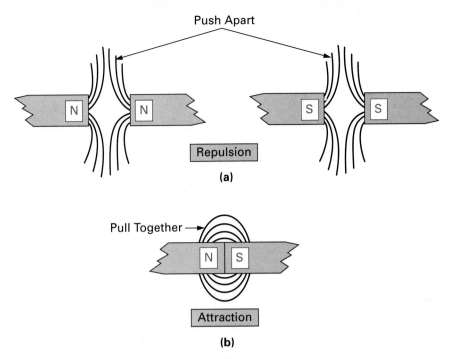

FIGURE 6-15 **Attraction and Repulsion. (a) Like Poles Repel Each Other. (b) Unlike Poles Attract Each Other.**

a north pole will result in a force of attraction between the two unlike poles [Figure 6-15(b)].

Artificial Magnets

There are three types of artificial magnets: (1) permanent magnets, (2) temporary magnets, and (3) electromagnets.

Permanent magnets. A permanent magnet is a piece of hardened steel or other special alloy that, when magnetized by the presence of a magnetic field, will retain its magnetism indefinitely (high retentivity), even when the material is removed from the magnetizing force or the magnetizing force is no longer present, as shown in Figure 6-16.

Temporary magnets. A temporary magnet is a piece of soft iron that will be magnetized while in the presence of a magnetizing field, but will be demagnetized the moment the soft iron is taken away from the magnetizing force or the magnetizing force is no longer present, as shown in Figure 6-17. Soft iron possesses low retentivity, which means it will not retain a magnetic field.

Electromagnets. An electromagnet is a temporary magnet that consists of an insulated wire wrapped around a soft-iron cylinder (core) to form a coil, as shown in Figure 6-18(a). A magnetic field is produced only while current flows through the coil, as shown in Figure 6-18(b). If a larger current is passed through the coil, a corresponding stronger magnetic field is generated, as shown in Figure 6-18(c).

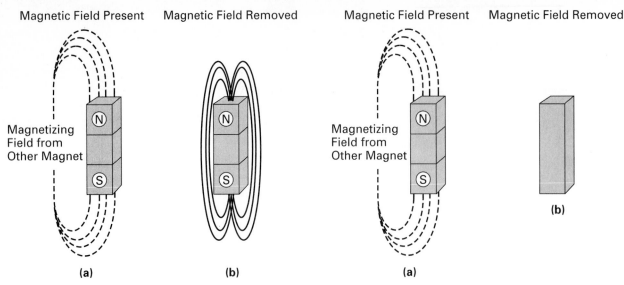

FIGURE 6-16   Permanent Magnets.
(a) Permanent magnet is magnetized when a
magnetizing field is present (domains aligned).
(b) Permanent magnet retains the magnetic energy
even when the magnetizing field is no longer present
(domains remain aligned).

FIGURE 6-17   Temporary Magnets.
(a) Temporary magnet is magnetized when the
magnetizing field is present (domains aligned).
(b) Temporary magnet loses magnetic energy when
the magnetizing field is removed (domains no
longer aligned).

Converting Magnetic Energy into Electrical Energy

A magnet like the one in Figure 6-19(a) can be used to generate electrical
current by movement of the magnet past a conductor, as shown in Figure
6-19(b). When the magnet is stationary, no voltage is induced into the wire and
therefore the light bulb is OFF. If the magnet is moved so that the magnetic
lines of force cut through the wire conductor, a voltage is induced in the conductor, and this voltage produces a current in the circuit causing the light bulb to
emit light. To achieve an energy conversion from magnetic to electric, you can
either move the magnet past the wire or move the wire past the magnet.

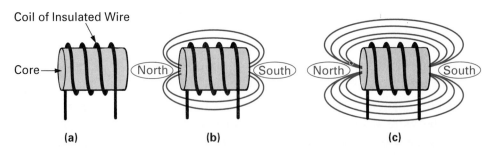

FIGURE 6-18   Electromagnets. (a) No Current, No Magnetic Field. (b) Small
Current, Small Magnetic Field. (c) Large Current, Large Magnetic Field.

(a)　　　　　　　　　　　**(b)**

**FIGURE 6-19   Direct Current from Magnetism. (a) Stationary Magnet.
(b) Moving Magnet.**

FIGURE 6-20   Generator.

To produce a continuous supply of electric current, the magnet or wire must be constantly in motion. The device that achieves this is the dc electric generator, illustrated in Figure 6-20.

SELF-TEST REVIEW QUESTIONS FOR SECTION 6-2-4

1. What is the name given to the natural magnetic rock?
2. How do the laws of attraction and repulsion apply to magnetism?
3. List the three types of artificial magnets.
4. What device is used to convert magnetic energy into electrical energy?

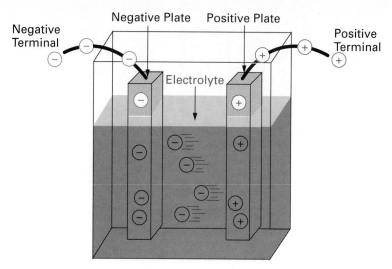

FIGURE 6-21 Basic Battery.

6-2-5   Chemically Generated DC: Chemical ——————→ Electrical

The chemical voltage source (the battery) is called the **voltaic cell.** Its principle of operation was discovered by Alessandro Volta, an Italian physicist, in 1800.

A car uses a lead–acid battery, a portable radio has a carbon–zinc or alkaline battery, and, in some instances, you may be using the rechargeable nickel–cadmium batteries. No matter what type of battery is used, they all produce voltage by the same basic principle and all have three basic components, shown in Figure 6-21: (1) a negative plate, (2) a positive plate, and (3) an electrolyte.

Battery Operation

Two dissimilar, separated metal plates such as copper and zinc are placed within a container that is filled with a liquid known as the **electrolyte,** usually an acid. A chemical reaction causes electrons to be repelled from one plate and attracted to the other, passing through the electrolyte. A large number of electrons collect on one plate (negative plate), while an absence or deficiency of electrons exists on the opposite, positive plate. The electrolyte acts on the two plates and transforms chemical energy into electrical energy, which can be taken from the cell at its two output terminals as an electrical current flow. If nothing is connected across the battery, as shown in Figure 6-22(a), a chemical reaction between the electrolyte and the negative electrode produces free electrons that travel from atom to atom and are then held in the negative electrode.

In Figure 6-22(b), a light bulb has been connected between the negative and positive electrodes. The mutual repulsion of the free electrons at the nega-

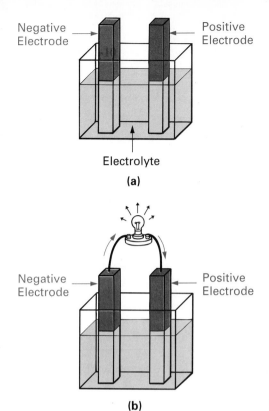

Negative Electrode · Positive Electrode · Electrolyte

(a)

Negative Electrode · Positive Electrode

(b)

FIGURE 6-22 **Battery Operation.**

tive electrode combined with the attraction of the positive electrode results in a migration of free electrons (current flow) through the light bulb, causing its filament to produce light.

Primary Cells

The chemical reaction within the cell will actually dissolve the negative plate until eventually it will be eaten away completely. This discharging process also results in hydrogen gas bubbles forming around the positive plate, causing a resistance between the two plates known as the battery's internal resistance to increase (called *polarization*). To counteract this problem, all dry cells have a chemical within them known as a depolarizer agent, which reduces the buildup of these gas bubbles. With time, however, the depolarizer's effectiveness will be reduced, and the battery's internal resistance will increase as the battery reaches a completely discharged condition. These types of batteries are known as **primary** (first time, last time) **cells** because they discharge once and must then be discarded. Almost all primary cells have their electrolyte in paste form which is why they are also referred to as **dry cells.** Wet cells have an electrolyte that is in liquid form.

Primary cell types.   Figure 6-23 illustrates some of the various types of primary cells on the market today. The shelf life of a battery is the length of time a battery can remain on the shelf (in storage) and still retain its usability. Most pri-

Primary Cell
Cell that produces electrical energy out through an internal electrochemical action; once discharged, it cannot be reused.

Dry Cell
Dc voltage-generating chemical cell using a nonliquid (paste) type of electrolyte.

FIGURE 6-23   Primary Rectangular, Cylindrical, and Button Cells and Batteries.

mary cells will deteriorate in a three-year period to approximately 80% of their original capacity.

Of all the types of primary cells, five dominate the market as far as applications and sales are concerned. These are the carbon–zinc, alkaline–manganese, mercury, silver oxide, and lithium, all of which are described and illustrated in Table 6-1. The cell voltage in Table 6-1 describes the voltage produced by one cell (one set of plates). If two or more sets of plates are installed in one package, the component is called a battery.

Secondary Cells

Secondary Cells
An electrolytic cell for generating electric energy. Once discharged the cell may be restored or recharged by sending an electric current through the cell in the opposite direction to that of the discharge current.

Secondary cells operate on the same basic principle as that of primary cells. However, in this case the plates are not eaten away or dissolved; they only undergo a chemical change during discharge. Once discharged, the secondary cell can have the chemical change that occurred during discharge reversed by recharging, resulting once more in a fully charged cell. Restoring a secondary cell to the charged condition is achieved by sending a current through the cell in the direction opposite to that of the discharge current, as illustrated in Figure 6-24.

In Figure 6-24(a), we have first connected a light bulb across the battery, and free electrons are being supplied by the negative electrode and moving through the light bulb or load and back to the positive electrode.

After a certain amount of time, any secondary cell will run down or will have discharged to such an extent that no usable value of current can be produced. The surfaces of the plates are changed, and the electrolyte lacks the necessary chemicals.

In Figure 6-24(b), during the recharging process, we use a battery charger, which reverses the chemical process of discharge by forcing electrons back into the cell and restoring the battery to its charged condition. The battery charger

TABLE 6-1 Primary Cell Types

Battery type	Features	Cell Voltage (V)	Applications	Construction
Carbon–zinc (also called Leclanché cell in honor of its inventor)	Most popular due to its low cost. Cylindrical D and C cells are most commonly used. Flat cells are stacked in series to obtain voltages greater than 1.5 V, as in the case of the 9 V battery	1.5	Portable radios, tape players, televisions, toys.	Has a zinc case (negative), a carbon center rod (positive), with ammonium chloride as the electrolyte and manganese dioxide as the depolarizing agent
Alkaline–maganese (also called alkaline cell)	Has three times the shelf life and capacity of carbon zinc. Cylindrical or miniature cell sizes are available	1.4	Portable radios, tape players, televisions, toys; large capacity is worth investment	The negative plate is granular zinc mixed with an electrolyte (alkaline); the positive plate is a polarizer in contact with the outer metal can
Mercury (also called mercuric–oxide)	Higher energy density than preceding two with good shelf life and small size; used in low-power applications. Flat or cylindrical cells are available	1.35 and 1.4	Watches, hearing aids, pacemakers, cameras, test equipment	A zinc anode and a mercuric oxide cathode in a potassium hydroxide electrolyte
Silver–oxide	High capacity; however, the material used (silver) makes it the most costly; can supply high currents for short periods of time; used in low-power applications	1.5	Watches, hearing aids, pacemakers, cameras	Contains a cathode of silver oxide, an alkaline electrolyte, and zinc anode
Lithium	High discharge rate and long shelf life; light weight and higher output voltage; the familiar cylindrical and flat cell are available	1.9	Liquid crystal watches, semi-conductor memories, hand-held calculators, sensor circuits	Contains a lithium anode, porous carbon cathode, and sulfur dioxide electrolyte

Metal Cap (+ Terminal)
Carbon Rod (+ Electrode)
Electrolyte (Paste)
Zinc Can (– Electrode)
(a) Carbon–Zinc

9 V Flat cell has 6 stacked 1.5 V cells

Alkaline Miniature Cell

Anode
Cathode
(b) Alkaline–Manganese

Amalgamated Zinc Anode
Mercuric Oxide Depolarizing Cathode
Potassium Hydroxide Electrolyte
(c) Mercury

Anode
Cathode
(d) Silver-Oxide

Cathode
Solid Electrolyte
Anode

voltage is normally set to about 115% of the battery voltage. (For example, a battery charger connected across a 12-V battery would be set to 115% of 12 V, or 13.8 V.) If this voltage is set too high, an excessive current can result, causing the battery to overheat.

Secondary cells are often referred to as **wet cells** because the electrolyte is not normally in paste form (dry) as in the primary cell, but is in liquid (wet) form and is free to move and flow within the container. Figure 6-25(a) gives more information on the discharge process, and Figure 6-25(b) on the charge process, that occur to the plates and electrolyte of a lead–acid secondary cell.

Wet Cell
Cell using a liquid electrolyte.

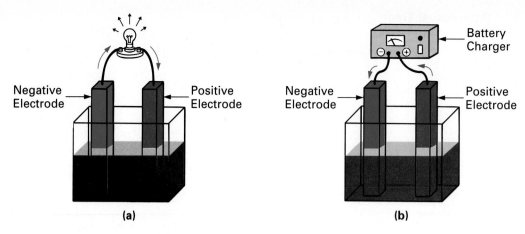

FIGURE 6-24 **Battery Operation. (a) Discharge. (b) Charge.**

Secondary cell types. Figure 6-26 illustrates the physical appearance of some of the types of secondary cells on the market today. The lead–acid and nickel–cadmium are the two most popular and are illustrated and described in Table 6-2 (p. 245). The two other types of secondary cells are the silver–zinc and silver–cadmium. Both have a large silver content and are therefore quite costly; however, they have the highest energy of all secondary cells and are used in specialized applications such as videotape recorders, portable televisions, and missiles.

Secondary cell capacity. Capacity (C) is measured by the amount of **ampere-hours** (Ah) a battery can supply during discharge. If, for example, a battery has a discharge capacity of 10 ampere-hours, the battery could supply 1 ampere for 10 hours, 10 amperes for 1 hour, 5 amperes for 2 hours, and so on, during discharge. Automobile batteries (12 V) will typically have an ampere-hour rating of between 100 and 300 Ah.

Ampere-hours
If you multiply the amount of current, in amperes, that can be supplied for a given time frame in hours, a value of ampere-hours is obtained.

Ampere-hour units are actually specifying the coulombs of charge in the battery. If, for example, a lead–acid car battery was rated at 150 Ah, this value would have to be converted to ampere-seconds to determine coulombs of charge, as 1 ampere-second (As) is equal to 1 coulomb. Ampere-hours are easily converted to ampere-seconds simply by multiplying ampere-hours by 3600, which is the number of seconds in 1 hour. In our example, a 150 Ah battery will have a charge of 54×10^4 coulombs.

■ **EXAMPLE:**

How many coulombs of charge does a 3.8-Ah lead–acid gel battery store?

■ *Solution:*

A 3.8-Ah battery can supply 3.8 A in 1 hour. This would be equivalent to 13,680 A in a second (3.8×3600). Since current is measured in coulombs/second, this battery when fully charged can store 13.68×10^3 C.

(a) Discharging

Charged EMF = 2 V

Spongy Lead Negative Terminal

Lead Peroxide Positive Terminal

Electrolyte: Sulfuric Acid, Water

Discharging

Decreasing Spongy Lead Increasing Lead Sulfate

Decreasing Lead Peroxide Increasing Lead Sulfate

Sulfuric Acid Decreasing, Water Increasing

Discharged

Minimum Spongy Lead Maximum Lead Sulfate

Minimum Lead Peroxide Maximum Lead Sulfate

Minimum Sulfuric Acid, Maximum Water

(b) Charging:

Battery Charger

Charging

Increasing Spongy Lead Decreasing Lead Sulfate

Increasing Lead Peroxide Decreasing Lead Sulfate

Increasing Sulfuric Acid, Decreasing Water

FIGURE 6-25   The Chemical Change of a Lead–Acid Cell During (a) Discharge and (b) Charge.

EXAMPLE:

A Ni-Cd battery is rated at 300 Ah and is fully discharged.

   a. How many coulombs of charge must the battery charger put into the battery to restore it to full charge?

   b. If the charger is supplying a charging current of 3 A, how long will it take the battery to fully charge?

Nickel-Metal Hydride Batteries

FIGURE 6-26 **Secondary Rectangular, Cylindrical, and Button Cells and Batteries.**

    c. Once fully charged, the battery is connected across a load that is pulling a current of 30 A. How long will it take until the battery is fully discharged?

    ■ *Solution:*

    a. The same amount that was taken out: $300\ \text{Ah} \times 3600 = 1080 \times 10^3\ \text{C}$

    b. $\dfrac{300\ \text{Ah}}{3\ \text{A}} = 100$ hours until charged

    c. $\dfrac{300\ \text{Ah}}{30\ \text{A}} = 10$ hours until discharged

TABLE 6-2 Secondary Cell Types.

Battery type	Features	Applications	Construction
Lead-acid (lead cell)	High cycle (discharge–charge) life and high current capacity; 2.1 V per cell; must be operated upright if a wet lead–acid. Lead–acid gell cells can be used in any position	*Gell Cell Batteries* televisions, recorders, robots, alarms, and tools *Wet Cell Batteries* Starting source for automobiles and power for robotics equipment	The electrodes are lead oxide immersed in an electrolyte of dilute sulfuric acid Wet-type lead–acid battery with 6 internal 2.1 V cells making 12.6 V car battery Positive Plate / Insulating Separators (porous rubber) / Negative Plate / Car Battery
Nickel-cadmium (Ni–Cd)	High capacity and high cost; three times discharge current of lead–acid for same amp-hour rating; can be sealed and operated in any position, so ideal for drills and other such portable equipment; 1.2 V per cell. Nickel cadmium batteries suffer from "memory effect." Short discharge and charge cycles cause the batteries' capacity to drop, because the cell appears to "remember" the lower capacity.	*Ni–Cd Gell Cell* televisions, radios, toys, recorders, shavers, toothbrushes *Ni–Cd Wet Cell Batteries* Starter source for jet engines (resembles lead–acid car battery in appearance). Robotics equipment	The positive plate is made of cadmium, the negative plate of nickel hydroxide, and the electrolyte used is potassium hydroxide Negative Sintered Plate / Separator / Nickel-plated Steel Jacket / Positive Sintered (applied in powered form) Plate
Nickel-metal Hydride (Ni–MH)	Discharge characteristics are very similar to those of the nickel–cadmium cell. The cell's nominal voltage is 1.2 V, and they do not suffer from memory effect.	Biggest applications include cellular phones, portable computers, camcorders, and other portable electronic systems	The positive plate is made of nickel oxyhydride, the negative plate of metal hydride, and the electrolyte used is potassium hydroxide (+) Positive Terminal / Positive Electrode / Separator / Negative Electrode / (–) Negative Terminal

245

Battery A   Battery B

<u>Series</u>
Voltage (V) = 2 x Cell Voltage
Current (I) = 1 Cell

FIGURE 6-27   Batteries in Series.

<u>Parallel</u>
Voltage (V) = 1 Cell
Current (I) = 2 x Cell Current

FIGURE 6-28   Batteries in Parallel.

Batteries in Series and Parallel

Batteries are often connected together to gain a higher voltage or current than can be obtained from one cell. Let's first analyze what can be obtained by connecting two cells in series, as shown in Figure 6-27.

When batteries are connected in series, the negative terminal of A is connected to the positive terminal of B. The total voltage across the bulb will be the sum of the two cell voltages, 18 V.

When batteries are connected in parallel, the negative terminal connects to the other negative terminal and the positive terminal connects to the positive. As can be seen in Figure 6-28, the voltage remains the same as for one cell; however, the current demand is now shared, and so the combined parallel batteries are able to supply a higher value of current compared to one battery. Each cell therefore provides half the total current flow through the load.

Internal Resistance

Figure 6-29 illustrates a typical lead–acid secondary cell on the left and its symbol on the right. When a circuit or piece of equipment is connected across the battery, as shown in Figure 6-29(a), the circuit or piece of equipment can be represented by its equivalent value of resistance, called the *load resistance* (R_L). In Figure 6-29(a) the switch is open, so the battery is not loaded and no load cur-

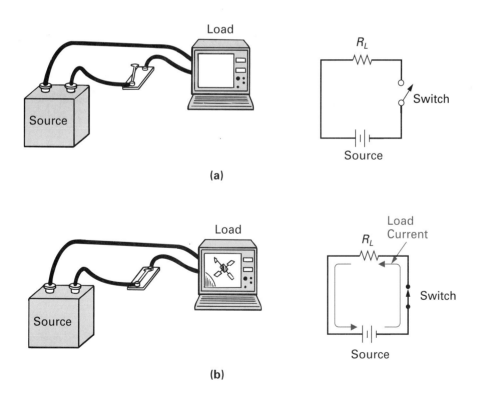

FIGURE 6-29 Loading a Battery.

FIGURE 6-30   Source and Load Resistance Divide the Voltage.

rent flows; in Figure 6-29(b) the battery has a completed current path, as the switch is now closed, so a load current will flow, the value of which depends on the load resistance and battery voltage.

The battery just discussed is known as an *ideal voltage source;* however, in reality, there is no such thing as an ideal voltage source. Batteries or any other type of voltage source are not 100% efficient; they all possess some form of internal resistance. This internal resistance is represented by a resistor connected in series with the battery symbol as seen in Figure 6-30. As you can see in the schematic circuit illustrated in this figure, R_L and R_S together form the total resistive load.

Very little voltage will be dropped across R_S, as it is normally small with respect to R_L. Internal inefficiencies of batteries and all voltage sources must always be kept small because a large R_S would drop a greater amount of the voltage, resulting in less output voltage to the load and therefore a waste of electrical power. For example, a lead–acid cell will typically have an internal resistance of 0.01 Ω (10 mΩ).

Maximum Power Transfer

A theorem that states maximum power will be transferred from source to load when the source resistance is equal to the load resistance.

Maximum Power Transfer

Maximum power will be delivered to the load when the resistance of the load (R_L) is equal to the resistance of the source (R_S). The best way to see if this theorem is correct is to apply it to a series of examples and then make a comparison.

■ **EXAMPLE:**

Figure 6-31(a) illustrates a 10-V battery with a 5-Ω internal resistance connected across a load. The load in this case is a light bulb, which has a load resistance of 1 Ω.

$$\text{circuit current, } I = \frac{V}{R}$$

$$= \frac{10\text{ V}}{R_S + R_L}$$

$$= \frac{10\text{ V}}{6\ \Omega}$$

$$= 1.66\text{ A}$$

The power supplied to the load is $P = I^2 \times R = 1.66^2 \times 1\ \Omega = 2.8\text{ W.}$

(b) Example ②

$R_{total} = R_{source} + R_{load}$

(a) Example ①

(c) Example ③

(d) Curve

FIGURE 6-31　**Maximum Power Transfer.**

■　**EXAMPLE:**

Figure 6-31(b) illustrates the same battery and R_S connected across a 5-Ω light bulb.

$$I = \frac{V}{R}$$

$$= \frac{10\,V}{R_S + R_L}$$

$$= \frac{10\,V}{10\,\Omega}$$

$$= 1\,A$$

Power supplied is $P = I^2 \times R = 1^2 \times 5\,\Omega = 5\,W$.

■　**EXAMPLE:**

Figure 6-31(c) illustrates the same battery again, and in this case a 10-Ω light bulb is connected in the circuit.

$$I = \frac{V}{R}$$

$$= \frac{10 \text{ V}}{R_S + R_L}$$

$$= \frac{10 \text{ V}}{15 \text{ }\Omega}$$

$$= 0.67 \text{ A}$$

Thus $P = I^2 \times R = 0.67^2 \times 10 \text{ }\Omega = 4.5 \text{ W}$.

As can be seen by the graph in Figure 6-31(d), which plots the power supplied to the load against load resistance, maximum power is delivered to the load (5 W) when the load resistance is equal to the source resistance.

The maximum power transfer condition is only used in special cases such as the automobile starter, where the load resistance remains constant. In most other cases, where load resistance can vary over a range of values, circuits are designed with a load resistance that will cause the best amount of power to be delivered. This is known as **optimum power transfer.**

Optimum Power Transfer
Since the ideal maximum power transfer conditions cannot always be achieved, most designers try to achieve optimum power transfer and have the source resistance and load resistance as close in value as possible.

SELF-TEST REVIEW QUESTIONS FOR SECTION 6-2-5

1. List the three basic components in a battery.
2. List all the various primary cell types.
3. List the various types of secondary cells available.
4. True or false: Two 1.5-V batteries in series would produce a total voltage of 1.5 V.
5. How does a battery's internal resistance relate to the maximum power transfer theorem?
6. Why do batteries have an internal resistance?

6-2-6 Electrically Generated DC

Up until now we have only discussed electrical energy in the form of direct current (dc); however, as we will discover in Chapter 11, electrical energy is more readily available in the form of alternating current (ac); in fact, that is what arrives at the wall receptacle at your home and workplace. We can use a dc power supply, like the one seen in Figure 6-32, to convert the ac electrical energy at the wall outlet to dc electrical energy for experiments.

Power in the laboratory or repair shop can be obtained from a battery, but since ac power is so accessible, dc power supplies are used. The power supply has advantages over the battery as a source of voltage in that it can quickly provide an accurate voltage that can easily be varied by a control on the front panel, and it never runs down.

FIGURE 6-32 **Direct Current from Alternating Current.**

During your electronic studies, you will use a power supply in the lab to provide different voltages for various experiments. In Figure 6-32 you can see that the power supply is being used to supply 12 V to a 12 kΩ resistor.

To review the operation of the dc power supply, which was discussed in the previous chapter, the ON/OFF switch is used to turn on or off the power supply. The VOLTS/AMPS switch is used to switch the meter between a voltmeter or an ammeter. In the VOLTS position, the meter indicates whatever voltage has been set by the voltage adjustment control, in this example, 12 V. If the VOLTS/AMPS switch is placed in the AMPS position, the meter will now monitor the amount of current flowing out of the power supply, which in this example is 1 mA.

SELF-TEST REVIEW QUESTIONS FOR SECTION 6-2-6

1. An electric company supplies what form of electrical energy?

2. True or false: A dc power supply converts dc electrical energy into ac.

3. True or false: A volts/amperes switch on a typical power supply will determine whether the output is voltage or current.

4. What advantages does the dc power supply have over a battery?

6-3 EQUIPMENT PROTECTION

Current must be monitored and not be allowed to exceed a safe level so as to protect users from shock, to protect the equipment from damage, and to prevent fire hazards. There are two basic types of protective devices: fuses and circuit breakers. Let's now discuss each of these in more detail.

6-3-1 Fuses

Fuse
This circuit- or
equipment-
protecting device
consists of a short,
thin piece of wire
that melts and breaks
the current path if
the current exceeds a
rated damaging level.

A **fuse** is a type of metal resistor that consists of a wire link or element of low-melting point inside a casing, as illustrated in Figure 6-33. Fuses have a current rating that indicates the maximum amount of current they will allow to pass through the fuse to the equipment. Figure 6-34, for example, illustrates a battery supplying power to a radio with a 2-A fuse connected between the two. When current passes through the fuse, some of the electrical energy is transformed into heat. If the current through the thin metal element exceeds the current rating of this fuse, in our example 2-A, the excessive current will create enough heat to melt the element and "open" or "blow" the fuse, thus disconnecting the radio and protecting it from damage.

One important point to remember is that if the current increases to a damaging level, the fuse will open and protect the radio. This implies that something went wrong with the battery and it started to supply too much current. This is almost never the case. What actually happens is that, as with all equipment, eventually something internally breaks down. This can cause the overall load resistance of the piece of equipment to increase or decrease. An increase in load resistance $(R_L\uparrow)$ means that the battery sees a higher resistance in the current path and therefore a small current would flow from the battery $(I\downarrow)$, and the user would be aware of the problem due to the nonoperation of the equipment. If an internal equipment breakdown causes the equipment's load resistance to decrease $(R_L\downarrow)$, the battery will see a smaller circuit resistance and will supply a heavier circuit current, $(I\uparrow)$, which could severely damage the equipment before the user had time to turn it off. Fuse protection is needed to disconnect the current automatically in this situation, to protect the equipment from damage.

Fuse elements come in various shapes and sizes so as to produce either quick heating and then melting (fast blow) or delayed heating and then melting (slow blow), as illustrated in Figure 6-35. The reason for the variety is based in the application differences. When turned on, some pieces of equipment will be of such low resistance that a short momentary current surge, sometimes in the region of four times the fuse's current rating, will result in the first couple of seconds. If a fast-blow fuse were placed in the circuit, it would blow at the instant the equipment was turned on, even though the equipment did not need to be protected; it merely needs a large amount of current initially to start up. A slow-blow fuse would be ideal in this application, as it would permit the initial heavy current and would begin to heat up and yet not blow due to the delay; then the

(a)   **(b)**

FIGURE 6-33   **Fuse. (a) Physical Appearance. (b) Schematic Symbol.**

FIGURE 6-34 **Application of a Fuse.**

surge would end, the current would decrease, and the fuse would still be intact (some slow-blow fuses will allow a 400% overload current to flow for a few seconds).

When on, some equipment cannot take any increase in current; otherwise, damage will occur. The slow blow would not be good in this application, as it would allow an increase of current to pass to the equipment for too long a

Type AG Fuses	Diameter and Length	
1 AG	1/4 in. ,	5/8 in.
3 AG	1/4 in. ,	1–1/4 in.
4 AG	9/32 in. ,	1–1/4 in.
5 AG	13/32 in. ,	1–1/2 in.
7 AG	1/4 in. ,	7/8 in.
8 AG	1/4 in. ,	1 in.
9 AG	1/4 in. ,	1–7/16 in.

FIGURE 6-35 **Fuse Types, Sizings, and Casings.**

FIGURE 6-36 **Voltage Rating of Fuses.**

period of time, whereas the fast blow would disconnect the current instantly if the current rating of the fuse were exceeded, consequently preventing equipment damage. The automobile was the first application for a fuse in a glass holder, and a size standard named AG (automobile glass) was established (see Figure 6-35).

Fuses also have a voltage rating that indicates the maximum circuit voltage that can be applied across the fuse by the circuit in which the fuse resides. This rating, which is important after the fuse has blown, prevents arcing across the blown fuse contacts, as illustrated in Figure 6-36. Once the fuse has blown, the circuit's positive and negative voltages are now connected across the fuse contacts; and if the voltage is too great, an arc can jump across the gap, causing a sudden surge of current, damaging the equipment connected.

(a)　　　　　　　　　　　　**(b)**

FIGURE 6-37 **Fuse Holders. (a) Holders. (b) Fuse and Holder in Equipment.**

Fuses are mounted within fuse holders and normally placed at the back of the equipment for easy access. Figure 6-37 illustrates fuse holders and how they are installed in a piece of test equipment. When replacing fuses, you must be sure that the power is off, because a potential is across it, and that a fuse with the correct current and voltage rating is used. Since it is merely a piece of wire, a good fuse should have a resistance of 0 Ω; that is, no voltage drop can be measured across it. A blown or burned-out fuse will, when removed, read infinite ohms, and when in its holder and power on, will have the full applied voltage across its two terminals.

6-3-2 Circuit Breakers

In many appliances and in a home electrical system, **circuit breakers** are used in place of fuses to prevent damaging current. A circuit breaker can open a current-carrying circuit without damaging itself, then be manually reset and used repeatedly, unlike a fuse, which must be replaced when it blows. To summarize, we can say that a circuit breaker is a reusable fuse. Figure 6-38 illustrates a circuit breaker's symbol, appearance, and typical application.

There are three types of circuit breakers: (1) thermal type, (2) magnetic type, and (3) thermomagnetic type. Let's now discuss each of these in more detail.

Thermal Circuit Breakers

The operation of this type of circuit breaker depends on temperature expansion due to electrical heating. Figure 6-39 illustrates the construction of a thermal circuit breaker. A U-shaped bimetallic (two-metal) strip is attached to the housing of the circuit breaker; it is composed of a layer of brass on one side and a layer of steel on the other. The current arrives at terminal *A*, enters the right side of the bimetallic U-shaped strip, leaves the left side, and travels to the upper contact, which is engaging the bottom contact, where it leaves and exits the circuit breaker from terminal *B*.

If current were to begin to increase beyond the circuit breaker's current rating, heating of the bimetallic strip would occur. As with all metals, heat causes the strip to expand. Some metals expand more than others, and this mea-

Circuit Breaker
Reusable fuse. This device will open a current-carrying path without damaging itself once the current value exceeds its maximum current rating.

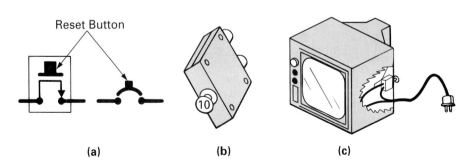

(a) (b) (c)

FIGURE 6-38 Circuit Breaker. (a) Symbol. (b) Appearance. (c) Typical Application.

FIGURE 6-39 **Thermal-Type Circuit Breaker.**

surement is called the *thermal expansion coefficient*. In our situation, brass will expand more than steel, resulting in the lower end of the shaped bimetallic strip bending to the right. This allows a catch to release a pivoted arm and to be pulled down by the tension of a spring. This action lifts the left side of the arm and the attached top contact. The result is to separate the top contact surface from the bottom contact surface, thereby opening the current path and protecting the circuit from the excessive current. This is also known as *tripping* the breaker. The reset button must now be engaged to close the contacts; however, if the problem still exists, the breaker will just trip once more, due to the excessive current.

Magnetic Circuit Breakers

This type of circuit breaker depends on the response of an electromagnet to break the circuit for protection. Figure 6-40 illustrates the construction of a magnetic circuit breaker.

A small level of current flows through the coil of the electromagnet, providing a small amount of magnetic pull on the iron arm, pulling it to the left; however, this magnetic force cannot overcome the pull to the right being generated by the spring *A*. This safe value of current will therefore pass from the *A* terminal, through the coil to the top contact, out of the bottom contact, and then exit the breaker at terminal *B*.

If the current exceeds the current rating of the circuit breaker, an increase in current causes an increase in the current through the coil of the electromagnet, which generates a greater magnetic force on the vertical arm, which pulls the top half of the vertical arm to the left and the lower half below the pivot to the right. This releases the catch holding the horizontal arm, allowing spring *B* to pull the right side of the lower arm down and, consequently, open the contact and disconnect or trip the breaker. The reset button must now be engaged to

FIGURE 6-40 Magnetic-Type Circuit Breaker.

close the contacts; however, if the problem still exists, the breaker will trip once again, as the excessive circuit current still exists.

Thermomagnetic Circuit Breakers

Figure 6-41(a) illustrates the typical home- or residential-type thermomagnetic circuit breaker. The thermal circuit breaker is similar to a slow-blow fuse in that it is ideal for passing momentary surges without tripping because of the delay caused by heating of the bimetallic strip. The magnetic circuit breaker,

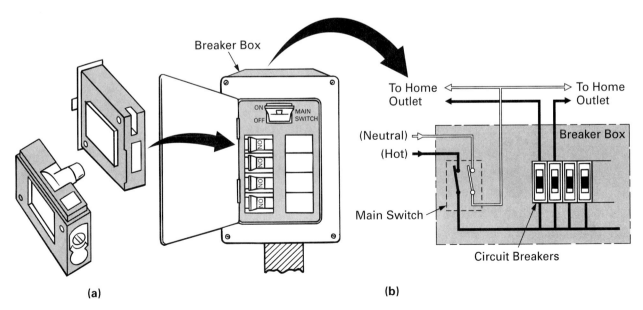

(a)   **(b)**

FIGURE 6-41 Thermomagnetic-Type Circuit Breaker.

however, is most similar to a fast-blow fuse in that it is tripped immediately when an increase of current occurs.

The thermomagnetic circuit breaker combines the advantages of both previously mentioned circuit breakers by incorporating both a bimetallic strip and an electromagnet as an actuating mechanism, and it operates as follows: For currents at and below the current rating, the circuit breaker is a short circuit and connects the current source to the load. For momentary overcurrent surges, the electromagnet is activated, but does not have enough force to trip the breaker. However, if overcurrent continues, the bimetallic strip will have heated and the combined forces of the bimetallic strip and electromagnet will trip the breaker and disconnect the current source from the load.

For a very large surge of current, the electromagnet receives enough current to trip the breaker independently of the bimetallic strip and disconnect the current source from the load. Figure 6-41(b) shows how a thermomagnetic circuit breaker is used in a typical residential circuit breaker cabinet.

6-4 SWITCHES

A *switch* is a device that completes (closes) or breaks (opens) the path of current, as seen in Figure 6-42. All mechanical switches can be classified into one of eight categories, illustrated in Figure 6-43. Many different variations of these eight different classifications exist. Figure 6-44 illustrates some of the different types on the market; but remember that they are all just basically a switch that merely opens or closes connections.

SELF-TEST REVIEW QUESTIONS FOR SECTIONS 6-3 AND 6-4

1. What do the current and voltage ratings of a fuse indicate?
2. What is the difference between a slow-blow and a fast-blow fuse?
3. List the three types of circuit breakers.
4. List the eight basic types of switches.

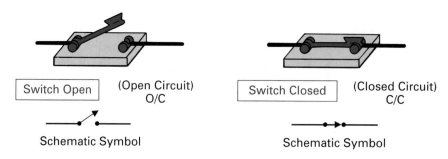

FIGURE 6-42   Open and Closed Switch.

(a) Single-pole, single-throw (SPST). A two-terminal switch with only one pole or moving contact, which can be cast in one direction only (single throw).

(b) Single-pole, double-throw (SPDT). A three-terminal switch for connecting one terminal to either of two other terminals.

(c) Double-pole, single-throw (DPST). A four-terminal switch that is used to connect or disconnect two pairs of terminals simultaneously.

(d) Double-pole, double-throw (DPDT). A switch that has six terminals and is used to connect one pair of terminals to either of the other two pairs.

(e) Normally open push button (NOPB). A switch that will make contact and pass current when it is pressed.

(f) Normally closed push button (NCPB). A switch that normally makes contact and passes current, but disconnects or opens when pressed.

(g) Rotary. An electromechanical device that is capable of making (closing contacts) or breaking (opening contacts) in a circuit.

(h) Dual-in-line package (DIP) switch. A dip switch is a group of separate miniature switches within two (dual) rows of external connecting pins or terminals.

Single pole: one moving contact.
Double pole: two moving contacts.
Single throw: pole can be thrown or cast in only one direction.
Double throw: pole can be thrown or cast in two directions.

FIGURE 6-43 Eight Basic Types of Switches.

(a)

(b)

(c)

(d)

(e)

(f)

(g)

FIGURE 6-44 **Switches. (Courtesy of ITW Switches.)**

SUMMARY

Source and Load

1. The phrase *load resistance* describes the device or equipment's circuit resistance, whereas the phrase *load current* describes the amount of current drawn by the device or equipment.

2. A device that causes a large load current to flow, due to its small load resistance, is called a large or heavy load because it is heavily loading down or working the supply or battery. On the other hand, a device that causes a small load current to flow, due to its large load resistance, is referred to as a small or light load because it is only lightly loading down or working the supply or battery.

3. Load current and load resistance are therefore inversely proportional to one another in that a large load resistance ($R_L\uparrow$) causes a small load current ($I_L\downarrow$), referred to as a light load, and vice versa.

4. The six basic forms of energy are mechanical, heat, light, magnetic, chemical, and electrical. To produce direct current (electrical energy), one of the other forms of energy must be used as a source and then converted or transformed into electrical energy in the form of a dc voltage that can be used to produce direct current.

Mechanically Generated DC

5. Positive and negative charges are known as *static* (stationary) *charges* or *static electricity*. Attraction and repulsion occur with positive and negative charges.

6. Friction is a form of mechanical energy.

7. Electrons in a glass rod (insulator) can be forced out of their orbits and captured by another material, such as wool, by the work of rubbing. A static negative charge now exists in the wool, while a static positive charge exists in the glass. These mechanically generated static charges can be discharged in one of three ways, producing electrical energy.

8. Once a state of equilibrium is achieved, two static charges will be neutralized and no further current will flow. If the connection has a small resistance, the discharge current will be brief and intense; on the other hand, if the connection has a large resistance, the discharge current will be prolonged and less intense.

9. Moving clouds in a storm cause friction between the molecules in the air. The buildup of a natural static charge in a cloud, if large enough, can cause a discharge or release of the electrical energy to the ground.

10. Mechanical energy in the form of pressure can also be used to generate electrical energy.

11. Quartz and Rochelle salts are solid compounds that will produce electricity when a pressure is applied.

12. Quartz is a natural or artificially grown crystal composed of silicon dioxide, from which thin slabs or plates are carefully cut and ground. Rochelle salt is a crystal of sodium potassium tartrate. They are both piezoelectric crystals. Piezoelectric effect is the generation of a voltage between the opposite faces of a crystal as a result of pressure being applied.

13. A voltage is produced between the opposite sides of a crystal (usually a crystalline lead compound) when it is either struck, pressured, or twisted.

14. Piezoelectric effect also describes the reverse of what has just been discussed, as it was discovered that crystals would not only generate a voltage when mechanical force was applied, but would generate mechanical force when a voltage was applied.

15. Piezoresistance of a semiconductor is described as a change in resistance due to a change in the applied strain, and therefore semiconductor resistors can be used as pressure sensors.

16. Unlike quartz, which generates a voltage, the piezoresistive sensor needs an excitation voltage applied, and its value will determine the value of output voltage.

Thermally Generated DC

17. Heat can be used to generate an electric charge.

18. When two dissimilar metals, such as copper and iron, are welded together and heated, an electrical charge is produced. This junction is called a ther-

mocouple, and the heat causes the release of electrons from their parent atoms.

19. The size of the charge is proportional to the temperature difference between the two metals.

20. Thermocouples are normally used to indicate temperature on a display calibrated in degrees.

Optically Generated DC

21. A photovoltaic cell also makes use of two dissimilar metals or semiconductor materials in its transformation of light to electrical energy.

22. A photoelectric cell is also called a solar cell.

23. When the light illuminates the light-sensitive material, a charge is generated, causing current to flow.

24. This phenomenon, known as photovoltaic action, finds application as a light meter for photographic purposes and as an electrical power source for use by the satellite.

25. Electric power companies make use of solar cells to convert sunlight into electrical energy for the consumer.

Magnetically Generated DC

26. In the ancient city of Magnesia in Asia Minor, stones composed of iron oxide were found to have the strange ability to attract and repel one another. It was also noted that when suspended from a string, these stones would always point in only one direction.

27. Historians are not sure whether the Chinese or Europeans first began using these "leading stones," or lodestones as they are now known, for long-distance land and sea travel, but the original compass was probably a lodestone suspended from a string.

28. Later it was found that a metal needle could be magnetized by stroking it across a lodestone, and so the next form of compass consisted of a magnetized needle attached to a piece of wood floating in a bowl of water; hence the origin of the term *needle of a compass*.

29. The lodestone is an example of a natural magnet, unlike the needle, which is called an artificial magnet because it must be artificially magnetized in order to possess a magnetic field, but they both have the ability to point in only one direction by interacting with the earth's natural magnetic field.

30. Magnetism was recorded by the Greeks as early as 800 B.C., but it was not until 1269 that Petrus Peregrinus de Maricout found that the invisible magnetic lines produced by lodestones intersected at two points, which he called *poles*.

31. In 1600, William Gilbert discovered that the magnetic field of a lodestone was lost when it was heated, but reappeared when the lodestone cooled.

32. In 1819, Hans Christian Oersted discovered that any current-carrying conductor generated a magnetic field, and this interrelationship between electricity and magnetism became known as electromagnetism.

33. In 1838, a German mathematician and physicist, Johann Gauss, published a paper on the earth's natural magnetic field, and in honor of his contribution the flux density of a magnetic field is measured in *gauss*.

34. A magnetic flux density of 1 gauss (G) is produced 1 centimeter away from a wire carrying a current of 5 A. To give some examples, the field strength of a magnetized needle may only be a fraction of a gauss, whereas a lodestone 1 inch long may have a strength of 100 G. The earth's magnetic field has a magnetic field strength of 0.3 G at the equator and 0.7 G at either pole.

35. An atom consists of electrons spinning around a nucleus; some atoms have few, while others have many electrons orbiting the atom's nucleus. Each electron is in fact a miniature magnet spinning around the nucleus. An electron at rest will not generate a magnetic field; however, orbiting electrons produce a magnetic field because of their high orbital velocity.

36. Tungsten has 74 electrons (will not magnetize) and iron has 26 electrons (will magnetize). With tungsten, 37 electrons rotate around the nucleus in the clockwise direction, while 37 electrons orbit in the counterclockwise direction. The clockwise magnetic field will cancel with the counterclockwise magnetic field, so the net magnetic field is zero, which is why tungsten cannot be magnetized. Iron has fewer electrons (26); however, 10 of the iron electrons spin clockwise and 16 spin counterclockwise, producing a net counterclockwise magnetic field of 6 electrons, which is why iron can be magnetized.

37. If approximately 100 trillion atoms of the same magnetic polarity are all arranged in one direction, they are called a domain. Domains within unmagnetized materials are all misaligned while aligned domains produce magnetized materials.

38. The magnetic field produced by a magnet is an invisible force concentrated at two opposite points called the magnetic poles. The south pole of a magnet is the pole that would point north toward the earth's north magnetic pole if the magnet were free to rotate.

39. The magnetic field is made up of invisible flux lines that leave the north pole and return at the south.

40. The law of attraction and repulsion could be applied to magnetic poles, just as it can be to positive and negative charges. A north against a north pole and south against a south pole will cause a force of repulsion between the two like poles while a south pole against a north pole will result in a force of attraction between the two unlike poles.

41. There are three types of artificial magnets:

    a. A permanent magnet is a piece of hardened steel or other special alloy that when magnetized by the presence of a magnetic field, will retain its magnetism indefinitely (high retentivity), even when the material is removed from the magnetizing force or the magnetizing force is no longer present.

    b. A temporary magnet is a piece of soft iron that will be magnetized while in the presence of a magnetizing field, but will be demagnetized the

FIGURE 6-45   Batteries.

Primary Cell Types

Battery type	Features	Cell Voltage (V)	Applications	Construction
Carbon–zinc (also called Leclanché cell in honor of its inventor)	Most popular due to its low cost. Cylindrical D and C cells are most commonly used. Flat cells are stacked in series to obtain voltages greater than 1.5 V, as in the case of the 9 V battery	1.5	Portable radios, tape players, televisions, toys.	
Alkaline–manganese (also called alkaline cell)	Has three times the shelf life and capacity of carbon zinc. Cylindrical or miniature cell sizes are available	1.4	Portable radios, tape players, televisions, toys; large capacity is worth investment	
Mercury (also called mercuric–oxide)	Higher energy density than preceding two with good shelf life and small size; used in low-power applications. Flat or cylindrical cells are available	1.35 and 1.4	Watches, hearing aids, pacemakers, cameras, test equipment	
Silver–oxide	High capacity; however, the material used (silver) makes it the most costly; can supply high currents for short periods of time; used in low-power applications	1.5	Watches, hearing aids, pacemakers, cameras	
Lithium	High discharge rate and long shelf life; light weight and higher output voltage; the familiar cylindrical and flat cell are available	1.9	Liquid crystal watches, semiconductor memories, hand-held calculators, sensor circuits	

Secondary Cell Types

Battery type	Features	Applications	Construction
Lead–acid (lead cell)	High cycle (discharge–charge) life and high current capacity; 2.1 V per cell; must be operated upright if a wet lead–acid. Lead–acid gell cells can be used in any position	*Gell Cell Batteries* televisions, recorders, robots, alarms, and tools *Wet Cell Batteries* Starting source for automobiles and power for robotics equipment	
Nickel–cadmium (Ni–Cd)	High capacity and high cost; three times discharge current of lead–acid for same amp-hour rating; can be sealed and operated in any position, so ideal for drills and other such portable equipment; 1.2 V per cell. Nickel cadmium batteries suffer from "memory effect." Short discharge and charge cycles cause the batteries capacity to drop, because the cell appears to "remember" the lower capacity.	*Ni–Cd Gell Cell* televisions, radios, toys, recorders, shavers, toothbrushes *Ni–Cd Wet Cell Batteries* Starter source for jet engines (resembles lead–acid car battery in appearance). Robotics equipment	
Nickel–metal Hydride (Ni–MH)	Discharge characteristics are very similar to those of the nickel–cadmium cell. The cell's nominal voltage is 1.2 V, and they do not suffer from memory effect.	Biggest applications include cellular phones, portable computers, camcorders, and other portable electronic systems	

FIGURE 6–45 (continued).

Identical cells

(a) In series:
$$V_T = V_1 + V_2 + V_3 + \ldots$$
(Total voltage is equal to the sum of all the cell voltages.)

(b) In parallel:
$$V_T = V_1 \text{ or } V_2 \text{ or } V_3 \text{ or } \ldots$$
(Total voltage is the same as for one cell voltage.)

moment the soft iron is taken away from the magnetizing force or the magnetizing force is no longer present. Soft iron possesses low retentivity which means it will not retain a magnetic field.

c. An electromagnet is a temporary magnet that consists of an insulated wire wrapped around a soft-iron cylinder (core) to form a coil. A magnetic field is produced only while current flows through the coil. If a larger current is passed through the coil, a corresponding stronger magnetic field is generated.

42. A magnet can be used to generate electrical current by movement of the magnet past a conductor. When the magnet is stationary, no current flows through the circuit, so the current meter indicates this zero current. If the magnet is moved so that the magnetic lines of force cross the wire conductor, a voltage is induced in the conductor.

43. To achieve an energy conversion from magnetic to electric, you can either move the magnet past the wire or move the wire past the magnet.

44. To produce a continuous supply of electric current, the magnet or wire must be constantly in motion. The device that achieves this is the dc electric generator.

Chemically Generated DC (Figure 6-45)

45. The chemical voltage source (the battery) is called the voltaic cell. Its principle of operation was discovered by Alessandro Volta, an Italian physicist, in 1800.

46. A battery has three basic components: (a) a negative plate, (b) a positive plate, and (c) an electrolyte.

47. Two dissimilar, separated metal plates such as copper and zinc are placed within a container that is filled with a liquid known as the electrolyte, usually an acid. A chemical reaction causes electrons to be repelled from one plate and attracted to the other, passing through the electrolyte. A large number of electrons collect on one plate (negative plate), while an absence or deficiency of electrons exists on the opposite, positive plate.

48. The electrolyte acts on the two plates and transforms chemical energy into electrical energy, which can be taken from the cell at its two output terminals as an electrical current flow.

49. The mutual repulsion of the free electrons at the negative electrode combined with the attraction of the positive electrode results in a migration of free electrons (current flow) through the light bulb, causing its filament to produce light.

50. The chemical reaction within the cell will actually dissolve the negative plate until eventually it will be eaten away completely.

51. These types of batteries are known as primary (first time, last time) cells because they discharge once and must then be discarded. Almost all primary cells have their electrolyte in paste form and are therefore referred to as dry cells, as opposed to wet cells, whose electrolyte is in liquid form.

52. The shelf life of a battery is the length of time a battery can remain on the shelf (in storage) and still retain its usability.

53. Of all the types of primary cells, five dominate the market as far as applications and sales are concerned. These are the carbon–zinc, alkaline–manganese, mercury, silver oxide, and lithium.

54. The cell voltage describes the voltage produced by one cell (one set of plates). If two or more sets of plates are installed in one package, the component is called a battery.

55. Secondary cells operate on the same basic principle as that of primary cells. However, in this case the plates are not eaten away or dissolved; they only undergo a chemical change during discharge. Once discharged, the secondary cell can have the chemical change that occurred during discharge reversed by recharging, resulting once more in a fully charged cell.

56. Restoring a secondary cell to the charged condition is achieved by sending a current through the cell in the direction opposite to that of the discharge current.

57. During the recharging process, we use a battery charger, which reverses the chemical process of discharge by forcing electrons back into the cell and restoring the battery to its charged condition. The battery charger voltage is normally set to about 115% of the battery voltage.

58. Secondary cells are often referred to as wet cells because the electrolyte is not normally in paste form (dry) as in the primary cell, but is in liquid (wet) form and is free to move and flow within the container. The lead–acid and nickel–cadmium are the two most popular. The two other types of secondary cells are the silver–zinc and silver–cadmium.

59. Capacity (C) is measured by the amount of ampere-hours (Ah) a battery can supply during discharge.

60. Ampere-hour units are actually specifying the coulombs of charge in the battery.

61. Batteries are often connected together to gain a higher voltage or current than can be obtained from one cell.

62. When batteries are connected in series, the negative terminal of battery A is connected to the positive terminal of battery B. The total voltage across the bulb will be the sum of the two cell voltages.

63. When batteries are connected in parallel, the negative terminal connects to the other negative terminal and the positive terminal connects to the positive. The voltage remains the same as for one cell; however, the current demand can now be shared.

64. Batteries or any other type of voltage source are not 100% efficient; they all possess some form of internal resistance. This internal resistance is represented by a resistor connected in series with the battery symbol.

65. Internal inefficiencies of batteries and all voltage sources must always be kept small because a large R_S would drop a greater amount of the voltage, resulting in less output voltage to the load and therefore a waste of electrical power.

66. Maximum power will be delivered to the load when the resistance of the load (R_L) is equal to the resistance of the source (R_S).

67. The maximum power transfer condition is only used in special cases such as the automobile starter, where the load resistance remains constant. In most other cases, where load resistance can vary over a range of values, circuits are designed with a load resistance that will cause the best amount of power to be delivered. This is known as optimum power transfer.

Electrically Generated DC

68. We can use ac electrical energy to generate dc electrical energy by use of a piece of equipment called a dc power supply.

69. The power supply has advantages over the battery as a source of voltage in that it can quickly provide an accurate voltage that can easily be varied by a control on the front panel, and it never runs down.

Fuses (Figure 6-46)

70. Current must be monitored and not be allowed to exceed a safe level so as to protect users from shock, to protect the equipment from damage, and to prevent fire hazards.

71. There are two basic types of protective devices: fuses and circuit breakers.

72. A fuse is a type of metal resistor that consists of a wire link or element of low-melting point inside a casing. Fuses have a current rating that indicates the maximum amount of current they will allow to pass through the fuse to the equipment.

73. When current passes through the fuse, some of the electrical energy is transformed into heat. If the current through the thin metal element exceeds the current rating of this fuse, the excessive current will create enough heat to melt the element and "open" or "blow" the fuse, thus disconnecting the equipment and protecting it from damage.

74. An increase in load resistance ($R_L\uparrow$) means that the battery sees a higher resistance in the current path and therefore a small current would flow from the battery ($I\downarrow$), and the user would be aware of the problem due to the nonoperation of the equipment.

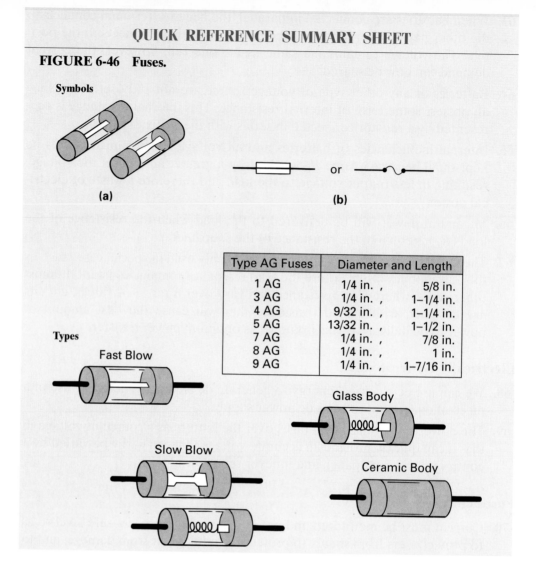

FIGURE 6-46 Fuses.

Symbols

(a)

(b)

Types

Fast Blow

Slow Blow

Type AG Fuses	Diameter and Length	
1 AG	1/4 in. ,	5/8 in.
3 AG	1/4 in. ,	1–1/4 in.
4 AG	9/32 in. ,	1–1/4 in.
5 AG	13/32 in. ,	1–1/2 in.
7 AG	1/4 in. ,	7/8 in.
8 AG	1/4 in. ,	1 in.
9 AG	1/4 in. ,	1–7/16 in.

Glass Body

Ceramic Body

75. If an internal equipment breakdown causes the equipment's load resistance to decrease ($R_L\downarrow$), the battery will see a smaller circuit resistance and will supply a heavier circuit current, ($I\uparrow$), which could severely damage the equipment before the user had time to turn it off. Fuse protection is needed to disconnect the current automatically in this situation, to protect the equipment from damage.

76. Fuse elements come in various shapes and sizes so as to produce either quick heating and then melting (fast blow) or delayed heating and then melting (slow blow).

77. Fuses also have a voltage rating that indicates the maximum circuit voltage that can be applied across the fuse by the circuit in which the fuse resides. This rating, which is important after the fuse has blown, prevents arcing across the blown fuse contacts.

78. Fuses are mounted within fuse holders and normally placed at the back of the equipment for easy access.

79. When replacing fuses, you must be sure that the power is off, because a potential is across it, and that a fuse with the correct current and voltage rating is used.

Circuit Breakers

80. In many appliances and in a home electrical system, circuit breakers are used in place of fuses to prevent damaging current.

81. A circuit breaker can open a current-carrying circuit without damaging itself, then be manually reset and used repeatedly, unlike a fuse, which must be replaced when it blows. There are three types of circuit breakers: (a) thermal type, (b) magnetic type, and (c) thermomagnetic type.

82. The operation of thermal circuit breakers depends on temperature expansion due to electrical heating. If current were to begin to increase beyond the circuit breaker's current rating, heating of the bimetallic strip would occur. The result is to separate the top contact surface from the bottom contact surface, thereby opening the current path and protecting the circuit from the excessive current. This is also known as *tripping* the breaker. The reset button must now be engaged to close the contacts.

83. Magnetic circuit breakers depend on the response of an electromagnet to break the circuit for protection. If the current exceeds the current rating of the circuit breaker, an increase in current causes an increase in the current through the coil of the electromagnet, which generates a greater magnetic force on the vertical arm, which opens the contact and disconnects or trips the breaker. The reset button must now be engaged to close the contacts.

84. The thermal circuit breaker is similar to a slow-blow fuse in that it is ideal for passing momentary surges without tripping because of the delay caused by heating of the bimetallic strip.

85. The magnetic circuit breaker, however, is most similar to a fast-blow fuse in that it is tripped immediately when an increase of current occurs.

86. The thermomagnetic circuit breaker combines the advantages of both previously mentioned circuit breakers by incorporating both a bimetallic strip and an electromagnet as an actuating mechanism. For currents at and below the current rating, the circuit breaker is a short circuit and connects the current source to the load. For momentary overcurrent surges, the electromagnet is activated, but does not have enough force to trip the breaker. However, if overcurrent continues, the bimetallic strip will have heated and the combined forces of the bimetallic strip and electromagnet will trip the breaker and disconnect the current source from the load. For a very large surge of current, the electromagnet receives enough current to trip the breaker independently of the bimetallic strip and disconnect the current source from the load.

FIGURE 6-47   Switches.

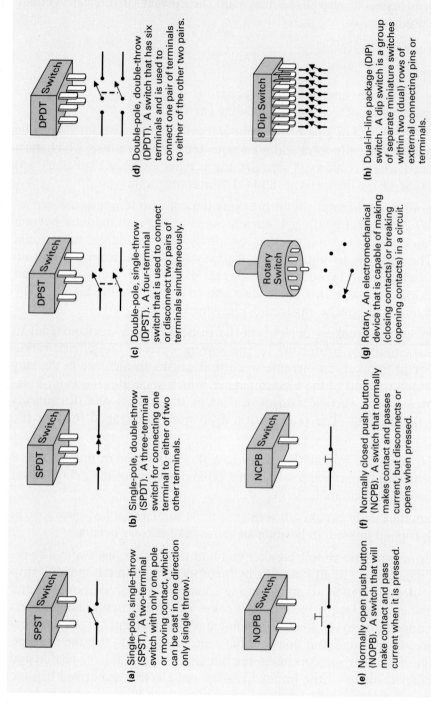

(a) Single-pole, single-throw (SPST). A two-terminal switch with only one pole or moving contact, which can be cast in one direction only (single throw).

(b) Single-pole, double-throw (SPDT). A three-terminal switch for connecting one terminal to either of two other terminals.

(c) Double-pole, single-throw (DPST). A four-terminal switch that is used to connect or disconnect two pairs of terminals simultaneously.

(d) Double-pole, double-throw (DPDT). A switch that has six terminals and is used to connect one pair of terminals to either of the other two pairs.

(e) Normally open push button (NOPB). A switch that will make contact and pass current when it is pressed.

(f) Normally closed push button (NCPB). A switch that normally makes contact and passes current, but disconnects or opens when pressed.

(g) Rotary. An electromechanical device that is capable of making (closing contacts) or breaking (opening contacts) in a circuit.

(h) Dual-in-line package (DIP) switch. A dip switch is a group of separate miniature switches within two (dual) rows of external connecting pins or terminals.

Single pole: one moving contact.
Double pole: two moving contacts.
Single throw: pole can be thrown or cast in only one direction.
Double throw: pole can be thrown or cast in two directions.

Switches (Figure 6-47)

87. A *switch* is a device that completes (closes) or breaks (opens) the path of current. All mechanical switches can be classified into one of eight categories, although many different variations of these eight different classifications exist.

Alkaline–manganese cell

Alternating current (ac)

Ampere-hour

Ampere-second

Artificial magnet

Battery

Battery charger

Blown fuse

Capacity

Carbon–zinc cell

Cell

Circuit breaker

Coil

Core

Dc generator

Depolarizer

Direct current (dc)

Discharge

Domain

Double-pole, double-throw (DPDT) switch

Double-pole, single-throw (DPST) switch

Dry cell

Dual in-line package (DIP) switch

Electrolyte

Electromagnet

Electromagnetism

Fast-blow fuse

Flux line

Friction

Fuse

Gauss

Generator

Lead–acid cell

Lithium cell

Load

Load current

Load resistance

Lodestone

Magnetic flux

Magnetic poles

Magnetic-type circuit breaker

Maximum power transfer

Mercuric oxide cell

Natural magnet

Nickel–cadmium cell

Normally closed pushbutton (NCPB) switch

Normally open push-button (NOPB) switch

Permanent magnet

Photocell

Photovoltaic action

Piezoelectric effect

Piezoresistance

Pole

Power supply

Pressure

Primary cell

Quartz

Rating

Retentivity

Rochelle salts

Rotary switch

Secondary cell

Shelf life

Silver–cadmium cell

Silver–oxide cell

Silver–zinc cell

Single-pole, double-throw (SPDT) switch

Single-pole, single-throw (SPST) switch

Slow-blow fuse

Solar cell

Solar panels

Solenoid

Source

Source current

Source resistance

Source voltage
Static charge
Supply
Supply voltage
Switch

Temporary magnet
Thermal-type circuit breaker
Thermocouple

Thermomagnetic-type circuit breaker
Voltaic cell
Wet cell

REVIEW QUESTIONS

Multiple-Choice Questions

1. Direct current is:
   a. A reversing of current continually in a circuit
   b. A flow of current in only one direction
   c. Produced by an ac voltage
   d. None of the above

2. If load resistance were doubled, load current will _____ .
   a. Halve   b. Double   c. Triple   d. Remain the same

3. Quartz is a solid compound that will produce electricity when _____ is applied.
   a. Friction   b. An electrolyte   c. A magnetic field   d. Pressure

4. Direct current is generated thermally by use of a:
   a. Crystal   b. Thermocouple   c. Thermistor   d. None of the above

5. An application of a photovoltaic cell would be:
   a. A satellite power source   c. Both (a) and (b)
   b. To turn on and off security lights   d. None of the above

6. The atomic theory of magnetism states that every _____ in motion is a miniature magnet.
   a. Electron   b. Proton   c. Neutron   d. Photon

7. Magnetic flux lines _____ the north pole and _____ at the south pole.
   a. Arrive, leave   c. Are never present, always present
   b. Leave, arrive   d. None of the above

8. Like magnetic poles _____, while unlike magnetic poles _____ .
   a. Attract, repel   b. Repel, attract   c. Repel, repel   d. Attract, attract

9. Of the three artificial magnets, the _____ holds its magnetic force even after the magnetizing force has been removed.
   a. Permanent magnet   c. Electromagnet
   b. Temporary magnet   d. Magnetite

10. Which type of battery can be used for only one discharge (and cannot be rejuvenated)?
    a. A lead–acid cell   c. A nickel–cadmium cell
    b. A secondary cell   d. A primary cell

11. A battery with a capacity of 12 ampere-hours could supply:
    a. 12 A for 1 hour   c. 6 A for 2 hours
    b. 3 A for 4 hours   d. All of the above

12. Batteries are normally connected in series to obtain a higher total:
    a. Voltage   c. Resistance
    b. Current   d. None of the above

13. The dc power supply converts:
    a. ac to dc   c. High dc to a low dc
    b. dc to ac   d. Low ac to a high ac

14. A fuse's current rating states the:
    a. Maximum amount of current allowed to pass
    b. Minimum amount of current allowed to pass
    c. Maximum permissible circuit voltage
    d. None of the above

15. A thermal-type circuit breaker is equivalent to a _____ fuse, whereas a magnetic-type circuit breaker is equivalent to a _____ fuse.
    a. Slow-blow, slow-blow   c. Slow-blow, fast-blow
    b. Fast-blow, slow-blow   d. Glass case, ceramic case

16. The type of circuit breaker found in the home is the:
    a. Thermal type   c. Magnetic type
    b. Thermomagnetic type   d. Carbon–zinc type

17. A single-pole, double-throw switch would have _____ terminals.
    a. Two   b. Three   c. Four   d. Five

18. Maximum power will be transferred from the source to the load when the source resistance is equal to:
    a. 25 Ω   b. Load current   c. Load resistance   d. Source voltage

19. A switch will very simply _____ (open) or _____ (close) a path of current.
    a. Break, make   b. make, break   c. s/c, o/c   d. o/c, break

20. With switches, the word *pole* describes a:
    a. Stationary contact   c. Path of current
    b. Moving contact   d. None of the above

Essay Questions

21. Describe briefly how direct current can be generated:
    a. Mechanically with pressure and friction (6-2-1)   d. Magnetically (6-2-4)
    b. Thermally (6-2-2)   e. Chemically (6-2-5)
    c. Optically (6-2-3)   f. Electrically (6-2-6)

22. List the three artificial magnets (6-2-4)

23. Briefly describe the operation of a battery. (6-2-5)

24. Simply state the difference between a primary and a secondary cell. (6-2-5)

25. Describe what can be gained by connecting batteries in series and parallel. (6-2-5)

26. Describe a fuse's: (6-3-1)

   a. Current rating **b.** Voltage rating

27. Briefly describe the operation of the following types of circuit breakers: (6-3-2)

   a. Thermal type **b.** Magnetic type

28. List the eight different switch classifications. (6-4)

29. List the four popular primary cells. (6-2-5)

30. List the two popular secondary cells. (6-2-5)

Practice Problems

31. If a 50 W light bulb is connected across a 120 V source and the following fuses are available, which should be used to protect the circuit? (6-3-1)

   a. 0.5 A/120 V (fast blow) **c.** 1 A/12 V (slow blow)
   b. ¾ A/120 V (fast blow) **d.** 0.5 A/120 V (slow blow)

32. Draw a diagram and indicate the total source voltage of: (6-2-5)

   a. Six 1.5-V cells connected in series with one another
   b. Six 1.5-V cells connected in parallel with one another

33. Draw a diagram and indicate the polarities of a 12 V lead–acid battery being charged by a 15 V battery charger. If this battery is rated at 150 Ah: (6-2-5)

   a. How many coulombs of charge will be stored in the fully charged condition?
   b. How long will it take the battery to charge if a charging current of 5 A is flowing?

34. Show how to connect two 12 V batteries to increase (a) the voltage and (b) the current. (6-2-5)

35. If a 6 V nickel–cadmium battery discharges at a rate of 3.5 A in 4 hours:

   a. Calculate its ampere-hour and ampere-second rating.
   b. How long will it take the battery to discharge if a current of 2 A is drawn? (6-2-5)

After completing this chapter, you will be able to:

1. Describe a series circuit.
2. Identify series circuits.
3. Connect components so that they are in series with one another.
4. Describe why current remains the same throughout a series circuit.
5. Explain how to calculate total resistance in a series circuit.
6. Explain how Ohm's law can be applied to calculate current, voltage, and resistance.

7. Describe why the series circuit is known as a voltage divider.
8. Describe a fixed- and a variable-voltage divider.
9. Explain how to calculate power in a series circuit.
10. Describe how to troubleshoot and recognize:
    a. An open component
    b. A component value variation
    c. A short circuit in a series circuit

Series DC Circuits

Hitler's Computer Mistake

In the early days of World War II, German scientist Konrad Zuse, who designed and built the first general-purpose computer, proposed constructing a computer that would operate 1000 times faster than anything else at that time. He was going to redesign his Z3 computer, which was being used at that time to solve engineering problems in aircraft and missile design, and include vacuum tubes instead of electromechanical relay switches. This proposal was rejected by Hitler, who was not interested in this long-term, two-year project, as he was sure that the war was going to be, for him, a certain, quick victory. Due to Hitler's shortsightedness, this powerful computer, which could have been used to break British communication codes, was never developed. However, unknown to both Hitler and Zuse, the British code-breaking computer project, called Ultra, had highest priority and was moving rapidly toward completion.

A series circuit, by definition, is the connecting of components end to end in a circuit to provide a single path for the current. This is true not only for resistors, but also for other components that can be connected in series; in all cases the components are connected in succession or strung together one after another so that only one path for current exists between the negative (−) and positive (+) terminals of the supply.

7-1 COMPONENTS IN SERIES

Series Circuit
Circuit in which the components are connected end to end so that current has only one path to follow throughout the circuit.

Figure 7-1 illustrates five examples of **series** resistive **circuits.** In all five examples, you will notice that the resistors are connected "in-line" with one another so that the current through the first resistor must pass through the second resistor, and the current through the second resistor must pass through the third, and so on.

■ **EXAMPLE:**

In Figure 7-2(a), seven resistors are laid out on a table top. With connecting wire, string all the resistors in series, starting at $R1$, and proceeding in numerical order through the resistors until reaching $R7$.

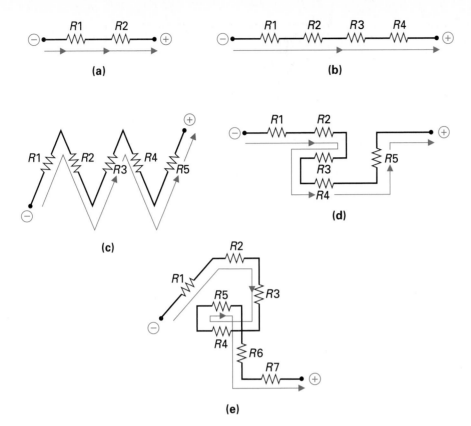

FIGURE 7-1   **Five Series Resistive Circuits.**

■ *Solution:*

In Figure 7-2(b), you can see that all the resistors are now connected in series, and the current has only one path to follow from negative to positive.

EXAMPLE:

Figure 7-3(a) shows four 1.5 V cells and three lamps. Using wires, connect all of the cells in series to create a 6 V battery source. Then connect all of the three lamps in series with one another, and finally, connect the 6 V battery source across the three-series-connected-lamp load.

■ *Solution:*

In Figure 7-3(b) you can see the final circuit containing a source, made up of four series-connected 1.5 V cells, and a load, consisting of three series-connected lamps. As explained in Chapter 6, when cells are connected in series, the total voltage (V_T) will be equal to the sum of all the cell voltages:

$$V_T = V_1 + V_2 + V_3 + V_4 = 1.5 \text{ V} + 1.5 \text{ V} + 1.5 \text{ V} + 1.5 \text{ V} = 6 \text{ V}$$

(b) Solution

FIGURE 7-2   Connecting Resistors in Series. (a) Problem. (b) Solution.

FIGURE 7-3   Series-Connected Cells and Lamps.

7-2 CURRENT IN A SERIES CIRCUIT

The current in a series circuit has only one path to follow and cannot divert in any other direction; consequently, the current through a series circuit is the same throughout that circuit.

Returning once again to the water analogy, you can see in Figure 7-4(a) that, if 2 gallons of water per second are being supplied by the pump, 2 gallons per second must be pulled into the pump; and if the rate at which water is leaving and arriving at the pump is the same, 2 gallons of water per second must be flowing throughout the circuit. The same value of water flow exists throughout a series-connected fluid system. If the values were adjusted to double the opposition to flow, then half, or 1 gallon per second, would be leaving the pump; but that same value of 1 gallon/second would be flowing throughout the system.

Similarly, with the electronic series circuit, shown in Figure 7-4(b), there is a total of 2 A leaving and 2 A arriving at the battery, so the same value of current exists throughout the series-connected electronic circuit. If the resistance of the circuit is doubled, then half, or 1 A of current, will leave the battery, but that same value of 1 A will flow throughout the entire circuit. Current is, therefore, exactly the same value at every point in a series circuit. This can be stated mathematically as

FIGURE 7-4 Series Circuit Current. (a) Fluid System. (b) Electric System.

(a) Schematic. (b) Pictorial

FIGURE 7-5 (a) Schematic. (b) Pictorial.

$$I_T = I_1 = I_2 = I_3 = \cdots$$

(Total current = current through R_1 = current through R_2 = current through R_3, and so on).

◼ **EXAMPLE:**

In Figure 7-5, a total current (I_T) of 1 A is flowing out of a battery, through two end-to-end resistors R_1 and R_2, and returning to the battery. Calculate:

   a. The current through R_1 (I_1)

   b. The current through R_2 (I_2)

◼ *Solution:*

Since R_1 and R_2 are connected in series, the current through both will be the same as the circuit current, which is equal to 1 A.

$$I_T = I_1 = I_2$$
$$1\,A = 1\,A = 1\,A$$

SELF-TEST REVIEW QUESTIONS FOR SECTIONS 7-1 AND 7-2

1. What is a series circuit?

2. What is the current flow through each of eight series-connected 8-Ω resistors if 8 A total current is flowing out of a battery?

7-3 RESISTANCE IN A SERIES CIRCUIT

Resistance is the opposition to current flow, and in a series circuit every resistor in series offers opposition to the current flow. In the water analogy of Figure 7-4, the total resistance or opposition to water flow is the sum of the two individual valve opposition values. Like the battery, the pump senses the total opposition in the circuit offered by all the valves or resistors, and the amount of current that flows is dependent on this resistance or opposition.

The total resistance in a series-connected electronic resistive circuit is thus equal to the sum of all the individual resistances, as shown in Figure 7-6(a)

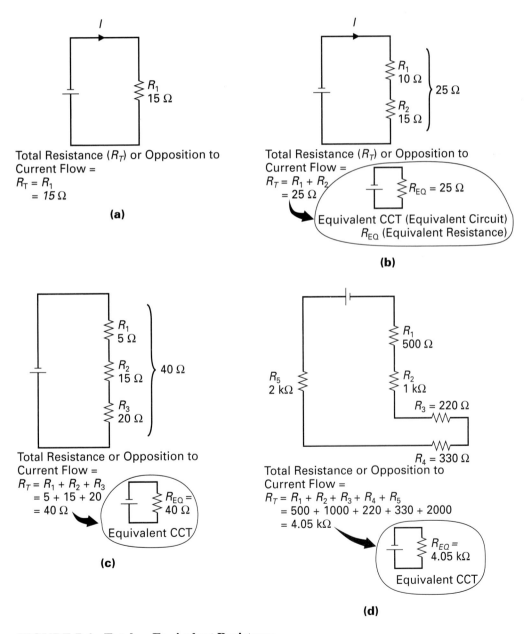

Total Resistance (R_T) or Opposition to Current Flow =
$R_T = R_1$
$\quad = 15\,\Omega$

(a)

Total Resistance (R_T) or Opposition to Current Flow =
$R_T = R_1 + R_2$
$\quad = 25\,\Omega$

Equivalent CCT (Equivalent Circuit)
R_{EQ} (Equivalent Resistance)

$R_{EQ} = 25\,\Omega$

(b)

Total Resistance or Opposition to Current Flow =
$R_T = R_1 + R_2 + R_3$
$\quad = 5 + 15 + 20$
$\quad = 40\,\Omega$

$R_{EQ} = 40\,\Omega$

Equivalent CCT

(c)

Total Resistance or Opposition to Current Flow =
$R_T = R_1 + R_2 + R_3 + R_4 + R_5$
$\quad = 500 + 1000 + 220 + 330 + 2000$
$\quad = 4.05\,k\Omega$

$R_{EQ} = 4.05\,k\Omega$

Equivalent CCT

(d)

FIGURE 7-6   Total or Equivalent Resistance.

through (d). An equivalent circuit could therefore be drawn for each of the circuits in Figure 7-6(b), (c), and (d) with one resistor, of a value equal to the sum of all the series resistance values.

No matter how many resistors are connected in series, the total resistance or opposition to current flow is always equal to the sum of all the resistor values. This formula can be stated mathematically as

$$R_T = R_1 + R_2 + R_3 + \cdots$$

(Total resistance = value of R_1 + value of R_2 + value of R_3, and so on.)

Equivalent Resistance (R_{eq})
Total resistance of all the individual resistances in a circuit.

Total resistance (R_T) is the only opposition the battery can sense; it does not see the individual separate resistors, but one **equivalent resistance.** Based on its voltage and this total resistance, a value of current will be produced to flow through the circuit (Ohm's law, $I = V/R$).

■ **EXAMPLE:**

Referring to Figure 7-7, calculate

    a. The circuit's total resistance

    b. The current flowing through R_2

■ *Solution:*

    a. $R_T = R_1 + R_2 + R_3 + R_4$

        $= 25\ \Omega + 20\ \Omega + 33\ \Omega + 10\ \Omega$

        $= 88\ \Omega$

    b. $I_T = I_1 = I_2 = I_3 = I_4$. Therefore, $I_2 = I_T = 3$ A.

FIGURE 7-7 (a) Schematic. (b) Pictorial.

FIGURE 7-8 **Three-Position Switch Controlling Lamp Brightness.**

▦ **EXAMPLE:**

Figure 7-8 shows how a single-pole three-position switch is being used to provide three different lamp brightness levels. In position ① R_1 is placed in series with the lamp, in position ② R_2 is placed in series with the lamp, and in position ③ R_3 is placed in series with the lamp. If the lamp has a resistance of 75 Ω, calculate the three values of current for each switch position.

▦ *Solution:*

$$\text{Position 1: } R_T = R_1 + R_{\text{lamp}}$$

$$= 25\ \Omega + 75\ \Omega = 100\ \Omega$$

$$I_T = \frac{V_T}{R_T} + \frac{12\ \text{V}}{100\ \Omega} = 120\ \text{mA}$$

$$\text{Position 2: } R_T = R_2 + R_{\text{lamp}} = 50\ \Omega + 75\ \Omega = 125\ \Omega$$

$$I_T = \frac{V_T}{R_T} = \frac{12\ \text{V}}{125\ \Omega} = 96\ \text{mA}$$

$$\text{Position 3: } R_T = R_3 + R_{\text{lamp}} = 75\ \Omega + 75\ \Omega = 150\ \Omega$$

$$I_T = \frac{V_T}{R_T} = \frac{12\ \text{V}}{150\ \Omega} = 80\ \text{mA}$$

SELF-TEST REVIEW QUESTIONS FOR SECTION 7-3

1. State the total resistance formula for a series circuit.
2. Calculate R_T if $R_1 = 2\ \text{k}\Omega$, $R_2 = 3\ \text{k}\Omega$, and $R_3 = 4700\ \Omega$.

FIGURE 7-9 **Series Circuit Voltage. (a) Fluid Analogy of Potentional Difference. (b) Electrical Potential Difference.**

7-4 VOLTAGE IN A SERIES CIRCUIT

A potential difference or voltage drop will occur across each resistor in a series circuit when current is flowing. The amount of voltage drop is dependent on the value of the resistor and the amount of current flow.

Let's examine this idea of potential difference or voltage drop by returning, once more, to the water analogy. In Figure 7-9(a), you can see that the high pressure from the pump's outlet is present on the left side of the valve; however, on the right side of the valve the high pressure is no longer present. The high potential that exists on the left of the valve is not present on the right, so a potential or pressure difference is said to exist across the valve.

Similarly, with the electronic circuit in Figure 7-9(b), the battery produces a high voltage or potential that is present at the top of the resistor; however, the high voltage that exists at the top of the resistor is not present at the bottom. Therefore, a potential difference or voltage drop is said to occur across the resistor.

The voltage drop across resistors can be found by utilizing Ohm's law: $V = I \times R$.

▨ **EXAMPLE:**

Referring to Figure 7-10, calculate:

  a. Total resistance (R_T)

  b. Amount of series current flowing throughout the circuit (I_T)

FIGURE 7-10   **Series Circuit Example.**

c. Voltage drop across R_1
d. Voltage drop across R_2
e. Voltage drop across R_3

■ *Solution:*

a. Total resistance $(R_T) = R_1 + R_2 + R_3$

$$= 20\ \Omega + 30\ \Omega + 50\ \Omega$$

$$= 100\ \Omega$$

b. Total current $(I_T) = \dfrac{V_{Source}}{R_T}$

$$= \dfrac{100\ V}{100\ \Omega}$$

$$= 1\ A$$

The same current will flow through the complete series circuit, so the current through R_1 will equal 1 A, the current through R_2 will equal 1 A, and the current through R_3 will equal 1 A.

c. Voltage across R_1 $(V_{R1}) = I_1 \times R_1$

$$= 1\ A \times 20\ \Omega$$

$$= 20\ V$$

d. Voltage across R_2 $(V_{R2}) = I_2 \times R_2$

$$= 1\ A \times 30\ \Omega$$

$$= 30\ V$$

e. Voltage across R_3 $(V_{R3}) = I_3 \times R_3$

$$= 1\ A \times 50\ \Omega$$

$$= 50\ V$$

Figure 7-11 shows the result of this example, and as you can see, the 20 Ω resistor drops 20 V, the 30 Ω resistor has 30 V across it, and the 50 Ω resistor has dropped 50 V. From this example, you will notice that the larger the resistor

FIGURE 7-11   (a) Schematic. (b) Pictorial.

value, the larger the voltage drop. Resistance and voltage drops are consequently proportional to one another.

$$V_{\text{drop}} \uparrow = I(\text{constant}) \times R \uparrow$$

$$V_{\text{drop}} \downarrow = I(\text{constant}) \times R \downarrow$$

Another interesting point you may have noticed from Figure 7-11 is that, if you were to add up all the voltage drops around a series circuit, they would equal the source (V_S) applied:

---

total voltage applied (V_S or V_T) = $V_{R1} + V_{R2} + V_{R3} + \cdots$

In the example, this is true, since

$$100 \text{ V} = 20 \text{ V} + 30 \text{ V} + 50 \text{ V}$$

$$100 \text{ V} = 100 \text{ V}$$

Kirchhoff's voltage law
The algebraic sum of the voltage drops in a closed path circuit is equal to the algebraic sum of the source voltage applied.

The series circuit has in fact divided up the applied voltage, and it appears proportionally across all the individual resistors. This characteristic was first observed by Gustav Kirchhoff in 1847; in honor of his discovery, this effect is known as **Kirchhoff's voltage law,** which states: The sum of the voltage drops in a series circuit is equal to the total voltage applied.

To summarize the effects of current, resistance, and voltage in a series circuit so far, we can say that:

1. The current in a series circuit has only one path to follow.

2. The value of current in a series circuit is the same throughout the entire circuit.

3. The total resistance in a series circuit is equal to the sum of all the resistances.

4. Resistance and voltage drops in a series circuit are proportional to one another, so a large resistance will have a large voltage drop and a small resistance will have a small voltage drop.

5. The sum of the voltage drops in a series circuit is equal to the total voltage applied.

■ **EXAMPLE:**

Calculate the voltage drop across the resistor R_1 in the circuit in Figure 7-12(a) for a resistance of 4 Ω or 2 Ω with a constant source of 4V.

■ *Solution:*

The voltage across R_1 can be calculated by using Ohm's law [Figure 7-12(b)].

$$V_{R1} = I_1 \times R_1$$
$$V_{R1} = 1 \text{ A} \times 4 \text{ Ω}$$
$$= 4 \text{ V}$$

If the resistance is changed to 2 Ω, the current flow within the circuit would be equal to [Figure 7-12(c)]

$$I = \frac{V_S}{R} = \frac{4 \text{ V}}{2 \text{ Ω}} = 2 \text{ A}$$

The voltage dropped across the resistor, however, would still be equal to

$$V_{R1} = I_1 \times R_1$$
$$= 2 \text{ A} \times 2 \text{ Ω}$$
$$= 4 \text{ V}$$

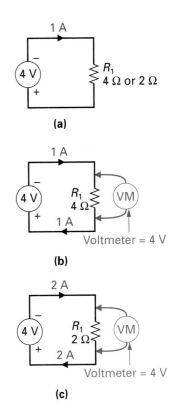

FIGURE 7-12 Single Resistor Circuit.

If only one resistor is connected in a series circuit, the entire applied voltage appears across this resistor. The amount of current flow is determined by the value of this single resistor and remains the same throughout the circuit.

■ **EXAMPLE:**

Referring to Figure 7-13(a), calculate:

a. Total circuit resistance

b. Value of current (I_T)

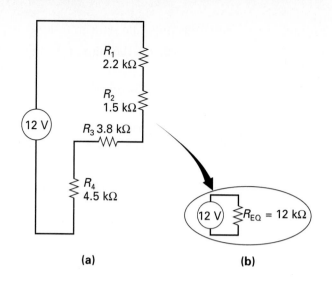

FIGURE 7-13 Series Circuit Example.

c. Voltage drop across each resistor

d. Then draw the circuit with respective voltages, resistances, and current inserted.

▓ *Solution:*

a. $R_T = R_1 + R_2 + R_3 + R_4$

$= 2.2 \text{ k}\Omega + 1.5 \text{ k}\Omega + 3.8 \text{ k}\Omega + 4.5 \text{ k}\Omega$

$= (2.2 \times 10^3) + (1.5 \times 10^3) + (3.8 \times 10^3) + (4.5 \times 10^3)$

$= 12 \text{ k}\Omega$ [Figure 7-13(b)]

b. $I_T = \dfrac{V_S}{R_T} = \dfrac{12 \text{ V}}{12 \text{ k}\Omega} = 1 \text{ mA}$

c. Voltage drop across each resistor:

$$V_{R1} = I_T \times R_1$$
$$= 1 \text{ mA} \times 2.2 \text{ k}\Omega$$
$$= 2.2 \text{ V}$$
$$V_{R2} = I_T \times R_2$$
$$= 1 \text{ mA} \times 1.5 \text{ k}\Omega$$
$$= 1.5 \text{ V}$$
$$V_{R3} = I_T \times R_3$$
$$= 1 \text{ mA} \times 3.8 \text{ k}\Omega$$
$$= 3.8 \text{ V}$$
$$V_{R4} = I_T \times R_4$$
$$= 1 \text{ mA} \times 4.5 \text{ k}\Omega$$
$$= 4.5 \text{ V}$$

d. See Figure 7-14.

FIGURE 7-14 Series Circuit Example with
All Values Inserted.

■ **EXAMPLE:**

Figure 7-15 shows a 12 V battery that has an internal source resistance (R_S or
R_{int}) of 0.5 Ω (500 mΩ). If the battery was to supply its maximum safe current,
which in this example is 2.5 A, what would be the output terminal voltage of the
battery?

■ *Solution:*

If the battery was supplying its maximum safe current of 2.5 A, the output voltage (V_{out}) would equal the source voltage ($V_S = 12$ V) minus the voltage drop
across the internal battery resistance (V_{Rint}). First let us calculate V_{Rint}:

**FIGURE 7-15   Voltage Drop across a Battery's
Internal Series Resistance.**

$$V_{Rint} = I \times R_{int}$$
$$= 2.5 \text{ A} \times 0.5 \text{ } \Omega = 1.25 \text{ V}$$

The output voltage therefore under full load will be

$$V_{out} = V_S - V_{Rint} = 12 \text{ V} - 1.25 \text{ V} = 10.75 \text{ V}$$

7-4-1 *Fixed Voltage Divider*

A series-connected circuit is often referred to as a *voltage-divider circuit*. The total voltage applied (V_T) or source voltage (V_S) is divided and dropped across the resistors in the series circuit. The amount of voltage dropped across a resistor is proportional to the value of resistance, and so a larger resistance causes a larger voltage drop across that resistor, while a smaller resistance causes a smaller voltage drop.

The voltage dropped across a resistor is normally a factor that needs to be calculated. The voltage-divider formula allows you to calculate the voltage drop across any resistor without having to work out current. This formula is stated as

$$V_X = \frac{R_X}{R_T} \times V_S$$

where V_X = voltage dropped across selected resistor

R_X = selected resistor's value

R_T = total series circuit resistance

V_S = source or applied voltage

Figure 7-16 illustrates a circuit from a previous problem. Normally, it would be necessary to:

$R_T = 12 \text{ k}\Omega$ **FIGURE 7-16   Series Circuit Example.**

1. Calculate the total resistance by adding up all the resistance values.

2. Once we have the total resistance and source voltage (V_S), we could then calculate current.

3. Having calculated the current flowing through each resistor, we use the current value to calculate the voltage dropped across any one of the four resistors merely by multiplying current by the individual resistance value.

The voltage-divider formula allows us to bypass the last two steps. If we know total resistance, supply voltage, and the resistance value, we can calculate the voltage drop across the resistor without having to calculate steps 2 and 3. For example, what would be the voltage dropped across R_2 and R_4 in Figure 7-16? The voltage dropped across R_2 is

$$V_{R2} = \frac{R_2}{R_T} \times V_S$$

$$= \frac{1.5 \text{ k}\Omega}{12 \text{ k}\Omega} \times 12 \text{ V}$$

$$= 1.5 \text{ V}$$

The voltage dropped across R_4 is

$$V_{R4} = \frac{R_4}{R_T} \times V_S$$

$$= \frac{4.5 \text{ k}\Omega}{12 \text{ k}\Omega} \times 12 \text{ V}$$

$$= 4.5 \text{ V}$$

CALCULATOR SEQUENCE

Step	Keypad Entry	Display Response
1.	① . ⑤ Ⓔ ③	1.5E3
2.	÷	
3.	① ② Ⓔ ③	12E3
4.	×	0.125
5.	① ②	12
6.	=	1.5

The voltage-divider formula could also be used to find the voltage drop across multiple resistors. For example, what would be the voltage dropped across R_2 and R_3? (This is illustrated in Figure 7-17.) The voltage across $R_2 + R_3$ is

FIGURE 7-17   Series Circuit Example.

$$V_{R2} \text{ and } V_{R3} = \frac{R_2 + R_3}{R_T} \times V_S$$

$$= \frac{5.3 \text{ k}\Omega}{12 \text{ k}\Omega} \times 12 \text{ V}$$

$$= 5.3 \text{ V}$$

EXAMPLE:

Calculate the voltage drop across (see Figure 7-18):

   a. R_1, R_2, and R_3 separately

   b. R_2 and R_3 together

   c. R_1, R_2, and R_3 together

Solution:

   a. The voltage drop across a resistor is proportional to the resistance value. The total resistance (R_T) in this circuit is 100 Ω or 100% of R_T. R_1 is 20% of the total resistance, so 20% of the source voltage will appear across R_1. R_2 is 30% of the total resistance, so 30% of the source voltage will appear across R_2. R_3 is 50% of the total resistance, so 50% of the source voltage will appear across R_3. This was a very simple problem in which the figures worked out very neatly. The voltage-divider formula achieves the very same thing by calculating the ratio of the resistance value to the total resistance. This percentage is then multiplied by the source voltage in order to find the desired resistor's voltage drop:

$$V_{R1} = \frac{R_1}{R_T} \times V_S$$

$$= \frac{20 \text{ }\Omega}{100 \text{ }\Omega} \times V_S$$

$$= 0.2 \times 100 \text{ V}$$

$$= 20 \text{ V}$$

FIGURE 7-18   **Series Circuit Example.**

$$V_{R2} = \frac{R_2}{R_T} \times V_S$$

$$= \frac{30\ \Omega}{100\ \Omega} \times V_S$$

$$= 0.3 \times 100\ \text{V}$$

$$= 30\ \text{V}$$

$$V_{R3} = \frac{R_3}{R_T} \times V_S$$

$$= \frac{50\ \Omega}{100\ \Omega} \times V_S$$

$$= 0.5 \times 100\ \text{V}$$

$$= 50\ \text{V}$$

b. Voltage dropped across R_2 and $R_3 = 30 + 50 = 80$ V.

c. Voltage dropped across R_1, R_2, and $R_3 = 20 + 30 + 50 = 100$ V.

The voltage-divider formula can be summarized by saying that: The voltage drop across a resistor or multiple resistors in a series circuit is equal to the ratio of that resistance (R_X) to the total resistance (R_T) multiplied by the source voltage (V_S).

If three voltages of 50, 80, and 100 V were required by a piece of equipment, we could use three individual power sources, which would be very expensive, or use one 100 V voltage source connected across the resistors from Example 7-10 to divide up the 100 V and supply the three required fixed voltages, as seen in Figure 7-19.

EXAMPLE:

Figure 7-20(a) shows a simplified circuit of an oscilloscope's cathode ray tube (CRT). In this example, a three-resistor series circuit and a −600 V supply voltage are being used to produce the needed three supply voltages for the CRT's three electrodes, called the focusing anode, the control grid, and the cathode. The heated cathode emits electrons that are collected and concentrated into a

FIGURE 7-19 Series Circuit Example.

beam by the combined electrostatic effect of the control grid and focusing anode. This beam of electrons passes through the apertures of the control grid and focusing anode and strikes the inner surface of the CRT screen. This inner surface is coated with a phosphorescent material that emits light when it is struck by electrons. The voltage between the cathode (K) and grid (G) of the CRT (V_{KG}) determines the intensity of the electron beam and therefore the brightness of the trace seen on the screen. The voltage between the grid (G) and anode (A) of the CRT (V_{GA}) determines the sharpness of the electron beam and therefore the focus of the trace seen on the screen. For the resistance values given, calculate the voltages on each of the three CRT electrodes.

■ **Solution:**

Referring to the illustration and calculations in Figure 7-20(b) you can see how the voltage-divider formula can be used to calculate the voltage drop across R_1

FIGURE 7-20   Fixed Voltage-Divider Circuit for Supplying Voltages to the Electrodes of a CRT.

$(V_{R1} = 300 \text{ V})$, R_2 $(V_{R2} = 255 \text{ V})$, and R_3 $(V_{R3} = 45 \text{ V})$. Since the grid of the CRT is connected directly to the -600 V supply, the grid voltage (V_G) will be

$$V_G = V_T = -600 \text{ V}$$

The cathode voltage (V_K) will be equal to the total supply voltage (V_T) minus the voltage drop across R_3 (V_{R3}), so

$$V_K = V_T - V_{R3} = (-600) - (-45) = -555 \text{ V}$$

The anode voltage (V_A) will be equal to the total supply voltage (V_T) minus the voltage drops across R_3 (V_{R3}) and R_2 (V_{R2}), and therefore

$$V_A = V_T - (V_{R3} + V_{R2}) = (-600) - (-45 + -225)$$
$$= -600 - 300 \text{ V} = -300 \text{ V}$$

The V_A, V_K, and V_G voltages are all negative voltage with respect to 0 V. The voltage V_{GA} and V_{KG}, shown on the left of Figure 7-20(b), will be the potential difference between the two electrodes. Therefore, V_{KG} is equal to the difference between V_K and V_G, and is equal to -45 V ($V_{KG} = -45 \text{ V} = V_{R3}$). The voltage between the CRT's grid and anode (V_{GA}) is equal to the difference between V_G and V_A, which will be -300 V ($V_{GA} = -300 \text{ V} = V_{R1}$).

■ **EXAMPLE:**

Figure 7-21(a) shows a 24 V voltage source driving a 10 Ω resistor that is located 1000 ft from the battery. If two 1000 ft lengths of AWG No. 13 wire are used to connect the source to the load, what will be the voltage applied across the load?

■ *Solution:*

Referring back to Table 3-2, you can see that AWG No. 13 copper cable has a resistance of 2.003 Ω for every 1000 ft. This means that our circuit should be redrawn as shown in Figure 7-21(b) to show the series resistances of wire 1 and wire 2. Using the voltage-divider formula, we can calculate the voltage drop across wire 1 and wire 2.

$$V_{W1} = \frac{R_{W1}}{R_T} \times V_T$$

$$= \frac{2 \text{ } \Omega}{14 \text{ } \Omega} \times 24 \text{ V} = 3.43 \text{ V}$$

Since the voltage drop across wire 2 will also be 3.43 V, the total voltage drop across both wires will be 6.86 V. The remainder, 17.14 V (24 V – 6.86 V = 17.14 V), will appear across the load resistor, R_L.

7-4-2 *Variable Voltage Divider*

When discussing variable-value resistors in Chapter 4, we talked about a potentiometer, or variable voltage divider, which consists of a fixed value of resistance between two terminals and a wiper that can be adjusted to vary resistance between its terminal and one of the other two. Figure 7-22(a) and (b) illustrate the potentiometer's schematic symbol and physical appearance.

Wire ① – 1000 Feet of No. 13 AWG Copper Wire

24 V

$R_L = 10 \, \Omega$

Wire ② – 1000 Feet of No. 13 AWG Copper Wire

(a)

Wire ①
Resistance = 2 Ω

V_{W1} = 3.43 V

24 V

Load Resistance,
$R_L = 10 \, \Omega$

V_{RL} = 17.14 V

Wire ②
Resistance = 2 Ω

V_{W2} = 3.43 V

(b)

FIGURE 7-21   **Series Wire Resistance.**

If the wiper is moved down, as seen in Figure 7-23(a), the resistance between terminals *A* and *B* will increase, while the resistance between *B* and *C* will decrease. If the wiper is moved up, as seen in Figure 7-23(b), the resistance between terminals *A* and *B* will decrease, while the resistance between *B* and *C* will increase. The resistances between *A* and *B* and *B* and *C* are inversely proportional to one another in that if one were to increase the other would decrease, and vice versa, and can be thought of as two separate resistors, as seen in Figure 7-23(c). However, the total of the resistance between *A* and *B* (R_{AB})

Terminal Ⓐ

Terminal Ⓑ

Terminal Ⓒ

(a)

(b)

FIGURE 7-22   **Potentiometer. (a) Schematic Symbol. (b) Physical Appearance.**

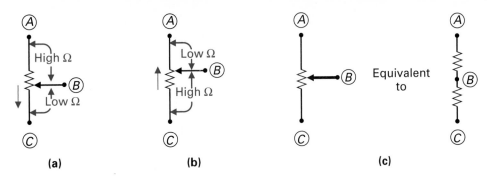

FIGURE 7-23   Potentiometer Operation.

and the resistance between B and C (R_{BC}) will always be equal to the rated value of the potentiometer, and equal to the resistance between A and C (R_{AC}).

$$R_{AB} + R_{BC} = R_{AC}$$

Figure 7-24(a) illustrates a 10 kΩ potentiometer that has been hooked up across a 10 V dc source with a voltmeter between terminals B and C. If the wiper terminal is positioned midway between A and C, the voltmeter should read 5 V, and the potentiometer will be equivalent to two 5 kΩ resistors in series, as shown in Figure 7-24(b). Kirchhoff's voltage law states that the entire source voltage will be dropped across the resistances in the circuit, and since the resistance values are equal, each will drop half of the source voltage, that is, 5 V.

In Figure 7-25(a), the wiper has been moved down so that the resistance between A and B is equal to 8 kΩ, and the resistance between B and C equals 2 kΩ. This will produce 2V on the voltmeter, as shown in Figure 7-25(b). The amount of voltage drop is proportional to the resistance, so a larger voltage will be dropped across the larger resistance. Using the voltage-divider formula, you can calculate that the 8 kΩ is 80% of the total resistance and therefore will drop 80% of the voltage:

$$V_{AB} = \frac{R_{AB}}{R_{\text{total}}} \times V_S = \frac{8\text{ kΩ}}{10\text{ kΩ}} \times 10\text{ V} = 8\text{ V}$$

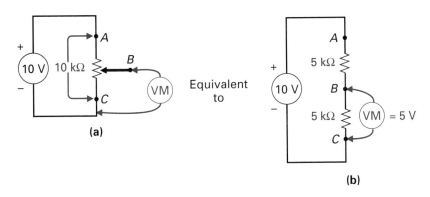

FIGURE 7-24   Potentiometer Wiper in Mid-position.

FIGURE 7-25   **Potentiometer Wiper in Lower Position.**

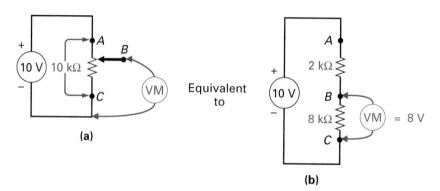

FIGURE 7-26   **Potentiometer Wiper in Upper Position.**

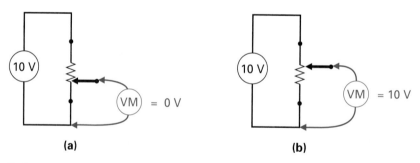

FIGURE 7-27   **Minimum and Maximum Settings of a Potentiometer.**

The 2 kΩ resistance between B and C is 20% of the total resistance and consequently will drop 20% of the total voltage:

$$V_{BC} = \frac{R_{BC}}{R_{\text{total}}} \times V_S = \frac{2 \text{ k}\Omega}{10 \text{ k}\Omega} \times 10 \text{ V} = 2 \text{ V}$$

which will be indicated on the voltmeter.

In Figure 7-26(a), the wiper has been moved up and now 2 kΩ exists between A and B, and 8 kΩ is present between B and C. In this situation, 2 V will be dropped across the 2 kΩ between A and B, and 8 V will be dropped across the 8 kΩ between B and C, as shown in Figure 7-26(b).

From this discussion, you can see that the potentiometer can be adjusted to supply different voltages on the wiper. This voltage can be decreased by moving the wiper down to supply a minimum of 0 V [Figure 7-27(a)], or the wiper can be moved up to supply a maximum of 10 V [Figure 7-27(b)], or it can be placed anywhere in between, which is why the potentiometer is known as a variable voltage divider.

■ **EXAMPLE:**

Figure 7-28 illustrates how a potentiometer can be used to control the output volume of an amplifier that is being driven by a compact disk (CD) player. The preamplifier is producing an output of 2 V, which is developed across a 50 kΩ potentiometer. If the wiper of the potentiometer is in its upper position, the full 2 V from the preamp will be applied into the input of the power amplifier. The power amplifier has a fixed voltage gain (A_V) of 12, and therefore the power amplifier's output is always 12 times larger than the input voltage. An input of 2 V (V_{in}) will therefore produce an output voltage (V_{out}) of 24 V ($V_{out} = V_{in} \times A_V = 2\,V \times 12 = 24\,V$). As the wiper is moved down, less of the 2 V from the preamp will be applied to the power amplifier, and therefore the output of the power amplifier and volume of the music heard will decrease. If the wiper of the potentiometer is adjusted so that a resistance of 20 kΩ exists between the wiper and the lower end of the potentiometer, what will be the input voltage to the power amplifier and output voltage to the speaker?

■ *Solution:*

By using the voltage-divider formula, we can determine the voltage developed across the potentiometer with 20 kΩ of resistance between the wiper (B) and lower end (C).

$$V_{in} = \frac{R_{BC}}{R_{AC}} \times V_{Pre}$$

$$= \frac{20\ k\Omega}{50\ k\Omega} \times 2\,V = 0.8\,V\ (800\,mV)$$

FIGURE 7-28   The Potentiometer as a Volume Control.

The voltage, V_{in}, is applied to the input of the power amplifier, which has an amplification factor or gain of 12, and therefore the output voltage will be

$$V_{out} = V_{in} \times A_V$$
$$= 0.8 \text{ V} \times 12 = 9.6 \text{ V}$$

SELF-TEST REVIEW QUESTIONS FOR SECTION 7-4

1. True or false: A series circuit is also known as a voltage-divider circuit.
2. True or false: The voltage drop across a series resistor is proportional to the value of the resistor.
3. If 6 Ω and 12 Ω resistors are connected across a dc 18 V supply, calculate I_T and the voltage drop across each.
4. State the voltage-divider formula.
5. Which component can be used as a variable voltage divider?
6. Could a rheostat be used in place of a potentiometer?

7-5 POWER IN A SERIES CIRCUIT

Power is the rate at which work is done. Work is said to have been done when energy is converted, in this case from electrical energy to heat energy. Resistors dissipate heat energy, and the rate at which they dissipate energy is called *power* and is measured in *watts* (joules per second). Resistors all have a resistive value, a tolerance, and a wattage rating. The wattage of a resistor is the amount of heat energy a resistor will dissipate per second, and this wattage is directly proportional to the resistor's size; a larger resistor will be able to dissipate more heat than a smaller resistor. Figure 7-29 reviews the size versus wattage rating of commercially available resistors. As you know, any of the power formulas can be used to calculate wattage. Resistors are manufactured in several different physical sizes, and if, for example, it is calculated that for a certain value of current and voltage a 5 W resistor is needed, and a ½ W resistor is put in its place, the ½ W resistor will burn out, because it is generating heat (5 W) faster than it can dissipate heat (½ W). A 10 W or 25 W resistor or greater could be used to replace a 5 W resistor; however, anything less than a 5 W resistor will burn out.

$\frac{1}{8}$ $\frac{1}{4}$ $\frac{1}{2}$ 1   2
Watt Watt Watt Watt Watts   FIGURE 7-29   **Resistor Wattage Ratings.**

It is, therefore, necessary that we have some way of calculating which wattage rating is necessary for each specific application.

A question you may be asking is why not just use large-wattage resistors everywhere. The two disadvantages with this are:

1. The larger the wattage, the greater the cost.
2. The larger the wattage, the greater the size and area the resistor occupies within the equipment.

EXAMPLE:

Figure 7-30 illustrates a 20 V battery driving a 12 V/ 1 A television set. R_1 is in series with the TV set and is being used to drop 8 V of the 20 V supply, so 12 V will be applied across the set.

a. What is the wattage rating for R_1?

b. What is the series load resistance of the TV set?

c. What is the amount of power being consumed by the TV set?

FIGURE 7-30   **Series Circuit Example.**

Solution:

a. Everything is known about R_1. Its resistance is 8 Ω, it has 1 A of current flowing through it, and 8 V is being dropped across it. Consequently, any one of the three power formulas can be used to calculate the wattage rating of R_1.

$$\text{power } (P) = I \times V = 1 \text{ A} \times 8 \text{ V} = 8 \text{ W}$$

or

$$P = I^2 \times R = 1^2 \times 8 = 8 \text{ W}$$

or

$$P = \frac{V^2}{R} = \frac{8^2}{8} = 8 \text{ W}$$

The nearest commercially available device would be a 10 W resistor; however, if size is not a consideration, it is ideal to double the wattage needed and use a 16 W resistor.

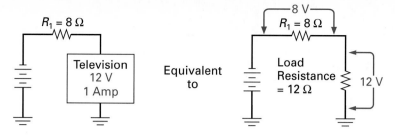

FIGURE 7-31   **Series Circuit Example with Values Inserted.**

b. You may recall that any piece of equipment is equivalent to a load resistance. The TV set has 12 V across it and is pulling 1 A of current; its load resistance can be calculated simply by using Ohm's law and deriving an equivalent circuit, as shown in Figure 7-31.

$$R_L \text{ (load resistance)} = \frac{V}{I}$$

$$= \frac{12 \text{ V}}{1 \text{ A}}$$

$$= 12 \text{ }\Omega$$

c. The amount of power being consumed by the TV set is

$$P = V \times I = 12 \text{ V} \times 1 \text{ A} = 12 \text{ W}$$

■ **EXAMPLE:**

Calculate the total amount of power dissipated in the series circuit in Figure 7-32.

FIGURE 7-32   **Series Circuit Example.**

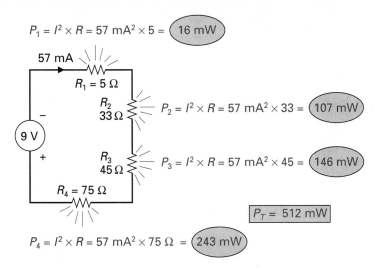

$P_1 = I^2 \times R = 57 \text{ mA}^2 \times 5 = \boxed{16 \text{ mW}}$

$P_2 = I^2 \times R = 57 \text{ mA}^2 \times 33 = \boxed{107 \text{ mW}}$

$P_3 = I^2 \times R = 57 \text{ mA}^2 \times 45 = \boxed{146 \text{ mW}}$

$P_T = 512 \text{ mW}$

$P_4 = I^2 \times R = 57 \text{ mA}^2 \times 75 \, \Omega = \boxed{243 \text{ mW}}$

FIGURE 7-33   Series Circuit Example with Values Inserted.

■ *Solution:*

The total power dissipated in a series circuit is equal to the sum of all the power dissipated by all the resistors. The easiest way to calculate the total power is to simplify the circuit to one resistance.

$$R_T = R_1 + R_2 + R_3 + R_4$$
$$= 5 \, \Omega + 33 \, \Omega + 45 \, \Omega + 75 \, \Omega$$
$$= 158 \, \Omega$$

We now have total resistance and total voltage, so we can calculate the total power:

$$P_T = \frac{V_S^2}{R_T} = \frac{9 \text{ V}^2}{158 \, \Omega} = \frac{81}{158} = 512.7 \text{ milliwatts (mW)}$$

The longer method would have been to calculate the current through the series current:

$$I = \frac{V_S}{R_T} = \frac{9 \text{ V}}{158 \, \Omega} = 56.96 \text{ mA} \quad \text{or} \quad 57 \text{ mA}$$

We could then calculate the power dissipated by each separate resistor and add up all the individual values to gain a total power figure. This is illustrated in Figure 7-33.

total power = addition of all the individual power losses

$$P_T = P_1 + P_2 + P_3 + P_4 + \cdots$$

$$= 16 \text{ mW} + 107 \text{ mW} + 146 \text{ mW} + 243 \text{ mW}$$
$$= 512 \text{ mW}$$

1. State the power formula.
2. Calculate the power dissipated by a 12 Ω resistor connected across a 12 V supply.
3. What fixed resistor type should probably be used for Question 2, and what would be a safe wattage rating?
4. What would be the total power dissipated if R_1 dissipates 25 W and R_2 dissipates 3800 mW?

7-6 TROUBLESHOOTING A SERIES CIRCUIT

A resistor will usually burn out and result in an open between its two leads when an excessive current flow occurs. This can normally, but not always, be noticed by a visual check of the resistor, which will appear charred due to the excessive heat. In some cases you will need to use your multimeter (combined ammeter, voltmeter, and ohmmeter) to check the circuit components to determine where a problem exists.

The two basic problems that normally exist in a series circuit are opens and shorts; however, a problem is not always as drastic as a short or an open, but may be a variation in a component's value over a long period of time, which will eventually cause a problem.

To summarize, then, we can say that one of three problems can occur to components in a series circuit:

1. A component will open (infinite resistance).
2. A component's value will change over a period of time.
3. A component will short (zero resistance).

The voltmeter is the most useful tool when checking series circuits as it can be used to measure voltage drops by connecting the meter leads across the component or resistors. Let's now analyze a problem and see if we can solve it by logically **troubleshooting** the circuit and isolating the faulty component. To begin with, let us take a look at the effects of an open component.

Troubleshooting
The process of locating and diagnosing malfunctions or breakdowns in equipment by means of systematic checking or analysis.

7-6-1 *Open Component in a Series Circuit*

A component is open when its resistance is the maximum possible (infinity). Let us see with a few examples what effect this will have.

▪ **EXAMPLE:**

Figure 7-34(a) illustrates a TV set with a load resistance of 3 Ω. The TV set is off because R_2 has burned out and become an open circuit. How would you determine that the problem is R_2?

FIGURE 7-34   Troubleshooting an Open in a Series Circuit.

■ *Solution:*

If an open circuit ever occurs in a series circuit, due in this case to R_2 having burned out, there can be no current flow, because series circuits have only one path for current to flow and that path has been broken ($I = 0$ A). Using the voltmeter to check the amount of voltage drop across each resistor, two results will be obtained:

1. The voltage drop across a good resistor will be zero volts.
2. The voltage drop across an open resistor will be equal to the source voltage V_S.

No voltage will be dropped across a good resistor because current is zero, and if $I = 0$, the voltage drop, which is the product of I and R, must be zero ($V = I \times R = 0 \times R = 0$ V). If no voltage is being dropped across the good resistor R_1 and the TV set resistance of 3 Ω, the entire source voltage will appear across the open resistor, R_2, in order that this series circuit comply with Kirchhoff's voltage law: V_S (9 V) $= V_{R1}$ (0 V) $+ V_{R2}$ (9 V) $+ V_L$ (0 V).

   To emphasize this point further, refer to Figure 7-34(b), which illustrates the fluid analogy. Like R_2, valve 2 has completely blocked any form of flow (water flow = 0), and looking at the pressure differences across all three valves, you can see that no pressure difference occurs across valves 1 and 3, but the entire pump pressure is appearing across valve 2, the component that has opened the circuit.

■ EXAMPLE:

Figure 7-35 illustrates a set of three lights connected across a 9 V battery. Bulb 3 is open and therefore there is no current flow; with no current flow, all three bulbs are off, and we need to isolate which of the three is faulty.

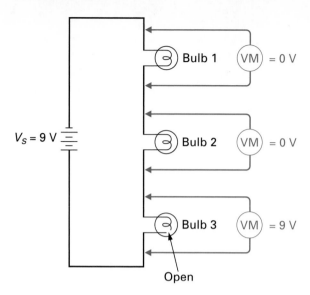

FIGURE 7-35 **Troubleshooting an Open Bulb in a Series Circuit.**

▨ *Solution:*

Using the voltmeter, there will be:

1. Zero volts dropped across bulb 1; 9 V appears on both sides of bulb 1, so the potential difference or voltage drop across bulb 1 is zero. Bulb 1 is OK.
2. Zero volts dropped across bulb 2; bulb 2 is OK.
3. Nine volts dropped across bulb 3; the entire source voltage is being dropped across bulb 3. Bulb 3 is open and needs to be replaced.

7-6-2 Component Value Variation in a Series Circuit

Resistors will rarely go completely open unless severely stressed because of excessive current flow. With age, resistors will normally change their resistance value. This occurs slowly and will generally always cause a decrease in the resistor's resistance and eventually cause a circuit problem. This lowering of resistance will cause an increase of current, which will cause an increase in the power dissipated; if the wattage of the resistor is exceeded, it can burn out. If they do not burn out but merely blow the circuit fuse due to this increase in current, the problem can be found by measuring the resistance values of each resistor or by measuring how much voltage is dropped across each resistor and comparing these to the calculated voltage, based on the parts list supplied by the manufacturer.

FIGURE 7-36 **Troubleshooting a Short in a Series Circuit.**

7-6-3 Shorted Component in a Series Circuit

A component has gone short when its resistance is 0 Ω. Let's work out a problem using the same example of the three bulbs across the 9 V battery, as shown in Figure 7-36.

■ **EXAMPLE:**

Bulbs 1 and 3 in Figure 7-36 are on and bulb 2 is off. A piece of wire exists between the two terminals of bulb 2, and the current is taking the lowest resistance path through the wire rather than through the filament resistor of the bulb. Bulb 2 therefore has no current flow through it, and light cannot be generated without current. How would you determine what the problem is?

■ *Solution:*

If bulb 2 was open (burned out), there could be no current flow in the circuit, so bulbs 1 and 2 would be off. Bulb 2 must therefore be shorted. Using the voltmeter to investigate this further, you find that bulb 2 drops 0 V across it because it has no resistance except the very small wire resistance of the bypass, which means that 4.5 V must be dropped across each working bulb (1 and 3) in order to comply with Kirchhoff's voltage law. The loss of bulb 2's resistance causes the overall circuit resistance offered by bulbs 1 and 3 to decrease, which will cause an increase in current; so bulbs 1 and 3 should glow more brightly.

To summarize opens and shorts in series circuits, we can say that:

1. The supply voltage appears across an open component.
2. Zero volts appears across a shorted component.

1. List the three basic problems that can occur with components in a series circuit.
2. How can an open component be detected in a series circuit?
3. True or false: If a series-connected resistor's value were to decrease, the voltage drop across that same resistor would increase.
4. How could a shorted component be detected in a series circuit?

SUMMARY

Series Circuits (Figure 7-37)

1. A series circuit is the connecting of components end to end in a circuit to provide a single path for the current. In all cases the components are connected in succession or strung together one after another so that only one path for current exists between the negative (−) and positive (+) terminals of the supply.

Current in a Series Circuit

2. The current in a series circuit has only one path to follow and cannot divert in any other direction; consequently, the current through a series circuit is the same throughout that circuit.

Resistance in a Series Circuit

3. Resistance is the opposition to current flow, and in a series circuit every resistor in series offers opposition to the current flow.
4. The total resistance in a series-connected electronic resistive circuit is thus equal to the sum of all the individual resistances.
5. Total resistance (R_T) is the only opposition the voltage source can sense; it does not see the individual separate resistors, but one equivalent resistance. Based on its voltage and this total resistance, a value of currrent will be produced to flow through the circuit (Ohm's law, $I = V/R$).

Voltage in a Series Circuit

6. A potential difference or voltage drop will occur across each resistor in a series circuit when current is flowing. The amount of voltage drop is dependent on the value of the resistor and the amount of current flow.
7. The voltage drop across resistors can be found by utilizing Ohm's law: $V = I \times R$.
8. The series circuit divides up the applied voltage, and it appears proportionally across all the individual resistors. This characteristic was first observed by Gustav Kirchhoff in 1847; in honor of his discovery, this effect is known as Kirchhoff's voltage law, which states: The sum of the voltage drops in a series circuit is equal to the total voltage applied.

FIGURE 7-37   Series Circuits.

Example Circuit (Resistors)

R_1 5 Ω
R_2 33 Ω
R_3 45 Ω
R_4 33 Ω

9 v

Example Circuit (Batteries)

V_1
V_2
V_3
V_4

$$V_T = V_1 + V_2 + V_3 + V_4$$

$$\boxed{I = I_1 = I_2 = I_3 = \cdots}$$

$$\boxed{R_T = R_1 + R_2 + R_3 + \cdots}$$

$$\boxed{V_S = V_{R1} + V_{R2} + V_{R3} + \cdots}$$

$$I_T = \frac{V_S}{R_T}$$

$V_{R1} = I_1 \times R_1,$
$V_{R2} = I_2 \times R_2,$
$V_{R3} = I_3 \times R_3,$ and so on

Voltage-divider formula $\boxed{V_X = \left(\dfrac{R_X}{R_T}\right) \times V_S}$

$$\boxed{P_T = P_1 + P_2 + P_3 + \cdots}$$

$$P_T = \frac{V_S^2}{R_T}$$

$$P_T = I^2 \times R_T$$

$$P_T = V_S \times I$$

I = series current in amperes
I_1 = current through R_1, in amperes
I_2 = current through R_2, in amperes
I_3 = current through R_3, in amperes
R_T = total circuit resistance, in ohms
R_1 = resistance of R_1, in ohms
R_2 = resistance of R_2, in ohms
R_3 = resistance of R_3, in ohms
V_S = source voltage, in volts
V_{R1}= voltage drop across R_1, in volts
V_{R2}= voltage drop across R_2, in volts
V_{R3}= voltage drop across R_3, in volts
V_X = voltage drop across selected resistor(s)
R_X = selected resistor(s) ohmic value
P_T = total power dissipated, in watts
P_1 = power dissipated by R_1, in watts
P_2 = power dissipated by R_2, in watts
P_3 = power dissipated by R_3, in watts

9. If only one resistor is connected in a series circuit, the entire applied voltage appears across this resistor. The amount of current flow is determined by the value of this single resistor and remains the same throughout the circuit.

10. A series-connected circuit is often referred to as a *voltage-divider circuit.* The total voltage applied (V_T) or source voltage (V_S) is divided and dropped across the resistors in the series circuit. The amount of voltage dropped across a resistor is proportional to the value of resistance, and so a larger resistance causes a larger voltage drop across that resistor, while a smaller resistance causes a smaller voltage drop.

11. The voltage-divider formula allows you to calculate the voltage drop across any resistor without having to work out current.

12. The voltage-divider formula can be summarized by saying that: The voltage drop across a resistor or multiple resistors in a series circuit is equal to the ratio of that resistance (R_X) to the total resistance (R_T) multiplied by the source voltage (V_S).

13. A potentiometer, or variable voltage divider, consists of a fixed value of resistance between two terminals and a wiper that can be adjusted to vary resistance between its terminal and one of the other two. If the wiper is moved down the resistance between terminals A and B will increase, while the resistance between B and C will decrease. If the wiper is moved up the resistance between terminals A and B will decrease, while the resistance between B and C will increase. The resistances between A and B and B and C are inversely proportional to one another to that if one were to increase the other would decrease, and vice versa.

14. The potentiometer can be adjusted to supply different voltages on the wiper. This voltage can be decreased by moving the wiper down to supply a minimum or the wiper can be moved up to supply a maximum.

Power in a Series Circuit

15. Power is the rate at which work is done. Work is said to have been done when energy is converted, in this case from electrical energy to heat energy.

16. Resistors dissipate heat energy, and the rate at which they dissipate energy is called *power* and is measured in *watts* (joules per second).

17. Resistors all have a resistive value, a tolerance, and a wattage rating. The wattage of a resistor is the amount of heat energy a resistor will dissipate per second, and this wattage is directly proportional to the resistor's size; a larger resistor will be able to dissipate more heat than a smaller resistor.

18. The larger the wattage, the greater the cost and the greater the size and area the resistor occupies within the equipment.

Troubleshooting a Series Circuit

19. A resistor will usually burn out and result in an open between its two leads when an excessive current flow occurs. This can normally, but not always, be noticed by a visual check of the resistor, which will appear charred due

to the excessive heat. In some cases you will need to use your multimeter to check the circuit components to determine where a problem exists.

20. The two basic problems that normally exist in a series circuit are opens and shorts; however, a problem is not always as drastic as a short or an open, but may be a variation in a component's value over a long period of time, which will eventually cause a problem.

21. The voltmeter is the most useful tool when checking series circuits as it can be used to measure voltage drops by connecting the meter leads across the component or resistors.

22. A component is open when its resistance is the maximum possible (infinity).

23. Using the voltmeter to check the amount of voltage drop across each resistor, two results will be obtained.

    a. The voltage drop across a good resistor will be zero volts.

    b. The voltage drop across an open resistor will be equal to the source voltage V_S.

24. With age, resistors will normally change their resistance value. This occurs slowly and will generally always cause a decrease in the resistor's resistance and eventually cause a circuit problem. This lowering of resistance will cause an increase of current, which will cause an increase in the power dissipated; if the wattage of the resistor is exceeded, it can burn out. If they do not burn out but merely blow the circuit fuse due to this increase in current, the problem can be found by measuring the resistance values of each resistor or by measuring how much voltage is dropped across each resistor and comparing these to the calculated voltage, based on the parts list supplied by the manufacturer.

25. A component has gone short when its resistance is $0 \, \Omega$.

26. To summarize opens and shorts in series circuits, we can say that:

    a. The supply voltage appears across an open component.

    b. Zero volts appears across a shorted component.

			NEW TERMS
Equivalent resistance	Series circuit	Variable voltage divider	
Fixed voltage dividers	Troubleshooting series circuits	Voltage divider	
Kirchhoff's voltage law			

Multiple-Choice Questions

REVIEW QUESTIONS

1. A series circuit:
   a. Is the connecting of components end to end
   b. Provides a single path for current
   c. Functions as a voltage divider
   d. All of the above

2. The total current in a series circuit is equal to:
   a. $I_1 + I_2 + I_3 + \cdots$ c. $I_1 = I_2 = I_3 = \cdots$
   b. $I_1 - I_2$ d. All of the above

3. If R_1 and R_2 are connected in series with a total current of 2 A, what will be the current flowing through R_1 and R_2, respectively?

   a. 1 A, 1 A
   b. 2 A, 1 A
   c. 2 A, 2 A
   d. All of the above could be true on some occasions.

4. The total resistance in a series circuit is equal to:

   a. The total voltage divided by the total current
   b. The sum of all the individual resistor values
   c. $R_1 + R_2 + R_3 + \cdots$
   d. All of the above
   e. None of the above are even remotely true.

5. Which of Kirchhoff's laws applies to series circuits:

   a. His voltage law
   b. His current law
   c. His power law
   d. None of them apply to series circuits, only parallel.

6. The amount of voltage dropped across a resistor is proportional to:

   a. The value of the resistor
   b. The current flow in the circuit
   c. Both (a) and (b)
   d. None of the above

7. If three resistors of 6 kΩ, 4.7 kΩ, and 330 Ω are connected in series with one another, what total resistance will the battery sense?

   a. 11.03 MΩ **c.** 6 kΩ
   b. 11.03 Ω **d.** 11.03 kΩ

8. The voltage-divider formula states that the voltage drop across a resistor or multiple resistors in a series circuit is equal to the ratio of that _____ to the _____ multiplied by the _____.

   a. Resistance, source voltage, total resistance
   b. Resistance, total resistance, source voltage
   c. Total current, resistance, total voltage
   d. Total voltage, total current, resistance

9. The _____ can be used as a variable voltage divider.

   a. Potentiometer   **c.** SPDT switch
   b. Fixed resistor   **d.** None of the above

10. A resistor of larger physical size will be able to dissipate _____ heat than a small resistor.

   a. More   **c.** About the same
   b. Less   **d.** None of the above

11. The _____ is the most useful tool when checking series circuits.

   a. Ammeter   **c.** Voltmeter
   b. Wattmeter **d.** Both (a) and (b)

12. When an open component occurs in a series circuit, it can be noticed because:
    a. Zero volts appears across it
    b. The supply voltage appears across it
    c. 1.3 V appears across it
    d. None of the above

13. Power can be calculated by:
    a. The addition of all the individual power figures
    b. The product of the total current and the total voltage
    c. The square of the total voltage divided by the total resistance
    d. All of the above

14. A series circuit is known as a:
    a. Current divider   c. Current subtractor
    b. Voltage divider   d. All of the above.

15. In a series circuit only _____ path(s) exists for current flow, while the voltage applied is distributed across all the individual resistors.
    a. Three   c. Four
    b. Several   d. One

Essay Questions

16. Describe a series-connected circuit. (7-1)
17. Describe and state mathematically what happens to current flow in a series circuit. (7-2)
18. Describe how total resistance can be calculated in a series circuit. (7-3)
19. Describe why voltage is dropped around a series circuit and how each voltage drop can be calculated. (7-4)
20. Briefly describe why resistance and voltage drops are proportional to one another. (7-4)
21. Describe a fixed and a variable voltage divider. (7-4-1 and 7-4-2)
22. How can individual and total power be calculated in a series circuit? (7-5)
23. How can you recognize shorts and opens in a series circuit when troubleshooting with a voltmeter? (7-6)
24. List the three problems that can occur in a series circuit. (7-6)
25. State Kirchhoff's voltage (series circuit) law. (7-4)

Practice Problems

26. If three resistors of 1.7 kΩ, 3.3 kΩ, and 14.4 kΩ are connected in series with one another across a 24 V source as shown in Figure 7-38, calculate:
    a. Total resistance (R_T)
    b. Circuit current
    c. Individual voltage drops
    d. Individual and total power dissipated

27. If 40 Ω and 35 Ω resistors are connected across a 24 V source, what would be the current flow through the resistors, and what resistance would cause half the current to flow?

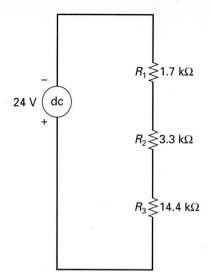

FIGURE 7-38

28. Calculate the total resistance (R_T) of the following series-connected resistors: 2.7 kΩ, 3.4 MΩ, 370 Ω, and 4.6 MΩ.

29. Calculate the value of resistors needed to divide up a 90 V source to produce 45 V and 60 V outputs, with a divider circuit current of 1 A.

30. If R_1 = 4.7 kΩ and R_2 = 6.4 kΩ and both are connected across a 9 V source, how much voltage will be dropped across R_2?

31. What current would flow through R_1 if it were one-third the ohmic value of R_2 and R_3, and all were connected in series with a total current of 6.5 mA flowing out of V_S?

32. Draw a circuit showing R_1 = 2.7 kΩ, R_2 = 3.3 kΩ, and R_3 = 0.027 MΩ in series with one another across a 20 V source. Calculate:

    a. I_T **d.** P_2 **g.** V_{R2}
    b. P_T **e.** P_3 **h.** V_{R3}
    c. P_1 **f.** V_{R1} **i.** I_{R1}

33. Calculate the current flowing through three light bulbs that are dissipating 120 W, 60 W, and 200 W when they are connected in series across a 120 V source. How is the voltage divided around the series circuit?

34. If three equal-value resistors are series connected across a dc power supply adjusted to supply 10 V, what percentage of the source voltage will appear across R_1?

FIGURE 7-39

FIGURE 7-40

$I = 100$ mA

$V_S = ?$ (dc)

$V_{R1} = ?$ $R_1 = 5\,\Omega$ $P_1 = ?$

$V_{R2} = ?$ $R_2 = 10\,\Omega$ $P_2 = ?$

$V_{R3} = ?$ $R_3 = 6\,\Omega$ $P_3 = ?$
$V_{R4} = ?$

$R_4 = 4\,\Omega$
$P_4 = ?$

FIGURE 7-41

35. Refer to the following figures and calculate:

   a. I (Figure 7-39)
   b. R_T and P_T (Figure 7-40)
   c. V_S, V_{R1}, V_{R2}, V_{R3}, V_{R4}, P_1, P_2, P_3, and P_4 (Figure 7-41)
   d. P_T, I, R_1, R_2, R_3, and R_4 (Figure 7-42)

Troubleshooting Questions

36. If three bulbs are connected across a 9 V battery in series, and the filament in one of the bulbs burned out, causing an open in the bulb, would the other lamps be on? Explain why.

37. Using a voltmeter, how would a short be recognized in a series circuit?

38. If one of three series-connected bulbs is shorted, will the other two bulbs be on? Explain why.

39. When one resistor in a series string is open, explain what would happen to the circuit's:

   a. Current
   b. Resistance
   c. Voltage across the open component
   d. Voltage across the other components

40. When one resistor in a series string is shorted, explain what would happen to the circuit's:

   a. Current
   b. Resistance
   c. Voltage across the shorted component
   d. Voltage across the other components

$P_T = ?$

$I = ?$

$P_1 = 12$ mW $P_2 = 7$ mW $P_3 = 16$ mW $P_4 = 3$ mW
$R_1 = ?$ $R_2 = ?$ $R_3 = ?$ $R_4 = ?$

12.5 V

FIGURE 7-42

After completing this chapter, you will be able to:

1. Describe the difference between a series and a parallel circuit.
2. Be able to recognize and determine whether circuit components are connected in series or parallel.
3. Explain why voltage measures the same across parallel-connected components.
4. State Kirchhoff's current law.
5. Describe why branch current and resistance are inversely proportional to one another.

6. Determine the total resistance of any parallel-connected resistive circuit.
7. Describe and be able to apply all formulas associated with the calculation of voltage, current, resistance, and power in a parallel circuit.
8. Describe how a short, open, or component variation will affect a parallel circuit's operation and how it can be recognized.

Parallel DC Circuits

OUTLINE

Bringing Home the Bacon

Sociable and athletic looking, Robert Noyce was the son of a small-town Iowa Congregational minister. As a college freshman in the spring of 1948, he decided to throw a Hawaiian luau for his friends. This was not a mistake, but stealing a pig and roasting it was, as the theft was uncovered. Noyce, a very promising student, was expelled and went to work for an insurance company.

After a short time, Noyce realized that he wanted to return to his studies, which proved not only profitable for him but also for the electronics industry. He graduated an honor student from MIT, became a brilliant physicist, invented the semiconductor integrated circuit, won the National Medal of Science Award, and founded Intel Corporation.

By tracing the path of current, we can determine whether a circuit has series-connected or parallel-connected components. In a series circuit, there is only one path for current, whereas in parallel circuits the current has two or more paths. These paths are known as *branches*. A parallel circuit, by definition, is when two or more components are connected to the same voltage source so that the current can branch out over two or more paths. In a parallel resistive circuit, two or more resistors are connected to the same source, so current splits to travel through each separate branch resistance.

8-1 COMPONENTS IN PARALLEL

Parallel Circuit
Also called shunt; circuit having two or more paths for current flow.

Many components, other than resistors, can be connected in parallel, and a **parallel circuit** can easily be identified because current is split into two or more paths. Being able to identify a parallel connection requires some practice, because they can come in many different shapes and sizes. The means for recognizing series circuits is that if you can place your pencil at the negative terminal of the voltage source (battery) and follow the wire connections through components to the positive side of the battery and only have one path to follow, the circuit is connected in series. If, however, you can place your pencil at the negative terminal of the voltage source and follow the wire and at some point have a choice of two or more routes, the circuit is connected with two or more parallel

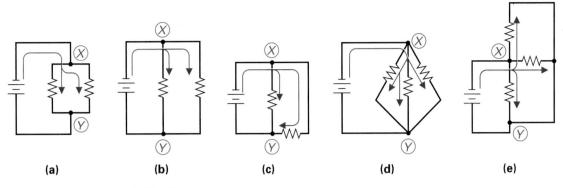

| (a) | (b) | (c) | (d) | (e) |

FIGURE 8-1 **Parallel Circuits.**

branches; the number of routes determines the number of parallel branches. Figure 8-1 illustrates five examples of parallel resistive circuits.

▪ EXAMPLE:

Figure 8-2(a) illustrates four resistors laid out on a table top.

   a. With wire leads, connect all four resistors in parallel between the negative and positive terminals of the battery.

   b. Draw the schematic diagram of the parallel-connected circuit.

▪ *Solution:*

Figure 8-2(b) shows how to connect the resistors in parallel.

▪ EXAMPLE:

Figure 8-3(a) shows four 1.5 V cells and three lamps. Using wires, connect all of the cells in parallel to create a 1.5 V source. Then connect all of the three lamps in parallel with one another and finally connect the 1.5 V source across the three parallel-connected lamp load.

▪ *Solution:*

In Figure 8-3(b) you can see the final circuit containing a source consisting of four parallel-connected 1.5 V cells, and a load consisting of three parallel-connected lamps. As explained in Chapter 6, when cells are connected in parallel, the total voltage remains the same as for one cell; however, the current demand can now be shared.

8-2 VOLTAGE IN A PARALLEL CIRCUIT

Figure 8-4(a) illustrates a simple circuit with four resistors connected in parallel across the voltage source of a 9 V battery. The current from the negative side of the battery will split between the four different paths or branches, yet the volt-

**FIGURE 8-2   Connecting Resistors in Parallel.
(a) Problem. (b) Solution.**

Four Parallel-Connected
1.5-V Cells = 1.5-V Battery

Three Parallel-
Connected Lamps

(b)

FIGURE 8-3   Parallel-Connected Cells and Lamps.

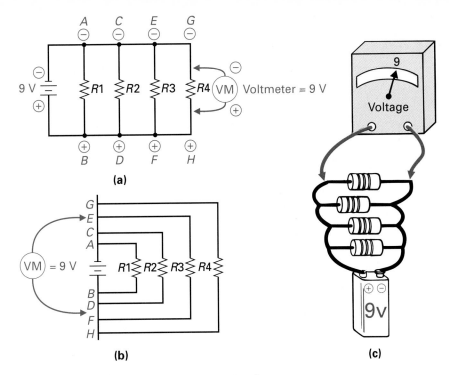

FIGURE 8-4 Voltage in a Parallel Circuit.

age drop across each branch of a parallel circuit is equal to the voltage drop across all the other branches in parallel. This means that if the voltmeter were to measure the voltage across A and B or C and D or E and F or G and H, they would all be the same or, in this example, would all drop 9 V.

It is quite easy to imagine why there will be the same voltage drop across all the resistors, seeing that points A, C, E, and G are all one connection and points B, D, F, and H are all one connection. Measuring the voltage drop with the voltmeter across any of the resistors is the same as measuring the voltage across the battery, as shown in Figure 8-4(b). As long as the voltage source remains constant, the voltage drop will always be common (9 V) across the parallel resistors, no matter what value or how many resistors are connected in parallel. The voltmeter is therefore measuring the voltage between two common points that are directly connected to the battery, and the voltage dropped across all these parallel resistors will be equal to the source voltage.

In Figure 8-5(a) and (b), the same circuit is shown in two different ways, so you can see how the same circuit can look completely different. In both examples, the voltage drop across any of the resistors will always be the same and, as long as the voltage source is not heavily loaded, equal to the source voltage. Just as you can trace the positive side of the battery to all four resistors, you can also trace the negative side to all four resistors.

Mathematically stated, we can say that in a parallel circuit

$$V_{R1} = V_{R2} = V_{R3} = V_{R4} = V_S$$

FIGURE 8-5   Parallel-Circuit Voltage Drop.

Voltage drop across R_1 = voltage drop across R_2 = voltage drop across R_3 (etc.) = source voltage

For the sake of discussion, we will relate this to a water analogy, as seen in Figure 8-6. The pressure across valves A and B will always be the same, even if one offers more opposition than the other. This is because the pressure measured across either valve will be the same as checking the pressure difference between piping X and Y, which run directly back to the pump, so the pressure across A and B is the same as the pressure difference across the pump.

■ **EXAMPLE:**

Refer to Figure 8-7 and calculate:

   a. Voltage drop across R_1

   b. Voltage drop across R_2

   c. Voltage drop across R_3

FIGURE 8-6   Fluid Analogy of Parallel-Circuit Pressure.

FIGURE 8-7   (a) Schematic. (b) Pictorial.

■ *Solution:*

Since all these resistors are connected in parallel, the voltage across every branch will be the same and equal to the source voltage applied. Therefore,

$$V_{R1} = V_{R2} = V_{R3} = V_S$$
$$7.5\text{ V} = 7.5\text{ V} = 7.5\text{ V} = 7.5\text{ V}$$

SELF-TEST REVIEW QUESTIONS FOR SECTIONS 8-1 AND 8-2

1. Describe a parallel circuit.
2. True or false: A parallel circuit is also known as a voltage-divider circuit.
3. What would be the voltage drop across R_1 if $V_S = 12$ V and R_1 and R_2 are both in parallel with one another and equal to 24 Ω each?
4. Can Kirchhoff's voltage law be applied to parallel circuits?

8-3   CURRENT IN A PARALLEL CIRCUIT

In addition to providing the voltage law for series circuits, Gustav Kirchhoff (in 1847) was the first to observe and prove that the sum of all the **branch currents** in a parallel circuit ($I_1 + I_2 + I_3$, etc.) was equal to the total current (I_T). In honor of his second discovery, this phenomenon is known as **Kirchhoff's current law.** It states that the sum of all the current entering a junction is equal to the sum of all the currents leaving that same junction.

Figure 8-8(a) and (b) illustrate two examples of how this law applies. In both examples, the sum of the currents entering a junction is equal to the sum of the currents leaving that same junction. In Figure 8-8(a) the total current arrives at a junction X and splits to produce three branch currents, I_1, I_2, and I_3, which cumulatively equal the total current (I_T) that arrived at the junction X. The same three branch currents combine at junction Y, and the total current (I_T) leaving that junction is equal to the sum of the three branch currents arriving at junction Y.

Branch Current
A portion of the total current that is present is one path of a parallel circuit.

Kirchhoff's Current Law
The sum of the currents flowing into a point in a circuit is equal to the sum of the currents flowing out of that same point.

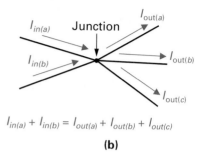

$$I_T = I_1 + I_2 + I_3 \quad \text{or} \quad I_1 + I_2 + I_3 = I_T$$

$$I_{in(a)} + I_{in(b)} = I_{out(a)} + I_{out(b)} + I_{out(c)}$$

(a)　　　　　　　　　　　　　　　　　　　　**(b)**

FIGURE 8-8　Kirchhoff's Current Law.

In Figure 8-8(b) you can see that there are two branch currents entering a junction [$I_{in(a)}$ and $I_{in(b)}$] and three branch currents leaving that same junction [$I_{out(a)}$, $I_{out(b)}$, and $I_{out(c)}$], and the sum of the input currents will equal the sum of the output currents: $I_{in(a)} + I_{in(b)} = I_{out(a)} + I_{out(b)} + I_{out(c)}$.

■ **EXAMPLE:**

Refer to Figure 8-9 and calculate the value of I_1.

■ *Solution:*

By Kirchhoff's current law,

$$I_T = I_1 + I_2$$
$$7\,\text{A} = ? + 3\,\text{A}$$
$$I_1 = 4\,\text{A}$$

■ **EXAMPLE:**

Refer to Figure 8-10 and calculate the value of I_T.

■ *Solution:*

By Kirchhoff's current law,

$$I_T = I_1 + I_2 + I_3 + I_4$$
$$? = 2\,\text{mA} + 17\,\text{mA} + 7\,\text{mA} + 37\,\text{mA}$$
$$I_T = 63\,\text{mA}$$

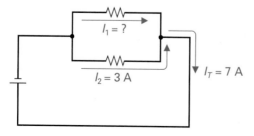

FIGURE 8-9　Calculating Branch Current.

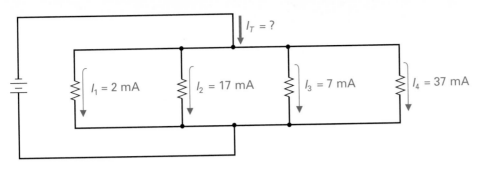

FIGURE 8-10   Calculating Total Current.

As with series circuits, to find out how much current will flow through a parallel circuit, we need to find out how much opposition or resistance is being connected across the voltage source.

$$I_T = \frac{V_S}{R_T}$$

$$\text{total current} = \frac{\text{source voltage}}{\text{total resistance}}$$

When we connect resistors in parallel, the total resistance in the circuit will actually decrease. In fact, the total resistance in a parallel circuit will always be less than the value of the smallest resistor in the circuit.

Let us now examine this point further. If two sets of identical resistors (R_1, R_2, and R_3) were used to build both a series and a parallel circuit, as seen in Figure 8-11, the total current flow in the parallel circuit would be larger than the

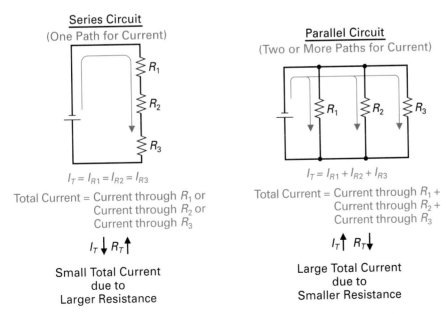

FIGURE 8-11   Series Circuit and Parallel Circuit Current Comparison.

FIGURE 8-12   The Parallel Circuit Current Divider.

total current in the series circuit, because the parallel circuit has two or more paths for current to flow, while the series circuit only has one. To explain why the total current will be larger in a parallel circuit, let us take the analogy of a freeway with only one path for traffic to flow. The freeway is equivalent to a series circuit, and a certain amount of traffic is allowed to flow along this freeway. If the freeway is expanded to accommodate two lanes, a greater amount of traffic can flow along the freeway in the same amount of time; with more lanes, you have a greater total amount of traffic flow and, in parallel circuits, more branches will allow a greater total amount of current flow because there is less resistance in more paths than there is with only one path. This concept is summarized in Figure 8-11.

Current Divider
A parallel network designed to proportionally divide the circuit's total current.

Just as a series circuit is often referred to as a voltage-divider circuit, a parallel circuit is often referred to as a **current-divider** circuit, because the total current arriving at a junction will divide or split into branch currents (Kirchhoff's law), as shown in Figure 8-12.

The current division is inversely proportional to the resistance in the branch seeing that the voltage across both resistors is constant and equal to the source voltage (V_S). This means that a large branch resistance will cause a small branch current ($I\downarrow = V/R\uparrow$), and a small branch resistance will cause a large branch current ($I\uparrow = V/R\downarrow$).

■ **EXAMPLE:**

FIGURE 8-13   **Parallel Circuit Example.**

Calculate the following for Figure 8-13:

   a. I_1

   b. I_2

   c. I_T

■ **Solution:**

Since R_1 and R_2 are connected in parallel across the 10 V source, the voltage across both resistors will be 10 V.

   a. $I_1 = \dfrac{V_{R1}}{R_1} = \dfrac{10\,\text{V}}{6\,\text{k}\Omega} = 1.6\,\text{mA}$ (smaller branch current through larger branch resistance).

b. $I_2 = \dfrac{V_{R2}}{R_2} = \dfrac{10\text{ V}}{3\text{ k}\Omega} = 3.3\text{ mA}$ (larger branch current through smaller branch resistance).

c. By Kirchhoff's current law,

$$I_T = I_1 + I_2$$
$$= 1.6\text{ mA} + 3.3\text{ mA}$$
$$= 4.9\text{ mA}$$

By rearranging Ohm's law, we can arrive at another formula, which is called the *current-divider formula* and can be used to calculate the current through any branch of a multiple-branch parallel circuit.

$$I_x = \dfrac{R_T}{R_x} \times I_T$$

where I_x = branch current desired
    R_T = total resistance
    R_x = resistance in branch
    I_T = total current

EXAMPLE:

Refer to Figure 8-14 and calculate the following if the total circuit resistance (R_T) is equal to 1 kΩ:

  a. $I_1 =$

  b. $I_2 =$

  c. $I_3 =$

FIGURE 8-14 **Parallel Example Circuit.**

Solution:

Since the source and therefore the voltage across each branch resistor are not known, we will use the current-divider formula to calculate I_1, I_2, and I_3.

  a. $I_1 = \dfrac{R_T}{R_1} \times I_T = \dfrac{1\text{ k}\Omega}{2\text{ k}\Omega} \times 10\text{ mA} = 5\text{ mA}$

    (smallest branch resistance has largest branch current)

  b. $I_2 = \dfrac{R_T}{R_2} \times I_T = \dfrac{1\text{ k}\Omega}{3\text{ k}\Omega} \times 10\text{ mA}$

    $= 3.33\text{ mA}$

c. $I_3 = \dfrac{R_T}{R_3} \times I_T = \dfrac{1\ \text{k}\Omega}{6\ \text{k}\Omega} \times 10\ \text{mA}$

$= 1.67\ \text{mA}$

(largest branch resistance has smallest branch current)

To double-check that the values for I_1, I_2, and I_3 are correct, you can apply Kirchhoff's current law, which is

$$I_T = I_1 + I_2 + I_3$$
$$10\ \text{mA} = 5\ \text{mA} + 3.33\ \text{mA} + 1.67\ \text{mA}$$
$$= 10\ \text{mA}$$

Step	Keypad Entry	Display Response
1.	① Ⓔ ③	1.E3
2.	÷	
3.	② Ⓔ ③	2.E3
4.	×	0.5
5.	① ⓪ Ⓔ ③ +/−	10E-3
6.	=	5.-03

FIGURE 8-15 Parallel Home Electrical System.

EXAMPLE:

A common use of parallel circuits is in the residential electrical system. All of the household lights and appliances are wired in parallel, as seen in the typical room wiring circuit in Figure 8-15(a). If it is a cold winter morning, and lamps 1 and 2 are switched ON, together with the space heater and hair dryer, what will the individual branch currents be, and what will be the total current drawn from the source?

Solution:

Figure 8-15(b) shows the schematic of the pictorial in Figure 8-15(a). Since all resistances are connected in parallel across a 120 V source, the voltage across all devices will be 120 V. Using Ohm's law we can calculate the four branch currents:

$$I_1 = \frac{V_{lamp1}}{R_{lamp1}} = \frac{120\ V}{125\ \Omega} = 960\ mA$$

$$I_2 = \frac{V_{lamp2}}{R_{lamp2}} = \frac{120\ V}{125\ \Omega} = 960\ mA$$

$$I_3 = \frac{V_{hairdryer}}{R_{hairdryer}} = \frac{120\ V}{40\ \Omega} = 3\ A$$

$$I_4 = \frac{V_{heater}}{R_{heater}} = \frac{120\ V}{12\ \Omega} = 10\ A$$

By Kirchhoff's current law,

$$I_T = I_1 + I_2 + I_3 + I_4$$
$$= 960\ mA + 960\ mA + 3\ A + 10\ A$$
$$= 14.92\ A$$

SELF-TEST REVIEW QUESTIONS FOR SECTION 8-3

1. State Kirchhoff's current law.
2. If $I_T = 4$ A and $I_1 = 2.7$ A in a two-resistor parallel circuit, what would be the value of I_2?
3. State the current-divider formula.
4. Calculate I_1 if $R_T = 1$ kΩ, $R_1 = 2$ kΩ, and $V_T = 12$ V.

8-4   RESISTANCE IN A PARALLEL CIRCUIT

We now know that parallel circuits will have a larger current flow than a series circuit containing the same resistors due to the smaller total resistance, as already illustrated in Figure 8-11. To calculate exactly how much total current will flow, we need to calculate the total resistance that the parallel circuit develops or contains.

The ability of a circuit to conduct current is a measure of that circuit's conductance, and you will remember from Chapter 2 that conductance (G) is equal to the reciprocal of resistance and is measured in siemens.

$$G = \frac{1}{R} \quad \text{(siemens)}$$

Every resistor in a parallel circuit will have a conductance figure that is equal to the reciprocal of its resistance, and the total conductance (G_T) of the circuit will be equal to the sum of all the individual resistor conductances; therefore,

$$G_T = G_{R1} + G_{R2} + G_{R3} + \cdots$$

(Total conductance is equal to the conductance of R_1 + the conductance of R_2 + the conductance of $R_3 + \cdots$.) Once you have calculated total conductance, the reciprocal of this figure will give you total resistance. If, for example, we have two resistors in parallel, as shown in Figure 8-16, the conductance for R_1 will equal

$$G_{R1} = \frac{1}{R_1} = \frac{1}{20\ \Omega} = 0.05\ \text{S}$$

The conductance for R_2 will equal

$$G_{R2} = \frac{1}{R_2} = \frac{1}{40\ \Omega} = 0.025\ \text{S}$$

The total conductance will therefore equal

$$G_{\text{total}} = G_{R1} + G_{R2}$$
$$= 0.05 + 0.025$$
$$= 0.075\ \text{S}$$

Since total resistance is equal to the reciprocal of total conductance, total resistance for the parallel circuit in Figure 8-16 will be

$$R_{\text{total}} = \frac{1}{G_{\text{total}}} = \frac{1}{0.075\ \text{S}} = 13.3\ \Omega$$

Combining these three steps (first calculate individual conductances, total conductance and then total resistance) we can arrive at the following *reciprocal formula:*

$$R_{\text{total}} = \frac{1}{(1/R_1) + (1/R_2)}$$

this formula states that the conductance of R_1 (G_{R1}) + conductance of R_2 (G_{R2}) = total conductance (G_T), and the reciprocal of total conductance is equal to total resistance

In the example for Figure 8-16, this combined general formula for total resistance can be verified by plugging in the example values.

FIGURE 8-16 Parallel Circuit Conductance and Resistance.

$$R_T = \frac{1}{(1/R_1) + (1/R_2)}$$

$$= \frac{1}{(1/20) + (1/40)}$$

$$= \frac{1}{0.05 + 0.025}$$

$$= \frac{1}{0.075}$$

$$= 13.3 \ \Omega$$

CALCULATOR SEQUENCE

Step	Keypad Entry	Display Response
1.	(Clear Memory)	
2.	2 0	20.
3.	1/x	5.E-2
4.	+	5.E-2
5.	4 0	40
6.	1/x	2.5E-2
7.	=	7.5E-2
8.	1/x	13.33333

The *reciprocal formula* for calculating total parallel circuit resistance for any number of resistors then is

$$R_T = \frac{1}{(1/R_1) + (1/R_2) + (1/R_3) + (1/R_4) + \cdots}$$

■ **EXAMPLE:**

Referring to Figure 8-17, calculate:

a. Total resistance

b. Voltage drop across R_2

c. Voltage drop across R_3

(a) (b)

FIGURE 8-17 (a) Parallel Example Circuit. (b) Equivalent Circuit.

■ Solution:

a. $R_T = \dfrac{1}{(1/R_1) + (1/R_2) + (1/R_3)}$

$= \dfrac{1}{(1/25\ \Omega) + (1/73\ \Omega) + (1/33\ \Omega)}$

$= \dfrac{1}{0.04 + 0.014 + 0.03}$

$= 11.9\ \Omega$

With parallel resistance circuits, the total resistance is always smaller than the smallest branch resistance. In this example the total opposition of this circuit is equivalent to 11.9 Ω, as shown in Figure 8-17(b).

With parallel resistive circuits, the voltage drop across any branch is equal to the voltage drop across each of the other branches and is equal to the source voltage, in this example 3.9 V.

8-4-1 Two Resistors in Parallel

If only two resistors are connected in parallel, a quick and easy formula called the *product-over-sum formula* can be used to calculate total resistance.

$$R_T = \dfrac{R_1 \times R_2}{R_1 + R_2}$$

$$\text{total resistance} = \dfrac{\text{product of both resistance values}}{\text{sum of both resistance values}}$$

Using this formula with the example in Figure 8-18, the total resistance can be calculated by the use of either parallel resistance formulas, (a) or (b), to make a comparison between the two:

(a) $R_T = \dfrac{R_1 \times R_2}{R_1 + R_2}$ (b) $R_T = \dfrac{1}{(1/R_1) + (1/R_2)}$

$= \dfrac{3.7\ \text{k}\Omega \times 2.2\ \text{k}\Omega}{3.7\ \text{k}\Omega + 2.2\ \text{k}\Omega}$ $= \dfrac{1}{(1/3.7\ \text{k}\Omega) + (1/2.2\ \text{k}\Omega)}$

$= \dfrac{8.14\ \text{k}\Omega}{5.9\ \text{k}\Omega}$ $= \dfrac{1}{(270.2 \times 10^{-6}) + (454.5 \times 10^{-6})}$

6.5 V R_1 3.7 kΩ R_2 2.2 kΩ

FIGURE 8-18   **Two Resistors in Parallel.**

$$= 1.38 \text{ k}\Omega$$

$$= \frac{1}{724.7 \times 10^{-6}}$$

$$= 1.38 \text{ k}\Omega$$

8-4-2 More Than Two Resistors in Parallel

You can see that the advantage of the product-over-sum parallel resistance formula (a) is its ease of use. Its disadvantage is that it can only be used for two resistors in parallel. In circuits containing more than two resistors, the previously discussed *reciprocal formula* must be used.

$$R_T = \frac{1}{(1/R_1) + (1/R_2) + (1/R_3) + (1/R_4) + \cdots}$$

8-4-3 Equal-Value Resistors in Parallel

If resistors of equal value are connected in parallel, a special case *equal-value formula* can be used to calculate the total resistance.

$$R_T = \frac{\text{value of one resistor } (R)}{\text{number of parallel resistors } (n)}$$

EXAMPLE:

Figure 8-19(a) shows how a stereo music amplifier is connected to drive two 8 Ω speakers, which are connected in parallel with one another. What is the total resistance connected across the amplifier's output terminals?

Solution:

Since both parallel-connected speakers have the same resistance [Figure 8-19(b)], the total resistance is most easily calculated by using the equal-value formula:

$$R_T = \frac{R}{n} = \frac{8 \, \Omega}{2} = 4 \, \Omega$$

EXAMPLE:

Refer to Figure 8-20 and calculate:

a. Total resistance in part (a)
b. Total resistance in part (b)

(a)

(b)

FIGURE 8-19 **Parallel-Connected Speakers.**

■ *Solution:*

a. Figure 8-20(a) has only two resistors in parallel and therefore the two parallel resistor formula can be used; the equivalent circuit is seen in Figure 8-20(c)

$$R_T = \frac{R_1 \times R_2}{R_1 + R_2}$$

$$= \frac{4.5\ \text{M}\Omega \times 3.2\ \text{M}\Omega}{4.5\ \text{M}\Omega + 3.2\ \text{M}\Omega}$$

$$= \frac{14.4\ \text{M}\Omega}{7.7\ \text{M}\Omega}$$

$$= 1.9\ \text{M}\Omega$$

b. Figure 8-20(b) has more than two resistors in parallel, and therefore the sum of conductances reciprocated formula must be used. The equivalent circuit can be seen in Figure 8-20(d).

$$R_T = \frac{1}{(1/R_1) + (1/R_2) + (1/R_3)}$$

$$= \frac{1}{(1/27\ \text{k}\Omega) + (1/10\ \text{k}\Omega) + (1/3.3\ \text{k}\Omega)}$$

$$= \frac{1}{440.0 \times 10^{-6}}$$

$$= 2.27\ \text{k}\Omega$$

CALCULATOR SEQUENCE FOR (A)

Step	Keypad Entry	Display Response
1.	4 . 5 E 6	4.5E6
2.	×	
3.	3 . 2 E 6	3.2E6
4.	=	1.44E13
5.	STO (store in memory)	
6.	4 . 5 E 6	4.5E6
7.	+	
8.	3 . 2 E 6	3.2E6
9.	=	7.7E6
10.	C/CE	0.
11.	RM (Recall memory)	1.44E13
12.	÷	
13.	7 . 7 E 6	7.7E6
14.	=	1.87E6

FIGURE 8-20 **Parallel Circuit Examples.**

■ **EXAMPLE:**

Find the total resistance of the parallel circuit in Figure 8-21.

■ *Solution:*

Since all four resistors are connected in parallel and are all of the same value, the equal-value resistors in parallel formula can be used.

$$R_T = \frac{R}{n} = \frac{2 \text{ k}\Omega}{4} = 500 \ \Omega$$

To summarize the effects of current, voltage, and resistance in parallel circuits, so far we could say:

1. Components are said to be connected in parallel when the current has to travel two or more paths between the negative and positive side of the voltage source.

2. The voltage across all the components in parallel is always the same.

3. The total current from the source is equal to the sum of all the branch currents (Kirchhoff's current law).

4. The amount of current flowing through each branch is inversely proportional to the resistance value in that branch.

5. The total resistance of a parallel circuit is always less than the value of the smallest branch resistor.

FIGURE 8-21 **Parallel Circuit Example.**

8-5 POWER IN A PARALLEL CIRCUIT

As with series circuits, the total power in a parallel resistive circuit is equal to the sum of all the power losses for each of the resistors in parallel.

$$P_T = P_1 + P_2 + P_3 + P_4 + \cdots$$

total power = addition of all the power losses

The formulas for calculating the amount of power dissipated are

$$P = I \times V$$
$$P = \frac{V^2}{R}$$
$$P = I^2 \times R$$

■ **EXAMPLE:**

Calculate the total amount of power dissipated in Figure 8-22.

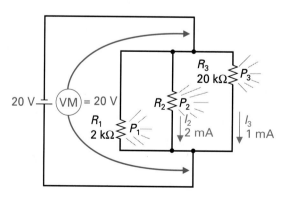

FIGURE 8-22 **Parallel Circuit Example.**

Solution:

The total power dissipated in a parallel circuit is equal to the sum of all the power dissipated by all the resistors. With P_{R1}, we only know voltage and resistance and therefore we can use the formula

$$P_{R1} = \frac{V^2}{R} = \frac{20^2}{2\ k\Omega} = 0.2\ W \quad \text{or} \quad 200\ mW$$

With P_{R2}, we only know current and voltage, and therefore we can use the formula

$$P_{R2} = I \times V = 2\ mA \times 20\ V = 40\ mW$$

With P_{R3}, we know V, I, and R; however, we will use the third power formula:

$$P_{R3} = I^2 \times R = 1\ mA^2 \times 20\ k\Omega = 20\ mW$$

Total power (P_T) equals the sum of all the power or wattage losses for each resistor:

$$P_T = P_{R1} + P_{R2} + P_{R3}$$
$$= 200\ mW + 40\ mW + 20\ mW$$
$$= 260\ mW$$

Step	Keypad Entry	Display Response
1.	2 0	20.
2.	x^2	400.
3.	÷	
4.	2 E 3	2.E3
5.	=	0.2

Step	Keypad Entry	Display Response
1.	2 E 3 +/−	2.E-3
2.	×	
3.	2 0	20.
4.	=	40E-3

Step	Keypad Entry	Display Response
1.	1 E 3 +/− x^2	1.E-6
2.	×	1.E-6
3.	2 0 E 3	20E3
4.	=	20.E-3

EXAMPLE:

In Figure 8-23, there are two ½ W (0.5 W) resistors connected in parallel. Should the wattage rating for each of these resistors be increased or decreased, or can they remain the same?

Solution:

Since current and voltage are known for both branches, the power formula used in both cases will be $P = I \times V$.

FIGURE 8-23 Parallel Circuit Example.

$$P_1 = I \times V \qquad\qquad P_2 = I \times V$$
$$= 20\,\text{mA} \times 5\,\text{V} \qquad = 200\,\text{mA} \times 5\,\text{V}$$
$$= 0.1\,\text{W} \qquad\qquad\quad = 1\,\text{W}$$

R_1 is dissipating 0.1 W and has been designed to dissipate up to 0.5 W; it is therefore safe in this application. However, R_2 is dissipating 1 W and is only designed to dissipate 0.5 W. R_2 will consequently overheat unless it is replaced with a resistor of the same ohmic value with a 1 W or greater rating.

■ **EXAMPLE:**

Figure 8-24(a) shows a simplified diagram of an automobile external light system. A 12 V lead–acid battery is used as a source and is connected across eight parallel-connected lamps. The left and right brake lights are controlled by the brake switch, which is attached to the brake pedal. When the light switch is turned on, both the rear taillights and the low-beam headlights are brought into circuit and turned on. The high-beam set of headlights are activated only if the

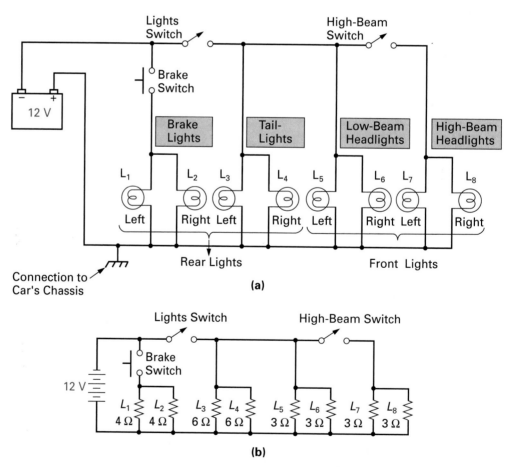

FIGURE 8-24 Parallel Automobile External Light System.

high-beam switch is closed. For the lamp resistances given, calculate the output power of each lamp, when in use.

■ *Solution:*

Figure 8-24(b) shows the schematic diagram of the pictorial in Figure 8-24(a). Since both V and R are known, we can use the V^2/R power formula.

Brake lights: Left lamp wattage $(P) = \dfrac{V^2}{R} = \dfrac{12\,V^2}{4\Omega} = 36\ W$

Right lamp wattage is the same as left.

Taillights: Left lamp wattage $(P) = \dfrac{V^2}{R} = \dfrac{12\,V^2}{6\Omega} = 24\ W$

Right lamp wattage is the same as left.

Low-beam headlights: $P = \dfrac{V^2}{R} = \dfrac{12\,V^2}{3\Omega} = 48\ W$

Each low-beam headlight is a 48-W lamp.

High-beam headlights: $P = \dfrac{V^2}{R} = \dfrac{12\,V^2}{3\Omega} = 48\ W$

Each high-beam headlight is a 48 W lamp.

SELF-TEST REVIEW QUESTIONS FOR SECTION 8-5

1. True or false: Total power in a parallel circuit can be obtained by using the same total power formula as for series circuits.
2. If $I_1 = 2$ mA and $V = 24$ V, calculate P_1.
3. If $P_1 = 22$ mW and $P_2 = 6400\ \mu W$, $P_T = ?$
4. Is it important to observe the correct wattage ratings of resistors when they are connected in parallel?

8-6 TROUBLESHOOTING A PARALLEL CIRCUIT

In the troubleshooting discussion on series circuits, we mentioned that one of three problems can occur, and these three also apply to parallel circuits:

1. A component will open.
2. A component will short.
3. A component's value will change over a period of time.

As a technician, it is important that you know how to recognize these circuit malfunctions and know how to isolate the cause and make the repair. Since you could be encountering opens, shorts, and component value changes in the parallel circuits you construct in lab, let us step through the troubleshooting procedure you should use.

Figure 8-25(a) indicates the normal readings that should be obtained from a typical parallel circuit if it is operating correctly. Keep these normal readings in mind, since they will change as circuit problems are introduced in the following section.

8-6-1   Open Component in a Parallel Circuit

In Figure 8-25(b), R_1 has opened, so there can be no current flow through the R_1 branch. An open is equivalent to the maximum possible resistance (infinite ohms), and therefore the total resistance of this parallel circuit will increase, causing the total current flow to decrease. The total current flow, which was 7 mA, will actually decrease by the amount that was flowing through the now open branch (1 mA) to a total current flow of 6 mA. If you had constructed this circuit in lab, you would probably not have noticed that there was anything wrong with the circuit until you started to make some multimeter measurements. A voltmeter measurement across the parallel circuit, as seen in Figure 8-25(b), would not have revealed any problem since the voltage measure across each resistor will always be the same and equal to the source voltage. A total current measurement, on the other hand, would indicate that there is something wrong since the measured reading does not equal the expected calculated value. Using your ammeter, you could isolate the branch fault by one of the following two methods:

1. If you measure each branch current, you will isolate the problem of R_1, because there will be no current flow through R_1. However, this could take three checks.

2. If you just measure total current (one check), you will notice that total current has decreased by 1 mA, and after making a few calculations you realize that the only path that had a branch current of 1 mA was through R_1, so R_1 must have gone open.

If R_1 opens, total current decreases by 1 mA to 6 mA ($I_T = I_1 + I_2 + I_3 =$ 0 mA + 4 mA + 2 mA = 6 mA).

If R_2 opens, total current decreases by 4 mA to 3 mA ($I_T = I_1 + I_2 + I_3 =$ 1 mA + 0 mA + 2 mA = 3 mA).

If R_3 opens, total current decreases by 2 mA to 5 mA ($I_T = I_1 + I_2 + I_3 =$ 1 mA + 4 mA + 0 mA = 5 mA).

Method 2 can be used as long as the resistors are unequal. If they are equal, method 1 may be used to locate the 0 current branch.

In most cases, the ohmmeter is ideal for locating open circuits. To make a resistance measurement, you should disconnect power from the circuit, isolate the device from the circuit, and then connect the ohmmeter across the device. Figure 8-25(c) shows why it is necessary for you to isolate the device to be tested from the circuit. With R_1 in circuit, the ohmmeter will be measuring the parallel resistance of R_2 and R_3's, even though the ohmmeter is connected across R_1. This will lead to a false reading and confuse the troubleshooting process. Disconnecting the device to be tested from the circuit, as shown in Figure 8-25(d), will give an accurate reading of R_1, R_2, and R_3's resistance and lead you to the cause of the problem.

$$I_{R1} = \frac{V_{R1}}{R_1} = \frac{12\ V}{12\ k\Omega} = 1\ mA$$

$$I_{R2} = \frac{V_{R2}}{R_2} = \frac{12\ V}{3\ k\Omega} = 4\ mA$$

$$I_{R3} = \frac{V_{R3}}{R_3} = \frac{12\ V}{6\ k\Omega} = 2\ mA$$

$$I_T = I_{R1} + I_{R2} + I_{R3}$$
$$= 1\ mA + 4\ mA + 2\ mA$$
$$= 7\ mA$$

(a)

(b)

With Resistor in Circuit, Branch Currents Will Give Incorrect Reading

Over-range (Or) Display Indicates Infinite Ohms (Open Circuit)

Disconnect Each Resister from Circuit and Check Their Individual Resistance Values

(c)

(d)

FIGURE 8-25   Troubleshooting an Open in a Parallel Circuit.

(a)   **(b)**

FIGURE 8-26   **Open Bulb in a Parallel Circuit. (a) Schematic. (b) Pictorial.**

■ **EXAMPLE:**

Figure 8-26 illustrates a set of three light bulbs connected in parallel across a 9 V battery. The filament in bulb 2 has burned out, causing an open, so there is no current flow through that path. However, there will be current flow through the other two, so bulbs 1 and 3 will be on. How would the faulty bulb be located?

■ *Solution:*

As always, a visual inspection would be your first approach. Since current is flowing through both B_1 and B_3, these two bulbs will be on. On the other hand, since B_2 is off, it would be easy for you to determine that bulb B_2 has blown.

Opens in parallel paths are generally quite easy to isolate since the device in the parallel path will not operate because its power source has been disconnected.

8-6-2 Shorted Component in a Parallel Circuit

In Figure 8-27(a), R_2, which has a resistance of 3 kΩ, has been shorted out. The only resistance in this center branch is the resistance of the wire, which would typically be a fraction of an ohm. In this example, let us assume that the resistance in the branch is 1 Ω, and therefore the current in this branch will attempt to increase to:

$$I_2 = \frac{12\text{ V}}{1\text{ }\Omega} = 12\text{ A}$$

Before the current even gets close to 12 A, the fuse in the dc power supply will blow, as shown in the inset in Figure 8-27(a), and disconnect power from the cir-

FIGURE 8-27 **Troubleshooting a Short in a Parallel Circuit.**

cuit. This is typically the effect you will see from a circuit short. Replacing the fuse will be a waste of time and money since the fuse will simply blow again once power is connected across the circuit. As before, the tool to use to isolate the cause is the ohmmeter, as shown in Figure 8-27(b). To check the resistance of each branch, disconnect power from the circuit, disconnect the devices to be tested from the circuit to prevent false readings, and then test each branch until the shorted branch is located. Once you have corrected the fault, you will need to change the power supply fuse or reset its circuit breaker.

8-6-3 *Component Value Variation in a Parallel Circuit*

The resistive values of resistors will change with age. This increase or decrease in resistance will cause a corresponding decrease or increase in branch current and therefore total current. This deviation from the desired value can also be checked with the ohmmeter. Be careful to take into account the resistor's tolerance because this can make you believe that the resistor's value has changed, when it is, in fact, within the tolerance rating.

8-6-4 *Summary of Parallel Circuit Troubleshooting*

1. An open component will cause no current flow within that branch. The total current will decrease, and the voltage across the component will be the same as the source voltage.
2. A shorted component will cause maximum current through that branch. The total current will increase and the voltage source fuse will normally blow.
3. A change in resistance value will cause a corresponding opposite change in branch current and total current.

SELF-TEST REVIEW QUESTIONS FOR SECTION 8-6

How would problems 1 through 3 be recognized in a parallel circuit?

1. An open component
2. A shorted component
3. A component's value variation
4. True or false: An ammeter is typically used to troubleshoot series circuits, whereas a voltmeter is typically used to troubleshoot parallel circuits.

SUMMARY   **Parallel Circuits (Figure 8-28)**

1. Series or parallel circuits can be determined by current.
2. In a series circuit, there is only one path for current, whereas in parallel circuits the current has two or more paths. These paths are known as *branches*.
3. A parallel circuit, by definition, is when two or more components are connected to the same voltage source so that the current can branch out over two or more paths.

FIGURE 8-28   Parallel Circuits.

Example Circuit (Resistors)

Example Circuit (Batteries)

$$V_T = V_1 = V_2 = V_3$$

$$I_T \uparrow$$

$$V_S = V_{R1} = V_{R2} = V_{R3} = \cdots$$

$$R_T = \frac{V_S}{I_T}$$

$$I_T = I_1 + I_2 + I_3 + \cdots$$

$$I_x = \frac{R_T}{R_x} \times I_T$$

$$R_T = \frac{1}{(1/R_1) + (1/R_2) + (1/R_3) + \cdots}$$

$$R_T = \frac{R_1 \times R_2}{R_1 + R_2}$$

$$R_T = \frac{R}{n}$$

$$P_T = P_1 + P_2 + P_3 + \cdots$$

$$P_T = \frac{V_S^2}{R_T} = I_T^2 \times R_T = V_S \times I_T$$

V_S = source voltage, in volts
V_{R1} = voltage across R_1, in volts
V_{R2} = voltage across R_2, in volts
V_{R3} = voltage across R_3, in volts
I_T = total current, in amperes
R_T = total resistance, in ohms
I_1 = current through R_1, in amperes
I_2 = current through R_2, in amperes
I_3 = current through R_3, in amperes
I_x = branch current desired, in amperes
R_x = resistance in branch
R_1 = resistance of R_1, in ohms
R_2 = resistance of R_2, in ohms
R_3 = resistance of R_3, in ohms
R_{EV} = resistance of one of the equivalent-value resistors
n = number of parallel resistors
P_T = total power dissipated, in watts
P_1 = power dissipated by R_1, in watts
P_2 = power dissipated by R_2, in watts
P_3 = power dissipated by R_3, in watts

4. In a parallel resistive circuit, two or more resistors are connected to the same source, so current splits to travel through each separate branch resistance.

5. Many components, other than resistors, can be connected in parallel, and a parallel circuit can easily be identified because current is split into two or more paths.

6. Measuring the voltage drop with the voltmeter across any of the resistors in a parallel circuit is the same as measuring the voltage across the voltage source.

7. Gustav Kirchhoff was the first to observe and prove that the sum of all the branch currents ($I_1 + I_2 + I_3$, etc.) was equal to the total current (I_T).

8. In honor of his second discovery, this phenomenon is known as Kirchhoff's current law. It states that the sum of all the current entering a junction is equal to the sum of all the currents leaving that same junction.

9. When we connect resistors in parallel, the total resistance in the circuit will decrease. In fact, the total resistance in a parallel circuit will always be less than the value of the smallest resistor in the circuit.

10. Just as a series circuit is often referred to as a voltage-divider circuit, a parallel circuit is often referred to as a current-divider circuit, because the total current arriving at a junction will divide or split into branch currents (Kirchhoff's law).

11. The current division is inversely proportional to the resistance in the branch seeing that the voltage across both resistors is constant and equal to the source voltage (V_S). This means that a large branch resistance will cause a small branch current ($I\downarrow = V/R\uparrow$), and a small branch resistance will cause a large branch current ($I\uparrow = V/R\downarrow$).

12. By rearranging Ohm's law, we can arrive at another formula, which is called the *current-divider formula* and can be used to calculate the current through any branch of a multiple-branch parallel circuit.

13. Parallel circuits will have a larger current flow than a series circuit containing the same resistors due to the smaller total resistance.

14. To calculate exactly how much total current will flow, we need to calculate the total resistance that the parallel circuit develops or contains.

15. The reciprocal formula allows us to calculate total parallel circuit resistance for any number of resistors.

16. With parallel resistance circuits, the total resistance is always smaller than the smallest branch resistance.

17. With parallel resistive circuits, the voltage drop across any branch is equal to the voltage drop across each of the other branches and is equal to the source voltage.

18. If only two resistors are connected in parallel, a quick and easy formula called the *product-over-sum formula* can be used to calculate total resistance.

19. If resistors of equal value are connected in parallel, a special case *equal-value formula* can be used to calculate the total resistance.

20. As with series circuits, the total power in a parallel resistive circuit is equal to the sum of all the power losses for each of the resistors in parallel.

21. An open component will cause no current flow within that branch. The total current will decrease, and the voltage across the component will be the same as the source voltage.

22. A shorted component will cause maximum current through that branch. The total current will increase and the voltage source fuse will normally blow.

23. A change in resistance value will cause a corresponding opposite change in branch current and total current.

			NEW TERMS
Branch current	Fixed current divider	Parallel circuit	
Current divider	Kirchhoff's current law		

Multiple-Choice Questions

REVIEW QUESTIONS

1. A parallel circuit has _____ path(s) for current to flow.

   a. One   **c.** Only three
   b. Two or more   **d.** None of the above

2. If a source voltage of 12 V is applied across four resistors of equal value in parallel, the voltage drops across each resistor would be equal to:

   a. 12 V   **c.** 4 V
   b. 3 V   **d.** 48 V

3. What would be the voltage drop across two 25 Ω resistors in parallel if the source voltage were equal to 9 V?

   a. 50 V   **c.** 12 V
   b. 25 V   **d.** None of the above

4. If a four-branch parallel circuit has 15 mA flowing through each branch, the total current into the parallel circuit will be equal to:

   a. 15 mA   **c.** 30 mA
   b. 60 mA   **d.** 45 mA

5. If the total three-branch parallel circuit current is equal to 500 mA, and 207 mA is flowing through one branch and 153 mA through another, what would be the current flow through the third branch?

   a. 707 mA   **c.** 140 mA
   b. 653 mA   **d.** None of the above

6. A large branch resistance will cause a _____ branch current.

   a. Large   **c.** Medium
   b. Small   **d.** None of the above are true

7. What would be the conductance of a 1 kΩ resistor?

   a. 10 mS   **c.** 2 kΩ
   b. 1 mS   **d.** All of the above

8. If only two resistors are connected in parallel, the total resistance equals:
   a. The sum of the resistance values
   b. Three times the value of one resistor
   c. The product over the sum
   d. All of the above

9. If resistors of equal value are connected in parallel, the total resistance can be calculated by:
   a. One resistor value divided by the number of parallel resistors
   b. The sum of the resistor values
   c. The number of parallel resistors divided by one resistor value
   d. All of the above could be true.

10. The total power in a parallel circuit is equal to the:
    a. Product of total current and total voltage
    b. Reciprocal of the individual power losses
    c. Sum of the individual power losses
    d. Both (a) and (b)
    e. Both (a) and (c)

Essay Questions

11. Describe the difference between a series and a parallel circuit. (8-1)
12. Explain and state mathematically the situation regarding voltage in a parallel circuit. (8-2)
13. State Kirchhoff's current law for parallel circuits. (8-3)
14. What is the current-divider formula? (8-3)
15. List the formulas for calculating the following total resistances: (8-4)
    a. Two resistors of different values
    b. More than two resistors of different values
    c. Equal-value resistors
16. Describe the relationship between branch current and branch resistance. (8-3)
17. Briefly describe total and individual power in a parallel resistive circuit. (8-5)
18. Discuss troubleshooting parallel circuits as applied to: (8-6)
    a. A shorted component   b. An open component
19. Explain why parallel circuits have a smaller total resistance and larger total current than series circuits. (8-3)
20. Briefly describe why Kirchhoff's voltage law applies to series circuits and why Kirchhoff's current law relates to parallel circuits. (8-3)

Practice Problems

21. Calculate the total resistance of four 30 kΩ resistors in parallel.
22. Find the total resistance for each of the following parallel circuits:
    a. 330 Ω and 560 Ω   c. 2.2 MΩ, 3 kΩ, and 220 Ω
    b. 47 kΩ, 33 kΩ, and 22 kΩ

23. If 10 V is connected across three 25 Ω resistors in parallel, what will be the total and individual branch currents?

24. If a four-branch parallel circuit has branch currents equal to 25 mA, 37 mA, 220 mA, and 0.2 A, what is the total circuit current?

25. If three resistors of equal value are connected across a 14 V supply and the total resistance is equal to 700 Ω, what is the value of each branch current?

26. If three 75 W light bulbs are connected in parallel across a 110 V supply, what is the value of each branch current? What is the branch current through the other two light bulbs if one burns out?

27. If 33 kΩ and 22 kΩ resistors are connected in parallel across a 20 V source, calculate:

    a. Total resistance   **d.** Total power dissipated
    b. Total current   **e.** Individual resistor power
    c. Branch currents   dissipated

28. If four parallel-connected resistors are each dissipating 75 mW, what is the total power being dissipated?

29. Calculate the branch currents through the following parallel resistor circuits when they are connected across a 10 V supply:

    a. 22 kΩ and 33 kΩ   **b.** 220 Ω, 330 Ω, and 470 Ω

30. If 30 Ω and 40 Ω resistors are connected in parallel, which resistor will generate the greatest amount of heat?

31. Calculate the total conductance and resistance of the following parallel circuits:

    a. Three 5 Ω resistors   **c.** 1 MΩ, 500 MΩ, 3.3 MΩ
    b. Two 200 Ω resistors **d.** 5 Ω, 3 Ω, 2 Ω

32. Connect the three resistors in Figure 8-29 in parallel across a 12 V battery and then calculate the following:

    a. V_{R1}, V_{R2}, V_{R3} **d.** P_T
    b. I_1, I_2, I_3 **e.** P_1, P_2, P_3
    c. I_T **f.** G_{R1}, G_{R2}, G_{R3}

33. Calculate R_T in Figure 8-30 (a), (b), (c), and (d).

34. Calculate the branch currents through four 60 W bulbs connected in parallel across 110 V. How much is the total current, and what would happen to the total current if one of the bulbs were to burn out? What change would occur in the remaining branch currents?

FIGURE 8-29   **Connect in Parallel Across a 12-volt Source.**

FIGURE 8-30   Calculate Total Resistance.

35. Calculate the following in Figure 8-31:

    a. I_2 **c.** V_S, I_1, I_2
    b. I_T **d.** R_2, I_1, I_2, P_T

Troubleshooting Questions

36. An open component in a parallel circuit will cause _____ current flow within that branch, which will cause the total current to

    _____.

    a. Maximum, increase   **c.** Maximum, decrease
    b. Zero, decrease   **d.** Zero, increase

37. A shorted component in a parallel circuit will cause _____ current through a branch, and consequently the total current will _____.

    a. Maximum, increase   **c.** Maximum, decrease
    b. Zero, decrease   **d.** Zero, increase

38. If a 10 kΩ and two 20 kΩ resistors are connected in parallel across a 20 V supply, and the total current measured is 2 mA, determine whether a problem exists in the circuit and, if it does, isolate the problem.

39. What situation would occur and how would we recognize the problem if one of the 20 kΩ resistors in Question 38 were to short?

40. With age, the resistance of a resistor will _____, resulting in a corresponding but opposite change in _____.

    a. Increase, branch current   **c.** Decrease, source resistance
    b. Change, source voltage   **d.** Change, branch current

FIGURE 8-31 Calculate the Unknown.

After completing this chapter, you will be able to:

1. Identify the difference between a series, a parallel, and a series–parallel circuit.

2. Describe how to use a three-step procedure to determine total resistance.

3. Describe for the series–parallel circuit how to use a five-step procedure to calculate:
   a. Total resistance.
   b. Total current.
   c. Voltage division.
   d. Branch current.
   e. Total power dissipated.

4. Explain what loading effect a piece of equipment will have when connected to a voltage divider.

5. Identify and describe the Wheatstone bridge circuit in both the balanced and unbalanced condition.

6. Describe the R–$2R$ ladder circuit used for digital-to-analog conversion.

7. Explain how to identify the following problems in a series–parallel circuit:
   a. Open series resistor.
   b. Open parallel resistor.
   c. Shorted series resistor.
   d. Shorted parallel resistor.
   e. Resistor value variation.

8. Describe the differences between a voltage and a current source.

9. Analyze series–parallel networks using:
   a. The superposition theorem.
   b. Thévenin's theorem.
   c. Norton's theorem.

Series–Parallel DC Circuits

The Christie Bridge Circuit

Who invented the Wheatstone bridge circuit? It was obviously Sir Charles Wheatstone. Or was it?

The Wheatstone bridge was actually invented by S.H. Christie of the Royal Military Academy at Woolwich, England. He described the circuit in detail in the *Philosophical Transactions* paper dated February 28, 1833. Christie's name, however, was unknown and his invention was ignored.

Ten years later, Sir Charles Wheatstone called attention to Christie's circuit. Sir Charles was very well known, and from that point on, and even to this day, the circuit is known as a Wheatstone bridge. Later, Werner Siemens would modify Christie's circuit and invent the variable-resistance arm bridge circuit, which would also be called a Wheatstone bridge.

No one has given full credit to the real inventors of these bridge circuits, until now!

The Christie Bridge

The Siemens Bridge

Very rarely are we lucky enough to run across straightforward series or parallel circuits. In general, all electronic equipment is composed of many components that are interconnected to form a combination of series and parallel circuits. In this chapter, we will be combining our knowledge of the series and parallel circuits discussed in the previous two chapters.

9-1 SERIES- AND PARALLEL-CONNECTED COMPONENTS

Figure 9-1(a) through (f) illustrates six different examples of **series–parallel resistive circuits.** The most important point to learn is how to distinguish between the resistors that are connected in series and the resistors that are connected in parallel, which will take a little practice.

One thing that you may not have noticed when examining Figure 9-1 is that:

Series–Parallel Circuit
Network or circuit that contains components that are connected in both series and parallel.

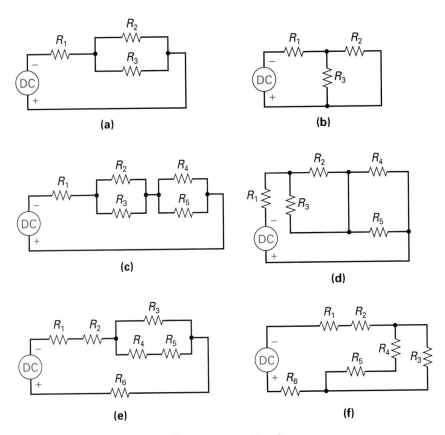

FIGURE 9-1   **Series–Parallel Resistive Circuits.**

Circuit 9-1(a) is equivalent to 9-1(b)

Circuit 9-1(c) is equivalent to 9-1(d)

Circuit 9-1(e) is equivalent to 9-1(f)

When analyzing these series–parallel circuits, always remember that current flow determines whether the resistor is connected in series or parallel. Begin at the negative side of the battery and apply these two rules:

1. If the total current has only one path to follow through a component, that component is connected in series.

2. If the total current has two or more paths to follow through two or more components, those components are connected in parallel.

Referring again to Figure 9-1, you can see that series or parallel resistor networks are easier to identify in parts (a), (c), and (e) than in parts (b), (d), and (f). Redrawing the circuit so that the components are arranged from left to right or from top to bottom is your first line of attack in your quest to identify series- and parallel-connected components.

▨ **EXAMPLE:**

Refer to Figure 9-2 and identify which resistors are connected in series and which are in parallel.

▨ *Solution:*

First, let's redraw the circuit so that the components are aligned either from left to right [Figure 9-3(a)] or from top to bottom [Figure 9-3(b)]. Placing your pencil at the negative terminal of the battery on whichever figure you prefer, either Figure 9-3(a) or (b), trace the current paths through the circuit toward the positive side of the battery, as illustrated in Figure 9-4.

The total current arrives first at R_1. There is only one path for current to flow, which is through R_1, and therefore R_1 is connected in series. The total current proceeds on past R_1 and arrives at a junction where current divides and travels through two branches, R_2 and R_3. Since current had to split into two paths, R_2 and

FIGURE 9-2
Series–Parallel Circuit Example.

(a)

(b)

FIGURE 9-3   Redrawn Series–Parallel Circuit. (a) Left to Right. (b) Top to Bottom.

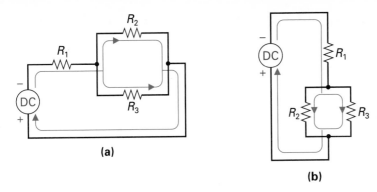

(a)

(b)

**FIGURE 9-4 Tracing Current Through a Series–Parallel
Circuit.**

R_3 are therefore connected in parallel. After the parallel connection of R_2 and R_3,
total current combines and travels to the positive side of the battery.

 To summarize, R_1 is in series with the parallel combination of R_2 and R_3.

EXAMPLE:

Refer to Figure 9-5 and identify which resistors are connected in series and
which are connected in parallel.

Solution:

Figure 9-6 illustrates the simplified, redrawn schematic of Figure 9-5. Total current leaves the negative terminal of the battery, and all of this current has to
travel through R_1, which is therefore a series-connected resistor. Total current
will split at junction A, and consequently, R_3 and R_4 with R_2 make up a parallel

**FIGURE 9-5   Series–Parallel
Circuit Example.**

**FIGURE 9-6
Redrawn Series–
Parallel Circuit
Example.**

combination. The current that flows through R_3 (I_2) will also flow through R_4 and have only one path to follow; therefore, R_3 is in series with R_4. I_1 and I_2 branch currents combine at junction B to produce total current, which has only one path to follow through the series resistor R_5, and finally, to the positive side of the battery.

To summarize, R_3 and R_4 are in series with one another and both are in parallel with R_2, and this combination is in series with R_1 and R_5.

9-2 TOTAL RESISTANCE IN A SERIES–PARALLEL CIRCUIT

No matter how complex or involved the series–parallel circuit, there is a simple three-step method to simplify the circuit to a single equivalent total resistance. Figure 9-7 illustrates an example of a series–parallel circuit. Once you have analyzed and determined the series–parallel relationship, we can proceed to solve for total resistance.

The three-step method is:

Step A: Determine the equivalent resistances of all branch series-connected resistors.

Step B: Determine the equivalent resistances of all parallel-connected combinations.

Step C: Determine the equivalent resistance of the remaining series-connected resistances.

Let's now put our theory to work with the example circuit in Figure 9-7.

STEP A Solve for all branch series-connected resistors. In our examples, this applies only to R_3 and R_4, and since this is a series connection, we have to use the series resistance formula.

$$R_{3,4} = R_3 + R_4 = 8 + 2 = 10 \ \Omega \qquad \text{(series resistance formula)}$$

With R_3 and R_4 solved, the circuit appears as indicated in Figure 9-8.

STEP B Solve for all parallel combinations. In this example, they are the two parallel combinations of (a) R_2 and $R_{3,4}$ and (b) R_5 and R_6 and R_7. Since these are parallel connections, use the parallel resistance formulas.

FIGURE 9-7 **Total Series–Parallel Circuit Resistance.**

FIGURE 9-8   After Completing Step A.

$$R_{2,3,4} = \frac{R_2 \times R_{3,4}}{R_2 + R_{3,4}} = \frac{12 \times 10}{12 + 10} = 5.5\ \Omega \qquad \text{(product-over-sum formula)}$$

$$R_{5,6,7} = \frac{1}{(1/R_5) + (1/R_6) + (1/R_7)} = 5.8\ \Omega \qquad \text{(reciprocal formula)}$$

With $R_{2,3,4}$ and $R_{5,6,7}$ solved, the circuit now appears as illustrated in Figure 9-9.

STEP C   Solve for the remaining series resistances. There are now four remaining series resistances, which can be reduced to one equivalent resistance (R_{eq}) or total resistance (R_T), as seen in Figure 9-10. By using the series resistance formula,

$$\begin{aligned} R_{eq} &= R_1 + R_{2,3,4} + R_{5,6,7} + R_8 \\ &= 4\ \Omega + 5.5\ \Omega + 5.8\ \Omega + 24\ \Omega \\ &= 39.3\ \Omega \end{aligned}$$

■ **EXAMPLE:**

Find the total resistance of the circuit in Figure 9-11.

■ *Solution:*

STEP A   Solve for all branch series-connected resistors. This applies to R_2 and R_3 (series connection):

$$R_{2,3} = R_2 + R_3 = 1\ k\Omega + 3\ k\Omega = 4\ k\Omega$$

The resulting circuit, after completing step A, is illustrated in Figure 9-12(a).

FIGURE 9-9   After Completing Step B.

FIGURE 9-10
After Completing
Step C.

FIGURE 9-11   Calculate Total Resistance.

STEP B   Solve for all parallel combinations. Observing the circuit resulting from step A [Figure 9-12(a)], you can see that current branches into three paths, so the parallel reciprocal formula can be used:

$$R_{2,3,4,5} = \frac{1}{(1/R_{2,3}) + (1/R_4) + (1/R_5)}$$

$$= \frac{1}{(1/4 \text{ k}\Omega) + (1/4 \text{ k}\Omega) + (1/8 \text{ k}\Omega)} = 1.6 \text{ k}\Omega$$

The resulting circuit, after completing step B, is illustrated in Figure 9-12(b).

STEP C   Solve for the remaining series resistances. Observing the circuit resulting from step B [Figure 9-12(b)], you can see that there are two remaining series resistances. The equivalent resistance (R_{eq}) is equal to

$$R_{eq} = R_1 + R_{2,3,4,5} = 2 \text{ k}\Omega + 1.6 \text{ k}\Omega = 3.6 \text{ k}\Omega$$

The total equivalent resistance, after completing all three steps, is illustrated in Figure 9-12(c).

(a)   **(b)**   **(c)**

FIGURE 9-12   Calculating Total Resistance. (a) Step A. (b) Step B. (c) Step C.

1. Given a series–parallel circuit, how can we determine which resistors are connected in series and which are connected in parallel?
2. Calculate the total resistance if two series-connected 12 kΩ resistors are connected in parallel with a 6 kΩ resistor.
3. State the three-step procedure used to determine total resistance in a circuit made up of both series and parallel resistors.
4. Sketch the following series–parallel resistor network made up of three resistors. R_1 and R_2 are in series with each other and are connected in parallel with R_3. If $R_1 = 470\ \Omega$, $R_2 = 330\ \Omega$, and $R_3 = 270\ \Omega$, what is R_T?

9-3 VOLTAGE DIVISION IN A SERIES–PARALLEL CIRCUIT

There is a simple three-step procedure for finding the voltage drop across each part of the series–parallel circuit. Figure 9-13 illustrates an example of a series–parallel circuit to which we will apply the three-step method for determining voltage drop.

STEP 1 Determine the circuit's total resistance. This is achieved by following the three-step method used previously for calculating total resistance.

Step A: $$R_{3,4} = 4 + 8 = 12\ \Omega$$

Step B: $$R_{2,3,4} = \frac{1}{(1/R_2) + (1/R_{3,4})} = 6\ \Omega$$

$$R_{5,6,7} = \frac{1}{(1/R_5) + (1/R_6) + (1/R_7)} = 12\ \Omega$$

Figure 9-14 illustrates the equivalent circuit up to this point. We end up with one series resistor (R_1) and two series equivalent resistors ($R_{2,3,4}$ and $R_{5,6,7}$). R_T is therefore equal to 28 Ω.

FIGURE 9-13 Series–Parallel Circuit Example.

FIGURE 9-14   After Completing Step 1.

STEP 2   Determine the circuit's total current. This step is achieved simply by utilizing Ohm's law.

$$I_T = \frac{V_T}{R_T} = \frac{84\text{ V}}{28\text{ }\Omega} = 3\text{ A}$$

STEP 3   Determine the voltage across each series resistor and each parallel combination (series equivalent resistor) in Figure 9-14. Since these are all in series, the same current (I_T) will flow through all three.

$$V_{R1} = I_T \times R_1 = 3\text{ A} \times 10\text{ }\Omega = 30\text{ V}$$
$$V_{R2,3,4} = I_T \times R_{2,3,4} = 3\text{ A} \times 6\text{ }\Omega = 18\text{ V}$$
$$V_{R5,6,7} = I_T \times R_{5,6,7} = 3\text{ A} \times 12\text{ }\Omega = 36\text{ V}$$

The voltage drops across the series resistor (R_1) and series equivalent resistors $(R_{2,3,4}$ and $R_{5,6,7})$ are illustrated in Figure 9-15.

Kirchhoff's voltage law states that the sum of all the voltage drops is equal to the source voltage applied. This law can be used to confirm that our calculations are all correct:

$$V_T = V_{R1} + V_{R2,3,4} + V_{R5,6,7}$$
$$= 30\text{ V} + 18\text{ V} + 36\text{ V}$$
$$= 84\text{ V}$$

To summarize, refer to Figure 9-16, which shows these voltage drops inserted into our original circuit. As you can see from this illustration:

FIGURE 9-15   After Completing Steps 2 and 3.

FIGURE 9-16   Detail of Step 3.

30 V is dropped across R_1.

18 V is dropped across R_2.

18 V is dropped across both R_3 and R_4.

36 V is dropped across R_5.

36 V is dropped across R_6.

36 V is dropped across R_7.

SELF-TEST REVIEW QUESTIONS FOR SECTION 9-3

---

1. State the three-step procedure used to calculate the voltage drop across each part of a series–parallel circuit.

2. Referring to Figure 9-13, double the values of all the resistors. Would the voltage drops calculated previously change, and if so, what would they be?

9-4 BRANCH CURRENTS IN A SERIES–PARALLEL CIRCUIT

In the preceding example, step 2 calculated the total current flowing in a series–parallel circuit. The next step is to find out exactly how much current is flowing through each parallel branch. This will be called step 4. Figure 9-17 shows the previously calculated data inserted in the appropriate places in our example circuit.

STEP 4   Total current (I_T) will exist at points A, B, C, and D. Between A and B, current has only one path to flow, which is through R_1. R_1 is therefore a series resistor, so $I_{R1} = I_T = 3$ A. Between points B and C, current has two paths: through R_2 (12 Ω) and through R_3 and R_4 (12 Ω).

FIGURE 9-17 **Series–Parallel Circuit Example with Previously Calculated Data.**

$$I_{R2} = \frac{V_{R2}}{R_2} = \frac{18 \text{ V}}{12 \text{ }\Omega} = 1.5 \text{ A}$$

$$I_{R3,4} = \frac{V_{R3,4}}{R_{3,4}} = \frac{18 \text{ V}}{12 \text{ }\Omega} = 1.5 \text{ A}$$

Not surprisingly, the total current of 3 A is split equally due to both branches having equal resistance.

The two 1.5-A branch currents will combine at point C to produce once again the total current of 3 A. Between points C and D, current has three paths to flow through, R_5, R_6, and R_7.

$$I_{R5} = \frac{V_{R5}}{R_5} = \frac{36 \text{ V}}{60 \text{ }\Omega} = 0.6 \text{ A}$$

$$I_{R6} = \frac{V_{R6}}{R_6} = \frac{36 \text{ V}}{24 \text{ }\Omega} = 1.5 \text{ A}$$

$$I_{R7} = \frac{V_{R7}}{R_7} = \frac{36 \text{ V}}{40 \text{ }\Omega} = 0.9 \text{ A}$$

All three branch currents will combine at point D to produce the total current of 3 A ($I_T = I_{R5} + I_{R6} + I_{R7} = 0.6 + 1.5 + 0.9 = 3$ A), proving Kirchhoff's current law.

9-5 POWER IN A SERIES–PARALLEL CIRCUIT

Whether resistors are connected in series or in parallel, the total power in a series–parallel circuit is equal to the sum of all the individual power losses.

$$P_t = P_1 + P_2 + P_3 + P_4 + \cdots$$

total power = addition of all power losses

The formulas for calculating the amount of power lost by each resistor are

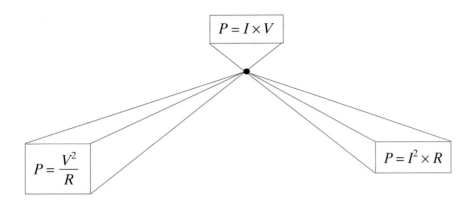

Let us calculate the power dissipated by each resistor; this final calculation will be called step 5. Since resistance, voltage, and current are known, either of the three formulas for power can be used to determine power.

$$P_{R1} = \frac{V_{R1}^2}{R_1} = \frac{30\ V^2}{10\ \Omega} = 90\ W$$

$$P_{R2} = \frac{V_{R2}^2}{R_2} = \frac{18\ V^2}{12\ \Omega} = 27\ W$$

$$P_{R3} = I_{R3}^2 \times R_3 = 1.5\ A^2 \times 4\ \Omega = 9\ W$$

$$P_{R4} = I_{R4}^2 \times R_4 = 1.5\ A^2 \times 8\ \Omega = 18\ W$$

$$P_{R5} = \frac{V_{R5}^2}{R_5} = \frac{36\ V^2}{60\ \Omega} = 21.6\ W$$

$$P_{R6} = \frac{V_{R6}^2}{R_6} = \frac{36\ V^2}{24\ \Omega} = 54\ W$$

$$P_{R7} = \frac{V_{R7}^2}{R_7} = \frac{36\ V^2}{40\ \Omega} = 32.4\ W$$

$$P_T = P_{R1} + P_{R2} + P_{R3} + P_{R4} + P_{R5} + P_{R6} + P_{R7}$$
$$= 90 + 27 + 9 + 18 + 21.6 + 54 + 32.4$$
$$= 252\ W$$

or

$$P_T = \frac{V_T^2}{R_T} = \frac{84\ V^2}{28\ \Omega}$$
$$= 252\ W$$

The total power dissipated in this example circuit is 252 W. All the information can now be inserted in a final diagram for the example, as shown in Figure 9-18.

FIGURE 9-18   **Series–Parallel Circuit Example with All Information Inserted.**

9-6 FIVE-STEP METHOD FOR SERIES–PARALLEL CIRCUIT ANALYSIS

Let's now combine and summarize all the steps for calculating resistance, voltage, current, and power in a series–parallel circuit by solving another problem. Before we begin, however, let us review the five-step procedure.

Solving for Resistance, Voltage, Current, and Power in a Series–Parallel Circuit

STEP 1   Determine the circuit's total resistance.

    Step A   Solve for series-connected resistors in all parallel combinations.

    Step B   Solve for all parallel combinations.

    Step C   Solve for remaining series resistances.

STEP 2   Determine the circuit's total current.

STEP 3   Determine the voltage across each series resistor and each parallel combination (series equivalent resistor).

STEP 4   Determine the value of current through each parallel resistor in every parallel combination.

STEP 5   Determine the total and individual power dissipated by the circuit.

■ **EXAMPLE:**

Referring to Figure 9-19, calculate:

  a. Total resistance

  b. Voltage drop across all resistors

c. Current through each resistor

d. Total power dissipated by the circuit

■ *Solution:*

The problem has basically asked us to calculate everything about the series–parallel circuit shown in Figure 9-19.

Applying the five-step procedure to our problem in Figure 9-19, we will first proceed with step 1, calculating the circuit's total resistance.

STEP 1   Determine the circuit's total resistance.

Step A: There are no series resistors within parallel combinations.

Step B: There are two-resistor (R_2, R_3) and three-resistor (R_5, R_6, R_7) parallel combinations in this circuit.

$$R_{2,3} = \frac{1}{(1/R_2) + (1/R_3)} = 222.2 \ \Omega$$

$$R_{5,6,7} = \frac{1}{(1/R_5) + (1/R_6) + (1/R_7)} = 500 \ \Omega$$

Figure 9-20 illustrates the circuit resulting after step B.

Step C: Solve for the remaining four resistances to gain the circuit's total resistance (R_T) or equivalent resistance (R_{eq}).

$$R_{eq} = R_1 + R_{2,3} + R_4 + R_{5,6,7}$$
$$= 1000 \ \Omega + 222.2 \ \Omega + 777.8 \ \Omega + 500 \ \Omega$$
$$= 2500 \ \Omega \quad \text{or} \quad 2.5 \ k\Omega$$

Figure 9-21 illustrates the circuit resulting after step C.

FIGURE 9-19   Apply the Five-Step Procedure to this Series–Parallel Circuit Example.

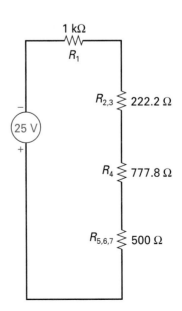

FIGURE 9-20   Circuit Resulting after Step 1B.

FIGURE 9-21   Circuit Resulting after Step 1C.

STEP 2   Determine the circuit's total current.

$$I_T = \frac{V_S}{R_T} = \frac{25\ \text{V}}{2.5\ \text{k}\Omega} = 10\ \text{mA}$$

STEP 3   Determine the voltage across each series resistor and each series equivalent resistor. To achieve this, we utilize the diagram obtained after completing step B (Figure 9-20):

$$V_{R1} = I_T \times R_1 = 10\ \text{mA} \times 1\ \text{k}\Omega = 10\ \text{V}$$

$$V_{R2,3} = I_T \times R_{2,3} = 10\ \text{mA} \times 222.2\ \Omega = 2.222\ \text{V}$$

$$V_{R4} = I_T \times R_4 = 10\ \text{mA} \times 777.8\ \Omega = 7.778\ \text{V}$$

$$V_{R5,6,7} = I_T \times R_{5,6,7} = 10\ \text{mA} \times 500\ \Omega = 5\ \text{V}$$

Figure 9-22 illustrates the results after step 3.

FIGURE 9-22   Circuit Resulting after Step 3.

STEP 4 Determine the value of current through each parallel resistor (Figure 9-23). R_1 and R_4 are series-connected resistors, and therefore their current will equal 10 mA.

$$I_{R1} = 10 \text{ mA}$$

$$I_{R4} = 10 \text{ mA}$$

The current through the parallel resistors is calculated by Ohm's law.

$$I_{R2} = \frac{V_{R2}}{R_2} = \frac{2.222 \text{ V}}{500 \text{ }\Omega} = 4.4 \text{ mA}$$

$$I_{R3} = \frac{V_{R3}}{R_3} = \frac{2.222 \text{ V}}{400 \text{ }\Omega} = 5.6 \text{ mA}$$

$$\left.\begin{array}{l} I_T = I_{R2} + I_{R3} \\ 10 \text{ mA} = 4.4 \text{ mA} + 5.6 \text{ mA} \end{array}\right\} \text{Kirchhoff's current law}$$

$$I_{R5} = \frac{V_{R5}}{R_5} = \frac{5 \text{ V}}{2 \text{ k}\Omega} = 2.5 \text{ mA}$$

$$I_{R6} = \frac{V_{R6}}{R_6} = \frac{5 \text{ V}}{2 \text{ k}\Omega} = 2.5 \text{ mA}$$

$$I_{R7} = \frac{V_{R7}}{R_7} = \frac{5 \text{ V}}{1 \text{ k}\Omega} = 5 \text{ mA}$$

$$\left.\begin{array}{l} I_T = I_{R5} + I_{R6} + I_{R7} \\ 10 \text{ mA} = 2.5 \text{ mA} + 2.5 \text{ mA} + 5 \text{ mA} \end{array}\right\} \text{Kirchhoff's current law}$$

STEP 5 Determine the total power dissipated by the circuit.

$$P_T = P_{R1} + P_{R2} + P_{R3} + P_{R4} + P_{R5} + P_{R6} + P_{R7}$$

FIGURE 9-23 Series–Parallel Circuit Example with Step 1, 2, and 3 Data Inserted.

or

$$P_T = \frac{V_T^2}{R_T}$$

Each resistor's power figure can be calculated and the sum would be the total power dissipated by the circuit. Since the problem does not ask for the power dissipated by each individual resistor, but for the total power dissipated, it will be easier to use the formula:

$$P_T = \frac{V_T^2}{R_T}$$

$$= \frac{25 \text{ V}^2}{2.5 \text{ k}\Omega}$$

$$= 0.25 \text{ W}$$

SELF-TEST REVIEW QUESTIONS FOR SECTIONS 9-4, 9-5, AND 9-6

1. State the five-step method used for series–parallel circuit analysis.
2. Design your own five-resistor series–parallel circuit, apply values, and use the five-step analysis method.

9-7 SERIES–PARALLEL CIRCUITS

9-7-1 *Loading of Voltage-Divider Circuits*

Loading
The adding of a load to a source.

The straightforward voltage divider was discussed in Chapter 7, but at that point we did not explore some changes that will occur if a load resistance is connected to the voltage divider's output. Figure 9-24 illustrates a voltage divider, and as you can see, the advantage of a voltage-divider circuit is that it can be used to produce several different voltages from one main voltage source by the use of a few chosen resistor values.

In our discussion on load resistance, we mentioned that every circuit or piece of equipment offers a certain amount of resistance, and this resistance represents how much a circuit or piece of equipment will load down the source supply.

Figure 9-25 illustrates an example voltage-divider circuit that is being used to develop a 10 V source from a 20 V battery supply. Figure 9-25(a) illustrates this circuit in the unloaded condition, and by making a few calculations you can understand everything about the circuit.

STEP 1 $R_T = R_1 + R_2 = 1 \text{ k}\Omega + 1 \text{ k}\Omega = 2 \text{ k}\Omega$

STEP 2 $I_T = \dfrac{V_T}{R_T} = \dfrac{20 \text{ V}}{2 \text{ k}\Omega} = 10 \text{ mA}$

FIGURE 9-24 **Voltage-Divider Circuit.**

The current that flows through a voltage divider, without a load connected, is called the **bleeder current.** In this example the current is equal to 10 mA. It is called the bleeder current because it is continually drawing or bleeding this current from the voltage source.

Bleeder Current Current drawn continuously from a voltage source. A bleeder resistor is generally added to lessen the effect of load changes or provide a voltage drop across a resistor.

STEP 3 $V_{R1} = V_{R2}$ (as resistors are the same value)

$$V_{R1} = 10 \text{ V}$$

$$V_{R2} = 10 \text{ V}$$

In Figure 9-25(b) we have connected a piece of equipment represented as a resistance (R_3) across the 10-V supply. This automatically turns the previous series circuit of R_1 and R_2 into a series–parallel circuit made up of R_1, R_2, and the 100-kΩ load resistance. By making a few more calculations, we can discover the changes that have occurred by connecting this load resistance.

STEP 1   Total resistance (R_T)

Step B: $R_{2,3} = \dfrac{R_2 \times R_3}{R_2 + R_3} = \dfrac{1 \text{ kΩ} \times 100 \text{ kΩ}}{1 \text{ kΩ} + 100 \text{ kΩ}} = 990.1 \text{ Ω}$

(a)   **(b)**

FIGURE 9-25 **Voltage-Divider Circuit. (a) Unloaded Output Voltage. (b) Loaded Output Voltage.**

Step C:
$$R_{1,2,3} = R_1 + R_{2,3}$$
$$= 1 \text{ k}\Omega + 990.1$$
$$= 1.99 \text{ k}\Omega$$

STEP 2 Total current (I_T)
$$I_T = \frac{20 \text{ V}}{1.99 \text{ k}\Omega} = 10.05 \text{ mA}$$

STEP 3 $V_{R1} = I_T \times R_1 = 10.05 \text{ mA} \times 1 \text{ k}\Omega = 10.05 \text{ V}$
$$V_{R2,3} = I_T \times R_{2,3} = 10.05 \text{ mA} \times 990.1 \text{ }\Omega = 9.95 \text{ V}$$

STEP 4 $I_{R1} = I_T = 10.05 \text{ mA}$
$$I_{R2} = \frac{V_{R2}}{R_2} = \frac{9.95 \text{ V}}{1 \text{ k}\Omega} = 9.95 \text{ mA}$$
$$I_{R3} = \frac{V_{R3}}{R_3} = \frac{9.95 \text{ V}}{100 \text{ k}\Omega} = 99.5 \text{ }\mu\text{A}$$

$$\left. \begin{array}{c} I_{R2} + I_{R3} = I_T \\ 9.95 \text{ mA} + 99.5 \text{ }\mu\text{A} = 10.05 \text{ mA} \end{array} \right\} \text{Kirchhoff's current law}$$

As you can see, the load resistance is pulling 99.5 μA, and this pulls the voltage down to 9.95 V from the required 10 V that was desired and is normally present in the unloaded condition.

When designing a voltage divider, design engineers need to calculate how much current a particular load will pull and then alter the voltage-divider resistor values to offset the loading effect when the load is connected.

9-7-2 The Wheatstone Bridge

Wheatstone Bridge
A four-arm, generally resistive, bridge that is used to measure resistance.

In 1850, Charles Wheatstone developed a circuit to measure resistance. This circuit, which is still widely used today, is called the **Wheatstone bridge** in his honor and is illustrated in Figure 9-26(a). In Figure 9-26(b), the same circuit has been drawn differently, yet the circuit functions exactly the same.

(a)　　　　　(b)

FIGURE 9-26 **Wheatstone Bridge. (a) Actual Circuit. (b) Redrawn Simplified Circuit.**

Balanced Bridge

Figure 9-27 illustrates an example circuit in which four resistors are connected together to form a series–parallel arrangement. Let us now use the five-step procedure to find out exactly what resistance, current, voltage, and power exist throughout the circuit.

STEP 1 Total resistance (R_T)

Step A:
$$R_{1,3} = R_1 + R_3 = 10 + 20 = 30$$
$$R_{2,4} = R_2 + R_4 = 10 + 20 = 30$$

Step B:
$$R_{Total}: (R_{1,2,3,4}) = \frac{R_{1,3} \times R_{2,4}}{R_{1,3} + R_{2,4}}$$

$$= \frac{30 \times 30}{30 + 30} = 15\ \Omega$$

Total resistance = 15 Ω

STEP 2 Total current (I_T)

$$I_T = \frac{V_T}{R_T} = \frac{30\ \text{V}}{15\ \Omega} = 2\ \text{A}$$

STEP 3 Since $R_{1,3}$ is in parallel with $R_{2,4}$, 30 V will appear across both $R_{1,3}$ and $R_{2,4}$.

$$V_T = V_{R1,3} = V_{R2,4} = 30\ \text{V}$$

voltage-divider formula
$$V_{R1} = \frac{R_1}{R_{1,3}} \times V_T$$

$$= \frac{10}{30} \times 30 = 10\ \text{V}$$

$$V_{R3} = \frac{R_3}{R_{1,3}} \times V_T = 20\ \text{V}$$

$$V_{R2} = \frac{R_2}{R_{2,4}} \times V_T = 10\ \text{V}$$

$$V_{R4} = \frac{R_4}{R_{2,4}} \times V_T = 20\ \text{V}$$

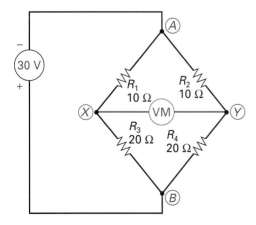

FIGURE 9-27 **Wheatstone Bridge Example Circuit.**

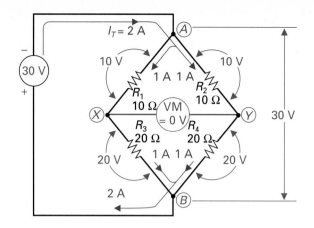

FIGURE 9-28   Balanced Wheatstone Bridge.

STEP 4 $I_{R1} = \dfrac{V_{R1,3}}{R_{1,3}} = \dfrac{30\text{ V}}{30\ \Omega} = 1\text{ A}$

$I_{R2,4} = \dfrac{V_{R2,4}}{R_{2,4}} = \dfrac{30\text{ V}}{30\ \Omega} = 1\text{ A}$

$\left.\begin{array}{l} I_{R1,3} + I_{R2,4} = I_T \\ 1\text{ A} + 1\text{ A} = 2\text{ A} \end{array}\right\}$ Kirchhoff's current law

STEP 5   Total power dissipated $(P_T) = I_T^2 \times R_T$

$$= 2\text{ A}^2 \times 15\ \Omega$$
$$= 4 \times 15\ \Omega$$
$$= 60\text{ W}$$

Figure 9-28 illustrates all of the step results inserted in the Wheatstone bridge example schematic. The Wheatstone bridge is said to be in the balanced condition since the voltage at point X will equal the voltage at point Y ($V_{R3} = V_{R4}$, 20 V = 20 V). This same voltage exists across R_3 and R_4, so the voltmeter, which is measuring the voltage difference between X and Y, will indicate 0 V potential difference.

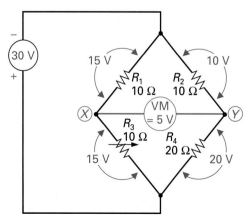

FIGURE 9-29   Unbalanced Wheatstone Bridge.

Unbalanced Bridge

In Figure 9-29 we have replaced R_3 with a variable resistor and set it to 10 Ω. The R_2 and R_4 resistor combination will not change its voltage drop; however, R_1 and R_3, which are now equal, will each split the 30 V supply, producing 15 V across R_3. The voltmeter will indicate the difference in potential (5 V) from the voltage across R_3 at point X (15 V) and across R_4 at point Y (20 V). The voltmeter is actually measuring the imbalance in the circuit, which is why this circuit in this condition is known as an *unbalanced bridge.*

Determining Unknown Resistance

Figure 9-30 illustrates a Wheatstone bridge into which an unknown resistor (R_{un}) and variable resistor (R_{va}) have been inserted. The variable value resistor is a calibrated resistor (resistance has been checked against a known, accurate resistance) in which the resistance can be adjusted and read from a calibrated dial.

The procedure to follow to find the value of the unknown resistor is as follows:

1. Adjust the variable value resistor until the voltmeter indicates that the Wheatstone bridge is balanced (voltmeter indicates 0 V).

2. Read the value of the variable value resistor. As long as $R_1 = R_2$, the variable resistance value will be the same as the unknown resistance value.

$$R_{va} = R_{un}$$

FIGURE 9-30   Using a Wheatstone Bridge to Determine an Unknown Resistance. (a) Schematic. (b) Pictorial.

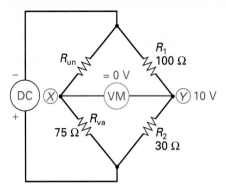

FIGURE 9-31 Wheatstone Bridge Circuit Example.

Since R_1 and R_2 are equal to one another, the voltage will be split across the two resistors, producing 10 V at point Y. The variable value resistor must therefore be adjusted so that it equals the unknown resistance, and therefore the same situation will occur, in that the 20 V source will be split, producing 10 V at point X, indicating a balanced zero-volt condition on the voltmeter. For example, if the unknown resistance is equal to 5 Ω, then only when the variable value resistor is adjusted and equal to 5 Ω would 10 V appear at point X and allow the circuit meter to read zero volts, indicating a balance. The variable value resistor resistance could be read (5 Ω) and the unknown resistor resistance would be known (5 Ω).

■ **EXAMPLE:**

What is the unknown resistance in Figure 9-31?

■ *Solution:*

The bridge is in a balanced condition as the voltmeter is reading a 0 V difference between points X and Y. In the previous section we discovered that if $R_1 = R_2$, then:

$$R_{va} = R_{un}$$

In this case, R_1 does not equal R_2, so a variation in the formula must be applied to take into account the ratio of R_1 and R_2.

$$R_{un} = R_{va} \times \frac{R_1}{R_2}$$

$$= 75\ \Omega \times \frac{100}{30}$$

$$= 75\ \Omega \times 3.33 = 250\ \Omega$$

Since R_1 is 3.33 times greater than R_2, then R_{un} must be 3.33 times greater than R_{va} if the Wheatstone bridge is in the balanced condition.

CALCULATOR SEQUENCE

Step	Keypad Entry	Display Response
1.	⒈ ⓪ ⓪	100
2.	÷	
3.	③ ⓪	30
4.	=	3.333
5.	×	
6.	⑦ ⑤	75
7.	=	250

FIGURE 9-32 *R–2R* Ladder Circuit.

9-7-3 *The* **R–2R** *Ladder Circuit*

Figure 9-32 illustrates a **R–2R ladder circuit**, which is a series–parallel circuit used for digital-to-analog conversion. To fully understand this circuit, our first step should be to find out exactly which branches will have which values of current. This can be obtained by finding out what value of resistance the current sees when it arrives at the three junctions *A, B,* and *C.* Let us begin with point *C* first and simplify the circuit. This is illustrated in Figure 9-33(a). No specific resistance value has been chosen, but in all cases 2*R* resistors (2 × *R*) are twice the resistance of an *R* resistor.

In Figure 9-33(a), if 2 mA of current arrives at point *C,* it sees 2*R* of resistance in parallel with a 2*R* resistance, and consequently the 2 mA of current splits and 1 mA flows through each branch. Two 2*R* resistors in parallel with one another would consequently be equivalent to one *R,* as seen in Figure 9-33(b).

In Figure 9-34(a), if 4 mA of current arrives at point *B,* it sees two series resistances to the right, which is equivalent to 2*R* [Figure 9-34(b)] and one resistance down of 2*R.* The 4 mA therefore splits, causing 2 mA down one path and 2 mA down the other. The two 2*R* resistors in parallel [Figure 9-34(b)] are equivalent to one *R,* as shown in Figure 9-34(c).

In Figure 9-35(a), if 8 mA of current arrives at point *A,* it sees two series resistors to the right, which is equivalent to 2*R* [Figure 9-35(b)], and one resis-

R–2R **Ladder Circuit**
A network or circuit composed of a sequence of *L* networks connected in tandem. This *R–2R* circuit is used in digital-to-analog converters.

(a)   **(b)**

FIGURE 9-33 *R–2R* **Equivalent Circuit at Junction C.**

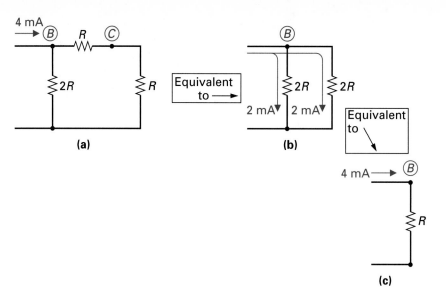

FIGURE 9-34 *R–2R Equivalent Circuit at Junction B.*

tance down of $2R$. The 8 mA of current therefore splits equally, causing 4 mA down one path and 4 mA down the other. The two $2R$ resistors in parallel [Figure 9-35(b)] are equivalent to one resistance R [Figure 9-35(c)].

Figure 9-36(a) through (f) summarizes the step-by-step simplification of this circuit.

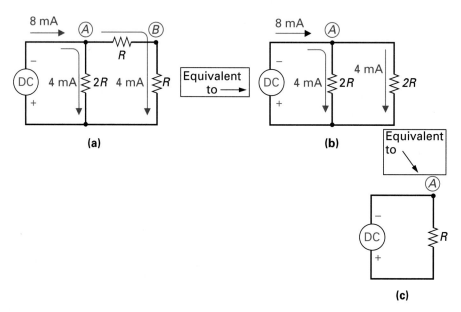

FIGURE 9-35 *R–2R Equivalent Circuit at Junction A.*

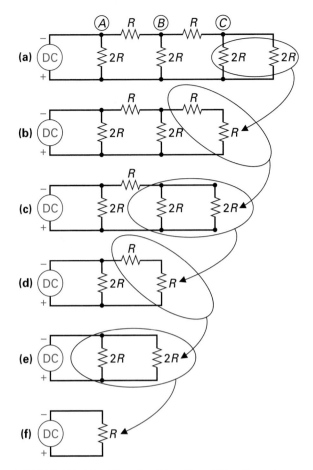

FIGURE 9-36 Step-by-Step Simplification of an
***R–2R* Ladder Circuit.**

The question you may have at this point is: What is the primary application of this circuit? The answer is as a current divider, as seen in Figure 9-37(a). The 8 mA of reference current is repeatedly divided by 2 as it moves from left to right, producing currents of 4 mA, 2 mA, and 1 mA. The result of this *R–2R* current division can be used in a circuit known as a *digital-to-analog converter* (DAC), which is illustrated in Figure 9-37(b).

Digital data or information exists within a computer, and this system expresses numbers and letters in two discrete steps; for example, on–off, high–low, open–closed, or 0–1. Only two conditions exist within the computer, and all information is represented by this two-state system.

Analog data or information exists outside a computer in our environment, and in this system, data are expressed in many different levels, as opposed to just two with digital information. Numbers are expressed in one of ten different levels (0–9), as opposed to only two (0–1) in digital.

Due to these differences, a device is needed that will interface (convert or link two different elements) the digital information within a computer to the

(b)

FIGURE 9-37 *R–2R* **Ladder in a Digital-to-Analog Converter.**

analog information that you and I understand. This device, called a digital-to-analog converter, uses the *R–2R* ladder circuit that we just discussed.

▪ **EXAMPLE:**

Determine the reference current and branch currents for the circuit in Figure 9-38.

▪ *Solution:*

In our simplification of the ladder circuit, we discussed previously that any *R–2R* ladder circuit can be simplified to one resistor equal to *R*, as seen in Figure 9-39(a). The reference or total current supplied will therefore equal

$$I_T = \frac{V_T}{R_T} = \frac{24\text{ V}}{1\text{ k}\Omega} = 24\text{ mA}$$

and the current will split through each branch, as shown in Figure 9-39(b).

FIGURE 9-38 **Calculate *R–2R* Circuit Reference Current and Branch Currents.**

FIGURE 9-39 *R–2R* Circuit with Data Inserted.

SELF-TEST REVIEW QUESTIONS FOR SECTION 9-7

1. What is meant by loading of a voltage-divider circuit?
2. Sketch a Wheatstone bridge circuit and list an application of this circuit.
3. What value will the total resistance of an *R–2R* ladder circuit always equal?
4. For what application could the *R–2R* ladder be used?

9-8 TROUBLESHOOTING SERIES–PARALLEL CIRCUITS

Troubleshooting is defined as the process of locating and diagnosing malfunctions or breakdowns in equipment by means of systematic checking or analysis. As discussed in previous resistive-circuit troubleshooting procedures, there are basically only three problems that can occur:

1. A component will open. This usually occurs if a resistor burns out or a wire or switch contact breaks.
2. A component will short. This usually occurs if a conductor, such as solder, wire, or some other conducting material, is dropped or left in the equipment, making or connecting two points that should not be connected.
3. There is a variation in a component's value. This occurs with age in resistors over a long period of time and can eventually cause a malfunction of the equipment.

Using the example circuit in Figure 9-40, we will step through a few problems, beginning with an open component. Throughout the troubleshooting, we will use the voltmeter whenever possible, as it can measure voltage by just connecting the leads across the component, rather than the ammeter, which has to be placed in the circuit, in which case the circuit path has to be opened. In some instances, ammeter use can be difficult to accomplish.

To begin, let's calculate the voltage drops and branch current obtained when the circuit is operating normally.

FIGURE 9-40 **Series–Parallel Circuit Example.**

STEP 1 (A) $R_{3,4} = R_3 + R_4 = 3\,\Omega + 9\,\Omega = 12\,\Omega$

(B) $R_{2,3,4} = \dfrac{R_2 \times R_{3,4}}{R_2 + R_{3,4}} = \dfrac{6\,\Omega \times 12\,\Omega}{6\,\Omega + 12\,\Omega} = 4\,\Omega$

(C) $R_{1,2,3,4} = R_T = R_1 + R_{2,3,4}$
$= 2\,\Omega + 4\,\Omega = 6\,\Omega$

STEP 2 $I_T = \dfrac{V_T}{R_T} = \dfrac{24\text{ V}}{6\,\Omega} = 4\text{ A}$

STEP 3 $V_{R1} = I_{R1} \times R_1 = 4\text{ A} \times 2\,\Omega = 8\text{ V}$

$V_{R2,3,4} = I_{R2,3,4} \times R_{2,3,4} = 4\text{ A} \times 4\,\Omega = 16\text{ V}$

(Kirchhoff's voltage law)

STEP 4 $I_{R1} = 4\text{ A}$ (series resistor)

$I_{R2} = \dfrac{V_{R2}}{R_2} = \dfrac{16\text{ V}}{6\,\Omega} = 2.7\text{ A}$

$I_{R3,4} = \dfrac{V_{R3,4}}{R_{3,4}} = \dfrac{16\text{ V}}{12\,\Omega} = 1.3\text{ A}$

All these results have been inserted in Figure 9-40.

9-8-1 Open Component

R_1 Open (Figure 9-41)

With R_1 open, there cannot be any current flow through the circuit as there is not a path from one side of the battery to the other. This fault can be recognized easily because approximately all of the applied 24 V will be measured across the open resistor (R_1), and 0 V will appear across all the other resistors.

FIGURE 9-41   Open Series-Connected Resistor in a Series–Parallel Circuit.

R_3 Open (Figure 9-42)

With R_3 open, there will be no current through the branch made up of R_3 and R_4. The current path will be through R_1 and R_2, and therefore the total resistance will now increase $(R_T\uparrow)$ from 6 Ω to

$$R_T = R_1 + R_2$$
$$= 2\,\Omega + 6\,\Omega = 8\,\Omega$$

FIGURE 9-42   Open Parallel-Connected Resistor in a Series–Parallel Circuit.

This 8 Ω is an increase in circuit resistance from the typical resistance, which was 6 Ω, which implies that an open has occurred to increase resistance. The total current will decrease ($I_T\downarrow$) from 4 A to

$$I_T = \frac{V_T}{R_T} = \frac{24\text{ V}}{8\text{ }\Omega} = 3\text{ A}$$

The voltage drop across the resistors will be

$$V_{R1} = I_T \times R_1 = 3\text{ A} \times 2\text{ }\Omega = 6\text{ V}$$
$$V_{R2} = I_T \times R_2 = 3\text{ A} \times 6\text{ }\Omega = 18\text{ V}$$

If one of the parallel branches is opened, the overall circuit resistance will always increase. This increase in the total resistance will cause an increase in the voltage dropped across the parallel branch (the greater the resistance, the greater the voltage drop), which enables the technician to localize the fault area and also to determine that the fault is an open.

The voltage measured with a voltmeter will be

$$R_1 = 6\text{ V}$$
$$R_2 = 18\text{ V}$$
$$R_3 = 18\text{ V (open)}$$
$$R_4 = 0\text{ V}$$

This identifies the problem as R_3 being open, as it drops the entire parallel circuit voltage (18 V) across itself, whereas normally the voltage would be dropped proportionally across R_3 and R_4, which are in series with one another.

9-8-2 *Shorted Component*

R_1 Shorted (Figure 9-43)

With R_1 shorted, the total circuit resistance will decrease ($R_T\downarrow$), causing an increase in circuit current ($I_T\uparrow$). This increase in current will cause an increase in the voltage dropped across the parallel branch; however, the fault can be located once you measure the voltage across R_1, which will read 0 V, indicating that this resistor has almost no resistance, as it has no voltage drop across it.

R_3 Shorted (Figure 9-44)

With R_3 shorted, there will be a decrease in the circuit's total resistance from 6 Ω to

$$R_{2,3,4} = \frac{R_2 \times R_{3,4}}{R_2 + R_{3,4}} = \frac{6 \times 9}{6 + 9} = 3.6$$
$$R_{1,2,3,4} = R_1 + R_{2,3,4}$$
$$= 2\text{ }\Omega + 3.6\text{ }\Omega$$
$$= 5.6\text{ }\Omega$$

This decrease in total resistance ($R_T\downarrow$) will cause an increase in total current ($I_T\uparrow$), which implies that a short has occurred to decrease resistance. The total current will now increase from 4 A to

FIGURE 9-43 Shorted Series-Connected Resistor in a Series–Parallel Circuit.

$$I_T = \frac{V_T}{R_T} = \frac{24 \text{ V}}{5.6 \text{ }\Omega} = 4.3 \text{ A}$$

The voltage drops across the resistors will be

$$V_{R1} = I_T \times R_1 = 4.3 \text{ A} \times 2 \text{ }\Omega = 8.6 \text{ V}$$

$$V_{R2,3,4} = I_T \times R_{2,3,4} = 4.3 \text{ A} \times 3.6 \text{ }\Omega = 15.4 \text{ V}$$

FIGURE 9-44 Shorted Parallel-Connected Resistor in a Series–Parallel Circuit.

FIGURE 9-45 **Summary of Symptoms for Opens and Shorts.**

When one of the parallel branch resistors is shorted, the overall circuit resistance will always decrease. This decrease in total resistance will cause a decrease in the voltage dropped across the parallel branch (the smaller the resistance, the smaller the voltage drop), and this enables the technician to localize the faulty area and also to determine that the fault is a short.

The voltage measured with the voltmeter (VM) will be

$$R_1 = 8.6 \text{ V}$$
$$R_2 = 15.4 \text{ V}$$
$$R_3 = 0 \text{ V (short)}$$
$$R_4 = 15.4 \text{ V}$$

This identifies the problem as R_3 being a short, as 0 V is being dropped across it. In summary, you may have noticed that (Figure 9-45):

1. An open component causes total resistance to increase ($R_T\uparrow$) and, therefore, total current to decrease ($I_T\downarrow$), and the open component, if in series, has the supply voltage across it, and if in a parallel branch, has the parallel branch voltage across it.

2. A shorted component causes total resistance to decrease ($R_T\downarrow$) and, therefore, total current to increase ($I_T\uparrow$), and the shorted component, if in series or parallel branches, will have 0 V across it.

9-8-3 Resistor Value Variation

R_2 Resistance Decreases (Figure 9-46)

If the resistance of R_2 decreases, the total circuit resistance (R_T) will decrease and the total circuit current (I_T) will increase. The result of this problem and the way in which the fault can be located is that when the voltage drop

FIGURE 9-46 **Parallel-Connected Resistor Value Decrease in Series–Parallel Circuit.**

across R_1 and the parallel branch is tested, there will be an increased voltage drop across R_1 due to the increased current flow, and a decrease in the voltage drop across the parallel branch due to a decrease in the parallel branch resistance.

With open and shorted components, the large voltage (open) or small voltage (short) drop across a component enables the technician to identify the faulty component. A variation in a component's value, however, will vary the circuit's behavior; but with this example, the symptoms could have been caused by a combination of variations. So once the area of the problem has been localized, the next troubleshooting step is to remove each of the resistors in the suspected faulty area and verify that their resistance values are correct by measuring their resistances with an ohmmeter. In this problem we had an increase in voltage across R_1 and a decrease in voltage across the parallel branch when measuring with a voltmeter, which was caused by R_2 decreasing. The same swing in voltage readings could also have been obtained by an increase in the resistance of R_1.

R_2 Resistance Increases (Figure 9-47)

If the resistance of R_2 increases, the total circuit resistance (R_T) will increase and the total circuit current (I_T) will decrease. The voltage drop across R_1 will decrease and the voltage across the branch will increase in value. Once again, these measured voltage changes could be caused by the resistance of R_2 increasing or the resistance of R_1 decreasing.

To reinforce your understanding let's work out a few examples of troubleshooting other series–parallel circuits.

FIGURE 9-47 Series-Connected Resistor Value
Decrease in Series–Parallel Circuit.

■ **EXAMPLE:**

If bulb 1 in Figure 9-48 goes open, what effect will it have, and how will the fault
be recognized?

▨ *Solution:*

A visual inspection of the circuit shows all bulbs off, as there is no current path
from one side of the battery to the other, since bulb 1 is connected in series and
has opened the only path. Since all bulbs are off, the faulty bulb cannot be visu-
ally isolated; however, you can easily localize the faulty bulb by using one of two
methods:

1. Use the voltmeter and check the voltage across each bulb. Bulb 1 would
   have 12 V across it, while 2 and 3 would have 0 V across them. This isolates
   the faulty component to bulb 1, since all the supply voltage is being measured across it.

2. By analyzing the circuit diagram, you can see that only one bulb can open
   and cause all the bulbs to go out, and that is bulb 1. If bulb 2 opens, 1 and 3

FIGURE 9-48 Series-Connected Open Bulb in
Series–Parallel Circuit.

would still be on, and if bulb 3 opens, 1 and 2 would still remain on. With power off, the ohmmeter could verify this open.

■ **EXAMPLE:**

One resistor in Figure 9-49 has shorted. From the voltmeter reading shown, determine which one.

■ *Solution:*

If the supply voltage is being measured across the parallel branch of R_2 and R_3, there cannot be any other resistance in circuit, so R_1 must have shorted. The next step would be to locate the component, R_1, and determine what has caused it to short. If we were not told that a resistor had shorted, the same symptom could have been caused if R_2 and R_3 were both open, and therefore the open parallel branch would allow no current to flow and maximum supply voltage would appear across it. The individual component resistance, when checked, will isolate the problem.

■ **EXAMPLE:**

Determine if there is an open or short in Figure 9-50. If so, isolate it by the two voltage readings that are shown in the circuit diagram.

■ *Solution:*

Performing a few calculations, you should come up with a normal total circuit resistance of 14 kΩ and a total circuit current of 1 mA. This should cause 12 V across R_1 and 2 V across the parallel branch under no-fault conditions. The decrease in the voltage drop across the series resistor R_1 leads you to believe that there has been a decrease in total circuit current, which must have been caused by a total resistance increase, which points to an open component (assuming that only an open or short can occur and not a component value variation).

FIGURE 9-49   Find the Shorted Resistor.

FIGURE 9-50   Does a Problem Exist?

If R_1 was open, all the 14 V would have been measured across R_1, which did not occur.

If R_3 opened,

$$\text{total resistance} = 15\text{ k}\Omega$$

$$\text{total current} = 0.93\text{ mA} \left(\frac{V_T}{R_T} = \frac{14\text{ V}}{15\text{ k}\Omega} \right)$$

$$V_{R1} = I_T \times R_1 = 11.16\text{ V}$$

Since the voltage dropped across R_1 was 9.3 V, R_3 is not the open.

If R_2 opened,

$$\text{total resistance} = 18\text{ k}\Omega$$

$$\text{total current} = 0.78\text{ mA} \left(\frac{V}{R} = \frac{14\text{ V}}{18\text{ k}\Omega} \right)$$

$$V_{R1} = I_T \times R_1 = 9.36\text{ V}$$

This circuit's problem is resistor R_2, which has burned out, causing an open circuit.

SELF-TEST REVIEW QUESTIONS FOR SECTION 9-8

Describe how to isolate the following problems in a series–parallel circuit.

1. An open component
2. A shorted component
3. A resistor value variation

9-9 THEOREMS FOR DC CIRCUITS

Series–parallel circuits can become very complex in some applications, and the more help you have in simplifying and analyzing these networks, the better. The following theorems can be used as powerful analytical tools for evaluating circuits. To begin, let's discuss the differences between voltage and current sources.

9-9-1 *Voltage and Current Sources*

The easiest way to understand a current source is to compare its feature to a voltage source, so let's begin by discussing voltage sources.

Voltage Source

Voltage Source
The circuit or device that supplies voltage to a load circuit.

The battery is an example of a **voltage source** that in the ideal condition will produce a fixed output voltage regardless of what load resistance is connected across its terminals. This means that even if a large load current is drawn from the battery (heavy load, due to a small load resistance) or if a small load

FIGURE 9-51 Ideal Voltage Source. (a) Heavy Load. (b) Light Load. (c) Symbol.

current results (light load, due to a large load resistance), the battery will always produce a constant output voltage, as seen in Figure 9-51.

In reality, every voltage source, whether a battery, power supply, or generator, will have some level of inefficiency and not only generate an output electrical dc voltage but also generate heat. This inefficiency is represented as an internal resistance, as seen in Figure 9-52(a), and in most cases this internal source resistance (R_{int}) is very low (several ohms) compared to the load resistance (R_L). In Figure 9-52(a), no load has been connected, so the output or open-circuit voltage will be equal to the source voltage, V_S. When a load is connected across the battery, as shown in Figure 9-52(b), R_{int} and R_L form a

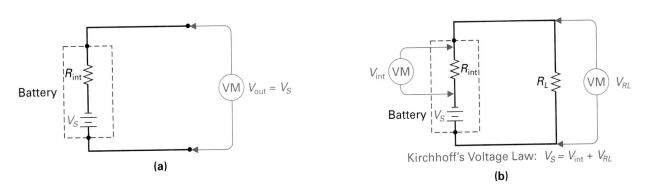

FIGURE 9-52 Realistic Voltage Source. (a) Unloaded. (b) Loaded.

series circuit, and some of the source voltage appears across R_{int}; so the output or load voltage is always less than V_S. Since R_{int} is normally quite small compared to R_L, the voltage source approaches ideal, as almost all the source voltage (V_S) appears across R_L.

In conclusion, a voltage source should have the smallest possible internal resistance, so that the output voltage (V_{out}) will remain constant and approximately equal to V_S independent of whether a light load (large R_L, small I_L) or heavy load (small R_L, large I_L) is connected across its output terminals.

EXAMPLE:

Calculate the output voltage in Figure 9-53 if R_L is equal to:

   a. 100 Ω

   b. 1 kΩ

   c. 100 kΩ

Solution:

(Voltage-divider formula)

   a. $R_L = 100$ Ω:

$$V_{out} = \frac{R_L}{R_T} \times V_S$$

$$= \frac{100\ \Omega}{110\ \Omega} \times 100\ \text{V} = 90.9\ \text{V}$$

   b. $R_L = 1$ kΩ (1000 Ω)

$$V_{out} = \frac{1000\ \Omega}{1010\ \Omega} \times 100\ \text{V} = 99.0\ \text{V}$$

   c. $R_L = 100$ kΩ (100,000 Ω)

$$V_{out} = \frac{100,000\ \Omega}{100,010\ \Omega} \times 100\ \text{V} = 99.99\ \text{V}$$

From this example you can see that the larger the load resistance, the greater the output voltage (V_{out} or V_{RL}). To explain this in a little more detail, we can say that a large R_L is considered a light load for the voltage source, as it only has to produce a small load current ($R_L\uparrow$, $I_L\downarrow$), and consequently, the heat generated by the source is small ($P_{R_{int}}\downarrow = I^2\downarrow \times R$) and the voltage source is more efficient (V_{out} almost equals V_S), approaching ideal ($V_{out} = V_S$). The voltage source in this

FIGURE 9-53   **Voltage Source Circuit Example.**

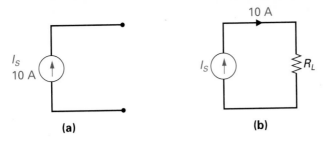

(a)

(b)

FIGURE 9-54   The Current Source Maintains a Constant Output Current Whether (a) Unloaded or (b) Loaded.

example, however, produced an almost constant output voltage (within 10% of V_S) despite the very large changes in R_L.

Current Source

Just as a voltage source has a certain voltage rating, the **current source** has a certain current rating; and just as a voltage source should deliver a constant output voltage, an ideal current source should deliver its constant rated current, regardless of what value of load resistance is connected across its output terminals, as seen in Figure 9-54.

A current source can be thought of as a voltage source with an extremely large internal resistance, as seen in Figure 9-55(a) and symbolized in Figure 9-55(b).

In conclusion, a current source should have a large internal resistance, so that, whatever the load resistance connected across the output, it will have very little effect on the total resistance and the load current will remain constant. The symbol for a constant current source has an arrow within a circle and this arrow points in the direction of current flow.

Current Source
The circuit or device that supplies current to a load circuit.

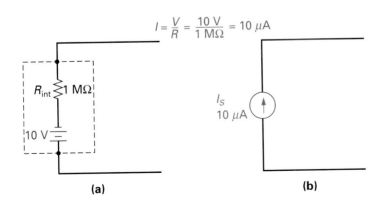

$$I = \frac{V}{R} = \frac{10\ V}{1\ M\Omega} = 10\ \mu A$$

(a)

(b)

FIGURE 9-55   The Large Internal Resistance of the Current Source.

FIGURE 9-56 Current Source Circuit Example.

■ **EXAMPLE:**

Calculate the load current supplied by the current source in Figure 9-56 if the following values of R_L are connected across the output terminals:

   a. 100 Ω

   b. 1 kΩ

   c. 100 kΩ

■ *Solution:*

   a. $R_L = 100\ \Omega\ (R_T = R_{int} + R_L = 1{,}000{,}000\ \Omega + 100\ \Omega = 1{,}000{,}100\ \Omega)$

$$I_L = \frac{V_S}{R_T}$$

$$= \frac{10\ \text{V}}{1{,}000{,}100\ \Omega} = 9.999\ \mu\text{A}$$

   b. $R_L = 1\ \text{k}\Omega\ (R_T = R_{int} + R_L = 1{,}001{,}000\ \Omega)$

$$I_L = \frac{10\ \text{V}}{1{,}001{,}000\ \Omega} = 9.99\ \mu\text{A}$$

   c. $R_L = 100\ \text{k}\Omega\ (R_T = R_{int} + R_L = 1{,}100{,}000\ \Omega$

$$I_L = \frac{10\ \text{V}}{1{,}100{,}000\ \Omega} = 9.09\ \mu\text{A}$$

From this example you can see that the current source delivered an almost constant output current regardless of the large load resistance change.

Superposition Theorem
In a network or circuit containing two or more voltage sources, the current at any point is equal to the algebraic sum of the individual source currents produced by each source acting separately.

9-9-2 Superposition Theorem

The **superposition theorem** is used not only in electronics, but also in physics and even economics. It is used to determine the net effect in a circuit that has two or more sources connected. The basic idea behind this theorem is that if two voltage sources are both producing a current through the same circuit, the net current can be determined by first finding the individual currents and then adding them together. Stated formally: *In a network containing two or more voltage sources, the current at any point is equal to the algebraic sum of the individual source currents produced by each source acting separately.*

The best way to fully understand the theorem is to apply it to a few examples. Figure 9-57(a) illustrates a simple series circuit with two resistors and two voltage sources. The 12 V source (V_1) is trying to produce a current in a clockwise direction, while the 24 V source (V_2) is trying to force current in a counterclockwise direction. What will be the resulting net current in this circuit?

STEP 1   To begin, let's consider what current would be produced in this circuit if only V_1 were connected, as shown in Figure 9-57(b).

$$I_1 = \frac{V_1}{R_T} = \frac{12 \text{ V}}{9 \ \Omega} = 1.33 \text{ A}$$

STEP 2   The next step is to determine how much current V_2 would produce if V_1 were not connected in the circuit, as shown in Figure 9-57(c).

$$I_2 = \frac{V_2}{R_T} = \frac{24 \text{ V}}{9 \Omega} = 2.66 \text{ A}$$

V_1 is attempting to produce 1.33 A in the clockwise direction, while V_2 is trying to produce 2.66 A in the counterclockwise direction. The net current will consequently be 1.33 A in the counterclockwise direction.

FIGURE 9-57   Superposition.

FIGURE 9-58   Superposition Circuit Example.

■ **EXAMPLE:**

Calculate the current through R_2 in Figure 9-58 using the superposition theorem.

■ **Solution:**

The first step is to calculate the current through R_2 due only to the voltage source V_1. This is shown in Figure 9-59(a). R_1 is a series-connected resistor, while R_2 and R_3 are connected in parallel with one another. So:

$$R_{2,3} = \frac{R_{val}}{n} = \frac{20\ \Omega}{2} = 10\ \Omega \qquad \text{(same-value parallel resistor formula)}$$

$$R_T = R_1 + R_{2,3} = 20\ \Omega + 10\ \Omega = 30\ \Omega \qquad \text{(total resistance)}$$

$$I_T = \frac{V_1}{R_T} = \frac{12\ V}{30\ \Omega} = 400\ \text{mA} \qquad \text{(total current)}$$

$$I_{R2} = \frac{R_{2,3}}{R_2} \times I_T = \frac{10\ \Omega}{20\ \Omega} \times 400\ \text{mA} = 200\ \text{mA} \qquad \text{(current-divider formula)}$$

This 200 mA of current is flowing down through R_2.

(a)

(b)

FIGURE 9-59   Superposition Circuit Solution.

The next step is to find the current flow through R_2 due only to the voltage source V_2. This is shown in Figure 9-59(b). In this instance, R_3 is a series-connected resistor, and R_1 and R_2 make up a parallel circuit. So:

$$R_{1,2} = 10 \ \Omega$$

$$R_T = R_{1,2} + R_3 = 10 \ \Omega + 20 \ \Omega = 30 \ \Omega$$

$$I_T = \frac{V_2}{R_T} = \frac{30 \text{ V}}{30 \ \Omega} = 1 \text{ A} \qquad (1000 \text{ mA})$$

$$I_{R2} = \frac{R_{1,2}}{R_2} \times I_T = \frac{10 \ \Omega}{20 \ \Omega} \times 1000 \text{ mA} = 500 \text{ mA}$$

This 500 mA of current will flow down through R_2.

Since both V_1 and V_2 produce a current flow down through R_2, so I_{R2} and I_{R3} have the same algebraic sign, the total current through R_2 is equal to the sum of the two currents produced by V_1 and V_2.

$$I_{R2} \text{ (total)} = I_{R2} \text{ due to } V_1 + I_{R2} \text{ due to } V_2$$

$$= 200 \text{ mA} + 500 \text{ mA}$$

$$= 700 \text{ mA}$$

■ **EXAMPLE:**

Calculate the current flow through R_1 in Figure 9-60.

■ *Solution:*

With the superposition theorem, current sources are treated differently from voltage sources in that each *current source is removed from the circuit and replaced with an open,* as illustrated in the first step of the solution shown in Figure 9-61(a). In this instance you can see that current has only one path (series circuit), so the current through R_1 is counterclockwise and equal to the I_1 source current, 100 μA.

In Figure 9-61(b), the current source I_1 has been removed and replaced with an open. R_1 and R_2 form a series circuit, so the total source current from I_2 flows through R_1 (I_{R1} =500 μA) in a clockwise direction.

Since I_1 is producing 100 μA of current through R_1 in a counterclockwise direction and I_2 is producing 500 μA of current in a clockwise direction, the resulting current through R_1 will equal

$$I_{R1} = I_2 \text{ (cw)} - I_1 \text{ (ccw)}$$

$$= 500 \text{ μA} - 100 \text{ μA} = 400 \text{ μA clockwise}$$

FIGURE 9-60 Superposition Circuit Example.

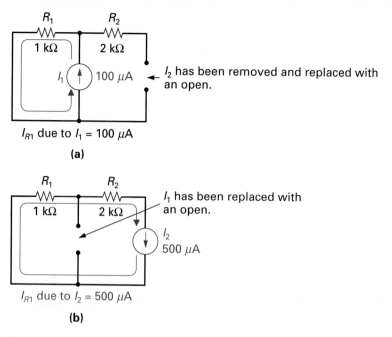

I_{R1} due to $I_1 = 100 \ \mu A$

(a)

I_{R1} due to $I_2 = 500 \ \mu A$

(b)

FIGURE 9-61   **Superposition Circuit Solution.**

9-9-3 Thévenin's Theorem

Thévenin's Theorem
Any network of voltage sources and resistors can be replaced by a single equivalent voltage source (V_{TH}) in series with a single equivalent resistance (R_{TH}).

Thévenin's theorem allows us to replace the complex networks in Figure 9-62(a) with an equivalent circuit containing just one source voltage (V_{TH}) and one series-connected resistance (R_{TH}), as in Figure 9-62(b). Stated formally: *Any network of voltage sources and resistors can be replaced by a single equivalent voltage source (V_{TH}) in series with a single equivalent resistance (R_{TH}).*

Figure 9-63(a) illustrates an example circuit. As with any theorem, a few rules must be followed to obtain an equivalent V_{TH} and R_{TH}.

STEP 1   The first step is to disconnect the load (R_L) and calculate the voltage that will appear across points A and B, as in Figure 9-63(b). This open-circuit voltage will be the same as the voltage drop across R_2 (V_{R2}) and is called the *Thévenin equivalent voltage* (V_{TH}). First, let's calculate current:

$$I_T = \frac{V_S}{R_T} = \frac{12 \ V}{9 \ \Omega} = 1.333 \ A$$

Therefore, V_{R2} or V_{TH} will equal

$$V_{R2} = I_T \times R_2 = 1.333 \times 6 \ \Omega = 8 \ V$$

so $V_{TH} = 8 \ V$.

STEP 2   Now that the Thévenin equivalent voltage has been calculated, the next step is to calculate the Thévenin equivalent resistance. In this step, the source voltage is removed and replaced with a short, as seen in Figure 9-63(c), and the Thévenin equivalent resistance is equal to whatever resistance exists

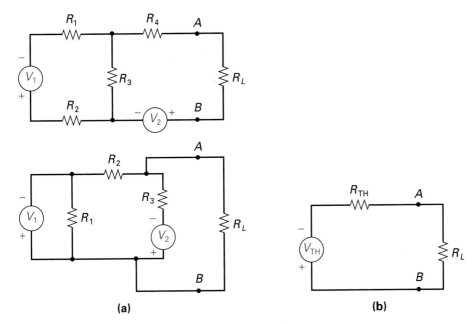

FIGURE 9-62 Thévenin's Theorem. (a) Complex Multiple Resistors and Source Networks Are Replaced by (b) One Source Voltage (V_{TH}) and One Series-Connected Resistance (R_{TH}).

between points A and B. In this example, R_1 and R_2 form a parallel circuit, the total resistance of which is equal to

$$R_T = \frac{R_1 \times R_2}{R_1 + R_2} = \frac{3 \times 6}{3 + 6} = \frac{18}{9} = 2\,\Omega$$

so $R_{TH} = 2\,\Omega$. The circuit to be Thévenized in Figure 9-63(a) can be represented by the Thévenin equivalent circuit shown in Figure 9-63(d).

The question you may have at this time is why we would need to simplify such a basic circuit, when Ohm's law could have been used just as easily to analyze the network. Thévenin's theorem has the following advantages:

1. If you had to calculate load current and load voltage (I_{RL} and V_{RL}) for 20 different values of R_L, it would be far easier to use the Thévenin equivalent circuit with the series-connected resistors R_{TH} and R_L, rather than applying Ohm's law to the series–parallel circuit made up of R_1, R_2, and R_L.

2. Thévenin's theorem permits you to solve complex circuits that could not easily be analyzed using Ohm's law.

■ **EXAMPLE:**

Determine V_{TH} and R_{TH} for the circuit in Figure 9-64.

■ *Solution:*

The first step is to remove the load resistor R_L and calculate what voltage will appear between points A and B, as shown in Figure 9-65(a). Removing R_L will

FIGURE 9-63 Thévenin's Theorem. (a) Example Circuit. (b) Obtaining Thévenin Voltage (V_{TH}). (c) Obtaining Thévenin Resistance. (d) Thévenin Equivalent Circuit.

FIGURE 9-64 Thévenin Circuit Example.

open the path for current to flow through R_4, which will consequently have no voltage drop across it. The voltage between points A and B therefore will be equal to the voltage dropped across R_3, and since R_1, R_2, and R_3 form a series circuit, the voltage-divider formula can be used:

$$V_{R3} = \frac{R_3}{R_T} \times V_S$$

$$= \frac{10\ \Omega}{40\ \Omega} \times 10\ V = 2.5\ V$$

Therefore,

$$V_{AB} = V_{R3} = V_{TH} = 2.5\ V$$

The next step is to calculate the Thévenin resistance, which will equal whatever resistance appears across the terminals A and B with the voltage source having been removed and replaced with a short, as shown in Figure 9-65(b). In Figure 9-65(c), the circuit has been redrawn so that the relationship between the resistors can be seen in more detail. As you can see, R_1 and R_2 are in series with one another and both are in parallel with R_3, and this combination is in series

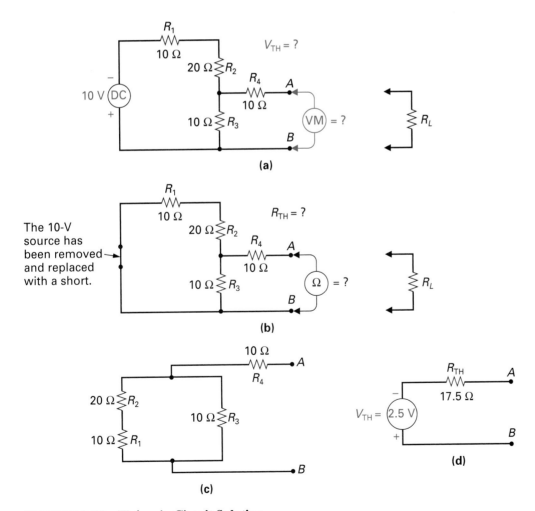

FIGURE 9-65 Thévenin Circuit Solution.

with R_4. Using our three-step procedure for calculating total resistance in a series–parallel circuit, the following results are obtained:

1. $R_{1,2} = R_1 + R_2 = 10\ \Omega + 20\ \Omega = 30\ \Omega$

2. $R_{1,2,3} = \dfrac{R_{1,2} \times R_3}{R_{1,2} + R_3} = \dfrac{30\ \Omega \times 10\ \Omega}{30\ \Omega + 10\ \Omega} = \dfrac{300\ \Omega}{40\ \Omega} = 7.5\ \Omega$

3. $R_T = R_{1,2,3} + R_4 = 7.5\ \Omega + 10\ \Omega = 17.5\ \Omega$

Figure 9-65(d) illustrates the Thévenin equivalent circuit.

9-9-4 Norton's Theorem

Norton's Theorem
Any network of voltage sources and resistors can be replaced by a single equivalent current source (T_N) in parallel with a single equivalent resistance (R_N).

Norton's theorem, like Thévenin's theorem, is a tool for simplifying a complex circuit into a more manageable one. Figure 9-66 illustrates the difference between a Thévenin equivalent and Norton equivalent circuit. Thévenin's theorem simplifies a complex network and uses an equivalent voltage source (V_{TH}) and an equivalent series resistance (R_{TH}). Norton's theorem, on the other hand, simplifies a complex circuit and represents it with an equivalent current source (I_N) in parallel with an equivalent Norton resistance (R_N), as shown in Figure 9-66.

FIGURE 9-66   Comparison of Thévenin's and Norton's Circuits.

As with any theorem, a set of steps has to be carried out to arrive at an equivalent circuit. The example we will use is shown in Figure 9-67(a) and is the same example used in Figure 9-66.

STEP 1 Calculate the Norton equivalent current source, which will be equal to the current that would flow between terminals A and B if the load resistor was removed and replaced with a short, as seen in Figure 9-67(b). Placing a short between terminals A and B will short out the resistor R_2, so the only resistance in the circuit will be R_1. The Norton equivalent current source in this example will therefore be equal to

$$I_N = \frac{V_S}{R_T} = \frac{12\ V}{3\ \Omega} = 4\ A$$

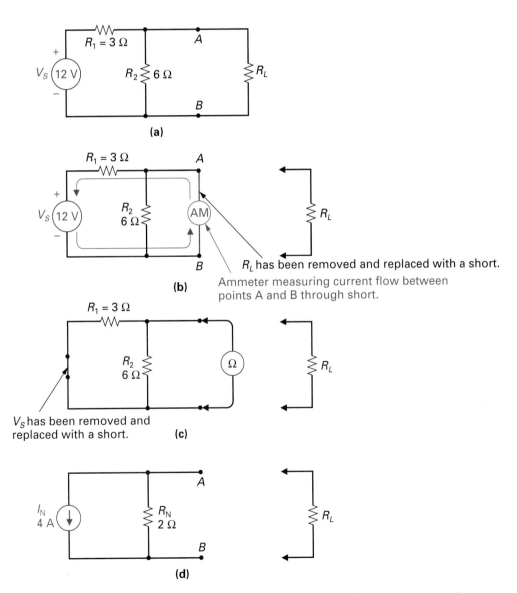

FIGURE 9-67 Norton's Theorem. **(a)** Example Circuit. **(b)** Obtaining Norton Current. **(c)** Obtaining Norton Resistance (R_N). **(d)** Norton Equivalent Circuit.

FIGURE 9-68 Norton Circuit Example.

STEP 2 The next step is to determine the value of the Norton equivalent resistance that will be placed in parallel with the current source, unlike Thévenin's equivalent resistance, which was placed in series. Like Thévenin's theorem, though, Norton's equivalent resistance (R_N) is equal to the resistance between terminals A and B when the voltage source is removed and replaced with a short, as shown in Figure 9-67(c). Since R_1 and R_2 form a parallel circuit, the Norton equivalent resistance will be equal to

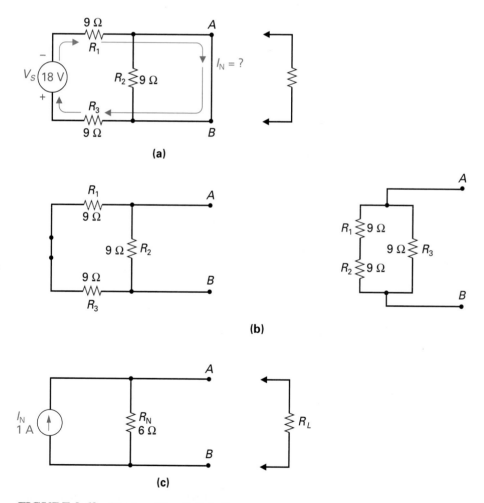

FIGURE 9-69 Norton Circuit Solution.

$$R_N = \frac{R_1 \times R_2}{R_1 + R_2} = \frac{3\,\Omega \times 6\,\Omega}{3\,\Omega + 6\,\Omega} = \frac{18\,\Omega}{9\,\Omega} = 2\,\Omega$$

The Norton equivalent circuit has been determined simply by carrying out these two steps and is illustrated in Figure 9-67(d).

■ **EXAMPLE:**

Determine I_N and R_N for the circuit in Figure 9-68.

■ *Solution:*

If a short is placed between terminals A and B, R_2 will be shorted out, so the current between points A and B, and therefore the Norton equivalent current, will be limited by only R_1 and R_3 and will equal [Figure 9-69(a)]

$$I_N = \frac{V_S}{R_T} = \frac{V_S}{R_1 + R_3} = \frac{18\,\text{V}}{9\,\Omega + 9\,\Omega} = 1\,\text{A}$$

Replacing V_S with a short, you can see that our Norton equivalent resistance between terminals A and B is made up of R_1 and R_3 in series with one another, and both are in parallel with R_2. R_N will therefore be equal to [Figure 9-69(b)]

$$R_{1,3} = R_1 + R_3 = 9\,\Omega + 9\,\Omega = 18\,\Omega$$

$$R_T = \frac{R_{1,3} \times R_2}{R_{1,3} + R_2} = \frac{18\,\Omega \times 9\,\Omega}{18\,\Omega + 9\,\Omega} = 6\,\Omega$$

The Norton equivalent circuit is shown in Figure 9-69(c).

SELF-TEST REVIEW QUESTIONS FOR SECTION 9-9

1. A constant voltage source will have a _____ internal resistance, while a constant current source will have a _____ internal resistance.   (large or small)
2. The superposition theorem is a logical way of analyzing networks with more than one _____.
3. Thévenin's theorem represents a complex two-terminal network as a single _____ source with a series-connected single _____.
4. Norton's theorem also allows you to analyze complex two-terminal networks as a single _____ source in parallel with a single resistor.

Series–Parallel Circuits (Figure 9-70)   **SUMMARY**

1. If current has only one path to follow through a component, that component is connected in series.
2. If the total current has two or more paths to follow, these components are connected in parallel.

FIGURE 9-70 Series–Parallel Circuits.

Example Circuit

Solving for Resistance, Voltage, Current, and Power in a Series–Parallel Circuit

STEP 1 Determine the circuit's total resistance.

> *Step A* Solve for series-connected resistors in all parallel combinations.
>
> *Step B* Solve for all parallel combinations.
>
> *Step C* Solve for remaining series resistances.

STEP 2 Determine the circuit's total current.

STEP 3 Determine the voltage across each series resistor and each parallel combination (series equivalent resistor).

STEP 4 Determine the value of current through each parallel resistor in every parallel combination.

STEP 5 Determine the total and individual power dissipated by the circuit.

The Wheatstone Bridge Circuit

Balanced

Unbalanced

FIGURE 9-70 (continued)

Wheatstone bridge

$$R_{\text{variable}} = R_{\text{unknown}}$$

Branch resistances are equal.

$$R_{\text{unknown}} = R_{\text{variable}} \times \left(\frac{R_1}{R_2}\right)$$

Branch resistors, R_1 and R_2, are not equal.

3. All electronic equipment is composed of many components interconnected to form a combination of series and parallel (series–parallel) circuits.

4. The total resistance in a series–parallel circuit can be calculated by:

Step A: Determining an equivalent resistance of all series-connected resistors in parallel combinations.

Step B: Determining the equivalent resistance of all parallel-connected combinations.

Step C: Determining the equivalent resistance of the remaining series-connected resistances.

5. a. The voltage division in a series–parallel circuit can be calculated by:

Step 1: Determining the circuit's total resistance by applying steps A, B, and C.

Step 2: Determining the circuit's total current.

Step 3: Determining the voltage drop across each series resistor and each parallel combination (series equivalent resistor).

b. The fourth step is to:

Step 4: Determine the amount of current flowing through each of the parallel combinations.

c. The fifth and final step is to:

Step 5: Determine the total power and individual power figures.

6. The Wheatstone bridge is a series–parallel circuit that was first developed in 1850 by Charles Wheatstone to measure resistance.

7. The R–$2R$ ladder circuit is a series–parallel circuit used for digital-to-analog conversion.

8. Troubleshooting is the process of locating and diagnosing malfunctions or breakdowns in equipment by means of systematic checking or analysis.

9. a. If a series-connected resistor in a series–parallel circuit opens, there cannot be any current flow, and the source voltage will appear across the open and 0 V will appear across all the others.

b. If a parallel branch in a series–parallel circuit opens, the overall circuit resistance will increase and the total current will decrease. A greater parallel resistance will cause a greater voltage drop across that parallel circuit.

10. a. If a series-connected resistor in a series–parallel circuit shorts, the total circuit resistance will decrease, resulting in an increase in total current. The fault can be located by measuring the voltage across the shorted resistor, which will be 0 V.

b. If a parallel branch in a series–parallel circuit shorts, the total resistance will decrease, resulting in a total current increase. This smaller parallel circuit resistance will result in a smaller voltage drop across the parallel circuit.

11. A resistor's resistance will typically change with age, resulting in a total resistance change and therefore a total current change. The faulty component can be located due to its abnormal voltage drop.

12. An ideal voltage source will have an internal resistance of $0 \ \Omega$ and provide a constant output voltage regardless of what load resistance is connected across its output.

Theorems for DC Circuits

13. In reality, every practical voltage source will have some level of inefficiency, and this is represented as an internal resistance, and the output voltage will remain relatively constant despite variations in load resistance.

14. An ideal current source will have an infinite internal resistance and provide a constant output current, regardless of what load resistance is connected across its output.

15. In reality, every practical current source has a relatively high internal resistance, and the output current will remain relatively constant despite variations in load resistance.

16. The superposition theorem is a useful tool when analyzing circuits with more than one voltage source.

17. Thévenin's theorem is another handy tool that can be used to represent complex series–parallel networks as a single voltage source (V_{TH}) in series with a single resistor (R_{TH}).

18. Norton's theorem can also simplify a complex series–parallel network to an equivalent form consisting of a single current source (I_N) in parallel with a single resistor (R_N).

NEW TERMS

Analog data	Current source	Loading
Balanced bridge	Digital data	Network
Bleeder current	Digital-to-analog converter	Norton resistance
Calibrated	Interface	Norton's theorem
Computer	Ladder circuit	Norton voltage

Series–parallel circuit   Thévenin's theorem   Voltage source

Superposition theorem   Thévenin voltage   Wheatstone bridge

Thévenin resistance   Unbalanced bridge

Multiple-Choice Questions

1. A series–parallel circuit is a combination of:
   a. Components connected end to end
   b. Series (one current path) circuits
   c. Both series and parallel circuits
   d. Parallel (two or more current path) circuits

2. Total resistance in a series–parallel circuit is calculated by applying the _____ resistance formula to series-connected resistors and the _____ resistance formula to resistors connected in parallel.
   a. Series, parallel   c. Series, series
   b. Parallel, series   d. Parallel, parallel

3. Total current in a series–parallel circuit is determined by dividing the total _____ by the total _____.
   a. Power, current   c. Current, resistance
   b. Voltage, resistance   d. Voltage, power

4. Branch current within series–parallel circuits can be calculated by:
   a. Ohm's law   c. Kirchhoff's current law
   b. The current-divider formula   d. All of the above

5. All voltages on a circuit diagram are with respect to _____ unless otherwise stated.
   a. The other side of the component   c. Ground
   b. The high-voltage source   d. All of the above

6. A _____ ground has the negative side of the source voltage connected to ground, while a _____ ground has the positive side of the source voltage connected to ground.
   a. Positive, negative   c. Positive, earth
   b. Chassis, earth   d. Negative, positive

7. The output voltage will always _____ when a load or voltmeter is connected across a voltage divider.
   a. Decrease   c. Increase
   b. Remain the same   d. All of the above could be considered true.

8. A Wheatstone bridge was originally designed to measure:
   a. An unknown voltage   c. An unknown power
   b. An unknown current   d. An unknown resistance

9. A balanced bridge has an output voltage:
   a. Equal to the supply voltage   c. Of 0 V
   b. Equal to half the supply voltage   d. Of 5 V

10. The total resistance of a ladder circuit is best found by starting at the point _____ the source.

    a. Nearest to   **c.** Midway between
    b. Farthest from   **d.** All of the above

11. The R–$2R$ ladder circuit finds its main application as a(an):

    a. Analog-to-digital converter
    b. Digital-to-analog converter
    c. Device to determine unknown resistance values
    d. All of the above

12. The Norton equivalent resistance (R_N) is always in _____ with the Norton equivalent current source (I_N).

    a. Proportion   **c.** Parallel
    b. Series   **d.** Either series or parallel

13. An ideal current source has _____ internal resistance, while an ideal voltage source has _____ internal resistance.

    a. An infinite, $0 \, \Omega$ of   **c.** $0 \, \Omega$ of, an infinite
    b. No, a large amount   **d.** No, an infinite

14. The Thévenin equivalent resistance (R_{TH}) is always in _____ with the Thévenin equivalent voltage (V_{TH}).

    a. Proportion   **c.** Parallel
    b. Series   **d.** Either series or parallel

15. The superposition theorem is useful for analyzing circuits with:

    a. Two or more voltage sources   **c.** Only two voltage sources
    b. A single voltage source   **d.** A single current source

16. A resistor, when it burns out, will generally:

    a. Decrease slightly in value   **c.** Short
    b. Increase slightly in value   **d.** Open

17. In a series–parallel resistive circuit, an open series-connected resistor will cause _____ current, whereas an open parallel-connected resistor will result in a total current _____.

    a. An increase in, decrease   **c.** Zero, decrease
    b. A decrease in, increase   **d.** None of the above

18. In a series–parallel resistive circuit, a shorted series-connected resistor will cause _____ current, whereas a shorted parallel-connected resistor will result in a total current _____.

    a. An increase in, increase   **c.** An increase in, decrease
    b. A decrease in, decrease   **d.** A decrease in, increase

19. A resistor's resistance will typically _____ with age, resulting in a total circuit current _____.

    a. Decrease, decrease   **c.** Decrease, increase
    b. Increase, increase   **d.** All of the above

20. The maximum power transfer theorem states that maximum power is transferred from source to a load when load resistance is equal to source resistance.

   a. True   **b.** False

Essay Questions

21. State the five-step method for determining a series–parallel circuit's resistance, voltage, current, and power values. (9-6)

22. Illustrate the following series–parallel circuits:

   a. R_1 in series with a parallel combination R_2, R_3, and R_4.
   b. R_1 in series with a two-branch parallel combination consisting of R_2 and R_3 in series and R_4 in parallel.
   c. R_1 in parallel with R_2, which is in series with a three-resistor parallel combination, R_3, R_4, and R_5.

23. Using the example in Question 22(c), apply values of your choice and apply the five-step procedure. (9-6)

24. Describe what is meant by "loading of a voltage-divider circuit." (9-7-1)

25. Illustrate and describe the Wheatstone bridge in the: (9-7-2)

   a. Balanced condition
   b. Unbalanced condition
   c. Application of measuring unknown resistances

26. Describe how the ladder circuit acts as a current divider. (9-7-3)

27. Briefly describe the difference between a voltage source and a current source. (9-9-1)

28. Briefly describe the following theorems: (9-9)

   a. Superposition   **c.** Norton's
   b. Thévenin's   **d.** Maximum power transfer

29. What would be the advantages of Thévenin's and Norton's theorems to obtain an equivalent circuit? (9-9)

30. Draw the components that would exist in a: (9-9)

   a. Thévenin equivalent circuit   **b.** Norton equivalent circuit

31. Describe the steps involved in obtaining a: (9-9)

   a. Thévenin equivalent circuit   **b.** Norton equivalent circuit

32. When troubleshooting series–parallel circuits, describe what effect (9-8)

   a. An open series
   b. An open parallel-connected resistor

   would have on total current and resistance, and how the opened resistor could be isolated.

33. When troubleshooting series–parallel circuits, describe what effect (9-8-2)

   a. A shorted series
   b. A shorted parallel-connected resistor

   would have on total current and resistance, and how the shorted resistors could be isolated.

FIGURE 9-71

FIGURE 9-72

34. Describe what effect a resistor's value variation would have and how it could be recognized. (9-8-3)

35. Give the divider formula, Ohm's law, and Kirchhoff's laws used to determine: (9-6)

   a. Branch currents
   b. Voltage drops in a series–parallel circuit

  (List all six.)

Practice Problems

36. R_3 and R_4 are in series with one another and are both in parallel with R_5. This parallel combination is in series with two series-connected resistors, R_1 and R_2. $R_1 = 2.5$ kΩ, $R_2 = 10$ kΩ, $R_3 = 7.5$ kΩ, $R_4 = 2.5$ kΩ, $R_5 = 2.5$ MΩ, and $V_S = 100$ V. For these values, calculate:

   a. Total resistance
   b. Total current
   c. Voltage across series resistors and parallel combinations
   d. Current through each resistor
   e. Total and individual power figures

37. Referring to the example in Question 36, calculate the voltage at every point of the circuit with respect to ground.

38. A 10 V source is connected across a series–parallel circuit made up of R_1 in parallel with a branch made up of R_2 in series with a parallel combination of R_3 and R_4. $R_1 = 100$ Ω, $R_2 = 100$ Ω, $R_3 = 200$ Ω, and $R_4 = 300$ Ω. For these values, apply the five-step procedure, and also determine the voltage at every point of the circuit with respect to ground.

FIGURE 9-73

FIGURE 9-74

39. Calculate the output voltage (V_{RL}) in Figure 9-71 if R_L is equal to:
    a. 25 Ω   **b.** 2.5 kΩ   **c.** 2.5 MΩ
40. What load current will be supplied by the current source in Figure 9-72 if R_L is equal to:
    a. 25 Ω   **b.** 2.5 kΩ   **c.** 2.5 MΩ
41. Use the superposition theorem to calculate total current:
    a. Through R_2 in Figure 9-73(a)
    b. Through R_3 in Figure 9-73(b)
42. Convert the following voltage sources to equivalent current sources:
    a. $V_S = 10$ V, $R_{int} = 15$ Ω
    b. $V_S = 36$ V, $R_{int} = 18$ Ω
    c. $V_S = 110$ V, $R_{int} = 7$ Ω
43. Use Thévenin's theorem to calculate the current through R_L in Figure 9-74. Sketch the Thévenin and Norton equivalent circuits for Figure 9-74.
44. Sketch the Thévenin and Norton equivalent circuits for the networks in Figure 9-75.
45. Convert the following current sources to equivalent voltage sources:
    a. $I_S = 5$ mA, $R_{int} = 5$ MΩ   **c.** $I_S = 0.0001$ A, $R_{int} = 2.5$ kΩ
    b. $I_S = 10$ A, $R_{int} = 10$ kΩ

(a)   **(b)**

FIGURE 9-75

Troubleshooting Questions

46. Referring to the example circuit in Figure 9-49, describe the effects you would get if a resistor were to short, and if a resistor were to open.

47. Design a simple five-resistor series–parallel circuit and insert a source voltage and resistance values. Apply the five-step series–parallel circuit procedure, and then theoretically open and short all the resistors and calculate what effect would occur and how you would recognize the problem.

48. Carbon composition resistors tend to increase in resistance with age, while most other types generally decrease in resistance. What effects would resistance changes have on their respective voltage drops?

49. Use Thévenin's theorem to simplify the circuit in Figure 9-76. What effect would the following faults have on the Thévenin equivalent circuit?

   a. R_2 is shorted.   **b.** R_2 is open.

50. Calculate the Norton equivalent for Figure 9-76 and describe what circuit differences will occur for the same faults listed in Question 49.

FIGURE 9-76

O B J E C T I V E S

After completing this chapter, you will be able to:

1. State the difference between an analog and digital readout multimeter.

2. Explain the construction and operation of a D'Arsonval or moving coil meter movement.

3. Describe meter resistance and sensitivity.

4. Describe why, when using the ammeter to measure current:
   a. Shunt resistors are used to achieve different range scales.
   b. Five rules must be applied.
   c. It affects the circuit in which it is connected.

5. Describe why, when using the voltmeter to measure voltage:
   a. Multiplier resistors are used to achieve different range scales.

   b. Voltmeter sensitivity determines meter accuracy.
   c. Five rules must be applied.
   d. It affects the circuit in which it is connected.

6. State the difference between an earth and chassis ground.

7. Explain why voltage measurements are usually measured with respect to ground.

8. Describe why, when using the ohmmeter to measure resistance:
   a. An internal battery source is used to supply current.
   b. The ohmmeter scale is nonlinear.
   c. The range scales of an ohmmeter are interpreted differently from those of the voltmeter and ammeter.
   d. Four rules must be applied.

The Series–Parallel Analog Multimeter

The Frog Lost!

The first meter movement that indicated values of current was called a galvanometer and therefore must have been invented by the Italian physiologist Luigi Galvani. This current meter was actually invented by Frenchman Claude Pouillet in 1837; however, the inventor fell into obscurity. Later his invention was named the galvanometer after Galvani, whose name was well known. One of Galvani's students discovered that a frog's legs twitched when it was touched by metal with a static machine operating nearby. Galvani followed up this observation, and discovered that a frog's leg also twitched when two different metals were connected at one end, and the other ends touched to the dissected leg. His theory was that electricity from the nerves was discharging through the metal circuit. Although he explained the frog-leg reaction incorrectly, this experiment gained him much fame; enough, in fact, to have Pouillet's invention attached to his name.

In the preceding three chapters we examined series circuits, parallel circuits, and finally a combination of the two: series–parallel circuits. In this chapter you will analyze the analog multimeter, which is used to measure or determine the magnitude of voltage, current, and resistance. Although the digital multimeter is more commonly used to measure these three electrical quantities, the analog multimeter is still in use and contains many good examples of practical series–parallel circuits. These applications of series–parallel circuits, therefore, will help reinforce your understanding of series–parallel networks.

Multimeter
Piece of electronic test equipment that can perform multiple tasks in that it can be used to measure voltage, current, or resistance.

10-1 THE ANALOG METER MOVEMENT

The **multimeter** is a device used for measurement and troubleshooting in electronic equipment. It is called a multimeter because it is like having three meters all rolled into one, as shown in Figure 10-1(a). The multimeter's three units are selected by the main function switch:

(a)

(b)

FIGURE 10-1   Multimeters. (a) Analog. (b) Digital.

1. The **ammeter** unit measures current.
2. The **voltmeter** unit measures voltage.
3. The **ohmmeter** unit measures resistance.

For many years, the analog readout type of multimeter shown in Figure 10-1(a) was the only type available. At present, these analog readout multimeters are still being used, but are slowly being replaced by digital readout multimeters, similar to the type shown in Figure 10-1(b).

Analog is a word that means "similar to." For example, a ball is an analog of the world. With an analog readout multimeter, the amount of pointer deflection across the scale is an analog, or similar to, the magnitude of the electrical property being measured **(analog data).** This type of indicator is therefore called an **analog display,** as opposed to a **digital display,** which makes use of deci-

Ammeter
Meter placed in the path of current flow to measure the amount.

Voltmeter
Instrument designed to measure the voltage or potential difference. Its scale can be graduated in kilovolts, volts, or millivolts.

Ohmmeter
Measurement device used to measure electrical resistance.

Analog Data
Information that is represented in a continuously varying form, as opposed to digital data, which have distinct or discrete values.

Analog Display
Display in which a moving pointer or some other analog device is used to display the data.

Digital Display
Display in which either light-emitting diodes (LEDs) or liquid crystal diodes (LCDs) form digits which are used to display the data.

FIGURE 10-2 **Moving Coil Meter Movement.**

mal digits to display the magnitude of the electrical property **(digital data).** Let us now examine the internal workings of the analog multimeter.

10-1-1 *Construction*

The D'Arsonval or moving coil type of meter movement illustrated in Figure 10-2 was invented by Jacques Arsène D'Arsonval and can be used in instruments to measure current, voltage, or resistance. A coil of copper wire is mounted on a moving core known as an **armature,** which is free to rotate between the two poles of a stationary permanent magnet. A spiral spring is attached to the moving coil and holds the armature in a rest position so that the pointer, which is also attached to the armature, is pointing to 0 on the scale.

10-1-2 *Operation*

This moving coil type of meter movement is used in the analog multimeter, and whether we are measuring current, voltage, or resistance, a current will always flow in the moving coil to represent the amount of current, voltage, or resistance being measured with our two test leads. Let's examine now what will actually happen when a current flows through the moving coil.

When a current is passed through the moving coil, a magnetic field is set up by the electromagnet, producing a north pole on the right side, and a south pole on the left side of the moving coil, as seen in Figure 10-3(a) [left-hand rule for

FIGURE 10-3 **Moving Coil Meter Movement Operation.**

electromagnets, Figure 10-3(b)]. The magnetic field of the electromagnet will interact with the magnetic field of the permanent magnet. This will force the armature to rotate in a clockwise direction, since like poles repel one another, as summarized in Figure 10-4.

The south pole generated in the electromagnet is repulsed by the south pole of the permanent magnet and attracted by the north pole of the permanent magnet, while the north pole generated in the electromagnet is repulsed by the north pole of the permanent magnet and attracted by the south pole of the permanent magnet. The overall effect is to overcome the tension created by the spring and move the pointer in a clockwise direction. If a greater current is passed through the electromagnet, a larger magnetic field will be generated, causing a stronger magnetic interaction between the electromagnet and permanent magnet, thus increasing the repulsion and attraction, resulting in an increased movement of the pointer across the scale in the clockwise direction. A greater current, therefore, causes a greater meter deflection.

If current flows through the electromagnet in the opposite direction, it will generate a north pole in the left side of the electromagnet and a south pole in the right side of the electromagnet. This will result in a counterclockwise movement of the pointer so that the pointer will strike the physical pointer stop, as shown in Figure 10-5. Too large a current could damage the meter, and this is why the meter leads are said to be polarity (positive and negative) sensitive. Therefore, when measuring a potential difference, the red positive test lead must always be placed on the point that is positive with respect to the other terminal, or, said another way, the negative black test lead must always probe the terminal that is negative with respect to the other terminal. Another point to remember is

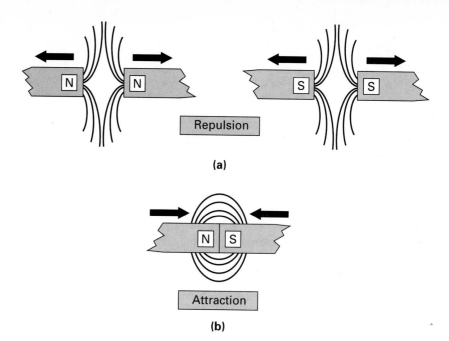

Repulsion

(a)

Attraction

(b)

FIGURE 10-4   Magnetic Law of Attraction and Repulsion. (a) Like
Poles Repulse One Another. (b) Unlike Poles Attract One Another.

Pointer Stop

Black

Red

FIGURE 10-5   Counterclockwise Meter
Movement.

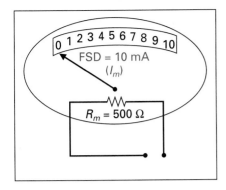

FIGURE 10-6 **Simplified Diagram of Meter Movement.**

that since gravity will have an effect on the pointer, the meter should be placed flat on its back so as to obtain the most accurate readings.

10-1-3 *Meter Resistance and Sensitivity*

The coil of the armature in a moving coil meter possesses some value of resistance and is called the movement's internal resistance (R_m). This internal **meter resistance** (R_m) is small and normally in the range 1 to 500 Ω.

    The wire used to make the coil is extremely thin (about the thickness of a human hair) and can therefore not, without damage, carry very much current. The typical current rating of a meter movement is therefore small, ranging from 10 μA to 10 mA. This value specifies the **meter FSD current**, the amount of current needed to cause the meter movement to deflect the needle to its full-scale deflection (FSD) position at the far right side of the scale. The maximum current of a meter movement (I_m) will therefore tell you how sensitive a meter is; for example, if a meter movement needs only 10 μA to deflect the pointer to FSD, it is much more sensitive than a meter movement that requires 10 mA. A simplified diagram of a meter movement showing R_m and I_m is illustrated in Figure 10-6.

Meter Resistance
The resistance of a meter's armature coil.

Meter FSD Current
The value of current needed to cause the meter movement to deflect the needle to its full-scale deflection (FSD) position.

SELF-TEST REVIEW QUESTIONS FOR SECTION 10-1

1. An _____ is used to measure current, a _____ is used to measure voltage, and an _____ is used to measure _____.

2. Define:
  a. R_m
  b. I_m

10-2 THE AMMETER

Figure 10-7(a) illustrates a meter with a maximum meter movement current (I_m) of 1 mA. This meter will function perfectly for any current value in the range from 0 to 1 mA. If the current being measured exceeds 1 mA, which is the cur-

(a)

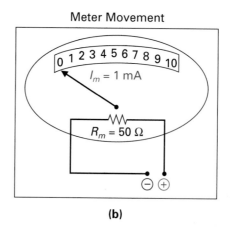

(b)

FIGURE 10-7   1-mA Ammeter. (a) Physical View. (b) Simplified Diagram.

rent rating of the fuse, the fuse will blow and protect the meter from damage. Some modifications are necessary if we wish to measure any current greater than the FSD current (I_m). To make this modification, we must first know the meter's internal resistance (R_m) and FSD current (I_m). In this example, we will use a 1 mA, 50 Ω meter movement shown in the simplified diagram in Figure 10-7(b).

10-2-1 Ammeter Range Scales

The maximum possible current that the meter shown in Figure 10-7(a) can measure as it is presently constructed is 1 mA; however, if we wish to measure current in the range from 0 to 10 mA, we will need some other path for the 9 mA to travel through so that only 1 mA will flow through the meter movement.

A resistor is connected in parallel or shunt with the meter movement resistance (R_m) to shunt the damaging current away from the meter movement, as shown in Figure 10-8. The next question is: What will be the value of the **shunt resistor** (R_{sh}) so that it will shunt 9 mA away from the meter movement when we are measuring a current value of 10 mA? This can be calculated by first finding the voltage across R_{sh}.

Shunt Resistor
A resistor connected in parallel or shunt with the meter movement of an ammeter.

$$V_{Rm} = I_{Rm} \times R_m = 1 \text{ mA} \times 50 \ \Omega = 50 \text{ mV}$$

Since $V_{Rm} = V_{Rsh}$ (the resistors are connected in parallel), the value of R_{sh} can therefore be calculated by applying Ohm's law:

$$R_{sh} = \frac{V_{Rsh}}{I_{Rsh}} = \frac{50 \text{ mV}}{9 \text{ mA}} = 5.6 \ \Omega$$

If a maximum current of 10 mA is measured by the meter, the 10 mA will travel into the meter, where it will split through a parallel circuit made up of R_m and R_{sh}. The current split will be as follows:

$$I_{Rm} = \frac{V_{Rm}}{R_m} = \frac{50 \text{ mV}}{50 \ \Omega} = 1 \text{ mA}$$

$$I_{Rsh} = \frac{V_{Rsh}}{R_{sh}} = \frac{50 \text{ mV}}{5.6 \ \Omega} = 9 \text{ mA}$$

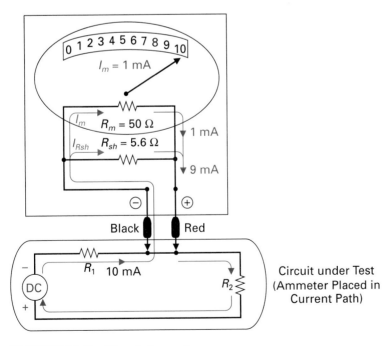

FIGURE 10-8   10-mA Ammeter.

For a current of 10 mA, you will have 1 mA flowing through the moving coil and this will cause FSD.

To have a multirange ammeter, there will have to be some way of switching in different values of shunt resistors for different current ranges. Figure 10-9 shows how this can be achieved. Figure 10-9(a) illustrates the different shunt resistors needed for different ranges.

If the rotary switch is placed in position A (0 to 1 mA range), there is no need for a shunt resistor because the FSD current for the moving coil is 1 mA. With the rotary switch in position B (0 to 10 mA range), we have switched in a

Rotary Switch

FIGURE 10-9   Multirange Ammeter. (a) Internal Circuit. (b) External Appearance. (c) Ammeter Symbol in Circuit.

shunt resistance of 5.6 Ω, and, as calculated previously, this 5.6 Ω resistor will shunt 9 mA and allow 1 mA through R_m when a maximum of 10 mA is being measured.

In position C (0 to 100 mA), the rotary switch has switched in the shunt resistor R_{sh2}, whose job it is to shunt 99 mA away from the moving coil if 100 mA is being measured. Since we already know that 50 mV will appear across all shunt resistors,

$$R_{sh2} = \frac{V_{Rsh2}}{I_{Rsh2}} = \frac{50\text{ mV}}{99\text{ mA}} = 0.5\ \Omega$$

If the rotary switch is placed in position D (0 to 1 A or 1000 mA), R_{sh3} is switched in parallel with R_m to shunt 999 mA away from the moving coil. R_{sh3} can be calculated and is equal to

$$R_{sh3} = \frac{V_{sh3}}{I_{sh3}} = \frac{50\text{ mV}}{999\text{ mA}} = 0.05\ \Omega$$

Figure 10-9(b) shows the external appearance of the rotary switch that is used by the technician to select the ammeter range scale required. Figure 10-9(c) illustrates the schematic symbol of the ammeter (AM) inserted in a circuit to measure current.

With multiple ranges, the scale has to be interpreted by the technician differently, depending on the range scale selected. For example, if the 1-mA range is selected, the 10 on the scale is now equivalent to 1 mA, the 9 on the scale is equivalent to 0.9 mA, the 8 to 0.8 mA, and so on. If the 10-mA range is selected, 10 on the scale is equal to 10 mA, 9 to 9 mA, 8 to 8 mA, and so on. If the 100-mA range is selected, a 2 on the scale is equivalent to 20 mA, all the way up to the 10 on the scale, which in this case is equal to 100 mA. On the 1-A (1000-mA) range scale, 6 on the scale is equivalent to 0.6 A or 600 mA, 7 on the scale is equal to 0.7 A or 700 mA, and so on.

Some range scales are not oriented only around 10, but can incorporate a scale of 30, as well as the 10, as seen in Figure 10-10. If the 100-mA or 1-A scales are selected, the upper scale from 0 to 10 is used, with 10 on the scale representing 100 mA or 1 A. If the 300-mA or 3-A scales are selected, the lower scale

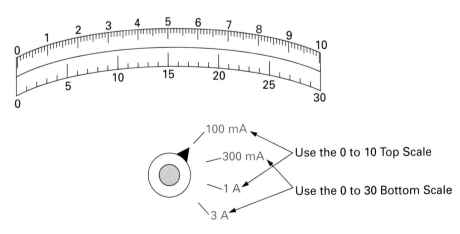

FIGURE 10-10 Ammeter Scales.

from 0 to 30 is used, with 30 on the scale representing 300 mA or 3 A, respectively.

10-2-2 Measuring Current

There are some important points to remember when using the ammeter to measure current in a circuit, as shown in Figure 10-11.

1. Always set the selector to the higher scale first (amperes) and then reduce as needed to the milliampere or microampere ranges, just in case a larger current than anticipated is within the circuit, in which case the ammeter could be damaged.

2. The positive lead of the ammeter (red) should be connected in the circuit so that it can be traced back to the positive side of the voltage supply, and the negative lead of the meter (black) should be connected in the circuit so that it can be traced back to the negative side of the voltage supply.

3. The ammeter must always be connected in the path of the current flow and, therefore, the circuit path must be opened and the ammeter inserted into the circuit.

4. Most analog meters have an accuracy of ±3% of full scale. For this reason, current readings should be as close to full scale as possible. For example, if 7 mA was measured on the 10-mA scale, the maximum error is ±0.3 mA. If 7 mA was measured on the 100-mA scale, however, the maximum error is ±3% of 100 or 3 mA. This means that on the 10-mA scale, a 7-mA reading could read from 6.7 to 7.3 mA, whereas on the 100-mA scale a 7-mA reading could read from 4 to 10 mA.

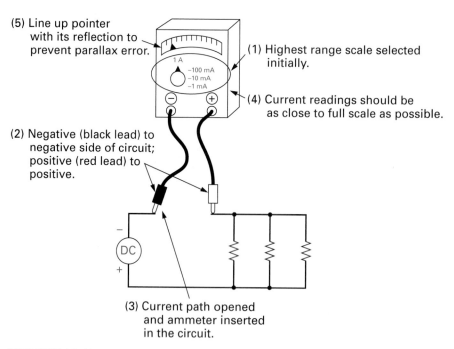

FIGURE 10-11   Connecting an Ammeter in Circuit to Measure Current.

5. Most analog meters include a mirror behind the scale and pointer. A reflection of the pointer will therefore appear in the mirror, and only when you are directly in front of the scale will the pointer line up with its reflection and enable you to read off an accurate value from the scale. The mirror is included to prevent a **parallax error.**

Parallax Error
The apparent displacement of an object's position caused by a shift in the point of observation of the object.

10-2-3 *Ammeter Loading Effect*

The ammeter is placed in a circuit to measure the current flowing in that circuit. The meter should ideally offer no resistance to the circuit, as any resistance in an ammeter will cause the overall circuit resistance to change, and this will change the current flow that we are trying to measure.

Figure 10-12 illustrates an example of an ammeter set up for the 10-mA range and, as you can see, the total resistance offered by the ammeter is the value of the shunt resistance in parallel with the coil resistance, a total of 5 Ω. The resistance of the circuit (10,000 Ω) is so large compared to the internal resistance of the meter (5 Ω) that this meter resistance can be disregarded, and the measured current will be relatively accurate. In this example, the total circuit resistance will equal 10,000 Ω (10 kΩ) and, therefore, the circuit current will equal 1.2 mA (12 V/10 kΩ). With the ammeter in the circuit, the total resistance will now equal 10,005 Ω and, therefore, the circuit current will equal 1.199 mA (12 V/10,005 Ω). It can be seen therefore that the ammeter has no significant effect on the circuit current it measures.

$$R_T = \frac{R_m \times R_{sh}}{R_m + R_{sh}} = \frac{50 \times 5.6}{50 + 5.6} = 5.0 \ \Omega$$

FIGURE 10-12 **Effect of Ammeter in Circuit.**

1. To have a multirange ammeter, different values of _____ resistors are connected in _____ with the meter's coil resistance.

2. If $I_m = 1$ mA and $R_m = 50\ \Omega$, what value of R_{sh} would be needed to measure 30 mA?

10-3 THE VOLTMETER

A voltage increase causes a current increase and this relationship between voltage and current means that the moving coil type of meter movement discussed previously can also be used to measure voltage, as seen in Figure 10-13(a). In this explanation we will, for variety, use a different movement, where $I_m = 1$ mA and $R_m = 10\ \Omega$. If 10 mV is connected across 10 Ω of moving coil resistance, a current of 1 mA will flow:

$$I = \frac{V}{R} = \frac{10\ \text{mV}}{10\ \Omega} = 1\ \text{mA}$$

This will cause full-scale deflection of the meter. With this setup, the meter can be used to measure any voltage from 0 to 10 mV.

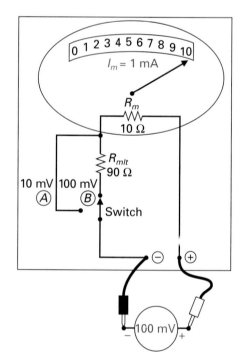

FIGURE 10-13   Voltmeter. (a) 10 mV Voltmeter. (b) 100 mV Voltmeter.

10-3-1 Voltmeter Range Scales

To measure a voltage greater than 10 mV, we will utilize a resistor, referred to as a **multiplier resistor,** to drop the extra voltage. For example, if we wished to measure a voltage in the range of 0 to 100 mV, we are already aware that a 10-mV drop across the moving coil resistance ($R_m = 10\ \Omega$) is all that is necessary to cause a full-scale deflection current of 1 mA ($I_m = 10\ \text{mV}/10 = 1\ \text{mA}$). To measure 100 mV, a multiplier resistor is connected in series with R_m to drop the extra 90 mV. The value of the multiplier resistor can be calculated by simply applying Ohm's law.

Multiplier Resistor
A resistor connected in series with the meter movement of a voltmeter.

$$R_{mlt} = \frac{V_{mlt}}{I_{mlt}} = \frac{90\ \text{mV}}{1\ \text{mA}} = 90\ \Omega$$

This modification with the 90 Ω multiplier resistor connected in series with the moving coil resistance will allow us to measure a voltage range from 0 to 100 mV. When 100 mV is being measured across the positive (+) and negative (−) terminals of the meter, 90 mV will be dropped across the multiplier and 10 mV across the moving coil resistor (R_m), causing full-scale deflection.

If a switch is included, as seen in Figure 10-13(b), we can either (1) switch out the multiplier (position A) to convert the meter to a 10-mV range, or (2) if position B is selected, the multiplier resistor can be brought in series with R_m to obtain a 100-mV range.

Figure 10-14(a) illustrates a multirange voltmeter in which four multiplier resistors can be switched into circuit. The voltage drop across the moving coil

(a)

(b)

(c)

FIGURE 10-14   Multirange Voltmeter. (a) Internal Circuit. (b) External Appearance. (c) Voltmeter Symbol in Circuit.

resistance must always be 10 mV to cause FSD. The remaining voltage on whichever range has been selected will be dropped across the series-connected multiplier resistor.

Position A. If the 10-mV range is selected, there is no need for a multiplier as 5 mV across the (+) and (−) terminals will cause one-half FSD and 10 mV across the terminals will cause FSD.

Position B. If the 100-mV range is selected, the difference in voltage, which in this case is 90 mV (100 mV − 10 mV), must be dropped across a multiplier resistor (R_{mlt1}) whose resistance will equal

$$R_{mlt1} = \frac{V_{mlt}}{I_{mlt}} = \frac{90 \text{ mV}}{1 \text{ mA}} = 90 \text{ }\Omega$$

Position C. When the 1-V (1000-mV) range is selected, the voltage difference that will have to be dropped across the multiplier resistor (R_{mlt2}) will be 990 mV (1000 mV − 10 mV). The value of R_{mlt2} will therefore be

$$R_{mlt2} = \frac{V_{mlt2}}{I_{mlt2}} = \frac{990 \text{ mV}}{1 \text{ mA}} = 990 \text{ }\Omega$$

Position D. [10-V range (10,000 mV)] The voltage difference is equal to 10,000 mV − 10 mV = 9990 mV.

$$R_{mlt3} = \frac{V_{mlt3}}{I_{mlt3}} = \frac{9990 \text{ mV}}{1 \text{ mA}} = 9990 \text{ }\Omega$$

Position E. [100-V range (100,000 mV)] The voltage difference is equal to 100,000 mV − 10 mV = 99,990 mV.

$$R_{mlt4} = \frac{V_{mlt4}}{I_{mlt4}} = \frac{99,990 \text{ mV}}{1 \text{ mA}} = 99,990 \text{ }\Omega$$

Figure 10-14(b) shows the external appearance of the rotary switch used by the technician to select the voltage range scale required. Figure 10-14(c) illustrates the schematic symbol of the voltmeter being used to measure the voltage drop across a resistor.

FIGURE 10-15   Voltmeter Scales and Ranges.

The range scales of the voltmeter are like the ammeter in that they may not only be oriented around 10, such as 10 mV, 100 mV, 1 V, 10 V, and 100 V, but also around, for example, 3, 30, or 300 V.

Figure 10-15 illustrates a multimeter with six range scales of 100 mV, 300 mV, 1 V, 3 V, 10 V, and 30 V. The top scale of 0 through 10 is used for the 100 mV, 1 V, and 10 V range scales, and the 0 through 30 scale is used for the 300 mV, 3 V, and 30 V range scales. The range scale selected refers to the FSD voltage, or maximum voltage that can be measured.

10-3-2 Voltmeter Sensitivity (Ω/V)

The sensitivity of a voltmeter is measured in a quantity known as the **ohms-per-volt rating** (Ω/V). The Ω/V rating can be calculated by taking the reciprocal of the full-scale deflection current of the moving coil (I_m). For example, a 1 mA meter will have an Ω/V rating of 1000 Ω/V.

Ohms-per-Volt Rating
Value that indicates the sensitivity of a voltmeter. The higher the ohms-per-volt rating, the more sensitive the meter.

$$\text{voltmeter sensitivity (Ω/V)} = \frac{1 \text{ V}}{\text{meter coil current } (I_m)}$$

$$= \frac{1 \text{ V}}{1 \text{ mA}} = 1000 \text{ Ω/V}$$

A meter that only requires 100 μA (I_m) for FSD for the moving coil will have a sensitivity of

$$V_m \text{ sensitivity} = \frac{1 \text{ V}}{I_m}$$

$$= \frac{1 \text{ V}}{100 \text{ μA}}$$

$$= 10 \text{ kΩ/V} \quad \text{or} \quad 10,000 \text{ Ω/V}$$

It can be said, therefore, that if a meter only needs a small current to cause FSD it will have a large Ω/V rating and will be classified as a more sensitive meter.

10-3-3 Measuring Voltage

There are some important points to remember and apply when measuring voltage with a voltmeter, as shown in Figure 10-16.

1. Always set the range selector to the highest range scale first (100 V) and then reduce as needed to the lower ranges (volts or millivolts).
2. Always connect the positive lead (+, red) of the voltmeter to the component end that is connected back to the positive side of the battery or source, and the negative lead (−, black) of the meter to the component end that is connected back to the negative side of the battery or source.

(5) Line up pointer with its reflection to prevent parallax error.

(3) Voltmeter across component to measure voltage drop.

(4) Voltage readings should be as close to full scale as possible.

Black

10 mV
100 mV
1 V
10 V
100 V

(1) Highest range scale selected initially.

DC

Red

(2) Negative lead (−, black) to negative side of component; positive lead (+, red) to positive side of component.

FIGURE 10-16   Connecting a Voltmeter to Measure Voltage.

3. The voltmeter must be connected across or in parallel with the component voltage to be measured.

4. Since most analog meters have an accuracy of ±3% of full scale, voltage readings should be as close to full scale as possible. For example, if a 7-V reading is taken on the 10-volt scale the maximum error is ±3% of 10 V, or 0.3 V. If 7 V was measured on the 100-V scale, however, a maximum error of ±3% of 100 V, or 3 V, could occur. This means that on the 10-V scale a 7 V measurement could read between 6.7 and 7.3 V, while on the 100-V scale a 7 V measurement could read between 4 and 10 V.

5. Always line the pointer up with its reflection in the mirror and then take a reading to prevent parallax error.

10-3-4 *Voltmeter Loading Effect*

A voltmeter is connected in parallel with the component to measure the voltage developed across that component. The voltmeter is, therefore, much easier to use for troubleshooting or testing purposes, as the circuit path does not have to be opened and the meter inserted in the current path, as with the ammeter.

When a voltmeter is connected across a component for voltage measurement, some of the circuit current flows through the voltmeter to deflect the meter movement. Figure 10-17(a) illustrates a voltmeter that has an internal resistance of 100 kΩ being used to measure the voltage drop across R_2. Without the voltmeter connected, the voltage will be dropped equally across R_1 and R_2, causing 5 V across R_2. If the voltmeter is now connected across R_2, the voltmeter resistance of 100 kΩ is now in parallel with the resistance of R_2, causing an equivalent resistance of 4.76 kΩ. The meter has loaded down the circuit and caused this change in resistance, which means that R_2 will not drop 5 V as nor-

FIGURE 10-17 Voltmeter Loading Effect.

mal, but will in fact drop 4.88 V. This error of 0.12 V caused by the loading effect of the meter is very small. As long as the internal voltmeter resistance is large with respect to the resistance of the component across which the voltage is being measured, the reading will be reasonably accurate. In the first example, Figure 10-17(a), the meter's resistance is 100 kΩ, which is very large with respect to the component's resistance of 5 kΩ, so the loading effect of the meter is negligible, causing a small error of 0.12 V.

In Figure 10-17(b), the same voltmeter, with a resistance of 100 kΩ, is being used to measure the voltage drop across R_2, whose resistance is also equal to 100 kΩ. Once again, the voltage of 10 V will be divided equally across R_1 and R_2 when the meter is not connected, causing a 5 V drop across R_2.

If the voltmeter is now connected across R_2 so as to measure the voltage drop across the resistor, the voltmeter's resistance of 100 kΩ is now in parallel with the resistance of R_2, causing an equivalent resistance of 50 kΩ. The voltage drop across R_2, which is normally 5 V, is now pulled down to 3.33 V due to the loading effect of the meter, which is a 1.67 V error.

These two examples illustrate the loading effect of a voltmeter on a circuit under test. However, if a voltmeter has a large internal resistance with respect to the component's resistance (10 or more times greater), the loading effect of the meter can be disregarded and the voltmeter will be more accurate.

10-3-5 *Voltage Reference*

A voltmeter will always measure voltage with respect to some other point. If a resistor has a voltage drop of 30 V, for example, this means that one side of the resistor is 30 V more positive or negative than the other. This could occur if one side of the resistor was at +30 V and the other at +60 V (30 V difference in potential) or if one side was −100 V and the other was −70 V (30 V difference). Generally, when troubleshooting, a voltage measurement is taken with respect to 0 V or **ground,** of which there are two basic types:

1. *Earth ground:* The power lines from your local electric company have a reference point referred to as earth ground; this is because one side of the voltage source is attached to a metal rod and inserted in the ground (Figure 10-18).

Ground
An intentional or accidental conducting path between an electrical circuit or system and the earth, or some conducting body acting in place of the earth.

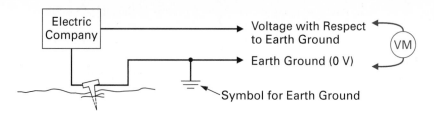

FIGURE 10-18   Earth Ground.

Positive Ground
A system whereby the positive terminal of the voltage source is connected to the system's conducting chassis or body.

Negative Ground
A system whereby the negative terminal of the voltage source is connected to the system's conducting chassis or body.

2. *Circuit or chassis ground:* In electronic equipment, the metal chassis or housing of the equipment is used as a common reference point. One side of the voltage supply is connected to the metal frame or chassis of the equipment and voltage is measured with respect to this common ground reference point, referred to as a circuit, chassis, or frame ground (Figure 10-19).

Another term that is used with all dc voltage sources is **positive ground** or **negative ground.** This refers to which polarity of the battery or source is connected to the chassis; for example, if the negative side of the battery is connected to the chassis, then positive voltages will be measured throughout the circuit with respect to the negative ground [Figure 10-20(a)]. On the other hand, if the positive side of the battery is connected to the chassis, then negative voltages

Symbol for Chassis or Circuit Ground

Metal Frame or Chassis

No need to run cable from bottom of resistor back to positive side of battery, as metal chassis of equipment makes the connection.

Voltages are measured by placing the negative lead on top of the resistor and the positive lead anywhere on the chassis. This will be the same as measuring directly across the resistor, as the chassis is a common positive reference point.

FIGURE 10-19   Chassis Ground.

(a)   (b)

FIGURE 10-20 **Voltage Reference. (a) Negative Ground, Positive Voltage Measurements. (b) Positive Ground, Negative Voltage Measurements.**

with respect to a positive ground will be present throughout the circuit [Figure 10-20(b)]. To obtain positive readings on the voltmeter in this instance, the meter leads will have to be reversed.

If a voltmeter is normally always used to measure voltage with respect to chassis ground (0 V), how could you find the voltage dropped across R_1 in Figure 10-21? The voltmeter will measure 50 V at point A, but this is the voltage dropped across both R_1 and R_2. The voltage dropped across R_1 can be obtained by measuring the voltage at point A with respect to chassis ground (+50 V) and then at point B with respect to chassis ground (+25 V). The difference between these two readings on your voltmeter will be the potential difference between A and B; in this case, 25 V is dropped across R_1.

The next question you are probably asking is: Why would we make two voltmeter checks to find out that we have dropped 25 V across R_1 when we could have just hooked up the voltmeter across R_1 and measured 25 V [Figure 10-22(a) and (b)]. The reason why voltage measurements are almost always measured with respect to chassis ground rather than the other side of the resistor or component is convenience and fault prevention. If voltages are measured with respect to chassis ground, the negative lead of the voltmeter can be permanently attached to the chassis, and you only have to worry about moving the positive meter lead to probe the desired testing points [Figure 10-22(a)].

If you measure the voltage dropped across a resistor, as shown in Figure 10-22(b), both the positive and negative leads must be held by the technician to probe in two places, which means:

1. The technician does not have a free hand because he or she is probing with both voltmeter leads.

FIGURE 10-21 **Voltmeter Probing.**

(a)   (b)

FIGURE 10-22   Measuring Voltage Drop. (a) Two Checks to Find the Voltage Drop across R_1. (b) One Check to Find the Voltage Drop across R_1.

2. If the technician is trying to position two probes at once in two different points, it becomes twice as difficult, and the risk of slipping and causing a short and then possibly an additional problem in the circuit under test is present.

For safety reasons, it is often recommended that the test leads be attached for voltage measurements with the equipment off; then turn it on to observe the reading. This is especially important in high-voltage systems of 300 V or more.

■ **EXAMPLE:**

Using the voltmeter with the negative lead tied to the chassis, the voltages at points *A, B,* and *C* are as shown in Figure 10-23. From these readings, calculate the voltage drops across $R_1, R_2, R_3,$ and R_4.

■ *Solution:*

The difference between *A* and *B* = 40 V − 15 V = 25 V; therefore, 25 V is dropped across R_1.

The difference between *B* and *C* = 15 V − 4 V = 11 V; therefore, 11 V is dropped across R_2 and R_3 (resistors are in parallel).

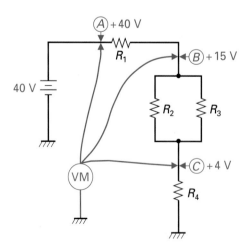

FIGURE 10-23   Calculating Voltage Drop.

The difference between C and chassis ground = 4 V − 0 V = 4 V; therefore, 4 V is dropped across R_4.

SELF-TEST REVIEW QUESTIONS FOR SECTION 10-3

1. To have a multirange voltmeter, different values of _____ resistors are connected in _____ with the meter's coil resistance.
2. Define ohms/volt and how it relates to a meter's sensitivity.

10-4 THE OHMMETER

Resistance can also be measured using the same moving coil meter movement. Figure 10-24(a) illustrates the meter configuration for measuring resistance. The only difference is that a series 1.5 V battery source is used to supply current. When the meter leads are connected together (shorted), the zero-ohms adjust resistor is adjusted to cause full-scale deflection (1 mA); this will occur when the meter's total resistance is equal to

$$R = \frac{V}{I} = \frac{1.5\ V}{1\ mA} = 1500\ \Omega$$

(a)

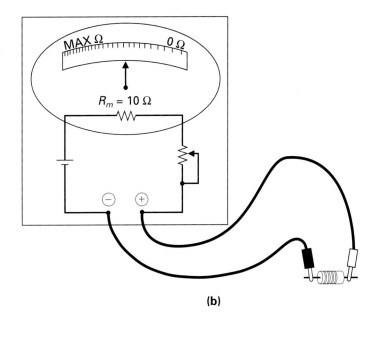

(b)

FIGURE 10-24 Ohmmeter. (a) Zeroing the Meter. (b) Measuring Resistance.

Since the moving coil resistance is equal to 10 Ω (R_m), the zero-ohms adjust resistance will need to be adjusted to equal 1490 Ω (1500 Ω − 10 Ω = 1490 Ω) in order to cause a total circuit resistance of 1500 Ω and, therefore, achieve full-scale deflection current (I_m) of 1 mA. Adjustment of the zero-ohms adjust resistor to obtain FSD is called **zeroing** the meter. When the meter leads are shorted together, the meter will deflect to the full-scale position (right side), indicating that there is no resistance (0 Ω) between the two meter leads. As different values of resistance are inserted between the two meter probes, different values of current will result, causing different levels of pointer deflection. A higher value of resistance will cause a smaller value of current to flow through the meter circuit, resulting in a smaller deflection of the pointer, as shown in Figure 10-24(b).

Zeroing
Calibrating a meter so that it shows a value of zero when zero is being measured.

The reason for the ohms adjust control is due to the discharge of the internal 1.5 V ohmmeter battery. As its voltage falls after months of use, the resistance of the zero-ohms adjust resistor is decreased to still cause an FSD current of 1 mA when the meter leads are shorted together. If, for example, the battery in the circuit of Figure 10-24(a) has discharged to 1 V, the zero-ohms adjust resistor needs to be varied to a value of 990 Ω, and therefore the total meter circuit resistance will equal

$$\text{zero-ohms resistance} + \text{moving coil resistance}$$
$$990\ \Omega + 10\ \Omega = 1000\ \Omega$$

This will cause a meter circuit current of

$$I_m = \frac{V}{R} = \frac{1\ \text{V}}{1000\ \Omega} = 1\ \text{mA}$$

The meter is now said to be zeroed, because 1 mA of meter circuit current is flowing through the moving coil, causing FSD.

When the ohmmeter's probes are shorted together, it causes FSD, which indicates to the technician that 0 Ω (a short circuit) exists between the two probes. On the other hand, if the two probes are open, the meter needle remains in its resting position, indicating that infinite (∞) resistance (maximum possible resistance or open circuit) exists between the two probes.

Figure 10-25(a) illustrates an example ohmmeter, which has a 1 mA, 10-Ω meter movement. Let's now proceed to take three examples of measured resistance and see how the ohmmeter responds to indicate these resistances.

1. If the two leads are shorted together, the total meter circuit resistance will equal

$$R_m + R_{\text{zero}} + \text{measured resistance} = 10\ \Omega + 1490\ \Omega + 0\ \Omega$$
$$= 1500\ \Omega$$

The moving coil current will consequently equal

$$I_m = \frac{V}{R} = \frac{1.5\ \text{V}}{1500\ \Omega} = 1\ \text{mA}$$

and this will cause full-scale deflection [Figure 10-25(b)].

2. If the two meter leads are connected across a 1500 Ω resistor, the total circuit resistance will equal

FIGURE 10-25   Ohmmeter's Response to Resistance. (a) Ohmmeter. (b) 0 Ω. (c) 1500 Ω. (d) 4500 Ω.

$$R_m + R_{\text{zero}} + \text{measured resistance} = 10\ \Omega + 1490\ \Omega + 1500\ \Omega$$
$$= 3000\ \Omega$$

The moving coil current will consequently equal

$$I_m = \frac{V}{R} = \frac{1.5\ \text{V}}{3000\ \Omega} = 0.5\ \text{mA}$$

Since 1 mA causes full-scale deflection, 0.5 mA will cause one-half full-scale deflection [Figure 10-25(c)].

3. If we connect our ohmmeter leads across a resistor of 4500 Ω, the total meter circuit resistance will equal

$$R_m + R_{\text{zero}} + \text{measured resistance} = 10\ \Omega + 1490\ \Omega + 4500\ \Omega$$
$$= 6000\ \Omega$$

The moving coil current consequently equals

$$I_m = \frac{V}{R} = \frac{1.5\ \text{V}}{6000\ \Omega} = 0.25\ \text{mA}$$

Since 1 mA causes full-scale deflection, 0.25 mA will cause one-quarter full-scale deflection [Figure 10-25(d)].

Nonlinear Scale
A scale whose divisions are not uniformly spaced.

Linear Scale
A scale whose divisions are uniformly spaced.

The ohmmeter scale is referred to as a **nonlinear scale,** as opposed to the voltmeter and ammeter scales, which are **linear.** Figure 10-26(a) illustrates the nonlinearity of the ohmmeter, while Figure 10-26(b) shows how the voltmeter and ammeter scales follow a linear variation.

Maybe the best way to clearly see the difference between the two types of scales is first to analyze the ohmmeter's nonlinear scale. The difference with the ohmmeter, first, is that 0 begins on the right side. Between three-quarters FSD and one-half FSD on the ohmmeter's output scale is a change of 500 to 1500 on the input resistance being measured, which is a 1000 Ω change. From one-quarter FSD to one-half FSD is an input resistance change of 3000 Ω. From one-half FSD to three-quarters FSD is the same amount of distance on the output scale, yet the input resistance change is three times greater between one-quarter and one-half FSD. The output scale, therefore, does not vary in direct proportion or at the same rate as the input resistance being measured and is consequently referred to as a nonlinear scale (the divisions are not uniformly spaced).

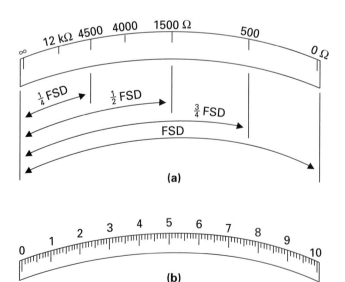

FIGURE 10-26   Nonlinear versus Linear Scale.
(a) Nonlinear Ohmmeter Scale. (b) Linear Voltmeter and Ammeter Scale.

If the output scale varies in direct proportion to the input being measured, as with the voltmeter and ammeter scales, the scale is said to be linear. This can be further explained by using the same simple example.

Between one-quarter and one-half FSD is a change of 2.5 to 5 V on the output scale and is in direct proportion to the input change of 2.5 to 5 V (or amperes). Between one-half and three-quarters FSD is exactly the same amount of change as one-quarter to one-half FSD, both on the output scale and the input being measured, which is why Figure 10-26(b) is referred to as a linear scale (all divisions are evenly spaced).

10-4-1 Ohmmeter Range Scales

Figure 10-27(a) illustrates the external appearance of an ohmmeter with multiple ranges, and Figure 10-27(b) illustrates the internal series and shunt resistors that allow different range settings to be attained. The range scales of an ohmmeter are interpreted differently from those of the ammeter and voltmeter. The ohmmeter's range scales in Figure 10-27(a) of $R \times 1$, $R \times 10$, and $R \times 100$ are used as multipliers; for example, if the pointer is at 500 and the range scale selected is $R \times 10$, the resistor's resistance will be equal to

$$R \times 10 = 500 \times 10 = 5000 \text{ or } 5 \text{ k}\Omega$$

FIGURE 10-27   Multirange Ohmmeter. (a) External Appearance. (b) Internal Circuit. (c) Ohmmeter Symbol in Circuit.

A typical ohmmeter will usually have a small battery of 1.5 V, which is used for the $R \times 1$, $R \times 10$, and $R \times 100$ ranges, while a larger battery of typically 9 or 15 V will be used for the $R \times 1000$ range and higher.

Figure 10-27(b) shows the internal circuit of an ohmmeter. When the $R \times 1$ range is selected, no series multiplier resistor or parallel shunt resistor is connected, and the meter will operate as follows: If the meter leads are shorted together, a resistance of 0 Ω exists between them, and the meter should cause full-scale deflection of the pointer (1 mA of moving coil current through R_m) to the 0 Ω position on the far-right side of the scale.

R_m and R_{zero} are in series with one another, giving a combined resistance of 1500 Ω and causing a total circuit resistance of 1500 Ω. Meter circuit current will therefore equal

$$I = \frac{V}{R} = \frac{1.5 \text{ V}}{1500 \text{ }\Omega} = 1 \text{ mA}$$

which will flow through R_m, causing FSD of the pointer to the 0 Ω position of the scale.

If a 1500-Ω resistor is connected between the two meter leads, the meter should cause the pointer to deflect to the one-half FSD position (0.5 mA of moving coil current through R_m). With the 1500 Ω resistor between the two meter leads, a total circuit resistance of

$$R_t = R_m + R_{\text{zero}} + R_{\text{measured}}$$

$$= 50 \text{ }\Omega + 1450 \text{ }\Omega + 1500 \text{ }\Omega$$

$$= 3000 \text{ }\Omega$$

This will cause a circuit current of $I = V/R = 1.5 \text{ V}/3000 \text{ }\Omega = 0.5 \text{ mA}$, which will cause one-half FSD of the pointer to the 1500 position on the scale.

For the $R \times 10$ and $R \times 100$ scales, an ohmmeter will switch in a different value of series and shunt resistors in order to obtain the $R \times 10$ and $R \times 100$ ohmmeter positions.

Figure 10-27(c) shows the schematic symbol for an ohmmeter.

10-4-2 *Measuring Resistance*

(See Figure 10-28.) There are a few points to remember when measuring resistance with an ohmmeter:

1. Connect (short) the positive and negative leads together and adjust the zero adjust control so that the pointer is at 0 Ω on all scales. This zeroing of the ohmmeter is also known as a confidence test since it lets you know if the meter is functioning.

2. Ensure that no power is connected to the component to be measured, as the circuit voltage and internal ohmmeter battery voltage could aid one another, causing an excessive current, which could damage the meter movement.

3. Connect the leads across the component and read off the resistance indicated by the scale, multiplying by 1, 10, 100, 1000, or more, depending on the ohms range selected.

(4) Lift one lead of component to disconnect it from the circuit, when resistance of component is being measured while component is in circuit.

(3) Read off value and multiply by range.

(1) Zero adjust with meter leads touching.

(2) Ensure that no power is connected to component.

FIGURE 10-28 **Measuring Resistance.**

4. If the resistance of a component is measured while the component is connected in a circuit, an invalid reading may be obtained due to the parallel effects of other resistances in the circuit. This problem can be overcome by disconnecting one end of the component from the circuit to obtain a true reading.

SELF-TEST REVIEW QUESTIONS FOR SECTION 10-4

1. What are the major differences between an ammeter/voltmeter and an ohmmeter?
2. State the four points that should be remembered when measuring resistance.

10-5 THE ANALOG MULTIMETER

After discussing the ammeter, voltmeter, and ohmmeter separately, we will now combine all three to end up with a multipurpose meter (multimeter), or, as they are sometimes called, a **VOM meter** (volt-ohm-milliamp meter), as shown in Figure 10-29. Having all three meters in one package is both convenient and cheaper than having to buy three meters.

VOM Meter
Abbreviation for volt-ohm-milliamp meter.

FIGURE 10-29 Multimeter.

Voltmeter: Has linear scale, uses multiplier resistors to drop excess voltage.

Range switch selects ranges for measuring either current, resistance, or voltage.

Meter jacks: Plug black test lead into negative jack and red lead into positive jack.

Ammeter: Has linear scale, uses shunt resistors to shunt away excess current.

Ohmmeter: Has nonlinear scale, uses battery to generate current and series and shunt resistors for different ranges.

448

1. Which meter has:
   a. A nonlinear scale?
   b. Only shunt resistors?
   c. Only multiplier resistors?
   d. An internal voltage source?
   e. A linear scale?
2. VOM is an abbreviation for _____.

The Analog Readout Volt-Ohm-Milliamp (VOM) Multimeter (Figure 10-30)   SUMMARY

1. The multimeter is a device used for the measurement of current, voltage, and resistance and is used for experimentation with and troubleshooting of electronic equipment.

2. The multipurpose meter has three meters all rolled into one package: the ammeter measures current, the voltmeter measures voltage, and the ohmmeter measures resistance.

3. The analog readout multimeter utilizes a moving coil, which is deflected by a current when measuring volts, ohms, or amperes.

4. The ammeter uses shunt resistors in parallel with the meter movement and has a very low internal resistance when placed in circuit to minimize the loading effect.

5. The voltmeter uses multiplier resistors in series with the meter movement and has a very large internal resistance when connected across a component to minimize the loading effect.

6. The ohmmeter has a combination of series and parallel resistors in and across the meter movement, and uses an internal current source in the form of a battery to deflect the pointer.

NEW TERMS

Analog data
Analog display
Analog readout
Armature
Chassis ground
D'Arsonval meter movement
Digital data
Digital display
Digital readout

Earth ground
Full-scale deflection (FSD)
Linear scale
Meter movement
Meter resistance
Meter sensitivity
Multimeter
Multiplier resistor
Multirange meter

Negative ground
Nonlinear scale
Ohms per volt (Ω/V)
Parallax error
Polarity sensitive
Positive ground
Sensitivity
Shunt resistor
Zeroing

FIGURE 10-30   The Volt-Ohm-Milliamp Multimeter.

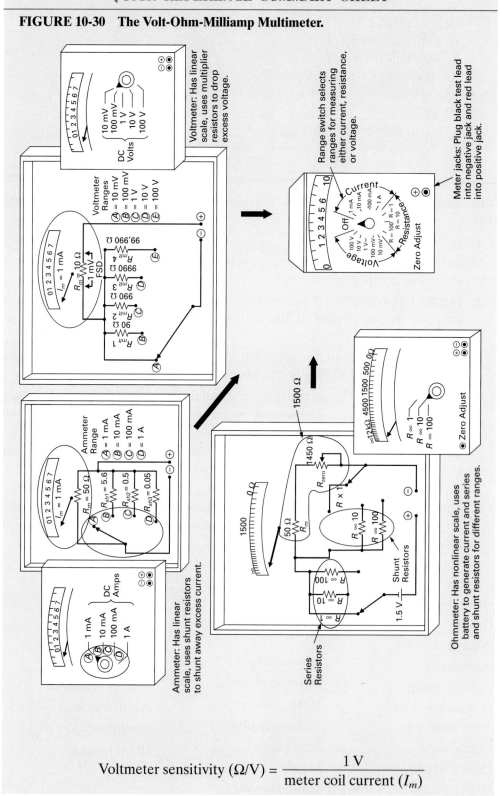

$$\text{Voltmeter sensitivity } (\Omega/\text{V}) = \frac{1 \text{ V}}{\text{meter coil current } (I_m)}$$

Multiple-Choice Questions

1. The device used to measure the number of coulombs per second passing a given point is called

   a. An ammeter   **c.** An ohmmeter
   b. A voltmeter   **d.** A wattmeter

2. The device used to measure the potential difference between two points is called:

   a. An ammeter   **c.** An ohmmeter
   b. A voltmeter   **d.** A wattmeter

3. The device used to measure the amount of opposition to current flow offered by a particular component is called:

   a. An ammeter   **c.** An ohmmeter
   b. A voltmeter   **d.** A wattmeter

4. The _____ or moving coil type of meter movement was named after its inventor.

   a. Joule   **c.** D'Arsonval
   b. Wattmeter   **d.** Guglielmo

5. A meter movement that requires only 10 µA to deflect the pointer to FSD is _____ than a meter that requires 10 mA.

   a. More sensitive   **c.** Insensitive
   b. Less sensitive   **d.** All of the above could be true.

6. Ammeter ranges are accomplished by _____ resistors connected in _____ with the meter movement.

   a. Multiplier, series   **c.** Multiplier, parallel
   b. Shunt, parallel   **d.** Shunt, series

7. Voltmeter ranges are accomplished by _____ resistors connected in _____ with the meter movement.

   a. Multiplier, series   **c.** Multiplier, parallel
   b. Shunt, parallel   **d.** Shunt, series

8. An ammeter should ideally offer _____ resistance to the circuit in which it is connected.

   a. $100 \text{ k}\Omega$ of   **c.** A very low
   b. A very large   **d.** $150 \ \Omega$ of

9. A voltmeter should ideally offer_____ resistance to the circuit in which it is connected.

   a. $100 \text{ k}\Omega$ of   **c.** A very low
   b. A very large   **d.** $150 \ \Omega$ of

10. The sensitivity of a voltmeter is measured in a quantity known as the _____ rating.

    a. Ohms/cm   **c.** Ohms/volt
    b. Coulombs/second   **d.** Amperes/volt

11. The only difference between the ohmmeter and the voltmeter or ammeter is that the ohmmeter uses a/an _____ to supply current.

    a. External voltage source   **d.** Both (b) and (c)
    b. Battery   **e.** Both (a) and (b)
    c. Internal voltage source

12. The two advantages of the digital multimeter are that it is:

    a. Easy to read and cheap in price
    b. Accurate and cheap in price
    c. Accurate and easy to interpret
    d. Inexpensive and has a quick response

13. The two advantages of the analog multimeter are that it is:

    a. Easy to read and cheap in price
    b. Accurate and cheap in price
    c. Accurate and easy to interpret
    d. Inexpensive and has a quick response

14. Which meter uses a nonlinear scale?

    a. Ammeter   **c.** Ohmmeter
    b. Voltmeter **d.** Both (a) and (b)

15. A meter containing an ammeter, voltmeter, and ohmmeter is generally referred to as a:

    a. VOM   **c.** Multimeter
    b. Wattmeter **d.** Both (a) and (c)

Essay Questions

16. List the four steps that should be remembered when using the ohmmeter. (10-4-2)

17. List the five steps that should be remembered when using the voltmeter. (10-3-3)

18. List the five steps that should be remembered when using the ammeter. (10-2-2)

19. Briefly describe the construction and operation of the D'Arsonval or moving coil type of meter movement. (10-1-1 and 10-1-2)

20. Briefly describe how ammeters have multiple range scales by the use of shunt resistors. (10-2-1)

21. Briefly describe how voltmeters have multiple range scales by the use of multiplier resistors (10-3-1)

22. Describe what loading effect an ammeter and voltmeter will have when measuring current and voltage in a circuit. (10-2-3 and 10-3-4)

23. How is the ohms-per-volt rating of a voltmeter obtained? (10-3-2)

24. What is the difference between an earth ground and a chassis ground? (10-3-5)

25. What is a multimeter or VOM? (10-5)

Practice Problems

26. At what position would the pointer be (1 to 10 scale) if 15 μA flows through a 60 μA meter movement?

27. If a voltmeter on the 1000-V range has a sensitivity of 60,000 Ω/V, what is its internal resistance?

28. For a 50 μA, 100-Ω ammeter meter movement, determine the value of shunt resistors for the:

    a. 100 mA range   **b.** 100 μA range

29. What is the value of multiplier resistors for a 50,000 Ω/V voltmeter on the:

    a. 10-V range?   **b.** 100-V range?

30. If an ohmmeter's pointer points to 60 on the $R \times 100$ range, the resistance of the resistor being measured is equal to how many ohms?

31. A voltmeter has a sensitivity of 20,000 Ω/V. Calculate the resistance of the voltmeter if the following voltage scales are being used:

    a. 25 V **b.** 100 V **c.** 1.5 V

32. Calculate the shunt resistances needed for the ranges B, C, and D of the ammeter shown in Figure 10-9 if $I_m = 100$ μA and $R_m = 1$ kΩ.

33. Calculate the multiplier resistances needed for the voltmeter shown in Figure 10-14 if $I_m = 50$ μA and $R_m = 2000$ Ω.

34. Sketch the schematic for the following meters and indicate resistor values.

    a. Ammeter $I_m = 50$ μA, $R_m = 1500$ Ω. Four range scales: 100 μA, 100 mA, 1 A, and 10 A.

    b. Voltmeter $I_m = 0.5$ mA, $R_m = 1500$ Ω. Four range scales: 1 V, 3 V, 10 V, 30 V.

    c. Ohmmeter $I_m = 1$ mA, $R_m = 150$ Ω, battery = 1.5 V, $R_{zero} = $?

35. To double the current and voltage ranges of the meters in Question 34(a) and (b), what changes would have to occur?

CHAPTER

11

O B J E C T I V E S

After completing this chapter, you will be able to:

1. Explain the difference between alternating current and direct current.

2. Describe how ac is used to deliver power and to represent information.

3. Give the three advantages that ac has over dc from a power point of view.

4. Describe basically the ac power distribution system from the electric power plant to the home or industry.

5. Describe the three waves used to carry information between two points.

6. Explain how the three basic information carriers are used to carry many forms of information on different frequencies within the frequency spectrum.

7. Explain the different characteristics of the five basic wave shapes.

8. Describe fundamental and harmonic frequencies.

9. Define the differences between electrical equipment and electronic equipment.

Alternating Current (AC)

The Laser

In 1898, H. G. Wells's famous book, *The War of the Worlds,* had Martian invaders with laserlike death rays blasting bricks, firing trees, and piercing iron as if it were paper. In 1917, Albert Einstein stated that, under certain conditions, atoms or molecules could absorb light and then be stimulated to release this borrowed energy. In 1954, Charles H. Townes, a professor at Columbia University, conceived and constructed with his students the first "maser" (acronym for "microwave amplification by stimulated emission of radiation"). In 1958, Townes and Arthur L. Shawlow wrote a paper showing how stimulated emission could be used to amplify light waves as well as microwaves, and the race was on to develop the first "laser." In 1960, Theodore H. Maiman, a scientist at Hughes Aircraft Company, directed a beam of light from a flash lamp into a rod of synthetic crystal, which responded with a burst of crimson light so bright that it outshone the sun.

An avalanche of new lasers emerged, some as large as football fields, while others were no bigger than a pinhead. They can be made to produce invisible infrared or ultraviolet light or any visible color in the rainbow, and the high-power lasers can vaporize any material a million times faster and more intensely than a nuclear blast, while the low-power lasers are safe to use in children's toys.

At present, the laser is being used by the FBI to detect fingerprints that are 40 years old, in defense programs, in compact disk players, in underground fiber optic communication to transmit hundreds of telephone conversations, to weld car bodies, to drill holes in baby-bottle nipples, to create three-dimensional images called holograms, and as a surgeon's scalpel in the operating room. Not a bad beginning for a device that when first developed was called "a solution looking for a problem."

In this chapter you will be introduced to alternating current (ac). While direct current (dc) can be used in many instances to deliver power or represent information, there are certain instances in which ac is preferred. For example, ac is easier to generate and is more efficiently transmitted as a source of electrical power for the home or business. Audio (speech and music) and video (picture) information are generally always represented in electronic equipment as an alternating current or alternating voltage signal.

In this chapter, we will begin by describing the difference between dc and ac, and then examine where ac is used. Following this we will discuss all the characteristics of ac waveform shapes, and finally, the differences between electricity and electronics.

11-1 THE DIFFERENCE BETWEEN DC AND AC

One of the best ways to describe anything new is to begin by redescribing something known and then discuss the unknown. The known topic is **direct current.** Direct current (dc) is the flow of electrons in one DIRECTion and one direction

Direct Current (dc)
Current flow in only one direction.

**FIGURE 11-1   Direct Current. (a) DC Sources. (b) DC Flow.
(c) Graphical Representation of DC.**

only. Dc voltage is nonvarying and normally obtained from a battery or power supply unit, as seen in Figure 11-1(a). The only variation in voltage from a battery occurs due to the battery's discharge, but even then, the current will still flow in only one direction, as seen in Figure 11-1(b).

A dc voltage of 9 or 6 V could be illustrated graphically as shown in Figure 11-1(c). Whether 9 or 6 V, the voltage can be seen to be constant or the same at any time.

Some power supplies supply a form of dc known as pulsating dc, which varies periodically from zero to a maximum, back to zero, and then repeats. Figure 11-2(a) illustrates the physical appearance and schematic diagram of a battery charger that is connected across two series resistors. The battery charger is generating a waveform, as shown in Figure 11-2(b), known as pulsating dc. At time 1 [Figure 11-2(c)], the power supply is generating 9 V and direct current is

FIGURE 11-2 Pulsating DC.

FIGURE 11-3   Alternating Current.

flowing from negative to positive. At time 2 [Figure 11-2(d)], the power supply is producing 0 V, and therefore no current is being produced. In between time 1 and time 2, the voltage out of the power supply will decrease from 9 V to 0 V; however, no matter what the voltage, whether 8, 7, 6, 5, 4, 3, 2, or 1 V, current will be flowing in only one direction (unidirectional) and is therefore referred to as dc.

Pulsating dc is normally supplied by a battery charger and is used to charge secondary batteries. It is also used to operate motors that convert the pulsating dc electrical energy into a mechanical rotation output. Whether steady or pulsating, direct current is current in only one DIRECTion.

Alternating current (ac) flows first in one direction and then in the opposite direction. This reversing current is produced by an alternating voltage source, as shown in Figure 11-3(a), which reaches a maximum in one direction (positive), decreases to zero, and then reverses itself and reaches a maximum in the opposite direction (negative). This is graphically illustrated in Figure 11-3(b).

During the time of the positive voltage alternation, the polarity of the voltage will be as shown in Figure 11-3(c), so current will flow from negative to positive in a counterclockwise direction.

During the time of the negative voltage alternation, the polarity of the voltage will reverse, as shown in Figure 11-3(d), causing current to flow once again from negative to positive, but in this case, in the opposite clockwise direction.

Alternating Current
Electric current that rises from zero to a maximum in one direction, falls to zero, and then rises to a maximum in the opposite direction, and then repeats another cycle, the positive and negative alternations being equal.

11-2 WHY ALTERNATING CURRENT?

The question that may be troubling you at this point is: If we have been managing fine for the past chapters with dc, why do we need ac?

There are two main applications for ac:

1. *Power transfer:* to supply electrical power for lighting, heating, cooling, appliances, and machinery in both home and industry
2. *Information transfer:* to communicate or carry information, such as radio music and television pictures, between two points

To begin with, let us discuss the first of these applications, power transfer.

11-2-1 *Power Transfer*

There are three advantages that ac has over dc from a power point of view. Let us now discuss these in detail.

Generator
Device used to convert a mechanical energy input into an electrical energy output.

1. Flashlights, radios, and portable televisions all use batteries (dc) as a source of power. In these applications where a small current is required, batteries will last a good length of time before there is a need to recharge or replace them. Many appliances and most industrial equipment need a large supply of current, and in this situation a **generator** would have to be used to generate this large amount of current. Generators operate in the opposite way to motors, in that a generator converts a mechanical rotation input into an electrical output. Generators can be used to generate either dc or ac, but ac generators can be larger, less complex internally, and cheaper to operate, which are the first reasons why we use ac instead of dc for supplying power. Figure 11-4 illustrates a typical ac generator.

2. From a power point of view, ac is always used by electric companies when transporting power over long distances to supply both the home and industry with electrical energy. Recalling the power formula, you will remember

FIGURE 11-4   AC Generator.

that power is proportional to either current or voltage squared ($P \propto I^2$ or $P \propto V^2$), which means that to supply power to the home or industry, we would supply either a large current or voltage. As you can see in Figure 11-5, between the electric power plant and home or industry are power lines carrying the power. The amount of power lost (heat) in these power lines can be calculated by using the formula $P = I^2 \times R$, where I is the current flowing through the line and R is the resistance of the power lines. This means that the larger the current, the greater the amount of power lost in the lines in the form of heat and therefore the less the amount of power supplied to the home or industry. For this reason, power companies transport electric energy at a very high voltage between 200,000 and 600,000 V. Since the voltage is high, the current can be low and provide the same amount of power to the consumer. Yet, by keeping the current low, the amount of heat loss generated in the power lines is minimal.

Now that we have discovered why it is more efficient over a long distance to transport high voltages than high current, what does this have to do with ac? An ac voltage can easily and efficiently be transformed up or down to a higher or lower voltage by utilizing a device known as a **transformer**, and even though dc voltages can be stepped up and down, the method is inefficient and more complex.

Transformer
Device consisting of two or more coils that are used to couple electric energy from one circuit to another, yet maintain electrical isolation between the two.

FIGURE 11-5   Power Lines Connect Power from the Electric Company to Home and Industry.

3. Nearly all electronic circuits and equipment are powered by dc voltages, which means that once the ac power arrives at the home or industry, in most cases it will have to be converted into dc power to operate electronic equipment. It is a relatively simple process to convert ac to dc; however, conversion from dc to ac is a complex and comparatively inefficient process.

Figure 11-6 illustrates ac power distribution from the electric power plant to the home and industry. The ac power distribution system begins at the electric power plant, which has the powerful large generators driven by turbines to generate large ac voltages. The turbines can be driven by either falling water (hydroelectric), or from steam, which is produced with intense heat by burning either coal, gas, or oil or from a nuclear reactor (thermoelectric). The turbine supplies the mechanical energy to the generator, to be transformed into ac electrical energy.

FIGURE 11-6   AC Power Distribution.

The generator generates an ac voltage of approximately 22,000 V, which is stepped up by transformers to approximately 500,000 V. This voltage is applied to the long-distance transmission lines, which connect the power plant to the city or town. At each city or town, the voltage is tapped off the long-distance transmission lines and stepped down to approximately 66,000 V and is distributed to large-scale industrial customers. The 66,000 V is stepped down again to approxi-

(a)

(b)

(c)

(d)

(e)

(f)

FIGURE 11-7   AC Power Distribution. (a) Hydroelectric Dam. (b) Electric Power Plant. (c) Generator. (d) Control Room. (e) Step-down Transformer. (f) Utility Pole.

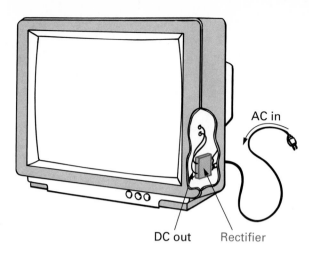

AC in

DC out   Rectifier

FIGURE 11-8   Rectification (Converting AC to DC) by Appliances.

mately 4800 V and distributed throughout the city or town by short-distance transmission lines. This 4800 V is used by small-scale industrial customers and residential customers who receive the ac power via step-down transformers on utility poles, which step down the 4800 V to 240 V and 120 V. The appearance of some of the devices that make up the power distribution system can be seen in Figure 11-7.

A large amount of equipment and devices within industry and the home will run directly from the ac power, such as heating, lighting, and cooling. Some equipment that runs on dc, such as televisions and computers, will accept the 120 V ac and internally convert it to the dc voltages required. Figure 11-8 illustrates a TV set that is operating from the 120 V ac from the wall outlet and internally converting it to dc so that it can power the circuits necessary to produce both audio (sound) and video (picture) information. This unit that converts ac to dc is called a **rectifier.**

Rectifier
Device that achieves rectification.

11-2-2 *Information Transfer*

Information, by definition, is the property of a signal or message that conveys something meaningful to the recipient. **Communication,** which is the transfer of information between two points, began with speech and progressed to handwritten words in letters and printed words in newspapers and books. To achieve greater distances of communication, face-to-face communications evolved into telephone and radio communications.

A simple communication system can be seen in Figure 11-9(a). The audio or **sound waves** generated when the person on the left speaks are converted to an equivalent electrical signal or wave by the microphone. This electrical wave is fed into an amplifier, which boosts the magnitude of the electrical signal. The enlarged electrical signal emerging out of the amplifier circuit is then converted back to the original sound waves of larger amplitude by the speaker, where they are then received by the recipient.

Communication
Transmission of information between two points.

Sound Wave
Traveling wave propogated in an elastic medium that travels at a speed of approximately 1133 ft/s.

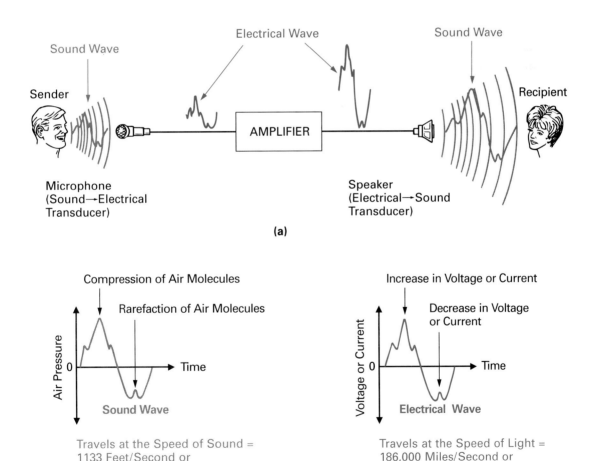

(a)

(b)

(c)

FIGURE 11-9 **Information Transfer. (a) Wire Communication System. (b) Sound Wave. (c) Electrical Wave.**

The voice information or sound wave produced by the sender is a variation in air pressure, as shown in Figure 11-9(b), and travels at the speed of sound. Sound waves or sounds are normally generated by a vibrating reed or plucked string in the case of musical instruments. In this example the sender's vocal cords vibrate backward and forward, producing a rarefaction or decreased air pressure, where few air molecules exist, and a compression or increased air pressure, where many air molecules exist. Like the ripples produced by a stone falling in a pond, the sound waves produced by the sender are constantly expanding and traveling outward.

The microphone is in fact a **transducer** (energy converter), because it converts the sound wave (which is a form of mechanical energy) into electrical energy in the form of voltage and current, which varies in the same manner as the sound wave and therefore contains the sender's message or information.

The **electrical wave,** shown in Figure 11-9(c), is a variation in voltage or current and can only exist in a wire conductor or circuit. This electrical signal

Transducer
Any device that converts energy from one form to another.

Electrical Wave
Traveling wave propagated in a conductive medium that is a variation in voltage or current and travels at slightly less than the speed of light.

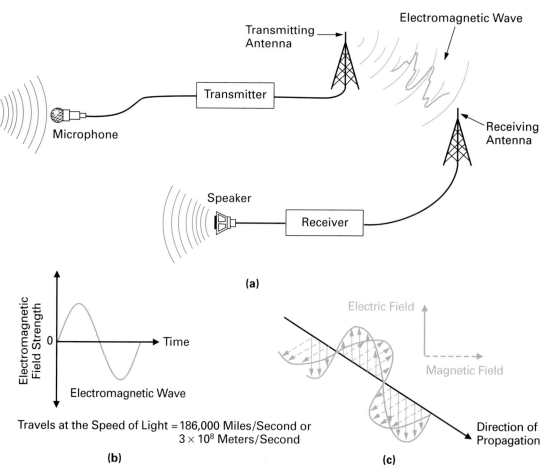

FIGURE 11-10   Information Transfer. (a) Wireless Communication System. (b) Simplified Electromagnetic Wave. (c) Components of an Electromagnetic Wave.

travels at the speed of light, and after being increased in magnitude by the amplifier, it is applied to a loudspeaker.

The speaker, like the microphone, is also an electroacoustical transducer that converts the electrical energy input into a mechanical soundwave output. These sound waves strike the outer eardrum, causing the ear diaphragm to vibrate, and these mechanical vibrations actuate nerve endings in the ear, which convert the mechanical vibrations into electrochemical impulses that are sent to the brain. The brain decodes this information by comparing these impulses with a library of previous sounds and so provides the sensation of hearing.

To communicate between two distant points, a wire must be connected between the microphone and speaker. However, if an electrical wave is applied to an antenna, the electrical wave is converted into a radio or electromagnetic wave, as shown in Figure 11-10(a), and communication is established without the need of a connecting wire—hence the term **wireless communication.** Antennas are designed to radiate and receive electromagnetic waves, which vary in field strength, as shown in Figure 11-10(b), and can exist in either air or space. These radio waves, as they are also known, travel at the speed of light and allow us to achieve great distances of communication.

More specifically, they are composed of two basic components, as shown in Figure 11-10(c). The electrical voltage applied to the antenna is converted into an electric field and the electrical current into a magnetic field. This **electromagnetic** (electric–magnetic) **wave** is used to carry a variety of information, such as speech, radio broadcasts, television signals, and so on.

In summary, the sound wave is a variation in air pressure, the electrical wave is a variation of voltage or current, and the electromagnetic wave is a variation of electric and magnetic field strength, as seen in Figure 11-11.

Wireless Communication
Term describing radio communication that requires no wires between the two communicating points.

Electromagnetic (radio) wave
Wave that consists of both an electric and magnetic variation, and travels at the speed of light.

EXAMPLE:

How long will it take the sound wave produced by a rifle shot to travel 9630.5 feet?

Solution:

This problem makes use of the following formula:

$$\text{Distance} = \text{velocity} \times \text{time}$$

or

$D = $ distance
$v = $ velocity
$t = $ time

If someone travels at 20 mph for 2 hours, the person will travel 40 miles ($D = v \times t = 20$ mph $\times 2$ hours $= 40$ miles). In this problem, the distance (9630.5) and the

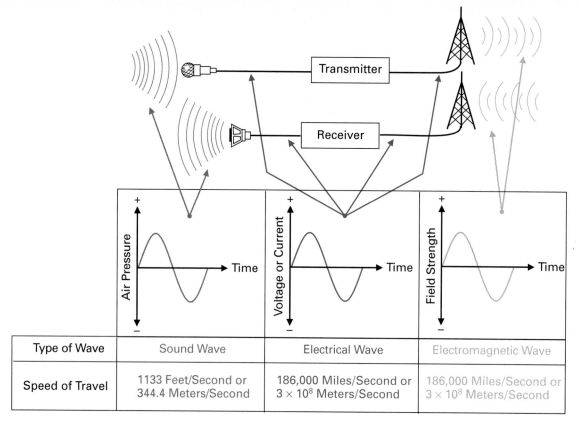

Type of Wave	Sound Wave	Electrical Wave	Electromagnetic Wave
Speed of Travel	1133 Feet/Second or 344.4 Meters/Second	186,000 Miles/Second or 3×10^8 Meters/Second	186,000 Miles/Second or 3×10^8 Meters/Second

FIGURE 11-11   **Summary of the Sound, Electrical, and Electromagnetic Waves.**

sound wave's velocity (1130 ft/sec) are known. By rearranging the formula, we can find time:

$$\text{time} = \frac{\text{distance}}{\text{velocity}} = \frac{9630.5 \text{ ft}}{1130 \text{ ft/s}} = 8.5 \text{ s}$$

■ **EXAMPLE:**

How long will it take an electromagnetic (radio) wave to reach a receiving antenna that is 2000 miles away from the transmitting antenna?

■ *Solution:*

In this problem, both distance (2000 miles) and velocity (186,000 miles/s) are known, and time has to be calculated:

$$\text{time} = \frac{\text{distance}}{\text{velocity}} = \frac{2000 \text{ miles}}{186,000 \text{ miles/s}}$$

$$= 1.075 \times 10^{-2} \quad \text{or} \quad 10.8 \text{ ms}$$

SELF-TEST REVIEW QUESTIONS FOR SECTION 11-2-2

1. Define *information* and *communication*.
2. The _____ wave is a variation in air pressure, the _____ wave is a variation in *field* strength, and the _____ wave is a variation of voltage or current.
3. Which of the three waves described in Question 2:
   a. Can only exist in the air?
   b. Can exist in either air or a vacuum?
   c. Exists in a wire conductor?
4. Sound waves travel at the speed of sound, which is _____, while electrical and electromagnetic waves travel at the speed of light, which is _____.
5. A human ear is designed to receive _____ waves, an antenna is designed to transmit or receive _____ waves, and an electronic circuit is designed to pass only _____ waves.
6. Give the names of the following energy converters or transducers:
   a. Sound wave (mechanical energy) to electrical wave
   b. Electrical wave to sound wave
   c. Electrical wave to electromagnetic wave
   d. Sound wave to electrochemical impulses

MINI-MATH REVIEW

This mini-math review is designed to remind you of the mathematical details relating to **trigonometry**, which will be used in following sections and chapters.

Trigonometry
The study of the properties of triangles and trigonometric functions and of their applications.

Introduction to Trigonometry

Electronic technicians, electronic engineers, mechanical engineers, electricians, carpenters, architects, navigators, and many other people in technical trades make use of trigonometry on a daily basis. Although volumes of theory are available about this subject, most of it is not that useful to us for everyday applications. In this mini-math review we will concentrate on the more practical part of trigonometry—the **right-angle triangle.**

Right-Angle Triangle
A triangle having a 90° or square corner.

The 3, 4, 5 Right-Angle Triangle

Like all triangles, the right-angle triangle or right triangle has three sides and three corners. Its distinguishing feature, however, is that two of the sides of this triangle are at right angles (at 90°) to one another, as shown in Figure

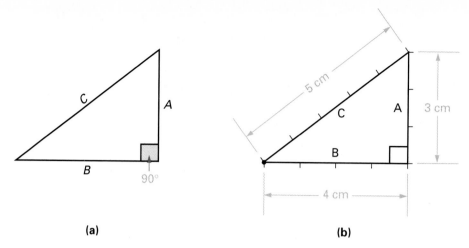

(a)   (b)

FIGURE 11-12   Right-Angle (3, 4, 5) Triangle.

11-12(a). The small square box within the triangle is placed in the corner to show that sides A and B are *square* or at *right angles* to one another.

If you study this right triangle you may notice two interesting facts about the relative lengths of sides A, B, and C. These observations are:

1. Side C is always longer than side A or side B.

2. The total length of side A and side B combined is always longer than side C.

The right triangle in Figure 11-12(b) has been drawn to scale to demonstrate another interesting fact about this triangle. If side A were to equal 3 cm and side B was made to equal 4 cm, side C would equal 5 cm. This demonstrates a basic relationship among the three sides and accounts for why this right triangle is sometimes referred to as a *3,4,5 triangle*. As long as the relative lengths remain the same, it makes no difference whether the sides are 3 cm, 4 cm, and 5 cm, or 30 km, 40 km, and 50 km. Unfortunately, in most applications our lengths of A, B, and C will not work out as easy as 3, 4, 5.

When a relationship exists among three quantities, we can develop a formula to calculate an unknown when two of the quantities are known. It was Pythagoras who first developed this basic formula or equation (known as the **Pythagorean Theorem**), which states:

Pythagorean Theorem
A theorem in geometry: the square of the length of the hypotenuse of a right triangle equals the sum of the squares of the lengths of the other two sides.

$$\boxed{C^2 = A^2 + B^2} \qquad \left(\begin{array}{l} 5^2 = 3^2 + 4^2 \\ 25 = 9\ + 16 \end{array}\right)$$

By using the rules of algebra, we can transpose this formula to derive formulas for C, B, and A.

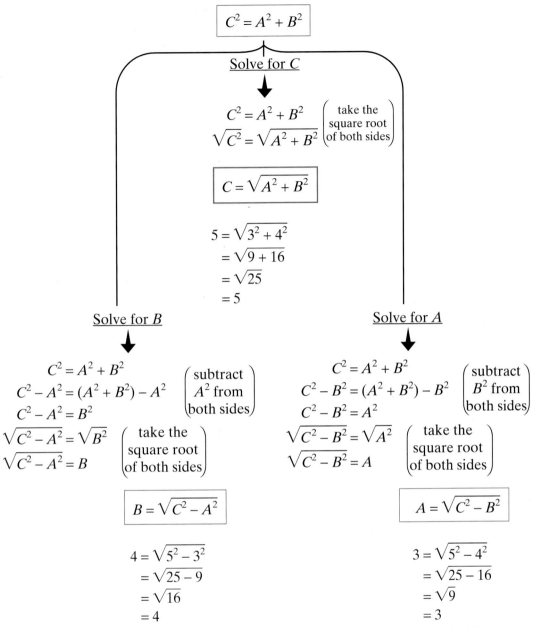

$$C^2 = A^2 + B^2$$

Solve for C

$$C^2 = A^2 + B^2$$
$$\sqrt{C^2} = \sqrt{A^2 + B^2} \quad \left(\begin{array}{l}\text{take the}\\ \text{square root}\\ \text{of both sides}\end{array}\right)$$

$$\boxed{C = \sqrt{A^2 + B^2}}$$

$$5 = \sqrt{3^2 + 4^2}$$
$$= \sqrt{9 + 16}$$
$$= \sqrt{25}$$
$$= 5$$

Solve for B

$$C^2 = A^2 + B^2$$
$$C^2 - A^2 = (A^2 + B^2) - A^2 \quad \left(\begin{array}{l}\text{subtract}\\ A^2 \text{ from}\\ \text{both sides}\end{array}\right)$$
$$C^2 - A^2 = B^2$$
$$\sqrt{C^2 - A^2} = \sqrt{B^2} \quad \left(\begin{array}{l}\text{take the}\\ \text{square root}\\ \text{of both sides}\end{array}\right)$$
$$\sqrt{C^2 - A^2} = B$$

$$\boxed{B = \sqrt{C^2 - A^2}}$$

$$4 = \sqrt{5^2 - 3^2}$$
$$= \sqrt{25 - 9}$$
$$= \sqrt{16}$$
$$= 4$$

Solve for A

$$C^2 = A^2 + B^2$$
$$C^2 - B^2 = (A^2 + B^2) - B^2 \quad \left(\begin{array}{l}\text{subtract}\\ B^2 \text{ from}\\ \text{both sides}\end{array}\right)$$
$$C^2 - B^2 = A^2$$
$$\sqrt{C^2 - B^2} = \sqrt{A^2} \quad \left(\begin{array}{l}\text{take the}\\ \text{square root}\\ \text{of both sides}\end{array}\right)$$
$$\sqrt{C^2 - B^2} = A$$

$$\boxed{A = \sqrt{C^2 - B^2}}$$

$$3 = \sqrt{5^2 - 4^2}$$
$$= \sqrt{25 - 16}$$
$$= \sqrt{9}$$
$$= 3$$

Let us now test these formulas with a few examples. In Figure 11-13(a), a 4 ft ladder has been placed in a position 2 ft from a wall. How far up the wall will the ladder reach?

In this example, A is unknown, and therefore

$$A = \sqrt{C^2 - B^2}$$
$$= \sqrt{4^2 - 2^2}$$
$$= \sqrt{16 - 4}$$
$$= \sqrt{12}$$
$$= 3.46 \quad \text{or} \quad 3.5 \text{ ft}$$

FIGURE 11-13   Right-Triangle Examples.

In Figure 11-13(b), a 16 ft flagpole is casting a 12 ft shadow. What would be the distance from the end of the shadow to the top of the flagpole?

In this example, C is unknown, and therefore

$$C = \sqrt{A^2 + B^2}$$
$$= \sqrt{16^2 + 12^2}$$
$$= \sqrt{256 + 144}$$
$$= \sqrt{400}$$
$$= 20 \text{ ft}$$

In Figure 11-13(c), a 200 ft piece of string is stretched from the top of a 125 ft cliff to a point on the beach. What would be the distance from this point to the cliff?

In this example, B is unknown, and therefore

$$B = \sqrt{C^2 - A^2}$$
$$= \sqrt{200^2 - 125^2}$$
$$= \sqrt{40,000 - 15,625}$$
$$= \sqrt{24,375}$$
$$= 156 \text{ ft}$$

Opposite, Adjacent, Hypotenuse, and Theta

Up until this point we have called the three sides of our right triangle A, B, and C. Figure 11-14 shows the more common names given to the three sides of a right-angle triangle. Side C, called the *hypotenuse side*, is always the longest of the three sides. Side B, called the *adjacent side*, always extends between the hypotenuse and the vertical side. An angle called *theta* (θ, a Greek letter) is formed between the hypotenuse side and the adjacent side. The side that is always opposite to the angle θ is called the *opposite side*.

Angles are always measured in degrees since all angles are part of a circle, like the one shown in Figure 11-15(a). A circle is divided into 360 small sections called degrees. In Figure 11-15(b) our right triangle has been placed within the circle. In this example the triangle occupies a 45 degree (45°) section of the circle and therefore theta equals 45 degrees ($\theta = 45°$). Moving from left to right in Figure 11-15(c) you will notice that angle θ has been increased from 5° to 85°. The length of the hypotenuse (H) in all of these examples remains the same; however, the adjacent side's length decreases ($A\downarrow$) and the opposite side's length increases ($O\uparrow$) as angle θ is increased ($\theta\uparrow$). This relationship between the relative length of a triangle's sides and theta means that we do not have to know the length of two sides to calculate a third. If we have the value of just one side and the angle theta, we can calculate the length of the other two sides.

Sine of Theta (Sin θ)

In the preceding section we discovered that a relationship exists between the relative length of a triangle's sides and the angle of theta. *Sine* is a comparison between the length of the opposite side and the length of the hypotenuse. Expressed mathematically,

FIGURE 11-14   Names Given to the Sides of a Right Triangle.

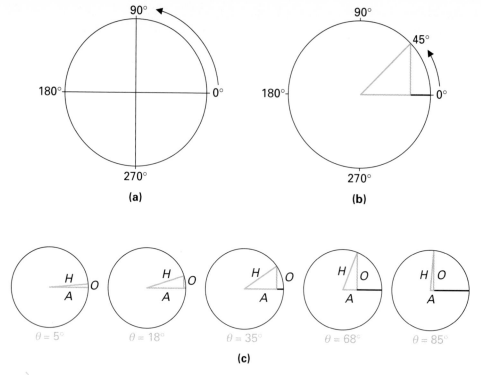

(a)

(b)

$\theta = 5°$ $\theta = 18°$ $\theta = 35°$ $\theta = 68°$ $\theta = 85°$

(c)

FIGURE 11-15   **Angle Theta (θ).**

$$\text{sine of theta (sin } \theta) = \frac{\text{opposite side } (O)}{\text{hypotenuse side } (H)}$$

Since the hypotenuse is always larger than the opposite side, the result will always be less than 1 (a fraction).

Let us use this formula on a few examples to see how it works. In Figure 11-16(a), angle theta is equal to 41° while the opposite side is equal to 20 cm in length. Calculate the length of the hypotenuse.

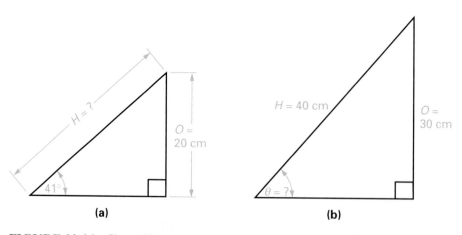

(a)   (b)

FIGURE 11-16   **Sine of Theta.**

Inserting these values in our formula, we obtain the following:

$$\text{Sin } \theta = \frac{O}{H}$$

$$\text{Sin } 41° = \frac{20 \text{ cm}}{H}$$

Looking up 41° in a sine trigonometry table or by using a scientific calculator that has all of the trigonometry tables stored permanently in its memory, you will find that the sine of 41° is 0.656. This value describes the fact that when $\theta = 41°$, the opposite side will be 0.656, or 65.6%, as long as the hypotenuse side. By inserting this value into our formula and transposing the formula according to the rules of algebra, we can determine the length of the hypotenuse.

$$\text{Sin } 41° = \frac{20 \text{ cm}}{H} \quad \left(\begin{array}{c} \text{calculator sequence:} \\ \boxed{4}\boxed{1}\boxed{\text{SIN}} \end{array} \right)$$

$$0.656 = \frac{20 \text{ cm}}{H}$$

$$0.656 \times H = \frac{20 \text{ cm} \times \cancel{H}}{\cancel{H}} \quad \text{(multiply both sides by } H\text{)}$$

$$\frac{\cancel{0.656} \times H}{\cancel{0.656}} = \frac{20 \text{ cm}}{0.656} \quad \text{(divide both sides by 0.656)}$$

$$H = \frac{20 \text{ cm}}{0.656} = 30.5 \text{ cm}$$

Figure 11-16(b) illustrates another example; however, in this case the lengths of sides H and O are known but θ is not.

$$\text{Sin } \theta = \frac{O}{H}$$

$$= \frac{30 \text{ cm}}{40 \text{ cm}}$$

$$= 0.75$$

The ratio of side O to side H is 0.75, or 75%, which means that the opposite side is 75%, or 0.75, as long as the hypotenuse side. To calculate angle θ we must isolate it on one side of the equation. To achieve this we must multiply both sides of the equation by *arcsin,* or *inverse sine* (invsin), which does the opposite of sine.

$$\text{Sin } \theta = 0.75$$

$$\text{invsin (sin } \theta) = \text{invsin } 0.75$$

$$\theta = \text{invsin } 0.75 \quad \left(\begin{array}{c} \text{calculator sequence:} \\ \boxed{.}\boxed{7}\boxed{5}\boxed{\text{INV}}\boxed{\text{SIN}} \end{array} \right)$$

$$= 48.6°$$

In summary, therefore, the sin trig functions take an angle θ and give you a number x. The inverse sin (arcsin) trig functions take a number x and give you an angle θ. In both cases, the number x is a ratio and describes what fraction the opposite side is in comparison to the hypotenuse.

Sine: angle $\theta \rightarrow$ number x

Inverse sine: number $x \rightarrow$ angle θ

Cosine of Theta (Cos θ)

As sine is a comparison between the opposite and the hypotenuse, cosine is a comparison between the adjacent and hypotenuse sides.

$$\text{cosine of theta (Cos }\theta) = \frac{\text{adjacent }(A)}{\text{hypotenuse }(H)}$$

Figure 11-17(a) illustrates an example in which the angle θ and the length of the hypotenuse are known. From this information calculate the length of the adjacent side.

$$\text{Cos }\theta = \frac{A}{H}$$

$$\text{Cos }30° = \frac{A}{40 \text{ cm}} \quad \left(\begin{array}{l}\text{calculator sequence:}\\ \boxed{3}\,\boxed{0}\,\boxed{\text{COS}}\end{array}\right)$$

$$0.866 = \frac{A}{40 \text{ cm}}$$

Looking up the cosine of 30° you will obtain the fraction 0.866. This value states that when θ = 30°, the adjacent side will always be 0.866, or 86.6%, as long as the hypotenuse. By transposing this equation we can calculate the length of the adjacent side:

$$0.866 = \frac{A}{40 \text{ cm}}$$

$$40 \times 0.866 = \frac{A}{40 \text{ cm}} \times 40 \quad \text{(multiply both sides by 40)}$$

$$A = 0.866 \times 40$$

$$= 34.64 \text{ cm}$$

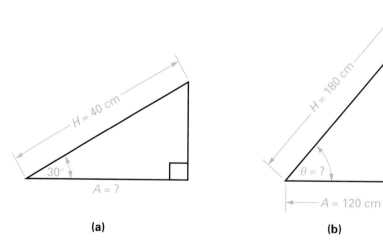

(a)   (b)

FIGURE 11-17   Cosine of Theta.

As another example, calculate the angle θ in Figure 11-17(b) if $A = 120$ cm and $H = 180$ cm.

$$\text{Cos } \theta = \frac{A}{H}$$

$$= \frac{120 \text{ cm}}{180 \text{ cm}}$$

$$= 0.667 \quad (A \text{ is } 66.7\% \text{ as long as } H)$$

$$\text{invcos}(\cos \theta) = \text{invcos } 0.667 \quad (\text{multiply both sides by invcos})$$

$$\theta = \text{invcos } 0.667 \quad \left(\begin{array}{l}\text{calculator sequence:} \\ \boxed{0}\boxed{.}\boxed{6}\boxed{6}\boxed{7}\boxed{\text{INV}}\boxed{\text{COS}}\end{array}\right)$$

$$= 48.2°$$

Once again the inverse cosine trig function does the reverse operation of the cosine function, in that

Cosine: angle $\theta \rightarrow$ number x

Inverse cosine: number $x \rightarrow$ angle θ

Tangent of Theta (Tan θ)

Tangent is a comparison between the opposite side of a right triangle and the adjacent side.

> $$\text{tangent of theta (Tan } \theta) = \frac{\text{opposite } (O)}{\text{adjacent } (A)}$$

Figure 11-18(a) illustrates an example in which $\theta = 65°$ and the opposite side $= 43$ cm. Calculate the length of the adjacent side.

$$\tan \theta = \frac{O}{A}$$

$$\tan 65° = \frac{43 \text{ cm}}{A} \quad (\text{calculator sequence: } \boxed{6}\boxed{5}\boxed{\text{TAN}})$$

$$2.14 = \frac{43 \text{ cm}}{A} \quad \left(\begin{array}{l}\text{whenever } \theta = 65°, \text{ the opposite} \\ \text{side will be 2.14 times longer} \\ \text{than the adjacent side}\end{array}\right)$$

$$2.14 \times A = \frac{43 \text{ cm}}{A} \times A \quad (\text{multiply both sides by } A)$$

$$\frac{2.14 \times A}{2.14} = \frac{43 \text{ cm}}{2.14} \quad (\text{divide both sides by } 2.14)$$

$$A = \frac{43 \text{ cm}}{2.14}$$

$$= 20.1 \text{ cm}$$

As another example, calculate the angle θ in Figure 11-18(b) if $O = 30$ cm and $A = 130$ cm.

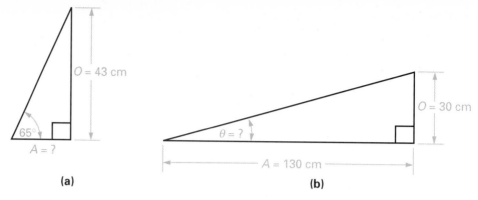

FIGURE 11-18   **Tangent of Theta.**

$$\tan \theta = \frac{O}{A}$$

$$= \frac{30 \text{ cm}}{130 \text{ cm}}$$

$$= 0.231 \quad (O \text{ is } 0.231 \text{ or } 23.1\% \text{ as long as } A)$$

$$\text{invtan} (\tan \theta) = \text{invtan } 0.231 \quad (\text{multiply both sides by invtan})$$

$$\theta = \text{invtan } 0.231 \quad \left(\begin{array}{c} \text{calculator sequence:} \\ \boxed{0}\boxed{.}\boxed{2}\boxed{3}\boxed{1}\boxed{\text{INV}}\boxed{\text{TAN}} \end{array}\right)$$

$$= 12.99 \text{ or } 13°$$

Summary

As you have seen, trigonometry involves the study of the relationship between the three sides (O, H, A) of the right triangle, and it also relates to the relationship between the sides of the right triangle and the number of degrees contained in the angle theta (θ).

If the length of two sides of a right triangle are known, and the length of the third side is needed, remember that

$$C^2 = A^2 + B^2$$

If the angle θ is known along with the length of one side, or if angle θ is needed and the lengths of the two sides are known, one of the three formulas can be chosen based on what variables are known, and what is needed.

$$\sin \theta = \frac{O}{H} \qquad \cos \theta = \frac{A}{H} \qquad \tan \theta = \frac{O}{A}$$

When I was introduced to trigonometry, my mathematics professor spent 15 minutes having the whole class practice what he described as an old oriental war cry which went like this: "SOH CAH TOA". After he explained that it wasn't a war cry but in fact a memory aid to help us remember that SOH was in fact Sin $\theta = O/H$, CAH was Cos $\theta = A/H$, and TOA was Tan $\theta = O/A$, we understood the method in his madness.

11-3 AC WAVE SHAPES

In all fields of electronics, whether medical, industrial, consumer, or data processing, different types of information are being conveyed between two points, and electronic equipment is managing the flow of this information.

Let's now discuss the basic types of ac wave shapes. The way in which a wave varies in magnitude with respect to time describes its wave shape. All ac waves can be classified into one of six groups, and these are illustrated in Figure 11-19.

11-3-1 The Sine Wave

The **sine wave** is the most common type of waveform. It is the natural output of a generator that converts a mechanical input, in the form of a rotating shaft, into an electrical output in the form of a sine wave. In fact, for one cycle of the input shaft, the generator will produce one sinusoidal ac voltage waveform, as shown in Figure 11-20. When the input shaft of the generator is at 0°, the ac output is

Sine Wave
Wave whose amplitude is the sine of a linear function of time. It is drawn on a graph that plots amplitude against time or radial degrees relative to the angular rotation of an alternator.

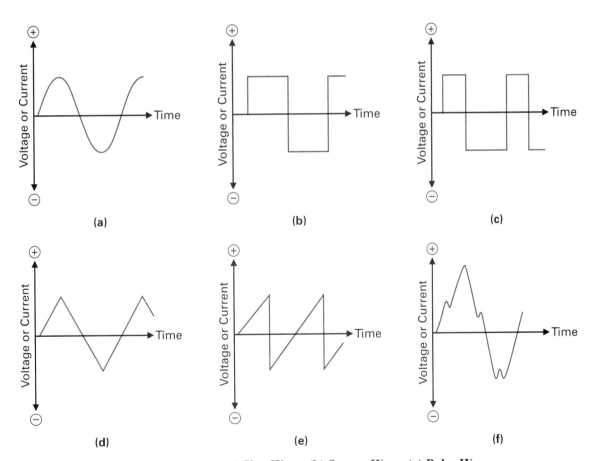

FIGURE 11-19 AC Wave Shapes. (a) Sine Wave. (b) Square Wave. (c) Pulse Wave. (d) Triangular Wave. (e) Sawtooth Wave. (f) Irregular Wave.

FIGURE 11-20 **Degrees of a Sine Wave.**

0 V. As the shaft is rotated through 360°, the ac output voltage will rise to a maximum positive voltage at 90°, fall back to 0 V at 180°, and then reach a maximum negative voltage at 270°, and finally return to 0 V at 360°. If this ac voltage is applied across a closed circuit, it produces a current that continually reverses or alternates in each direction.

Figure 11-21 illustrates the sine wave, with all the characteristic information inserted, which at first glance looks a bit ominous. Let's analyze and discuss each piece of information individually, beginning with the sine wave's amplitude.

Amplitude
Magnitude or size an alternation varies from zero.

Amplitude

Figure 11-22 plots direction and amplitude against time. The **amplitude** or magnitude of a wave is often represented by a **vector** arrow, also illustrated in Figure 11-22. The vector's length indicates the magnitude of the current or voltage, while the arrow's point is used to show the direction, or polarity.

Vector
Quantity that has both magnitude and direction. They are normally represented as a line, the length of which indicates magnitude and the orientation of which, due to the arrowhead on one end, indicates direction.

Peak Value

The peak of an ac wave occurs on both the positive and negative alternation, but is only at the peak (maximum) for an instant. Figure 11-23(a) illustrates an ac current waveform rising to a positive peak of 10 A, falling to zero, and then reaching a negative peak of 10 A in the reverse direction; Figure 11-23(b) shows an ac voltage waveform reaching positive and negative peaks of 9 V.

Peak Value
Maximum or highest-amplitude level.

480   CHAPTER 11 / ALTERNATING CURRENT (AC)

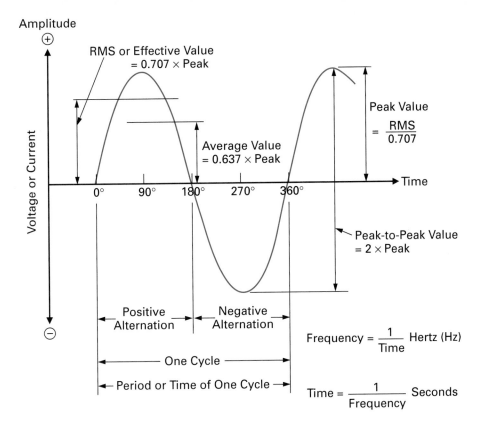

FIGURE 11-21 Sine Wave.

Peak-to-Peak Value

The **peak-to-peak value** of a sine wave is the value of voltage or current between the positive and negative maximum values of a waveform. For example, the peak-to-peak value of the current waveform in Figure 11-23(a) is equal to $I_{\text{p-p}} = 2 \times I_\text{p} = 20$ A. In Figure 11-23(b), it would be equal to $V_{\text{p-p}} = 2 \times V_\text{p} = 18$ V.

Peak-to-Peak Value
Difference between the maximum positive and maximum negative values.

FIGURE 11-22 Sine-Wave Amplitude.

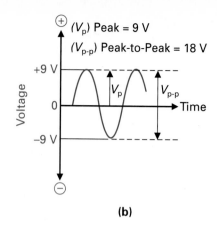

FIGURE 11-23 Peak and Peak-to-Peak of a Sine Wave.

$$p - p = 2 \times peak$$

RMS Value

Rms value of an ac voltage, current, or power waveform is equal to 0.707 times the peak value. The rms value is the effective or dc value equivalent of the ac wave.

RMS or Effective Value

Both the positive and negative alternation of a sine wave can accomplish the same amount of work; but the ac waveform is only at its maximum value for an instant in time, spending most of its time between peak currents. Our examples in Figure 11-23(a) and (b), therefore, cannot supply the same amount of power as a dc value of 10 A or 9 V.

The effective value of a sine wave is equal to 0.707 of the peak value. Let's now see how this value was obtained. Power is equal to either $P = I^2 \times R$ or $P = V^2/R$; said another way, power is proportional to the voltage or current squared. If every instantaneous value of either the positive or negative half-cycle of any voltage or current sinusoidal waveform is squared, as shown in Figure 11-24, and then averaged out to obtain the mean value, the square root of this mean value would be equal to 0.707 of the peak.

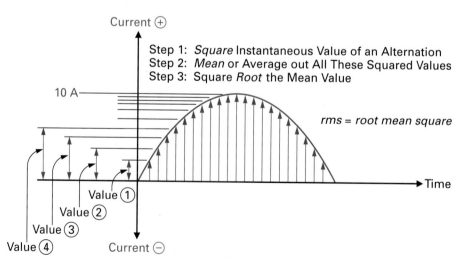

FIGURE 11-24 Obtaining the RMS Value of 0.707 for a Sine Wave.

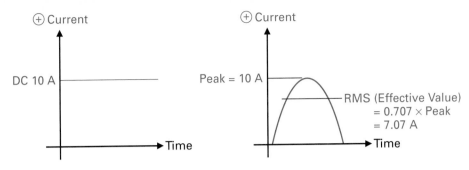

FIGURE 11-25   Effective Equivalent.

For example, if the process is carried out on the 10 A current waveform considered previously, the result would equal 7.07 A (0.707 × 10 A = 7.07 A), which is 0.707 of the peak value of 10 A.

$$\text{rms} = 0.707 \times \text{peak}$$

This root-mean-square (rms) result of 0.707 can always be used to tell us how effective an ac sine wave will be. For example, a 10 A dc source would be 10 A effective because it is continually present and always delivering power to the circuit to which it is connected, while a 10 A ac source would only be 7.07 A effective, as seen in Figure 11-25, because it is at 10 A for only a short period of time. As another example, a 10 V ac sine-wave alternation would be as effective or supply the same amount of power to a circuit as a 7.07 V dc source.

Unless otherwise stated, ac values of voltage or current are always given in rms, so the peak value can be calculated by transposing the original rms formula of rms = peak × 0.707, and ending up with

$$\text{peak} = \frac{\text{rms}}{0.707}$$

Since 1/0.707 = 1.414, the peak can also be calculated by

$$\text{peak} = \text{rms} \times 1.414$$

Average Value

The **average value** of the positive or negative alternation is found by taking either the positive or negative alternation, and listing the amplitude or vector length of current or voltages at 1° intervals, as shown in Figure 11-26(a). The sum of all these values is then divided by the total number of values (averaging), which for all sine waves will calculate out to be 0.637 of the peak voltage or cur-

Average Value
Mean value found when the area of a wave above a line is equal to the area of the wave below the line.

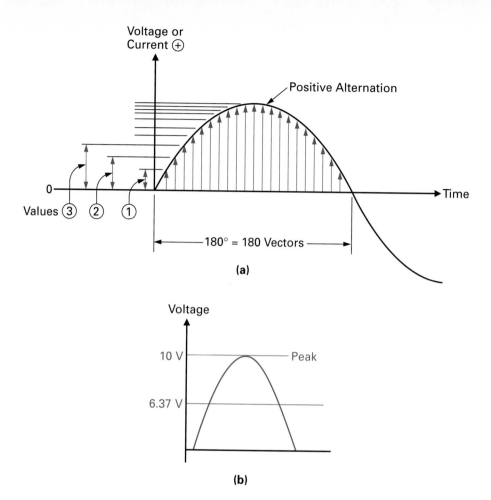

FIGURE 11-26 **Average Value of a Sine-Wave Alternation = 0.637 × Peak.**

rent. For example, the average value of a sine-wave alternation with a peak of 10 V, as seen in Figure 11-26(b), is equal to

$$\text{average} = 0.637 \times \text{peak}$$

$$= 0.637 \times 10 \text{ V}$$
$$= 6.37 \text{ V}$$

The average of a positive or negative alternation (half-cycle) is equal to 0.637 × peak; however, the average of the complete cycle, including both the positive and negative half-cycles, is mathematically zero as the amount of voltage or current above the zero line is equal but opposite to the amount of voltage or current below the zero line, as shown in Figure 11-27.

■ **EXAMPLE:**

Calculate V_p, V_{p-p}, V_{rms}, and V_{avg} of a 16 V peak sine wave.

Positive Alternation = Negative Alternation

Area = Area

and Mathematically Cancel to Equal Zero

**FIGURE 11-27 Average Value of a
Complete Sine-Wave Cycle = 0.**

Solution:

$$V_p = 16 \text{ V}$$
$$V_{p-p} = 2 \times V_p = 2 \times 16 \text{ V} = 32 \text{ V}$$
$$V_{rms} = 0.707 \times V_p = 0.707 \times 16 \text{ V} = 11.3 \text{ V}$$
$$V_{avg} = 0.637 \times V_p = 0.637 \times 16 \text{ V} = 10.2 \text{ V}$$

EXAMPLE:

Calculate V_p, V_{p-p}, and V_{avg} of a 120 V (rms) ac main supply.

Solution:

$$V_p = \text{rms} \times 1.414 = 120 \text{ V} \times 1.414 = 169.68 \text{ V}$$
$$V_{p-p} = 2 \times V_p = 2 \times 169.68 \text{ V} = 339.36 \text{ V}$$
$$V_{avg} = 0.637 \times V_p = 0.637 \times 169.68 \text{ V} = 108.09 \text{ V}$$

The 120 V (rms) that is delivered to every home and business has a peak of 169.68 V; however, it will deliver the same power as 120 V dc.

Frequency and Period

The **period** (*t*) is the time required for one complete cycle (positive and negative alternation) of the sinusoidal current or voltage waveform. A *cycle*, by definition, is the change of an alternating wave from zero to a positive peak, to zero, then to a negative peak, and finally, back to zero (see Figure 11-28).

Frequency is the number of repetitions of a periodic wave in a unit of time. It is symbolized by *f* and is given the unit hertz (cycles per second), in honor of a German physicist, Heinrich Hertz.

Period
Time taken to complete one complete cycle of a periodic or repeating waveform.

Frequency
Rate or recurrences of a periodic wave normally within a unit of one second, measured in hertz (cycles/second).

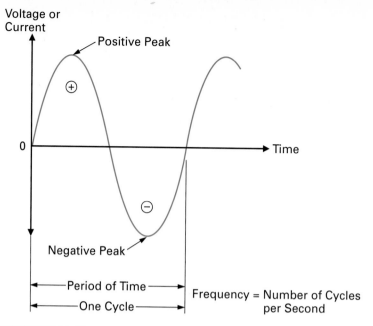

FIGURE 11-28 **Frequency and Period.**

Sinusoidal waves can take a long or a short amount of time to complete one cycle. This time is related to frequency in that period is equal to the reciprocal of frequency, and vice versa.

$$f\,(\text{hertz}) = \frac{1}{t}$$

$$t\,(\text{seconds}) = \frac{1}{f}$$

where t = period

f = frequency

For example, the ac voltage of 120 V (rms) arrives at the household electrical outlet alternating at a frequency of 60 hertz (Hz). This means that 60 cycles arrive at the household electrical outlet in 1 second. If 60 cycles occur in 1 second, as seen in Figure 11-29(a), it is actually taking ¹⁄₆₀ of a second for one of the 60 cycles to complete its cycle, which calculates out to be

$$\text{¹⁄₆₀ of 1 second} = \frac{1\ \text{second}}{60\ \text{cycles}} \times 1\ \text{second} = 16.67\ \text{milliseconds (ms)}$$

So the time or period of one cycle can be calculated by using the formula period $(t) = 1/f = 1/60$ Hz = 16.67 ms, as shown in Figure 11-29(b).

If the period or time of a cycle is known, the frequency can be calculated; for example,

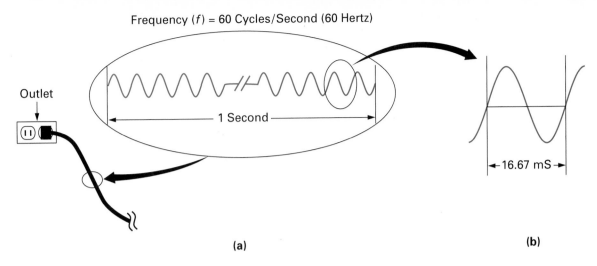

Frequency (f) = 60 Cycles/Second (60 Hertz)

Outlet

1 Second

16.67 mS

(a)   (b)

FIGURE 11-29   120-V, 60-Hz AC Supply.

$$\text{frequency } (f) = \frac{1}{\text{period}} = 1/16.67 \text{ ms} = 60 \text{ Hz}$$

As illustrated in Figure 11-6, all homes in the United States receive at their wall outlets an ac voltage of 120 V rms at a frequency of 60 Hz. This frequency was chosen for convenience, as a lower frequency would require larger transformers, and if the frequency were too low, the slow switching (alternating) current through the light bulb would cause it to flicker. A higher frequency than 60 Hz was found to cause an increase in the amount of heat generated in the core of all power distribution transformers due to eddy currents and hysteresis losses. Consequently, 60 Hz was chosen in the United States; however, other countries, such as England and most of Europe, use an ac power line frequency of 50 Hz (240 V).

EXAMPLE:

If a sine wave has a period of 400 μs, what is its frequency?

Solution:

$$\text{frequency } (f) = \frac{1}{\text{time } (t)} = \frac{1}{400 \text{ μs}} = 2.5 \text{ kHz}$$

EXAMPLE:

If it takes a sine wave 25 ms to complete two cycles, how many of the cycles will be received in 1 s?

Solution:

If the period of two cycles is 25 ms, one cycle period will equal 12.5 ms. The number of cycles per second or frequency will equal

$$f = \frac{1}{t} = \frac{1}{12.5 \text{ ms}} = 80 \text{ Hz} \quad \text{or} \quad 80 \text{ cycles/second}$$

■ **EXAMPLE:**

Calculate the period of the following:

    a. 100 MHz
    b. 40 cycles every 5 seconds
    c. 4.2 kilocycles/second
    d. 500 kHz

■ *Solution:*

$$f = \frac{1}{t} \qquad \text{therefore, } t = \frac{1}{f}$$

a. $t = \dfrac{1}{100 \text{ MHz}} = 10$ nanoseconds (ns)

b. 40 cycles/5 s = 8 cycles/second (8 Hz)

$t = \dfrac{1}{8 \text{ Hz}} = 125$ ms

c. $t = \dfrac{1}{4.2 \text{ kHz}} = 238$ µs

d. $t = \dfrac{1}{500 \text{ kHz}} = 2$ µs

Wavelength

Wavelength
Distance between two points of corresponding phase and is equal to waveform velocity or speed divided by frequency.

    Wavelength, as its name states, is the physical length of one complete cycle and is generally measured in meters. The wavelength (λ, lambda) of a complete cycle is dependent on the frequency and velocity of the transmission:

$$\lambda = \frac{\text{velocity}}{\text{frequency}}$$

    Electromagnetic waves. Radio waves travel at the speed of light in air or a vacuum, which is 3×10^8 meters/second or 3×10^{10} cm/second.

$$\lambda \text{ (m)} = \frac{3 \times 10^8 \text{ m/s}}{f \text{ (Hz)}} \qquad \text{or} \qquad \lambda \text{ (cm)} = \frac{3 \times 10^{10} \text{ cm/s}}{\text{frequency (Hz)}}$$

(There are 100 centimeters [cm] in 1 meter [m], therefore cm = 10^{-2}, and m = 10^0 or 1.) Subsequently, the higher the frequency, the shorter the wavelength, which is why a shortwave radio receiver is designed to receive high frequencies ($\lambda\downarrow = 3 \times 10^8/f\uparrow$).

■ EXAMPLE:

Calculate the wavelength of the electromagnetic waves illustrated in Figure 11-30.

■ *Solution:*

a. $\lambda = \dfrac{3 \times 10^8}{f \,(\text{Hz})}$ m/s $= \dfrac{3 \times 10^8}{10 \text{ kHz}} = 30{,}000$ m   or   30 km

b. $\lambda = \dfrac{3 \times 10^{10}}{f \,(\text{Hz})}$ cm/s $= \dfrac{3 \times 10^{10}}{2182 \text{ kHz}} = 13{,}748.9$ cm   or   137.489 m

c. $\lambda = \dfrac{3 \times 10^{10}}{f \,(\text{Hz})}$ cm/s $= \dfrac{3 \times 10^{10}}{4.0 \text{ GHz}} = \dfrac{3 \times 10^{10}}{4 \times 10^9} = 7.5$ cm   or   0.075 m

FIGURE 11-30   Electromagnetic Wavelength Examples.

Sound waves.   Sound waves travel at a slower speed than electromagnetic waves, as their mechanical vibrations depend on air molecules, which offer resistance to the traveling wave. For sound waves, the wavelength formula will be equal to

$$\lambda \ (m) = \frac{344.4 \ m/s}{f \ (Hz)}$$

▦ **EXAMPLE:**

Calculate the wavelength of the sound waves illustrated in Figure 11-31.

▦ *Solution:*

a. $\lambda \ (m) = \dfrac{344.4 \ m/s}{f \ (Hz)} = \dfrac{344.4}{35 \ kHz} = 9.8 \times 10^{-3} \ m = 9.8 \ mm$

b. 300 Hz: $\lambda \ (m) = \dfrac{344.4 \ m/s}{300 \ Hz} = 1.15 \ m$

3000 Hz: $\lambda \ (m) = \dfrac{344.4 \ m/s}{3000 \ Hz} = 0.115 \ m \quad or \quad 11.5 \ cm$

(a)

Frequency Range = 300 Hz to 3 kHz
Wavelength Range = ? to ?

(b)

FIGURE 11-31   Sound Wavelength Examples.

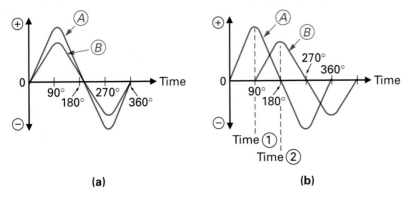

(a) **(b)**

FIGURE 11-32   Phase Relationship. (a) In Phase. (b) Out of Phase.

Phase Relationships

The **phase** of a sine wave is always relative to another sine wave of the same frequency. Figure 11-32(a) illustrates two sine waves that are in phase with one another, while Figure 11-32(b) shows two sine waves that are out of phase with one another. Sine wave A is our reference, since the positive-going zero crossing is at $0°$, its positive peak is at $90°$, its negative-going zero crossing is at $180°$, its negative peak is at $270°$, and the cycle completes at $360°$. In Figure 11-32(a), sine wave B is in phase with A since its peaks and zero crossings occur at the same time as sine wave A's. In Figure 11-32(b), sine wave B has been shifted to the right by $90°$ with respect to the reference sine wave A. This **phase shift** or **phase angle** of $90°$ means that sine wave A *leads* B by $90°$, or B *lags* A by $90°$. Sine wave A is said to lead B as its positive peak, for example, occurs first at time 1, while the positive peak of B occurs later at time 2.

Phase
Angular relationship between two waves, normally between current and voltage in an ac circuit.

Phase Shift or Angle
Change in phase of a waveform between two points, given in degrees of lead or lag. Phase difference between two waves, normally expressed in degrees.

▢ **EXAMPLE:**

What are the phase relationships between the two waveforms illustrated in Figure 11-33(a) and (b)?

▢ *Solution:*

a. The phase shift or angle is $90°$. Sine wave B leads sine wave A by $90°$, or A lags B by $90°$.

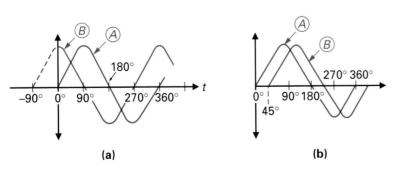

(a) **(b)**

FIGURE 11-33   Phase Relationship Examples.

b. The phase shift or angle is 45°. Sine wave A leads sine wave B by 45°, or B lags A by 45°.

The Meaning of Sine

The square, rectangular, triangular, and sawtooth waveform shapes are given their names because of their waveform shapes. The sine wave is the most common type of waveform shape, and why the name *sine* was given to this wave needs to be explained further.

Figure 11-34(a) shows the correlation between the 360° of a circle and the 360° of a sine wave. Within the circle a triangle has been drawn to represent the right-angle triangle formed at 30°. The hypotenuse side will always remain at the same length throughout 360°, and will equal the peak of the sine wave. The opposite side of the triangle is equal to the amplitude vector of the sine wave at 30°. To calculate the amplitude of the opposite side, and therefore the amplitude of the sine wave at 30°, we use the sine of theta formula discussed previously in the mini-math review on trigonometry.

$$\sin \theta = \frac{\text{opposite}}{\text{hypotenuse}}$$

$$\sin 30° = \frac{O}{H} \quad \left(\begin{array}{c} \text{calculator sequence:} \\ \boxed{3}\,\boxed{0}\,\boxed{\text{SIN}} \end{array} \right)$$

$$0.500 = \frac{O}{H}$$

This tells us that at 30°, the opposite side is 0.500 or half the size of the hypotenuse. At 30°, therefore, the amplitude of the sine wave will be 0.5 (50%) of the peak value.

Figure 11-34(b) lists the sine values at 15° increments. Figure 11-34(c) shows an example of a 10-V-peak sine wave. At 15°, a sine wave will always be at 0.259 (sine of 15°) of the peak value, which for a 10 V sine wave will be 2.59 V (0.259×10 V = 2.59 V). At 30°, a sine wave will have increased to 0.500 (sine of 30°) of the peak value. At 45°, a sine wave will be at 0.707 of the peak, and so on. The sine wave is consequently called a sine wave, as it changes in amplitude at the same rate as the sine trigonometric function.

11-3-2 *The Square Wave*

Square Wave
Wave that alternates between two fixed values for an equal amount of time.

The **square wave** is a periodic (repeating) wave that alternates from a positive peak value to a negative peak value, and vice versa, for equal lengths of time.

In Figure 11-35 you can see an example of a square wave that is at a frequency of 1 kHz and has a peak of 10 V. If the frequency of a wave is known, its period or time of one cycle can be calculated by using the formula $T = 1/f = 1/1$ kHz = 1 ms or $\frac{1}{1000}$ of a second. One complete cycle will take 1 ms to complete, so the positive and negative alternations will each last for 0.5 ms.

If the peak of the square wave is equal to 10 V, the peak-to-peak value of this square wave will equal $V_{p-p} = 2 \times V_p = 20$ V.

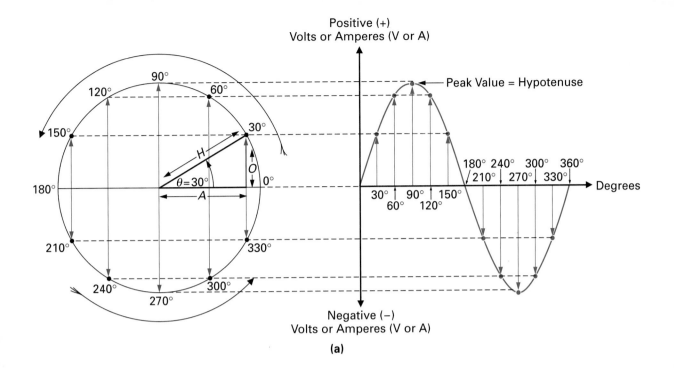

(a)

Angle	Sine
0°	0.000
15	0.259
30	0.500
45	0.707
60	0.866
75	0.966
90	1.000

(b)

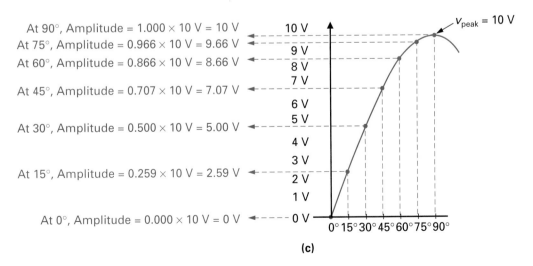

At 90°, Amplitude = 1.000 × 10 V = 10 V
At 75°, Amplitude = 0.966 × 10 V = 9.66 V
At 60°, Amplitude = 0.866 × 10 V = 8.66 V
At 45°, Amplitude = 0.707 × 10 V = 7.07 V
At 30°, Amplitude = 0.500 × 10 V = 5.00 V
At 15°, Amplitude = 0.259 × 10 V = 2.59 V
At 0°, Amplitude = 0.000 × 10 V = 0 V

V_{peak} = 10 V

(c)

FIGURE 11-34 **Meaning of Sine.**

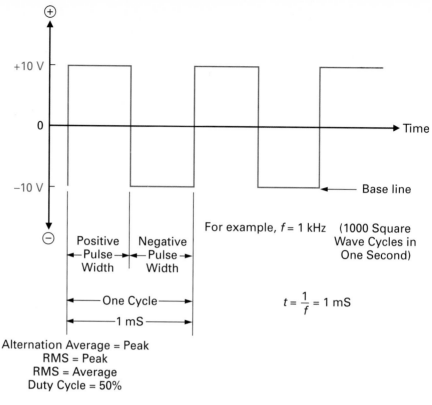

Alternation Average = Peak
RMS = Peak
RMS = Average
Duty Cycle = 50%

FIGURE 11-35 Square Wave.

To summarize this example, the square wave alternates from a positive peak value of +10 V to a negative peak value of −10 V for equal time lengths (half-cycles) of 0.5 ms.

Duty Cycle

Duty cycle is an important relationship, which has to be considered when discussing square waveforms. The **duty cycle** is the ratio of a pulse width (positive or negative pulse or cycle) to the overall period or time of the wave and is normally given as a percentage.

Duty Cycle
A term used to describe the amount of ON time versus OFF time. ON time is usually expressed as a percentage.

$$\text{duty cycle } (\%) = \frac{\text{pulse width } (P_w)}{\text{period } (t)} \times 100\%$$

The duty cycle of the example square wave in Figure 11-35 will equal

$$\text{duty cycle } (\%) = \frac{\text{pulse width } (P_w)}{\text{period } (p)} \times 100\%$$

$$= \frac{0.5 \text{ ms}}{1 \text{ ms}} \times 100\%$$

$$= 50\%$$

FIGURE 11-36   2- to 18-V Square Wave.

Since a square wave always has a positive and a negative alternation that are equal in time, the duty cycle of all square waves is equal to 50%, which actually means that the positive cycle lasts for 50% of the time of one cycle.

Average

The average or mean value of a square wave can be calculated by using the formula

$$V \text{ or } I \text{ average} = \text{baseline} + (\text{duty cycle} \times \text{peak to peak})$$

The average of the complete square wave cycle in Figure 11-35 should calculate out to be zero, as the amount above the line equals the amount below. If we apply the formula to this example, you can see that

$$V_{avg} = \text{baseline} + (\text{duty cycle} \times \text{peak to peak})$$
$$= -10 \text{ V} + (0.5 \times 20 \text{ V})$$
$$= -10 \text{ V} + 10 \text{ V}$$
$$= 0 \text{ V}$$

However, a square wave does not always alternate about 0; for example, Figure 11-36 illustrates a 16 V_{p-p} square wave that rests on a baseline of 2 V. The average value of this square wave is equal to

$$V_{avg} = \text{baseline} + (\text{duty cycle} \times \text{peak to peak})$$
$$= 2 \text{ V} \times (0.5 \times 16 \text{ V})$$
$$= (+2) + (+8 \text{ V})$$
$$= 10 \text{ V}$$

EXAMPLE:

Calculate the duty cycle and V_{avg} of a square wave of 0 to 5 V.

■ Solution:

The duty cycle of a square wave is always 0.5 or 50%. The average is:

$$V_{avg} = \text{baseline} + (\text{duty cycle} \times V_{p-p})$$
$$= 0 \text{ V} + (0.5 \times 5 \text{ V})$$
$$= 0 \text{ V} + 2.5 \text{ V} = 2.5 \text{ V}$$

Up to this point, we have seen the ideal square wave, which has instantaneous transition from the negative to the positive values, and vice versa, as shown in Figure 11-37(a). In fact, the transitions from negative to positive (positive or leading edge) and from positive to negative (negative or trailing edge) are not as ideal as shown here. It takes a small amount of time for the wave to increase to its positive value (the **rise time**) and consequently an equal amount of time for a wave to decrease to its negative value (the **fall time**). Rise time (T_R), by definition, is the time it takes for an edge to rise from 10% to 90% of its full amplitude, while fall time (T_F) is the time it takes for an edge to fall from 90% to 10% of its full amplitude, as shown in Figure 11-37(b).

With a waveform such as that in Figure 11-37(b), it is difficult, unless a standard is used, to know exactly what points to use when measuring the width of either the positive or negative alternation. The standard width is always measured between the two 50% amplitude points, as shown in Figure 11-38.

Rise Time
Time it takes a positive edge of a pulse to rise from 10% to 90% of its peak value.

Fall Time
Time it takes a negative edge of a pulse to fall from 90% to 10% of its peak value.

(a)

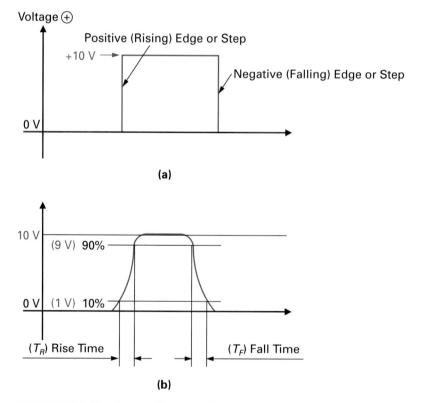

(b)

FIGURE 11-37   Square Wave's Rise and Fall Times. (a) Ideal. (b) Actual.

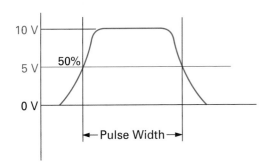

FIGURE 11-38   **Pulse Width of a Square Wave.**

Time-Domain Analysis
A method of representing a waveform by plotting its amplitude versus time.

Frequency-Domain Analysis
A method of representing a waveform by plotting its amplitude versus frequency.

Fundamental Frequency
This sine wave is always the lowest frequency and largest amplitude component of any waveform shape and is used as a reference.

Harmonic Frequency
Sine wave that is smaller in amplitude and is some multiple of the fundamental frequency.

Frequency-Domain Analysis

A *periodic wave* is a wave that repeats the same wave shape from one cycle to the next. Figure 11-39(a) is a **time-domain** representation of a periodic sine wave, which is the same way it would appear on an oscilloscope display as it plots the sine wave's amplitude against time. Figure 11-39(b) is a **frequency-domain** representation of the same periodic sine wave, and this graph, which shows the wave as it would appear on a spectrum analyzer, plots the sine wave's amplitude against frequency instead of time. This graph shows all the frequency components contained within a wave, and since, in this example, the sine wave has a period of 1 ms and therefore a frequency of 1 kHz, there is one bar at the 1 kHz point of the graph and its size represents the sine wave's amplitude. Pure sine waves have no other frequency components.

Other periodic wave shapes, such as square, pulse, triangular, sawtooth, or irregular, are actually made up of a number of sine waves having a particular frequency, amplitude, and phase. To produce a square wave, for instance, you would start with a sine wave, as shown in Figure 11-40(a), whose frequency is equal to the resulting square-wave frequency. This sine wave is called the **fundamental frequency,** and all the other sine waves that will be added to this fundamental are called **harmonics** and will always be lower in amplitude and higher in frequency. These harmonics or multiples are harmonically related to the funda-

FIGURE 11-39   **Analysis of a 1-kHz Sine Wave. (a) Time Domain. (b) Frequency Domain.**

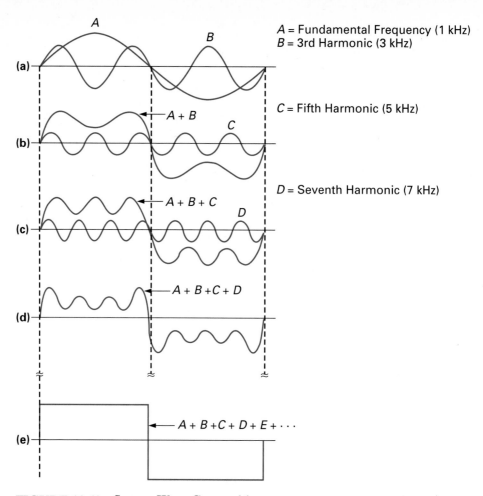

FIGURE 11-40   Square-Wave Composition.

mental, in that the second harmonic is twice the fundamental frequency, the third harmonic is three times the fundamental frequency, and so on.

Square waves are composed of a fundamental frequency and an infinite number of odd harmonics (third, fifth, seventh, and so on). If you look at the progression in Figure 11-40(a) through (d), you see that by continually adding

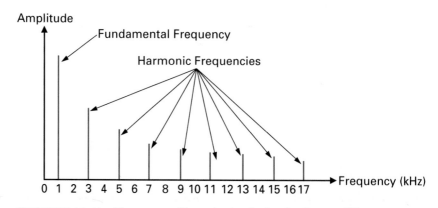

FIGURE 11-41   Frequency-Domain Analysis of a Square Wave.

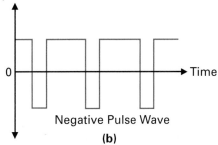

Average = Duty Cycle × Peak
RMS   = Average

Positive Pulse Wave
(a)

Negative Pulse Wave
(b)

FIGURE 11-42   Rectangular or Pulse Wave.

these odd harmonics the waveform comes closer to a perfect square wave, as shown in Figure 11-40(e).

Figure 11-41 plots the frequency domain of a square wave, with the bars representing the odd harmonics of decreasing amplitude.

11-3-3   The Rectangular or Pulse Wave

The **rectangular wave** is similar to the square wave in many respects, in that it is a periodic wave that alternately switches between one of two fixed values. The difference in the rectangular wave is that it does not remain at the two peak values for equal lengths of time, as shown in the examples in Figure 11-42(a) and (b).

In Figure 11-42(a), the rectangular wave remains at its negative level for a longer period than its positive, while the rectangular wave in Figure 11-42(b) stays at its positive value for the longer period of time and is only momentarily at its negative value.

PRF, PRT, and Pulse Length

When discussing a rectangular wave, a few terms change. Instead of stating the cycles per second as frequency, it is called **pulse repetition frequency** (PRF), which is far more descriptive. The reciprocal of frequency is time, and with rectangular pulse waveforms the reciprocal of the PRF is **pulse repetition time** (PRT), as summarized in Table 11-1. Frequency is equivalent to PRF and time to PRT; the only difference is the name.

Let us look at the example in Figure 11-43 of a 5 V rectangular wave at a frequency of 1 kHz and a pulse width of 1 μs, and practice with these new terms. With a pulse repetition frequency of 1 kHz, the time between the leading edges of pulses (PRT) will be 1/1 kHz = 1 ms. **Pulse width** (P_w), **pulse duration** (P_d), or **pulse length** (P_l) are all terms that describe the length of time for which the pulse lasts, and in this example it is equal to 1 μs, which means that 999 μs exists between the end of one pulse and the beginning of the next.

Rectangular (Pulse) Wave
Also known as a pulse wave; it is a repeating wave that only alternates between two levels or values and remains at one of these values for a small amount of time relative to the other.

Pulse Repetition Frequency
The number of times per second that a pulse is transmitted.

Pulse Repetition Time
The time interval between the start of two consecutive pulses.

Pulse Width, Pulse Length, or Pulse Duration
The time interval between the leading edge and trailing edge of a pulse at which the amplitude reaches 50% of the peak pulse amplitude.

TABLE 11-1

SQUARE AND SINE WAVE		RECTANGULAR WAVE
$\text{Frequency } (f) = \dfrac{1}{\text{time } (t)}$		$\dfrac{\text{pulse repetition}}{\text{frequency (PRF)}} = \dfrac{1}{\substack{\text{pulse repetition} \\ \text{time (PRT)}}}$
	Equivalent to	
$\text{Time } (t) = \dfrac{1}{\text{frequency } (f)}$		$\dfrac{\text{pulse repetition}}{\text{time (PRT)}} = \dfrac{1}{\substack{\text{pulse repetition} \\ \text{frequency (PRF)}}}$

Duty Cycle

The duty cycle is calculated in exactly the same way as for the square wave and is a ratio of the pulse width to the overall time (PRT). In our example in Figure 11-43 the duty cycle will be equal to

$$\text{duty cycle } (\%) = \frac{\text{pulse width } (P_w)}{\text{PRT}} \times 100\%$$

$$= \frac{1\,\mu s}{1000\,\mu s}$$

$$= \text{duty cycle figure of } 0.001$$

$$= 0.001 \times 100\%$$

$$= 0.1\%$$

The result tells us that the positive pulse lasts for 0.1% of the total time (PRT).

Average

The average or mean value of this waveform is calculated by using the same square-wave formula. The average of the pulse wave in Figure 11-43 will be

$$V \text{ or } I \text{ average} = \text{baseline} + (\text{duty cycle} \times \text{peak to peak})$$

$$V_{\text{avg}} = 0\text{ V} + (0.001 \times 5\text{ V})$$

$$= 0\text{ V} + (5\text{ mV})$$

$$= 5\text{ mV}$$

FIGURE 11-43 PRF and PRT of a Pulse Wave.

FIGURE 11-44 Average of a Pulse Wave.

Figure 11-44 illustrates the average value of this rectangular waveform. If the voltage and width of the positive pulse are taken and spread out over the entire PRT, they will have a mean level equal, in this example, to 5 mV.

■ **EXAMPLE:**

Calculate the duty cycle and average voltage of the following radar pulse waveform:

$$\text{peak voltage, } V_p = 20 \text{ kV}$$
$$\text{pulse length, } P_l = 1 \text{ μs}$$
$$\text{baseline voltage} = 0\text{V}$$
$$\text{PRF} = 3300 \text{ pulses per second (pps)}$$

■ *Solution:*

$$\text{duty cycle} = \frac{\text{pulse length } (P_l)}{\text{PRT}} \times 100\%$$

$$= \frac{1 \text{ μs}}{303 \text{ μs}} \times 100\% \left(\text{PRT} = \frac{1}{\text{PRF}} = \frac{1}{3300} = 303 \text{ μs} \right)$$

$$= (3.3 \times 10^{-3}) \times 100\%$$

$$= 0.33\%$$

$$V_{\text{avg}} = \text{baseline} + (\text{duty cycle} \times V_{\text{p-p}})$$

$$= 0 \text{ V} + [(3.3 \times 10^{-3}) \times 20 \times 10^{3}]$$

$$= 66 \text{ V}$$

Frequency-Domain Analysis

The pulse or rectangular wave is closely related to the square wave, as shown in Figure 11-45; however, some changes occur in its harmonic content. One is that even-number harmonics are present and their amplitudes do not fall

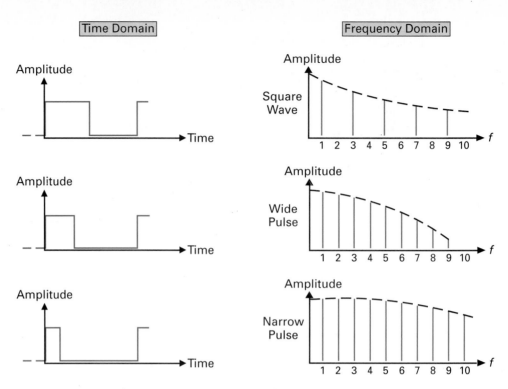

FIGURE 11-45   **Time- and Frequency-Domain Analysis of a Pulse Waveform.**

off as quickly as do those of the square wave. The amplitude and phase of these sine-wave harmonics are determined by the pulse width and pulse repetition frequency, and the narrower the pulse, the greater the number of harmonics present.

11-3-4   The Triangular Wave

Triangular Wave
A repeating wave that has equal positive and negative ramps that have linear rates of change with time.

Linear
Relationship between input and output in which the output varies in direct proportion to the input.

A **triangular wave** consists of a positive and negative ramp of equal values, as shown in Figure 11-46. Both the positive and negative ramps have a linear increase and decrease, respectively. Linear, by definition, is the relationship between two quantities that exists when a change in a second quantity is directly proportional to a change in the first quantity. The two quantities in this case are voltage or current and time. As shown in Figure 11-46, if the increment of change of voltage ΔV (pronounced "delta vee") is changing at the same rate as the increment of time, Δt ("delta tee"), the ramp is said to be **linear.**

With Figure 11-47(a), the voltage has risen 1 V in 1 second (time 1) and maintains that rise through to time 2 and, consequently, is known as a linearly rising slope. In Figure 11-47(b), the voltage is falling first from 6 to 5 V, which is a 1 V drop in 1 second, and in time 2 from 6 V to 2 V, which is a 4 V drop in 4 s. The rate of fall still remains the same, so the waveform is referred to as a linearly falling slope.

FIGURE 11-46   Triangular Wave.

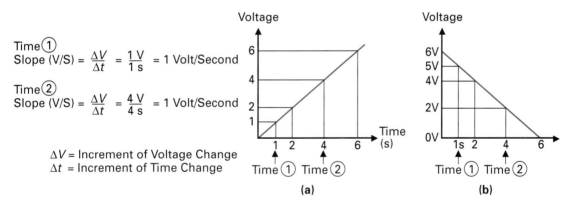

Time ① Slope (V/S) = $\frac{\Delta V}{\Delta t}$ = $\frac{1\text{ V}}{1\text{ s}}$ = 1 Volt/Second

Time ② Slope (V/S) = $\frac{\Delta V}{\Delta t}$ = $\frac{4\text{ V}}{4\text{ s}}$ = 1 Volt/Second

ΔV = Increment of Voltage Change
Δt = Increment of Time Change

FIGURE 11-47   Linear Triangular-Wave Rise and Fall.

This formula for slope will also apply to a current waveform, and is

$$\text{slope (A/s)} = \frac{\Delta I}{\Delta t}$$

where ΔI = increment of current change

Δt = increment of time change

With triangular waves, frequency and time apply as usual, as seen in Figure 11-48, with

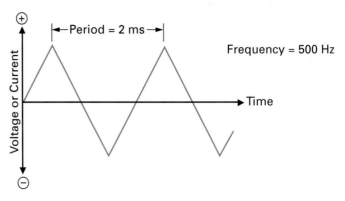

FIGURE 11-48 **Triangular-Wave Period and Frequency.**

$$\text{frequency} = \frac{1}{\text{time}} \quad (\text{Hz})$$

$$\text{time} = \frac{1}{\text{frequency}} \quad (\text{s})$$

Frequency-Domain Analysis

The time domain of a triangular wave is shown in Figure 11-49(a), while the frequency domain of a triangular wave is shown in Figure 11-49(b). Frequency domain analysis is often used to test electronic circuits as it tends to highlight problems that would normally not show up when a sine-wave signal is applied.

11-3-5 *The Sawtooth Wave*

Sawtooth Wave
Repeating waveform that rises from zero to a maximum value linearly and then falls to zero and repeats.

On an oscilloscope display (time-domain presentation), the **sawtooth wave** is very similar to a triangular wave, in that a sawtooth wave has a linear ramp. However, unlike the triangular wave, which reverses and has an equal but oppo-

(a)

Triangle Wave Spectrum

(b)

FIGURE 11-49 **Analysis of a Triangular Wave. (a) Time Domain. (b) Frequency Domain.**

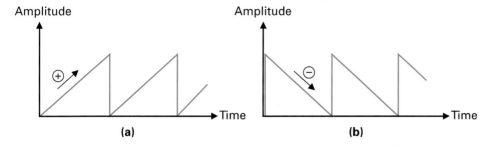

FIGURE 11-50   Sawtooth Wave. (a) Positive Ramp. (b) Negative Ramp.

site ramp back to its starting level, the sawtooth "flies" back to its starting point immediately and then repeats the previous ramp, as seen in Figure 11-50, which shows both a positive and a negative ramp sawtooth.

Figure 11-51 shows the odd and even harmonics contained in the frequency domain analysis of a negative-going ramp.

11-3-6   Other Waveforms

The waveforms discussed so far are some of the more common types; however, since every waveform shape (except a pure sine wave) is composed of a large number of sine waves combined in an infinite number of ways, any waveform shape is possible. Figure 11-52 illustrates a variety of waveforms that can be found in all fields of electronics, and the appendix contains a detailed frequency spectrum chart detailing the applications of many electromagnetic frequencies.

SELF-TEST REVIEW QUESTIONS FOR SECTION 11-3

1. Sketch the following waveforms:
   a. Sine wave
   b. Square wave
   c. Rectangular wave
   d. Triangular wave
   e. Sawtooth wave

2. An oscilloscope gives a _____-domain representation of a periodic wave, while a spectrum analyzer gives a _____-domain representation.

3. What are the wavelength formulas for sound waves and electromagnetic waves, and why are they different?

4. A square wave is composed of an infinite number of _____ harmonics.

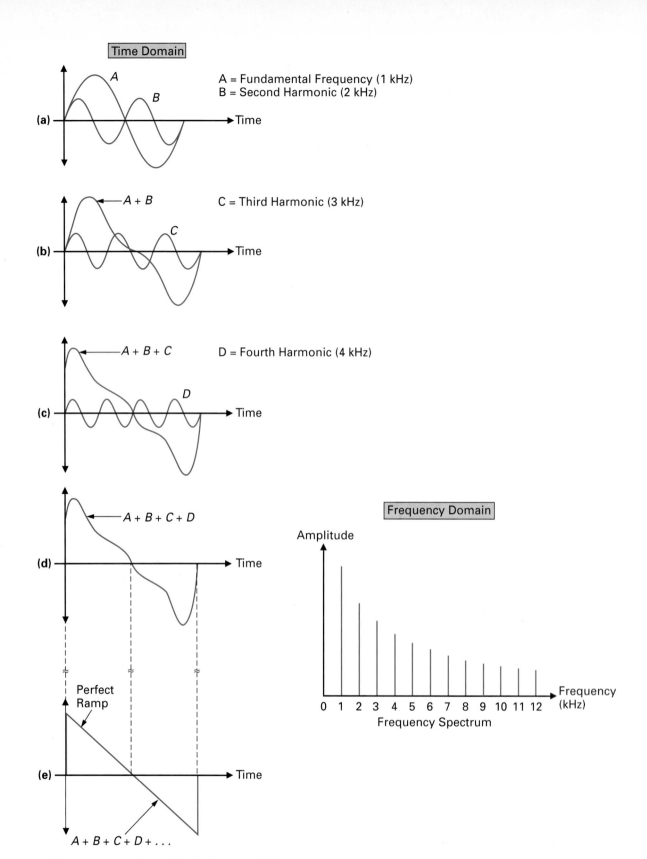

FIGURE 11-51 Analysis of a Sawtooth Wave.

FIGURE 11-52   Waveforms. (a) Telephone Communications. (b) Radio Broadcast. (c) Medical. (d) Radar/Sonar. (e) Communications. (f) Industrial.

11-4   ELECTRICITY AND ELECTRONICS

In the beginning of this chapter, it was stated that ac is basically used in two applications, for (1) power transfer and (2) information transfer. These two uses for ac help define the difference between electricity and electronics. Electronic equipment manages the flow of information, while electrical equipment manages the flow of power. In summary:

EQUIPMENT	MANAGES
Electrical	Power (large values of V and I)
Electronic	Information (small values of V and I)

To use a few examples, we can say that a dc power supply is a piece of electrical equipment since it is designed to manage the flow of power. A TV set, however, is an electronic system since its electronic circuits are designed to manage the flow of audio (sound) and video (picture) information.

Since most electronic systems include a dc power supply, we can say that the electrical circuits manage the flow of power, and this power supply enables the electronic circuits to manage the flow of information.

SELF-TEST REVIEW QUESTIONS FOR SECTION 11-4

1. _____ equipment manages the flow of information, and these ac waveforms normally have small values of current and voltage.
2. _____ equipment manages the flow of power, and these ac waveforms normally have large values of current and voltage.

SUMMARY

Direct Current (DC) and Alternating Current (AC)

1. While direct current (dc) can be used in many instances to deliver power or represent information, there are certain instances in which ac is preferred. For example, ac is easier to generate and is more efficiently transmitted as a source of electrical power for the home or business. Audio (speech and music) and video (picture) information are generally always represented in electronic equipment as an alternating current or alternating voltage signal.

2. Direct current (dc) is the flow of electrons in one DIRECTion and one direction only. Dc voltage is nonvarying and normally obtained from a battery or power supply unit. The only variation in voltage from a battery occurs due to the battery's discharge, but even then, the current will still flow in only one direction.

3. Some power supplies supply a form of dc known as pulsating dc, which varies periodically from zero to a maximum, back to zero, and then repeats. Pulsating dc is normally supplied by a battery charger and is used to charge secondary batteries. It is also used to operate motors that convert the pulsating dc electrical energy into a mechanical rotation output. Whether steady or pulsating, direct current is current in only one DIRECTion.

4. Alternating current (ac) flows first in one direction and then in the opposite direction. This reversing current is produced by an alternating voltage source, which reaches a maximum in one direction (positive), decreases to zero, and then reverses itself and reaches a maximum in the opposite direction (negative). During the time of the positive voltage alternation, the polarity of the voltage will cause current to flow from negative to positive in one direction. During the time of the negative voltage alternation, the polarity of the voltage will reverse, causing current to flow once again from negative to positive, but in this case, in the opposite direction.

5. There are two main applications for ac:
   a. *Power transfer:* to supply electrical power for lighting, heating, cooling, appliances, and machinery in both home and industry

b. *Information transfer:* to communicate or carry information, such as radio music and television pictures, between two points.

6. There are three advantages that ac has over dc from a power point of view.

a. Many appliances and most industrial equipment need a large supply of current, and in this situation a generator would have to be used to generate this large amount of current. Generators can be used to generate either dc or ac, but ac generators can be larger, less complex internally, and cheaper to operate, which are the first reasons why we use ac instead of dc for supplying power.

b. From a power point of view, ac is always used by electric companies when transporting power over long distances to supply both the home and industry with electrical energy. The amount of power lost (heat) in these power lines can be calculated by using the formula $P = I^2 \times R$, where I is the current flowing through the line and R is the resistance of the power lines. This means that the larger the current, the greater the amount of power lost in the lines in the form of heat and therefore the less the amount of power supplied to the home or industry. For this reason, power companies transport electric energy at a very high voltage between 200,000 and 600,000 V. Since the voltage is high, the current can be low and provide the same amount of power to the consumer. An ac voltage can easily and efficiently be transformed up or down to a higher or lower voltage by utilizing a device known as a transformer, and even though dc voltages can be stepped up and down, the method is inefficient and more complex.

c. Nearly all electronic circuits and equipment are powered by dc voltages, which means that once the ac power arrives at the home or industry, in most cases it will have to be converted into dc power to operate electronic equipment. It is a relatively simple process to convert ac to dc; however, conversion from dc to ac is a complex and comparatively inefficient process.

Power Transfer

7. The ac power distribution system begins at the electric power plant, which has the powerful large generators driven by turbines to generate large ac voltages.

8. The turbines can be driven by either falling water (hydroelectric), or from steam, which is produced with intense heat by burning either coal, gas, or oil or from a nuclear reactor (thermoelectric).

9. The turbine supplies the mechanical energy to the generator, to be transformed into ac electrical energy.

10. The generator generates an ac voltage of approximately 22,000 V, which is stepped up by transformers to approximately 500,000 V.

11. At each city or town, the voltage is tapped off the long-distance transmission lines and stepped down to approximately 66,000 V and is distributed to large-scale industrial customers.

12. The 66,000 V is stepped down again to approximately 4800 V and distributed throughout the city or town by short-distance transmission lines. This 4800 V is used by small-scale industrial customers and residential customers who receive the ac power via step-down transformers on utility poles, which step down the 4800 V to 240 V and 120 V.

13. A large amount of equipment and devices within industry and the home will run directly from the ac power, such as heating, lighting, and cooling.

14. Some equipment that runs on dc, such as televisions and computers, will accept the 120 V ac and internally convert it to the dc voltages required.

15. The unit that converts ac to dc is called a rectifier.

Information Transfer

16. Information, by definition, is the property of a signal or message that conveys something meaningful to the recipient.

17. Communication, which is the transfer of information between two points, began with speech and progressed to handwritten words in letters and printed words in newspapers and books. To achieve greater distances of communication, face-to-face communications evolved into telephone and radio communications.

18. Sound waves or sounds are normally generated by a vibrating reed or plucked string or a person's vocal cords.

19. Like the ripples produced by a stone falling in a pond, the sound waves are constantly expanding and traveling outward.

20. The microphone is in fact a transducer (energy converter), because it converts the sound wave (which is a form of mechanical energy) into electrical energy in the form of voltage and current, which varies in the same manner as the sound wave and therefore contains the sender's message or information.

21. The electrical wave is a variation in voltage or current and can only exist in a wire conductor or circuit. This electrical signal travels at the speed of light.

22. The speaker, like the microphone, is also an electroacoustical transducer that converts the electrical energy input into a mechanical sound-wave output.

23. To communicate between two distant points, a wire must be connected between the microphone and speaker. However, if an electrical wave is applied to an antenna, the electrical wave is converted into a radio or electromagnetic wave.

24. Antennas are designed to radiate and receive electromagnetic waves, which vary in field strength, and can exist in either air or space. These radio waves, as they are also known, travel at the speed of light and allow us to achieve great distances of communication.

25. More specifically, radio waves are composed of two basic components. The electrical voltage applied to the antenna is converted into an electric field and the electrical current into a magnetic field. This electromagnetic (elec-

QUICK REFERENCE SUMMARY SHEET

FIGURE 11-53 AC Wave Shapes.

The Sine Wave

$Frequency = \dfrac{1}{Time}$ Hertz (Hz)

$Time = \dfrac{1}{Frequency}$ Seconds

The Square Wave

Alternation Average = Peak
RMS = Peak
RMS = Average
Duty Cycle = 50%

For example, f = 1 kHz (1000 Square Wave Cycles in One Second)

$t = \dfrac{1}{f} = 1$ mS

The Rectangular or Pulse Wave

Average = Duty Cycle × Peak
RMS = Average

The Triangular Wave

Period = 2 ms
Frequency = 500 Hz

The Sawtooth Wave

Distance = velocity × time

$$C^2 = A^2 + B^2$$

$$C = \sqrt{A^2 + B^2} \qquad B = \sqrt{C^2 - A^2} \qquad A = \sqrt{C^2 - B^2}$$

$$\sin\theta = \frac{O}{H} \qquad \cos\theta = \frac{A}{H} \qquad \tan\theta = \frac{O}{A}$$

$$f = \frac{1}{t} \qquad t = \frac{1}{f}$$

f = frequency, in hertz (Hz)
t = period or time, in seconds (s)

Sine wave: $\qquad V_{p-p} = 2 \times V_p \qquad V_{p-p}$ = peak-to-peak voltage

V_p = peak voltage

$I_{p-p} = 2 \times I_p \qquad I_{p-p}$ = peak-to-peak current

I_p = peak current

$$\text{rms} = 0.707 \times \text{peak}$$

$$\text{Peak} = \frac{\text{rms}}{0.707} \quad \text{or} \quad 1.414 \times \text{rms}$$

$$\text{Average} = 0.637 \times \text{peak}$$

Square wave: $\qquad$ Duty cycle (%) = $\dfrac{\text{pulse width } (P_w)}{\text{period } (t)} \times 100\%$

$$V \text{ or } I \text{ average} = \text{baseline} + \begin{pmatrix} \text{duty cycle} \times \\ \text{peak to peak} \end{pmatrix}$$

Rectangular wave: PRF = $\dfrac{1}{\text{PRT}}$

PRF = pulse repetition frequency in pulses per second (pps)

PRT = $\dfrac{1}{\text{PRF}}$

PRT = pulse repetition time, in seconds(s)

Triangular wave: Slope (A/s) = $\dfrac{\Delta I}{\Delta t}$

delta I (ΔI) = increment of current change

delta t (Δt) = increment of time

Wavelength (λ) = $\dfrac{\text{velocity}}{\text{frequency (Hz)}}$ $\qquad$ *Sound wave:* λ (m) = $\dfrac{344.4 \text{ m/s}}{f}$

Electromagnetic wave: λ (cm) = $\dfrac{3 \times 10^{10} \text{ cm/s}}{f}$ $\quad$ or $\quad$ λ (m) = $\dfrac{3 \times 10^8 \text{ m/s}}{f}$

Sound, Electrical, and Electromagnetic Waves

Type of Wave	Sound Wave	Electrical Wave	Electromagnetic Wave
Speed of Travel	1133 Feet/Second or 344.4 Meters/Second	186,000 Miles/Second or 3 × 10⁸ Meters/Second	186,000 Miles/Second or 3 × 10⁸ Meters/Second

tric–magnetic) wave is used to carry a variety of information, such as speech, radio broadcasts, television signals, and so on.

AC Wave Shapes (Figure 11-53)

26. If the length of two sides of a right triangle are known, and the length of the third side is needed, remember that

$$C^2 = A^2 + B^2$$

27. If the angle θ is known along with the length of one side, or if angle θ is needed and the lengths of the two sides are known, one of the three formulas can be chosen based on what variables are known, and what is needed.

$$\sin \theta = \frac{O}{H} \qquad \cos \theta = \frac{A}{H} \qquad \tan \theta = \frac{O}{A}$$

28. The sine wave is the most common type of waveform. It is the natural output of a generator. If this ac voltage is applied across a closed circuit, it produces a current that continually reverses or alternates in each direction.

29. The amplitude or magnitude of a wave is often represented by a vector arrow. The vector's length indicates the magnitude of the current or voltage, while the arrow's point is used to show the direction, or polarity.

30. The peak of an ac wave occurs on both the positive and negative alternation, but is only at the peak (maximum) for an instant.

31. The peak-to-peak value of a sine wave is the value of voltage or current between the positive and negative maximum values of a waveform.

32. Both the positive and negative alternation of a sine wave can accomplish the same amount of work; but the ac waveform is only at its maximum value for an instant in time, spending most of its time between peak currents. The effective value of a sine wave is equal to 0.707 of the peak value.

33. This root-mean-square (rms) result of 0.707 can always be used to tell us how effective an ac sine wave will be.

34. Unless otherwise stated, ac values of voltage or current are always given in rms.

35. The average value of the positive or negative alternation is found by taking either the positive or negative alternation and listing the amplitude or vector length of current or voltages at 1° intervals. The sum of all these values is then divided by the total number of values (averaging), which for all sine waves will calculate out to be 0.637 of the peak voltage or current.

36. The average of a positive or negative alternation (half-cycle) is equal to 0.637 × peak; however, the average of the complete cycle, including both the positive and negative half-cycles, is mathematically zero as the amount of voltage or current above the zero line is equal but opposite to the amount of voltage or current below the zero line.

37. The period (t) is the time required for one complete cycle (positive and negative alternation) of the sinusoidal current or voltage waveform. A *cycle,* by definition, is the change of an alternating wave from zero to a positive peak, to zero, then to a negative peak, and finally, back to zero.

38. Frequency is the number of repetitions of a periodic wave in a unit of time. It is symbolized by f and is given the unit hertz (cycles per second), in honor of a German physicist, Heinrich Hertz.

39. Sinusoidal waves can take a long or a short amount of time to complete one cycle. This time is related to frequency in that period is equal to the reciprocal of frequency.

40. All homes in the United States receive at their wall outlets an ac voltage of 120 V rms at a frequency of 60 Hz. This frequency was chosen for convenience, as a lower frequency would require larger transformers, and if the frequency were too low, the slow switching (alternating) current through the light bulb would cause it to flicker.

41. A higher frequency than 60 Hz was found to cause an increase in the amount of heat generated in the core of all power distribution transformers due to eddy currents and hysteresis losses.

42. Wavelength, as its name states, is the physical length of one complete cycle and is generally measured in meters. The wavelength (λ, lambda) of a complete cycle is dependent on the frequency and velocity of the transmission.

43. Electromagnetic waves or radio waves travel at the speed of light in air or a vacuum, which is 3×10^{8} meters/second or 3×10^{10} cm/second.

44. Sound waves travel at a slower speed than electromagnetic waves, as their mechanical vibrations depend on air molecules, which offer resistance to the traveling wave.

45. The phase of a sine wave is always relative to another sine wave of the same frequency.

46. The sine wave is called a sine wave because it changes in amplitude at the same rate as the sine trigonometric function.

47. The square wave is a periodic (repeating) wave that alternates from a positive peak value to a negative peak value, and vice versa, for equal lengths of time.

48. The duty cycle is the ratio of a pulse width (positive or negative pulse or cycle) to the overall period or time of the wave and is normally given as a percentage.

49. Since a square wave always has a positive and a negative alternation that are equal in time, the duty cycle of all square waves is equal to 50%, which actually means that the positive cycle lasts for 50% of the time of one cycle.

50. The average of the complete square wave cycle should calculate out to be zero, as the amount above the line equals the amount below.

51. It takes a small amount of time for the wave to increase to its positive value (the rise time) and consequently an equal amount of time for a wave to decrease to its negative value (the fall time). Rise time (T_R), by definition, is the time it takes for an edge to rise from 10% to 90% of its full amplitude, while fall time (T_F) is the time it takes for an edge to fall from 90% to 10% of its full amplitude.

52. A *periodic wave* is a wave that repeats the same wave shape from one cycle to the next.

53. A time-domain representation of a periodic sine wave, which is the same way it would appear on an oscilloscope display, plots the sine wave's amplitude against time.

54. A frequency-domain representation of the same periodic sine wave, which shows the wave as it would appear on a spectrum analyzer, plots the sine wave's amplitude against frequency instead of time. Pure sine waves have no other frequency components.

55. Other periodic wave shapes, such as square, pulse, triangular, sawtooth, or irregular, are actually made up of a number of sine waves having a particular frequency, amplitude, and phase.

56. The fundamental frequency sine wave is always the lowest frequency and largest amplitude component of any waveform shape and is used as a reference.

57. Sine wave harmonic frequencies are smaller in amplitude and are some multiple of the fundamental frequency.

58. Square waves are composed of a fundamental frequency and an infinite number of odd harmonics (third, fifth, seventh, and so on).

59. The rectangular wave is similar to the square wave in many respects, in that it is a periodic wave that alternately switches between one of two fixed values. The difference in the rectangular wave is that it does not remain at the two peak values for equal lengths of time.

60. Pulse repetition frequency is the number of times per second that a pulse is transmitted.

61. Pulse repetition time is the time interval between the start of two consecutive pulses.

62. Pulse width, pulse length, or pulse duration is the time interval between the leading edge and trailing edge of a pulse at which the amplitude reaches 50% of the peak pulse amplitude.

63. The triangular wave is a repeating wave that has equal positive and negative ramps that have linear rates of change with time.

64. The term linear describes a relationship between input and output in which the output varies in direct proportion to the input.

65. The sawtooth wave is a repeating waveform that rises from zero to a maximum value linearly and then falls to zero and repeats.

Electricity and Electronics

66. AC is basically used in two applications, for (1) power transfer and (2) information transfer.

67. Electronic equipment manages the flow of information, while electrical equipment manages the flow of power.

68. A dc power supply is a piece of electrical equipment since it is designed to manage the flow of power. A TV set, however, is an electronic system since

its electronic circuits are designed to manage the flow of audio (sound) and video (picture) information.

69. Since most electronic systems include a dc power supply, we can say that the electrical circuits manage the flow of power, and this power supply enables the electronic circuits to manage the flow of information.

Adjacent

Alternating current (ac)

Amplifier

Amplitude

Antenna

Audio

Average value

Battery charger

Communication

Cosine

Delta (Δ)

Direct current

Duty cycle

Effective value

Electrical wave

Electric field

Electroacoustical transducer

Electromagnetic wave

Fall time

Frequency

Frequency domain

Fundamental frequency

Harmonics

Hypotenuse

Information

Lag

Lead

Leading edge

Microphone

Opposite

Oscilloscope

Peak-to-peak value (p–p)

Peak value (p)

Period (t)

Periodic

Phase shift

Power transfer

Pulsating dc

Pulse duration (P_d)

Pulse length (P_l)

Pulse repetition frequency (PRF)

Pulse repetition time (PRT)

Pulse waveform

Pulse width (P_w)

Pythagorean Theorem

Rectangular waveform

Rectifier

Right-angle triangle

Rise time

Root mean square (rms)

Sawtooth waveform

Sine

Sine wave

Sinusoidal

Slope

Sound wave

Speaker

Spectrum analyzer

Speed of light

Speed of sound

Square wave

Tangent

Time domain

Trailing edge

Transducer

Transformer

Transmission lines

Triangular wave

Trigonometry

Utility pole

Vector

Video

Wavelength

Multiple-Choice Questions

REVIEW QUESTIONS

1. A current that rises from zero to maximum positive, returns to zero, and then repeats is known as:
   - **a.** Alternating current
   - **b.** Ac
   - **c.** Pulsating dc
   - **d.** Steady dc

2. A current that rises from zero to maximum positive, decreases to zero, and then reverses to reach a maximum in the opposite direction (negative) is known as:

   a. Alternating current   **c.** Steady direct current
   b. Pulsating direct current   **d.** All of the above

3. The advantage(s) of ac over dc from a power distribution point of view is/are:

   a. Generators can supply more power than batteries
   b. Ac can be transformed to a high or low voltage easily, minimizing power loss
   c. Ac can easily be converted into dc
   d. All of the above
   e. Only (a) and (c)

4. The approximate voltage appearing on long-distance transmission lines in the ac distribution system is:

   a. 250 V   **c.** 500,000 V
   b. 2500 V   **d.** 250,000 V

5. The most common type of alternating wave shape is the:

   a. Square wave   **c.** Rectangular wave
   b. Sine wave   **d.** Triangular wave

6. _____ equipment manages the flow of information.

   a. Electronic   **c.** Discrete
   b. Electrical   **d.** Integrated

7. _____ equipment manages the flow of power.

   a. Electronic   **c.** Discrete
   b. Electrical   **d.** Integrated

8. The peak-to-peak value of a sine wave is equal to:

   a. Twice the rms value   **c.** Twice the peak value
   b. 0.707 times the rms value   **d.** $1.14 \times$ the average value

9. The rms value of a sine wave is also known as the:

   a. Effective value   **c.** Peak value
   b. Average value   **d.** All of the above

10. The peak value of a 115 V (rms) sine wave is:

    a. 115 V   **c.** 162.7 V
    b. 230 V   **d.** Two of the above could be true.

11. The mathematical average value of a sine wave cycle is:

    a. $0.637 \times$ peak   **c.** $1.414 \times$ rms
    b. $0.707 \times$ peak   **d.** Zero

12. The frequency of a sine wave is equal to the reciprocal of _____.

    a. The period   **d.** Both (a) and (b)
    b. One cycle time   **e.** None of the above
    c. One alternation

13. What is the period of a 1-MHz sine wave?

a. 1 ms   **c.** 10 ms
b. One millionth of a second   **d.** 100 μs

14. The sine of 90° is:

a. 0   **c.** 1
b. 0.5   **d.** Any of the above

15. What is the frequency of a sine wave that has a cycle time of 1 ms?

a. 1 MHz   **c.** 200 m
b. 1 kHz   **d.** 10 kHz

16. The pulse width (P_w) is the time between the _____ points on the positive and negative edges of a pulse.

a. 10%   **c.** 50%
b. 90%   **d.** All of the above

17. The duty cycle is the ratio of _____ to period.

a. Peak   **c.** Pulse length
b. Average power   **d.** Both (a) and (c)

18. With a pulse waveform, PRF can be calculated by taking the reciprocal of:

a. The duty cycle   **c.** P_d
b. PRT   **d.** P_l

19. The sound wave exists in _____ and travels at approximately _____.

a. Space, 1130 ft/s   **c.** Air, 3×10^6 m/s
b. Wires, 186,282.397 miles/s   **d.** None of the above

20. The electrical and electromagnetic waves travel at a speed of:

a. 186,000 miles/s   **c.** 162,000 nautical miles/s
b. 3×10^8 meters/s   **d.** All of the above

Essay Questions

21. Describe the three advantages that ac has over dc from a power point of view. (11-2-1)

22. Describe briefly the ac power distribution system. (11-2-1)

23. What is the difference between the words electricity and electronics? (11-4)

24. What are the five basic ac information wave shapes? (11-3)

25. Describe briefly the following terms as they relate to the sine wave: (11-3-1)

a. RMS   **d.** Average   **g.** Period
b. Peak   **e.** The name sine   **h.** Wavelength
c. Peak to peak   **f.** Frequency   **i.** Phase

26. Describe briefly the following terms as they relate to the square wave: (11-3-2)

a. Duty cycle   **b.** Average

27. Describe briefly the following terms as they relate to the rectangular wave: (11-3-3)

    a. PRT   **c.** Duty cycle
    b. PRF   **d.** Average

28. Briefly describe the meaning of the terms *fundamental frequency* and *harmonics*. (11-3-2)

29. List and describe all the pertinent information relating to the following information carriers: (11-3)

    a. Sound wave   **c.** Electromagnetic wave
    b. Electrical wave

30. Describe the difference between frequency- and time-domain analysis. (11-3-2)

Practice Problems

31. Calculate the periods of the following sine-wave frequencies:

    a. 27 kHZ   **d.** 365 Hz
    b. 3.4 MHz   **e.** 60 Hz
    c. 25 Hz   **f.** 200 kHz

32. Calculate the frequency for each of the following values of time:

    a. 16 ms   **d.** 0.05 s
    b. 1 s   **e.** 200 μs
    c. 15 μs   **f.** 350 ms

33. A 22 V peak sine wave will have the following values:

    a. Rms voltage =   **c.** Peak-to-peak voltage =
    b. Average voltage =

34. A 40 mA rms sine wave will have the following values:

    a. Peak current =   **c.** Average current =
    b. Peak-to-peak current =

35. How long would it take an electromagnetic wave to travel 60 miles?

36. An 11 kHz rectangular pulse, with a pulse width of 10 μs, will have a duty cycle of _____%.

37. Calculate the PRT of a 400 kHz pulse waveform.

38. Calculate the average current of the pulse waveform in Question 36 if its peak current is equal to 15 A.

39. What is the duty cycle of a 10 V peak square wave at a frequency of 1 kHz?

40. Considering a fundamental frequency of 1 kHz, calculate the frequency of its:

    a. Third harmonic   **c.** Seventh harmonic
    b. Second harmonic

OBJECTIVES

After completing this chapter, you will be able to:

1. Describe how the multimeter is able to measure ac as well as dc.

2. Explain some of the multimeter accessories, such as:
   a. Current clamps
   b. RF probes
   c. High-voltage probes

3. Compare the advantages and disadvantage of the analog multimeter and digital multimeter.

4. Define the function and basic operation of;
   a. A frequency counter
   b. An audio-frequency generator
   c. A radio-frequency generator

5. Describe how to troubleshoot electronic equipment by applying a six-step troubleshooting procedure.

6. Explain some frequently used troubleshooting techniques.

AC Test Equipment and Troubleshooting Techniques

Logging On

During the seventeenth century, European thinkers were obsessed with any device that could help in mathematical calculation. Scottish mathematician John Napier decided to meet this need, and in 1614 he published his new discovery of logarithms. In this book, consisting mostly of tediously computed tables, Napier stated that a logarithm is the exponent of a base number, for example, 100 is 10^2, 27 is $10^{1.43136}$, 10 is 10^1, 6 is $10^{0.77815}$, and any number no matter how large or small can be represented in this manner. He also outlined how the multiplication of two numbers could be achieved by simple addition. For example, if the logarithm of one number, 2, which is 0.30103, is added to the logarithm of another number, 4, which is 0.60206, the result is equal to the logarithm of 8, which is 0.90309. Therefore, the multiplication of two large numbers could now be achieved by looking up the logarithms of the two numbers in a log table, adding them together, and then finding the number that corresponds to that sum in an antilog (reverse log) table. In this example, the antilog of 0.9039 is 8.

Napier's table of logarithms were used by William Oughtred, who, just ten years after Napier's death in 1617, developed a hand mechanical device that could be used for rapid calculation. This device, considered the first pocket calculator, was the slide rule.

As well as being a brilliant mathematician, Napier was also interested in designing military weapons. One such unfinished project was a death ray system consisting of an arrangement of mirrors and lenses that, when aligned, would produce a concentrated lethal beam of sunlight.

I N T R O D U C T I O N

As a technician or engineer, you are going to be required to diagnose and repair a problem in the shortest amount of time possible. To aid in the efficiency of this fault finding and repair process, you can make use of certain pieces of test equipment. Humans have five kinds of sensory systems: touch, taste, sight, sound, and smell. Four can be used for electronic troubleshooting: sight, sound, touch, and smell.

Electronic test equipment can be used either to *sense* a circuit's condition or to *generate* a signal to see the response of the component or circuit to that signal. For example, the dc multimeter can be used to measure voltage, current, or resistance. When voltage or current is selected, the meter senses the power present in the circuit and gives an indication of that power as voltage or current on either an analog or digital readout display. When resistance is being measured, the multimeter utilizes an internal battery to generate a current that flows out of the meter. The amount of current flow determines the amount of meter deflection, and this current is dependent on the amount of resistance connected to the meter to be measured.

In Chapter 5 you were introduced to the four basic test instruments: the dc power supply, the multimeter, the function generator, and the oscilloscope. In this chapter we extend our studies of test instruments, and examine a six-step troubleshooting procedure that will help you to establish good troubleshooting techniques.

12-1 ADDITIONAL TEST INSTRUMENTS

12-1-1 *The AC Meter*

In Chapter 5 we discussed how a multimeter, like the one shown in Figure 12-1, could be used to measure direct current and voltage. Most multimeters can be used to measure either dc or ac. When the technician or engineer wishes to measure ac, the ac current or voltage is converted to dc internally by a circuit known as a **rectifier,** as shown in Figure 12-2, before passing on to the meter.

The dc produced by the rectification process is in fact pulsating, as shown in Figure 12-3, so the current through the meter is a series of pulses rising from zero to maximum (peak) and from maximum back to zero. Frequencies below 10 Hz (lower frequency limit) will cause the digits on a digital multimeter to continually increase in value and then decrease in value as the meter follows the

Rectifier
A device that converts alternating current into a unidirectional or dc current.

Ohms(Ω), Amps (A), Volts (V) Indication

LSI CMOS Analog/ Digital Processor

Audible Beeper Signals Continuity with 3500–Hz Tone.

Rotary Switch Provides Function Selection.

Metal Oxide Varistor (MOV) Protects Against High Transient Voltages.

Nine-volt Battery Clip

Display Window
Liquid Crystal Display (LCD)

High Impact Case

Voltage Ranges

Resistance Ranges

Current Ranges Protected by Fuse

Silicon Seal Makes Meter Water, Dirt, and Grime Resistant

FIGURE 12-1   Multimeter.

FIGURE 12-2   AC Meter Uses a Rectifier to Produce DC.

pulsating dc. This makes it difficult to read the meter. This effect will also occur with an analog multimeter which, from 10 Hz to approximately 2 kHz, will not be able to follow the fluctuation, causing the needle of the meter movement to remain in a position equal to the average value of the pulsating dc from the rectifier (0.637 of peak). Most meters are normally calibrated internally to indicate rms values (0.707 of peak) rather than average values, because this effective value is most commonly used when expressing ac voltage or current. The upper frequency limit of the ac meter is approximately 2 to 8 kHz. Beyond this limit the meter becomes progressively inaccurate due to the reactance of the capacitance in the rectifier of both types of meters, and the reactance of the moving coil in the analog multimeter. This reactance will result in inaccurate indications due to the change in opposition at different ac input frequencies.

FIGURE 12-3   Rectifier.

Current Clamps

The voltmeter is probably the most frequently used setting on the multimeter. A meter reading can be obtained by just connecting the probes across the component or source to be measured, unlike the ammeter, which requires that the circuit current path be opened, and the ammeter inserted in the path of current flow. If a current measurement is required, a clamp can be used, as seen in Figure 12-4, which allows us to sense the amount of current flow through the conductor without opening the current path.

(a)

(b)

FIGURE 12-4   **Current Clamp.**

The alternating current flowing through the conductor produces an expanding and collapsing magnetic field, which cuts across the coil of wire wound around the core of the clamp, and induces an alternating voltage in the coil (1 mA induces 1 mV). The induced alternating voltage causes an alternating current to flow, which is converted to dc by the rectifier and used to operate the meter. The larger the current flowing in the conductor, the larger the magnetic field surrounding the conductor, which results in a greater induced voltage, current, and consequent meter movement. These clamps are generally ineffective at measuring smaller currents (microamperes), because the magnetic field produced by the current is too weak.

Radio-Frequency Probes

Radio-Frequency Probe
A probe used in conjunction with an ac meter to measure high-frequency RF signals.

As mentioned previously, the meter has a high range limit of approximately 2 kHz. If higher-frequency (radio frequencies) electrical waves are to be measured, a **radio-frequency** (RF) **probe,** as shown in Figure 12-5, can be used. The probe picks up the high-frequency ac voltage from a conducting point on the circuit and passes or couples it to a capacitor, which blocks any dc that could be present at the test point, as we only want to measure the high-frequency ac. The rectifier within the probe will convert this ac input into a dc output, which will be displayed as an rms value on the meter display.

High-Voltage Probe

High-Voltage Probe
Accessory to the voltmeter that has added multiplier resistors within the probe to divide up the large potential being measured by the probe.

The typical multimeter can handle voltages up to approximately 1000 V. If you wish to measure voltages higher than this, another component, known as a **high-voltage** (HV) **probe,** such as the one shown in Figure 12-6, can be used. The high-voltage probe has additional multiplier resistors to drop the extra voltage. Most high-voltage probes are designed so that $1/100$ of voltage at the probe tip

FIGURE 12-5   Radio-Frequency Probe.

Additional Multiplier Resistors

FIGURE 12-6 High-Voltage Probe.

from the test point will appear out of the probe and be applied to the meter. For example, if 10 kV is being measured, 100 V will appear out of the probe and be applied to the meter ($\frac{1}{100} \times 10,000$ V = 100 V). This probe would be called a × 100 probe because the 100 V shown on the meter display would now have to be multiplied by 100 for the operator to determine the voltage (100 V × 100 = 10 kV). The high-voltage probe is especially well insulated to protect its user, who should apply all safety precautions and exercise extreme caution.

EXAMPLE:

A DMM indicates 3.9 V on its display when a test point is probed by a × 1000 high-voltage probe. What is the voltage at this test point?

Solution:

A × 1000 probe will divide the voltage by 1000, so the displayed voltage must be multiplied by 1000 to obtain the correct value:

$$3.9 \text{ V} \times 1000 = 3.9 \text{ kV}$$

Analog and Digital Multimeters

The digital meter is superior to the analog meter in two basic ways:

1. The digital multimeter (DMM) has an easy-to-read display, with decimal points and polarity, while with the analog multimeter the value has to be interpreted by the needle position and the range selected, which may result in human errors.

2. Some of the best analog multimeters have accuracies of 1%, whereas a DMM will typically have an accuracy of 0.01%.

The analog multimeter, however, has one advantage over the DMM: When measuring low-frequency ac signals of several hertz, the analog multimeter will deflect between zero and some value due to the pulsating dc. The operator will see that the signal is pulsating at a low frequency due to the back-and-forth meter needle movement, whereas with the DMM this continual change will cause the digits on the display to change continually and therefore not allow the operator to take a reading.

When it comes to a choice between an analog or a digital multimeter, it seems that most people prefer the DMM because of its easy-to-read display and accuracy.

12-1-2 The Oscilloscope

Oscilloscope
Instrument used to view signal amplitude, frequency, and shape at different points throughout a circuit.

Figure 12-7 illustrates a typical **oscilloscope** (sometimes abbreviated to *scope*), which is used primarily to display the shape and spacing of electrical signals. The oscilloscope displays the actual sine, square, rectangular, triangular, or sawtooth wave shape that is occurring at any point in a circuit. This display is made on a cathode-ray tube (CRT), which is also used in television sets and computers to display video information. From the display on the CRT, we can measure or calculate time, frequency, and amplitude characteristics such as rms, average, peak, and peak-to-peak.

The oscilloscope allows us to see what is happening at every point through a circuit. In Figure 12-8 you can see the different waveforms at different points on the circuit. There are also voltage test points that can be tested with a voltmeter if a scope is not available. A voltmeter, as you can well imagine, does not supply the technician or engineer with as much information as the oscilloscope.

(a)

(b)

FIGURE 12-7   **Typical Oscilloscope. (a) Oscilloscope. (b) Oscilloscope Probe.**

FIGURE 12-8 *Schematic Diagram with Voltage and Waveform Test Points.*

Controls

Oscilloscopes come with a wide variety of features and functions; however, the basic operational features are almost identical. Figure 12-9 illustrates the front panel of a typical oscilloscope; we will now discuss the various control functions. Some of these controls are difficult to understand without practice and experience, so practical experimentation is essential if you hope to gain a clear understanding of how to operate an oscilloscope.

General controls (see Figure 12-9)

Intensity control: Controls the brightness of the trace, which is the pattern produced on the screen of a CRT.

Focus control: Used to focus the trace.

Power OFF/ON: Switch will turn on oscilloscope while indicator shows when oscilloscope is turned on.

FIGURE 12-9   Oscilloscope Controls.

Some oscilloscopes have the ability to display more than one pattern or trace on the CRT screen, as seen by the examples in Figure 12-10. A dual-trace oscilloscope can produce two traces or patterns on the CRT screen at the same time, whereas a single-trace oscilloscope can trace out only one pattern on the screen. The dual-trace oscilloscope is very useful, as it allows us to make comparisons between the phase, amplitude, shape, and timing of two signals from two separate test points. One signal or waveform is applied to the channel *A* input of the oscilloscope, while the other waveform is applied to the channel *B* input.

Channel selection (see Figure 12-9)

Mode switch: This switch allows us to select which channel input should be displayed on the CRT screen.

CHA: The input arriving at channel *A*'s jack is displayed on the screen as a single trace.

CHB: The input arriving at channel *B*'s jack is displayed on the screen as a single trace.

Dual: The inputs arriving at jacks *A* and *B* are both displayed on the screen, as a dual trace.

Calibration
To determine, by measurement or comparison with a standard, the correct value of each scale reading.

(a) **Calibration** Output   This output connection provides a point where a fixed 1-V peak-to-peak square-wave signal can be obtained at a frequency of 1 kHz. This signal is normally fed into either channel *A* or *B*'s input to test probes and the oscilloscope operation.

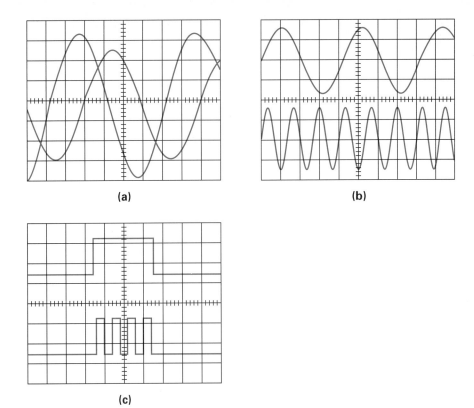

FIGURE 12-10   Sample of Dual-Trace Oscilloscope Displays for Comparison.

(b) Channel *A* and *B* Horizontal Controls

↔ *Position control:* This control will move the position of the one (single-trace) or two (dual-trace) waveforms horizontally (left or right) on the CRT screen.

Sweep time/cm switch: The oscilloscope contains circuits that produce a beam of light that is swept continually from the left to the right of the CRT screen. When no input signal is applied, this sweep will produce a straight horizontal line in the center of the screen. When an input signal is present, this horizontal sweep is influenced by the input signal, which moves it up and down to produce a pattern on the CRT screen the same as the input pattern (sine, square, sawtooth, and so on). This sweep time/cm switch selects the speed of the sweep from left to right, and it can be either fast (0.2 microseconds per centimeter; 0.2 μs/cm) or slow (0.5 second per centimeter; 0.5 s/cm). A low-frequency input signal (long cycle time or period) will require a long time setting (0.5 s/cm) so that the sweep can capture and display one or more cycles of the input. A number of settings are available, with lower time settings displaying fewer cycles and higher time settings showing more cycles of an input.

(c) **Triggering** Controls   These provide the internal timing control between the sweep across the screen and the input waveform.

Triggering
Initiation of an action in a circuit which then functions for a predetermined time, for example, the duration of one sweep in a cathode-ray tube.

Triggering level control. This determines the point where the sweep starts.

Slope switch (+): Sweep is triggered on positive-going slope.
     (−): Sweep is triggered on negative-going slope.

Source switch, CHA: The input arriving into channel *A* jack triggers the sweep.
    CHB: The input arriving into channel *B* jack triggers the sweep.
    EXT: The signal arriving at the external trigger jack is used to trigger the sweep.

(d) Channel *A* and *B* Vertical Controls The *A* and *B* channel controls are identical.

Volts/cm switch: This switch sets the number of volts to be displayed by each major division on the vertical scale of the screen.

↕*Position control:* Moves the trace up or down for easy measurement or viewing.

AC-DC-GND *switch:* In the AC position, a capacitor on the input will pass the ac component entering the input jack, but block any dc components.
     In the GND position, the input is grounded (0 V) so that the operator can establish a reference.
     In the DC position, both ac and dc components are allowed to pass on to and be displayed on the screen.

Measurements

The oscilloscope is probably the most versatile of test equipment, as it can be used to test:

Dc voltage

Ac voltage

Waveform duration

Waveform frequency

Waveform shape

(a) Voltage Measurement The screen is divided into eight vertical and ten horizontal divisions, as shown in Figure 12-11. This 8×10 cm grid is called the *graticule.* Every vertical division has a value depending on the setting of the volts/cm control. For example, if the volts/cm control is set to 5 V/div or 5 V/cm, the waveform shown in Figure 12-12(a), which rises up four major divisions, will have a peak positive alternation value of 20 V (4 div $\times$ 5 V/div = 20 V).

As another example, look at the positive alternation in Figure 12-12(b). The positive alternation rises up three major divisions and then extends another four subdivisions, which are each equal to 1 V because five subdivisions exist within one major division, and one major division is, in this example, equal to 5 V. The positive alternation shown in Figure 12-12(b) therefore has a peak of three major divisions (3 $\times$ 5 V/cm = 15 V), plus three subdivisions (3 $\times$ 1 V = 3 V), which equals 18 volts peak.

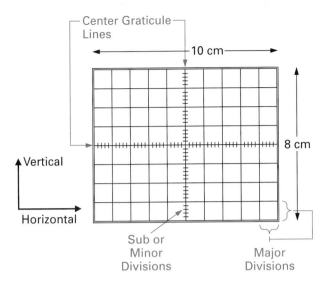

FIGURE 12-11 Oscilloscope Grid.

In Figure 12-12(c), we have selected the 10 volt/cm position, which means that each major division is equal to 10 V and each subdivision is equal to 2 V. In this example, the waveform peak will be equal to two major divisions ($2 \times 10 = 20$ V), plus four subdivisions (4×2 V $= 8$ V), which is equal to 28 V. Once the peak value of a sine wave is known, the peak to peak, average, and rms can be calculated mathematically.

When measuring a dc voltage with the oscilloscope, the volts/cm is applied in the same way, as shown in Figure 12-13. A positive dc voltage in this situation will cause deflection toward the top of the screen, whereas a negative voltage will cause deflection toward the bottom of the screen.

To determine the dc voltage, count the number of major divisions and to this add the number of minor divisions. In Figure 12-13, a major division equals 1 V/cm and, therefore, a minor division equals 0.2 V/cm, so the dc voltage being measured is interpreted as +2.6 V.

(b) Time and Frequency Measurement   The frequency of an alternating wave, such as that seen in Figure 12-14(a), is inversely proportional to the amount of time it takes to complete one cycle ($f = 1/t$). Consequently, if time can be measured, frequency can be determined.

The time/cm control relates to the horizontal line on the oscilloscope graticule and is used to determine the period of a cycle so that frequency can be calculated. For example, in Figure 12-14(b), a cycle lasts five major horizontal divisions, and since the 20 μs/division setting has been selected, the period of the cycle will equal 5×20 μs/division $= 100$ μs. If the period is equal to 100 μs, the frequency of the waveform will be equal to $f = 1/t = 1/100$ μs $= 10$ kHz.

■ **EXAMPLE:**

A complete sine-wave cycle occupies four horizontal divisions and four vertical divisions from peak to peak. If the oscilloscope is set on the 20 ms/cm and 500 mV/cm, calculate:

(a)

(b)

(c)

FIGURE 12-12   Measuring AC Peak Voltage.

a. $V_{p\text{-}p}$
b. t
c. f
d. V_p
e. V_{avg} of peak
f. V_{rms}

■ *Solution:*

4 horizontal divisions $\times$ 20 ms/div = 80 ms
4 vertical divisions $\times$ 500 mV/div = 2 V

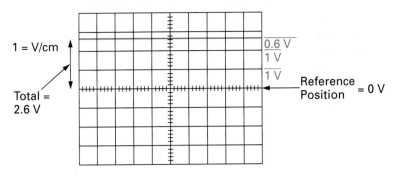

FIGURE 12-13   Measuring DC Voltage.

a. $V_{\text{p-p}} = 2\ \text{V}$

b. $t = 80\ \text{ms}$

c. $f = \dfrac{1}{t} = 12.5\ \text{Hz}$

d. $V_{\text{p}} = 0.5 \times V_{\text{p-p}} = 1\ \text{V}$

e. $V_{\text{avg}} = 0.637 \times V_{\text{p}} = 0.637\ \text{V}$

f. $V_{\text{rms}} = 0.707 \times V_{\text{p}} = 0.707\ \text{V}$

Hand-held Scopemeters

The hand-held **scopemeter**, shown in Figure 12-15, combines a multimeter, oscilloscope, frequency counter, and signal generator in one easy-to-carry battery-operated unit. For a technician, this test instrument is portable, easy to set up, and easy to use since it will automatically change its settings for the best operating mode and continue to adjust itself as the input changes. To explain this test instrument in more detail, let us examine some of its key functions.

Measure menu: The measurement menu button gives you direct access to a quick pop-up menu of more than 30 measurements including: V_{rms}, V_{mean} (arithmetic average), $V_{\text{peak-to-peak}}$, frequency, time delay, rise time, phase,

Scopemeter
A hand-held, battery-operated instrument that combines a multimeter, oscilloscope, frequency counter, and signal generator in one.

(a)

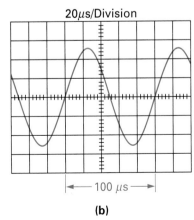

(b)

FIGURE 12-14   Time and Frequency Measurement.

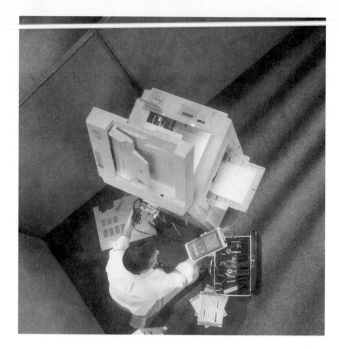

FIGURE 12-15   The Hand-Held Scopemeter. (Courtesy of Fluke Corporation. Reproduced with permission.)

current, and so on. Once the measurement is selected, the scopemeter automatically sets itself up and takes the measurement, as seen in the example in Figure 12-16(a).

Auto set: As you move from test point to test point, the scopemeter will handle the changing inputs automatically in this continuous autoset mode. Each time the signal input changes, the scopemeter will automatically search for the best trigger level, timebase (sweep time), and range scale to speed up measurements and reduce errors, as seen in Figure 12-16(b).

Min max trendplot: This function is used to display and record up to 40 days of signal trend which is sometimes needed in order to pinpoint intermittent problems that occur randomly. Figure 12-16(c) shows an example of how the display shows minimum, maximum, and average readings.

Save: Most scopemeters have large internal memories for saving screen images, setups, and waveforms from the field. These stored measurements can be recalled at any time, downloaded or sent to a personal computer for reports, or sent directly to a printer. Figure 12-16(d) shows how these waveforms and setups would appear on a computer screen.

Function Generator
Signal generator that can function as a sine, square, rectangular, triangular, or sawtooth waveform generator.

12-1-3   The Function Generator

In many cases you will wish to generate a waveform of a certain shape and frequency, and then apply this waveform to your newly constructed circuit. One of the most versatile waveform generators is the **function generator,** so called because it can function as a sine wave, square wave, rectangular wave, triangular

Choose a measurement mode.

Select the measurement

The measurement is taken

(a)

Regardless of changing speed and load...

...continuous Autoset gives a properly scaled signal display.

(b)

(c)

(d)

FIGURE 12-16 Scopemeter Operation. (Courtesy of Fluke Corporation. Reproduced with permission.)

(a)

Switches Used to
Select Waveform Shape

Keypad Used to
Enter Frequency
Desired

Unit Switch: Used to Select Either
    Hz, kHz, or MHz

TTL/600 Ω: Used to Select Either
    Digital (Transistor, Transistor Logic/TTL) or
    Analog (600 Ω-Output Impedance)

Cycle #: Selects Number of Cycles to
    Be Produced

(b)

FIGURE 12-17   Function Generators.

wave, or sawtooth wave generator. Figure 12-17(a) shows a photograph of a typical function generator, and Figure 12-17(b) shows a line drawing of a typical programmable function generator.

Frequency Counter
Meter used to
measure the
frequency or cycles
per second of a
periodic wave.

12-1-4   The Frequency Counter

It is very important that anyone involved in the design, manufacture, and servicing of electronic equipment be able to accurately measure the frequency of a periodic wave. Without the ability to measure frequency accurately, there

OFF/AUTO Switch: Used to Turn on and off
Equipment; When On, the Counter
Automatically Switches Ranges Dependent
on Input

Frequency Counter

OFF   AUTO

Over

654.321

KHz
MHz
GHz

6 Digit/Autorange

Input Jack

Overrange Indicator: Flashes When beyond
Range of Counter

kHz, MHz, and GHz Unit Indicators:
Used to Indicate That Display Is
Indicating the Frequency Input
in kHz, MHz, or GHz

(a)

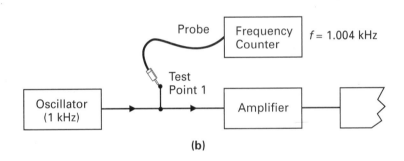

Probe

Frequency
Counter

f = 1.004 kHz

Test
Point 1

Oscillator
(1 kHz)

Amplifier

(b)

(c)

FIGURE 12-18   Frequency Counter.

FIGURE 12-19   Audio-Frequency Generator.

could be no communications, home entertainment, or a great number of other systems.

The frequency meter or counter, shown in Figure 12-18(a), analyzes the frequency of the periodic wave being applied to the input jack and provides a readout on the display of its frequency. In Figure 12-18(b), the frequency counter is being used to determine the accuracy of an oscillator's output of 1 kHz. Figure 12-18(c) shows a photograph of a typical frequency counter. Most frequency counters on the market today can measure frequencies anywhere from hertz to gigahertz.

12-1-5   The Audio-Frequency Generator

Audio-Frequency Generator
A signal generator that can be set to generate a sinusoidal AF signal voltage at any desired frequency in the audio spectrum.

Audio is a Greek word meaning "I hear," and although we can only hear sound-wave frequencies from about 20 to 20,000 Hz, audio-frequency generators will generally produce electrical sine or square waves from 1 Hz to 1 MHz. An audio-frequency (AF) generator or oscillator can be seen in Figure 12-19.

12-1-6   The Radio-Frequency Generator

Radio-Frequency Generator
A generator capable of supplying RF energy at any desired frequency in the radio spectrum.

Figure 12-20(a) shows a typical RF generator that can be used to generate any frequency from 3 kHz to 3 GHz. The electrical wave produced can be either a constant-amplitude continuous wave, shown in Figure 12-20(b), or a varying-amplitude signal, known as an amplitude-modulated (AM) wave, shown in Figure 12-20(c).

Scale Has Different Ranges:
Each Range Is Selected by
Range Switch

RF Frequency Gen.

Range Switch

Used to Select
Specific Frequency
within Range
Selected

Used to Vary
the Amplitude
of the Output
Voltage

Off On   Power   Ind

Ext
Mod
In

CW AM EXT

RF
Voltage
Out

Modulation

Gnd Out

Modulation Switch: Used to select
either continuous wave or amplitude-
modulated output. In the External position
an amplitude modulated output is produced
at the freqency of the external modulating
signal in.

Grounding
Terminal

Radio
Frequency
Output
Terminal

(a)

Continuous Wave
(CW)

Amplitude Modulated
(AM)

(b)

(c)

FIGURE 12-20   **Radio-Frequency Generator.**

1. True or false: The current clamp does not require the circuit current path to be broken in order to measure current. Can it be used to measure dc amps?

2. What is the difference between an RF and an HV probe?

3. The _____ is used primarily to display the shape and spacing of electrical signals.

4. Name the device within the oscilloscope on which the waveforms can be seen.

5. On the 20-μs/div time setting, a full cycle occupies four major divisions. What is the waveform's frequency and period?

6. On the 2-V/cm voltage setting, the waveform swings up and down two major divisions (a total of 4 cm of vertical swing). Calculate the waveform's peak and peak-to-peak voltage.

7. What is the advantage(s) of the dual-trace oscilloscope?

8. List some of the waveform shapes typically generated by a function generator.

9. What is the function of a frequency counter?

10. What is the difference between an audio- and a radio-frequency generator?

12-2 TROUBLESHOOTING TECHNIQUES

Troubleshooting
The process of locating and repairing malfunctions in equipment by means of systematic analysis and testing.

Troubleshooting is the process of locating and repairing malfunctions in equipment by means of systematic analysis and testing. For you to become a good electronics technician or troubleshooter, you will need to have a thorough understanding of *electronics, test equipment, troubleshooting techniques,* and *system repair.* Your understanding of electronics and test equipment will build on the foundation you have established so far in your studies.

Your knowledge of troubleshooting techniques and system repair will improve as you continue experimenting in lab since, as you have probably already discovered, very few circuits operate properly the first time. Try to view these circuit failures as opportunities rather than disasters, and benefit from the troubleshooting experience you can obtain.

In this section we step through a basic six-step troubleshooting procedure, which is illustrated in Figure 12-21. This procedure can be used as a guide to help you quickly isolate and repair basic circuit or complex system malfunctions.

12-2-1 *Step 1: Symptom Recognition*

Before even touching a piece of test equipment you should familiarize yourself with the operation of the circuit or system. This information can be obtained from operator's manuals, technical manuals, and other documentation. Many technicians bypass this step and proceed directly to the detailed testing, only to

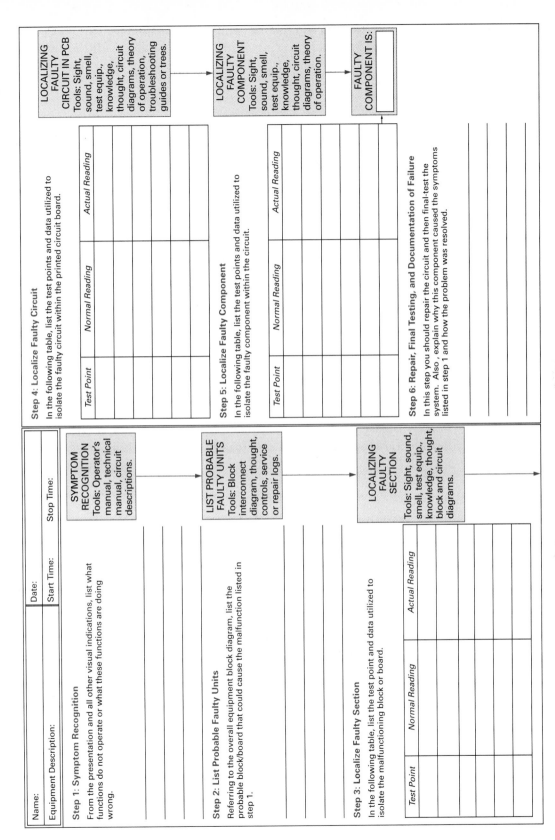

FIGURE 12-21 Six-Step Troubleshooting Procedure.

543

Circuit Failure
A malfunction or breakdown of a component in a circuit.

Operator Error
An incorrect use of the controls of a circuit or system.

discover that what appeared to be a complex **circuit failure** turned out to be a simple **operator error.** Once you are completely familiar with the operation of the circuit or system, operate the controls to determine what functions do not operate or what these functions are doing wrong.

12-2-2 Step 2: List Probable Faulty Units

Referring to the circuit's schematic diagram or the system's block interconnect diagram, list the probable block or board that could cause the malfunction that was listed in step 1.

12-2-3 Step 3: Localize Faulty Section

In this step you remove equipment covers and begin testing the entire system to isolate the malfunctioning block or section, as shown in Figure 12-22.

As you proceed with this step, keep several thoughts in mind:

1. Always look for obvious errors, such as that the power is not connected, the system or circuit is not turned on, and so on.

2. Use your senses of sight, sound, and smell to check for burning components, crackling connections, and so on.

Half-Split Method
A troubleshooting technique used to isolate a faulty block in a circuit or system. In this method a mid-point is chosen and tested to determine which half is malfunctioning.

3. Use a troubleshooting technique called the **half-split method** to help quickly isolate the area of malfunction. For example, if a circuit or system has a good input and a bad output, instead of checking at each and every point between input and output, test a point midway between input and

FIGURE 12-22 Localizing the Faulty Block or Board.

output. If the signal is good, the problem is in the second half of the circuit or system, whereas if the signal is bad, the problem is in the first half of the circuit or system.

4. If the entire circuit or system, or if a section of the circuit or system, is "dead," you can probably safely assume that you have a power problem. Use your voltmeter to measure the ac input to the power supply and the dc output supplies, and don't forget to check one of the most common malfunctions—a blown fuse or tripped circuit breaker.

12-2-4   Step 4: Localize Faulty Circuit

Once you have isolated the problem to a particular board, your next step is to localize the faulty circuit on the printed circuit board (PCB), as shown in Figure 12-23. Keep in mind to use your senses, check first for the obvious, and apply the half-split method. Many equipment manufacturers include PCB troubleshooting trees or guides in the system technical manual, which will help you quickly isolate the faulty section of a circuit.

12-2-5   Step 5: Localize Faulty Component

In most cases it is important to troubleshoot and repair a system quickly, since failures result in a loss of productivity, and therefore money. In some instances, therefore, it may only be necessary for you to locate the faulty PCB and install a replacement, and then isolate the faulty component at a later time. This technique is known as *board-level substitution.*

If a replacement board is not available, you will have to troubleshoot the circuit to find its faulty component. At this stage you may need to refer back to the detailed circuit description, use the troubleshooting guides, perform several tests on the circuit, and logically try to zero in on the faulty component.

FIGURE 12-23   **Localizing Faulty Circuit on Printed Circuit Board.**

FIGURE 12-24   Making the Repair.

Component-level substitution can also be used to speed up the troubleshooting process. For example, once you have located the problem in a general area containing several components, replace the main components one at a time with known good components until the failure is fixed. Although this seems time consuming, it can save time with hard-to-find, elusive circuit problems.

12-2-6 *Step 6: Repair, Final Testing, and Documentation of Failure*

Making the repair may be as simple as changing a fuse, reinstalling a connector, removing a short, or repairing an open wire. Generally, though, it involves the removal and replacement of a component, as shown in Figure 12-24. If the component is mounted in a protoboard, this process will be simple; however, if the component is soldered into a PCB, you will have to follow the desoldering steps outlined in Chapter 5. When a component is removed, make sure that you make a note of its orientation (if this is important) and be sure to handle all static-sensitive components with care.

Once the repair has been made, you should final-test the circuit or system to see that the symptoms listed in step 1 have been resolved, and then document the failure.

SELF-TEST REVIEW QUESTIONS FOR SECTION 12-2

---

1. To become a good troubleshooter you will need a thorough knowledge of _____, _____, _____, and _____.

2. List the six-step troubleshooting procedure.

SUMMARY

1. Test equipment can be used to either sense a circuit's condition or generate a signal to see the response of the component or circuit to that signal.

2. Most multimeters can be used to measure either ac or dc.

3. When multimeters are selected to measure ac, internally a rectifier is used to convert the ac input into a dc voltage.

4. Multimeters are normally calibrated to indicate rms values of the ac being measured.

5. A current clamp allows the technician or engineer to measure ac current without opening the current path.

6. The RF probe can be used with the multimeter to more accurately measure higher frequencies above 2 kHz.

7. The high-voltage probe can be used to measure voltages in the kilovolt range by connecting additional multiplier resistors.

8. The analog multimeter unlike the DMM can be used to measure low frequency ac.

9. The digital multimeter is generally more popular because of its easy-to-read display and accuracy.

10. The oscilloscope displays the shape and spacing of electrical signals, and can therefore be used to measure dc voltage, ac voltage, waveform duration, waveform frequency, and waveform shape.

11. The oscilloscope can be used to display waveform shapes, and from this presentation we can calculate the waveform's time, frequency, and amplitude characteristics.

12. A dual-trace oscilloscope can produce two traces or waveforms on the screen, which allows the technician or engineer to make comparisons between phase, amplitude, shape, or timing.

13. The function generator can produce a sine wave, square wave, rectangular wave, triangular wave, or sawtooth wave.

14. The frequency counter measures and displays the number of cycles per second (hertz) on a digital display.

15. The audio-frequency, radio-frequency, and function generators are all used to produce a wide range of electrical waves for testing or controlling the operation of different circuits.

16. A function generator may function or perform as a sine, square, triangular, pulse, or sawtooth waveform generator.

17. To become a good technician you have to have a thorough knowledge of electronics, test equipment, troubleshooting techniques, and system repair.

18. Troubleshooting is the process of locating and repairing malfunctions in equipment by means of systematic analysis and testing.

19. The six-step troubleshooting procedure is:
    a. Learn to recognize symptoms.
    b. List probable faulty units.

c. Localize faulty section.

d. Localize faulty circuit.

e. Localize faulty component.

f. Repair, final-test, and document failure.

NEW TERMS

AC meter

Audio-frequency (AF) generator

Circuit failure

Current clamp

Frequency counter

Half-split method

High-voltage (HV) probe

Instrument

Operator error

Radio-frequency (RF) generator

Radio-frequency (RF) probe

Rectifier

Signal generator

Substitution

Troubleshooting

REVIEW QUESTIONS

Multiple-Choice Questions

1. Which of the following would be considered a sensing instrument?
   a. Ohmmeter
   b. Voltmeter
   c. Audio oscillator
   d. Two of the above

2. Which of the following generates instead of sensing voltage or current?
   a. Voltmeter
   b. Ammeter
   c. Ohmmeter
   d. Two of the above

3. When the analog multimeter is selected to measure ac:
   a. A rectifier is connected internally in circuit.
   b. A pulsating dc waveform is applied to the meter movement.
   c. The meter will indicate the rms or dc equivalent of the ac input.
   d. All of the above.

4. What is the frequency limit associated with accurately measuring ac with a multimeter?
   a. 10 Hz to 2 kHz
   b. 1000 to 15,000 Hz
   c. 15 MHz to 10 GHz
   d. All three of the above

5. When connected to a multimeter, the current clamp:
   a. Allows you to measure ac voltage with the same ease as current
   b. Indicates current flow in a conductor based on magnetic field strength
   c. Allows you to measure ac current with the same ease as voltage
   d. Both (a) and (b)
   e. Both (b) and (c)

6. Radio-frequency probes are used to detect high-frequency _____ waves and display a rms value on the meter.
   a. Sound
   b. Electrical
   c. Electromagnetic
   d. None of the above

7. A × 1000 high-voltage probe means that _____ of the voltage at the probe tip will appear out of the probe and be applied to the meter.

    a. One tenth   **c.** One thousandth
    b. 1000%   **d.** Most

8. Which meter type has the highest accuracy?

    a. The AMM (analog multimeter)
    b. The DMM (digital multimeter)
    c. The VOM (volt-ohm-milliammeter)
    d. Both (a) and (c)

9. What are two advantages of the analog multimeter?

    a. Able to read low-frequency ac and low in cost
    b. Accuracy and ability to read low-frequency ac
    c. Easy-to-read display and accuracy
    d. Low cost and accuracy

10. The oscilloscope can be used to measure:

    a. Ac and dc voltage   **d.** Rise and fall times
    b. Frequency   **e.** All of the above
    c. Duration

Essay Questions

11. How can a multimeter be used to measure ac current or voltage? (12-1-1)

12. Briefly describe the following multimeter accessories: (12-1-1)

    a. The current clamp   **c.** The high-voltage probe
    b. The RF probe

13. Describe the pros and cons of analog and digital readout multimeters. (12-1-1)

14. Explain the purpose of the frequency counter. (12-1-4)

15. If a multimeter indicates the rms value of a sine wave, what does an oscilloscope indicate? (12-1-2)

16. How can the oscilloscope be used to measure: (12-1-2)

    a. Dc voltage   **b.** Frequency

17. What advantage does a dual-trace oscilloscope have over a single-trace oscilloscope? (12-1-2)

18. Briefly describe the function of:

    a. AF generator controls (12-1-5)   **b.** RF generator controls (12-1-6)

19. In relation to the oscilloscope, describe: (12-1-2)

    a. The sweep time/cm switch   **b.** The volts/cm switch

20. What is troubleshooting? (12-2)

21. What understanding is necessary to be an electronics technician? (12-2)

22. List the six-step troubleshooting guide. (12-2)

23. Describe the following troubleshooting techniques: (12-2)

    a. Half-split method   **b.** Substitution

24. What is the difference between component-level troubleshooting and board-level troubleshooting?

25. Briefly describe the six-step troubleshooting procedure. (12-2)

Practice Problems

26. If one cycle of a sine wave occupies 4 cm on the oscilloscope horizontal grid and 5 cm from peak to peak on the vertical grid, calculate frequency, period, rms, average, and peak for the following control settings:

   a. 0.5 V/cm, 20 µs/cm   **c.** 50 mV/cm, 0.2 µs/cm

   b. 10 V/cm, 10 ms/cm

27. Assuming the same graticule and switch settings of the oscilloscope in Figure 12-9, what would be the lowest setting of the volts/cm and time/division switches to fully view a 6 V rms, 350 kHz sine wave?

28. If the volts/cm switch is positioned to 10 V/cm and the waveform extends 3.5 divisions from peak to peak, what is the peak-to-peak value of this wave?

29. If a square wave occupies 5.5 horizontal cm on the 1 µs/cm position, what is its frequency?

30. Which settings would you use on an autoset scopemeter to fully view an 8 V rms, 20 kHz triangular wave?

After completing this chapter, you will be able to:

1. Define the term *capacitance*.
2. Describe the basic capacitor construction.
3. Explain the charging and discharging process and its relationship to electrostatics.
4. State the unit of capacitance and explain how it relates to charge and voltage.
5. List and explain the factors determining capacitance.
6. Describe capacitance breakdown and capacitor leakage.
7. Calculate total capacitance in parallel and series capacitance circuits.

8. Describe the advantages and differences between the five basic types of fixed capacitors.
9. Describe the advantages and differences between the four basic types of variable capacitors.
10. Explain the characteristics and new techniques used to create the one-farad capacitor.
11. Describe the coding of capacitor values on the body by use of alphanumerics or color.
12. Explain the capacitor time constant as it relates to dc charging and discharging.
13. Explain how the capacitor charges and discharges when ac is applied.

Capacitance and Capacitors

Sounds Like Electronics

Robert Moog, a self-employed engineer of Trumansburg, New York, was frequently commissioned by his close friend, composer Herbert Deutsch, to build voltage-controlled oscillators and amplifiers. Combining these devices and adding some versatility, Moog built an analog music synthesizer that allowed the operator to control many tones simultaneously and rapidly alter tones during performances.

Synthesizers, devices used to generate a number of frequencies, had been around for a number of years, but electronic music did not find wide acceptance with the popular music audience until a young musician named Walter Carlos used a Moog instrument to produce a totally synthesized hit record. The album *Switched-On Bach*, released in 1968, was an interpretation of music by eighteenth-century composer Johann Sebastian Bach, and both the album and instrument were such breakthroughs in electronic music that in no time at all composers were writing a wide variety of original music specifically for Moog synthesizers.

Up to this point we have concentrated on circuits containing only resistance, which oppose the flow of current and then convert or dissipate power in the form of heat. Capacitance and inductance are two circuit properties that act differently from resistance in that they will charge or store the supplied energy and then return almost all the stored energy back to the circuit, rather than lose it in wasted heat. Inductance will be discussed in a following chapter.

Capacitance is the ability of a circuit or device to store electrical charge. A device or component specifically designed to have this capacity or capacitance is called a *capacitor*. A capacitor stores an electrical charge similar to a bucket holding water. Using the analogy throughout this chapter that a capacitor is similar to a bucket will help you gain a clear understanding of the capacitor's operation. For example, the capacitor holds charge in the same way that a bucket holds water. A larger capacitor will hold more charge and will take longer to charge, just as a larger bucket will hold more water and take longer to fill. A larger circuit resistance means a smaller circuit current, and therefore a larger capacitor charge time. Similarly, a smaller hose will have a greater water resistance producing a smaller water flow, and therefore the bucket will take a

(b)

Two Metal Plates (Conductor)

Dielectric (insulator)

Two Connecting Leads

(c)

(a)

**FIGURE 13-1   Capacitor. (a) Physical Appearance (Courtesy of Sencore).
(b) Schematic Symbol. (c) Basic Construction.**

long time to fill. Capacitors store electrons, and basically, the amount of electrons stored is a measure of the capacitor's capacitance.

13-1 CAPACITOR CONSTRUCTION

Figure 13-1 illustrates the main parts and schematic symbol of the **capacitor.** Several years ago, capacitors were referred to as condensers; however, that term is very rarely used today. Two leads are connected to two parallel metal conductive plates, which are separated by an insulating material known as a *dielectric.* The conductive plates are normally made of metal foil, while the dielectric can be paper, air, glass, ceramic, mica, or some other form of insulator.

Capacitor
Device that stores electric energy in the form of an electric field that exists within a dielectric (insulator) between two conducting plates each of which is connected to a lead. This device was originally called a condensor.

SELF-TEST REVIEW QUESTIONS FOR SECTION 13-1

1. What are the principal parts of a capacitor?
2. What is the name given to the insulating material between the two conductive plates?

13-2 CHARGING AND DISCHARGING A CAPACITOR

Like a secondary battery, a capacitor can be made to charge or discharge, and when discharging it will return all of the energy it consumed during charge.

13-2-1 Charging a Capacitor

Capacitance is the ability of a capacitor to store an electrical charge. Figure 13-2 illustrates how the capacitor stores an electric charge. This capacitor is shown as two plates with air acting as the dielectric.

In Figure 13-2(a), the switch is open, so no circuit current results. An equal number of electrons exist on both plates, so the voltmeter (VM) indicates zero, or no potential difference exists across the capacitor.

In Figure 13-2(b), the switch is now closed and electrons travel to the positive side of the battery, away from the right-hand plate of the capacitor. This creates a positive right-hand capacitor plate, which results in an attraction of free electrons from the negative side of the battery to the left-hand plate of the capacitor. In fact, for every electron that leaves the right-hand capacitor plate and is attracted into the positive battery terminal, another electron leaves the

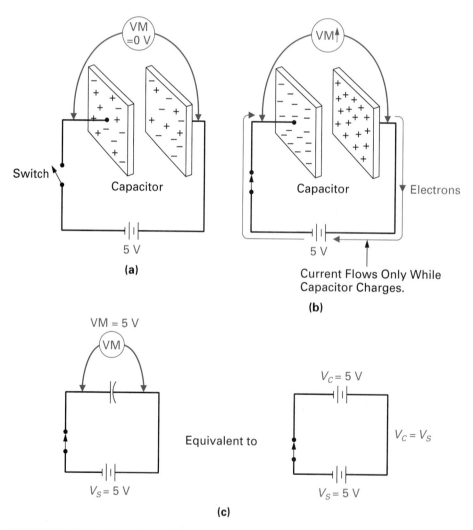

FIGURE 13-2   Charging a Capacitor. (a) Uncharged Capacitor. (b) Charging Capacitor. (c) Charged Capacitor.

negative side of the battery and travels to the left-hand plate of the capacitor. Current appears to be flowing from the negative side of the battery, around to the positive, through the capacitor. This is not really the case, because no electrons can flow through the insulator or dielectric. There appears to be one current flowing throughout the circuit, when in fact there are actually two separate currents—one from the battery to the capacitor and the other from the capacitor to the battery.

A voltmeter across the capacitor will indicate an increase in the potential difference between the plates, and the capacitor is said to be charging toward, in this example, 5 V. This potential difference builds up across the two plates until the potential difference (voltage) across the capacitor is equal to the potential difference (voltage) of the battery. In this example, when the capacitor reaches a potential difference or charge of 5 V, the capacitor will be equivalent to a 5 V battery, as seen in Figure 13-2(c). In this condition no potential difference exists between the battery and capacitor, and therefore no current flow can exist without a potential difference and the capacitor is said to be charged.

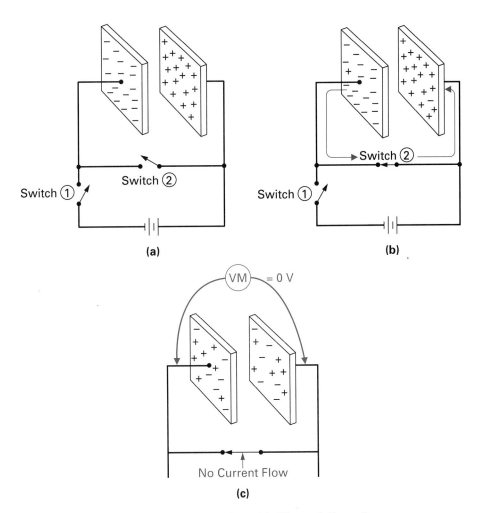

FIGURE 13-3   Discharging a Capacitor. (a) Charged Capacitor.
(b) Discharging Capacitor. (c) Discharged Capacitor.

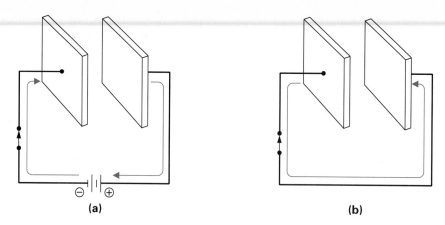

FIGURE 13-4 **Capacitor Charge and Discharge Current Paths.**
(a) Charging. (b) Discharging.

13-2-2 Discharging a Capacitor

If the capacitor is now disconnected from the circuit by opening switch 1, as shown in Figure 13-3(a), it will remain in its charged condition. If switch 2 is now closed, a path exists across the charged capacitor, as shown in Figure 13-3(b), and the excess of electrons on the left plate will flow through the conducting wire to the positive plate on the right side. The capacitor is now said to be *discharging.* When equal numbers of electrons exist on both sides, the capacitor is said to be *discharged;* and since both plates have equal charge, the potential difference across the capacitor will be zero, as shown in Figure 13-3(c).

Figure 13-4(a) and (b) illustrate the capacitor's charge and discharge currents, which flow in opposite directions. In both cases the current flow is always from one plate to the other and never exists through the dielectric insulator.

SELF-TEST REVIEW QUESTIONS FOR SECTION 13-2

1. True or false: When both plates of a capacitor have an equal charge, the capacitor is said to be charged.
2. If 2 μA of current flows into one plate of a capacitor and 2 μA flows out of the other plate, how much current is flowing through the dielectric of this working capacitor?

Electrostatic or Electric Field
Force field produced by static electrical charges. Also called a voltage field, it is a field or force that exists in the space between two different potentials or voltages.

13-3 ELECTROSTATICS

Just as a magnetic field is produced by the flow of current, an **electric field** is produced by voltage.

Current generates a magnetic field

Switch
Electrostatic Lines
of Flux or Force

FIGURE 13-5   Electrostatic (Electric) Field between the Plates of a Charged Capacitor.

Voltage generates an electric field

Figure 13-5 illustrates an example capacitor circuit with the capacitor charged and the switch open. In this condition, the capacitor retains its charge and an invisible electric or electrostatic (voltage) field will be produced by non-moving or static electrical (electrostatic) charges of different polarities. You will remember that like charges repel and unlike charges attract. Invisible electrostatic lines of flux or force can be illustrated to show this electrostatic force of attraction or repulsion. These lines are polarized away from a positive electrostatic (stationary electrical) charge and toward the negative electrostatic charge, as shown in Figure 13-6(a). If two like charges are in close proximity to one another, the electrostatic lines organize themselves into a pattern, as shown in Figure 13-6(b).

A charged capacitor has an electric or electrostatic field existing between the positively charged and negatively charged plates. The **strength of the electrostatic field** is proportional to the charge or potential difference on the plates and inversely proportional to the distance between the plates.

Field Strength
The strength of an electric, magnetic, or electromagnetic field at a given point.

(a)

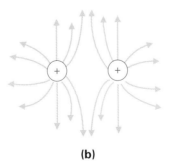

(b)

FIGURE 13-6   Electrostatic Field. (a) Electrostatic Lines of Attraction. (b) Electrostatic Lines of Repulsion.

$$\text{field strength (V/m)} = \frac{\text{charge difference } (V), \text{volts}}{\text{distance between plates } (d), \text{meters}}$$

The dielectric or insulator between the plates, like any other material, has its own individual atoms, and although the dielectric electrons are more tightly bound to their atoms than conductor electrons, stresses are placed on the atoms within the dielectric, as seen in Figure 13-7. The electrons in orbit around the dielectric atoms are displaced or distorted by the electric field existing between the positive and negative plate. If the charge potential across the capacitor is high enough and the distance between the plates is small enough, the attraction

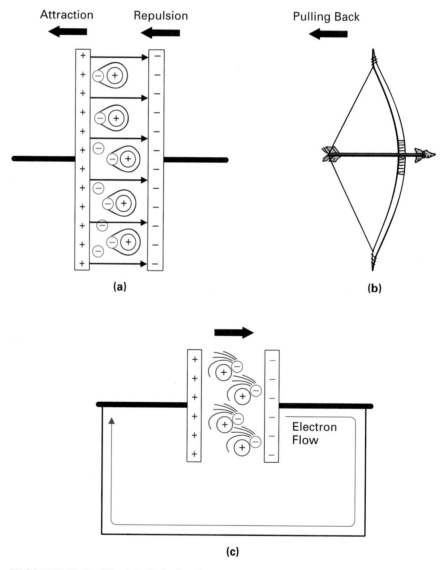

FIGURE 13-7   **Electric Polarization.**

and repulsion exerted on the dielectric atom can be large enough to free the dielectric atom's electrons. The material then becomes ionized, and a chain reaction of electrons jumping from one atom to the next in a right-to-left movement occurs. If this occurs, a large number of electrons will flow from the negative to the positive plate, and the dielectric is said to have broken down. This situation occurs if the capacitor is placed in a circuit where the voltages within the circuit exceed the voltage rating of the capacitor.

If the voltage rating of the capacitor is not exceeded, an electrostatic or electric field still exists between the two plates and causes this pulling of the atom's electrons within the dielectric toward the positive plate. This displacement, known as **electric polarization,** is similar to the pulling-back effect on a bow, as shown in Figure 13-7(b). When the capacitor is given a path for discharge, as shown in Figure 13-7(c), the electric field in the dielectric, which is causing the distortion, is the force field that drives the electrons, like the bow drives the arrow.

Electric Polarization
A displacement of bound charges in a dielectric when placed in an electric field.

To summarize electrostatics and capacitors, the charges on the plates of a capacitor produce an electric field, the electric field causes the distortion of the atoms known as *electric polarization,* and this pulling back or distortion, which is held there by the electric field, is the electron moving force (emf) that drives the electrons when a discharge path is provided. The energy in a capacitor is actually stored in the electric or electrostatic field within the dielectric.

Now that we understand these points, we can see where the word **dielectric** comes from. The dielectric is the insulating material that exists between two (di) plates and undergoes electric polarization when an electric field exists within it (dielectric).

Dielectric
Insulating material between two *(di)* plates in which the *electric* field exists.

SELF-TEST REVIEW QUESTIONS FOR SECTION 13-3

1. An _____ field is generated by the flow of current, while an _____ field is produced by voltage.
   a. Electric, magnetic
   b. Magnetic, electric
2. State the formula for calculating field strength.
3. What is electric polarization?
4. Describe why the term *dielectric* is used to indicate the insulating material between the plates of a capacitor.

13-4 THE UNIT OF CAPACITANCE

Capacitance is the ability of a capacitor to store an electrical charge, and the unit of capacitance is the **farad** (F), named in honor of Michael Faraday's work in 1831 in the field of capacitance. A capacitor with the capacity of 1 farad (1 F) can store 1 coulomb of electrical charge (6.24×10^{18} electrons) if 1 volt is applied across the capacitor's plates, as seen in Figure 13-8.

Farad
Unit of capacitance.

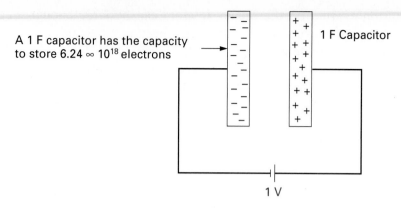

A 1 F capacitor has the capacity to store 6.24 ∞ 10^{18} electrons

1 F Capacitor

1 V

FIGURE 13-8   One Farad of Capacitance.

A 1 F capacitor is a very large value and not frequently found in electronic equipment. Most values of capacitance found in electronic equipment are in the units between the microfarad ($\mu F = 10^{-6}$) and picofarad ($pF = 10^{-12}$). A micro-farad is 1 millionth of a farad (10^{-6}). So if a 1 F capacitor can store 6.24×10^{18} electrons with 1 V applied, a 1 μF capacitor, which has 1 millionth the capacity of a 1 F capacitor, can store only 1 millionth of a coulomb, or $(6.24 \times 10^{18}) \times (1 \times 10^{-6}) = 6.24 \times 10^{12}$ electrons when 1 V is applied, as shown in Figure 13-9.

■ **EXAMPLE:**

Convert the following to either microfarads or picofarads (whichever is more appropriate):

a. 0.00002 F

b. 0.00000076 F

c. 0.00047×10^{-7} F

■ *Solution:*

a. 20 μF

b. 0.76 μF

c. 47 pF

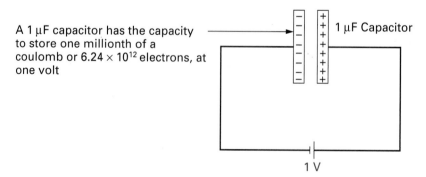

A 1 μF capacitor has the capacity to store one millionth of a coulomb or 6.24×10^{12} electrons, at one volt

1 μF Capacitor

1 V

FIGURE 13-9   One-millionth of a Farad.

Since there is a direct relationship between capacitance, charge, and voltage, there must be a way of expressing this relationship in a formula.

$$\text{capacitance, } C \text{ (farads)} = \frac{\text{charge, } Q \text{ (coulombs)}}{\text{voltage, } V \text{ (volts)}}$$

where C = capacitance, in farads

   Q = charge, in coulombs

   V = voltage, in volts

By transposition of the formula, we arrive at the following combinations for the same formula:

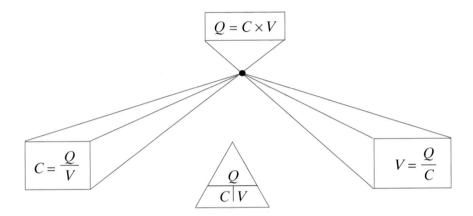

EXAMPLE:

If a capacitor has the capacity to hold 36 C ($36 \times 6.24 \times 10^{18} = 2.25 \times 10^{20}$ electrons) when 12 V is applied across its plates, what is the capacitance of the capacitor?

Solution:

$$C = \frac{Q}{V}$$

$$= \frac{36 \text{ C}}{12 \text{ V}}$$

$$= 3 \text{ F}$$

EXAMPLE:

How many electrons could a 3 µF capacitor store when 5 V is applied across it?

Solution:

$$Q = C \times V$$

$$= 3 \text{ µF} \times 5 \text{ V}$$

$$= 15 \text{ µC}$$

(15 microcoulombs is 15 millionths of a coulomb.) Since $1\text{ C} = 6.24 \times 10^{18}$ electrons, $15\ \mu\text{C} = (15 \times 10^{-6}) \times 6.24 \times 10^{18} = 9.36 \times 10^{13}$ electrons.

■ **EXAMPLE:**

If a capacitor of 2 F has stored 42 C of charge (2.63×10^{20} electrons), what is the voltage across the capacitor?

■ *Solution:*

$$V = \frac{Q}{C}$$
$$= \frac{42\text{ C}}{2\text{ F}}$$
$$= 21\text{ V}$$

SELF-TEST REVIEW QUESTIONS FOR SECTION 13-4

1. What is the unit of capacitance?
2. State the formula for capacitance in relation to charge and voltage.
3. Convert 30,000 µF to farads.
4. If a capacitor holds 17.5 C of charge when 9 V is applied, what is the capacitance of the capacitor?

13-5 FACTORS DETERMINING CAPACITANCE

The capacitance of a capacitor is determined by three factors:

1. The plate area of the capacitor
2. The distance between the plates
3. The type of dielectric used

Let's now discuss these three factors in more detail, beginning with the plate area.

13-5-1 *Plate Area (A)*

The capacitance of a capacitor is directly proportional to the plate area. This area in square centimeters is the area of only one plate and is calculated by multiplying length by width. This is illustrated in Figure 13-10(a) and (b). In these two examples, the (b) capacitor plate is twice as large as the (a) capacitor plate, and since capacitance is proportional to plate area ($C \propto A$), the capacitor in example (b) will have double the capacity or capacitance of the capacitor in example (a). Since the energy of a charged capacitor is in the electric field between the plates

Plate Area (m²) = Length × Width
            = 2 × 2
            = 4 Square Centimeters

Plate Area (m²) = Length × Width
            = 4 × 2
            = 8 Square Centimeters

Small Area = Small Capacitance
   (A↓)   = (C↓)

Large A = Large C
  (A↑)   = (C↑)

Example **(a)**

Example **(b)**

FIGURE 13-10   **Capacitance Is Proportional to Plate Area.**

and the plates of the capacitor (b) are double those of (a), there is twice as much area for the electric field to exist, and this doubles the capacitor's capacitance.

13-5-2   Distance between the Plates (d)

The distance or separation between the plates is dependent on the thickness of the dielectric used. The capacitance of a capacitor is inversely proportional to this distance between plates, in that an increase in the distance ($d\uparrow$) causes a

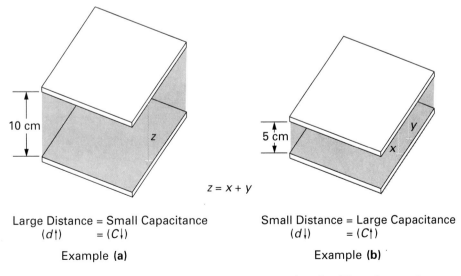

$z = x + y$

Large Distance = Small Capacitance
   (d↑)   = (C↓)

Small Distance = Large Capacitance
   (d↓)   = (C↑)

Example **(a)**

Example **(b)**

FIGURE 13-11   **Capacitance Is Inversely Proportional to Plate Separation or Distance (d).**

decrease in the capacitor's capacitance ($C\downarrow$). In Figure 13-11(a), a large distance between the capacitor plates results in a small capacitance, whereas in Figure 13-11(b) the dielectric thickness and the plate separation are half that of capacitor (a). This illustrates also how the capacitance of a capacitor can be doubled, in this case by halving the space between the plates. The gap across which the electric lines of force exist is halved in capacitor (b), and this doubles the strength of the electric field, which consequently doubles capacitance. Simply stated, an electric line of force (Z) in Figure 13-11(a) can be used to produce two electric lines of force (X and Y) in Figure 13-11(b) if the distance is half.

13-5-3 Dielectric Constant

Dielectric Constant (K)
The property of a material that determines how much electrostatic energy can be stored per unit volume when unit voltage is applied. Also called permittivity.

The insulating dielectric of a capacitor concentrates the electric lines of force between the two plates. Consequently, different dielectric materials can change the capacitance of a capacitor by being able to concentrate or establish an electric field with greater ease than other dielectric insulating materials. The **dielectric constant (K)** is the ease with which an insulating material can establish an electrostatic (electric) field. A vacuum is the least effective dielectric and has a dielectric constant of 1, as seen in Table 13-1. All the other insulators listed in this table will support electrostatic lines of force more easily than a vacuum. The vacuum is used as a reference. All the other materials have dielectric constant values that are relative to the vacuum dielectric constant of 1. For example, mica has a dielectric constant of 5.0, which means that mica can cause an electric field

TABLE 13-1 Dielectric Constants

MATERIAL	DIELECTRIC CONSTANT (K)[a]
Vacuum	1.0
Air	1.0006
Teflon	2.0
Wax	2.25
Paper	2.5
Amber	2.65
Rubber	3.0
Oil	4.0
Mica	5.0
Ceramic (low)	6.0
Bakelite	7.0
Glass	7.5
Water	78.0
Ceramic (high)	8000.

[a]The different material compositions can cause different values of K.

five times the intensity of a vacuum; and, since capacitance is proportional to the dielectric constant ($C \propto K$), the mica capacitor will have five times the capacity of the same-size vacuum dielectric capacitor. In another example, we can see that the capacitance of a capacitor can be increased by a factor of almost 8000 by merely using ceramic rather than air as a dielectric between the two plates.

13-5-4 *The Capacitance Formula*

Thus plate area, separation, and the dielectric used are the three factors that change the capacitance of a capacitor. The formula that combines these three factors is

$$C = \frac{(8.85 \times 10^{-12}) \times K \times A}{d}$$

where C = capacitance, in farads (F)

8.85×10^{-12} is a constant

K = dielectric constant

A = plate area, in square meters (m²)

d = distance between the plates, in meters (m)

This formula summarizes what has been said in relation to capacitance. The capacitance of a capacitor is directly proportional to the dielectric constant (K) and the plates' area (A) and is inversely proportional to the dielectric thickness or distance between the plates (d).

▨ EXAMPLE:

What is the capacitance of a ceramic capacitor with a 0.3 m² plate area and a dielectric thickness of 0.0003 m?

▨ *Solution:*

$$C = \frac{(8.85 \times 10^{-12}) \times K \times A}{d}$$

$$= \frac{(8.85 \times 10^{-12}) \times 6 \times 0.3 \text{ m}^2}{0.0003 \text{ m}}$$

$$= 5.31 \times 10^{-8}$$

$$= 0.0531 \text{ μF}$$

CALCULATOR SEQUENCE

Step	Keypad Entry	Display Response
1.	⑧ . ⑧ ⑤ Ⓔ ① ② +/−	8.85E-12
2.	×	
3.	⑥	6
4.	×	5.31E-11
5.	⓪ . ③	
6.	÷	1.59E-11
7.	⓪ . ⓪ ⓪ ⓪ ③	
8.	=	5.31E-8

1. List the three variable factors that determine the capacitance of a capacitor.
2. State the formula for capacitance.
3. If a capacitor's plate area is doubled, the capacitance will _____?
4. If a capacitor's dielectric thickness is halved, the capacitance will _____?

13-6 DIELECTRIC BREAKDOWN AND LEAKAGE

Capacitors store a charge just as a container or tank stores water. The amount of charge stored by a capacitor is proportional to the capacitor's capacitance and the voltage applied across the capacitor ($Q = C \times V$). The charge stored by a fixed-value capacitor (C is fixed) therefore can be increased by increasing the voltage across the plates, as shown in Figure 13-12(a). If the voltage across the capacitor is increased further, the charge held by the capacitor will increase until the dielectric between the two plates of the capacitor breaks down and a spark jumps or arcs between the plates.

Using the water analogy shown in Figure 13-12(b), if the pressure of the water being pumped in is increased, the amount of water stored in the tank will also increase until a time is reached when the tank's seams at the bottom of the tank cannot contain the large amount of pressure and break down under strain. The amount of water stored in the tank is proportional to the pressure applied,

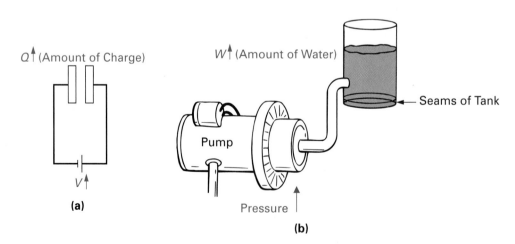

FIGURE 13-12   Breakdown. (a) Voltage Increase, Charge Increases until Dielectric Breakdown. (b) Pressure Increase, Water Increases until Seams Break Down.

TABLE 13-2 Dielectric Strengths

MATERIAL	DIELECTRIC STRENGTH (V/mm)
Air	787
Oil	12,764
Ceramic	39,370
Paper	49,213
Teflon	59,055
Mica	59,055
Glass	78,740

FIGURE 13-13 Capacitor Leakage.

just as the amount of charge stored in a capacitor is proportional to the amount of voltage applied.

The **breakdown voltage** of a capacitor is determined by the strength of the dielectric used. Table 13-2 illustrates some of the different strengths of many of the common dielectrics. As an example, let's consider a capacitor that uses 1 mm of air as a dielectric between its two plates. This particular capacitor can withstand any voltage up to 787 V. If the voltage is increased further, the dielectric will break down and current will flow between the plates, destroying the capacitor (air capacitors, however, can recover from ionization).

The ideal or perfect insulator should have a resistance equal to infinite ohms. Insulators or the dielectric used to isolate the two plates of a capacitor are not perfect and therefore have some very high values of resistance. This means that some value of resistance exists between the two plates, as shown in Figure 13-13; although this value of resistance is very large, it will still allow a small amount of current to flow between the two plates (in most applications a few nanoamperes or picoamperes). This small current, referred to as **leakage current,** causes any charge in a capacitor to slowly, over a long period of time, discharge between the two plates. A capacitor should have a large leakage resistance to ensure the smallest possible leakage current.

Breakdown Voltage
The voltage at which breakdown occurs in a dielectric or insulation.

Leakage Current
Small, undesirable flow of current through an insulator or dielectric.

SELF-TEST REVIEW QUESTIONS FOR SECTION 13-6

1. Which dielectric material has the best breakdown voltage figure?
2. True or false: Current flows through the dielectric at and below the breakdown voltage of the material.
3. A capacitor should have a _____ leakage resistance to ensure a _____ leakage current.
4. Leakage current is normally in:
   a. Milliamperes   c. Microamperes
   b. Nanoamperes   d. Kiloamperes

13-7 CAPACITORS IN COMBINATION

Like resistors, capacitors can be connected in either series or parallel. The rules for determining total capacitance for parallel- and series-connected capacitors are opposite to series- and parallel-connected resistors.

13-7-1 Capacitors in Parallel

In Figure 13-14(a), you can see a 2 μF and 4 μF capacitor connected in parallel with one another. As the top plate of capacitor A is connected to the top plate of capacitor B with a wire, and a similar situation occurs with the bottom plates, you can see that this is the same as if the top and bottom plates were touching one another, as shown in Figure 13-14(b). When drawn so that the respective plates are touching, the dielectric constant and plate separation is the same as shown in Figure 13-14(a); however, now we can easily see that the plate area is actually increased. Consequently, if capacitors are connected in parallel, the effective plate area is increased; and since capacitance is proportional to plate area $[C\uparrow = (8.85 \times 10^{-12}) \times K \times A\uparrow/d]$, the capacitance will also increase. Total capacitance is actually calculated by adding the plate areas, so total capacitance is equal to the sum of all the individual capacitances in parallel.

$$C_T = C_1 + C_2 + C_3 + C_4 + \cdots$$

(a)

$C_{Total} = 2 \ \mu F + 4 \ \mu F = 6 \ \mu F$

(b)   **FIGURE 13-14   Capacitors in Parallel.**

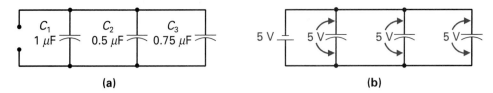

(a)　　　　　　　　　　　(b)

FIGURE 13-15　Example of Parallel-Connected Capacitors.

▦ **EXAMPLE:**

Determine the total capacitance of the circuit in Figure 13-15(a). What will be the voltage drop across each capacitor?

▦ *Solution:*

$$C_T = C_1 + C_2 + C_3$$
$$= 1 \ \mu F + 0.5 \ \mu F + 0.75 \ \mu F$$
$$= 2.25 \ \mu F$$

As with any parallel-connected circuit, the source voltage appears across all the components. If, for example, 5 V is connected to the circuit of Figure 13-15(b), all the capacitors will charge to the same voltage of 5 V because the same voltage always exists across each section of a parallel circuit.

13-7-2　Capacitors in Series

In Figure 13-16(a), we have taken the two capacitors of 2 μF and 4 μF and connected them in series. Since the bottom plate of the *A* capacitor is connected to the top plate of the *B* capacitor, they can be redrawn so that they are touching, as shown in Figure 13-16(b).

The top plate of the *A* capacitor is connected to a wire into the circuit, and the bottom plate of *B* is connected to a wire into the circuit. This connection creates two center plates that are isolated from the circuit and can therefore be disregarded, as shown in Figure 13-16(c). The first thing you will notice in this illustration is that the dielectric thickness ($d\uparrow$) has increased, causing a greater separation between the plates. The effective plate area of this capacitor has decreased, as it is just the area of the top plate only. Even though the bottom plate extends out further, the electric field can only exist between the two plates, so the surplus metal of the bottom plate has no metal plate opposite for the electric field to exist.

Consequently, when capacitors are connected in series the effective plate area is decreased ($A\downarrow$) and the dielectric thickness increased ($d\uparrow$), and both of these effects result in an overall capacitance decrease ($C\downarrow\downarrow = (8.85 \times 10^{-12}) \times K \times A\downarrow/d\uparrow$).

The plate area is actually decreased to the smallest individual capacitance connected in series, which in this example is the plate area of *A*. If the plate area were the only factor, then capacitance would always equal the smallest capacitor value. However, the dielectric thickness is always equal to the sum of all the

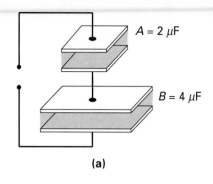

A = 2 μF

B = 4 μF

(a)

(b)

$$C_t = \frac{2\ \mu F \times 4\ \mu F}{2\ \mu F + 4\ \mu F} = 1.33\ \mu F$$

(c)

FIGURE 13-16   Capacitors in Series.

capacitor dielectrics, and this factor always causes the total capacitance (C_T) to be less than the smallest individual capacitance when capacitors are connected in series.

The total capacitance of two or more capacitors in series therefore is calculated by using the following formulas: For two capacitors in series,

$$C_T = \frac{C_1 \times C_2}{C_1 + C_2}$$

(product-over-sum formula)

For more than two capacitors in series,

$$C_T = \frac{1}{(1/C_1) + (1/C_2) + (1/C_3) + \cdots}$$

(reciprocal formula)

■ **EXAMPLE:**

Determine the total capacitance of the circuit in Figure 13-17.

■ *Solution:*

$$C_T = \frac{1}{(1/C_1) + (1/C_2) + (1/C_3)}$$

$$= \frac{1}{(1/4 \ \mu F) + (1/2 \ \mu F) + (1/1 \ \mu F)}$$

$$= \frac{1}{1.75 \times 10^6} = 5.7143 \times 10^{-7}$$

$$= 0.5714 \ \mu F \quad \text{or} \quad 0.6 \ \mu F$$

FIGURE 13-17 **Example of Series-Connected Capacitors.**

The total capacitance for capacitors in series is calculated in the same way as total resistance when resistors are in parallel.

As with series-connected resistors, all of the voltage drops across the series-connected resistors will equal the voltage applied (Kirchhoff's voltage law). With capacitors connected in series, the charged capacitors act as a voltage divider, and therefore the voltage-divider formula can be applied to capacitors in series.

$$V_{cx} = \frac{C_T}{C_x} \times V_T$$

where V_{cx} = voltage across desired capacitor

$\quad C_T$ = total capacitance

$\quad C_x$ = desired capacitor's value

$\quad V_T$ = total supplied voltage

■ **EXAMPLE:**

Using the voltage-divider formula, calculate the voltage dropped across each of the capacitors in Figure 13-17 if $V_S = 24$ V.

■ *Solution:*

$$V_{C1} = \frac{C_T}{C_1} \times V_S = \frac{0.5714 \ \mu F}{4 \ \mu F} \times 24 \ V = 3.4 \ V$$

$$V_{C2} = \frac{C_T}{C_2} \times V_S = \frac{0.5714 \ \mu F}{2 \ \mu F} \times 24 \ V = 6.9 \ V$$

$$V_{C3} = \frac{C_T}{C_3} \times V_S = \frac{0.5714 \ \mu F}{1 \ \mu F} \times 24 \ V = 13.7 \ V$$

$$V_S = V_{C1} + V_{C2} + V_{C3} = 3.4 + 6.9 + 13.7 = 24 \ V$$

$$\text{(Kirchhoff voltage law)}$$

If the capacitor values are the same, as seen in Figure 13-18(a), the voltage is divided equally across each capacitor, as each capacitor has an equal amount of charge and therefore has half of the applied voltage (in this example, 3 V across each capacitor).

When the capacitor values are different, the smaller value of capacitor will actually charge to a higher voltage than the larger capacitor. In the example in Figure 13-18(b), the smaller capacitor is actually half the size of the other capacitor, and it has charged to twice the voltage. Since Kirchhoff's voltage law has to apply to this and every series circuit, you can easily calculate that the voltage across C_1 will equal 4 V and is twice that of C_2, which is 2 V. To understand this fully, we must first understand that although the capacitance is different, both capacitors have an equal value of coulomb charge held within them, which in this example is 8 μC.

$$Q_1 = C_1 \times V_1$$
$$= 2 \ \mu F \times 4 \ V = 8 \ \mu C$$
$$Q_2 = C_2 \times V_2$$
$$= 4 \ \mu F \times 2 \ V = 8 \ \mu C$$

This equal charge occurs because the same amount of current flow exists throughout a series circuit, so both capacitors are being supplied with the same number or quantity of electrons. The charge held by C_1 is large with respect to its small capacitance, whereas the same charge held by C_2 is small with respect to its larger capacitance.

If the charge remains the same (Q is constant) and the capacitance is small, the voltage drop across the capacitor will be large, because the charge is large with respect to the capacitances.

$$V\uparrow = \frac{Q}{C\downarrow}$$

$$V_{C1} = \frac{C_T}{C_1} \times V_T = \frac{1 \ \mu F}{2 \ \mu F} \times 6 \ V = 3 \ V$$

$$V_{C2} = \frac{C_T}{C_2} \times V_T = \frac{1 \ \mu F}{2 \ \mu F} \times 6 \ V = 3 \ V$$

(a)

$$V_{C1} = \frac{C_T}{C_1} \times V_T = \frac{1.33 \ \mu F}{2 \ \mu F} \times 6 \ V = 4 \ V$$

$$V_{C2} = \frac{C_T}{C_2} \times V_T = \frac{1.33 \ \mu F}{4 \ \mu F} \times 6 \ V = 2 \ V$$

(b)

FIGURE 13-18 Voltage Drops across Series-Connected Capacitors.

On the other hand, for a constant charge, a large capacitance will have a small charge voltage because the charge is small with respect to the capacitance:

$$V\downarrow = \frac{Q}{C\uparrow}$$

We can apply the water analogy once more and imagine two series-connected buckets, one of which is twice the size of the other. Both are being supplied by the same series pipe, which has an equal flow of water throughout, and are consequently each holding an equal amount of water, for example, 1 gallon. The 1 gallon of water in the small bucket is large with respect to the size of the bucket, and a large amount of pressure exists within that bucket. The 1 gallon of water in the large bucket is small with respect to the size of the bucket, so a small amount of pressure exists within this bucket. The pressure within a bucket is similar to the voltage across a capacitor, and therefore a small bucket or capacitor will have a greater pressure or voltage associated with it, while a large bucket or capacitor will develop a small pressure or voltage.

To summarize capacitors in series, all the series-connected components will have the same charging current throughout the circuit, and because of this, two or more capacitors in series will always have equal amounts of coulomb charge. If the charge (Q) is equal, the voltage across the capacitor is determined by the value of the capacitor. A small capacitance will charge to a larger voltage ($V\uparrow = Q/C\downarrow$), whereas a large value of capacitance will charge to a smaller voltage ($V\downarrow = Q/C\uparrow$).

SELF-TEST REVIEW QUESTIONS FOR SECTION 13-7

1. If 2 µF, 3 µF, and 5 µF capacitors are connected in series, what will be the total circuit capacitance?
2. If 7 pF, 2 pF, and 14 pF capacitors are connected in parallel, what will be the total circuit capacitance?
3. State the voltage-divider formula as it applies to capacitance.
4. True or false: With resistors, the large value of resistor will drop a larger voltage, whereas with capacitors the smaller value of capacitor will actually charge to a higher voltage.

13-8 TYPES OF CAPACITORS

Capacitors come in a variety of shapes and sizes and can be either fixed or variable in their values of capacitance. Within these groups, capacitors are generally classified by the dielectric used between the plates.

A **fixed-value capacitor** is a capacitor whose capacitance value remains constant and cannot be altered. Fixed capacitors normally come in a disk or a tubular package, as seen in Figure 13-19, and consist of metal foil plates separated by

Fixed-Value Capacitor
A capacitor whose value is fixed and cannot be varied.

FIGURE 13-19   Fixed-Value Capacitors (Courtesy of Sencore).

one of the following types of insulators (dielectric), which is the means by which we classify them.

1. Mica
2. Ceramic
3. Paper
4. Plastic
5. Electrolytic

A *variable-value capacitor* is a capacitor whose capacitance value can be changed by rotating a shaft. The variable capacitor normally consists of one electrically connected movable plate and one electrically connected stationary plate. There are basically four types of variable-value capacitors, which are also classified by the dielectric used.

1. Air
2. Mica
3. Ceramic
4. Plastic

First, let's take a closer look at the different types of fixed-value capacitors.

13-8-1 *Fixed-Value Capacitors*

Mica Capacitor
Fixed capacitor that uses mica as the dielectric between its plates.

Mica

Figure 13-20 illustrates the physical appearance and construction of the **mica capacitor.** In Figure 13-20(a), which illustrates the construction of the mica capacitor, you can see that thin foil plates (normally aluminum) are alternately

(a)

(b)

FIGURE 13-20   Mica Capacitors. (a) Construction. (b) Physical Appearance (Courtesy of Sencore).

stacked to form the plates of the capacitor, all of which are isolated from one another by a thin layer of mica dielectric. Every other plate is connected to one lead on the right, while the other metal foil plates are attached to the other connecting lead on the left. This arrangement of stacked plates provides an increase in the capacitance due to the larger overall plate area.

The complete assembly of plates and mica is then sealed inside a protective casing. Figure 13-20(b) shows some examples of dipped mica capacitors. If molding equipment is used to protect and seal the assembly the capacitor is referred to as a *molded* mica capacitor; however, if the plate and mica assembly is sealed by a dipping process, the capacitor is referred to as a *dipped* mica capacitor.

Ceramic

Figure 13-21 illustrates the construction and physical appearance of the molded and dipped types of **ceramic capacitor.** In their construction, you will notice the stacking of plates and isolation by ceramic, similar to the mica capacitor. Generally, the low-dielectric-constant ceramic ($K = 6.0$) is placed within the molded package, while the high-dielectric-constant ceramic (different composition means that $K = 8000$) is placed within the dipped package to obtain higher voltage ratings and high values of capacitance in small packages.

Ceramic Capacitor
Capacitor in which the dielectric material used between the plates is ceramic.

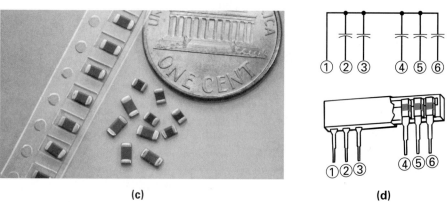

FIGURE 13-21 Ceramic Capacitors. (a) Construction. (b) Physical Appearance (Courtesy of Sencore). (c) Chip Ceramic Casing. (Product photo supplied as a courtesy of NIC Components Corp, Amityville, N.Y.) (d) SIP Casing.

Figure 13-21(c) shows the ceramic chip surface-mount capacitors that are now available on the market for both discrete and integrated types of circuits. Values typically range from 16 to 1600 pF. Different values of capacitance can be obtained by using different types of ceramic, thereby varying the dielectric constant.

Figure 13-21(d) shows how the chip ceramic capacitors can be mounted in a SIP (single in-line package) casing.

Lead to
Inner Foil Sheet

Inner Foil

Paper Dielectric

Outer Foil

Lead to
Outer Foil Sheet

(a)

(b)

(c)

FIGURE 13-22   **Paper Capacitors.**
(a) Construction. (b) Axial Lead Wrapped.
(c) Radial Lead Dipped.

Paper

Figure 13-22 illustrates the construction and physical appearance of tubular **paper capacitors.** Two long strips of foil are separated by a paper dielectric that has been saturated with paraffin to ensure that the paper dielectric is a good insulator. The two foil plates separated by the paper dielectric are rolled into a tubular shape and placed in either a molded or a dipped case.

The leads of the capacitor can be either axial lead, as shown in Figure 13-22(b) (leads coming out of either end), or radial lead, as shown in Figure 13-22(c) (leads coming out of a single end).

Plastic

Plastic film capacitors have almost completely replaced the older types of paper capacitors, and their method of construction is illustrated in Figure 13-23(a). The construction is identical to the paper type; however, in this instance a plastic film dielectric strip is used as the dielectric and then rolled into a cylinder. Mylar, polycarbonate, Teflon, and polypropylene are all examples of plastic that are used as a dielectric.

The final cylindrical package can be either encased in an outer plastic layer, as shown in Figure 13-23(b), or dipped, as shown in Figure 13-23(c).

Electrolytic

Figure 13-24 shows the construction and physical appearance of typical **electrolytic capacitors.** These types of capacitors are constructed in a similar manner as both paper and plastic types of capacitors, the difference being that

Paper Capacitors
Fixed capacitor using oiled or waxed paper as a dielectric.

Plastic Film Capacitor
A capacitor in which alternate layers of metal aluminum foil are separated by thin films of plastic dielectric.

Electrolytic Capacitor
Capacitor having an electrolyte between the two plates; due to chemical action, a very thin layer of oxide is deposited on only the positive plate, which accounts for why this type of capacitor is polarized.

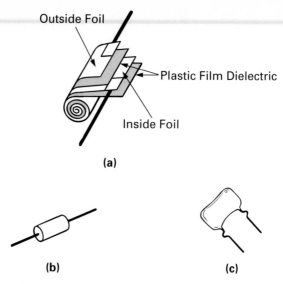

Outside Foil

Plastic Film Dielectric

Inside Foil

(a)

(b)　　　　　　**(c)**

**FIGURE 13-23　Plastic Capacitors.
(a) Construction. (b) Axial Lead. (c) Radial
Lead.**

the foil plates are separated by a strip of gauze that has been saturated with a conductive fluid known as an *electrolyte*. In the manufacturing process, a dc voltage causes a current flow in one direction, which causes the electrolyte to interact chemically with the aluminum foil and create a coating of aluminum oxide on the surface of the positive aluminum foil plate, as shown in Figure 13-24(a), which causes it to change chemically. This oxide, which is formed by an electrochemical reaction, becomes the dielectric for the electrolytic capacitor, and because it is extremely thin, the electrolytic capacitor can have a large capaci-

Positive Plate

Negative Plate

Aluminum Foil

Aluminum Oxide

Gauze
(Saturated with Electrolyte)

Aluminum Foil

(a)　　　　　　　　　　　**(b)**　　　　　　**(c)**

**FIGURE 13-24　Electrolytic Capacitors. (a) Construction. (b) Physical Appearance
(Courtesy of Sencore). (c) Schematic Symbol for Electrolytic.**

Name	Construction	Approximate Range of Values and Tolerances	Characteristics	
Mica		1 pF–0.1 μF ±1% to ±5%	Lower voltage rating than other capacitors of the same size	Small Capacitor Values
Ceramic		*Low Dielectric K*: 1 pF–0.01 μF ±0.5% to ±10% *High Dielectric K*: 1 pF–0.1 μF ±10% to ±80%	Most popular small valve capacitor due to lower cost than mica, and its ruggedness	
Paper		1 pF–1 μF ± 10%	Has a large plate area and therefore large capacitance for a small size	
Plastic		1 pF–10 μF ± 5% to ±10%	Has almost completely replaced paper capacitors; has large capacitance valves for small size and high voltage ratings	Large Capacitor Values
Electrolytic (Aluminum and Tantalum)		1 μF–1F ± 10% to ±50%	Most popular large valve capacitor: large capacitance into small area, wide range of valves. Disadvantages are: cannot be used in AC circuits as they are polarized; poor tolerances; low leakage resistance and so high leakage current. *Tantalum* advantages over aluminum include smaller size, longer life than aluminum, which has an approximate lifespan of 12 years. Disadvantages: 4 to 5 times the price.	

FIGURE 13-25   **Summary of Fixed-Value Capacitors.**

tance for a small size. The chemical change in the positive aluminum plate during this electrochemical process makes the electrolytic capacitor **polarized.** Thus the electrolytic capacitor must always have a positive charge applied to its positive plate and a negative charge to its negative plate. If this rule is not followed, the electrolytic capacitor becomes a safety hazard, because it can in the worst-case condition explode violently.

Figure 13-24(c) shows the schematic symbols for electrolytic capacitors, which are always marked with a (+) or (−) sign to indicate polarity. In Figure 13-24(b) you can see the manufacturer's method of polarity indication.

Figure 13-24(b) shows a "can" electrolytic capacitor. With this type of electrolytic capacitor, the negative outside foil plate is connected to the metal can casing, which acts as the negative capacitor lead.

Tantalum instead of aluminum is used in some electrolytic capacitors for the plates and has advantages over the aluminum type, which are:

1. Higher capacitance per volt for a given unit volume
2. Longer life and excellent shelf life (storage)
3. Able to operate in a wider temperature range
4. Temperature stability is better
5. Construction features make them more rugged

The cost of tantalum electrolytics, however, is almost four to five times that of aluminum electrolytics, and their operating voltages are much lower.

Figure 13-25 illustrates, reviews, and includes added information on the five basic fixed capacitors discussed previously.

13-8-2 *Variable-Value Capacitors*

Variable-value capacitors are the second basic type, and within this group, four types are commercially available. Like the fixed-value type of capacitors, they are classified by dielectric.

Polarized Electrolytic Capacitor
An electrolytic capacitor in which the dielectric is formed adjacent to one of the metal plates, creating a greater opposition to current in one direction only.

Tantalum Capacitor
Electrolytic capacitor having a tantalum foil anode.

Variable-Value Capacitor
A capacitor whose value can be varied.

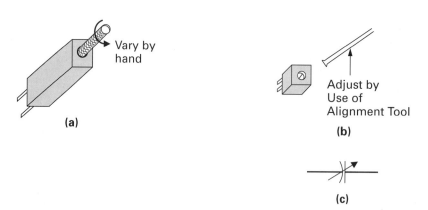

FIGURE 13-26   (a) Variable Capacitor. (b) Adjustable Capacitor.
(c) Schematic Symbol for Variable or Adjustable Capacitor.

Variable
1. Air
2. Mica

The variable-value type uses a hand-rotated shaft to vary the effective plate area, and so capacitance, as shown in Figure 13-26(a).

Adjustable
3. Ceramic
4. Plastic

The adjustable type uses screw-in or screw-out mechanical adjustment to vary the distance between the plates, and so capacitance, as shown in Figure 13-26(b).

Capacitance is dependent on effective plate area and can therefore be varied by either changing the effective plate area or distance between the plates. The dielectric constant is fixed and depends on the particular type being used. The schematic symbol for a variable capacitor is shown in Figure 13-26(c).

FIGURE 13-27 **Variable Air-Type Capacitors. (a) Construction. (b) Physical Appearance (Courtesy of Sencore). (c) Ganged.**

Air (Variable)

Figure 13-27(a) illustrates the construction of a typical air dielectric variable capacitor. With this type of capacitor, the effective plate area is adjusted to vary capacitance by causing a set of rotating plates (rotor) to mesh with a set of stationary plates (stator). When the rotor plates are fully out, the capacitance is minimum, and when the plates are fully in, the capacitance is maximum, because the maximum amount of rotor plate area is now opposite the stator plate, creating the maximum value of capacitance.

The plates are usually made of aluminum to prevent corrosion and the dielectric between the plates is air. The capacitor has to be carefully manufactured to ensure that the rotor plates do not touch the stator plates when the shaft is rotated and the plates interweave. Figure 13-27(b) illustrates a typical variable air capacitor.

In radio equipment, it is sometimes necessary to have two or more variable-value capacitors that have been constructed in such a way that a common shaft (rotor) runs through all the capacitors and varies their capacitance simultaneously, as shown in Figure 13-27(c). If you mechanically couple (gang) two or more variable capacitors so that they can all be operated from a single control, the component is known as a **ganged capacitor,** and the symbol for this arrangement is also shown in Figure 13-27(c).

Ganged Capacitor
Mechanical coupling of two or more capacitors, switches, potentiometers, or any other components so that the activation of one control will operate all sections.

Mica, Ceramic, and Plastic (Adjustable)

Figure 13-28 illustrates some of the typical packages for mica, ceramic, or plastic film types of adjustable capacitors, which are also referred to as *trimmers.* The adjustable capacitor generally has one stationary plate and one spring metal moving plate. The screw forces the spring metal plate closer or farther away from the stationary plate, varying the distance between the plates and so changing capacitance. The two plates are insulated from one another by either mica, ceramic, or plastic film, and the advantages of each are the same as for fixed-value capacitors.

These types of capacitors should only be adjusted with a plastic or non-metallic alignment tool, because a metal screwdriver may affect the capacitance of the capacitor when nearby, making it very difficult to adjust for a specific value of capacitance.

FIGURE 13-28   Adjustable Mica, Ceramic, or Plastic Capacitors. (a) Construction. (b) Physical Appearance (Courtesy of Sencore).

13-8-3 The One-Farad Capacitor

Most capacitor values are normally measured in either microfarads or pico-farads. The traditional picofarad and microfarad capacitors, which have been discussed, consist of two conductor plates separated by an insulator; when volt-age is applied to the capacitor, current flows and a charge builds up within the capacitor and is stored. The amount of charge stored is dependent on the plate area, plate separation, and the dielectric constant of the insulator used. All three of these factors can be varied separately or in combination to create the capaci-tors that we have today.

The ceramic capacitor has become the most popular small-value capacitor due to the high dielectric constants that can be obtained from ceramic (8000 or more). The electrolytic capacitor has become the most popular large-value capacitor because the extremely thin dielectric oxide layer enables us to obtain large values for a small size.

Many techniques have been tried to extend the capacitance value up toward the farad; however, the conductive surface area is the main factor that needs to be increased in order to gain high values of capacitance, as seen in Fig-ure 13-29(a). For example, if two pieces of aluminum the size of fingernails were held together, a fingernail thickness apart, a capacitance of approximately 1 pF would be created. This 1 pF capacitor would now have to be increased by a fac-tor of 1,000,000,000,000 to obtain a capacitor of 1 F.

To increase surface area, a process known as *double etch* has been used, as seen in Figure 13-29(b), where, instead of having a flat plate surface, as seen in Figure 13-29(a), large pits are etched onto the surface and then smaller pits are etched onto the larger pits to create a larger surface area. This method, however, still does not get us close to 1 F.

The 1 F capacitor uses a relatively new double-layer technique, the effect of which was first noticed more than a century ago by the German scientist Her-mann von Helmholtz. Helmholtz discovered that a double layer of charge will build up on the surface between a solid and a liquid.

The double-layer capacitor uses activated charcoal as the solid plates, and a liquid electrolyte between the two plates and an insulating separator allow a charge to build up within the capacitor due to the solid (charcoal) and liquid (electrolyte).

Activated charcoal is used as the plate because it has an almost infinite number of tiny particles making up its surface area, as seen in Figure 13-29(c), which yields a massive plate surface area. For example, 4 grams of activated charcoal would have an approximate surface area equivalent to a football field.

Using this technique a 1 F capacitor can be manufactured, like that seen in Figure 13-29(d). This 1 F capacitor package actually holds six 6 F capacitors con-nected in series, producing a total capacitance of

$$C_t \text{ (total series capacitance)} = \frac{1}{\frac{1}{6} + \frac{1}{6} + \frac{1}{6} + \frac{1}{6} + \frac{1}{6} + \frac{1}{6}}$$
$$= 1 \text{ F}$$

The reason why six 6 F capacitors are included is that the breakdown voltage of the double-layer capacitor is approximately 1 V, so by connecting six of these capacitors in series, the dielectric thickness is six times greater, increasing the

FIGURE 13-29 The 1-F Capacitor. (a) Traditional Plate Shape. (b) Double-Etched Plate Area. (c) Activated Charcoal Surface Area. (d) Large-Farad-Value Capacitors (Courtesy of Sencore).

breakdown voltage up to a more practical dc working voltage of 5 V. Due to the large value and the variations in activated charcoal, the tolerance of these capacitors can range from −20% to +80% of their value (0.8 to 1.8 F).

One major application for these capacitors is as a power backup for a computer's memory. When using a computer, it is important that you regularly save in a permanent memory the information you are working with in temporary memory. The reason for this is that the temporary memory will lose all of its information if there is a momentary power failure. In this situation, the 1 F capacitor could be used as a backup source of power to the computer's temporary memory during momentary power outages in order to prevent the loss of information.

1. List the five types of fixed-value capacitors.
2. Which fixed-value capacitor is the most popular in applications where:
   a. Large values are required?
   b. Small values are required?
3. List the four basic types of variable capacitors.
4. Which variable capacitor types are best suited for:
   a. Large value variations?
   b. Small value variations?
5. Which capacitor type is said to be polarity conscious?

13-9 CODING OF CAPACITANCE VALUES

The capacitor value, tolerance, and voltage rating need to be shown in some way on the capacitor's exterior. Presently, two methods are used: (1) alphanumeric labels and (2) color coding.

13-9-1 Alphanumeric Labels

Manufacturers today most commonly use letters of the alphabet and numbers (alphanumerics) printed on either the disk or tubular body to indicate the capacitor's specifications, as illustrated in Figure 13-30.

The tubular type, which tends to be larger in size, as seen in Figure 13-30(b), is the easier of the two since the information is basically uncoded. The value of capacitance and unit, typically the microfarad (μF or MF), tolerance figure (preceded by ± or followed by %), and voltage rating (followed by a V for

FIGURE 13-30   Alphanumeric Coding of Capacitors. (a) Disk Type. (b) Tubular.
(c) 0.47 μF, +80%, −20% Tolerance, 10V (Courtesy of Sencore).

voltage) are printed on all sizes of tubular cases. The remaining letters or numbers are merely manufacturer's codes for case size, series, and the like.

With disk capacitors (dipped or molded), as seen in Figure 13-30(a), certain rules have to be applied when decoding the notations. Many capacitors of this type do not define the unit of capacitance; in this situation, try to locate a decimal point. If a decimal point exists, for example 0.01 or 0.001, the value is in microfarads (10^{-6}). If no decimal point exists, for example 50 or 220, the value is in picofarads (10^{-12}) and you must analyze the number in a little more detail. Figure 13-30(b) and (c) give additional examples.

■ **EXAMPLE:**

What is the value of the following capacitor if it is labeled 50, 50 V, ±5?

■ *Solution:*

Since no decimal point is present, the unit is in picofarads:

$$50 \text{ pF}, \quad 50 \text{ V}, \quad \pm 5\%$$

If no decimal point is present and three digits exist and the last digit is a zero, the value is as stands and in picofarads. If the third digit is a number other than 0 (1 to 9), it is a multiplier and describes the number of zeros to be added to the picofarad value.

■ **EXAMPLE:**

If two capacitors are labeled with the following coded values, how should they be interpreted?

  a. 220

  b. 104

■ *Solution:*

Since no decimal point is present, they are both in picofarads.

  a. If the last of the three digits is a zero, the value is as it stands.

$$\underline{2\ 2\ 0} = 220 \text{ pF}$$

three-digit value

  b. If the third digit is a number from 1 to 9, it is a multiplier.

$$\underline{1\ 0}\ \underline{4}$$

two-digit value ———⌐ ⌐——— multiplier

So

$$10 \times 10^4 = 100,000 \text{ pF}$$

or

$$100,000 \times 10^{-12} = 0.1 \times 10^{-6} = 0.1 \text{ μF}$$

TABLE 13-13

Color	1st Digit	2nd Digit	Multiplier	Tolerance
Black		0	1	±20%
Brown	1	1	10^{-1}	±1%
Red	2	2	10^{-2}	±2%
Orange	3	3	10^{-3}	±3%
Yellow	4	4	10^{-4}	±4%
Green	5	5	10^{-5}	±5%
Blue	6	6	10^{-6}	
Violet	7	7	10^{-7}	
Gray	8	8	10^{-8}	
White	9	9	10^{-9}	±10%

The tolerance of the capacitor is sometimes clearly indicated, for example, ±5 or 10%; in other cases, a letter designation is used, such as

$$F = \pm 1\%$$

$$G = \pm 2\%$$

$$J = \pm 5\%$$

$$K = \pm 10\%$$

$$M = \pm 20\%$$

$$Z = -20\%, +80\%$$

Unfortunately, there does not seem to be a standard among capacitor manufacturers, which can cause confusion when trying to determine the value of capacitance. Therefore, if you are not completely sure, you should always consult technical data or information sheets from the manufacturer.

13-9-2 Color Coding

Table 13-3 illustrates the capacitor color code, which is almost identical to the resistor color code except for certain tolerances. Although the alphanumeric labels are now commonplace, the color code is still being used by some manufacturers.

SELF-TEST REVIEW QUESTIONS FOR SECTION 13-9

What are the following values of capacitance?

1. 470 ± 2
2. 0.47 ± 5

13-10 CAPACITIVE TIME CONSTANT

When a capacitor is connected across a dc voltage source, it will charge to a value equal to the voltage applied. If the charged capacitor was then connected across a load, the capacitor would then discharge through the load. The time it takes a capacitor to charge or discharge can be calculated if the circuit's resistance and capacitance are known. Let us now examine how we can calculate a capacitor's charge time and discharge time.

13-10-1 DC Charging

When a capacitor is connected across a dc voltage source, such as a battery or power supply, current will flow and the capacitor will charge up to a value equal to the dc source voltage, as shown in Figure 13-31. When the charge switch is first closed, as seen in Figure 13-31(a), there is no voltage across the capacitor at that instant and therefore a potential difference exists between the battery and capacitor. This causes current to flow and begin charging the capacitor.

Once the capacitor begins to charge, the voltage across the capacitor does not instantaneously rise to 100 V; it takes a certain amount of time before the capacitor voltage is equal to the battery voltage. When the capacitor is fully charged no potential difference exists between the voltage source and the capacitor. Consequently, no more current flows in the circuit as the capacitor has reached its full charge, as seen in Figure 13-31(b). The amount of time it takes for a capacitor to charge to the supplied voltage (in this example, 100 V) is dependent on the circuit's resistance and capacitance value. If the circuit's resistance is increased, the opposition to current flow will be increased, and it will take the capacitor a longer period of time to obtain the same amount of charge because the circuit current available to charge the capacitor is less.

If the value of capacitance is increased, it again takes a longer time to charge to 100 V because a greater amount of charge is required to build up the voltage across the capacitor to 100 V.

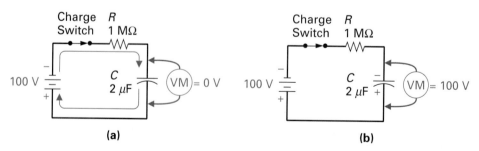

(a)　　　　　　　　　　　　**(b)**

FIGURE 13-31　Capacitor Charging. (a) Switch Is Closed and Capacitor Begins to Charge. (b) Capacitor Charged.

The circuit's resistance (R) and capacitance (C) are the two factors that determine the charge time (τ). Mathematically, this can be stated as

$$\tau = R \times C$$

where τ = **time constant** (s)

$\quad R$ = resistance (Ω)

$\quad C$ = capacitance (F)

In this example, we are using a resistance of 1 MΩ and a capacitance of 2 μF, which means that the time constant is equal to

$$\tau = R \times C$$
$$= 2\ \mu\text{F} \times 1\ \text{M}\Omega$$
$$= (2 \times 10^{-6}) \times (1 \times 10^{6})$$
$$= 2\ \text{s}$$

Time Constant
Time needed for either a voltage or current to rise to 63.2% of the maximum or fall to 36.8% of the initial value. The time constant of an RC circuit is equal to the product of R and C.

Two seconds is the time, so what is the constant? The constant value that should be remembered throughout this discussion is "**63.2**."

Figure 13-32 illustrates the rise in voltage across the capacitor from 0 to a maximum of 100 V in five time constants (5 × 2 s = 10 s). So where does 63.2 come into all this?

First time constant: In $1RC$ seconds (1 × R × C = 2 s), the capacitor will charge to 63.2% of the applied voltage (63.2% × 100 V = 63.2 V).

Second time constant: In $2RC$ seconds (2 × R × C = 4 s), the capacitor will charge to 63.2% of the remaining voltage. In the example, the capacitor will be charged to 63.2 V in one time constant, and therefore the voltage remaining is equal to 100 V − 63.2 V = 36.8 V. At the end of the second time constant, therefore, the capacitor will have charged to 63.2% of the remaining voltage (63.2% × 36.8 V = 23.3 V), which means that it will have reached 86.5 V (63.2 + 23.3 = 86.5 V) or 86.5% of the applied voltage.

Third time constant: In $3RC$ seconds (6 s), the capacitor will charge to 63.2% of the remaining voltage:

$$\text{remaining voltage} = 100\ \text{V} - 86.5\ \text{V}$$
$$= 13.5\ \text{V}$$
$$63.2\% \text{ of } 13.5\ \text{V} = 8.532\ \text{V}$$

At the end of the third time constant, therefore, the capacitor will have charged to 86.5 V + 8.532 = 95 V, or 95% of the applied voltage.

Fourth time constant: In $4RC$ seconds (8 s), the capacitor will have charged to 63.2% of the remaining voltage (100 V − 95 V = 5 V); therefore, 63.2% of 5 V = 3.2 V. So the capacitor will have charged to 95 V + 3.2 V = 98.2 V, or 98.2% of the applied voltage.

Fifth time constant: In $5RC$ seconds (10 s), the capacitor is considered to be fully charged since the capacitor will have reached 63.2% of the remaining

FIGURE 13-32 Charging Capacitor.

voltage (100 V − 98.2 V = 1.8 V); therefore, 63.2% of 1.8 V = 1.1 V. So the capacitor will have charged to 98.2 V + 1.1 V = 99.3 V, or 99.3% of the applied voltage.

The voltage waveform produced by the capacitor acquiring a charge is known as an *exponential* waveform, and the voltage across the capacitor is said to rise exponentially. An exponential rise is also referred to as a *natural increase*. There are many factors that exponentially rise and fall. For example, we grow exponentially, in that there is quite a dramatic change in our height in the early years and then this increase levels off and reaches a maximum.

Before the switch is closed and even at the instant the switch is closed, the capacitor is not charged, which means that there is no capacitor voltage to oppose the supply voltage and, therefore, a maximum current of *V*/*R*, 100 V/1 MΩ = 100 μA flows. This current begins to charge the capacitor, and a potential difference begins to build up across the plates of the capacitor, and this voltage opposes the supply voltage, causing a decrease in charging current. As the capacitor begins to charge, less of a potential difference exists between the supply voltage and capacitor voltage and so the current begins to decrease.

To calculate the current at any time, we can use the formula

$$i = \frac{V_S - V_C}{R}$$

where i = instantaneous current

V_S = source voltage

V_C = capacitor voltage

R = resistance

For example, the current flowing in the circuit after 1 time constant will equal the source voltage, 100 V, minus the capacitor's voltage, which in 1 time constant will be 63.2% of the source voltage or 63.2 V, divided by the resistance.

$$i = \frac{V_S - V_C}{R}$$
$$= \frac{100 \text{ V} - 63.2 \text{ V}}{1 \text{ M}\Omega}$$
$$= 36.8 \text{ }\mu\text{A}$$

As the charging continues, the potential difference across the plates exponentially rises to equal the supply voltage, as seen in Figure 13-33(a), while the current exponentially falls to zero, as shown in Figure 13-33(b). The constant of 63.2 can be applied to the exponential fall of current from 100 μA to 0 μA in 5RC seconds.

When the switch was closed to start charging of the capacitor, there was no charge on the capacitor; therefore, a maximum potential difference existed between the battery and capacitor, causing a maximum current flow of 100 μA ($I = V/R$).

FIGURE 13-33   Exponential Rise in Voltage and Fall in Current in a Charging Capacitive Circuit.

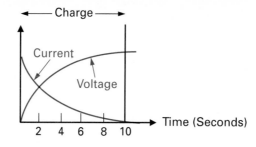

FIGURE 13-34 **Phase Difference between Voltage and Current in a Charging Capacitive Circuit.**

First time constant: In $1RC$ seconds, the current will have exponentially decayed to 63.2% of its maximum (63.2% of 100 μA = 63.2 μA) to a value of 36.8 μA (100 μA − 63.2 μA). In the example of 2 μF and 1 MΩ, this occurs in 2 s.

Second time constant: In $2RC$ seconds ($2 \times R \times C = 4$ s), the current will decrease 63.2% of the remaining current, which is

$$63.2\% \text{ of } 36.8 \text{ μA} = 23.26 \text{ μA}$$

The current will drop 23.26 μA from 36.8 μA and reach 13.5 μA or 13.5%.

Third time constant: In $3RC$ seconds (6 s), the capacitor's charge current will decrease 63.2% of the remaining current (13.5 μA) to 5 μA or 5%.

Fourth time constant: In $4RC$ seconds (8 s), the current will have decreased to 1.8 μA or 1.8%.

Fifth time constant: In $5RC$ seconds (10 s), the charge current is now 0.7 μA or 0.7%. At this time, the charge current is assumed to be zero and the capacitor is now charged to a voltage equal to the applied voltage.

Studying the exponential rise of the voltage and the exponential decay of current in a capacitive circuit, you will notice an interesting relationship. In a pure resistive circuit, the current flow through a resistor would be in step with the voltage across that same resistor, in that an increased current would cause a corresponding increase in voltage drop across the resistor. Voltage and current are consequently said to be *in step* or *in phase* with one another.

With the capacitive circuit, the current flow in the circuit and voltage across the capacitor are not in step or in phase with one another. When the switch is closed to charge the capacitor, the current is maximum (100 μA), while the voltage across the capacitor is zero. After five time constants (10 s), the capacitor's voltage is now maximum (100 V) and the circuit current is zero, as seen in Figure 13-34. The circuit current flow is out of phase with the capacitor voltage, and this difference is referred to as a *phase shift*. In any circuit containing capacitance, current will lead voltage.

13-10-2 DC Discharging

Figure 13-35 illustrates the circuit, voltage, and current waveforms that occur when a charged capacitor is discharged from 100 V to 0 V. The 2 μF capacitor, which was charged to 100 V in 10 s ($5RC$), is discharged from 100 to 0 V in the same amount of time.

FIGURE 13-35   Discharging Capacitor. (a) Voltage Waveform. (b) Current Waveform.

Looking at the voltage curve, you can see that the voltage across the capacitor decreases exponentially, dropping 63.2% to 36.8 V in $1RC$ seconds, another 63.2% to 13.5 V in $2RC$ seconds, another 63.2% to 5 V in $3RC$ seconds, and so on, until zero.

The current flow within the circuit is dependent on the voltage in the circuit, which is across the 2 μF capacitor. As the voltage decreases, the current will also decrease by the same amount ($I\downarrow = V\downarrow/R$).

$$\text{discharge switch closed: } I = \frac{V}{R} = \frac{100\text{ V}}{1\text{ M}\Omega} = 100\ \mu\text{A} \quad \text{maximum}$$

$$1RC\ (2)\ \text{seconds: } I = \frac{V}{R} = \frac{36.8\text{ V}}{1\text{ M}\Omega} = 36.8\ \mu\text{A}$$

$$2RC\ (4)\ \text{seconds: } I = \frac{V}{R} = \frac{13.5\text{ V}}{1\text{ M}\Omega} = 13.5\ \mu\text{A}$$

$$3RC\ (6)\ \text{seconds: } I = \frac{V}{R} = \frac{5\text{ V}}{1\text{ M}\Omega} = 5.0\ \mu\text{A}$$

$$4RC\ (8)\ \text{seconds: } I = \frac{V}{R} = \frac{1.8\text{ V}}{1\text{ M}\Omega} = 1.8\ \mu\text{A}$$

$$5RC\ (10)\ \text{seconds: } I = \frac{V}{R} = \frac{0.7\text{ V}}{1\text{ M}\Omega} = 0.7\ \mu\text{A} \quad \text{zero}$$

1. What is the capacitor time constant?
2. In one time constant, a capacitor will have charged to what percentage of the applied voltage?
3. In one time constant, a capacitor will have discharged to what percentage of its full charge?
4. True or false: The charge or discharge of a capacitor follows a linear rate of change.

13-11 AC CHARGE AND DISCHARGE

Let us begin by returning to our charged capacitor that was connected across a 100-V dc source, as shown in Figure 13-36(a). When a capacitor is connected across a dc source, it charges in five time constants and then the voltage across the capacitor will oppose the source voltage, causing current to stop. Since no current flows in this circuit, the capacitor is effectively acting as an open circuit when a constant dc voltage is applied.

If the 100 V source is reversed, as shown in Figure 13-36(b), the 100 V charge on the capacitor is now aiding the path of the battery instead of pushing or reacting against it. The discharge current will flow in the opposite direction to the charge current, until the capacitor voltage is equal to 0 V, at which time it will begin to charge in the reverse direction, as shown in Figure 13-36(c).

If the battery source is once more reversed, as shown in Figure 13-36(d), the charge current will discharge the capacitor to 0 V and then begin once more to charge it in the reverse direction, as shown in Figure 13-36(e).

The result of this switching or alternating procedure is that there is current flow in the circuit at all times (except for the instant when the battery source is

FIGURE 13-36 AC Charge and Discharge.

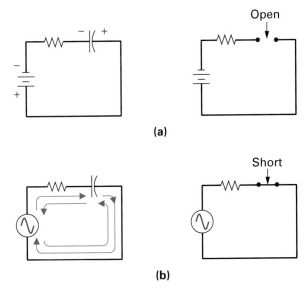

(a)

(b)

FIGURE 13-37 Capacitors' Reaction to DC and AC. (a) Capacitor Will Block DC. (b) Capacitor Will Pass AC.

removed and the polarity reversed). The switching of a dc voltage source in this way has almost the same effect as if we were applying an alternating voltage, as shown in Figure 13-36(f).

Figure 13-37 summarizes the capacitor's reaction to a dc source and an ac source. When a dc voltage is applied across a capacitive circuit, as shown in Figure 13-37(a), the capacitor will charge and then oppose the dc source, preventing current flow. Since no current can flow when the capacitor is charged, the capacitor acts like an open circuit, and it is this action that accounts for why capacitors are referred to as an "open to dc."

When an ac voltage is applied across a capacitive circuit, as shown in Figure 13-37(b), the continual reversal of the applied voltage will cause a continual charging and discharging of the capacitor, and therefore circuit current will always be present. Any type of ac source or in fact any fluctuating or changing dc source will cause a circuit current, and if current is present, the opposition is low. It is this action that accounts for why capacitors are referred to as a "short to ac."

This ability of a capacitor to block dc and seem to pass ac will be exploited and explained later in applications of capacitors.

SELF-TEST REVIEW QUESTIONS FOR SECTION 13-11

1. What differences occur when ac is applied to a capacitor rather than dc?
2. True or false: A capacitor's reaction to ac and dc accounts for why it is known as an ac short and dc block.

13-12 PHASE RELATIONSHIP BETWEEN CAPACITOR CURRENT AND VOLTAGE

Thinking of the applied alternating voltage as a dc source that is being continually reversed sometimes makes it easier to understand how a capacitor reacts to ac. Whether the applied voltage is dc or ac, the rule holds true in that a 90° phase shift or difference exists between circuit current and capacitor voltage. The exact relationship between ac current and voltage in a capacitive circuit is illustrated in Figure 13-38.

This phase shift that exists between the circuit current and the capacitor voltage is normally expressed in degrees. At 0°, the capacitor is fully discharged (0 V), and the source is supplying a maximum circuit charge current. From 0° to 90°, the capacitor will charge toward a maximum positive value, and this increase in capacitor voltage will oppose the source voltage, whose circuit charging current will slowly decrease to 0 A. At 90°, the capacitor is fully charged, and the circuit current is at 0 A, so the capacitor will return to the circuit the energy it consumed during the positive (+) charge cycle. This discharge circuit current is in the opposite direction to the positive charge current. At 180°, the capacitor is fully discharged (0 V), and the source is supplying a maximum circuit charge current. From 180° to 270°, the capacitor will charge toward a maximum negative value, and this negative increase in capacitor voltage will oppose the source voltage, whose circuit charging current will slowly decrease to 0 A. At 270°, the capacitor is fully charged and the circuit current is at 0 A, so the capacitor will discharge and return the energy it consumed during the negative (−) charge cycle. This discharge circuit current is in the opposite direction to the negative charge current.

Throughout this cycle, notice that the voltage across the capacitor follows the circuit current. This current, therefore, leads the voltage by 90°, and the 90° leading phase shift (current leading voltage) will occur only in a capacitive circuit.

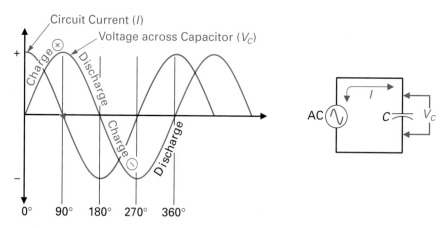

FIGURE 13-38   AC Current Leads Voltage in a Capacitive Circuit.

SELF-TEST REVIEW QUESTIONS FOR SECTION 13-12

1. What does a voltage-current phase shift mean?
2. True or false: In a resistive circuit, the current leads the voltage by exactly 90°.
3. True or false: In a purely capacitive circuit, a 90° phase difference occurs between current and voltage.
4. True or false: Phase differences between voltage and current only occur when ac is applied.

Capacitance and Capacitors (Figure 13-39)   **SUMMARY**

1. Resistance will oppose the flow of current and then convert or dissipate power in the form of heat.

2. Capacitance and inductance are two circuit properties that act differently from resistance in that they will charge or store the supplied energy and then return almost all the stored energy back to the circuit, rather than lose it in wasted heat.

3. Capacitance is the ability of a circuit or device to store electrical charge. A device or component specifically designed to have this capacity or capacitance is called a *capacitor*. A capacitor stores an electrical charge similar to a bucket holding water.

4. Capacitors store electrons, and basically, the amount of electrons stored is a measure of the capacitor's capacitance.

5. Capacitors, referred to in the past as condensers, have two leads connected to two parallel metal conductive plates, which are separated by an insulating material known as a *dielectric*. The conductive plates are normally made of metal foil, while the dielectric can be paper, air, glass, ceramic, mica, or some other form of insulator.

6. Like a secondary battery, a capacitor can be made to charge or discharge, and when discharging it will return all of the energy it consumed during charge.

7. Just as a magnetic field is produced by the flow of current, an electric field is produced by voltage.

8. A charged capacitor has an electric or electrostatic field existing between the positively charged and negatively charged plates. The strength of the electrostatic field is proportional to the charge or potential difference on the plates and inversely proportional to the distance between the plates.

9. The dielectric or insulator between the plates, like any other material, has its own individual atoms, and although the dielectric electrons are more tightly bound to their atoms than conductor electrons, stresses are placed on the atoms within the dielectric. The electrons in orbit around the dielectric atoms are displaced or distorted by the electric field existing between

FIGURE 13-39   Capacitance and Capacitors.

$$\text{Field strength (V/m)} = \frac{\text{charge difference } (V) \text{ in volts}}{\text{distance between plates } (d) \text{ in meters}}$$

$$\left(\text{V/m} = \text{volts per meter}\right)$$

$$Q = C \times V$$

Q = change, in coulombs
C = capacitance, in farads
V = voltage, in volts

$$C = \frac{Q}{V}$$

$$V = \frac{Q}{C}$$

$$C = \frac{(8.85 \times 10^{-12}) \times K \times A}{d}$$

C = capacitance in farads
8.85×10^{-12} is a constant
K = dielectric constant
A = plate area, in square meters
d = distance between the plates, in meters

Capacitors in parallel:

$$C_T = C_1 + C_2 + C_3 + C_4 + \cdots$$

C_T = total capacitance, in farads
C_1 = capacitance of C_1, in farads
C_2 = capacitance of C_2, in farads
C_3 = and so on

Capacitors in series:

Two capacitors

$$C_T = \frac{C_1 \times C_2}{C_1 + C_2}$$

More than two capacitors

$$C_T = \frac{1}{(1/C_1) + (1/C_2) + (1/C_3) + \cdots}$$

Voltage divider formula

$$V_{cx} = \frac{C_T}{C_x} \times V_T$$

V_{cx} = desired voltage
C_T = total capacitance
C_x = selected capacitor value
V_T = supplied voltage

$$\tau = R \times C$$

τ = time constant, in seconds
R = resistance, in ohms
C = capacitance, in farads

$$i = \frac{V_S - V_C}{R}$$

i = instantaneous current, in amperes
V_S = source voltage, in volts
V_C = capacitor voltage, in volts
R = resistance, in ohms

FIGURE 13-39 (continued)

Dielectric Constants

MATERIAL	DIELECTRIC CONSTANT (K)[a]
Vacuum	1.0
Air	1.0006
Teflon	2.0
Wax	2.25
Paper	2.5
Amber	2.65
Rubber	3.0
Oil	4.0
Mica	5.0
Ceramic (low)	6.0
Bakelite	7.0
Glass	7.5
Water	78.0
Ceramic (high)	8000.

[a]The different material compositions can cause different values of K.

Dielectric Strengths

MATERIAL	DIELECTRIC STRENGTH (V/mm)[a]
Air	787
Oil	12,764
Ceramic	39,370
Paper	49,213
Teflon	59,055
Mica	59,055
Glass	78,740

Fixed-Value Capacitors

Name	Construction	Approximate Range of Values and Tolerances	Characteristics	
Mica		1 pF–0.1 μF ±1% to ±5%	Lower voltage rating than other capacitors of the same size	Small Capacitor Values
Ceramic	Ceramic Dielectric / Lead Wire	*Low Dielectric K:* 1 pF–0.01 μF ±0.5% to ±10% *High Dielectric K:* 1 pF–0.1 μF ±10% to ±80%	Most popular small valve capacitor due to lower cost than mica, and its ruggedness	
Paper	Lead to Inner Foil Sheet / Inner Foil / Lead to Outer Foil Sheet / Outer Foil	1 pF–1 μF ± 10%	Has a large plate area and therefore large capacitance for a small size	
Plastic	Outside Foil / Inside Foil	1 pF–10 μF ± 5% to ±10%	Has almost completely replaced paper capacitors; has large capacitance valves for small size and high voltage ratings	Large Capacitor Values
Electrolytic (Aluminum and Tantalum)	(Tantalum has tantalum rather than aluminum foil plates) Aluminum Foil / Aluminum Oxide / Aluminum Foil / Gauze (Saturated with Electrolyte)	1 μF–1F ± 10% to ±50%	Most popular large valve capacitor: large capacitance into small area, wide range of valves. Disadvantages are: cannot be used in AC circuits as they are polarized; poor tolerances; low leakage resistance and so high leakage current. *Tantalum* advantages over aluminum include smaller size, longer life than aluminum, which has an approximate lifespan of 12 years. Disadvantages: 4 to 5 times the price.	

FIGURE 13-39 (continued)

Variable-Value Capacitors

Rotating (Rotor) Plates

Connection to Rotor Plates

Connection to Stator Plates

Stationary (Stator) Plates

Screw Adjustment

Spring Metal Movable Plate

Dielectric (Mica, Ceramic, or Plastic Film)

Stationary Plate

DC Charging

Charge Switch R 1 MΩ
100 V C 2 μF (VM) = 0 V

Charge Switch R 1 MΩ
100 V C 2 μF (VM) = 100 V

100 V
63.2%

1RC 2RC 3RC 4RC 5RC
Switch Closed

Voltage across Capacitor
(Exponential Rise)

100 μA
36.8%
13.5%
5% 1.8% 0.7%

1RC 2RC 3RC 4RC 5RC
2   4   6   8   10 Seconds
Switch Closed

Charge Current to Capacitor
(Exponential Fall)

DC Discharging

R 1 MΩ
100 V Discharge Switch C 2 μF (VM)

100 V
36.8 V
13.5 V
5 V 1.8 V 0.7 V

1RC 2RC 3RC 4RC 5RC Seconds
Discharge Switch Closed

R 1 MΩ
100 V Discharge Switch C 2 μF (AM)

100 μA
36.8 μA
13.5 μA
5 μA 1.8 μA 0.7 μA

1RC 2RC 3RC 4RC 5RC Seconds
(2) (4) (6) (8) (10)
Discharge Switch Closed

the positive and negative plate. If the charge potential across the capacitor is high enough and the distance between the plates is small enough, the attraction and repulsion exerted on the dielectric atom can be large enough to free the dielectric atom's electrons. The material then becomes ionized, and a chain reaction of electrons jumping from one atom to the next in a right-to-left movement occurs. If this occurs, a large number of electrons will flow from the negative to the positive plate, and the dielectric is said to have broken down. This situation occurs if the capacitor is placed in a circuit where the voltages within the circuit exceed the voltage rating of the capacitor.

10. If the voltage rating of the capacitor is not exceeded, an electrostatic or electric field still exists between the two plates and causes this pulling of the atom's electrons within the dielectric toward the negative plate. This displacement, known as electric polarization, is similar to the pulling-back effect on a bow. When the capacitor is given a path for discharge the electric field in the dielectric, which is causing the distortion, is the force field that drives the electrons, like the bow drives the arrow.

11. The energy in a capacitor is actually stored in the electric or electrostatic field within the dielectric.

12. *Capacitance* is the ability of a capacitor to store an electrical charge, and the unit of capacitance is the farad (F), named in honor of Michael Faraday's work in 1831 in the field of capacitance.

13. A capacitor with the capacity of 1 farad (1 F) can store 1 coulomb of electrical charge (6.24×10^{18} electrons) if 1 volt is applied across the capacitor's plates.

14. The dielectric constant (K) is the ease with which an insulating material can establish an electrostatic (electric) field. The vacuum is used as a reference. All the other materials have dielectric constant values that are relative to the vacuum dielectric constant of 1.

15. The capacitance of a capacitor is directly proportional to the dielectric constant (K) and the plates' area (A) and is inversely proportional to the dielectic thickness or distance between the plates (d).

16. Capacitors store a charge just as a container or tank stores water. The amount of charge stored by a capacitor is proportional to the capacitor's capacitance and the voltage applied across the capacitor ($Q = C \times V$).

17. If the voltage across a capacitor is increased, the charge held by the capacitor will increase until the dielectric between the two plates of the capacitor breaks down and a spark jumps or arcs between the plates.

18. The breakdown voltage of a capacitor is determined by the strength of the dielectric used.

19. The ideal or perfect insulator should have a resistance equal to infinite ohms. Insulators or the dielectric used to isolate the two plates of a capacitor are not perfect and therefore have some very high values of resistance. This means that some value of resistance exists between the two plates. Although this value of resistance is very large, it will still allow a small

amount of current to flow between the two plates (in most applications a few nanoamperes or picoamperes). This small current, referred to as leakage current, causes any charge in a capacitor to slowly, over a long period of time, discharge between the two plates. A capacitor should have a large leakage resistance to ensure the smallest possible leakage current.

20. Like resistors, capacitors can be connected in either series or parallel. The rules for determining total capacitance for parallel- and series-connected capacitors are opposite to series- and parallel-connected resistors.

21. If capacitors are connected in parallel, the effective plate area is increased; and since capacitance is proportional to plate area $[C\uparrow = (8.85 \times 10^{-12}) \times K \times A\uparrow/d]$, the capacitance will also increase.

22. The total capacitance for capacitors connected in parallel is actually calculated by adding the plate areas, so total capacitance is equal to the sum of all the individual capacitances in parallel.

23. When capacitors are connected in series the effective plate area is decreased $(A\downarrow)$ and the dielectric thickness increased $(d\uparrow)$, and both of these effects result in an overall capacitance decrease $[C\downarrow\downarrow = (8.85 \times 10^{-12}) \times K \times A\downarrow/d\uparrow]$. The plate area is actually decreased to the smallest individual capacitance connected in series. If the plate area were the only factor, then capacitance would always equal the smallest capacitor value. However, the dielectric thickness is always equal to the sum of all the capacitor dielectrics, and this factor always causes the total capacitance (C_T) to be less than the smallest individual capacitance when capacitors are connected in series.

24. The total capacitance of two or more capacitors in series therefore is calculated by using the product-over-sum formula, or the reciprocal formula.

25. As with series-connected resistors, all of the voltage drops across the series-connected resistors will equal the voltage applied (Kirchhoff's voltage law). With capacitors connected in series, the charged capacitors act as a voltage divider, and therefore the voltage-divider formula can be applied to capacitors in series.

26. Capacitors in series will have the same charging current throughout the circuit, and because of this, two or more capacitors in series will always have equal amounts of coulomb charge. If the charge (Q) is equal, the voltage across the capacitor is determined by the value of the capacitor. A small capacitance will charge to a larger voltage $(V\uparrow = Q/C\downarrow)$, whereas a large value of capacitance will charge to a smaller voltage $(V\downarrow = Q/C\uparrow)$.

27. A fixed-value capacitor is a capacitor whose capacitance value remains constant and cannot be altered. Fixed capacitors normally come in a disk or a tubular package and consist of metal foil plates separated by one of the following types of insulators (dielectric), which is the means by which we classify them.

    a. Mica   **d.** Plastic
    b. Ceramic **e.** Electrolytic
    c. Paper

28. A *variable-value capacitor* is a capacitor whose capacitance value can be changed by rotating a shaft. The variable capacitor normally consists of one electrically connected movable plate and one electrically connected stationary plate. There are basically four types of variable-value capacitors, which are also classified by the dielectric used.

a. Air   **c.** Ceramic
b. Mica **d.** Plastic

29. Ceramic chip surface-mount capacitors that are now available on the market for both discrete and integrated types of circuits. Values typically range from 16 to 1600 pF.

30. The electrolytic capacitor must always have a positive charge applied to its positive plate and a negative charge to its negative plate. If this rule is not followed, the electrolytic capacitor becomes a safety hazard, because it can in the worst-case condition explode violently.

31. Tantalum instead of aluminum is used in some electrolytic capacitors for the plates and has advantages over the aluminum type, which are:

a. Higher capacitance per volt for a given unit volume
b. Longer life and excellent shelf life (storage)
c. Able to operate in a wider temperature range
d. Temperature stability is better
e. Construction features make them more rugged

The cost of tantulum electrolytics, however, is almost four to five times that of aluminum electrolytics, and their operating voltages are much lower.

32. The variable-value type capacitors use a hand-rotated shaft to vary the effective plate area, and so capacitance.

33. The adjustable type capacitors use screw-in or screw-out mechanical adjustment to vary the distance between the plates, and so capacitance.

34. In radio equipment, it is sometimes necessary to have two or more variable-value capacitors that have been constructed in such a way that a common shaft (rotor) runs through all the capacitors and varies their capacitance simultaneously. If you mechanically couple (gang) two or more variable capacitors so that they can all be operated from a single control, the component is known as a ganged capacitor.

35. Capacitors should only be adjusted with a plastic or nonmetallic alignment tool, because a metal screwdriver may affect the capacitance of the capacitor when nearby, making it very difficult to adjust for a specific value of capacitance.

36. The 1 F capacitor uses a relatively new double-layer technique, the effect of which was first noticed more than a century ago by the German scientist Hermann von Helmholtz. Helmholtz discovered that a double layer of charge will build up on the surface between a solid and a liquid. The double-layer capacitor uses activated charcoal as the solid plates, and a liquid electrolyte between the two plates and an insulating separator allow a charge to build up within the capacitor due to the solid (charcoal) and liquid (electrolyte).

37. The capacitor value, tolerance, and voltage rating need to be shown in some way on the capacitor's exterior. Presently, two methods are used: (a) alphanumeric labels and (b) color coding.

38. When a capacitor is connected across a dc voltage source, it will charge to a value equal to the voltage applied. If the charged capacitor was then connected across a load, the capacitor would then discharge through the load. The time it takes a capacitor to charge or discharge can be calculated if the circuit's resistance and capacitance are known.

39. Time constant is the time needed for either a voltage or current to rise to 63.2% of the maximum or fall to 36.8% of the initial value. The time constant of an *RC* circuit is equal to the product of *R* and *C*.

40. The voltage waveform produced by the capacitor acquiring a charge is known as an *exponential* waveform, and the voltage across the capacitor is said to rise exponentially. An exponential rise is also referred to as a *natural increase*.

41. In a pure resistive circuit, the current flow through a resistor would be in step with the voltage across that same resistor, in that an increased current would cause a corresponding increase in voltage drop across the resistor. Voltage and current are consequently said to be *in step* or *in phase* with one another.

42. With the capacitive circuit, the current flow in the circuit and voltage across the capacitor are not in step or in phase with one another. When the switch is closed to charge the capacitor, the current is maximum while the voltage across the capacitor is zero. After five time constants the capacitor's voltage is now maximum and the circuit current is zero. The circuit current flow is out of phase with the capacitor voltage, and this difference is referred to as a *phase shift*. In any circuit containing capacitance, current will lead voltage.

43. When a dc voltage is applied across a capacitive circuit the capacitor will charge and then oppose the dc source, preventing current flow. Since no current can flow when the capacitor is charged, the capacitor acts like an open circuit, and it is this action that accounts for why capacitors are referred to as an "open to dc."

44. When an ac voltage is applied across a capacitive circuit, the continual reversal of the applied voltage will cause a continual charging and discharging of the capacitor, and therefore circuit current will always be present. Any type of ac source or in fact any fluctuating or changing dc source will cause a circuit current, and if current is present, the opposition is low. It is this action that accounts for why capacitors are referred to as a "short to ac."

45. This ability of a capacitor to block dc and seem to pass ac will be exploited in applications of capacitors.

46. Thinking of the applied alternating voltage as a dc source that is being continually reversed sometimes makes it easier to understand how a capacitor reacts to ac. Whether the applied voltage is dc or ac, the rule holds true in that a 90° phase shift or difference exists between circuit current and capacitor voltage.

47. This phase shift that exists between the circuit current and the capacitor voltage is normally expressed in degrees.

48. In a capacitive circuit, circuit current leads the capacitor voltage by 90°, and the 90° leading phase shift (current leading voltage) will occur only in a purely capacitive circuit.

REVIEW QUESTIONS

Multiple-Choice Questions

1. Capacitors were originally referred to as:
   a. Vacuum tubes   **b.** Condensers   **c.** Inductors   **d.** Suppressors

2. When a capacitor charges:
   a. The voltage across the plates rises exponentially
   b. The circuit current falls exponentially
   c. The capacitor charges to the source voltage in $5RC$ seconds
   d. All of the above

3. A _____ field is generated by the flow of current and an _____ field is generated by voltage.
   a. Magnetic, electrostatic   **c.** Electric, magnetic
   b. Electric, electrostatic   **d.** All of the above may be considered true.

4. The strength of an electric field in a capacitor is proportional to the _____ and inversely proportional to the _____ .
   a. Plate separation, charge
   b. Plate separation, potential plate difference
   c. Plate potential difference, plate separation
   d. Both (a) and (b)

5. The plates of a capacitor are generally made out of a:
   a. Resistive material   **c.** Conductive material
   b. Semiconductor material   **d.** Two of the above could be true.

6. The energy of a capacitor is stored in:
   a. The magnetic field within the dielectric
   b. The magnetic field around the capacitor leads
   c. The electric field within the plates
   d. The electric field within the dielectric

7. What is the capacitance of a capacitor if it can store 24 C of charge when 6 V is applied across the plates?
   a. 2 μF   **b.** 3 μF   **c.** 4.7 μF   **d.** None of the above

8. The capacitance of a capacitor is directly proportional to:
   a. The plate area   **d.** Both (a) and (c)
   b. The distance between the plates   **e.** Both (a) and (b)
   c. The constant of the dielectric used

9. The capacitance of a capacitor is inversely proportional to:
   a. The plate area   **d.** Both (a) and (c)
   b. The distance between the plates   **e.** Both (a) and (b)
   c. The dielectric used

10. The breakdown voltage of a capacitor is determined by:
   a. The type of dielectric used
   b. The size of the capacitor plates
   c. The wire gauge of the connecting leads
   d. Both (a) and (b)

11. A capacitor should have a _____ leakage resistance to ensure a _____ leakage current.
   a. Large, small   **b.** Large, medium   **c.** Small, large   **d.** Small, small

12. Total series capacitance of two capacitors is calculated by:
   a. Using the product over sum formula
   b. Using the voltage divider formula
   c. Using the series resistance formula on the capacitors
   d. Adding all the individual values

13. Total parallel capacitance is calculated by:
   a. Using the product over sum formula
   b. Using the voltage divider formula
   c. Using the parallel resistance formula on capacitors
   d. Adding all the individual values

14. The mica and ceramic fixed capacitors have:

    a. An arrangement of stacked plates
    b. An electrolyte substance between the plates
    c. An adjustable range
    d. All of the above

15. An electrolytic capacitor:

    a. Is the most popular large-value capacitor
    b. Is polarized
    c. Can have either aluminum or tantalum plates
    d. All of the above

16. Variable capacitors normally achieve a large variation in capacitance by varying _____, while adjustable trimmer capacitors only achieve a small capacitance range by varying _____ .

    a. Dielectric constant, plate area   **c.** Plate separation, dielectric constant
    b. Plate area, plate separation   **d.** Plate separation, plate area

17. In one time constant, a capacitor will charge to _____ of the source voltage.

    a. 86.5% **b.** 63.2% **c.** 99.3% **d.** 98.2%

18. When ac is applied across a capacitor, a _____ phase shift exists between circuit current and capacitor voltage.

    a. 45° **b.** 60° **c.** 90° **d.** 63.2°

19. A capacitor consists of:

    a. Two insulated plates separated by a conductor
    b. Two conductive plates separated by a conductor
    c. Two conductive plates separated by an insulator
    d. Two conductive plates separated by conductive spacer

20. A 47-μF capacitor charged to 6.3 V will have a stored charge of:

    a. 296.1 μC   **d.** All of the above
    b. 2.96×10^{-4} C **e.** Both (a) and (b)
    c. 0.296 mC

Essay Questions

21. Describe the construction and main parts of a capacitor. (13-1)

22. What happens to a capacitor during:

    a. Charge (13-2-1) **b.** Discharge (13-2-2)

23. Describe how electrostatics relates to capacitance and give the formula for electric field strength. (13-3)

24. Briefly explain the relationship between capacitance, charge, and voltage. (13-4)

25. Describe the three factors affecting the capacitance of a capacitor. (13-5)

26. Briefly describe:
    a. The term *capacitor breakdown* (13-6)
    b. The term *capacitor leakage* (13-6)
    c. Why the smaller-value capacitor in a two-capacitor series circuit has the larger voltage developed across it (13-7-2)

27. List the formula(s) used to calculate total capacitance when capacitors are connected in:
    a. Parallel (13-7-1) b. Series (13-7-2)

28. Describe the following types of capacitors: (13-8)
    Fixed Value
    a. Mica b. Ceramic c. Paper d. Plastic e. Electrolytic
    Variable Value
    a. Air b. Mica, ceramic, and plastic

29. Briefly explain the technique and materials used to create the 1 F capacitor. (13-8-3)

30. Explain how the constant 63.2 is used in relation to the charge and discharge of a capacitor. (13-10)

31. How is the coding of a capacitor's value, tolerance, and voltage rating indicated on a capacitor? (13-9)

32. Describe the differences between dc and ac capacitor charging and discharging. (13-11)

33. Would a mica dielectric concentrate on electrostatic field more or less than air; if so, by how much? (13-5)

34. Calculate the total capacitance of the circuits illustrated in Figure 13-40, and describe which capacitors are in parallel and which are in series. (13-7)

35. What advantages and disadvantages do tantalum electrolytic capacitors have over the aluminum type? (13-8-1)

36. Which of the fixed-value capacitor types is polarity sensitive, and what does this mean? (13-8-2)

37. What is a ganged capacitor, and what is its application? (13-8-2)

38. Describe 1 F of capacitance. (13-4)

(a)

(b)

FIGURE 13-40

39. Briefly describe why series-connected capacitors are treated like parallel-connected resistors, and why parallel-connected capacitors are treated like series-connected resistors when calculating total capacitance. (13-7)

Practice Problems

40. If a 10 µF capacitor is charged to 10 V, how many coulombs of charge has it stored?

41. Calculate the electric field strength within the dielectric of a capacitor that is charged to 6 V and the dielectric thickness is 32 µm (32 millionths of a meter).

42. If a 0.006-µF capacitor has stored 125×10^{-6} C of charge, what potential difference would appear across the plates?

43. Calculate the capacitance of the capacitor that has the following parameter values: $A = 0.008$ m^2; $d = 0.00095$ m; the dielectric used is paper.

44. Calculate the total capacitance if the following are connected in:
   a. Parallel: 1.7 µF, 2.6 µF, 0.03 µF, 1200 pF
   b. Series: 1.6 µF, 1.4 µF, 4 µF

45. If three capacitors of 0.025 µF, 0.04 µF, and 0.037 µF are connected in series across a 12-V source, as shown in Figure 13-41, what would be the voltage drop across each?

46. Give the value of the following alphanumeric capacitor value codes:
   a. 104   **b.** 125   **c.** 0.01   **d.** 220

47. Give the value of the following color-coded capacitor values:
   a. Yellow, violet, black   **b.** Orange, yellow, violet

48. What would be the time constant of the following RC circuits?
   a. $R = 6$ kΩ, $C = 14$ µF   **c.** $R = 170$ Ω, $C = 24$ µF
   b. $R = 12$ MΩ, $C = 1400$ pF   **d.** $R = 140$ kΩ, $C = 0.007$ µF

49. If 10 V were applied across all the RC circuits in Question 48, what would be the voltage across each capacitor after one time constant, and how much time would it take each capacitor to fully charge?

50. In one application, a capacitor is needed to store 25 µC of charge and will always have 125 V applied across its terminals. Calculate the capacitance value needed.

FIGURE 13-41

After completing this chapter, you will be able to:

1. Define and explain capacitive reactance.

2. Describe impedance, phase angle, power, and power factor as they relate to a series and parallel *RC* circuit.

3. Explain some of the more common capacitor failures and how to use an ohmmeter and capacitance analyzer to test them.

4. Describe how the capacitor and resistor in combination can be used:
   a. To combine ac and dc
   b. To act as a voltage divider
   c. As a filter
   d. As an integrator
   e. As a differentiator

Capacitive Circuits, Testing, and Applications

OUTLINE

Let's Toss for It!

In 1938, Bill Hewlett and Dave Packard, close friends and engineering graduates at Stanford University, set up shop in the one-car garage behind the Packard's rented home in Palo Alto, California. In the garage, the two worked on what was to be the first product of their lifetime business together, an electronic oscillating instrument that represented a breakthrough in technology and was specifically designed to test sound equipment.

Late in the year, the oscillator (designated the 200A "because the number sounded big") was presented at a West Coast meeting of the Institute of Radio Engineers (now the Institute of Electrical and Electronics Engineers—IEEE) and the orders began to roll in. Along with the first orders for the 200A was a letter from Walt Disney Studios asking the two to build a similar oscillator covering a different frequency range. The Model 200B was born shortly thereafter, and Disney purchased eight to help develop the unique sound system for the classic film *Fantasia*. By 1940, the young company had outgrown the garage and moved into a small rented building nearby.

Over the years, the company continued a steady growth, expanding its product line to more than 10,000, including computer systems and peripheral products, test and measuring instruments, hand-held calculators, medical electronic equipment, and systems for chemical analysis. Employees have increased from 2 to almost 82,000, and the company is one of the 100 largest industrial corporations in America, with a net revenue of $8.1 billion. Whose name should go first was decided by the toss of a coin on January 1, 1939, and the outcome was Hewlett-Packard or HP.

In Chapter 13 you discovered that when a capacitor was connected to a dc circuit, it charged up until it was at a value equal to the applied dc voltage, and then from that point on current was zero. When a capacitor was connected to an ac circuit, however, the continual reversal of the applied voltage resulted in a continuous charge and discharge of the capacitor, so circuit current was always present. This ability of the capacitor makes it an extremely useful device in ac circuits.

In this chapter we first study capacitive reactance, which is the opposition a capacitor offers to current flow. Then we study the ac circuit characteristics of a series *RC* circuit and parallel *RC* circuit. To complete our study of capacitors, in the final two sections in this chapter we discuss the testing of capacitors and then some of the more common applications of capacitors.

14-1 CAPACITIVE REACTANCE

Resistance (R), by definition, is the opposition to current flow with the dissipation of energy and is measured in ohms. Capacitors oppose current flow like a resistor; however, a resistor dissipates energy, whereas a capacitor stores energy (when it charges) and then gives back its energy into the circuit (when it discharges). Because of this difference, a new term had to be used to describe the opposition offered by a capacitor. **Capacitive reactance (X_C)**, by definition, is the opposition to current without the dissipation of energy and is also measured in ohms.

Capacitive Reactance (X_C) Measured in ohms, it is the ability of a capacitor to oppose current flow without the dissipation of energy.

If capacitive reactance is basically opposition, it is inversely proportional to the amount of current flow. If a large current is within a circuit, the opposition or reactance must be low ($I\uparrow$, $X_C\downarrow$). Conversely, a small circuit current will be the result of a large opposition or reactance ($I\downarrow$, $X_C\uparrow$).

When a dc source is connected across a capacitor, current will flow only for a short period of time ($5RC$ seconds) to charge the capacitor. After this time, there is no further current flow. Consequently, the capacitive reactance or opposition offered by a capacitor to dc is infinite (maximum).

Alternating current is continuously reversing in polarity, resulting in the capacitor continuously charging and discharging. This means that charge and discharge currents are always flowing around the circuit, and if we have a certain value of current, we must also have a certain value of reactance or opposition.

Initially, when the capacitor's plates are uncharged, they will not oppose or react against the charging current and therefore maximum current will flow ($I\uparrow$) and the reactance will be very low ($X_C\downarrow$). As the capacitor charges, it will oppose or react against the charge current, which will decrease ($I\downarrow$), so the reactance will increase ($X_C\uparrow$). The discharge current is also highest at the start of discharge ($I\uparrow$, $X_C\downarrow$) as the voltage of the charged capacitor is also high; but as the capacitor discharges, its voltage decreases and the discharge current will also decrease ($I\downarrow$, $X_C\uparrow$).

To summarize, at the start of a capacitor charge or discharge, the current is maximum, so the reactance is low. This value of current then begins to fall to zero, so the reactance increases.

If the applied alternating current is at a high frequency, as shown in Figure 14-1(a), it is switching polarity more rapidly than a lower frequency and there is very little time between the start of charge and discharge. As the charge and discharge currents are largest at the beginning of the charge and discharge of the capacitor, the reactance has very little time to build up and oppose the current, which is why the current is a high value and the capacitive reactance is small at higher frequencies. With lower frequencies, as shown in Figure 14-1(b), the applied alternating current is switching at a slower rate, and therefore the reactance, which is low at the beginning, has more time to build up and oppose the current.

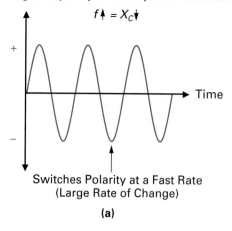

High Frequency = Low Capacitive Reactance

$$f\!\uparrow = X_C\!\downarrow$$

Time

Switches Polarity at a Fast Rate
(Large Rate of Change)

(a)

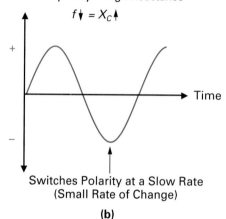

Low Frequency = High Reactance

$$f\!\downarrow = X_C\!\uparrow$$

Time

Switches Polarity at a Slow Rate
(Small Rate of Change)

(b)

**FIGURE 14-1   Capacitive Reactance Is
Inversely Proportional to Frequency.**

Capacitive reactance is therefore inversely proportional to frequency:

$$\text{capacitive reactance } (X_C) \propto \frac{1}{f \,(\text{frequency})}$$

Frequency, however, is not the only factor that determines capacitive reactance. Capacitive reactance is also inversely proportional to the value of capacitance. If a larger capacitor value is used a longer time is required to charge the capacitor ($t\!\uparrow = C\!\uparrow R$), which means that current will be present for a longer period of time, so the overall current will be large ($I\!\uparrow$); consequently, the reactance must be small ($X_C\!\downarrow$). On the other hand, a small capacitance value will charge in a small amount of time ($t\!\downarrow = C\!\downarrow R$) and the current is present for only a short period of time. The overall current will therefore be small ($I\!\downarrow$), indicating a large reactance ($X_C\!\uparrow$).

$$\text{capacitive reactance } (X_C) \propto \frac{1}{C \,(\text{capacitance})}$$

Capacitive reactance (X_C) is therefore inversely proportional to both frequency and capacitance and can be calculated by using the formula

$$X_C = \frac{1}{2\pi f C}$$

where X_C = capacitive reactance, in ohms

2π = constant

f = frequency, in hertz

C = capacitance, in farads

EXAMPLE:

Calculate the reactance of a 2 μF capacitor when a 10 kHz sine wave is applied.

Solution:

$$X_C = \frac{1}{2\pi f C}$$

$$= \frac{1}{2 \times \pi \times 10 \text{ kHz} \times 2 \text{ μF}} = 8 \text{ Ω}$$

SELF-TEST REVIEW QUESTIONS FOR SECTION 14-1

1. Define *capacitive reactance.*
2. State the formula for capacitive reactance.
3. Why is capacitive reactance inversely proportional to frequency and capacitance?
4. If $C = 4$ μF and $f = 4$ kHz, calculate X_C.

14-2 SERIES *RC* CIRCUIT

In a purely resistive circuit, as shown in Figure 14-2(a), the current flowing within the circuit and the voltage across the resistor are in phase with one another. In a purely capacitive circuit, as shown in Figure 14-2(b), the current flowing in the circuit leads the voltage across the capacitor by 90°.

Purely resistive: 0° phase shift (I is in phase with V)

Purely capacitive: 90° phase shift (I leads V by 90°)

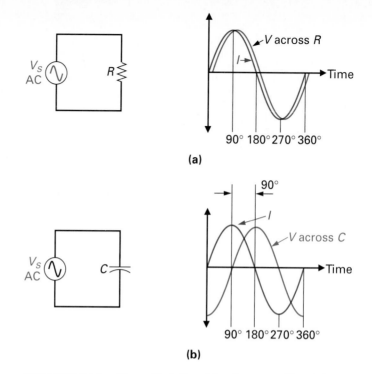

(a)

(b)

FIGURE 14-2   Phase Relationships between *V* and *I*.
(a) Resistive Circuit: Current and Voltage Are in Phase.
(b) Capacitive Circuit: Current Leads Voltage by 90°.

If we connect a resistor and capacitor in series, as shown in Figure 14-3(a), we have probably the most commonly used electronic circuit, which has many applications. The voltage across the resistor (V_R) is always in phase with the circuit current (I), as can be seen in Figure 14-3(b), because maximum points and zero crossover points occur at the same time. The voltage across the capacitor (V_C) lags the circuit current by 90°.

Since the capacitor and resistor are in series, the same current is supplied to both components; Kirchhoff's voltage law can be applied, which states that the

(a)

(b)

FIGURE 14-3   *RC* Series Circuit. (a) Circuit. (b) Waveforms.

sum of the voltage drops around a series circuit is equal to the voltage applied (V_S). The voltage drop across the resistor (V_R) and the voltage drop across the capacitor (V_C) are out of phase with one another, which means that their peaks occur at different times. The signal for the applied voltage (V_S) is therefore obtained by adding the values of V_C and V_R at each instant in time, plotting the results, and then connecting the points with a line; this is represented in Figure 14-3(b) by the shaded waveform.

Although the waveforms in Figure 14-3(b) indicate the phase relationship between I, V_S, V_R, and V_C, it seems difficult to understand clearly the relationship among all four because of the crisscrossing of waveforms. An easier method of representation is to return to the circle and vectors that were introduced in the trigonometry mini-math review in Chapter 11.

14-2-1 Vector Diagram

A vector (or phasor) is a quantity that has both magnitude and direction and is represented by a line terminated at the end by an arrowhead, as seen in Figure 14-4(a). Vectors are generally always used to represent a physical quantity that has two properties. For example, if you are traveling 60 miles per hour in a southeast direction, the size or magnitude of the vector would represent 60 mph, while the direction of the vector would point southeast. In an ac circuit containing a reactive component such as a capacitor, the vector is used to represent a voltage or current. The magnitude of the vector represents the value of voltage or current, while the direction of the vector represents the phase of the voltage or current.

A **vector diagram** is an arrangement of vectors to illustrate the magnitude and phase relationships between two or more quantities of the same frequency within an ac circuit. Figure 14-4(b) illustrates the basic parts of a vector diagram. As an example, the current (I) vector is at the 0° position and the size of the arrow represents the peak value of alternating current.

Vector Diagram
Arrangement of vectors showing the phase relationships between two or more ac quantities of the same frequency.

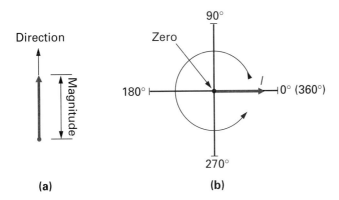

FIGURE 14-4   Vectors. (a) Vector. (b) Vector Diagram.

14-2-2 Voltage

Figure 14-5(a), (b), and (c) repeat our previous *RC* series circuit with waveforms and a vector diagram. In Figure 14-5(b), the current peak flowing in the series *RC* circuit occurs at 0° and will be used as a reference; therefore, the vector of current in Figure 14-5(c) is to the right in the 0° position. The voltage across the resistor (V_R) is in phase or coincident with the current (I), as shown in Figure 14-5(b), and the vector that represents the voltage across the resistor (V_R) over laps or coincides with the I vector, at 0°.

The voltage across the capacitor (V_C) is, as shown in Figure 14-5(b), 90° out of phase (lagging) with the circuit's current, so the V_C vector in Figure 14-5(c) is drawn at −90° (minus sign indicates lag) to the current vector, and the length of this vector represents the magnitude of the voltage across the capacitor. Since the ohmic values of the resistor (R) and the capacitor (X_C) are equal, the voltage drop across both components is the same. The V_R and V_C vectors are subsequently equal in length.

The source voltage (V_S) is, by Kirchhoff's voltage law, equal to the sum of the series voltage drops (V_C and V_R); however, since these voltages are not in phase with one another, we cannot simply add the two together. The source voltage (V_S) will be the sum of both V_C and V_R at a particular time. Studying the waveforms in Figure 14-5(b), you will notice that peak source voltage will occur at 45°. By vectorially adding the two voltages V_R and V_C in Figure 14-5(c), we obtain a resultant V_S vector that has both magnitude and phase. The angle theta (θ) formed between circuit current (I) and source voltage (V_S) will always be less than 90°, and in this example is equal to −45° because the voltage drops across R and C are equal due to R and X_C being of the same ohmic value.

If V_C and V_R are drawn to scale, the peak source voltage (V_S) can be calculated by using the same scale; however, a mathematical rather than graphical method can be used to save the drafting time.

In Figure 14-6(a), we have taken the three voltages (V_R, V_C, and V_S) and formed a right-angle triangle, as shown in Figure 14-6(b). The Pythagorean theo-

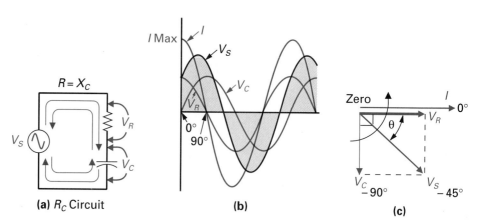

(a) R_C Circuit   **(b)**   **(c)**

FIGURE 14-5 *RC* **Series Circuit Analysis.**

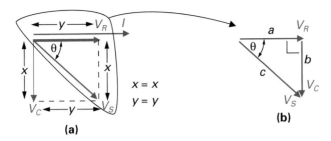

FIGURE 14-6 Voltage and Current Vector Diagram of a Series _RC_ Circuit.

rem for right-angle triangles states that if you take the square of a (V_R) and add it to the square of b (V_C), the square root of the result will equal c (the source voltage, V_S).

$$V_S = \sqrt{V_R^2 + V_C^2}$$

By transposing the formula according to the rules of algebra we can calculate any unknown if two variables are known.

$$V_C = \sqrt{V_S^2 - V_R^2}$$
$$V_R = \sqrt{V_S^2 - V_C^2}$$

■ **EXAMPLE:**

Calculate the source voltage applied across an _RC_ series circuit if $V_R = 12$ V and $V_C = 8$ V.

■ **_Solution:_**

$$V_S = \sqrt{V_R^2 + V_C^2}$$
$$= \sqrt{12\ V^2 + 8\ V^2}$$
$$= \sqrt{144 + 64}$$
$$= 14.42\ V$$

CALCULATOR SEQUENCE

Step	Keypad Entry	Display Response
1.	⓵ ⓶	12.0
2.	x^2	144.0
3.	+	
4.	8	8.0
5.	x^2	64.0
6.	=	208.0
7.	√x	14.42220

14-2-3 *Impedance*

Impedance (Z)
Measured in ohms, it
is the total
opposition a circuit
offers to current flow
(reactive and
resistive).

Since resistance is the opposition to current with the dissipation of heat, and reactance is the opposition to current without the dissipation of heat, a new term is needed to describe the total resistive and reactive opposition to current. **Impedance** (designated Z) is also measured in ohms and is the total circuit opposition to current flow. It is a combination of resistance (R) and reactance (X_C); however, in our capacitive and resistive circuit, a phase shift or difference exists, and just as V_C and V_R cannot be merely added together to obtain V_S, R and X_C cannot be simply summed to obtain Z.

If the current within a series circuit is constant (the same throughout the circuit), the resistance of a resistor (R) or reactance of a capacitor (X_C) will be directly proportional to the voltage across the resistor (V_R) or the capacitor (V_C).

$$V_R\updownarrow = I \times R\updownarrow, \qquad V_C\updownarrow = I \times X_C\updownarrow$$

A vector diagram can be drawn similarly to the voltage vector diagram to illustrate opposition, as shown in Figure 14-7(a). The current is used as a reference ($0°$); the resistance vector (R) is in phase with the current vector (I), since V_R is always in phase with I. The capacitive reactance (X_C) vector is at $-90°$ to the resistance vector, due to the $90°$ phase shift between a resistor and capacitor. The lengths of the resistance vector (R) and capacitive reactance vector (X_C) are equal in this example. By vectorially adding R and X_C, we have a resulting impedance (Z) vector.

By using the three variables, which have again formed a right-angle triangle [Figure 14-7(b)], we can apply the Pythagorean theorem to calculate the total opposition or impedance (Z) to current flow, taking into account both R and X_C.

$$Z = \sqrt{R^2 + X_C^2}$$

$$R = \sqrt{Z^2 - X_C^2}$$
$$X_C = \sqrt{Z^2 - R^2}$$

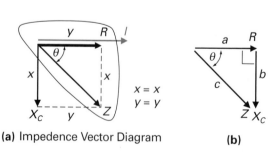

(a) Impedence Vector Diagram   **(b)**

FIGURE 14-7 **Resistance, Reactance, and Impedance Vector Diagram of a Series *RC* Circuit.**

EXAMPLE:

Calculate the total impedance of a series RC circuit if $R = 27\ \Omega$, $C = 0.005\ \mu F$, and the source frequency $= 1$ kHz.

Solution:

The total opposition (Z) or impedance is equal to

$$Z = \sqrt{R^2 + X_C^2}$$

R is known, but X_C will need to be calculated.

$$X_C = \frac{1}{2\pi f C}$$

$$= \frac{1}{2 \times \pi \times 1\ \text{kHz} \times 0.005\ \mu F}$$

$$= 31.8\ \text{k}\Omega$$

Since $R = 27\ \Omega$ and $X_C = 31.8$ kΩ, then

$$Z = \sqrt{R^2 + X_C^2}$$

$$= \sqrt{27\ \Omega^2 + 31.8\ \text{k}\Omega^2}$$

$$= \sqrt{729 + 1 \times 10^9}$$

$$= 31.8\ \text{k}\Omega$$

As you can see in this example, the small resistance of 27 Ω has very little effect on the circuit's total opposition or impedance, due to the relatively large capacitive reactance of 31,800 Ω.

EXAMPLE:

Calculate the total impedance of a series RC circuit if $R = 45$ kΩ and $X_C = 45\ \Omega$.

Solution:

$$Z = \sqrt{R^2 + X_C^2}$$

$$= \sqrt{45\ \text{k}\Omega^2 + 45^2}$$

$$= 45\ \text{k}\Omega$$

In this example, the relatively small value of X_C had very little effect on the circuit's opposition or impedance, due to the large circuit resistance.

EXAMPLE:

Calculate the total impedance of a series RC circuit if $X_C = 100\ \Omega$ and $R = 100\ \Omega$.

Solution:

$$Z = \sqrt{R^2 + X_C^2}$$

$$= \sqrt{100^2 + 100^2}$$

$$= 141.4\ \Omega$$

In this example R was equal to X_C.

We can define the total opposition or impedance in terms of Ohm's law, in the same way as we defined resistance.

$$I = \frac{V}{Z}$$ $$Z = \frac{V}{I}$$ $$V = I \times Z$$

$$\frac{V}{I \mid Z}$$

By transposition, we can arrive at the usual combinations of Ohm's law.

14-2-4 Phase Angle or Shift (θ)

In a purely resistive circuit, the total opposition (Z) is equal to the resistance of the resistor, so the phase shift (θ) is equal to 0° [Figure 14-8(a)].

In a purely capacitive circuit, the total opposition (Z) is equal to the capacitive reactance (X_C) of the capacitor, so the phase shift (θ) is equal to −90° [Figure 14-8(b)].

When a circuit contains both resistance and capacitive reactance, the total opposition or impedance has a phase shift that is between 0 and 90°. Referring back to the impedance vector diagram in Figure 14-7 and Example 14-5, you can see that we have used a simple example where R has equaled X_C, so the phase shift (θ) has always been −45°. If the resistance and reactance are different from one another, as was the case in Examples 14-3 and 14-4, the phase shift (θ) will change, as shown in Figure 14-9. As can be seen in this illustration, the phase shift (θ) is dependent on the ratio of capacitive reactance to resistance (X_C/R). A more resistive circuit will have a phase shift between 0 and 45°, while a more reactive circuit will have a phase shift between 45 and 90°. By the use of trigonometry (the science of triangles), we can derive a formula to calculate the degree of phase shift, since two quantities X_C and R are known.

FIGURE 14-8   Phase Angles. (a) Purely Resistive Circuit. (b) Purely Capacitive Circuit.

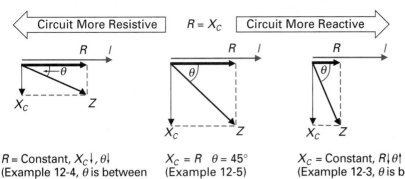

R = Constant, $X_C\downarrow$, $\theta\downarrow$
(Example 12-4, θ is between
0° and 45°)

$X_C = R$ $\theta = 45°$
(Example 12-5)

X_C = Constant, $R\downarrow\theta\uparrow$
(Example 12-3, θ is between
45° and 90°)

FIGURE 14-9   Phase Angle of a Series _RC_ Circuit.

The phase angle, θ, is equal to

$$\theta = \text{invtan}\ \frac{X_C}{R}$$

This formula will determine by what angle Z leads R. Since X_C/R is equal to V_C/V_R, the phase angle can also be calculated if V_R and V_C are known.

$$\theta = \text{invtan}\ \frac{V_C}{V_R}$$

This formula will determine by what angle V_R leads V_S.

■ **EXAMPLE:**

Calculate the phase shift or angle in two different series _RC_ circuits if:

　a. $V_R = 12\ \text{V}, V_C = 8\ \text{V}$

　b. $R = 27\ \Omega, X_C = 31.8\ \text{k}\Omega$

■ *Solution:*

　a. $\theta = \text{invtan}\dfrac{V_C}{V_R}$

　　$= \text{invtan}\dfrac{8\ \text{V}}{12\ \text{V}}$

　　$= 33.7°$ (V_R leads V_S by 33.7°)

　b. $\theta = \text{invtan}\dfrac{X_C}{R}$

　　$= \text{invtan}\dfrac{31.8\ \text{k}\Omega}{27\ \Omega}$

　　$= 89.95°$ (R leads Z by 89.95°)

14-2-5 Power

In this section we examine power in a series ac circuit. Let us begin with a simple resistive circuit and review the power formulas used previously.

Purely Resistive Circuit

In Figure 14-10 you can see the current, voltage, and power waveforms generated by applying an ac voltage across a purely resistive circuit. The applied voltage causes current to flow around the circuit, and the electrical energy is converted into heat energy. This heat or power is dissipated and lost and can be calculated by using the power formula.

$$P = V \times I$$
$$P = I^2 \times R$$
$$P = \frac{V^2}{R}$$

Voltage and current are in phase with one another in a resistive circuit, and instantaneous power is calculated by multiplying voltage by current at every instant through 360° ($P = V \times I$). The sinusoidal power waveform is totally positive, because a positive voltage multiplied by a positive current gives a positive

FIGURE 14-10   Power in a Purely Resistive Circuit.

value of power, and a negative voltage multiplied by a negative current will also produce a positive value of power. For these reasons, a resistor is said to generate a positive power waveform, which you may have noticed is twice the frequency of the voltage and current waveforms; two power cycles occur in the same time as one voltage and current cycle.

The power waveform has been split in half, and this line that exists between the maximum point (8 W) and zero point (0 W) is the average value of power (4 W) that is being dissipated by the resistor.

Purely Capacitive Circuit

In Figure 14-11, you can see the current, voltage, and power waveforms generated by applying an ac voltage source across a purely capacitive circuit. As expected, the current leads the voltage by 90°, and the power wave is calculated by multiplying voltage by current, as before, at every instant through 360°. The resulting power curve is both positive and negative. During the positive alternation of the power curve, the capacitor is taking power as the capacitor charges.

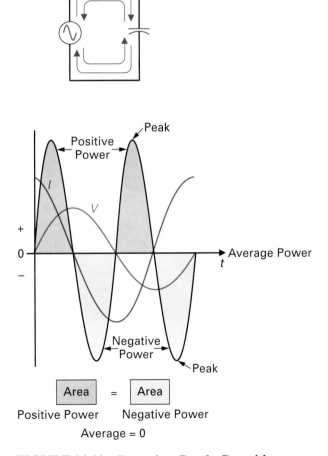

FIGURE 14-11 Power in a Purely Capacitive Circuit.

When the power alternation is negative, the capacitor is giving back the power it took as it discharges back into the circuit.

The average power dissipated is once again the value that exists between the maximum positive and maximum negative points and causes the area above this line to equal the area below. This average power level calculates out to be zero, which means that no power is dissipated in a purely capacitive circuit.

Resistive and Capacitive Circuit

In Figure 14-12 you can see the current, voltage, and power waveforms generated by applying an ac voltage source across a series-connected *RC* circuit. The current leads the voltage by some phase angle less than 90°, and the power waveform is once again determined by the product of voltage and current. The negative alternation of the power cycle indicates that the capacitor is discharging and giving back the power that it consumed during the charge.

The positive alternation of the power cycle is much larger than the negative alternation because it is the combination of both the capacitor taking power during charge and the resistor consuming and dissipating power in the form of heat. The average power being dissipated will be some positive value, due to the heat being generated by the resistor.

Power Factor

In a purely resistive circuit, all the energy supplied to the resistor from the source is dissipated in the form of heat. This form of power is referred to as **resistive power** (P_R) or **true power,** and is calculated with the formula.

Resistive Power or True Power
The average power consumed by a circuit during one complete cycle of alternating current.

$$P_R = I^2 \times R$$

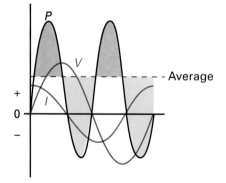

FIGURE 14-12   Power in a Resistive and Capacitive Circuit.

In a purely capacitive circuit, all the energy supplied to the capacitor is stored from the source and then returned to the source, without energy loss. This form of power is referred to as **reactive power** (P_X) or **imaginary power.**

Reactive Power or Imaginary Power Also called wattless power, it is the power value obtained by multiplying the effective value of current by the effective value of voltage and the sine of the angular phase difference between current and voltage.

$$P_X = I^2 \times X_C$$

When a circuit contains both capacitance and resistance, some of the supply is stored and returned by the capacitor and some of the energy is dissipated and lost by the resistor.

Figure 14-13(a) illustrates another vector diagram. Just as the voltage across a resistor is 90° out of phase with the voltage across a capacitor, and resistance is 90° out of phase with reactance, resistive power will be 90° out of phase with reactive power.

(a)

(b)

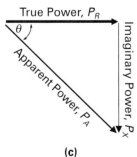

(c)

FIGURE 14-13   **Apparent Power.**

Apparent Power
The power value obtained in an ac circuit by multiplying together effective values of voltage and current, which reach their peaks at different times.

If we take the three variables from Figure 14-13(b) to form a right-angle triangle as in Figure 14-13(c), we can vectorially add true power and imaginary power to produce a resultant **apparent power** vector. Apparent power is the power that appears to be supplied to the load and includes both the true power dissipated by the resistance and the imaginary power delivered to the capacitor.

Applying the Pythagorean theorem once again, we can calculate apparent power by

$$P_A = \sqrt{P_R{}^2 + P_X{}^2}$$

where P_A = apparent power, in volt-amperes (VA)
P_R = true power, in watts (W)
P_X = reactive power, in volt-amperes reactive (VAR)

Power Factor
Ratio of actual power to apparent power. A pure resistor has a power factor of 1 or 100% while a capacitor has a power factor of 0 or 0%.

The **power factor** is a ratio of true power to apparent power and is a measure of the loss in a circuit. It can be calculated by using the formula

$$PF = \frac{\text{true power } (P_R)}{\text{apparent power } (P_A)}$$

Figure 14-14 helps explain what the power factor of a circuit actually indicates. In a purely resistive circuit, as shown in Figure 14-14(a), the apparent power will equal the true power ($P_A = \sqrt{P_R{}^2 + 0^2}$). The power factor will therefore equal 1 (PF = P_R/P_A). In a purely capacitive circuit, as shown in Figure 14-14(b), the true power will be zero, and therefore the power factor will equal 0 (PF = $0/P_A$). A power factor of 1 therefore indicates a maximum power loss (cir-

Purely Resistive

In a resistive circuit the reactive power is zero, and therefore the true power = apparent power.

The power factor is therefore equal to:

$$PF = \frac{P_R}{P_A} = 1 \quad \text{(maximum value)}$$

(a)

Purely Reactive

In a reactive circuit the resistive power (P_R) is zero.

The power factor is therefore equal to:

$$PF = \frac{P_R}{P_A} = 0 \quad \text{(minimum value)}$$

(b)

FIGURE 14-14   Circuit's Power Factor. (a) Purely Resistive, PF = 1. (b) Purely Reactive, PF = 0.

cuit is resistive), while a power factor of 0 indicates no power loss (circuit is capacitive). With circuits that contain both resistance and reactance, the power factor will be somewhere between zero (0 = reactive) and one (1 = resistive).

Since true power is determined by resistance and apparent power is dependent on impedance, as shown in Figure 14-13(a), the power factor can also be calculated by using the formula

$$PF = \frac{R}{Z}$$

As the ratio of true power (adjacent) to apparent power (hypotenuse) determines the angle θ, the power factor can also be determined by the cosine of angle θ.

$$PF = \cos \theta$$

EXAMPLE:

Calculate the following for a series RC circuit if $R = 2.2$ kΩ, $X_C = 3.3$ kΩ, and $V_S = 5$ V.

    a. Z
    b. I
    c. θ
    d. P_R
    e. P_X
    f. P_A
    g. PF

Solution:

a. $Z = \sqrt{R^2 + X_C^2}$

    $= \sqrt{2.2 \text{ k}\Omega^2 + 3.3 \text{ k}\Omega^2}$

    $= 3.96$ kΩ

b. $I = \dfrac{V_S}{Z} = \dfrac{5 \text{ V}}{3.96 \text{ k}\Omega} = 1.26$ mA

c. $\theta = \text{invtan} \dfrac{X_C}{R} = \text{invtan} \dfrac{3.3 \text{ k}\Omega}{2.2 \text{ k}\Omega}$

    $= \text{invtan } 1.5 = 56.3°$

d. True power $= I^2 \times R$

        $= 1.26 \text{ mA}^2 \times 2.2 \text{ k}\Omega$

        $= 3.49$ mW

e. Reactive power $= I^2 \times X_C$

        $= 1.26 \text{ mA}^2 \times 3.3 \text{ k}\Omega$

        $= 5.24 \times 10^{-3}$ or 5.24 mVAR

Step	Keypad Entry	Display Response
1.	③ . ③ Ⓔ ③	3.3E3
2.	÷	
3.	② . ② Ⓔ ③	2.2E3
4.	=	1.5
5.	inv tan	56.309932

f. Apparent power = $\sqrt{P_R^2 + P_X^2}$

$= \sqrt{3.49 \text{ mW}^2 + 5.24 \text{ mW}^2}$

$= 6.29 \times 10^{-3}$ or 6.29 mVA

g. Power factor $= \dfrac{R}{Z} = \dfrac{2.2 \text{ k}\Omega}{3.96 \text{ k}\Omega} = 0.55$

or

$= \dfrac{P_R}{P_A} = \dfrac{3.49 \text{ mW}}{6.29 \text{ mW}} = 0.55$

or

$= \cos\theta = \cos 56.3° = 0.55$

SELF-TEST REVIEW QUESTIONS FOR SECTION 14-2

1. What is the phase relationship between current and voltage in a series *RC* circuit?

2. What is a phasor diagram?

3. Define and state the formula for impedance.

4. What is the phase angle or shift in:

   a. A purely resistive circuit?

   b. A purely capacitive circuit?

   c. A series circuit consisting of *R* and *C*?

14-3 PARALLEL *RC* CIRCUIT

Now that we have analyzed the characteristics of a series *RC* circuit, let us connect a resistor and capacitor in parallel.

14-3-1 Voltage

As with any parallel circuit, the voltage across all components in parallel is equal to the source voltage; therefore,

$$V_R = V_C = V_S$$

14-3-2 Current

In Figure 14-15(a), you will see a parallel circuit containing a resistor and a capacitor. The current through the resistor and capacitor is simply calculated by applying Ohm's law

(a)

(b)

FIGURE 14-15 **Parallel *RC* Circuit.**

$$\text{(resistor current) } I_R = \frac{V_S}{R}$$

$$\text{(capacitor current) } I_C = \frac{V_S}{X_C}$$

Total current (I_T), however, is not as simply calculated. As expected, resistor current (I_R) is in phase with the applied voltage (V_S), as shown in the vector diagram in Figure 14-15(b). Capacitor current will always lead the applied voltage by 90°, and as the applied voltage is being used as our reference at 0° on the vector diagram, the capacitor current will have to be drawn at +90° in order to lead the applied voltage by 90°, since vector diagrams rotate in a counterclockwise direction.

Total current is therefore the vector sum of both the resistor and capacitor currents. Using the Pythagorean theorem, total current can be calculated by

$$I_T = \sqrt{I_R^2 + I_C^2}$$

14-3-3 *Phase Angle*

The angle by which the total current (I_T) leads the source voltage (V_S) can be determined with either of the following formulas:

$$\theta = \text{invtan } \frac{I_C \text{ (opposite)}}{I_R \text{ (adjacent)}} \qquad \theta = \text{invtan } \frac{R}{X_C}$$

14-3-4 Impedance

Since the circuit is both capacitive and resistive, the total opposition or impedance of the parallel *RC* circuit can be calculated by

$$Z = \frac{V_S}{I_T}$$

The impedance of a parallel *RC* circuit is equal to the total voltage divided by the total current. Using basic algebra, this basic formula can be rearranged to express impedence in terms of reactance and resistance.

$$Z = \frac{R \times X_C}{\sqrt{R^2 + X_C^2}}$$

14-3-5 Power

With respect to power, there is no difference between a series circuit and a parallel circuit. The true power or resistive power (P_R) dissipated by an *RC* circuit is calculated with the formula

$$P_R = I_R^2 \times R$$

The imaginary power or reactive power (P_X) of the circuit can be calculated with the formula

$$P_X = I_C^2 \times X_C$$

The apparent power is equal to the vector sum of the true power and the reactive power.

$$P_A = \sqrt{P_R^2 + P_X^2}$$

As with series RC circuits the power factor is calculated as

$$PF = \frac{P_R \text{ (resistive power)}}{P_A \text{ (apparent power)}}$$

A power factor of 1 indicates a purely resistive circuit, while a power factor of 0 indicates a purely reactive circuit.

EXAMPLE:

Calculate the following for a parallel RC circuit in which $R = 24\ \Omega$, $X_C = 14\ \Omega$, and $V_S = 10$ V.

a. I_R
b. I_C
c. I_T
d. Z
e. θ

Solution:

a. $I_R = \dfrac{V_S}{R} = \dfrac{10\text{ V}}{24\ \Omega} = 416.66$ mA

b. $I_C = \dfrac{V_S}{X_C} = \dfrac{10\text{ V}}{14\ \Omega} = 714.28$ mA

c. $I_T = \sqrt{I_R^2 + I_C^2}$

$\quad = \sqrt{416.66\text{ mA}^2 + 714.28\text{ mA}^2}$

$\quad = \sqrt{0.173 + 0.510}$

$\quad = 826.5$ mA

d. $Z = \dfrac{V_S}{I_T} = \dfrac{10\text{ V}}{826.5\text{ mA}} = 12\ \Omega$

or

$\quad = \dfrac{R \times X_C}{\sqrt{R^2 + X_C^2}} = \dfrac{24 \times 14}{\sqrt{24^2 + 14^2}}$

$\quad = 12\ \Omega$

e. $\theta = \arctan \dfrac{I_C}{I_R} = \arctan \dfrac{714.28\text{ mA}}{416.66\text{ mA}}$

$\quad = \arctan 1.714 = 59.7°$

or

$\theta = \arctan \dfrac{R}{X_C} = \arctan \dfrac{24\ \Omega}{14\ \Omega}$

$\quad = \arctan 1.714 = 59.7°$

CALCULATOR SEQUENCE

Step	Keypad Entry	Display Response
1.	[7][1][4][.][2][8][E][3][+/−]	714.28E-3
2.	[÷]	
3.	[4][1][6][.][6][6][E][3][+/−]	416.66E-3
4.	[=]	1.7142994
5.	[inv][tan]	59.743762

1. What is the phase relationship between current and voltage in a parallel *RC* circuit?

2. Could a parallel *RC* circuit be called a voltage divider?

3. State the formula used to calculate:

    a. I_T

    b. Z

4. Will capacitor current lead or lag resistor current in a parallel *RC* circuit?

14-4 TESTING CAPACITORS

Now that you have a good understanding as to how a capacitor should function, let us investigate how to diagnose a capacitor malfunction.

14-4-1 *The Ohmmeter*

A faulty capacitor may have one of three basic problems:

1. A short, which is easy to detect and is caused by a contact from plate to plate.

2. An open, which is again quite easy to detect and is normally caused by one of the leads becoming disconnected from its respective plate.

3. A leaky dielectric or capacitor breakdown, which is quite difficult to detect, as it may only short at a certain voltage. This problem is usually caused by the deterioration of the dielectric, which starts displaying a much lower dielectric resistance than it was designed for. The capacitor with this type of problem is referred to as a *leaky capacitor*.

Capacitors of 0.5 µF and larger can be checked by using an analog ohmmeter or a digital multimeter with a bar graph display by using the procedure shown in Figure 14-16.

Step 1: Ensure that the capacitor is discharged by shorting the leads together.

Step 2: Set the ohmmeter to the highest ohms range scale. Zero the ohmmeter.

Step 3: Connect the meter to the capacitor, observing the correct polarity if an electrolytic is being tested, and observe the meter pointer. The capacitor will initially be discharged, and therefore maximum current will flow from the meter battery to the capacitor; maximum current means low resistance, which is why the meter's pointer deflects to the far right to indicate 0 Ω.

Step 4: As the capacitor charges, it will cause current flow from the meter's battery to decrease, and consequently, the meter needle will move toward the left side of the scale.

Step 1: Discharge Capacitor

Step 2: Set Analog Ohmmeter to
High Ohms Range

Step 3: Meter Deflects Rapidly
to 0 Ohms Initially

Step 4: Meter Should Then Return to
Infinity as the Capacitor Charges

FIGURE 14-16   Testing a Capacitor of More Than 0.5 μF.

A good capacitor will cause the meter to react as just explained. A larger capacitance will cause the meter to move slowly to infinity (∞), as it will take a longer time to charge, while a smaller value of capacitance will charge at a much faster rate, causing the meter to deflect rapidly toward ∞. For this reason, the ohmmeter cannot be reliably used to check capacitors with values of less than 0.5 μF, because the capacitor charges up too quickly and the meter does not have enough time to respond.

A shorted capacitor will cause the meter to deflect to zero ohms and remain in that position. An open capacitor will cause no meter deflection (infinite resistance) because there is no path for current to flow.

A leaky capacitor will deflect to the right, as normal. If the meter pointer returns almost all the way back to ∞, only a small current is still flowing and the capacitor has a small dielectric leak. If the meter only comes back to halfway or a large distance away from infinity, a large amount of current is still flowing and the capacitor has a large dielectric leak (defect).

When using the analog ohmmeter to test capacitors, there are some other points that you should be aware of:

1. Electrolytics are noted for having a small yet noticeable amount of inherent leakage; therefore, do not expect the needle pointer to move all the way to the left (∞ ohms). Most electrolytic capacitors that are still functioning normally will show a resistance of 200 kΩ or more.

2. Some ohmmeters utilize internal batteries of up to 15 V, so be careful not to exceed the voltage rating of the capacitor.

14-4-2   The Capacitance Meter or Analyzer

The ohmmeter check tests the capacitor under a low-voltage condition. This may be adequate for some capacitor malfunctions; however, a problem that often occurs with capacitors is that they short or leak at a high voltage. The ohmmeter test is also adequate for capacitors of 0.5 µF or greater; however, a smaller capacitor cannot be tested because its charge time is too fast for the meter to respond. The ohmmeter cannot check for high-voltage failure, for small-value capacitance, or if the value of capacitance has changed through age or extreme thermal exposure.

Capacitance Meter
Instrument used to measure the capacitance of a capacitor or circuit.

A **capacitance meter** or analyzer, which is illustrated in Figure 14-17, can totally check all aspects of a capacitor in a range of values from approximately 1 pF to 20 F. The tests that are generally carried out include:

1. *Capacitor value change* [Figure 14-18(a)]: Capacitors will change their value over a period of time. Ceramic capacitors often change 10 to 15% within the first year as the ceramic material relaxes. Electrolytics change their value due to the electrolytic solution simply drying out. Some capacitors are simply labeled incorrectly by manufacturers or the technician cannot determine the correct value because of the labeling used. A value change accounts for approximately 25% of all defective capacitors.

2. *Capacitor leakage* [Figure 14-18(b)]: Leakage occurs due to an imperfection of the dielectric. Although the dielectric's resistance is very high, a small amount of leakage current will flow between the plates. This resistance, which is between the plates and therefore effectively in parallel with the capacitor, can become too low and cause the circuit that the capacitor is in

FIGURE 14-17   Capacitor Analyzer.

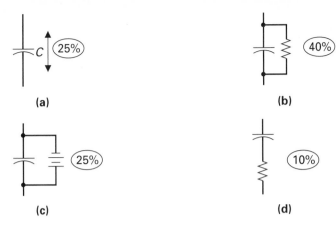

FIGURE 14-18 Problems with Capacitors. (a) Value Change. (b) Leakage. (c) Dielectric Absorption. (d) Equivalent Series Resistance.

to malfunction. Most capacitance meters will perform a leakage test with operating potentials up to 650 V. Leakage accounts for approximately 40% of all defective capacitors.

3. *Dielectric absorption* [Figure 14-18(c)]: This occurs mainly in electrolytics when they will take on a charge but will not fully discharge. This residual charge remains within the capacitor, similar to a small dc battery, and this changes the effective value of the capacitor once it is in circuit during operation. If this causes the value to change more than 15%, the capacitor should be rejected. Dielectric absorption accounts for approximately 25% of all defective capacitors.

4. *Equivalent series resistance* (ESR) [Figure 14-18(d)]: Series resistance is found in the capacitor leads, lead to plate connection, and electrolyte (almost always occurs in electrolytics) and causes the effective circuit capacitance value to change. Equivalent series resistance accounts for 10% of all defective capacitors.

SELF-TEST REVIEW QUESTIONS FOR SECTION 14-4

---

1. The analog ohmmeter can only be used to test capacitors that are _____ or more in value.
2. A _____ _____ tests a capacitor for value change, leakage, dielectric absorption, and its equivalent series resistance.

14-5 APPLICATIONS OF CAPACITORS

There are many applications of capacitors, some of which will be discussed now; others will be presented later and in your course of electronic studies.

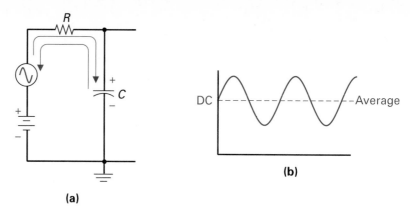

(a)

(b)

FIGURE 14-19   Superimposing AC on a DC Level.

14-5-1   Combining AC and DC

Figure 14-19(a) and (b) show how the capacitor can be used to combine ac and dc. The capacitor is large in value (electrolytic typically) and can be thought of as a very large bucket that will fill or charge to a dc voltage level. The ac voltage will charge and discharge the capacitor, which is similar to pouring in and pulling out (alternating) more and less water. The resulting waveform is a combination of ac and dc that varies above and below an average dc level. In this instance, the ac is said to be superimposed on a dc level.

14-5-2   The Capacitive Voltage Divider

In Chapter 13 we illustrated how capacitors could be connected in series across a dc voltage source to form a voltage divider. This circuit is repeated in Figure 14-20(a). Figure 14-20(b) shows how two capacitors can be used to divide up an ac voltage source of 30 V rms. Like a resistive voltage divider, voltage drop is proportional to current opposition, which with a capacitor is called *capacitive reactance* (X_C). The larger the capacitive reactance $(X_C\uparrow)$, the larger the voltage drop $(V\uparrow)$. Since capacitive reactance is inversely proportional to the capacitor value and the frequency of the ac source, a change in input frequency will cause a change in the capacitor's capacitive reactance, and this will change the voltage drop across the capacitor. Although a 90° phase shift is present between voltage and current in a purely capacitive circuit (I leads V by 90°), there is no phase difference between the input voltage and the output voltages.

Filter
Network composed of resistor, capacitor, and inductors used to pass certain frequencies yet block others through heavy attenuation.

14-5-3   RC Filters

A **filter** is a circuit that allows certain frequencies to pass but blocks other frequencies. In other words, it filters out the unwanted frequencies but passes the wanted or selected ones.

$$V_{C1} = \frac{C_T}{C_1} \times V_T = \frac{1.33 \ \mu F}{2 \ \mu F} \times 6 \ V = 4 \ V$$

$$V_{C2} = \frac{C_T}{C_2} \times V_T = \frac{133 \ \mu F}{4 \ \mu F} \times 6 \ V = 2 \ V$$

(a)

$$I = \frac{V}{X_C} = \frac{30 \ V}{300 \ \Omega} = 100 \ mA$$

$$V_{C1} = I \times X_{C1} = 100 \ mA \times 200 \ \Omega = 20 \ V$$

$$V_{C2} = I \times X_{C2} = 100 \ mA \times 100 \ \Omega = 10 \ V$$

(b)

FIGURE 14-20   Capacitive Voltage Divider. (a) DC Circuit.
(b) AC Circuit.

There are two basic RC filters:

1. The **low-pass filter,** which can be seen in Figure 14-21(a); as its name implies, it passes the low frequencies but heavily **attenuates** the higher frequencies.

2. The **high-pass filter,** which can be seen in Figure 14-21(b); as its name implies, it allows the high frequencies to pass but heavily attenuates the lower frequencies.

With either the low- or high-pass filter, it is important to remember that capacitive reactance is inversely proportional to frequency ($X_C \propto 1/f$) as stated in the center of Figure 14-21.

With the low-pass filter shown in Figure 14-21(a), the output is connected across the capacitor. As the frequency of the input increases, the amplitude of the output decreases. At dc (0 Hz) and low frequencies, the capacitive reactance is very large ($X_C \uparrow = 1/f \downarrow$) with respect to the resistor. All the input will appear across the capacitor, because the capacitor and resistor form a voltage divider, as shown in Figure 14-22(a). As with any voltage divider, the larger opposition to current flow will drop the larger voltage. Since the output voltage is determined by the voltage drop across the capacitor, almost all the input will appear across the capacitor and therefore be present at the output.

If the frequency of the input increases, the reactance of the capacitor will decrease ($X_C \downarrow = 1/f \uparrow$), and a larger amount of the signal will be dropped across the resistor. As frequency increases, the capacitor becomes more of a short circuit (lower reactance), and the output, which is across the capacitor, decreases, as shown in Figure 14-22(b).

Low-Pass Filter
Network or circuit designed to pass any frequencies below a critical or cutoff frequency and reject or heavily attenuate all frequencies above.

Attenuate
To reduce in amplitude an action or signal.

High-Pass Filter
Network or circuit designed to pass any frequencies above a critical or cutoff frequency and reject or heavily attenuate all frequencies below.

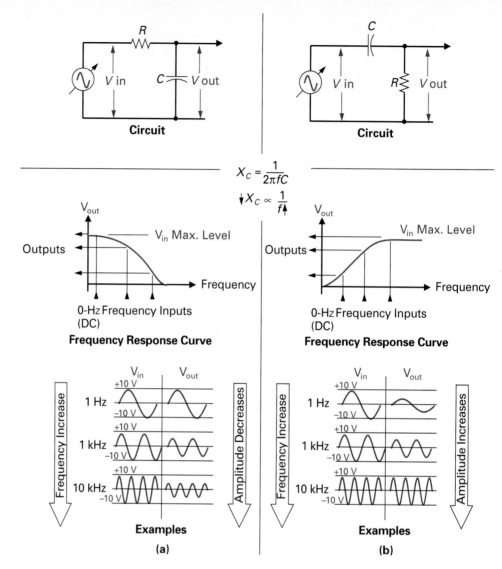

FIGURE 14-21 *RC* Filters. (a) Low-Pass Filter. (b) High-Pass Filter.

Frequency Response Curve
A graph indicating how effectively a circuit or device responds to the frequency spectrum.

Below the circuit of the low-pass filter in Figure 14-21(a), you will see a graph known as the **frequency response curve** for the low-pass filter. This curve illustrates that as the frequency of the input increases the voltage at the output will decrease.

With the high-pass filter, seen in Figure 14-21(b), the capacitor and resistor have traded positions to show how the opposite effect for the low-pass filter occurs.

At low frequencies, the reactance will be high and almost all of the signal in will be dropped across the capacitor. Very little signal appears across the resistor, and, consequently, the output, as shown in Figure 14-23(a). As the frequency of the input increases, the reactance of the capacitor decreases, allowing more of

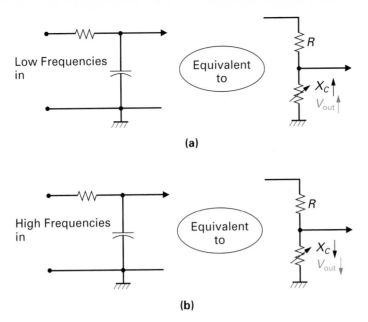

(a)

(b)

FIGURE 14-22 Low-Pass Filter. (a) Low-Pass Filter, High
X_C. (b) Low-Pass Filter, Low X_C.

the input signal to appear across the resistor and therefore appear at the output,
as shown in Figure 14-23(b).

Below the circuit of the high-pass filter in Figure 14-21(b), you will see the
frequency response curve for the high-pass filter. This curve illustrates that as
the frequency of the input increases the voltage at the output increases.

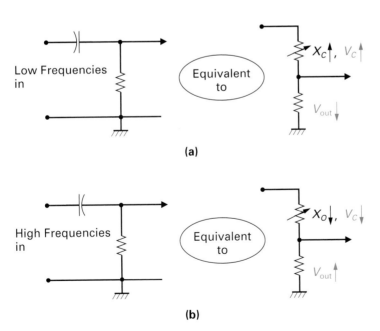

(a)

(b)

FIGURE 14-23 High-Pass Filter. (a) High X_C. (b) Low X_C.

14-5-4 The RC Integrator

Up until now we have analyzed the behavior of an *RC* circuit only when a sine-wave input signal was applied. In both this and the following sections we demonstrate how an *RC* circuit will react to a square-wave input, and show two other important applications of capacitors. The term **integrator** is derived from a mathematical function in calculus. This combination of *R* and *C*, in some situations, displays this mathematical function. Figure 14-24(a) illustrates an integrator circuit that can be recognized by the series connection of *R* and *C*, but mainly from the fact that the output is taken across the capacitor.

Integrator
Device that approximates and whose output is proportional to an integral of the input signal.

If a 10 V square wave is applied across the circuit, as shown in the waveforms in Figure 14-24(b), and the time constant of the *RC* combination calculates out to be 1 second, the capacitor will charge when the square wave input is positive toward the applied voltage (10 V) and reaches it in five time constants (5 seconds). Since the positive alternation of the square wave lasts for 6 seconds, the capacitor will be fully charged 1 second before the positive alternation ends, as shown in Figure 14-25.

When the positive half-cycle of the square-wave input ends after 6 seconds, the input falls to 0 V and the circuit is equivalent to that shown in Figure 14-26(a). The 10 V charged capacitor now has a path to discharge and in five time constants (5 seconds) is fully discharged, as shown in Figure 14-26(b).

If the same *RC* integrator circuit was connected to a square wave that has a 1-second positive half-cycle, the capacitor will not be able to charge fully toward 10 V. In fact, during the positive alternation of 1 second (one time constant),

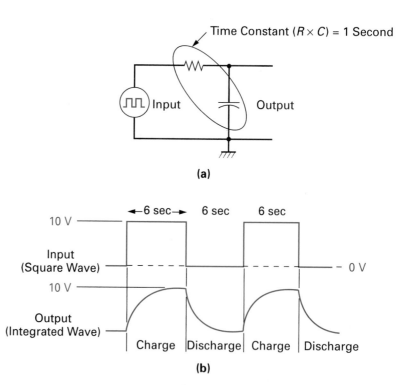

FIGURE 14-24 *RC* Integrator. (a) Circuit. (b) Waveforms.

**FIGURE 14-25   Integrator Response to Positive Step.
(a) Equivalent Circuit. (b) Waveforms for Charge.**

the capacitor will reach 63.2% of the applied voltage (6.32 V), and then during the 0 V half-cycle it will discharge to 63.2% of 6.32 V, to 2.33 V, as shown by the waveform in Figure 14-27. The voltage across the capacitor, and therefore the output voltage, will gradually build up and eventually level off to an average value of 5 V in about five time constants (5 seconds).

In summary, if the period or time of the square wave is decreased, or if the time constant is increased, the same effect results. The capacitor has less time to charge and discharge and reaches an average value of the input voltage, which for a square wave is half of its amplitude.

Figure 14-28 illustrates the circuit waveform produced when the time constant is long with respect to the period of the input waveform. If the time constant is further increased so that it is even longer with respect to the input waveform, the output will have an even smaller peak-to-peak variation.

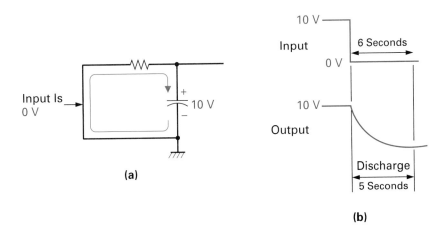

**FIGURE 14-26   Integrator Response to Negative Step. (a) Equivalent
Circuit. (b) Waveforms for Discharge.**

FIGURE 14-27   Integrator Response to Square Wave. (a) Circuit. (b) Waveforms.

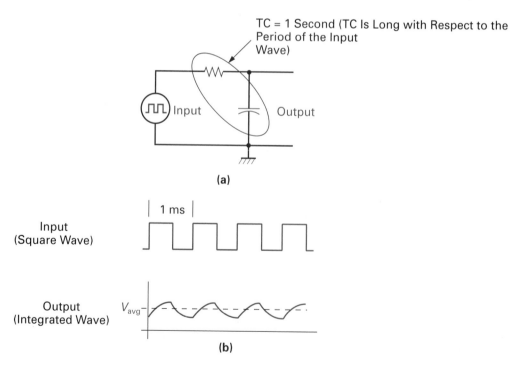

FIGURE 14-28   Integrator Response to Long Time Constant. (a) Circuit.
(b) Waveforms.

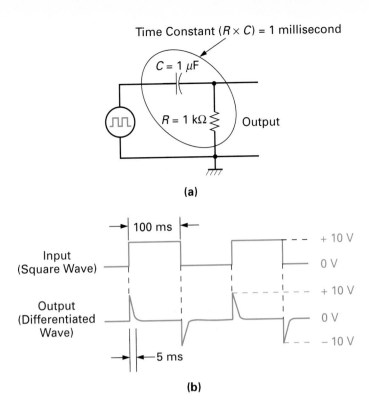

(a)

(b)

FIGURE 14-29 *RC* **Differentiator. (a) Circuit.**
(b) Waveforms.

14-5-5 *The* RC *Differentiator*

Figure 14-29(a) illustrates the **differentiator** circuit, which is the integrator's opposite. In this case the output is taken across the resistor instead of the capacitor, and the time constant is always short with respect to the input square wave period.

The differentiator output waveform, shown in Figure 14-29(b), is taken across the resistor and is the result of the capacitor's charge and discharge.

Differentiator
A circuit whose output voltage is proportional to the rate of change of the input voltage. The output waveform is then the time derivative of the input waveform, and the phase of the output waveform leads that of the input by 90°.

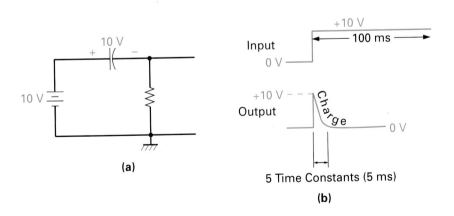

(a)

(b)

FIGURE 14-30 **Differentiator's Response to Positive Step.**
(a) Equivalent Circuit. (b) Waveforms.

(a)

(b)

FIGURE 14-31 **Differentiator's Response to a Negative Step. (a) Equivalent Circuit. (b) Waveforms.**

When the square wave swings positive, the equivalent circuit is that shown in Figure 14-30(a). When the 10 V is initially applied (positive step of the square wave), all the voltage is across the resistor, and therefore at the output, as the capacitor cannot charge instantly. As the capacitor begins to charge, more of the voltage is developed across the capacitor and less across the resistor. The voltage across the capacitor exponentially increases and reaches 10 V in five time constants (5×1 ms = 5 ms), while the voltage across the resistor, and therefore the output, exponentially falls from its initial 10 V to 0 V in five time constants as shown in Figure 14-30(b), at which time all the voltage is across the capacitor and no voltage will be across the resistor.

When the positive half-cycle of the square wave ends and the input falls to zero, the circuit is equivalent to that shown in Figure 14-31(a). The negative

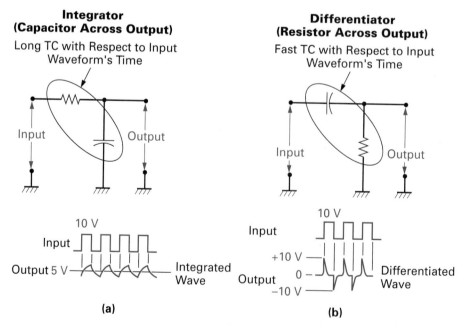

(a)

(b)

FIGURE 14-32 **Summary of Integrator and Differentiator.**

plate of the capacitor is now applied directly to the output. Since the capacitor cannot instantly discharge, the output drops suddenly down to –10 V as shown in Figure 14-31(b). This is the voltage across the resistor, and therefore the output. The capacitor is now in series with the resistor and therefore has a path through the resistor to discharge, which it does in five time constants to 0 V.

Figure 14-32(a) and (b) illustrate the integrator and differentiator circuits and waveforms, which are used extensively in many applications and equipment, such as computers, robots, lasers, and communications.

SELF-TEST REVIEW QUESTIONS FOR SECTION 14-5

1. Give three circuit applications for a capacitor.
2. With an *RC* low-pass filter, the _____ is connected across the output. Whereas with an *RC* high-pass filter, the _____ is connected across the output.
3. An integrator has a _____ time constant compared to the period of the input square wave.
4. A _____ circuit produces positive and negative spikes at the output when a square wave is applied at the input.

Capacitive Circuits, Testing, and Applications (Figure 14-33)

SUMMARY

1. Resistance (*R*), by definition, is the opposition to current flow with the dissipation of energy and is measured in ohms. Capacitors oppose current flow like a resistor; however, a resistor dissipates energy, whereas a capacitor stores energy (when it charges) and then gives back its energy into the circuit (when it discharges). Because of this difference, a new term had to be used to describe the opposition offered by a capacitor. Capacitive reactance (X_C), by definition, is the opposition to current without the dissipation of energy and is also measured in ohms.

2. If capacitive reactance is basically opposition, it is inversely proportional to the amount of current flow. If a large current is within a circuit, the opposition or reactance must be low ($I\uparrow$, $X_C\downarrow$). Conversely, a small circuit current will be the result of a large opposition or reactance ($I\downarrow$, $X_C\uparrow$).

3. When a dc source is connected across a capacitor, current will flow only for a short period of time (5*RC* seconds) to charge the capacitor. After this time, there is no further current flow. Consequently, the capacitive reactance or opposition offered by a capacitor to dc is infinite (maximum).

4. Alternating current is continuously reversing in polarity, resulting in the capacitor continuously charging and discharging. This means that charge and discharge currents are always flowing around the circuit, and if we have a certain value of current, we must also have a certain value of reactance or opposition.

FIGURE 14-33 **Capacitive Circuits, Testing, and Applications.**

$$X_C = \frac{1}{2\pi f C}$$

X_C = capacitive reactance, in ohms
2π = constant
f = frequency, in hertz
C = capacitance, in farads

Series RC Circuits:

$$V_S = \sqrt{V_R^2 + V_C^2}$$

V_S = source voltage
V_R = voltage drop across resistor
V_C = voltage drop across capacitor

$$V_C = \sqrt{V_S^2 - V_R^2}$$
$$V_R = \sqrt{V_S^2 - V_C^2}$$

$$Z = \sqrt{R^2 + X_C^2}$$

Z = impedance, in ohms
R = resistance, in ohms
X_C = capacitive reactance, in ohms

$$R = \sqrt{Z^2 - X_C^2}$$
$$X_C = \sqrt{Z^2 - R^2}$$

$$\theta = \text{invtan}\, \frac{X_C}{R}$$

θ = phase angle or shift, in degrees

$$\theta = \text{invtan}\, \frac{V_C}{V_R}$$

$$P_A = \sqrt{P_R^2 + P_X^2}$$

P_A = apparent power, in volt-amperes (VA)
P_R = true power, in watts (W)
P_X = reactive power, in volt-amperes reactive (VAR)

$$PF = \frac{P_R}{P_A} = \frac{R}{Z} = \cos\theta$$

PF = power factor

Parallel RC Circuits:

$$I_R = \frac{V_S}{R}$$

I_R = current through resistor
I_C = current through capacitor

$$I_C = \frac{V_S}{X_C}$$

$$I_T = \sqrt{I_R^2 + I_C^2}$$

I_T = total circuit current

$$Z = \frac{V}{I_T}$$

$$Z = \frac{R \times X_C}{\sqrt{R^2 + X_C^2}}$$

$$\theta = \text{invtan}\, \frac{I_C}{I_R} \qquad \text{or} \qquad \theta = \text{invtan}\, \frac{R}{X_C}$$

FIGURE 14-33 (continued)

RC Series Circuit

$R = X_C$

R_C Circuit

RC Parallel Circuit

$$I_R = \frac{V}{R}$$

$$I_C = \frac{V}{X_C}$$

RC Filter Circuits

Circuit

Circuit

$$X_C = \frac{1}{2\pi fC}$$

$$\downarrow X_C \propto \frac{1}{f} \uparrow$$

Frequency Response Curve

Frequency Response Curve

Examples

Examples

RC Integrator and Differentiator Circuit

**Integrator
(Capacitor Across Output)**

Long TC with Respect to Input
Waveform's Time

**Differentiator
(Resistor Across Output)**

Fast TC with Respect to Input
Waveform's Time

5. If the applied alternating current is at a high frequency it is switching polarity more rapidly than a lower frequency and there is very little time between the start of charge and discharge. As the charge and discharge currents are largest at the beginning of the charge and discharge of the capacitor, the reactance has very little time to build up and oppose the current, which is why the current is a high value and the capacitive reactance is small at higher frequencies. With lower frequencies, the applied alternating current is switching at a slower rate, and therefore the reactance, which is low at the beginning, has more time to build up and oppose the current. Capacitive reactance is therefore inversely proportional to frequency.

6. Capacitive reactance is also inversely proportional to the value of capacitance. If a larger capacitor value is used a longer time is required to charge the capacitor ($t\uparrow = C\uparrow R$), which means that current will be present for a longer period of time, so the overall current will be large ($I\uparrow$); consequently, the reactance must be small ($X_C\downarrow$). On the other hand, a small capacitance value will charge in a small amount of time ($t\downarrow = C\downarrow R$) and the current is present for only a short period of time. The overall current will therefore be small ($I\downarrow$), indicating a large reactance ($X_C\uparrow$). Capacitive reactance (X_C) is therefore inversely proportional to both frequency and capacitance.

Series *RC* Circuit

7. In a purely resistive circuit, the current flowing within the circuit and the voltage across the resistor are in phase with one another.

8. In a purely capacitive circuit, the current flowing in the circuit leads the voltage across the capacitor by 90°.

9. If we connect a resistor and capacitor in series, we have probably the most commonly used electronic circuit, which has many applications. The voltage across the resistor (V_R) is always in phase with the circuit current (I), because maximum points and zero crossover points occur at the same time. The voltage across the capacitor (V_C) lags the circuit current by 90°.

10. Since the capacitor and resistor are in series, the same current is supplied to both components; Kirchhoff's voltage law can be applied, which states that the sum of the voltage drops around a series circuit is equal to the voltage applied (V_S). The voltage drop across the resistor (V_R) and the voltage drop across the capacitor (V_C) are out of phase with one another, which means that their peaks occur at different times. The signal for the applied voltage (V_S) is therefore obtained by adding the values of V_C and V_R at each instant in time, plotting the results, and then connecting the points with a line.

11. A vector (or phasor) is a quantity that has both magnitude and direction and is represented by a line terminated at the end by an arrowhead.

12. Vectors are generally always used to represent a physical quantity that has two properties.

13. A vector diagram is an arrangement of vectors to illustrate the magnitude and phase relationships between two or more quantities of the same frequency within an ac circuit.

14. By vectorially adding the two voltages V_R and V_C we obtain a resultant V_S vector that has both magnitude and phase. The angle theta (θ) formed between circuit current (I) and source voltage (V_S) will always be less than 90°.

15. Since resistance is the opposition to current with the dissipation of heat, and reactance is the opposition to current without the dissipation of heat, a new term is needed to describe the total resistive and reactive opposition to current. Impedance (designated Z) is also measured in ohms and is the total circuit opposition to current flow. It is a combination of resistance (R) and reactance (X_C); however, in our capacitive and resistive circuit, a phase shift or difference exists, and just as V_C and V_R cannot be merely added together to obtain V_S, R and X_C cannot be simply summed to obtain Z.

16. By vectorially adding R and X_C, we have a resulting impedance (Z) vector.

17. In a purely resistive circuit, the total opposition (Z) is equal to the resistance of the resistor, so the phase shift (θ) is equal to 0°.

18. In a purely capacitive circuit, the total opposition (Z) is equal to the capacitive reactance (X_C) of the capacitor, so the phase shift (θ) is equal to –90°.

19. When a circuit contains both resistance and capacitive reactance, the total opposition or impedance has a phase shift that is between 0 and 90°.

20. Voltage and current are in phase with one another in a resistive circuit, and instantaneous power is calculated by multiplying voltage by current at every instant through 360° ($P = V \times I$). The sinusoidal power waveform is totally positive, because a positive voltage multiplied by a positive current gives a positive value of power, and a negative voltage multiplied by a negative current will also produce a positive value of power.

21. A resistor is said to generate a positive power waveform, which is twice the frequency of the voltage and current waveforms; two power cycles occur in the same time as one voltage and current cycle.

    The power waveform has been split in half, and the line that exists between the maximum point and zero point is the average value of power that is being dissipated by the resistor.

22. The current leads the voltage by 90° in a purely capacitive circuit, and the power wave is calculated by multiplying voltage by current, as before, at every instant through 360°. The resulting power curve is both positive and negative. During the positive alternation of the power curve, the capacitor is taking power as the capacitor charges. When the power alternation is negative, the capacitor is giving back the power it took as it discharges back into the circuit.

    The average power dissipated is once again the value that exists between the maximum positive and maximum negative points and causes the area above this line to equal the area below. This average power level calculates out to be zero, which means that no power is dissipated in a purely capacitive circuit.

23. When an ac voltage source is applied across a series-connected RC circuit, the current leads the voltage by some phase angle less than 90°, and the power waveform is once again determined by the product of voltage and current. The negative alternation of the power cycle indicates that the capac-

itor is discharging and giving back the power that it consumed during the charge.

24. The positive alternation of the power cycle is much larger than the negative alternation because it is the combination of both the capacitor taking power during charge and the resistor consuming and dissipating power in the form of heat. The average power being dissipated will be some positive value, due to the heat being generated by the resistor.

25. In a purely resistive circuit, all the energy supplied to the resistor from the source is dissipated in the form of heat. This form of power is referred to as resistive power (P_R) or true power.

26. In a purely capacitive circuit, all the energy supplied to the capacitor is stored from the source and then returned to the source, without energy loss. This form of power is referred to as reactive power (P_X) or imaginary power.

27. When a circuit contains both capacitance and resistance, some of the supply is stored and returned by the capacitor and some of the energy is dissipated and lost by the resistor.

28. Apparent power is the power that appears to be supplied to the load and includes both the true power dissipated by the resistance and the imaginary power delivered to the capacitor.

29. The power factor is a ratio of true power to apparent power and is a measure of the loss in a circuit.

Parallel *RC* Circuit

30. As with any parallel circuit, the voltage across all components in parallel is equal to the source voltage.

31. The current through the resistor and capacitor is simply calculated by applying Ohm's law.

32. Total current (I_T), however, is not as simply calculated. As expected, resistor current (I_R) is in phase with the applied voltage (V_S). Capacitor current will always lead the applied voltage by 90°, and as the applied voltage is being used as our reference at 0° on the vector diagram, the capacitor current will have to be drawn at +90° in order to lead the applied voltage by 90°, since vector diagrams rotate in a counterclockwise direction.

33. Total current is therefore the vector sum of both the resistor and capacitor currents.

34. The angle by which the total current (I_T) leads the source voltage (V_S) is call the phase angle.

35. The impedance of a parallel *RC* circuit is equal to the total voltage divided by the total current.

36. With respect to power, there is no difference between a series circuit and a parallel circuit.

Testing Capacitors

37. A faulty capacitor may have one of three basic problems:

a. A short, which is easy to detect and is caused by a contact from plate to plate.

b. An open, which is again quite easy to detect and is normally caused by one of the leads becoming disconnected from its respective plate.

c. A leaky dielectric or capacitor breakdown, which is quite difficult to detect, as it may only short at a certain voltage. This problem is usually caused by the deterioration of the dielectric, which starts displaying a much lower dielectric resistance than it was designed for. The capacitor with this type of problem is referred to as a *leaky capacitor.*

38. Capacitors of 0.5 µF and larger can be checked by using an analog ohmmeter or a digital multimeter with a bar graph display by using the following procedure:

a. Ensure that the capacitor is discharged by shorting the leads together.

b. Set the ohmmeter to the highest ohms range scale. Zero the ohmmeter.

c. Connect the meter to the capacitor, observing the correct polarity if an electrolytic is being tested, and observe the meter pointer. The capacitor will initially be discharged, and therefore maximum current will flow from the meter battery to the capacitor; maximum current means low resistance, which is why the meter's pointer deflects to the far right to indicate 0 Ω.

d. As the capacitor charges, it will cause current flow from the meter's battery to decrease, and consequently, the meter needle will move toward the left side of the scale.

39. The ohmmeter check tests the capacitor under a low-voltage condition. This may be adequate for some capacitor malfunctions; however, a problem that often occurs with capacitors is that they short or leak at a high voltage. A capacitance meter or analyzer can totally check all aspects of a capacitor in a range of values from approximately 1 pF to 20 F. The tests that are generally carried out include:

a. Capacitor value change   **c.** Dielectric absorption
b. Capacitor leakage   **d.** Equivalent series resistance

Applications of Capacitors

40. The capacitor can be used to combine ac and dc. The capacitor is large in value (electrolytic typically) and can be thought of as a very large bucket that will fill or charge to a dc voltage level. The ac voltage will charge and discharge the capacitor, which is similar to pouring in and pulling out (alternating) more and less water. The resulting waveform is a combination of ac and dc that varies above and below an average dc level. In this instance, the ac is said to be superimposed on a dc level.

41. Capacitors could be connected in series across a dc voltage source to form a voltage divider. Like a resistive voltage divider, voltage drop is proportional to current opposition, which with a capacitor is called *capacitive reactance* (X_C). The larger the capacitive reactance ($X_C\uparrow$), the larger the voltage drop ($V\uparrow$). Since capacitive reactance is inversely proportional to the capacitor value and the frequency of the ac source, a change in input frequency will cause a change in the capacitor's capacitive reactance, and this will change the voltage drop across the capacitor. Although a 90° phase shift is present between voltage and current in a purely capacitive circuit (I leads V by 90°), there is no phase difference between the input voltage and the output voltages.

42. A filter is a circuit that allows certain frequencies to pass but blocks other frequencies. In other words, it filters out the unwanted frequencies but passes the wanted or selected ones.

     There are two basic RC filters:

     a. The low-pass filter, which as its name implies, passes the low frequencies but heavily attenuates the higher frequencies.

     b. The high-pass filter, which as its name implies, allows the high frequencies to pass but heavily attenuates the lower frequencies.

43. The frequency response curve for the low-pass filter illustrates that as the frequency of the input increases the voltage at the output will decrease. The frequency response curve for the high-pass filter illustrates that as the frequency of the input increases the voltage at the output increases.

44. The term integrator is derived from a mathematical function in calculus. An integrator circuit can be recognized by the series connection of R and C, but mainly from the fact that the output is taken across the capacitor. The voltage across the capacitor of an integrator, and therefore the output voltage, will gradually build up and eventually level off to an average value in about five time constants. If the period or time of the square wave input is decreased, or if the time constant is increased, the capacitor has less time to charge and discharge and reaches an average value of the input voltage, which for a square wave is half of its amplitude.

45. With the differentiator circuit, the output is taken across the resistor instead of the capacitor, and the time constant is always short with respect to the input square wave period. The differentiator output waveform is taken across the resistor and is the result of the capacitor's charge and discharge.

NEW TERMS		
Apparent power	Imaginary power
Average power	Impedance
Capacitance meter	Integrator
Capacitive circuit	Leaky capacitor
Capacitive reactance	Low-pass filter
Dielectric absorption	Parallel RC circuit
Differentiator	Positive power
Equivalent series resistance	Power factor
Filter	Pythagorean theorem
Frequency response curve	Reactance
High-pass filter	Reactive power

Resistive circuit

Series *RC* circuit

Superimposed

Theta

True power

Vector

Vector diagram

Multiple-Choice Questions

1. Capacitive reactance is inversely proportional to:
   a. Capacitance and resistance   c. Capacitance and impedance
   b. Frequency and capacitance   d. Both (a) and (c)

2. The impedance of an *RC* series circuit is equal to:
   a. The sum of R and X_C
   b. The square root of the sum of R^2 and X_C^2
   c. The square of the sum of R and X_C
   d. The sum of the square root of R and X_C

3. In a purely resistive circuit:
   a. The current flowing in the circuit leads the voltage across the capacitor by 90°.
   b. The circuit current and resistor voltage are in phase with one another.
   c. The current leads the voltage by 45°.
   d. The current leads the voltage by a phase angle between 0 and 90°.

4. In a purely capacitive circuit:
   a. The current flowing in the circuit leads the voltage across the capacitor by 90°.
   b. The circuit current and resistor voltage are in phase with one another.
   c. The current leads the voltage by 45°.
   d. The current leads the voltage by a phase angle between 0 and 90°.

5. In a series circuit containing both capacitance and resistance:
   a. The current flowing in the circuit leads the voltage across the capacitor by 90°.
   b. The circuit current and resistor voltage are in phase with one another.
   c. The current leads the voltage by 45°.
   d. Both (a) and (b).

6. In a series *RC* circuit, the source voltage is equal to:
   a. The sum of V_R and V_C c. The vectoral sum of V_R and V_C
   b. The difference between V_R and V_C d. The sum of V_R and V_C squared.

7. As the source frequency is increased, the capacitive reactance will:
   a. Increase   c. Be unaffected
   b. Decrease   d. Increase, depending on harmonic content

8. The phase angle of a series *RC* circuit indicates by what angle V_S _____ V_R.
   a. Lags   b. Leads   c. Leads or lags   d. None of the above

9. In a series *RC* circuit, the vector combination of *R* and X_C is the circuit's
_____.

    a. Phase angle   **b.** Apparent power   **c.** Source voltage   **d.** Impedance

10. In a parallel *RC* circuit, the total current is equal to:

    a. The sum of I_R and I_C **c.** The vectoral sum of I_R and I_C
    b. The difference between I_R and I_C **d.** The sum of I_R and I_C squared

11. _____ is the opposition offered by a capacitor to current flow without the dissipation of energy.

    a. Capacitive reactance   **d.** Phase angle
    b. Resistance   **e.** The power factor
    c. Impedance

12. _____ is the total reactive and resistive circuit opposition to current flow.

    a. Capacitive reactance   **d.** Phase angle
    b. Resistance   **e.** The power factor
    c. Impedance

13. _____ is the ratio of true (resistive) power to apparent power and is therefore a measure of the loss in a circuit.

    a. Capacitive reactance   **d.** Phase angle
    b. Resistance   **e.** The power factor
    c. Impedance

14. In a series *RC* circuit, the leading voltage will be measured across the:

    a. Resistor   **c.** Source
    b. Capacitor **d.** Any of the choices are true.

15. In a series *RC* circuit, the lagging voltage will be measured across the:

    a. Resistor   **c.** Source
    b. Capacitor **d.** Any of the choices are true.

Essay Questions

16. Give the formula and define the term *capacitive reactance*. (14-1)
17. In a series *RC* circuit, give the formulas for calculating: (14-2)

    a. V_S, when V_R and V_C are known
    b. *Z*, when *R* and X_C are known
    c. *Z*, when *I* and *V* are known
    d. θ, when X_C and *R* are known
    e. θ, when V_C and V_R are known
    f. Power factor, when *R* and *Z* are known
    g. Power factor, when P_R and P_A are known

18. In a parallel *RC* circuit, give the formulas for calculating: (14-3)

    a. I_R, when *V* and *R* are known
    b. I_C, when *V* and X_C are known
    c. *I*, when I_R and I_C are known
    d. *Z*, when *V* and *I* are known
    e. *Z*, when *R* and X_C are known

19. What is meant by *long* or *short time constant,* and do large or small values of RC produce a long or a short time constant? (Chapter 13)

20. Describe how the inverse relationship between frequency and capacitive reactance can be used for the application of filtering. (14-5-3)

21. Illustrate the voltage and current waveforms across a resistor and capacitor in series when a dc voltage is applied during charge and when the same resistor and capacitor are connected to discharge. (Chapter 13)

22. Describe and illustrate how the capacitor can be used in the following applications:

   a. Combining ac and dc (14-5-1)
   b. A voltage divider (14-5-2)
   c. Filtering high and low frequencies (14-5-3)
   d. Integrating a square wave (14-5-4)
   e. Differentiating a square wave (14-5-5)

23. Describe the difference between reactance, resistance, and impedance. (14-2)

24. Sketch the phase relationships between: (14-2-3)

   a. V_R and I in a purely resistive circuit
   b. V_C and I in a purely capacitive circuit

25. What are positive power and negative power? (14-2-4)

Practice Problems

26. Calculate the capacitive reactance of the following capacitor circuits with the following parameters:

   a. $f = 1$ kHz, $C = 2$ μF
   b. $f = 100$ Hz, $C = 0.01$ μF
   c. $f = 17.3$ MHz, $C = 47$ μF

27. In a series RC circuit, the voltage across the capacitor is 12 V and the voltage across the resistor is 6 V. Calculate the source voltage.

28. Calculate the impedance for the following series RC circuits:

   a. 2.7 MΩ, 3.7 μF, 20 kHz
   b. 350 Ω, 0.005 μF, 3 MHz
   c. $R = 8.6$ kΩ, $X_C = 2.4$ Ω
   d. $R = 4700$ Ω, $X_C = 2$ kΩ

29. In a parallel RC circuit with parameters of $V_S = 12$ V, $R = 4$ MΩ, and $X_C = 1.3$ kΩ, calculate:

   a. I_R **b.** I_C **c.** I_T **d.** Z **e.** θ

30. Calculate the total reactance in:

   a. A series circuit where $X_{C1} = 200$ Ω, $X_{C2} = 300$ Ω, $X_{C3} = 400$ Ω
   b. A parallel circuit where $X_{C1} = 3.3$ kΩ, $X_{C2} = 2.7$ kΩ

31. Calculate the capacitance needed to produce 10 kΩ of reactance at 20 kHz.

32. At what frequency will a 4.7 μF capacitor have a reactance of 2000 Ω?

(a)

(b)

FIGURE 14-34

33. A series *RC* circuit contains a resistance of 40 Ω and a capacitive reactance of 33 Ω across a 24 V source.

    a. Sketch the schematic diagram.   **b.** Calculate Z, I, V_R, V_C, I_R, I_C, and θ.

34. A parallel *RC* circuit contains a resistance of 10 kΩ and a capacitive reactance of 5 kΩ across a 100-V source.

    a. Sketch the schematic diagram.   **b.** Calculate I_R, I_C, I_T, Z, V_R, V_C, and θ.

35. Calculate V_R and V_C for the circuits seen in Figure 14-34(a) and (b).

36. Calculate the impedance of the four circuits shown in Figure 14-35.

37. In Figure 14-36, the output voltage, since it is taken across the capacitor, will _____ the voltage across the resistor by _____ degrees.

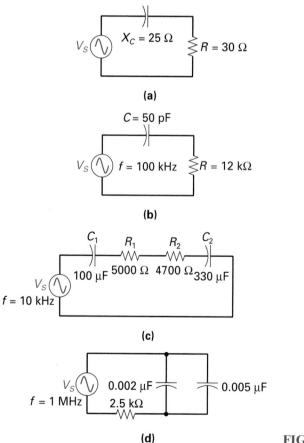

(a)

(b)

(c)

(d)

FIGURE 14-35

FIGURE 14-36

38. If the positions of the capacitor and resistor in Figure 14-36 are reversed, the output voltage, since it is now taken across the resistor, will _____ the voltage across the capacitor by _____ degrees.

39. Calculate the resistive power, reactive power, apparent power, and power factor for the circuit seen in Figure 14-36; $V_{in} = 24$ V and $f = 35$ kHz.

40. Refer to Figure 14-37 and calculate the following:
   a. [Figure 14-37(a)] X_C, I, Z, I_R, θ, V_R, V_C
   b. [Figure 14-37(b)] V_R, V_C, I_R, I_C, I_T, Z, θ

Troubleshooting Problems

41. Describe the three basic problems that normally occur with faulty capacitors.

42. Describe how to use the ohmmeter to check capacitors and also explain some of its limitations.

43. Describe the four basic tests performed by a capacitor meter or analyzer.

44. Which capacitor problem accounts for the largest percentage of defective capacitors? Explain exactly what this malfunction is.

45. If the needle of a meter goes to zero ohms and remains there, the capacitor is _____ . If the capacitor is _____, however, no charging will occur and the meter will indicate infinite ohms.

(a)

(b)

FIGURE 14-37

After completing this chapter, you will be able to:

1. Describe what is meant by the word *electromagnetism.*

2. Explain the atomic theory of electromagnetism.

3. Describe the left-hand rule of electromagnetism.

4. Describe how a magnetic field is generated by current flow through a:
   a. Conductor
   b. Coil

5. Describe the difference between dc and ac electromagnetism.

6. Explain the following magnetic terms:
   a. Magnetic flux
   b. Flux density
   c. Magnetizing force
   d. Magnetomotive force
   e. Reluctance
   f. Permeability (relative and absolute)

7. Explain the relationship between flux density and magnetizing force.

8. Describe the cycle known as the hysteresis loop.

9. Describe the following applications of electromagnetism:
   a. Magnetic-type circuit breaker
   b. Electric bell
   c. Relay and its applications
   d. Solenoid-type electromagnet and its application

10. Define *electromagnetic induction.*

11. State Faraday's and Lenz's laws relating to electromagnetic induction.

12. Describe the following applications of electromagnetic induction:
    a. AC generator
    b. Moving coil microphone

Electromagnetism and Electromagnetic Induction

Problem-Solver

Charles Proteus Steinmetz (1865–1923) was an outstanding electrical genius who specialized in mathematics, electrical engineering, and chemistry. His three greatest electrical contributions were his investigation and discovery of the law of hysteresis, his investigations in lightning, which resulted in his theory on traveling waves, and his discovery that complex numbers could be used to solve ac circuit problems. Solving problems was in fact his specialty, and on one occasion he was commissioned to troubleshoot a failure on a large company system that no one else had been able to repair. After studying the symptoms and schematics for a short time, he chalked an X on one of the metal cabinets, saying that this was where they would find the problem, and left. He was right, and the problem was remedied to the relief of the company executives; however, they were not pleased when they received a bill for $1000. When they demanded that Steinmetz itemize the charges, he replied—$1 for making the mark and $999 for knowing where to make the mark.

The strong message this vignette conveys is that you will get $1 for physical labor and $999 for mental labor, and this is a good example as to why you should continue in your pursuit of education.

It was during a classroom lecture in 1820 that Danish physicist Hans Christian Oersted accidentally stumbled on an interesting reaction. As he laid a compass down on a bench he noticed that the compass needle pointed to an adjacent conductor that was carrying a current, instead of pointing to the earth's north pole. It was this discovery that first proved that magnetism and electricity were very closely related to one another. This phenomenon is now called *electromagnetism* since it is now known that any conductor carrying an electro or electrical current will produce a magnetic field.

In 1831, the English physicist Michael Faraday explored further Oersted's discovery of electromagnetism and found that the process could be reversed. Faraday observed that if a conductor was passed through a magnetic field, a voltage would be induced in the conductor and cause a current to flow. This phenomenon is referred to as *electromagnetic induction.*

In this chapter we examine the terms and characteristics of electromagnetism (electricity to magnetism) and electromagnetic induction (magnetism to electricity).

15-1 ELECTROMAGNETISM

The electron plays a very important role in magnetism. However, the real key or link between electricity and a magnetic field is motion. Anytime a charged particle moves, a magnetic field is generated. Consequently, current flow, which is the movement of electrons, produces a magnetic field.

15-1-1 Atomic Theory of Electromagnetism

Every orbiting electron in motion around its nucleus generates a magnetic field. When electrons are forced to leave their parent atom by voltage and flow toward the positive polarity, they are all moving in the same direction and each electron's magnetic field will add to the next. The accumulation of all these electron fields will create the magnetic field around the conductor, as shown in Figure 15-1.

A simple experiment can be performed to prove that this invisible magnetic field does in fact exist around a conductor, the setup of which is illustrated in Figure 15-2. With the switch open, as shown in Figure 15-2(a), no current will flow, and therefore no magnetic field is generated around the conductor and the iron filings on the cardboard will be disorganized.

With the switch closed, a current of 3 A ($I = V/R = 9$ V/3 $\Omega = 3$ A) will flow through the circuit and a magnetic field will be set up around the conductor. The iron filings become organized in circles due to the influence of the magnetic field, as shown in Figure 15-2(b).

15-1-2 Electromagnetism's Left-Hand Rule

As we have seen so far, there is a relationship between the current in a conductor and the magnetic field around the conductor. The **left-hand rule** states that if an insulated conductor is held with your left hand so that your thumb is pointing in the direction of electron flow (to the positive potential), your fingers will be pointing in the direction of the magnetic force, as illustrated in Figure 15-3.

Electromagnetism
Relates to the magnetic field generated around a conductor when current is passed through it.

Left-Hand Rule
If the fingers of the left hand are placed around the wire so that the thumb points in the direction of the electron flow, the fingers will be pointing in the direction of the magnetic field being produced by the conductor.

FIGURE 15-1   Electron Magnetic Fields.

FIGURE 15-2   Magnetic Field Experiment. (a) Switch Open, No Current Flow, No Magnetic Field. (b) Switch Closed, Current Flow, Magnetic Field.

SELF-TEST REVIEW QUESTIONS FOR SECTIONS 15-1-1 AND 15-1-2

1. True or false: An electric field always encircles a current-carrying conductor.
2. What subatomic particle is said to have its own magnetic field?
3. For what purpose could the left-hand rule be used?
4. From which pole do the magnetic lines of force originate?

15-1-3   The Electromagnet

If our conductor is wound to form a coil, as illustrated in Figure 15-4(a), the conductor current will generate its own magnetic field, which will combine additively within the coil, and the net effect of all these conductor magnetic fields

FIGURE 15-3   Left-Hand Rule of Electromagnetism.

can be used to generate a strong coil magnetic field. This coil of conductor which is used to generate a strong magnetic field is called an **electromagnet.**

If a greater number of loops are added, a greater or stronger magnetic field is generated, as seen in Figure 15-4(b). If the coils were wound in an opposing opposite direction, the opposing magnetic fields would cancel and so no magnetic field would be generated, as seen in Figure 15-4(c).

The left-hand rule can also be applied to electromagnets to determine which of the poles will be the north end of the electromagnet. This is illustrated in Figure 15-5, which shows that if you wrap the fingers of your left hand around the **coil** so that your fingers are pointing in the direction of current flow, your thumb will be pointing to the north end of the electromagnet.

15-1-4   DC versus AC Electromagnetism

A magnetic field results whenever a current flows through any piece of conductor or wire, as shown in Figure 15-6(a). If a conductor is wound to form a spiral, as illustrated in Figure 15-6(b), the conductor, which is now referred to as a *coil,*

Electromagnet
A magnet consisting of a coil wound on a soft iron or steel core. When current is passed through the coil a magnetic field is generated and the core is strongly magnetized to concentrate the magnetic field.

Coil
Number of turns of wire wound around a core to produce magnetic flux (an electromagnet) or to react to a changing magnetic flux (an inductor).

FIGURE 15-4   Coils of an Electromagnet. (a) Small Number of Turns. (b) Large Number of Turns. (c) Opposing Turns.

will, as a result of current flow, develop a magnetic field, which will sum or intensify within the coil. If many coils are wound in the same direction, an electromagnet is formed, as shown in Figure 15-6(c), which will produce a concentrated magnetic field whenever a current is passed through its coils.

Up until now, we have only been discussing current flow through a coil in one direction (dc). A dc voltage produces a fixed current in one direction and

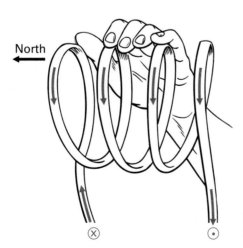

FIGURE 15-5   Left-Hand Rule for Electromagnets.

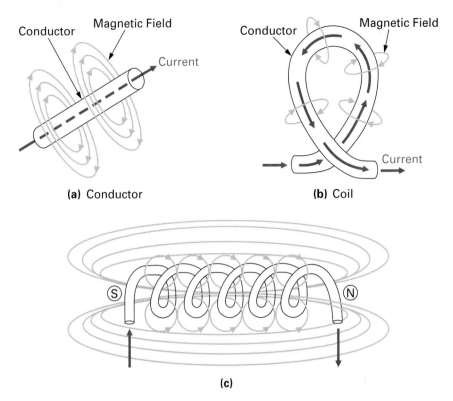

FIGURE 15-6 **Conductor and Coil Electromagnetism.**

therefore generates a magnetic field in a coil of fixed polarity, as shown in Figure 15-7(a), and as determined by the left-hand rule.

Alternating current (ac) is continually varying, and as the polarity of the magnetic field is dependent on the direction of current flow (left-hand rule), the magnetic field will also be alternating in polarity, as shown in Figure 15-7(b). Let us look at times 1 through 4 in Figure 15-7 in more detail.

Time 1: The alternating voltage has risen to a maximum positive level and causes current flow as seen in the circuit. This will cause a magnetic field with a south pole above and a north pole below.

Time 2: Between positions 1 and 2, the voltage, and therefore current, will decrease from a maximum positive value to zero. This will cause a corresponding collapse of the magnetic field from maximum (time 1) to zero (time 2).

Time 3: Voltage and consequently current increase from zero to maximum negative between positions 2 and 3. The increase in current flow causes a similar increase or buildup of magnetic flux, producing a north pole above and south pole below.

Time 4: From time 3 to time 4, the current within the circuit diminishes to zero, and the magnetic field once again collapses. The cycle then repeats.

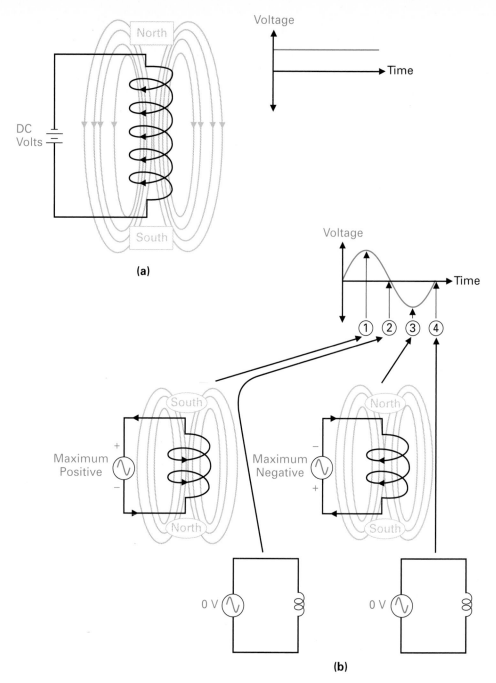

FIGURE 15-7 **Electromagnets. (a) DC Electromagnet. (b) AC Electromagnet.**

In summary:

1. Direct current (dc) produces a constant magnetic field of a fixed polarity, for example, north–south.

2. An alternating current (ac) produces an alternating magnetic field, which continuously switches polarity, for example, north–south, south–north, north–south, and so on.

1. What name could be used instead of *electromagnet?*

2. True or false: The greater the number of loops in an electromagnet, the greater or stronger the magnetic field.

3. What form of current flow produces a magnetic field that maintains a fixed magnetic polarity?

4. True or false: A magnetic field is produced only when ac is passed through a conductor.

5. Describe how the left-hand rule applies to electromagnetism.

6. What form of current flow produces a magnetic field that continuously switches in polarity?

15-1-5 *Magnetic Terms*

Magnetic Flux (ϕ) and Flux Density (B)

One magnetic line of flux or force is called a **maxwell,** in honor of James Maxwell's work in this field. However, this unit is too small and impractical and therefore the amount of **magnetic flux** (symbolized ϕ, phi) is measured in webers instead of maxwells. One *weber* is equal to 10^8 (100,000,000) magnetic lines of force or maxwells.

Maxwell
One magnetic line of force or flux is called a maxwell.

Magnetic Flux
The magnetic lines of force produced by a magnet.

> magnetic flux (ϕ) = number of lines of force (or maxwells),
> in webers

If 10^8 lines of force exist in 1 square meter, and 10^8 lines of force exist in 1 square centimeter, which is the stronger magnetic field? As far as magnetic flux (ϕ) is concerned, 1 weber exists in both. So some other way is needed to specify how many lines of force exist in a given area, and this will tell us if it is a strong or weak magnetic field.

Flux density (B) is equal to the number of magnetic lines of flux (ϕ) per square meter, and it is given in the unit of tesla (T).

Flux Density
A measure of the strength of a wave.

$$\text{flux density } (B) = \frac{\text{magnetic flux } (\phi)}{\text{area } (A)}$$

where B = flux density, in teslas (T)

ϕ = Magnetic flux, in webers (Wb)

A = area, in square meters (m^2)

EXAMPLE:

If magnet A produces 10^8 (100,000,000) lines of flux (1 weber) in 1 square centimeter, and magnet B produces 10^8 lines of flux in 1 square meter, which magnet is producing the more concentrated or intense magnetic field?

▨ *Solution:*

Since there are 10,000 square centimeters in 1 square meter, 1 square centimeter (1 cm^2) equals one ten-thousandth of a square meter (0.01 m^2).

$$\text{Magnet } A: \text{flux density} = \frac{1 \text{ weber}}{0.0001 \text{ (m}^2)}$$

$$= 10{,}000 \text{ tesla}$$

$$\text{Magnet } B: \text{flux density} = \frac{1 \text{ weber}}{1 \text{ (m}^2)}$$

$$= 1 \text{ tesla (T)}$$

Flux density determined that magnet A produced the more concentrated magnetic force.

Magnetomotive Force (MMF)

The magnetic flux produced by an electromagnet is produced by current flowing through a coil of wire. As discussed previously, electromotive force (emf) is the pressure or voltage that forces electrons to move. **Magnetomotive force** (mmf) is the magnetic pressure that produces the magnetic field. The formula for mmf is

Magnetomotive Force
Force that produces a magnetic field.

$$\boxed{\text{mmf} = I \times N \text{ (ampere-turns)}}$$

where mmf = magnetomotive force, in ampere-turns (At)

I = current, in amperes

N = number of turns in the coil

The formula basically says that if you increase the current through the coil or increase the number of turns in the coil, you will increase the magnetic pressure (magnetomotive force), and by increasing magnetic pressure, you will increase the magnetic field produced, as shown in Figure 15-8.

EXAMPLE:

What is the magnetomotive force produced when 3 A of current flows through 5 turns in a coil?

▨ *Solution:*

$$\text{mmf} = I \text{ (current)} \times N \text{ (number of turns)}$$

$$= 3A \times 5 \text{ turns}$$

$$= 15 \text{ At (ampere-turns)}$$

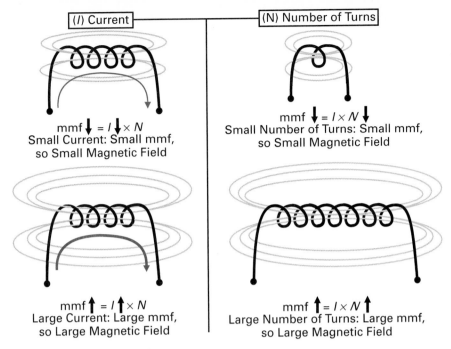

FIGURE 15-8 **Magnetomotive Force.**

Magnetizing Force (*H*)

Example 15-2 showed that magnetomotive force (mmf) is equal to the product of current and the number of turns; however, if a 20-turn coil or solenoid is stretched out to twice its length, the magnetic field or force will be half as strong, because there would be less of a reinforcing effect between the coils due to the greater distance between the coils. The length of the coil is consequently a factor that also determines the field intensity, and this new term, **magnetizing force** (*H*), is the reference we commonly use to describe the magnetic field intensity.

Magnetizing Force
Also called magnetic field strength, it is the magnetomotive force per unit length at any given point in a magnetic circuit.

$$\text{magnetizing force } (H) = \frac{I \times N}{l}$$

or

$$H = \frac{\text{mmf}}{l}$$

where H = magnetizing force in ampere-turns/meter (At/m)

$\quad I$ = current, in amperes (A)

$\quad N$ = number of coil turns

$\quad l$ = length of coil, in meters (m)

Reluctance (ℜ)

Reluctance
Resistance to the flow of magnetic lines of force.

Reluctance, in reference to magnetic energy, is equivalent to resistance in electrical energy. Reluctance is the opposition or resistance to the establishment of a magnetic field in an electromagnet. The formula for calculating reluctance is the magnetic energy equivalent of the electrical Ohm's law as shown:

Magnetic

$$\text{reluctance } (\mathfrak{R}) = \frac{\text{mmf (magnetomotive force)}}{\phi \text{ (magnetic flux)}}$$

(measured in ampere-turns/weber)

Electrical

$$\text{resistance } (R) = \frac{V \text{ (electromotive force)}}{I \text{ (current)}}$$

(measured in ohms)

Permeability (μ)

Permeability
Measure of how much better a material is as a path for magnetic lines of force with respect to air, which has a permeability of 1 (symbolized by the Greek lowercase letter mu, μ).

As magnetic reluctance is equivalent to electrical resistance, magnetic permeability is equivalent to electrical conductance. **Permeability** is a measure of how easily a material will allow a magnetic field to be set up within it. Permeability is symbolized by the Greek lowercase letter mu (μ) and is measured in henrys per meter (H/m). A high permeability figure (μ↑) indicates that a magnetic field can easily be established within a material and therefore that this material's reluctance must be low (ℜ↓). On the other hand, a low permeability figure (μ↓) indicates that there will be a large reluctance (ℜ↑) to establishing a magnetic field. This is mathematically stated as:

Magnetic

$$\text{permeability } (\mu) = \frac{1}{\mathfrak{R} \text{ (reluctance)}}$$

(Permeability is inversely proportional to reluctance and is measured in henrys per meter.)

Electric

$$\text{conductance } (G) = \frac{1}{R \text{ (resistance)}}$$

(Conductance is inversely proportional to resistance and is measured in siemens.)

TABLE 15-1   Permeabilities of Various Materials

MATERIAL	RELATIVE PERMEABILITY (μ_r)	PERMEABILITY (μ)
Air or vacuum	1	1.26×10^{-6}
Nickel	50	6.28×10^{-5}
Cobalt	60	7.56×10^{-5}
Cast iron	90	1.1×10^{-4}
Machine steel	450	5.65×10^{-4}
Transformer iron	5,500	6.9×10^{-3}
Silicon iron	7,000	8.8×10^{-3}
Permaloy	100,000	0.126
Supermalloy	1,000,000	1.26

The permeability figures in henrys/meter for different materials are listed in Table 15-1. **Relative** (with respect to) is a word that means that a comparison has to be made. Relative permeability (μ_r) is the measure of how well another given material will conduct magnetic lines of force with respect to, or relative to, our reference material, air, which has a relative permeability value of 1. Referring to the relative permeability column in Table 15-1, you can see that magnetic lines of flux will pass through nickel 50 times easier than through air. The relative permeability (μ_r) of air, which is equal to 1, should not be confused with the absolute permeability (μ_0) of free space or air, which is equal to $4\pi \times 10^{-7}$ or 1.26×10^{-6}.

Relative
Not independent; compared or with respect to some other value of a measured quantity.

permeability (μ) = relative permeability (μ_r)
$\times$ absolute permeability of air (μ_0)

Whether permeability (μ), relative permeability (μ_r), or absolute permeability (μ_0) is used as a standard makes little difference. A high permeability figure indicates a low reluctance ($\mu\uparrow$, $\Re\downarrow$), and vice versa.

EXAMPLE:

If the magnetic flux produced by a material is equal to 335 μWb and the mmf equals 15 At:

a. What is the reluctance of the material?

b. What is the material's permeability?

CALCULATOR SEQUENCE

Step	Keypad Entry	Display Response
1.	① ⑤	15
2.	÷	
3.	③ ③ ⑤ E ⑥ +/−	335E-6
4.	=	44776.1
5.	1/x	22.3E-6

a. Reluctance $(\Re) = \dfrac{\text{mmf (magnetomotive force)}}{\phi \text{ (magnetic flux)}}$

$\quad = \dfrac{15 \text{ At (ampere-turns)}}{335 \text{ } \mu\text{Wb}}$

$\quad = 44.8 \times 10^3 \text{ At/Wb}$

b. Permeability $(\mu) = \dfrac{1}{44.8 \times 10^3 \text{ (reluctance)}}$

$\quad = 22.3 \times 10^{-6} \text{ henrys/meter (H/m)}$

Summary

The magnetic field strength of an electromagnet can be increased by increasing the number of turns in its coil, increasing the current through the electromagnet, or decreasing the length of the coil ($H = I \times N/l$). The field strength can be increased further by placing an iron core within the electromagnet, as illustrated in Figure 15-9(a). An iron core has less reluctance or opposition to the magnetic lines of force than air, so the flux density (B) is increased. Another way of saying this is that the permeability or conductance of magnetic lines of force within iron is greater than that of air, and if the permeability of iron is large, its reluctance must be small. The symbol for an iron core electromagnet is seen in Figure 15-9(b).

15-1-6 Flux Density (B) versus Magnetizing Force (H)

B–H Curve
Curve plotted on a graph to show successive states during magnetization of a ferromagnetic material.

The **B–H curve** in Figure 15-10 illustrates the relationship between the two most important magnetic properties: flux density (B) and magnetizing force (H). Figure 15-10(a) illustrates the B–H curve, while Figure 15-10(b) illustrates the positive rising portion of the ac current that is being applied to the iron core electromagnetic circuit in Figure 15-10(c).

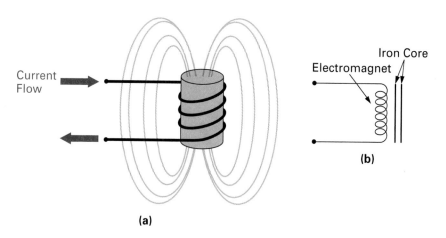

Current Flow

Iron Core

Electromagnet

(b)

(a)

FIGURE 15-9 Iron-Core Electromagnet. (a) Operation. (b) Symbol.

The magnetizing force is actually equal to $H = I$ (current) $\times N$ (number of turns)/l (length of coil); but since the number of turns and length of coil are fixed for the coil being used, the magnetizing force (H) is proportional to the current (I) applied, which is shown in Figure 15-10(b). The positive rise of the current from zero to maximum positive is applied through the electromagnet and will produce a corresponding bloom or buildup in magnetic flux at a rate indicated by the B–H curve shape in Figure 15-10(a).

It is important to note that as the magnetizing force (current) is increased, there are three distinct stages in the change of flux density or magnetic flux out.

Stage 1: Up to this point, the increase in flux density is slow, as a large amount of force is required to begin alignment of the molecule magnets.

Stage 2: Increase in flux density is now rapid and almost linear as the molecule magnets are aligning easily.

Stage 3: In this state they cannot be magnetized any further because all the molecule magnets are fully aligned and no more flux density can be easily obtained. This is called the **saturation point,** and it is the state of magnetism beyond which an electromagnet is incapable of further magnetic strength, that is, the point beyond which the B–H curve is a straight, horizontal line, indicating no change.

Saturation Point
The point beyond which an increase in one of two quantities produces no increase in the other.

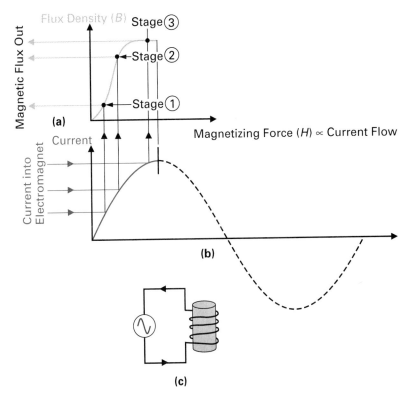

FIGURE 15-10 **(a) Flux Density versus Magnetizing Force Curve (*B*–*H* Curve). (b) Current Applied to Circuit. (c) Circuit.**

Saturation can easily be described by a simple analogy of a sponge. A dry sponge can only soak up a certain amount of water. As it continues to absorb water, a point will be reached where it will have soaked up the maximum amount of water possible. At this point, the sponge is said to be saturated with water, and no matter how much extra water you supply, it cannot hold any more.

The electromagnet is saturated at stage 3 and cannot produce any more magnetic flux, even though more magnetizing force is supplied as the sine wave continues on to its maximum positive level.

Hysteresis
A lag between cause and effect. With magnetism it is the amount that the magnetization of a material lags the magnetizing force due to molecular friction.

Looking at these three stages and the *B–H* curve that is produced, you can see that, in fact, the magnetization (setting up of the magnetic field, *B*) lags the magnetizing force (*H*) because of molecular friction. This lag or time difference between magnetizing force (*H*) and flux density (*B*) is known as **hysteresis.**

Figure 15-11(a) illustrates what is known as a *hysteresis loop,* which is formed when you plot magnetizing force (*H*) against flux density (*B*) through a complete cycle of alternating current, as seen in Figure 15-11(b). Initially, when the electric circuit switch is open, the iron core is unmagnetized; therefore, both

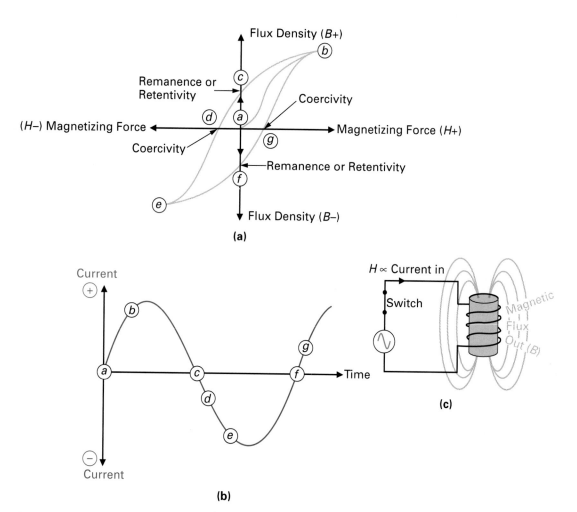

FIGURE 15-11   *B–H* **Hysteresis Loop.**

H and *B* are zero at point *a*. When the switch is closed, as seen in Figure 15-11(c), the current [Figure 15-11(b)] is increased and flux density [Figure 15-11(a)] increases until saturation point *b* is reached. This part of the waveform (from point *a* to *b*) is exactly the same as the *B–H* curve, discussed previously. The current continues on beyond saturation point *b*; however, the flux density cannot increase beyond saturation.

At point *c*, the magnetizing force (current) is zero and *B* (flux density) falls to a value *c* that is the positive magnetic flux remaining after the removal of the magnetizing force (*H*). This particular value of flux density is termed **remanence or retentivity.**

The current or magnetizing force now reverses, and the amount of current in the reverse direction that causes flux density to be brought down from *c* to zero (point *d*) after the core has been saturated is termed the **coercive force.** The current, and therefore magnetizing force, continues on toward a maximum negative until saturation in the opposite magnetic polarity occurs at point *e*.

At point *f,* the magnetizing force (current) is zero and *B* falls to remanence, which is the negative magnetic flux remaining after the removal of the magnetizing force *H*. The value of current between *f* and *g* is the coercive force needed in the reverse direction to bring flux density down to zero (point *g*).

In Figure 15-12 it is easier to see how the current variation corresponds to the magnetizing force variation.

Remanence or Retentivity
Amount a material remains magnetized after the magnetizing force has been removed.

Coercive Force
Magnetizing force needed to reduce the residual magnetism within a material to zero.

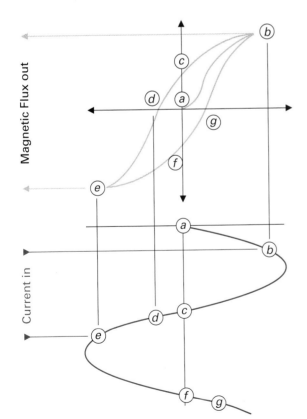

FIGURE 15-12   AC Current In, Magnetic Flux Out.

Define the following:

1. Magnetic flux

2. Flux density

3. Magnetomotive force

4. Magnetizing force

5. Reluctance

6. Permeability

7. True or false: The hysteresis loop is formed when you plot H against B through a complete cycle of ac.

8. What is magnetic saturation?

15-1-7 *Applications of Electromagnetism*

Magnetic-Type Circuit Breaker

The magnetic-type circuit breaker shown in Figure 15-13 was first discussed in Chapter 6 and is one application of an electromagnet. This and all other circuit breakers are used for current protection. If the rated current value or below is passing through the circuit breaker, the electromagnet will not generate a strong enough magnetic field to pull the iron arm to the left and release the catch so the horizontal arm will open the contacts. However, if the current rating of the circuit breaker is exceeded, the increase in current will cause a corresponding increase in magnetic flux, attracting the iron arm, opening the contacts, and protecting the equipment from the dangerous level of current.

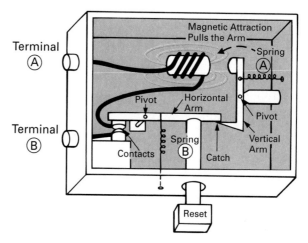

FIGURE 15-13   Magnetic-Type Circuit Breaker.

Electric Bell

The electric bell in Figure 15-14 utilizes a soft-iron core electromagnet, a striker, and a gong. When the bell push is pressed, a complete electrical circuit is made from the positive terminal to the electromagnet, the contact, the striker, and then back to the negative terminal. The electromagnet energizes (produces a magnetic field) and attracts the soft-iron striker, which strikes the gong and at the same time breaks the circuit, deenergizing the electromagnet. The striker is pulled back by the spring, and the circuit is reestablished, resulting in a continuous ringing of the bell.

Relay

A **relay** is an electromechanical device that either makes (closes) or breaks (opens) a circuit by moving contacts together or apart. Figure 15-15 illustrates the relay; Figure 15-15(a) shows the normally open (NO) relay and Figure 15-15(b) shows the normally closed (NC) relay.

Operation. In both cases, the relay consists basically of an electromagnet connected to lines x and y, a movable iron arm known as the armature, and some contacts. When current passes through x to y, the electromagnet generates a magnetic field, or is said to be **energized,** which attracts the armature toward the electromagnet. When this occurs, it closes or makes the normally open relay contacts and opens or breaks the normally closed relay contacts.

If the electromagnet is deenergized by stopping the current through the coil, the spring will pull back the armature to open the NO relay contacts or close the NC relay contacts between A and B.

Relay
Electromechanical device that opens or closes contacts when a current is passed through a coil.

Energized
Being electrically connected to a voltage source so that the device is activated.

FIGURE 15-14   Electric Bell.

FIGURE 15-15 relays (a) Single-Pole, Single-Throw (SPST), Normally Open (NO) Relay (Contacts Are Open until Activated). (b) Single-Pole, Single-Throw (SPST), Normally Closed (NC) Relay (Contacts Are Closed until Activated).

To summarize, then, you can see that the "normal" condition for the contacts between *A* and *B* is when the electromagnet is deenergized. In the deenergized condition, the normally open relay contacts are open and the normally closed relay contacts are closed.

The two relays discussed so far are actually single-pole, single-throw relays, as they have one movable contact (single pole) and one stationary contact that the pole can be thrown to, as shown in Figure 15-15. There are actually four basic configurations for relays and all are illustrated in Table 15-2. Variations on these basic four can come in all shapes and sizes, with one relay controlling sometimes several sets of contacts. Figure 15-16 illustrates several different styles and packages of relays that are available.

Applications of a relay. The relay is generally used in two basic applications:

1. To enable one master switch to operate several remote or difficultly placed contact switches, as illustrated in Figure 15-17. When the master switch is closed, the relay is energized, closing all its contacts and turning on all the lights. The advantage of this is twofold in that, first, the master switch can turn on three lights at one time, which saves time for the operator, and second, only one set of wires need be taken from the master switch to the lights, rather than three sets for all three lights.

TABLE 15-2 Relay Types

SPST Relay

Relay Type	De-energized	Energized
NO	A to B Open	A to B Closed
NC	A to B Closed	A to B Open

Normally Open (NO) Normally Closed (NC)

(a) Single-Pole, Single-Throw (SPST)

A relay with only one moving and one stationary contact. Available as either normally open (NO) or normally closed (NC).

Single pole: one moving contact.
Double pole: two moving contacts.
Single throw: pole can be thrown or cast in only one direction.
Double throw: pole can be thrown or cast in one of two directions.

SPDT Relay

Contacts	De-energized	Energized
A to B	Closed	Open
B to C	Open	Closed

(b) Single-Pole, Double-Throw (SPDT)

A relay with one moving and two stationary contacts. One set of contacts is normally closed (A and B), while the other set is normally open (B and C).

2 Normally Open   2 NNormally Closed   1 Normally Open/1 Normally Closed

(c) Double-Pole, Single-Throw (DPST)

A relay with two moving poles or armatures and two stationary contacts. Available as either both sets normally closed, both normally open, or one set normally open and the other set normally closed.

(d) Double-Pole, Double-Throw (DPDT)

A relay with two moving contacts and four stationary contacts. Two sets of contacts are normally closed (A and B/C and D), while the other two sets are normally open (B and D/E and F).

(a)

(b)

(c)

(d)

(e)

(f)

(g)

(h)

FIGURE 15-16   Relay Styles and Packages.

2. The second basic application of the relay is to enable a switch in a low-voltage circuit to operate relay contacts in a high-voltage circuit, as shown in Figure 15-18. The operator activates the switch in the low-voltage circuit, which will energize the relay, closing its contacts and connecting the high voltage to the motor. Figure 15-19 illustrates the circuit arrangement that would have to be used without a relay.

In a high-voltage circuit, a large current will be flowing, so a large or heavy-gauge wire will be needed to handle this current. With the relay circuit in Figure 15-18, large-gauge wire is needed only for the short distance between the relay and the motor, while in the circuit without the relay (Figure 15-19), heavy-gauge wire, which is more expensive than smaller-gauge wire, is needed for the long distance between the operator's switch and the motor.

An automobile starter circuit is illustrated in Figure 15-20. In this application of a relay, we will see how a relay can be used to supply the large dc current needed to activate a starter motor of an automobile.

Master Switch

Triple-Pole, Single-Throw
(3PST) NO

(a)

Bulb ①
Bulb ②
Bulb ③

3 P S T
Relay

(b)

FIGURE 15-17 **One Master Switch Operating Several Remotes.**
(a) Schematic. (b) Pictorial.

When the ignition switch is engaged in the passenger compartment by the driver, current flows through a light-gauge wire from the negative side of the battery, through the relay's electromagnet, through the ignition switch, and back to the positive side of the battery. This current flow through the electromagnet of the relay energizes the relay and closes the relay's contacts. Closing the relay's contacts makes a path for the current to flow through the heavy-gauge cable from the negative side of the battery, through the relay contacts and starter motor, and back to the positive side of the battery. The starter motor's output shaft spins the engine, causing it to start.

This application is a perfect example of how a relay can be used to close contacts in a heavy-current (heavy-gauge cable) circuit, while the driver has only to close contacts in a small-current (light-gauge cable) circuit.

If the relay were omitted, the driver's ignition switch would have to be used to connect the 12 V and large current to the starter motor. This would require that:

1. Heavy-gauge expensive cable be connected in a longer path between starter motor and passenger compartment.

2. The ignition switch would need to be larger to handle the heavier current.

(a)

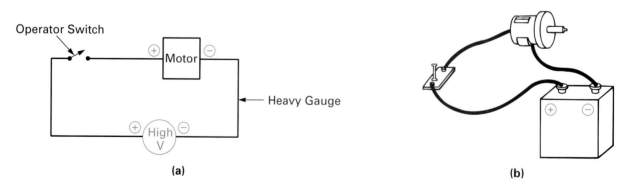

(b)

FIGURE 15-18   Low-Current Switch Enabling a High-Current Circuit. (a) Schematic. (b) Pictorial.

Reed Relay or Switch

Relay that consists of two thin magnetic strips within a glass envelope with a coil wrapped around the envelope so that when it is energized the relay's contacts or strips will snap together, making a connection between the two leads attached to each of the reed strips.

Reed relays and switches.   The **reed relay or switch** consists of two flat magnetic strips mounted inside a capsule, which is normally made of glass. The reed relay and reed switch are illustrated in Figure 15-21(a) and (b). The reed relay differs from the reed switch in that a reed relay has its own energizing coil, while the reed switch needs an external magnetic force to operate.

The reed relay and reed switch are both operated by a magnetic force that is provided by a coil for the reed relay and by a separate permanent magnet for the reed switch. When the magnetic force is present, opposite magnetic polarities are induced in the overlapping, high-permeability reed blades, causing them

(a)

(b)

FIGURE 15-19   Without Relay (More Expensive).

(a)

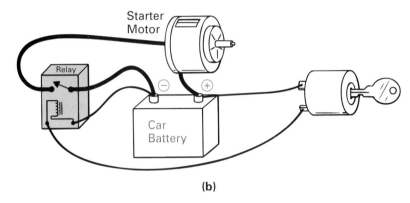

(b)

FIGURE 15-20   Automobile Starting Circuit.

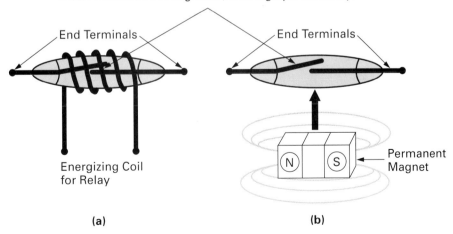

(a)   (b)

**FIGURE 15-21   Reed Relay and Switch. (a) Reed Relay: Contacts Close
When Electromagnetic Coil Is Energized. (b) Reed Switch: Contacts Close
When External Magnetic Field Comes in Close Proximity, Due to Induced
Magnetism in the Contacts.**

FIGURE 15-22   **Reed Switch as Part of a Home Security System.**

to attract one another by induced magnetism and snap together, thus closing the circuit. When the magnetic force is removed, the blades spring apart due to spring tension and open the circuit.

Referring to Figure 15-22 you will see a simple home security circuit. When the window is closed in the normal condition, the permanent magnet is directly adjacent to the reed switch, and therefore its contacts are closed, allowing current to flow from the negative side of the small 3-V battery, through the closed reed switch contacts, the relay's electromagnet, and back to the positive side of the 3-V battery. The relay is a normally closed (NC) SPST; but as current is flowing through the coil of the relay, the contacts are open, preventing the large positive and negative voltage from the 12-V battery from reaching the siren.

If the window is forced and opened by an intruder, the permanent magnet will no longer be in close proximity to the reed switch and the reed switch's contacts will open. When the reed switch's contacts open, the relay's coil will no longer have current flowing through it, and therefore the relay will deenergize and its contacts will return to their normal closed condition. A large current is now permitted to flow from the negative side of the large 12-V battery, through the relay's contacts, the siren, and back to the positive side of the battery. The siren will sound, as it now has 12 V applied to it, and alert the occupant of the home.

Solenoid-Type Electromagnet

Up to now, electromagnets have been used to close or open a set, or sets, of contacts to either make or break a current path. These electromagnets have used a stationary soft-iron core; however, some electromagnets are constructed with movable iron cores, as shown in Figure 15-23, which can be used to open or block the passage of a gas or liquid through a valve. These are known as **solenoid**-type electromagnets.

Solenoid
Coil and movable iron core that when energized by an alternating or direct current will pull the core into a central position.

When no current is flowing through the solenoid coil, no magnetic field is generated, so no magnetic force is exerted on the movable iron core, as shown in Figure 15-23(a), and therefore the compression spring maintains it in the up position, with the valve plug on the end of the core preventing the passage of either a liquid or gas through the valve (valve closed).

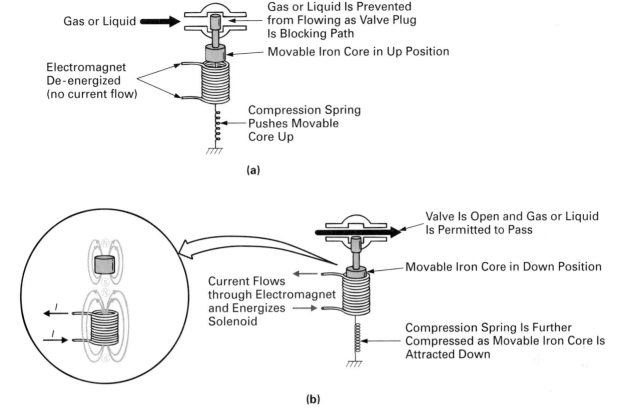

FIGURE 15-23 Solenoid-Type Electromagnet. (a) Deenergized. (b) Energized.

When a current flows through the electromagnet, the solenoid coil is energized, creating a magnetic field, as shown in Figure 15-23(b). Due to the influence of the coil's magnetic field, the movable soft-iron core will itself generate a magnetic field, as seen in the insert in Figure 15-23(b). This condition will create a north pole at the top of the solenoid coil and a south pole at the bottom of the movable core, and the resulting attraction will pull down the core (which is free to slide up and down), pulling with it the valve plug and opening the valve.

Solenoid-type electromagnets are actually constructed with the core partially in the coil and are used in washing machines to control water and in furnaces to control gas.

SELF-TEST REVIEW QUESTIONS FOR SECTION 15-1-7

1. List two applications of the electromagnet.
2. What is the difference between an NO and an NC relay?
3. What is the difference between a reed relay and a reed switch?
4. Give an application for a conventional relay and one for the reed switch.

Electromagnetic induction is the name given to the action that causes electrons to flow within a conductor when that conductor is moved through a magnetic field. Stated another way, electromagnetic induction is the voltage or emf induced or produced in a coil as the magnetic lines of force link with the turns of a coil. Since this phenomenon was first discovered by Michael Faraday, let us begin by studying **Faraday's Law.**

15-2-1   *Faraday's Law*

In 1831, Michael Faraday carried out an experiment including the use of a coil, a zero center ammeter (galvanometer), and a bar permanent magnet, as shown in Figure 15-24. Faraday discovered that [Figure 15-25(a) through (f)]:

a. When the magnet is moved into a coil the magnetic lines of flux cut the turns of the coil. This action that occurs when the magnetic lines of flux link with a conductor is known as *flux linkage*. Whenever flux linkage occurs, an emf is induced in the coil known as an induced voltage, which causes current to flow within the circuit and the meter to deflect in one direction, for example, to the right. Faraday discovered, in fact, that if the magnet was moved into the coil, or if the coil was moved over the magnet, an emf or voltage was induced within the coil. What actually occurs is that the electrons within the coil are pushed to one end of the coil by the magnetic field, creating an abundance of electrons at one end of the coil (negative charge) and an absence of electrons at the other end of the coil (positive charge). This potential difference across the coil will produce a current flow if a complete path for current (closed circuit) exists.

b. When the magnet is stationary within the coil, the magnetic lines are no longer cutting the turns of the coil, and so there is no induced voltage and the meter returns to zero.

FIGURE 15-24   Faraday's Electromagnetic Induction Experiment Components.

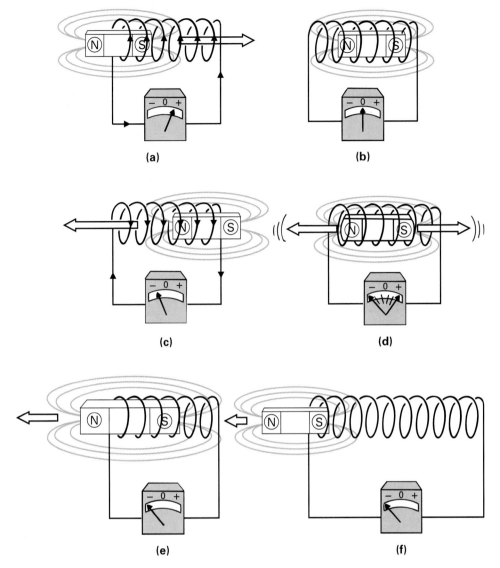

FIGURE 15-25 **Faraday's Electromagnetic Induction Discoveries.**

c. When the magnet is pulled back out of the coil, a voltage is induced that causes current to flow in the opposite direction to that of (a) and the meter deflects in the opposite direction, for example, to the left.

d. If the magnet is moved into or out of the coil at a greater speed, the voltage induced also increases, and therefore so does current.

e. If the size of the magnet and therefore the magnetic flux strength are increased, the induced voltage also increases.

f. If the number of turns in the coil is increased, the induced voltage also increases.

In summary, whenever there is relative motion or movement between the coil of a conductor and the magnetic lines of flux, a potential difference will be

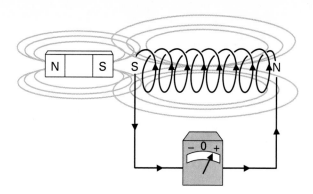

FIGURE 15-26 Lenz's Law.

induced and this action is called *electromagnetic induction*. The magnitude of the induced voltage depends on the number of turns in the coil, rate of change of flux linkage, and flux density.

15-2-2 Lenz's Law

Lenz's Law
The current induced in a circuit due to a change in the magnetic field is so directed as to oppose the change in flux, or to exert a mechanical force opposing the motion.

About the same time a German physicist, Heinrich Lenz, performed a similar experiment along the same lines as Faraday. His law states that the current induced in a coil due to the change in the magnetic flux is such as to oppose the cause producing it.

To explain this law further, refer to Figure 15-26. When the magnet moves into the coil, a voltage is induced such that the current flows in the coil and produces a pole at the face of the coil (left-hand rule), which opposes the entry of the magnet. In the example in Figure 15-26, we can see that as the permanent magnet moves into the coil, its magnetic lines of flux cut the turns of the coil and induce a voltage (electromagnetic induction), which causes current to flow in the coil as indicated by the meter movement. If you apply the left-hand rule to the coil, you can see that the current flow within the coil has produced a south pole on the left-hand side of the coil (electromagnetism) to oppose the entry or motion of the magnet that is producing the current.

15-2-3 The Weber

Weber (Wb)
Unit of magnetic flux. One weber is the amount of flux that when linked with a single turn of wire for an interval of 1 second, will induce an electromotive force of 1 V.

Consideration of Faraday's law enables us to take a closer look at the unit of flux (ϕ). The **weber** is equal to 10^8 magnetic lines of force, and from the electromagnetic induction point of view, if 1 weber of magnetic flux cuts a conductor for a period of 1 second, a voltage of 1 volt will be induced.

15-2-4 Applications of Electromagnetic Induction

Generator
Device used to convert a mechanical energy input into an electrical energy output.

AC Generator

The ac **generator** or alternator is an example of a device that uses electromagnetic induction to generate electricity. If you stroll around your city or town during the day or night and try to spot every piece of equipment, appliance, or device that is running from the ac electricity supplied by generators from the electric power plant, it begins to make you realize how inactive, dark, and difficult our modern society might become without ac power.

When discussing Faraday's discoveries of electromagnetic induction in Figure 15-25, the coil or conductor remained stationary and the magnet was moved. It can be operated in the opposite manner so that the magnetic field remains stationary and the coil or conductor is moved. As long as the magnetic lines of force have relative motion with respect to the conductor, an emf will be induced in the coil. Figure 15-27(a) illustrates a piece of wire wound to form a coil and attached to a galvanometer; its needle rests in the center position, indicating zero current. If the conductor remains stationary within the magnetic lines of flux being generated by the permanent magnet, there is no emf or voltage induced into the wire and so no current flow through the circuit and meter.

If the conductor is moved past the permanent magnet so that it cuts the magnetic lines of flux, as seen in Figure 15-27(b), an emf is generated within the conductor, which is known as an induced voltage, and this will cause current to flow through the wire in one direction and be indicated on the meter by the deflection of the needle to the left.

If the conductor is moved in the opposite direction past the magnetic field, it will induce a voltage of the opposite polarity and cause the meter to deflect to the right, as shown in Figure 15-27(c).

The value or amount of induced voltage is indicated by how far the meter deflects and this voltage is dependent on three factors:

1. The speed at which the conductor passes through the magnetic field
2. The strength or flux density of the magnetic field
3. The number of turns in the coil

FIGURE 15-27 (a) Stationary Conductor, No Induced Voltage. (b) Moving Conductor, Induced Voltage. (c) Moving Conductor, Induced Voltage.

If the speed at which the conductor passes through the magnetic field or the strength of the magnetic field or the number of turns of the coil is increased, then the induced voltage will also increase. This is merely a repetition of Faraday's law, but in this case we moved the conductor instead of the magnetic field; however, the results were the same.

Basic generator. The basic generator action makes use of Faraday's discoveries of electromagnetic induction. In Chapter 11 the physical appearance of a large 700,000-kW power plant generator was shown. Figure 15-28 illustrates some smaller generators that are mobile and can therefore be used in remote locations.

Figure 15-29 illustrates the basic generator's construction. The basic principle of operation is that the mechanical drive energy input will produce ac electrical energy out by means of electromagnetic induction.

Armature
Rotating or moving component of a magnetic circuit.

A loop of conductor, known as an **armature,** is rotated continually through 360° by a mechanical drive. This armature resides within a magnetic field produced by a dc electromagnet. Voltage will be induced into the armature and will appear on slip rings, which are also being rotated. A set of stationary brushes rides on the rotating slip rings and picks off the generated voltage and applies this voltage across the load. This voltage will cause current to flow within the circuit and be indicated by the zero center ammeter.

Let's now take a closer look at the armature as it sweeps through 360°, or one complete revolution. Figure 15-30 illustrates four positions of the armature as it rotates through 360° in the clockwise direction.

Position 1: At this instant, the armature is in a position such that it does not cut any magnetic lines of force. The induced voltage in the armature conductor is equal to 0 V and there is no current flow through the circuit.

(a)

(b)

FIGURE 15-28   AC Generator. (a) 250-kW Diesel Generator. (b) Small (2 kW) Camping Generator.

FIGURE 15-29   Basic Generator Construction.

Position 2: As the conducting armature moves from position 1 to position 2, you can see that more and more magnetic lines of flux will be cut, and the induced emf in the armature (being coupled off by the brushes from the slip rings) will also increase to a maximum value. The current flow throughout the circuit will rise to a maximum as the voltage increases, and this can be seen by the zero center ammeter deflection to the right. From the ac waveform in Figure 15-30, you can see the sinusoidal increase from zero to a maximum positive as the armature is rotated from 0 to 90°.

Position 3: The armature continues its rotation from 90 to 180°, cutting through a maximum quantity and then fewer magnetic lines of force. The induced voltage decreases from the maximum positive at 90° to 0 V at 180°. At the 180° position, as with the 0° position, the armature is once again perpendicular to the magnetic field, and so no lines are cut and the induced voltage is equal to 0 V.

Position 4: From 180 to 270°, the armature is still moving in a clockwise direction, and as it travels toward 270°, it cuts more and more magnetic lines of force. The direction of the cutting action between 0 and 90° causes a positive induced voltage; and since the cutting position between 180 and 270° is the reverse, a negative induced voltage will result in the armature causing current flow in the opposite direction, as indicated by the deflection of the zero center ammeter to the left. The voltage induced when the armature is at 270° will be equal but opposite in polarity to the voltage generated when the armature was at the 90° position. The current will therefore also be equal in value but opposite in its direction of flow.

From position 4 (270°), the armature turns to the 360° or 0° position, which is equivalent to position 1, and the induced voltage decreases from maximum negative to zero. The cycle then repeats.

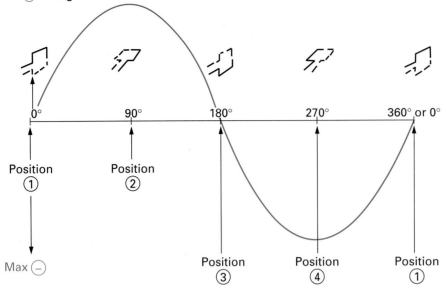

Max ⊕ Voltage and Current Produced

0°   90°   180°   270°   360° or 0°

Position ①   Position ②

Max ⊖

Position ③   Position ④   Position ①

Process Repeats Continuously

FIGURE 15-30   360° Generator Operation.

To summarize, in Figure 15-30, one complete revolution of the mechanical energy input causes one complete cycle of the ac electrical energy output, with a sine wave being generated as a result of circular motion.

Moving Coil Microphone

The moving coil **microphone** is an example of how electromagnetic induction is used to convert information-carrying sound waves to information-carrying electrical waves. Sound is the movement of pressure waves in the air. To create these pressure waves, a device such as a string, reed, or stretched membrane or the human vocal cords must be vibrated to compress and expand the nearby air molecules. Figure 15-31 illustrates a taut string that is vibrating back and forth and generating maximum (*A*) and minimum (*C*) pressure regions.

The frequency or pitch of the sound wave is determined by the number of complete vibrations per second (hertz or cycles/second), while the amplitude or intensity of the sound wave is determined by the amount the string shifts from left to right from its normal position (*B*).

A moving coil microphone converts mechanical sound waves into an electrical replica by use of electromagnetic induction. Figure 15-32 illustrates the physical appearance and construction of a moving coil type of microphone. A coil of wire is suspended in an air gap between magnetic poles and attached to a delicate diaphragm (flexible membrane). A strong magnetic field from a permanent magnet surrounds the coil, and a perforated protecting cover or shield is included to protect the delicate diaphragm.

Sound waves strike the diaphragm, causing it to vibrate back and forth. Since the coil is attached to the diaphragm, it will also be moved back and forth. This movement will cause the coil to cut the magnetic lines of force from the permanent magnet, and a resulting alternating voltage will be induced in the coil (electromagnetic induction). The electrical voltage produces an alternating current, which will have the same waveform shape (and consequently information) as the sound wave that generated it, as seen in Figure 15-32(c). This electrical

Microphone
Electroacoustic transducer that responds and converts a sound wave input into an equivalent electrical wave out.

FIGURE 15-31   Sound Wave.

(a)

Sound Signal Input

Diaphram   Shield

N   S   N

C

Permanent Magnet

Electrical Signal Output

(b)

Sound Wave
(Input)

Variation of
Air Pressure

Electrical Wave
(Output)

Variation of
Voltage and
Current

(c)

FIGURE 15-32   **Moving Coil Microphone. (a) Physical Appearance. (b) Construction. (c) Input/Output Waveforms.**

signal is often referred to as an *analog signal.* Analog is a term meaning "similar to," and as you can see in Figure 15-32(c), the electrical wave output of the microphone is an analog (similar to) of the sound wave input.

This microphone is sometimes called a dynamic or moving (dynamic is the opposite of static) coil microphone since it has a moving coil within it.

SELF-TEST REVIEW QUESTIONS FOR SECTION 15-2

1. Define *electromagnetic induction*.
2. Briefly describe Faraday's and Lenz's laws in relation to electromagnetic induction.
3. What waveform shape does the ac generator produce?
4. Why do you think a microphone is called an electroacoustical transducer?

SUMMARY

1. During a classroom lecture in 1820 Danish physicist Hans Christian Oersted accidentally stumbled on an interesting reaction. As he laid a compass down on a bench he noticed that the compass needle pointed to an adjacent conductor that was carrying a current, instead of pointing to the earth's north pole. It was this discovery that first proved that magnetism and electricity were very closely related to one another.

2. This phenomenon is now called *electromagnetism* since it is now known that any conductor carrying an electro or electrical current will produce a magnetic field.

3. In 1831, the English physicist Michael Faraday explored further Oersted's discovery of electromagnetism and found that the process could be reversed. Faraday observed that if a conductor was passed through a magnetic field, a voltage would be induced in the conductor and cause a current to flow. This phenomenon is referred to as *electromagnetic induction*.

Electromagnetism

4. Every orbiting electron in motion around its nucleus generates a magnetic field. When electrons are forced to leave their parent atom by voltage and flow toward the positive polarity, they are all moving in the same direction and each electron's magnetic field will add to the next. The accumulation of all these electron fields will create the magnetic field around the conductor.

5. The left-hand rule states that if an insulated conductor is held with your left hand so that your thumb is pointing in the direction of electron flow (to the positive potential), your fingers will be pointing in the direction of the magnetic force.

6. If our conductor is wound to form a coil the conductor current will generate its own magnetic field, which will combine additively within the coil, and the net effect of all these conductor magnetic fields can be used to generate a strong coil magnetic field. This coil of conductor which is used to generate a strong magnetic field is called an electromagnet.

7. If a greater number of loops are added, a greater or stronger magnetic field is generated. If the coils were wound in an opposing opposite direction, the opposing magnetic fields would cancel and so no magnetic field would be generated.

8. The left-hand rule can also be applied to electromagnets to determine which of the poles will be the north end of the electromagnet.

9. If a conductor is wound to form a spiral, the conductor, which is now referred to as a *coil,* will, as a result of current flow, develop a magnetic field, which will sum or intensify within the coil. If many coils are wound in the same direction, an electromagnet is formed, which will produce a concentrated magnetic field whenever a current is passed through its coils.

10. A dc voltage produces a fixed current in one direction and therefore generates a magnetic field in a coil of fixed polarity, and as determined by the left-hand rule.

11. Alternating current (ac) is continually varying, and as the polarity of the magnetic field is dependent on the direction of current flow (left-hand rule), the magnetic field will also be alternating in polarity.

Magnetic Terms (Figure 15-33)

12. One magnetic line of flux or force is called a maxwell, in honor of James Maxwell's work in this field. However, this unit is too small and impractical and therefore the amount of magnetic flux (symbolized ϕ, phi) is measured in webers instead of maxwells. One *weber* is equal to 10^8 (100,000,000) magnetic lines of force or maxwells.

13. Flux density (B) is equal to the number of magnetic lines of flux (ϕ) per square meter, and it is given in the unit of telsa (T).

14. The magnetic flux produced by an electromagnet is produced by current flowing through a coil of wire. Magnetomotive force (mmf) is the magnetic pressure that produces the magnetic field.

15. The length of the coil determines the field of intensity, and the new term, magnetizing force (H), is the reference we commonly use to describe the magnetic field intensity.

16. Reluctance, in reference to magnetic energy, is equivalent to resistance in electrical energy. Reluctance is the opposition or resistance to the establishment of a magnetic field in an electromagnet.

17. As magnetic reluctance is equivalent to electrical resistance, magnetic permeability is equivalent to electrical conductance. Permeability is a measure of how easily a material will allow a magnetic field to be set up within it. Permeability is symbolized by the Greek lowercase letter mu (μ) and is measured in henrys per meter (H/m).

18. Relative (with respect to) is a word that means that a comparison has to be made. Relative permeability (μ_r) is the measure of how well another given material will conduct magnetic lines of force with respect to, or relative to, our reference material, air, which has a relative permeability value of 1.

19. The relative permeability (μ_r) of air, which is equal to 1, should not be confused with the absolute permeability (μ_0) of free space or air, which is equal to $4\pi \times 10^{-7}$ or 1.26×10^{-6}.

FIGURE 15-33 Magnetic Terms.

1 magnetic line of force = 1 maxwell
10^8 lines of force or maxwells = 1 weber

$$B = \frac{\phi}{A}$$

B = flux density, in teslas (T)
ϕ = magnetic flux, in webers (Wb)
A = area, in square meters (m^2)

$$mmf = I \times N$$

mmf = magnetomotive force, in ampere turns (At)
I = current, in amperes
N = number of turns in coil

$$H = \frac{I \times N}{l}$$

H = magnetizing force, in ampere turns per meter (At/m)
l = length of coil, in meters

$$\mathfrak{R} = \frac{mmf}{\phi}$$

$\mathfrak{R}$ = reluctance, in ampere turns per weber (At/Wb)

$$\mu = \frac{1}{\mathfrak{R}}$$

μ = permeability, in henrys per meter (H/m)

$$\mu = \mu_r \times \mu_0$$

μ_r = relative permeability
μ_0 = absolute permeability = $4\pi \times 10^{-7}$

Permeabilities of Various Materials

MATERIAL	RELATIVE PERMEABILITY (μ_r)	PERMEABILITY (μ)
Air or vacuum	1	1.26×10^{-6}
Nickel	50	6.28×10^{-5}
Cobalt	60	7.56×10^{-5}
Cast iron	90	1.1×10^{-4}
Transformer iron	5,500	6.9×10^{-3}
Silicon iron	7,000	8.8×10^{-3}
Permalloy	100,000	0.126
Supermalloy	1,000,000	1.26

(a)

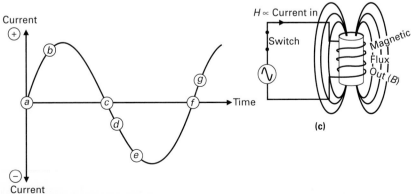

(b)

(c)

20. Whether permeability (μ), relative permeability (μ_r) or absolute permeability (μ_0) is used as a standard makes little difference. A high permeability figure indicates a low reluctance ($\mu \uparrow$, $\Re \downarrow$), and vice versa.

21. The magnetic field strength of an electromagnet can be increased by increasing the number of turns in its coil, increasing the current through the electromagnet, or decreasing the length of the coil ($H = I \times N/l$). The field strength can be increased further by placing an iron core within the electromagnet. An iron core has less reluctance or opposition to the magnetic lines of force than air, so the flux density (B) is increased. Another way of saying this is that the permeability or conductance of magnetic lines of force within iron is greater than that of air, and if the permeability of iron is large, its reluctance must be small.

22. The B–H curve illustrates the relationship between the two most important magnetic properties: flux density (B) and magnetizing force (H).

23. The saturation point is the state of magnetism beyond which an electromagnet is incapable of further magnetic strength, that is, the point beyond which the B–H curve is a straight, horizontal line, indicating no change.

24. Hysteresis is a lag between cause and effect. With magnetism it is the amount that the magnetization of a material lags the magnetizing force due to molecular friction.

25. Remanence or retentivity is the amount a material remains magnetized after the magnetizing force has been removed. Coercive force is the magnetizing force needed to reduce the residual magnetism within a material to zero.

Magnetic-Type Circuit Breaker

26. The magnetic-type circuit breaker is one application of an electromagnet. This and all other circuit breakers are used for current protection. If the rated current value or below is passing through the circuit breaker, the electromagnet will not generate a strong enough magnetic field to pull the iron arm to the left and release the catch so the horizontal arm will open the contacts. However, if the current rating of the circuit breaker is exceeded, the increase in current will cause a corresponding increase in magnetic flux, attracting the iron arm, opening the contacts, and protecting the equipment from the dangerous levels of current.

Electric Bell

27. The electric bell utilizes a soft-iron core electromagnet, a striker, and a gong. When the bell push is pressed, a complete electrical circuit is made from the positive terminal to the electromagnet, the contact, the striker, and then back to the negative terminal. The electromagnet energizes (produces a magnetic field) and attracts the soft-iron striker, which strikes the gong and at the same time breaks the circuit, deenergizing the electromagnet. The striker is pulled back by the spring, and the circuit is reestablished, resulting in a continuous ringing of the bell.

FIGURE 15-34   The Relay.

Relay Types

(a) Single-Pole, Single-Throw (SPST)

A relay with only one moving and one stationary contact. Available as either normally open (NO) or normally closed (NC).

(b) Single-Pole, Double-Throw (SPDT)

A relay with one moving and two stationary contacts. One set of contacts is normally closed (A and B), while the other set is normally open (B and C).

Single pole: one moving contact. *Double pole:* two moving contacts. *Single throw:* pole can be thrown or cast in only one direction. *Double throw:* pole can be thrown or cast in one of two directions.

(c) Double-Pole, Single-Throw (DPST)

A relay with two moving poles or armatures and two stationary contacts. Available as either both sets normally closed, both normally open, or one set normally open and the other set normally closed.

(d) Double-Pole, Double-Throw (DPDT)

A relay with two moving contacts and four stationary contacts. Two sets of contacts are normally closed (A and B/C and D), while the other two sets are normally open (B and D/E and F).

The Relay (Figure 15-34)

28. A relay is an electromechanical device that either makes (closes) or breaks (opens) a circuit by moving contacts together or apart.

29. The relay consists basically of an electromagnet connected to lines *x* and *y*, a movable iron arm known as the armature, and some contacts. When current passes through *x* to *y*, the electromagnet generates a magnetic field, or is said to be energized, which attracts the armature toward the electromagnet. When this occurs, it closes or makes the normally open relay contacts and opens or breaks the normally closed relay contacts.

If the electromagnet is deenergized by stopping the current through the coil, the spring will pull back the armature to open the NO relay contacts or close the NC relay contacts between *A* and *B*.

30. The relay is generally used in two basic applications:
    a. To enable one master switch to operate several remote or difficultly placed contact switches.
    b. To enable a switch in a low-voltage circuit to operate relay contacts in a high-voltage circuit.

31. The reed relay or switch consists of two flat magnetic strips mounted inside a capsule, which is normally made of glass. The reed relay differs from the reed switch in that a reed relay has its own energizing coil, while the reed switch needs an external magnetic force to operate.

Solenoid-Type Electromagnet

32. Some electromagnets are constructed with movable iron cores which can be used to open or block the passage of a gas or liquid through a valve. These are known as solenoid-type electromagnets.

Electromagnetic Induction

33. Electromagnetic induction is the name given to the action that causes electrons to flow within a conductor when that conductor is moved through a magnetic field. Stated another way, electromagnetic induction is the voltage or emf induced or produced in a coil as the magnetic lines of force link with the turns of a coil.

34. In 1831, Michael Faraday carried out an experiment including the use of a coil, a zero center ammeter (galvanometer), and a bar permanent magnet. Faraday discovered that:
    a. When the magnet is moved into a coil the magnetic lines of flux cut the turns of the coil. This action that occurs when the magnetic lines of flux link with a conductor is known as *flux linkage*. Whenever flux linkage occurs, an emf is induced in the coil known as an induced voltage, which causes current to flow within the circuit and the meter to deflect in one direction.
    b. When the magnet is stationary within the coil, the magnetic lines are no longer cutting the turns of the coil, and so there is no induced voltage and the meter returns to zero.
    c. When the magnet is pulled back out of the coil, a voltage is induced that causes current to flow in the opposite direction and the meter deflects in the opposite direction.
    d. If the magnet is moved into or out of the coil at a greater speed, the voltage induced also increases, and therefore so does current.
    e. If the size of the magnet and therefore the magnetic flux strength are increased, the induced voltage also increases.

f. If the number of turns in the coil is increased, the induced voltage also increases.

35. About the same time a German physicist, Heinrich Lenz, performed a similar experiment along the same lines as Faraday. His law states that the current induced in a coil due to the change in the magnetic flux is such as to oppose the cause producing it.

AC Generator (Figure 15-35)

36. The ac generator or alternator is an example of a device that uses electromagnetic induction to generate electricity.

37. The generator's basic principle of operation is that the mechanical drive energy input will produce ac electrical energy out by means of electromagnetic induction.

38. A loop of conductor, known as an armature, is rotated continually through 360° by a mechanical drive. This armature resides within a magnetic field produced by a dc electromagnet. Voltage will be induced into the armature and will appear on slip rings, which are also being rotated. A set of stationary brushes rides on the rotating slip rings and picks off the generated voltage and applies this voltage across the load. This voltage will cause current to flow within the circuit and be indicated by the zero center ammeter.

39. One complete revolution of the mechanical energy input causes one complete cycle of the ac electrical energy output, with a sine wave being generated as a result of circular motion.

Moving Coil Microphone

40. The moving coil microphone is an example of how electromagnetic induction is used to convert information carrying sound waves to information carrying electrical waves.

41. A moving coil microphone converts mechanical sound waves into an electrical replica.

42. The operation of the microphone is as follows. Sound waves strike the diaphragm, causing it to vibrate back and forth. Since the coil is attached to the diaphragm, it will also be moved back and forth. This movement will cause the coil to cut the magnetic lines of force from the permanent magnet, and a resulting alternating voltage will be induced in the coil (electromagnetic induction). The electrical voltage produces an alternating current, which will have the same waveform shape (and consequently information) as the sound wave that generated it. This electrical signal is often referred to as an *analog signal*. Analog is a term meaning "similar to," and the electrical wave output of the microphone is an analog (similar to) of the sound wave input.

43. This microphone is sometimes called a dynamic or moving (dynamic is the opposite of static) coil microphone since it has a moving coil within it.

FIGURE 15-35   The AC Generator.

Max ⊕ Voltage and Current Produced

0°   90°   270°   360° or 0°

Position ①   Position ②

Max ⊖

Position ③   Position ④   Position ①

Process Repeats Continuously

Position ①   Position ②   Position ③   Position ④

0°   90°   180°   270°

Rotary Mechanical Input

Rotating Armature

Electromagnet

Electromagnet

Rotating Slip Rings

Current Meter

Stationary Brushes Ride on Rotating Slip Rings

Load

Absolute permeability (μ_0)
Alternating magnetic field
Ampere turns per meter (At/m)
Ampere turns per weber (At/Wb)
Armature
B–H Curve
Coercive force
Coercivity
Coil
Deenergized
Diaphragm
Electromagnet
Electromagnetic induction
Electromagnetism
Energized
Faraday's law
Flux density (B)
Flux linkage
Galvanometer
Generator
Henrys per meter (H/m)

Hysteresis loop
Left-hand rule
Lenz's law
Magnetic field
Magnetic flux (ϕ)
Magnetizing force (H)
Magnetomotive force (mmf)
Maxwell (Mx)
Moving coil microphone
Permeability (μ)
Reed relay
Reed switch
Relative permeability (μ_r)
Relay
Reluctance
Remanence
Retentivity
Saturation
Solenoid
Tesla
Weber

Multiple-Choice Questions

1. Electromagnetism was first discovered by:
   a. Hans Christian Oersted   c. James Watt
   b. Heinrich Hertz   d. James Clark Maxwell

2. Every current-carrying conductor generates a(an):
   a. Electric field   c. Both (a) and (b)
   b. Magnetic field   d. None of the above

3. If a greater number of loops of a conductor in a coil are added, a _____ magnetic field is generated.
   a. Weaker   b. Stronger

4. If a smaller current is passed through a coil, a _____ magnetic field is generated.
   a. Weaker   b. Stronger

5. One magnetic line of flux is known as a _____.
   a. Weber   c. Maxwell
   b. Tesla   d. Oersted

6. 10^8 magnetic lines of flux are referred to as one _____.

   a. Weber   **c.** Maxwell
   b. Tesla   **d.** Oersted

7. Flux density is equal to the magnetic flux divided by area and is measured in _____.

   a. Webers   **c.** Maxwells
   b. Teslas   **d.** Oersteds

8. By increasing either current or the number of turns in a coil, the mmf, which is an abbreviation for _____, will increase.

   a. Multiple magnetic formulas   **c.** Electromotive force
   b. Magnetomotive force   **d.** None of the above

9. _____ is a term used to describe the magnetic field intensity and is equal to the mmf divided by the length of the coil.

   a. Flux density   **c.** Magnetizing force
   b. Magnetomotive force   **d.** Reluctance

10. Reluctance is equivalent to electrical:

    a. Current   **c.** Resistance
    b. Voltage   **d.** Power

11. Permeability is equivalent to electrical:

    a. Current   **c.** Resistance
    b. Conductance   **d.** Voltage

12. The absolute permeability of air is equal to:

    a. 1   **b.** 1.26×10^{-6} **c.** 4π **d.** None of the above

13. An electromagnet can be used in the application of:

    a. A relay   **c.** A circuit breaker
    b. An electric bell   **d.** All of the above

14. Relays can be used to:

    a. Allow one master switch to enable several others
    b. Allow several switches to enable one master
    c. Allow a switch in a low-current circuit to close contacts in a high-current circuit
    d. Both (a) and (b)
    e. Both (a) and (c)

15. The starter relay in an automobile is used to:

    a. Allow one master switch to enable several others
    b. Allow several switches to enable one master
    c. Allow a switch in a low-current circuit to close contacts in a high-current circuit
    d. Both (a) and (b)

16. The _____ uses an electromagnet around two flat magnetic strips mounted inside a glass capsule.

    a. Reed switch   **c.** Reed relay
    b. Magnetic circuit breaker   **d.** Starter relay

17. The reed switch could be used in a(an):
    a. Home security system   c. Magnetic-type circuit breaker
    b. Automobile starter   d. All of the above

18. The relative permeability of air or a vacuum is equal to:
    a. 4π b. 6.26×10^{-6} c. 1   d. None of the above

19. An electromagnet is also known as a:
    a. Coil   d. Both (a) and (c)
    b. Solenoid   e. Both (a) and (b)
    c. Resistor

20. A normally open relay (NO) will have:
    a. Contacts closed until activated   c. All contacts permanently open
    b. Contacts open until activated   d. All contacts permanently closed

21. Direct current produces a _____ magnetic field of _____ polarity.
    a. Alternating, unchanging   c. Constant, unchanging
    b. Constant, alternating   d. Both (a) and (c)

22. Magnetizing force (H) is equal to:
    a. $I \times N \times l$ c. $I \times N/l$
    b. $I \times N + l$ d. $N \times l/I$

23. Electromagnetism:
    a. Is the magnetism resulting from electrical current flow
    b. Is the electrical voltage resulting in a coil from the relative motion of a magnetic field
    c. Both (a) and (b)
    d. None of the above

24. Electromagnetic induction:
    a. Is the magnetism resulting from electrical current flow
    b. Is the electrical voltage resulting in a coil from the relative motion of a magnetic field
    c. Both (a) and (b)
    d. None of the above

25. The _____ plots magnetizing force against flux density through a complete cycle of alternating current.
    a. B–H curve   c. Power curve
    b. Coercive force   d. Hysteresis loop

26. When the magnetic flux linking a conductor is changing, an emf is induced, the magnitude of which depends on the number of coil turns, rate of change of flux linkage change, and flux density. This law was discovered by:
    a. Heinrich Lenz   c. Michael Faraday
    b. Guglielmo Marconi   d. Joseph Henry

27. The current induced in a coil due to the change in the magnetic flux is such as to oppose the cause producing it. This law was discovered by:
    a. Heinrich Lenz   c. Michael Faraday
    b. Guglielmo Marconi   d. Joseph Henry

28. An ac generator uses _____ to generate ac electricity.
  a. Electromagnetism   **c.** Magnetism
  b. Electromagnetic induction   **d.** None of the above

29. The generator converts _____ energy into _____ energy.
  a. Electrical, electrical   **c.** Chemical, electrical
  b. Mechanical, electrical   **d.** None of the above

30. The loop of conductor rotated through 360° in a generator is known as a(an):
  a. Electromagnet   **c.** Armature
  b. Field coil   **d.** Both (a) and (b)

31. Sound waves are a form of:
  a. Electrical energy   **c.** Magnetic energy
  b. Chemical energy   **d.** Mechanical energy

32. The moving coil microphone converts _____ waves into _____ waves.
  a. Sound, electrical   **c.** Electromagnetic, sound
  b. Electrical, sound   **d.** Sound, radio

33. There is a reciprocal relationship between permeability and _____.
  a. Flux density   **c.** Remanence
  b. Magnetizing force   **d.** Reluctance

34. The amount of current in the reverse direction needed to reduce the flux density (B) to zero is termed the:
  a. Coercive force   **c.** Electromotive force
  b. Magnetizing force   **d.** All of the above

Essay Questions

35. Describe what is meant by the word *electromagnetism*. (15-1)

36. Briefly describe how the left-hand rule of electromagnetism is applied to: (15-1-2)
  a. Conductors   **b.** Coils

37. How many maxwells make up 1 weber? (15-1-5)

38. Give the formulas for the following: (15-1-5)
  a. Flux density   **d.** Reluctance
  b. Magnetomotive force   **e.** Permeability
  c. Magnetizing force

39. Describe the operation of: (15-1-7)
  a. The magnetic-type circuit breaker
  b. The electric bell
  c. The NO and NC relays

40. Describe the difference and operation of the: (15-1-7)

    a. Reed relay **b.** Reed switch

41. Explain the operation and application of the solenoid-type electromagnet. (15-1-7)

42. Why is a wire looped many times to form an electromagnet more useful than a straight piece of wire? (15-1-3)

43. From which pole does the magnetic flux emerge and into which pole end does the flux return? (15-1-1)

44. What are the differences between magnetic flux and flux density? (15-1-5)

45. What effect does current have when it is passed through a coil of conductor? (15-1)

46. What effect does a magnet have when it is moved into and out of a coil? (15-2)

47. State Faraday's law. (15-2-1)

48. State Lenz's law. (15-2-2)

49. Describe the different effects when ac and dc are passed through a coil. (15-1-4)

50. Illustrate and describe all the different points on a hysteresis curve. (15-1-6)

51. Describe the meaning of the following terms: (15-1-6)

    a. Flux density   **d.** Coercive force
    b. Magnetizing   **e.** Electromagnetism
    c. Remanence   **f.** Electromagnetic induction

52. Briefly describe the operation of the generator. (15-2-4)

53. Briefly describe the operation of the moving coil microphone. (15-2-4)

54. Illustrate and describe the operation of the generator through 360°. (15-2-4)

55. List some other applications of ac electromagnetism. (15-1-7)

Practice Problems

56. If a magnet has a pole area of 6.4×10^{-3} m^2 and a 1200-μWb total flux, what flux density would the pole produce?

57. Calculate the magnetomotive force produced when 760 mA flow through 25 turns.

58. Calculate the magnetizing force (H) or field intensity if a 15-cm, 40-turn coil has a current of 1.2 A flowing through it.

59. Calculate the permeability (μ) of the following materials ($\mu_0 = 4\pi \times 10^{-7}$):

    a. Cast iron **b.** Nickel **c.** Machine steel

60. Calculate the reluctance of a magnetic circuit if the magnetic flux produced is equal to 2.3×10^{-4} Wb, and is produced by 3 A flowing through a solenoid of 36 turns.

61. If a coil of 50 turns is passing a current of 4 A, calculate the mmf.

62. Calculate the reluctance of an iron core when mmf = 150 At and ϕ = 360 µWb.

63. Calculate mmf when a 9-V battery is connected across a 50-turn, 23-Ω coil.

64. Calculate the permeability of a permalloy core.

65. Calculate the magnetizing force (H) of the coil in Question 63 if it were 0.7 m long.

After completing this chapter, you will be able to:

1. Describe self-induction.

2. List and explain the factors affecting inductance.

3. Give the formula for inductance.

4. Identify inductors in series and parallel and understand how to calculate total inductance when inductors are in combination.

5. List and explain the fixed and variable types of inductors.

6. Explain the inductive time constant.

7. Give the formula for inductive reactance.

8. Describe all aspects relating to a series and parallel *RL* circuit.

9. State the three typical malfunctions of inductors and explain how they can be recognized.

10. Describe how inductors can be used for the following applications:
    a. *RL* filter
    b. *RL* integrator
    c. *RL* differentiator

Inductance and Inductors

OUTLINE

What's Cooking!

As a boy, George R. Stibitz was a natural experimenter intrigued by any electrical or electronic gadget. His parents were always anxious about these tinkerings, but wishing not to discourage his imagination and keenness to learn, his father, a professor of theology, often gave him devices to occupy him. On one occasion at the age of 8, his latest experiment, in which he connected an electric motor to the home electrical outlet, caused a circuit overload and almost sent the family home in Dayton, Ohio, up in smoke.

In 1937, Stibitz, a young mathematician working at Bell Telephone Laboratories (the research division of AT&T), had another one of his brainstorms and in his spare time he began building the prototype, which consisted of telephone-system components, batteries, and other devices all interconnected by a mass of wires. The machine he was putting together would be the first system able to achieve binary arithmetic in the United States; since it was assembled on the kitchen table, he called it the Model K.

After developing a more ambitious system in 1940 with veteran Bell engineer Samuel B. Williams, the new digital calculator could handle the complex number problems that were needed for the design of long-distance telephone networks. The system was installed in Bell Labs' Manhattan, New York, headquarters and drew a large amount of attention from the American Mathematical Society, who invited Stibitz to give a presentation on the machine at Dartmouth College in Hanover, New Hampshire. Stibitz later wrote, "With my usual genius for making things more difficult for myself and others, I suggested direct telegraph operation of the complex number calculator from Hanover, and this was decided upon." Both Stibitz and Williams worked tirelessly setting up the demonstration, and on September 11, 1940, Stibitz typed in complex number problems on a keyboard in Hanover and within a minute a correct answer came racing back from New York. This was the first demonstration of long-distance computing.

Three mathematicians who would themselves greatly influence computer science sat in the audience during this presentation—John von Neumann, Norbert Weiner, and John W. Mauchly who in just a few years' time would help invent ENIAC, the world's first large-scale electronic digital computer.

In Chapter 15 two important rules were discussed that relate to this chapter on inductance. The first was that a magnetic field will build up around any current-carrying conductor (electromagnetism), and the second was that a voltage will be induced into a conductor when it is subjected to a moving magnetic field. These two rules form the basis for a phenomenon that is covered in this chapter and is called *self-inductance* or more commonly *inductance*. Inductance, by definition, is the ability of a device to oppose a change in current flow, and the device designed specifically to achieve this function is called the *inductor*. There is, in fact, no physical difference between an inductor and an electromagnet, since they are both coils. The two devices are given different names because they are used in different applications even though their construction and principle of operation are the same. An electromagnet is used to generate a magnetic field in response to current, while an inductor is used to oppose any change in current.

In this chapter we examine all aspects of inductance and inductors, including inductive circuits, types, reactance, testing, and applications.

16-1   SELF-INDUCTION

In Figure 16-1 an external voltage source has been connected across a coil, forcing a current through the coil. This current will generate a magnetic field that will increase in field strength in a short time from zero to maximum, expanding from the center of the conductor (electromagnetism). The expanding magnetic lines of force have relative motion with respect to the stationary conductor, so an induced voltage results (electromagnetic induction). The blooming magnetic field generated by the conductor is actually causing a voltage to be induced in the conductor that is generating the magnetic field. This effect of a current-carrying coil of conductor inducing a voltage within itself is known as **self-inductance.** This phenomenon was first discovered by Heinrich Lenz, who observed that the induced voltage causes an induced (bucking) current to flow in the coil, which opposes the source current producing it.

Figure 16-2(a) shows an inductor connected across a dc source. When the switch is closed, a circuit current will exist through the inductor and the resistor. As the current rises toward its maximum value, the magnetic field expands, and throughout this time of relative motion between field and conductor, an induced voltage will be present. This induced voltage will produce an induced current to oppose the change in the circuit current.

When the current reaches its maximum, the magnetic field, which is dependent on current, will also reach a maximum value and then no longer expand but remain stationary. When the current remains constant, no change will occur in the magnetic field and therefore no relative motion will exist between the conductor and magnetic field, resulting in no induced voltage or current to oppose circuit current, as shown in Figure 16-2(b). The coil has accepted electrical

Self-Inductance
The property that causes a counterelectromotive force to be produced in a conductor when the magnetic field expands or collapses with a change in current.

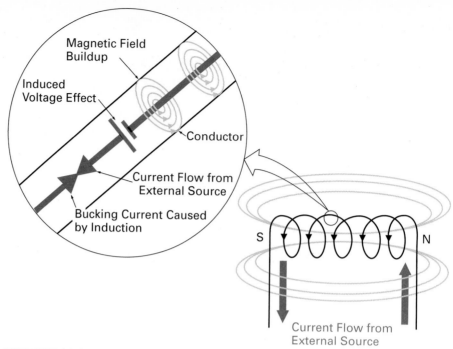

FIGURE 16-1   Self-Induction of a Coil.

Counter emf (Counter electromotive force)
Abbreviated "counter emf," or "back emf," it is the voltage generated in an inductor due to an alternating or pulsating current and is always of opposite polarity to that of the applied voltage.

Inductance
Property of a circuit or component to oppose any change in current as the magnetic field produced by the change in current causes an induced countercurrent to oppose the original change.

Henry
Unit of inductance.

energy and is storing it in the form of a magnetic energy field, just as the capacitor stored electrical energy in the form of an electric field.

If the switch is put in position *B,* as shown in Figure 16-2(c), the current from the battery will be zero, and the magnetic field will collapse as it no longer has circuit current to support it. As the magnetic lines of force collapse, they cut the conducting coils, causing relative motion between the conductor and magnetic field. A voltage is induced in the coil, which will produce an induced current to flow in the same direction as the circuit current was flowing before the switch was opened. The coil is now converting the magnetic field energy into electrical energy and returning the original energy that it stored.

After a short period of time, the magnetic field will have totally collapsed, the induced voltage will be zero, and the induced current within the circuit will therefore also no longer be present.

This induced voltage is called a **counter emf** or *back emf.* It opposes the applied emf (or battery voltage). The ability of a coil or conductor to induce or produce a counter emf within itself as a result of a change in current is called *self-inductance,* or more commonly **inductance** (symbolized *L*). The unit of inductance is the **henry** (H), named in honor of Joseph Henry, an American physicist, for his experimentation within this area of science. The inductance of an inductor is 1 henry when a current c°hange of 1 ampere per second causes an induced voltage of 1 volt. Inductance is therefore a measure of how much counter emf (induced voltage) can be generated by an inductor for a given amount of current change through that same inductor.

Expanding Magnetic Field

Stationary Magnetic Field

Collapsing Magnetic Field

(a)

(b)

(c)

FIGURE 16-2 **Self-Inductance. (a) Switch Closed. Increase in Circuit Current and, Therefore, Increase in Magnetic Field Induces Voltage in Coil. This Results in an Induced Current to Oppose Circuit Current. (b) Constant Circuit Current. Stationary Magnetic Field and, Therefore, No Induced Voltage or Current to Oppose Circuit Current. (c) Switch Opened. No Circuit Current and, Therefore, Magnetic Field Collapses and Induces Voltage in Coil. This Results in an Induced Current in Same Direction as Circuit Current.**

This counter emf or induced voltage can be calculated by the formula

$$V_{ind} = L \times \frac{\Delta I}{\Delta t}$$

where L = inductance, in henrys (H)

ΔI = increment of change of current (I)

Δt = increment of change with respect to time (t)

A larger inductance $(L\uparrow)$ will create a larger induced voltage $(V_{ind}\uparrow)$, and if the rate of change of current with respect to time is increased $(\Delta I/\Delta t\uparrow)$, the induced voltage or counter emf will also increase $(V_{ind}\uparrow)$.

EXAMPLE:

What voltage is induced across an inductor of 4 H when the current is changing at a rate of:

a. 1 A/s?

b. 4 A/s?

Solution:

a. $V_{ind} = L \times \dfrac{\Delta I}{\Delta t} = 4\ H \times 1\ A/s = 4\ V$

b. $V_{ind} = L \times \dfrac{\Delta I}{\Delta t} = 4\ H \times 4\ A/s = 16\ V$

The faster the coil current changes, the larger the induced voltage.

SELF-TEST REVIEW QUESTIONS FOR SECTION 16-1

---

1. Define self-induction.

2. What is counter emf, and how can it be calculated?

3. Calculate the voltage induced in a 2mH inductor if the current is increasing at a rate of 4 kA/s.

16-2   THE INDUCTOR

Inductor
Coil of conductor used to introduce inductance into a circuit.

An **inductor** is basically an electromagnet, as its construction and principle of operation are the same. We use the two different names because they have different applications. The purpose of the electromagnet or solenoid is to generate a magnetic field, while the purpose of an inductor or coil is to oppose any change of circuit current.

In Figure 16-3 a steady value of direct current is present within the circuit and the inductor is creating a steady or stationary magnetic field. If the current in the circuit is suddenly increased (by lowering the circuit resistance), the change in the expanding magnetic field will induce a counter emf within the inductor. This induced voltage will oppose the source voltage from the battery and attempt to hold current at its previous low level.

The counter emf cannot completely oppose the current increase, for if it did, the lack of current change would reduce the counter emf to zero. Current therefore incrementally increases up to a new maximum, which is determined by the applied voltage and the circuit resistance ($I = V/R$). Once the new higher level of current has been reached and remains constant, there will no longer be a change. This lack of relative motion between field and conductor will no longer generate a counter emf, so the current will remain at its new higher constant value.

This effect also happens in the opposite respect. If current decreases (by increasing circuit resistance), the magnetic lines of force will collapse because of the reduction of current and induce a voltage in the inductor, which will produce

Inductance (L) = 10 mH (millihenrys)

FIGURE 16-3   Inductor's Ability to Oppose Current Change.

an induced current in the same direction as the circuit current. These two combine and tend to maintain the current at the higher previous constant level. Circuit current will fall, however, as the induced voltage and current are only present during the change (in this case the decrease from the higher current level to the lower); and once the new lower level of current has been reached and remains constant, the lack of change will no longer induce a voltage or current. So the current will then remain at its new lower constant value.

The inductor is therefore an electronic component that will oppose any changes in circuit current, and this ability or behavior is referred to as *inductance*. Since the current change is opposed by a counter emf, inductance may also be defined as the ability of a device to induce a counter emf within itself for a change in current.

SELF-TEST REVIEW QUESTIONS FOR SECTION 16-2

1. What is the difference between an electromagnet and an inductor?
2. True or false: The inductor will oppose any changes in circuit current.

16-3 FACTORS DETERMINING INDUCTANCE

The inductance of an inductor is determined by four factors:

1. Number of turns
2. Area of the coil
3. Length of the coil
4. Core material used within the coil

Let's now discuss how these four factors can affect inductance, beginning with the number of turns.

16-3-1 Number of Turns (N) (Figure 16-4)

If an inductor has a greater number of turns, the magnetic field produced by passing current through the coil will have more magnetic force than an inductor with fewer turns. A greater magnetic field will cause a larger counter emf, because more magnetic lines of flux will cut more coils of the conductor, producing a larger inductance value. Inductance *(L)* is therefore proportional to the number of turns *(N)*:

$$L \propto N$$

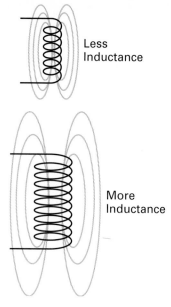

FIGURE 16-4
Inductance Is
Proportional to the
Number of Turns
($L \propto N$) in a Coil.

FIGURE 16-5   Inductance
Is Proportional to the Area
($L \propto A$) in a Coil.

16-3-2   *Area of Coil (A) (Figure 16-5)*

If the area of the coil is increased for a given number of turns, more magnetic lines of force will be produced, and if the magnetic field is increased, the inductance will also increase. Inductance *(L)* is therefore proportional to the cross-sectional area of the coil *(A):*

$$L \propto A$$

16-3-3   *Length of Coil (l) (Figure 16-6)*

If, for example, four turns are spaced out (long-length coil), the summation that occurs between all the individual coil magnetic fields will be small. On the other hand, if four turns are wound close to one another (short-length coil), all the

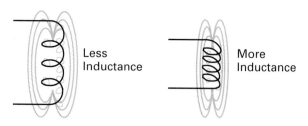

FIGURE 16-6   Inductance Is Inversely
Proportional to the Length ($L \propto 1/l$) of a Coil.

individual coil magnetic fields will easily interact and add together to produce a larger magnetic field and, therefore, greater inductance. Inductance is therefore inversely proportional to the length of the coil, in that a longer coil, for a given number of turns, produces a smaller inductance, and vice versa.

$$L \propto \frac{1}{l}$$

16-3-4 Core Material (μ) (Figure 16-7)

Most inductors have core materials such as nickel, cobalt, iron, steel, ferrite, or an alloy. These cores have magnetic properties that concentrate or intensify the magnetic field. Permeability is another factor that is proportional to inductance and the figures for various materials are shown in Table 16-1. The greater the permeability of the core material, the greater the inductance is.

$$L \propto \mu$$

Air Core

Less Inductance

Iron Core

Much More Inductance

TABLE 16-1 Permeabilities of Various Materials

MATERIAL	PERMEABILITY (μ)
Air or vacuum	1.26×10^{-6}
Nickel	6.28×10^{-5}
Cobalt	7.56×10^{-5}
Cast iron	1.1×10^{-4}
Machine steel	5.65×10^{-4}
Transformer iron	6.9×10^{-3}
Silicon iron	8.8×10^{-3}
Permalloy	0.126
Supermalloy	1.26

FIGURE 16-7
Inductance Is Proportional to the Permeability of the Coil's Material ($L \propto \mu$).

16-3-5 Formula for Inductance

All the four factors described can be placed in a formula to calculate inductance.

$$L = \frac{N^2 \times A \times \mu}{l}$$

where L = inductance, in henrys (H)

N = number of turns

A = cross-sectional area, in square meters (m²)

μ = permeability

l = length of core, in meters (m)

■ **EXAMPLE:**

Refer to Figure 16-8(a) and (b) and calculate the inductance of each.

■ *Solution:*

a. $L = \dfrac{5^2 \times 0.01 \times (6.28 \times 10^{-5})}{0.001} = 15.7 \text{ mH}$

b. $L = \dfrac{10^2 \times 0.1 \times (1.1 \times 10^{-4})}{0.1} = 11 \text{ mH}$

Length is 0.001 M

Area = 0.01 m²

N = 5 Turns

Nickel Core, so $\mu = 6.28 \times 10^{-5}$

(a)

Length is 0.1 M

Area = 0.1 m²

N = 10 Turns

Cast Iron Core, so $\mu = 1.1 \times 10^{-4}$

(b)

FIGURE 16-8 **Inductor Examples.**

1. List the four factors that determine the inductance of an inductor.
2. State the formula for inductance.

16-4 INDUCTORS IN COMBINATION

Inductors oppose the change of current in a circuit and so are treated in a manner similar to resistors connected in combination. Two or more inductors in series merely extend the coil length and increase inductance. Inductors in parallel are treated in a manner similar to resistors, with the total inductance being less than that of the smallest inductor's value.

16-4-1 *Inductors in Series*

When inductors are connected in series with one another, the total inductance is calculated by summing all the individual inductances.

$$L_T = L_1 + L_2 + L_3 + \ldots$$

■ **EXAMPLE:**

Calculate the total inductance of the circuit shown in Figure 16-9.

■ *Solution:*

$$
\begin{aligned}
L_T &= L_1 + L_2 + L_3 \\
&= 5 \text{ mH} + 7 \text{ mH} + 10 \text{ mH} \\
&= 22 \text{ mH}
\end{aligned}
$$

FIGURE 16-9 Inductors in Series.

16-4-2 *Inductors in Parallel*

When inductors are connected in parallel with one another, the reciprocal (two or more inductors) or product over sum (two inductors) formula can be used to find total inductance, which will always be less than the smallest inductor's value.

$$L_T = \frac{1}{(1/L_1) + (1/L_2) + (1/L_3) + \ldots}$$

$$L_T = \frac{L_1 \times L_2}{L_1 + L_2}$$

EXAMPLE:

Determine L_T for the circuits in Figure 16-10(a) and (b).

Solution:

a. General formula:

$$L_T = \frac{1}{(1/L_1) + (1/L_2) + (1/L_3)}$$

$$= \frac{1}{(1/10 \text{ mH}) + (1/5 \text{ mH}) + (1/20 \text{ mH})}$$

$$= 2.9 \text{ mH}$$

b. Product over sum:

$$L_T = \frac{L_1 \times L_2}{L_1 + L_2}$$

$$= \frac{10 \ \mu\text{H} \times 2 \ \mu\text{H}}{10 \ \mu\text{H} + 2 \ \mu\text{H}}$$

$$= \frac{20 \times 10^{-12} \text{H}}{12 \ \mu\text{H}} = 1.67 \ \mu\text{H}$$

L_1 L_2 L_3

10 mH   5 mH   20 mH

(a)

10 μH   2 μH

(b)

FIGURE 16-10   Inductors in Parallel.

1. True or false: To calculate total inductance, inductors can be treated in the same manner as capacitors.
2. State the formula for calculating total inductance in:
   a. A series circuit
   b. A parallel circuit
3. Calculate the total circuit inductance if 4 mH and 2 mH are connected:
   a. In series
   b. In parallel

16-5 TYPES OF INDUCTORS

As with resistors and capacitors, inductors are basically divided into the two categories of fixed-value and variable-value inductors, as shown by the symbols in Figure 16-11(a) and (b). Within these two categories, inductors are generally classified by the type of core material used. Figure 4-11(c) shows a variety of different types.

(a)   (b)

(c)

FIGURE 16-11   Inductor Types. (a) Fixed-Value Inductor Symbol. (b) Variable-Value Inductor Symbol. (c) Physical Appearance of Inductors. (Left photo courtesy of Sencore. Right photo courtesy of Coilcraft Inc., Cary, Illinois, 60013.)

16-5-1　Fixed-Value Inductors

Fixed-Value Inductor
An inductor whose value cannot be changed.

With **fixed-value inductors,** the inductance value remains constant and cannot be altered. It is usually made of solid copper wire with an insulating enamel around the conductor. This enamel or varnish on the conductor is needed since the turns of a coil are generally wound on top of one another. The three major types of fixed-value inductors on the market today are:

1. Air core　**2.** Iron core　**3.** Ferrite core

Air Core

Figure 16-12 illustrates some typical air-core fixed-value inductor types. The insulated copper wire can be wound on nonmagnetic materials such as plastic, ceramic, or Bakelite, which due to their nonmagnetic properties will have no effect on the inductance value. These materials act as a support or form for the inductor, whose wire size may be too small to support itself. If a heavier or more rigid wire size is used and it can support itself, there will be no need for a nonmagnetic form.

Air-Core Inductor
An inductor that has no metal core.

Inductors of this type, which make use of nonmagnetic form or do not have a form, are known as **air-core inductors.** Air-core inductors have the advantage that their inductance value does not vary with current, as do iron- and ferrite-core inductors. Their typical low values (below 10 μH) normally find application in high-frequency equipment, such as AM and FM radio, TV, and other communications transmitter and receiver circuits.

Iron Core

Iron-Core Inductor or Choke
Inductor used to impede the flow of alternating or pulsating current.

Figure 16-13 illustrates a typical **iron-core inductor.** The name **choke** is used interchangeably with the word *inductor,* because choke basically describes an inductor's behavior. As we discovered previously, the inductor will oppose any change of current flow, whether it is in the form of pulsating dc or ac, because of its self-inductance, which generates a counter emf to limit or choke the flow of current.

(a)

Copper Pattern Forming a Fixed Inductor

Printed Circuit Board (PCB)

(b)

(c)

FIGURE 16-12　Air-Core Fixed-Value Inductors. (a) Air-Core Inductor (Courtesy of Dale Electronics). (b) PBC Inductor. (c) Symbol.

E Piece   **I Piece**

(b)

(a)   **(c)**

FIGURE 16-13   Iron-Core Fixed-Value Inductors. (a) Physical Appearance (Courtesy of Coilcraft). (b) Construction. (c) Symbol.

An iron core has a higher permeability figure than air, which means that it will concentrate the magnetic lines of force and therefore increase inductance.

$$L\uparrow = \frac{N^2 \times A \times \mu\uparrow}{l}$$

The EI **ferrous** or compound of iron core [Figure 16-13(b)] has two sections, one of which resembles the letter E and the other I. The conductor is wrapped around the center section of the E to form the coil. These inductors or chokes can be made with the highest inductance values, up to the hundreds of henrys, and are used in dc and low-frequency ac circuits. Figure 16-13(c) shows the schematic symbol for an iron-core inductor.

Ferrous
Composed of and/or containing iron. A ferrous metal exhibits magnetic characteristics, as opposed to nonferrous metals.

Ferrite Core

Figure 16-14 illustrates two typical **ferrite-core inductors. Ferrite** is a chemical compound, which is basically powdered iron oxide and ceramic. Although its permeability is less than iron, it has a higher permeability than air and can therefore obtain a higher inductance value than air for the same number of turns.

The toroidal inductor (doughnut shaped) illustrated in Figure 16-14(a) has an advantage over the cylinder-shaped core in that the magnetic lines of force do not have to pass through air. With a cylindrical core, the magnetic lines of force merge out of one end of the cylinder, propagate or travel through air, and then enter in the other end of the cylinder. The nonmagnetic air, which has a low permeability, will reduce the strength of the field and therefore the inductance value:

$$L\downarrow = \frac{N^2 \times A \times \mu\downarrow}{l}$$

Ferrite-Core Inductor
An inductor containing a ferrite core.

Ferrite
A powdered, compressed, and sintered magnetic material having high resistivity. The high resistance makes eddy-current losses low at high frequencies.

(a)

(b)

(c)

FIGURE 16-14 **Ferrite-Core Fixed-Value Inductors. (a) Toroidal Inductor (Courtesy of Coilcraft). (b) Molded Inductor (Courtesy of Dale Electronics). (c) Schematic Symbol.**

If one could bend around either end of the cylinder to make a toroid, the magnetic lines of force emerging from one end of the inductor's core would pass from ferrite to ferrite, which has a high permeability.

Toroidal ferrite-core inductors have a greater inductance than the cylindrical-type inductors; however, they are more expensive to manufacture. They are used in both low- and high-frequency applications.

The molded inductor, shown in Figure 16-14(b), consists of a spiral metal film deposited on a cylindrical ferrite core. The entire assembly is encapsulated, and the available values range from 1.2 μH to 10 mH, with maximum currents of about 70 mA. Due to its small size, the value is often color coded onto the body instead of using alphanumeric or typographical labels, which can be used on all of the other larger inductors. Figure 16-14(c) shows the schematic symbol for an iron-core inductor.

16-5-2 Variable-Value Inductors

Variable-Value Inductor
An inductor whose value can be varied.

With **variable-value inductors,** the inductance value can be changed by adjusting the position of the movable core with respect to the stationary coil. There is basically only one type of variable-value inductor in wide use today: the ferrite core. Figure 16-15 illustrates a typical ferrite-core variable inductor. A screw

FIGURE 16-15 **Ferrite-Core Variable-Value Inductors. (a) Construction. (b) Physical Appearance (Courtesy of Sencore). (c) Symbol.**

adjustment moves a sliding ferrite core or slug farther in or out of the stationary coil. If the slug is all the way out, the permeability figure is low, because an air core is being used ($L\downarrow \propto \mu\downarrow$). If the slug is adjusted and screwed into the coil, the air is replaced by the ferrite slug, which has a higher permeability and, therefore, inductance value ($L\uparrow \propto \mu\uparrow$).

Variable inductors should only be adjusted with a plastic or nonmetallic alignment tool since any metal object in the vicinity of the inductor will interfere with the magnetic field and change the inductance of the inductor. Variable-value inductors are used extensively in radio circuits.

SELF-TEST REVIEW QUESTIONS FOR SECTION 16-5

1. List the three fixed-value inductor types.
2. Under which category would a fixed-value inductor with a nonmagnetic core come?
3. What is a ferrite compound?
4. What advantage does the toroid-shaped inductor have over a cylindrical inductor?
5. What type of variable-value inductor is the only one in wide use today?
6. What factor is varied to change the value of the variable inductor?

16-6 INDUCTIVE TIME CONSTANT

Inductors will not have any effect on a steady value of direct current (dc) from a dc voltage source. If, however, the dc is changing (pulsating), the inductor will oppose the change whether it is an increase or decrease in direct current, because a change in current causes the magnetic field to expand or contract, and in so doing it will cut the coil of the inductor and induce a voltage that will counter the applied emf.

16-6-1 DC Current Rise

Figure 16-16(a) illustrates an inductor *(L)* connected across a dc source (battery) through a switch and series-connected resistor. When the switch is closed, current will flow and the magnetic field will begin to expand around the inductor. This field cuts the coils of the inductor and induces a counter emf to oppose the rise in current. Current in an inductive circuit, therefore, cannot rise instantly to its maximum value, which is determined by Ohm's law $(I = V/R)$. Current will in fact take a time to rise to maximum, as graphed in Figure 16-16(b), due to the inductor's ability to oppose change.

It will actually take five time constants (5τ) for the current in an inductive circuit to reach maximum value. This time can be calculated by using the formula

$$\tau = \frac{L}{R} \quad \text{seconds}$$

The constant to remember is the same as before: 63.2%. In one time constant $(1 \times L/R)$ the current in the *RL* circuit will have reached 63.2% of its maximum value. In two time constants $(2 \times L/R)$, the current will have increased 63.2% of the remaining current, and so on, through five time constants.

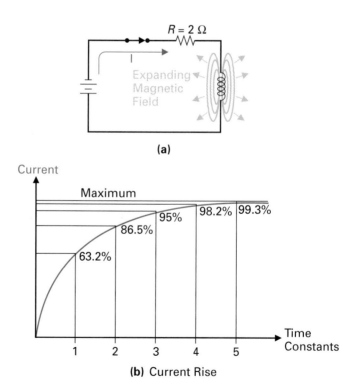

(a)

(b) Current Rise

FIGURE 16-16 **DC Inductor Current Rise.**

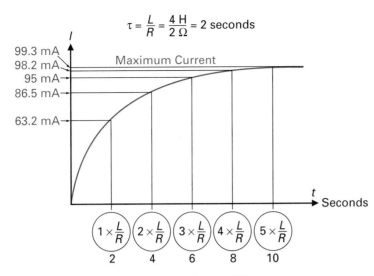

$$\tau = \frac{L}{R} = \frac{4\,H}{2\,\Omega} = 2 \text{ seconds}$$

FIGURE 16-17 **Exponential Current Rise.**

For example, if the maximum possible circuit current is 100 mA and an inductor of 4 H is connected in series with a resistor of 2 Ω, the current will increase as shown in Figure 16-17.

Referring back to the τ = L/R formula, you will notice that how quickly an inductor will allow the current to rise to its maximum value is proportional to the inductance and inversely proportional to the resistance. A larger inductance increases the strength of the magnetic field, so the opposition or counter emf increases, and the longer it takes for current to rise to a maximum ($\tau\uparrow = L\uparrow/R$). If the circuit resistance is increased, the maximum current will be smaller, and a smaller maximum is reached more quickly than a higher ($\tau\downarrow = L/R\uparrow$).

16-6-2 *DC Current Fall*

When the inductor's dc source of current is removed, as shown in Figure 16-18(a), by placing the switch in position *B*, the magnetic field will collapse and cut the coils of the inductor, inducing a voltage and causing a current to flow in the same direction as the original source current. This current will exponentially decay, or fall from the maximum to zero level, in five time constants ($5 \times L/R = 5 \times 4/2 = 10$ seconds), as shown in Figure 16-18(b).

EXAMPLE:

Calculate the circuit current at each of the five time constants if a 12 V dc source is connected across a series *RL* circuit, and *R* = 60 Ω and *L* = 24 mH. Plot the results on a graph showing current against time.

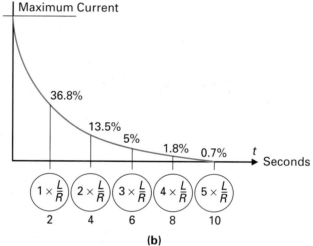

FIGURE 16-18 Exponential Current Fall.

■ *Solution:*

$$\text{maximum current, } I_{\max} = \frac{V_S}{R} = \frac{12 \text{ V}}{60 \text{ }\Omega} = 200 \text{ mA}$$

$$\text{time constant, } \tau = \frac{L}{R} = \frac{24 \text{ mH}}{60 \text{ }\Omega} = 400 \text{ }\mu\text{s}$$

At 1 time constant (400 µs after source voltage is applied), the current will be

$$I = 63.2\% \text{ of } I_{\max}$$
$$= 0.632 \times 200 \text{ mA} = 126.4 \text{ mA}$$

At two time constants (800 µs after source voltage is applied):

$$I = 86.5\% \text{ of } I_{\max}$$
$$= 0.865 \times 200 \text{ mA} = 173 \text{ mA}$$

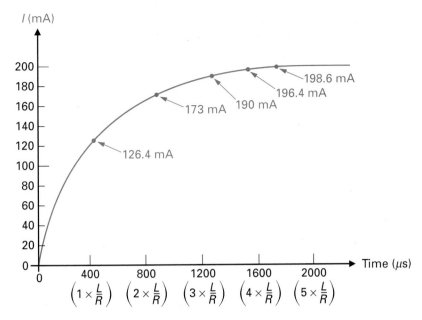

FIGURE 16-19   **Exponential Current Rise for Example 16-5.**

At three time constants (1200 μs or 1.2 ms):

$$I = 95\% \text{ of } I_{max}$$
$$= 0.95 \times 200 \text{ mA} = 190 \text{ mA}$$

At four time constants (1.6 ms):

$$I = 98.2\% \text{ of } I_{max}$$
$$= 0.982 \times 200 \text{ mA} = 196.4 \text{ mA}$$

At five time constants (2 ms):

$$I = 99.3\% \text{ of } I_{max}$$
$$= 0.993 \times 200 \text{ mA} = 198.6 \text{ mA, approximately maximum (200 mA)}$$

See Figure 16-19.

16-6-3 *AC Rise and Fall*

If an alternating (ac) voltage is applied across an inductor, as shown in Figure 16-20(a), the inductor will continuously oppose the alternating current because it is always changing. Figure 16-20(b) shows the phase relationship between the voltage across an inductor or counter emf and the circuit current.

The current in the circuit causes the magnetic field to expand and collapse and cut the conducting coils, resulting in an induced counter emf. At points *X* and *Y*, the steepness of the current waveform indicates that the current will be changing at its maximum rate, and therefore the opposition or counter emf will also be maximum. When the current is at its maximum positive or negative value, it has a very small or no rate of change (flat peaks). Therefore, the opposi-

(a)

(b)

FIGURE 16-20 **The Voltage across an Inductor Leads the Circuit Current by 90° in an Inductive Circuit.**

tion or counter emf should be very small or zero, as can be seen by the waveforms. The counter emf is, therefore, said to be 90° out of phase with the circuit current.

To summarize, we can say that the voltage across the inductor (V_L) or counter emf leads the circuit current (I) by 90°.

SELF-TEST REVIEW QUESTIONS FOR SECTION 16-6

1. How does the inductive time constant relate to the capacitive time constant?
2. True or false: The greater the value of the inductor, the longer it would take for current to rise to a maximum.
3. True or false: A constant dc level is opposed continuously by an inductor.
4. What reaction does an inductor have to ac?

16-7 INDUCTIVE REACTANCE

Inductive Reactance
Measured in ohms, it is the opposition to alternating or pulsating current flow without the dissipation of energy.

Reactance is the opposition to current flow without the dissipation of energy, as opposed to resistance, which is the opposition to current flow with the dissipation of energy.

Inductive reactance (X_L) is the opposition to current flow offered by an inductor without the dissipation of energy. It is measured in ohms and can be calculated by using the formula:

$$X_L = 2\pi \times f \times L$$

where X_L = inductive reactance, in ohms (Ω)

2π = 2π radians, 360° or 1 cycle

f = frequency, in hertz (Hz)

L = inductance, in henrys (H)

Inductive reactance is proportional to frequency ($X_L \propto f$) because a higher frequency (fast-switching current) will cause a greater amount of current change, and a greater change will generate a larger counter emf, which is an opposition or reactance against current flow. When 0 Hz is applied to a coil (dc), there exists no change, so the inductive reactance of an inductor to dc is zero ($X_L = 2\pi \times 0 \times L = 0$).

Inductive reactance is also proportional to inductance because a larger inductance will generate a greater magnetic field and subsequent counter emf, which is the opposition to current flow.

Ohm's law can be applied to inductive circuits just as it can be applied to resistive and capacitive circuits. The current flow in an inductive circuit *(I)* is proportional to the voltage applied *(V)*, and inversely proportional to the inductive reactance (X_L). Expressed mathematically,

$$I = \frac{V}{X_L}$$

■ **EXAMPLE:**

Calculate the current flowing in the circuit illustrated in Figure 16-21.

■ *Solution:*

The current can be calculated by Ohm's law and is a function of the voltage and opposition, which in this case is inductive reactance.

$$I = \frac{V}{X_L}$$

However, we must first calculate X_L:

$$X_L = 2\pi \times f \times L$$
$$= 6.28 \times 50 \text{ kHz} \times 15 \text{ mH}$$
$$= 4710 \ \Omega \text{ or } 4.71 \text{ k}\Omega$$

Current is therefore equal to

$$I = \frac{V}{X_L} = \frac{10 \text{ V}}{4.71 \text{ k}\Omega} = 2.12 \text{ mA}$$

FIGURE 16-21

■ **EXAMPLE:**

What opposition or inductive reactance will a motor winding or coil offer if $V = 12$ V and $I = 4.5$ A?

■ *Solution:*

$$X_L = \frac{V}{I} = \frac{12 \text{ V}}{4.5 \text{ A}} = 2.66 \text{ } \Omega$$

SELF-TEST REVIEW QUESTIONS FOR SECTION 16-7

1. Define and state the formula for inductive reactance.
2. Why is inductive reactance proportional to frequency and the inductance value?
3. How does inductive reactance relate to Ohm's law?
4. True or false: Inductive reactance is measured in henrys.

16-8 SERIES *RL* CIRCUIT

In a purely resistive circuit, as seen in Figure 16-22, the current flowing within the circuit and the voltage across the resistor are in phase with one another. In a purely inductive circuit, as shown in Figure 16-23, the current will lag the applied voltage by 90°. If we connect a resistor and inductor in series, as shown in Figure 16-24(a), we will have the most common combination of *R* and *L* used in electronic equipment.

16-8-1 *Voltage*

The voltage across the resistor and inductor shown in Figure 16-24(a) can be calculated by using Ohm's law:

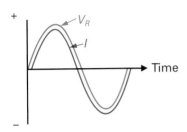

FIGURE 16-22 Purely Resistive Circuit: Current and Voltage Are in Phase.

FIGURE 16-23 Purely Inductive Circuit: Current Lags Applied Voltage by 90°.

$$V_R = I \times R$$

$$V_L = I \times X_L$$

The vector diagram in Figure 16-24(b) illustrates current (I) as the reference at $0°$ and, as expected, the voltage across the resistor (V_R) is in phase with the circuit current. The voltage across the inductor (V_L) leads the circuit current and the voltage across the resistor (V_R) by $90°$, or the circuit current vector lags the voltage across the inductor by $90°$.

As with any series circuit, we have to apply Kirchhoff's voltage law when calculating the value of applied or source voltage (V_S), which, due to the phase difference between V_R and V_L, is the vector sum of all the voltage drops. By cre-

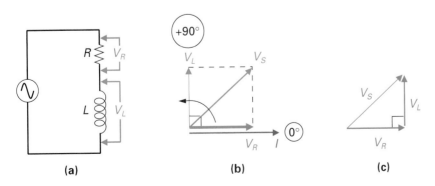

(a)   **(b)**   **(c)**

FIGURE 16-24 Series *RL* Circuit.

ating a right triangle from the three quantities, as shown in Figure 16-24(c), and applying the Pythagorean theorem, we arrive at a formula for source voltage.

$$V_S = \sqrt{V_R^2 + V_L^2}$$

As with any formula with three quantities, if two are known, the other can be calculated by simply rearranging the formula to

$$V_S = \sqrt{V_R^2 + V_L^2}$$

$$V_R = \sqrt{V_S^2 - V_L^2}$$

$$V_L = \sqrt{V_S^2 - V_R^2}$$

■ **EXAMPLE:**

FIGURE 16-25   **Voltage in a Series *RL* Circuit.**

Calculate V_R, V_L, and V_S for the circuit shown in Figure 16-25.

■ *Solution:*

$$V_R = I \times R$$
$$= 100 \text{ mA} \times 55 \text{ } \Omega$$
$$= 5.5 \text{ V}$$
$$V_L = I \times X_L$$
$$= 100 \text{ mA} \times 26 \text{ } \Omega$$
$$= 2.6 \text{ V}$$
$$V_S = \sqrt{V_R^2 + V_L^2}$$
$$= \sqrt{5.5 \text{ V}^2 + 2.6 \text{ V}^2}$$
$$= 6 \text{ V}$$

16-8-2 Impedance (Z)

Impedance is the total opposition to current flow offered by a circuit with both resistance and reactance. It is measured in ohms and can be calculated by using Ohm's law:

$$Z = \frac{V}{I}$$

FIGURE 16-26   Impedance in a Series *RL* Circuit.

Just as a phase shift or difference exists between V_R and V_L and they cannot be added to find applied voltage, the same phase difference exists between R and X_L, so impedance or total opposition cannot be simply the sum of the two, as shown in Figure 16-26.

The impedance of a series *RL* circuit is equal to the square root of the sum of the squares of resistance and reactance, and by rearrangement, X_L and R can also be calculated if the other two values are known:

$$Z = \sqrt{R^2 + X_L^2}$$

$$R = \sqrt{Z^2 - X_L^2}$$

$$X_L = \sqrt{Z^2 - R^2}$$

■ **EXAMPLE:**

Referring back to Figure 16-25, calculate Z.

■ *Solution:*

$$Z = \sqrt{R^2 + X_L^2}$$
$$= \sqrt{55^2 + 26^2}$$
$$= 60.8 \; \Omega$$

16-8-3   Phase Shift

If a circuit is purely resistive, the phase shift (θ) between the source voltage and circuit current is zero. If a circuit is purely inductive, voltage leads current by 90°; therefore, the phase shift is +90°. If the resistance and inductive reactance are equal, the phase shift will equal +45°, as shown in Figure 16-27.

The phase shift in an inductive and resistive circuit is the degrees of lead between the source voltage (V_S) and current (I), and by looking at the examples in Figure 16-27, you can see that the phase angle is proportional to

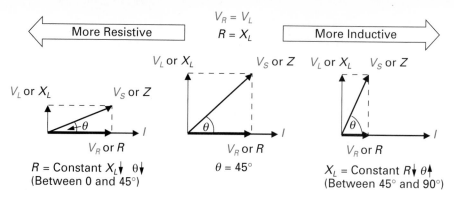

FIGURE 16-27 **Phase Shift (θ) in a Series *RL* Circuit.**

reactance and inversely proportional to resistance. Mathematically, it can be expressed as

$$\theta = \arctan \frac{X_L}{R}$$

As the current is the same in both the inductor and resistor in a series circuit, the voltage drops across the inductor and resistor are directly proportional to reactance and resistance:

$$V_R \updownarrow = I \,(\text{constant}) \times R \updownarrow, \quad V_L \updownarrow = I \,(\text{constant}) \times X_L \updownarrow$$

the phase shift can also be calculated by using the voltage drop across the inductor and resistor.

$$\theta = \arctan \frac{V_L}{V_R}$$

■ **EXAMPLE:**

Referring back to Figure 16-25, calculate the phase shift between source voltage and circuit current.

■ **Solution:**

Since the ratio of X_L/R and V_L/V_R are known for this example circuit, either of the phase shift formulas can be used.

$$\theta = \arctan \frac{X_L}{R}$$

$$= \arctan \frac{26\ \Omega}{55\ \Omega}$$

$$= \arctan 0.4727$$

$$= 25.3°$$

or

$$\theta = \arctan \frac{V_L}{V_R}$$

$$= \arctan \frac{2.6 \text{ V}}{5.5 \text{ V}}$$

$$= \arctan 0.4727$$

$$= 25.3°$$

The source voltage in this example circuit leads the circuit current by 25.3°.

16-8-4 *Power*

Purely Resistive Circuit

Figure 16-28 illustrates the current, voltage, and power waveforms produced when applying an ac voltage across a purely resistive circuit. Voltage and current are in phase, and true (P_R) power in watts can be calculated by multiplying current by voltage ($P = V \times I$).

The sinusoidal power waveform is totally positive, as a positive voltage multiplied by a positive current produces a positive value of power, and a negative voltage multiplied by a negative current also produces a positive value of

Voltage and Current
Are in RMS
(Effective) as Normal

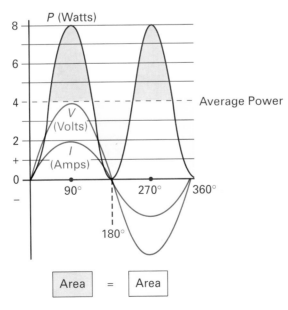

FIGURE 16-28 Purely Resistive Circuit.

power. For this reason, the resistor is said to develop a positive power waveform that is twice the frequency of the voltage or current waveform.

The average value of power dissipated by a purely resistive circuit is the halfway value between maximum and zero, in this example, 4 watts.

Purely Inductive Circuit

The pure inductor, like the capacitor, is a reactive component, which means that it will consume power without the dissipation of energy. The capacitor holds its energy in an electric field, while the inductor consumes and holds its energy in a magnetic field and then releases it back into the circuit.

The power curve alternates equally above and below the zero line, as seen in Figure 16-29. During the first positive power half-cycle, the circuit current is on the increase, to maximum (point A), the magnetic field is building up, and the inductor is storing electrical energy. When the circuit current is on the decline

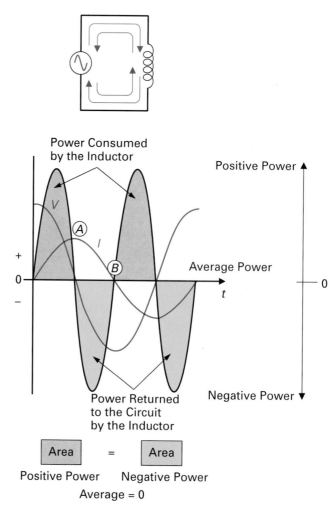

FIGURE 16-29 Purely Inductive Circuit.

between *A* and *B*, the magnetic field begins to collapse and self-induction occurs and returns electrical energy back into the circuit. The power alternation is both positive when the inductor is consuming power and negative when the inductor is returning the power back into the circuit. As the positive and negative power alternations are equal but opposite, the average power dissipated is zero.

Resistive and Inductive Circuit

An inductor is different from a capacitor in that it has a small amount of resistance no matter how pure the inductor. For this reason, inductors will never have an average power figure of zero, because even the best inductor will have some value of inductance and resistance within it, as seen in Figure 16-30. The reason an inductor has resistance is that it is simply a piece of wire, and any piece of wire has a certain value of resistance, as well as inductance. This coil resistance should be, and normally is, very small and can usually be ignored; however, in some applications even this small resistance can prevent the correct operation of a circuit, so a value or term had to be created to specify the differences in the quality of inductor available.

The **quality factor** *(Q)* of an inductor is the ratio of the energy stored in the coil by its inductance to the energy dissipated in the coil by the resistance; therefore, the higher the *Q*, the better the coil is at storing energy rather than dissipating it:

Quality Factor
Quality factor of an inductor or capacitor; it is the ratio of a component's reactance (energy stored) to its effective series resistance (energy dissipated).

$$\text{quality } (Q) = \frac{\text{energy stored}}{\text{energy dissipated}}$$

The energy stored is dependent on the inductive reactance (X_L) of the coil, and the energy dissipated is dependent on the resistance (R) of the coil. The quality factor of a coil or inductor can therefore also be calculated by using the formula

$$Q = \frac{X_L}{R}$$

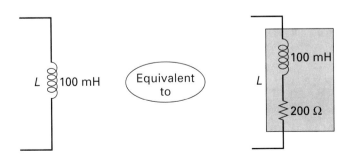

FIGURE 16-30 **Resistance within an Inductor.**

Calculate the quality factor Q of a 22 mH coil connected across a 2 kHz, 10 V source if its internal coil resistance is 27 Ω.

■ *Solution:*

$$Q = \frac{X_L}{R}$$

The reactance of the coil is not known but can be calculated by the formula

$$X_L = 2\pi f L$$
$$= 2\pi \times 2\ \text{kHz} \times 22\ \text{mH} = 276.5\ \Omega$$

Therefore,

$$Q = \frac{X_L}{R} = \frac{276.5\ \Omega}{27\ \Omega} = 10.24$$

Inductors, therefore, will never appear as pure inductance, but rather as an inductive and resistive *(RL)* circuit, and the resistance within the inductor will dissipate *true power.*

Figure 16-31 illustrates a circuit containing R and L and the power waveforms produced when $R = X_L$; the phase shift (θ) is equal to 45°.

The positive power alternation, which is above the zero line, is the combination of the power dissipated by the resistor and the power consumed by the inductor, while circuit current is on the rise. The negative power alternation is the power that was given back to the circuit by the inductor while the inductor's magnetic field was collapsing and returning the energy that was consumed.

FIGURE 16-31 Power in a Resistive and Inductive Circuit.

Power Factor

When a circuit contains both inductance and resistance, some of the energy is consumed and then returned by the inductor (reactive or imaginary power), and some of the energy is dissipated and lost by the resistor (resistive or true power).

Apparent power is the power that appears to be supplied to the load and is the vector sum of both the reactive and true power; it can be calculated by using the formula

$$P_A = \sqrt{P_R^2 + P_X^2}$$

where P_A = apparent power, in volt-amperes (VA)

P_R = true power, in watts (W)

P_X = reactive power, in volt-amperes reactive (VAR)

The power factor is a ratio of the true power to the apparent power and is therefore a measure of the loss in a circuit.

$$PF = \frac{\text{true power } (P_R)}{\text{apparent power } (P_A)}$$

or

$$PF = \frac{R}{Z}$$

or

$$PF = \cos\theta$$

EXAMPLE:

Calculate the following for a series RL circuit if $R = 40$ kΩ, $L = 450$ mH, $f = 20$ kHz, and $V_S = 6$ V:

a. X_L

b. Z

c. I

d. θ

e. Apparent power

f. PF

Solution:

a. $X_L = 2\pi f L = 2\pi \times 20 \text{ kHz} \times 450 \text{ mH}$

$\qquad = 56.5 \text{ k}\Omega$

b. $Z = \sqrt{R^2 + X_L^2}$

$\qquad = \sqrt{40 \text{ k}\Omega^2 + 56.5 \text{ k}\Omega^2} = 69.23 \text{ k}\Omega$

c. $I = \dfrac{V_S}{Z} = \dfrac{6 \text{ V}}{69.23 \text{ k}\Omega} = 86.6 \text{ µA}$

d. $\theta = \arctan \dfrac{X_L}{R} = \arctan \dfrac{56.5 \text{ k}\Omega}{40 \text{ k}\Omega} = 54.7°$

e. Apparent power $= \sqrt{(\text{true power})^2 + (\text{reactive power})^2}$

$\qquad P_R = I^2 \times R = 86.6 \text{ µA}^2 \times 40 \text{ k}\Omega = 300 \text{ µW}$

$\qquad P_X = I^2 \times X_L = 86.6 \text{ µA}^2 \times 56.5 \text{ k}\Omega = 423.7 \text{ µVAR}$

$\qquad P_A = \sqrt{P_R^2 + P_X^2}$

$\qquad\quad = \sqrt{300 \text{ µW}^2 + 423.7 \text{ µW}^2} = 519.2 \text{ µVA}$

f. $\text{PF} = \dfrac{P_R}{P_A} = \dfrac{300 \text{ µW}}{519.2 \text{ µW}} = 0.57$

$\qquad = \dfrac{R}{Z} = \dfrac{40 \text{ k}\Omega}{69.23 \text{ k}\Omega} = 0.57$

$\qquad = \cos \theta = \cos 54.7° = 0.57$

SELF-TEST REVIEW QUESTIONS FOR SECTION 16-8

1. True or false: In a purely inductive circuit, the current will lead the applied voltage by 90°.
2. Calculate the applied source voltage V_S in an *RL* circuit where $V_R = 4$ V and $V_L = 2$ V.
3. Define and state the formula for impedance when R and X_L are known.
4. If $R = X_L$, the phase shift will equal _____.
5. What is positive power?
6. Define Q and state the formula when X_L and R are known.
7. What is the difference between true and reactive power?
8. State the power factor formula.

16-9 PARALLEL *RL* CIRCUIT

Now that we have seen the behavior of resistors and inductors in series, let us analyze the *RL* parallel circuit.

16-9-1 Current

In Figure 16-32(a) you will see a parallel combination of a resistor and inductor. The voltage across both components is equal because of the parallel connection, and the current through each branch is calculated by applying Ohm's law.

$$\text{(resistor current) } I_R = \frac{V_S}{R}$$

$$\text{(inductor current) } I_L = \frac{V_S}{X_L}$$

Total current (I_T) is equal to the vector combination of the resistor current and inductor current, as shown in Figure 16-32(b). Figure 16-32(c) illustrates the current waveforms.

$$I_R = \frac{V_S}{R}$$

$$I_L = \frac{V_S}{R}$$

$X_L = R$
$I_L = I_R$

(a)

Phase Angle −45°

$$I_T = \sqrt{I_R^2 + I_L^2}$$

−45°

−90°

(b)

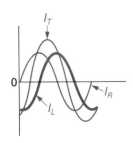

(c)

FIGURE 16-32   Parallel *RL* Circuit.

$$I_T = \sqrt{I_R^2 + I_L^2}$$

16-9-2 Phase Angle

The angle by which the total current (I_T) leads the source voltage (V_S) can be determined with either of the following formulas:

$$\theta = \arctan \frac{I_L}{I_R}$$

or

$$\theta = \arctan \frac{R}{X_L}$$

16-9-3 Impedance

The total opposition or impedance of a parallel *RL* circuit can be calculated by

$$Z = \frac{V_S}{I_T}$$

Using basic algebra, this formula can be rearranged to express impedance in terms of reactance and resistance.

$$Z = \frac{R \times X_L}{\sqrt{R^2 + X_L^2}}$$

16-9-4 Power

As with series *RL* circuits, resistive power and reactive power can be calculated by

$$P_R = I_R^2 \times R$$

$$P_X = I_L^2 \times X_L$$

The apparent power of the circuit is calculated by

$$P_A = \sqrt{P_R{}^2 + P_X{}^2}$$

and, finally, the power factor is equal to

$$PF = \cos \theta = \frac{P_R}{P_A}$$

EXAMPLE:

Calculate the following for a parallel RL circuit if $R = 45 \ \Omega$, $X_L = 1100 \ \Omega$, and $V_S = 24$ V:

a. I_R

b. I_L

c. I_T

d. Z

e. θ

Solution:

a. $I_R = \dfrac{V_S}{R} = \dfrac{24 \text{ V}}{45 \ \Omega} = 533.3$ mA

b. $I_L = \dfrac{V_S}{X_L} = \dfrac{24 \text{ V}}{1100 \ \Omega} = 21.8$ mA

c. $I_T = \sqrt{I_R{}^2 + I_L{}^2}$

$= \sqrt{533.3 \text{ mA}^2 + 21.8 \text{ mA}^2} = 533.7$ mA

d. $Z = \dfrac{R \times X_L}{\sqrt{R^2 + X_L{}^2}} = \dfrac{45 \ \Omega \times 1100 \ \Omega}{\sqrt{45 \ \Omega^2 + 1100 \ \Omega^2}}$

$= \dfrac{49.5 \text{ k}\Omega}{1100.9 \ \Omega} = 44.96 \ \Omega$

e. $\theta = \arctan \dfrac{R}{X_L} = \arctan \dfrac{45 \ \Omega}{1100 \ \Omega} = 2.34°$

Therefore, I_T lags V_S by 2.34°.

SELF-TEST REVIEW QUESTIONS FOR SECTION 16-9

1. True or false: When R and L are connected in parallel with one another, the voltage across each will be different and dependent on the values of R and X_L.

2. State the formula for calculating total current (I_T) when I_R and I_L are known.

FIGURE 16-33   Defective Inductors.

16-10   TESTING INDUCTORS

Basically, only three problems can occur with inductors:

1. An open
2. A complete short
3. A section short (value change)

16-10-1   *Open [Figure 16-33(a)]*

This problem can be isolated with an ohmmeter. Depending on the winding's resistance, the coil should be in the range of zero to a few hundred ohms. An open accounts for 75% of all defective inductors.

FIGURE 16-34   Capacitor and Inductor Analyzer.

16-10-2 *Complete or Section Short [Figure 16-33(b)]*

A coil with one or more shorted turns or a complete short can be checked with an ohmmeter and thought to be perfectly good because of the normally low resistance of a coil, as it is just a piece of wire. But if it is placed in a circuit with a complete or section short present, it will not function effectively as an inductor, if at all. For these checks, an **inductor analyzer** needs to be used like the one seen in Figure 16-34, which can be used to check capacitance and inductance. Complete or section shorts account for 25% of all defective inductors.

Inductor Analyzer
A test instrument designed to test inductors.

SELF-TEST REVIEW QUESTIONS FOR SECTION 16-10

1. How could the following inductor malfunctions be recognized?
   a. An open
   b. A complete or section short
2. Which inductor malfunction accounts for almost 75% of all failures?

16-11 APPLICATIONS OF INDUCTORS

16-11-1 RL *Filters*

The **RL filter** will achieve results similar to the *RC* filter in that it will pass some frequencies and block others, as seen in Figure 16-35. The inductive reactance of the coil and the resistance of the resistor form a voltage divider. Since inductive reactance is proportional to frequency ($X_L \propto f$), the inductor will drop less voltage at lower frequencies ($f\downarrow, X_L\downarrow, V_L\downarrow$) and more voltage at higher frequencies ($f\uparrow, X_L\uparrow, V_L\uparrow$).

RL **Filter**
A selective circuit of resistors and inductors which offers little or no opposition to certain frequencies while blocking or alternating other frequencies.

With the low-pass filter shown in Figure 16-35(a) the output is developed across the resistor. If the frequency of the input is low, the inductive reactance will be low, so almost all the input will be developed across the resistor and applied to the output. If the frequency of the input increases, the inductor's reactance will increase, resulting in almost all the input being dropped across the inductor and none across the resistor and therefore the output.

With the high-pass filter, shown in Figure 16-35(b), the inductor and resistor have been placed in opposite positions. If the frequency of the input is low, the inductive reactance will be low, so almost all the input will be developed across the resistor and very little will appear across the inductor and therefore at the output. If the frequency of the input is high, the inductive reactance will be high, resulting in almost all of the input being developed across the input and therefore appearing at the output.

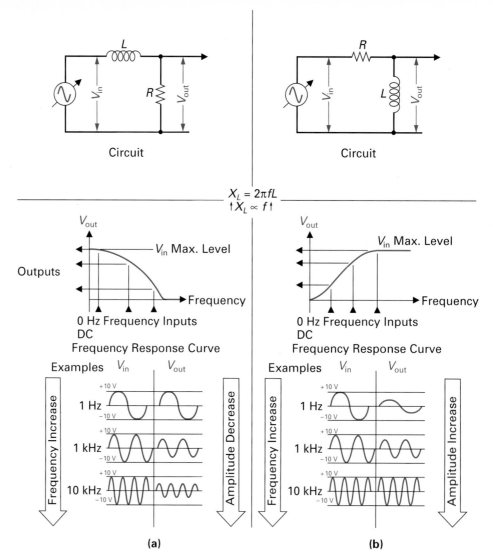

FIGURE 16-35 *RL* **Filter. (a) Low-Pass Filter. (b) High-Pass Filter.**

16-11-2 **RL** *Integrator*

RL Integrator
An *RL* circuit with an output proportionate to the integral of the input signal.

In the preceding section on *RL* filters we saw how an *RL* circuit reacted to a sine-wave input of different frequencies. In this section and the following we will see how an *RL* circuit will react to a square-wave input, and show two other important applications of inductors. In the **RL integrator,** the output is taken across the resistor, as seen in the circuit in Figure 16-36(a). The output shown in Figure 16-36(b) is the same as the previously described *RC* integrator's output. When the input rises from 0 to 10 V at the leading positive edge of the square wave input, the situation is as seen in Figure 16-37(a).

The inductor's 10 V counter emf opposes the sudden input change from 0 to 10 V, and if 10 V is across the inductor, 0 V must be across the resistor

(a)

Input

Output

Voltage
across
Resistor

(b)

FIGURE 16-36 *RL* **Integrator.**

Expanding Magnetic Field, Self-Induction

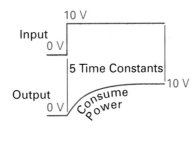

Input

5 Time Constants

Output

Consume Power

(a)

Stationary Magnetic Field, No Self-Induction

Input

Output

(b)

Collapsing Magnetic Field, Self-Induction

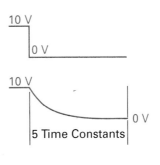

5 Time Constants

(c)

FIGURE 16-37 **Input/Output Analysis of *RL* Integrator.**

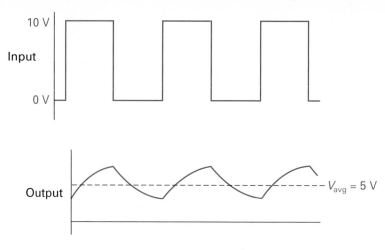

FIGURE 16-38

(Kirchhoff's voltage law), and therefore appearing at the output. After five time constants ($5 \times L/R$), the inductor's current in the circuit will have built up to maximum (V_{in}/R), and the inductor will be an equivalent short circuit, because no change and consequently no back emf exist, as shown in Figure 16-37(b). All the input voltage will now be across the resistor and therefore at the output.

When the square-wave input drops to zero, the circuit is equivalent to that seen in Figure 16-37(c), and the collapsing magnetic field will cause an induced voltage within the conductor, which will cause current to flow within the circuit for five time constants, whereupon it will reach 0.

As with the *RC* integrator, if the period of the square wave is decreased or if the time constant is increased, the output will reach an average value of half the input square wave's amplitude, as shown in Figure 16-38.

FIGURE 16-39   *RL* **Differentiator.**

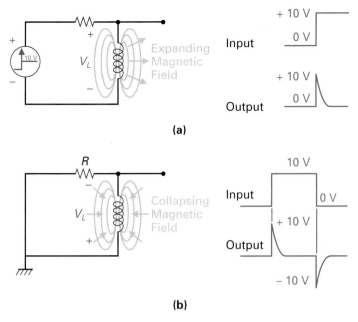

(a)

(b)

FIGURE 16-40   Input/Output Analysis of *RL* Differentiator.

16-11-3   RL *Differentiator*

With the ***RL differentiator,*** the output is taken across the inductor, as shown in Figure 16-39, and is the same as the *RC* differentiator's output. When the square-wave input rises from 0 to 10 V, the inductor will generate a 10 V counter emf across it, as shown in Figure 16-40(a), and this 10 V will all appear across the output. As the circuit current exponentially rises, the voltage across the inductor, and therefore at the output, will fall exponentially to 0 V after five time constants.

When the square wave input falls from 10 to 0 V, the circuit is equivalent to that shown in Figure 16-40(b), and the collapsing magnetic field induces a counter emf. The sudden −10 V impulse at the output will decrease to 0 V as the circuit current decreases in five time constants.

RL **Differentiator**
An *RL* circuit whose output voltage is proportional to the rate of change of the input voltage.

SELF-TEST REVIEW QUESTIONS FOR SECTION 16-11

1. List three circuit applications of the inductor.

2. Will an *RL* filter act as a low-pass filter or high-pass filter, or can it be made to function as either?

3. What are the waveform differences between an integrator and differentiator?

SUMMARY

1. Two important rules relate to inductance. The first was that a magnetic field will build up around any current-carrying conductor (electromagnetism), and the second was that a voltage will be induced into a conductor when it is subjected to a moving magnetic field. These two rules form the basis for a phenomenon called *self-inductance* or more commonly *inductance*.

2. Inductance, by definition, is the ability of a device to oppose a change in current flow, and the device designed specifically to achieve this function is called the *inductor*.

3. There is, in fact, no physical difference between an inductor and an electro-magnet, since they are both coils. The two devices are given different names because they are used in different applications even though their construction and principle of operation is the same. An electromagnet is used to generate a magnetic field in response to current, while an inductor is used to oppose any change in current.

Inductance (Figure 16-41)

4. When an external voltage source is connected across a coil, it will force a current through the coil. This current will generate a magnetic field that will increase in field strength in a short time from zero to maximum, expanding from the center of the conductor (electromagnetism). The expanding magnetic lines of force have relative motion with respect to the stationary conductor, so an induced voltage results (electromagnetic induction). The blooming magnetic field generated by the conductor is actually causing a voltage to be induced in the conductor that is generating the magnetic field. This effect of a current carrying coil of conductor inducing a voltage within itself is known as self-inductance.

5. This phenomenon was first discovered by Heinrich Lenz, who observed that the induced voltage causes an induced (bucking) current to flow in the coil, which opposes the source current producing it.

6. The induced voltage is called a counter emf or back emf. It opposes the applied emf (or battery voltage). The ability of a coil or conductor to induce or produce a counter emf within itself as a result of a change in current is called *self-inductance,* or more commonly inductance (symbolized *L*).

7. The unit of inductance is the henry (H), named in honor of Joseph Henry, an American physicist, for his experimentation within this area of science.

8. The inductance of an inductor is 1 henry when a current change of 1 ampere per second causes an induced voltage of 1 volt.

9. Inductance is therefore a measure of how much counter emf (induced voltage) can be generated by an inductor for a given amount of current change through that same inductor.

10. A larger inductance ($L\uparrow$) will create a larger induced voltage ($V_{ind}\uparrow$), and if the rate of change of current with respect to time is increased ($\Delta I/\Delta t\uparrow$), the induced voltage or counter emf will also increase ($V_{ind}\uparrow$).

QUICK REFERENCE SUMMARY SHEET

FIGURE 16-41   Inductance.

$$V_{ind} = L \times \frac{\Delta I}{\Delta t}$$

L = inductance, in henrys
ΔI = increment of change of current
Δt = increment of change of time

$$L = \frac{N^2 \times A \times \mu}{l}$$

N = number of turns
A = cross-sectional area, in square meters (m²)
μ = permeability
l = length of core, in meters (m)

Series *RL* Circuit:

- $L_T = L_1 + L_2 + L_3 + \ldots$ L_T = total inductance, in henrys
- $V_R = I \times R$
- $V_L = I \times X_L$
- $V_S = \sqrt{V_R^2 + V_L^2}$ $V_R = \sqrt{V_S^2 - V_L^2}$ $V_L = \sqrt{V_S^2 - V_R^2}$
- $Z = \frac{V}{I}$ $Z = \sqrt{R^2 + X_L^2}$ $R = \sqrt{Z^2 - X_L^2}$ $X_L = \sqrt{Z^2 - R^2}$
- $\theta = \arctan \frac{X_L}{R}$ $\theta = \arctan \frac{V_L}{V_R}$

Parallel *RL* Circuit:

- $I_R = \frac{V_S}{R}$
- $I_L = \sqrt{V_S \backslash X_L}$
- $I_T = \sqrt{I_R^2 + I_L^2}$
- $Z = \frac{R \times X_L}{\sqrt{R^2 + X_L^2}}$
- $\theta = \arctan \frac{R}{X_L}$
- $L_T = \frac{1}{(1/L_1) + (1/L_2) + (1/L_3) + \ldots}$
- $L_T = \frac{L_1 \times L_2}{L_1 + L_2}$
- $P_A = \sqrt{P_R^2 + P_X^2}$ P_A = apparent power, in volt-amperes (VA)

  P_R = true power, in watts (W)

  P_X = reactive power, in volt-amperes reactive (VAR)

- $PF = \frac{\text{true power}}{\text{apparent power}}$ $PF = \frac{Z}{R}$ $PF = \cos\theta$ PF = power factor

$$\tau = \frac{L}{R}$$ τ = inductive time constant, in seconds (s)

$$X_L = 2\pi f L$$

X_L = inductive reactance, in ohms (Ω)
2π = a constant
f = frequency, in hertz (Hz)
L = inductance, in henrys (H)

$$Q = \frac{\text{energy stored}}{\text{energy dissipated}}$$ Q = quality factor of an inductor $$Q = \frac{X_L}{R}$$

FIGURE 16-41 (continued)

Permeabilities of Various Materials

MATERIAL	PERMEABILITY (μ)
Air or vacuum	1.26×10^{-6}
Nickel	6.28×10^{-5}
Cobalt	7.56×10^{-5}
Cast iron	1.1×10^{-4}
Machine steel	5.65×10^{-4}
Transformer iron	6.9×10^{-3}
Silicon iron	8.8×10^{-3}
Permalloy	0.126
Supermalloy	1.26

DC Inductor Current Rise and Fall

Series RL Circuit

FIGURE 16-41 (continued)

Parallel RL Circuit

$$I_R = \frac{V_S}{R}$$

$$I_L = \frac{V_S}{R}$$

$$X_L = R$$
$$I_L = I_R$$

(a)

$$I_T = \sqrt{I_R^2 + I_L^2}$$

(b)

(c)

RL Filter

11. If the current in the circuit is suddenly increased (by lowering the circuit resistance), the change in the expanding magnetic field will induce a counter emf within the inductor. This induced voltage will oppose the source voltage from the battery and attempt to hold current at its previous low level.

12. The counter emf cannot completely oppose the current increase, for if it did, the lack of current change would reduce the counter emf to zero. Current therefore incrementally increases up to a new maximum, which is determined by the applied voltage and the circuit resistance $(I = V/R)$.

13. Once the new higher level of current has been reached and remains constant, there will no longer be a change. This lack of relative motion between field and conductor will no longer generate a counter emf, so the current will remain at its new higher constant value.

14. This effect also happens in the opposite respect. If current decreases (by increasing circuit resistance), the magnetic lines of force will collapse because of the reduction of current and induce a voltage in the inductor, which will produce an induced current in the same direction as the circuit current. These two combine and tend to maintain the current at the higher previous constant level. Circuit current will fall, however, as the induced voltage and current are only present during the change (in this case the decrease from the higher current level to the lower); and once the new lower level of current has been reached and remains constant, the lack of change will no longer induce a voltage or current. So the current will then remain at its new lower constant value.

15. The inductor is therefore an electronic component that will oppose any changes in circuit current, and this ability or behavior is referred to as *inductance*. Since the current change is opposed by a counter emf, inductance may also be defined as the ability of a device to induce a counter emf within itself for a change in current.

Factors Determining Inductance

16. The inductance of an inductor is determined by four factors:
    a. Number of turns
    b. Area of the coil
    c. Length of the coil
    d. Core material used within the coil

17. If an inductor has a greater number of turns, the magnetic field produced by passing current through the coil will have more magnetic force than an inductor with fewer turns. A greater magnetic field will cause a larger counter emf, because more magnetic lines of flux will cut more coils of the conductor, producing a larger inductance value. Inductance (L) is therefore proportional to the number of turns (N).

18. If the area of the coil is increased for a given number of turns, more magnetic lines of force will be produced, and if the magnetic field is increased, the inductance will also increase. Inductance (L) is therefore proportional to the cross-sectional area of the coil (A).

19. Inductance is inversely proportional to the length of the coil, in that a longer coil, for a given number of turns, produces a smaller inductance, and vice versa.

20. Most inductors have core materials such as nickel, cobalt, iron, steel, ferrite, or an alloy. These cores have magnetic properties that concentrate or intensify the magnetic field. Permeability is another factor that is proportional to inductance.

Inductors in Combination

21. Inductors oppose the change of current in a circuit and so are treated in a manner similar to resistors connected in combination.

22. Two or more inductors in series merely extend the coil length and increase inductance.

23. Inductors in parallel are treated in a manner similar to resistors, with the total inductance being less than that of the smallest inductor's value.

Types of Inductors

24. As with resistors and capacitors, inductors are basically divided into the two categories of fixed-value and variable-value inductors.

25. With fixed-value inductors, the inductance value remains constant and cannot be altered. It is usually made of solid copper wire with an insulating enamel around the conductor. This enamel or varnish on the conductor is needed since the turns of a coil are generally wound on top of one another. The three major types of fixed-value inductors on the market today are:

a. Air core
b. Iron core
c. Ferrite core

26. The name choke is used interchangeably with the word *inductor*, because choke basically describes an inductors' behavior, which generates a counter emf to limit or choke the flow of current.

27. Ferrite is a chemical compound, which is basically powdered iron oxide and ceramic. Although its permeability is less than iron, it has a higher permeability than air and can therefore obtain a higher inductance value than air for the same number of turns.

28. The toroidal inductor (doughnut shaped) has an advantage over the cylinder-shaped core in that the magnetic lines of force do not have to pass through air. With a cylindrical core, the magnetic lines of force merge out of one end of the cyclinder, propagate or travel through air, and then enter in the other end of the cylinder. The nonmagnetic air, which has a low permeability, will reduce the strength of the field and therefore the inductance value.

29. With variable-value inductors, the inductance value can be changed by adjusting the position of the movable core with respect to the stationary coil. A screw adjustment moves a sliding ferrite core or slug farther in or out of the stationary coil. If the slug is all the way out, the permeability figure is low, because an air core is being used ($L \downarrow \propto \mu \downarrow$). If the slug is adjusted and screwed into the coil, the air is replaced by the ferrite slug, which has a higher permeability and, therefore, inductance value ($L \uparrow \propto \mu \uparrow$).

30. Variable inductors should only be adjusted with a plastic or nonmetallic alignment tool since any metal object in the vicinity of the inductor will interfere with the magnetic field and change the inductance of the inductor.

Inductive Time Constant

31. Inductors will not have any effect on a steady value of direct current (dc) from a dc voltage source. If, however, the dc is changing (pulsating), the inductor will oppose the change whether it is an increase or decrease in direct current, because a change in current causes the magnetic field to expand or contract, and in so doing it will cut the coil of the inductor and induce a voltage that will counter the applied emf.

32. Current in an inductive circuit, therefore, cannot rise instantly to its maximum value, which is determined by Ohm's law ($I = V/R$). Current will in fact take a time to rise to maximum, due to the inductor's ability to oppose change.

33. It will actually take five time constants (5τ) for the current in an inductive circuit to reach maximum value.

34. The constant to remember is the same as before: 63.2%. In one time constant ($1 \times L/R$) the current in the LR circuit will have reached 63.2% of its maximum value. In two time constants ($2 \times L/R$), the current will have increased 63.2% of the remaining current, and so on, through five time constants.

35. When the inductor's dc source of current is removed, the magnetic field will collapse and cut the coils of the inductor, inducing a voltage and causing a current to flow in the same direction as the original source current. This current will exponentially decay, or fall from the maximum to zero level, in five time constants.

36. If an alternating (ac) voltage is applied across an inductor, the inductor will continuously oppose the alternating current because it is always changing. The current in the circuit causes the magnetic field to expand and collapse and cut the conducting coils, resulting in an induced counter emf.

37. The voltage across the inductor (V_L) or counter emf leads the circuit current (I) by 90°.

Inductive Reactance

38. Reactance is the opposition to current flow without the dissipation of energy, as opposed to resistance, which is the opposition to current flow with the dissipation of energy.

39. Inductive reactance (X_L) is the opposition to current flow offered by an inductor without the dissipation of energy.

40. Inductive reactance is proportional to frequency ($X_L \propto f$) because a higher frequency (fast switching current) will cause a greater amount of current change, and a greater change will generate a larger counter emf, which is an opposition or reactance against current flow. When 0 Hz is applied to a coil (dc), there exists no change, so the inductive reactance of an inductor to dc is zero ($X_L = 2\pi \times 0 \times L = 0$).

41. Inductive reactance is also proportional to inductance because a larger inductance will generate a greater magnetic field and subsequent counter emf, which is the opposition to current flow.

42. Ohm's law can be applied to inductive circuits just as it can be applied to resistive and capacitive circuits. The current flow in an inductive circuit (I) is proportional to the voltage applied (V), and inversely proportional to the inductive reactance (X_L).

Series RL Circuit

43. In a purely resistive circuit, the current flowing within the circuit and the voltage across the resistor are in phase with one another.

44. In a purely inductive circuit, the current will lag the applied voltage by 90°.

45. If we connect a resistor and inductor in series, we will have the most common combination of R and L used in electronic equipment.

46. The voltage across the resistor (V_R) is in phase with the circuit current. The voltage across the inductor (V_L) leads the circuit current and the voltage across the resistor (V_R) by 90°, or, the circuit current vector lags the voltage across the inductor by 90°.

47. As with any series circuit, we have to apply Kirchhoff's voltage law when calculating the value of applied or source voltage (V_S), which, due to the phase difference between V_R and V_L, is the vector sum of all the voltage drops.

48. Impedance is the total opposition to current flow offered by a circuit with both resistance and reactance. It is measured in ohms and can be calculated by using Ohm's law. The impedance of a series RL circuit is equal to the square root of the sum of the squares of resistance and reactance, and by rearrangement, X_L and R can also be calculated if the other two values are known.

49. If a circuit is purely resistive, the phase shift (θ) between the source voltage and circuit current is zero. If a circuit is purely inductive, voltage leads current by 90°; therefore, the phase shift is + 90°. If the resistance and inductive reactance are equal, the phase shift will equal + 45°.

50. The phase shift in an inductive and resistive circuit is the degrees of lead between the source voltage (V_S) and current (I). The phase angle is proportional to reactance and inversely proportional to resistance.

51. As the current is the same in both the inductor and resistor in a series circuit, the voltage drops across the inductor and resistor are directly proportional to reactance and resistance.

52. When applying an ac voltage across a purely resistive circuit, voltage and current are in phase, and true (P_R) power in watts can be calculated by multiplying current by voltage ($P = V \times I$).

53. The pure inductor, like the capacitor, is a reactive component, which means that it will consume power without the dissipation of energy. The capacitor holds its energy in an electric field, while the inductor consumes and holds its energy in a magnetic field and then releases it back into the circuit. The power alternation is both positive when the inductor is consuming power and negative when the inductor is returning the power back into the circuit. As the positive and negative power alternations are equal but opposite, the average power dissipated is zero.

54. An inductor is different from a capacitor in that it has a small amount of resistance no matter how pure the inductor. For this reason, inductors will never have an average power figure of zero, because even the best inductor will have some value of inductance and resistance within it.

55. The reason an inductor has resistance is that it is simply a piece of wire, and any piece of wire has a certain value of resistance, as well as inductance.

56. The quality factor (Q) of an inductor is the ratio of the energy stored in the coil by its inductance to the energy dissipated in the coil by the resistance;

therefore, the higher the Q, the better the coil at storing energy rather than dissipating it. The energy stored is dependent on the inductive reactance (X_L) of the coil, and the energy dissipated is dependent on the resistance (R) of the coil.

57. Inductors, therefore, will never appear as pure inductance, but rather as an inductive and resistive (RL) circuit, and the resistance within the inductor will dissipate *true power*.

58. When a circuit contains both inductance and resistance, some of the energy is consumed and then returned by the inductor (reactive or imaginary power), and some of the energy is dissipated and lost by the resistor (resistive or true power).

59. Apparent power is the power that appears to be supplied to the load and is the vector sum of both the reactive and true power.

Parallel *RL* Circuit

60. With a parallel combination of a resistor and inductor, the voltage across both components is equal because of the parallel connection, and the current through each branch is calculated by applying Ohm's law.

61. Total current (I_T) is equal to the vector combination of the resistor current and inductor current.

Testing Inductors

62. Basically, only three problems can occur with inductors:
   a. An open
   b. A complete short
   c. A section short (value change)

63. Depending on the winding's resistance, the coil should be in the range of zero to a few hundred ohms. An open accounts for 75% of all defective inductors.

64. A coil with one or more shorted turns or a complete short can be checked with an ohmmeter and thought to be perfectly good because of the normally low resistance of a coil, as it is just a piece of wire.

65. An inductor analyzer can be used to check capacitance and inductance. Complete or section shorts account for 25% of all defective inductors.

RL Filters

66. The *RL* filter will achieve results similar to the *RC* filter in that it will pass some frequencies and block others. The inductive reactance of the coil and the resistance of the resistor form a voltage divider. Since inductive reactance is proportional to frequency $(X_L \propto f)$, the inductor will drop less voltage at lower frequencies $(f\downarrow, X_L\downarrow, V_L\downarrow)$ and more voltage at higher frequencies $(f\uparrow, X_L\uparrow, V_L\uparrow)$.

RL Integrator

67. In the *RL* integrator, the output is taken across the resistor. The output is the same as the previously described *RC* integrator's output. As with the *RC* integrator, if the period of the square wave is decreased or if the time constant is increased, the output will reach an average value of half the input square wave's amplitude.

68. With the *RL* differentiator, the output is taken across the inductor.

Air-core inductor	Henry	*RL* differentiator
Choke	Inductance	*RL* filter
Counter emf	Inductive reactance	*RL* integrator
Delta	Inductive time constant	Self-inductance
Ferrite-core inductor	Inductor	Slug
Ferrous	Iron-core inductor	Toroidal inductor
Form	*Q* or quality factor	

Multiple-Choice Questions

1. Self-inductance is a process by which a coil will induce a voltage within _____.

   a. Another inductor   **c.** Itself
   b. Two or more close proximity inductors   **d.** Both (a) and (b)

2. Mutual inductance is a process by which a coil will induce a voltage within _____.

   a. Another inductor   **c.** Itself
   b. Two or more close proximity inductors   **d.** Both (a) and (b)

3. The inductor stores electrical energy in the form of a _____ field, just as a capacitor stores electrical energy in the form of a _____ field.

   a. Electric, magnetic   **b.** Magnetic, electric

4. The inductor is basically:

   a. An electromagnet
   b. A coil of wire
   c. A coil of conductor formed around a core material
   d. All of the above
   e. None of the above

5. The inductance of an inductor is proportional to _____, and inversely proportional to _____.

   a. $N, A, \mu; l$ **c.** $\mu, l, N; A$
   b. $A, \mu, l; N$ **d.** $N, A, l; \mu$

6. The total inductance of a series circuit is:
  a. Less than the value of the smallest inductor
  b. Equal to the sum of all the inductance values
  c. Equal to the product over sum
  d. All of the above

7. The total inductance of a parallel circuit can be calculated by:
  a. Using the product-over-sum formula
  b. Using L divided by N for equal-value inductors
  c. Using the reciprocal resistance formula
  d. All of the above

8. Air-core fixed-value inductors can use air or nonmagnetic forms, such as:
  a. Iron, cardboard   **c.** Ceramic, cardboard
  b. Ceramic, copper   **d.** Silicon, germanium

9. Ferrite is a chemical compound, which is basically powdered:
  a. Iron oxide and ceramic   **c.** Mylar and iron
  b. Iron and steel   **d.** Gauze and electrolyte

10. The ferrite-core variable inductor varies inductance by changing:
  a. μ **c.** N
  b. l **d.** A

11. It will actually take _____ time constants for the current in an inductive circuit to reach a maximum value.
  a. 63.2   **c.** 1.414
  b. 1   **d.** 5

12. The time constant for a series inductive/resistive circuit is equal to:
  a. $L \times R$ **c.** V/R
  b. L/R **d.** $2\pi \times f \times L$

13. Inductive reactance (X_L) is proportional to:
  a. Time or period of the ac applied
  b. Frequency of the ac applied
  c. The stray capacitance that occurs due to the air acting as a dielectric between two turns of a coil
  d. The value of inductance
  e. Two of the above are true.

14. In a series RL circuit, the source voltage (V_S) is equal to:
  a. The square root of the sum of $V_R{}^2$ and $V_L{}^2$
  b. The vector sum of V_R and V_L
  c. $I \times Z$
  d. Two of the above are partially true.
  e. Answers (a), (b), and (c) are correct.

15. In a purely resistive circuit, the phase shift is equal to _____, whereas in a purely inductive or capacitive circuit, the phase shift is _____ degrees.
  a. 45, 0   **c.** 45, 90
  b. 90, 0   **d.** None of the above

16. With an *RL* integrator, the output is taken across the:

    a. Inductor   **c.** Resistor
    b. Capacitor   **d.** Transformer's secondary

17. With an *RL* differentiator, the output is taken across the:

    a. Inductor   **c.** Resistor
    b. Capacitor   **d.** Transformer's secondary.

18. An inductor or choke between the input and output forms a:

    a. High-pass filter   **b.** Low-pass filter

19. An inductor or choke connected to ground or in shunt forms a:

    a. Low-pass filter   **b.** High-pass filter

20. Lenz's law states that when current is passed through a conductor a self-induced voltage in a coil will:

    a. Aid the applied source voltage
    b. Aid the increasing current from the source
    c. Produce an opposing current
    d. Both (a) and (c)

21. When tested with an ohmmeter, an open coil would show:

    a. Zero resistance   **c.** A 100 Ω to 200 Ω resistance
    b. An infinite resistance   **d.** Both (b) and (c)

22. Inductive reactance:

    a. Increases with frequency
    b. Is proportional to inductance
    c. Reduces the amplitude of alternating current
    d. All of the above

23. The current through an inductor _____ the voltage across the same inductor by _____.

    a. Lags, 90°   **c.** Leads, 90°
    b. Lags, 45°   **d.** Leads, 45°

24. The phasor combination of X_L and R is the circuit's:

    a. Reactance   **c.** Power factor
    b. Total resistance   **d.** Impedance

25. In a series *RL* circuit, where $V_R = 200$ mV and $V_L = 0.2$ V, $\theta = $:

    a. 45°   **c.** 0°
    b. 90°   **d.** 1°

Essay Questions

26. Briefly describe the terms: (Chapter 15, introduction)

    a. Electromagnetism   **b.** Electromagnetic induction

27. What is self-induction, and how does it relate to an inductor? (16-1)

28. Give the formula for inductance, and explain the four factors that determine inductance. (16-3)

29. List all the formulas for calculating total inductance when inductors are connected in: (16-4)

    a. Series **b.** Parallel

30. What are a fixed- and a variable-value inductor? (16-5)

31. List the important factors of the: (16-5)

    a. Air-, iron-, and ferrite-core fixed-value inductors
    b. Ferrite-core variable-value inductor

32. Describe the current rise and fall through an inductor when: (16-6)

    a. Dc is applied **b.** Ac is applied

33. Define the following terms:

    a. Inductive reactance (16-7) **d.** Q factor (16-8-4)
    b. Impedance (16-8-2)   **e.** Power factor (16-8-4)
    c. Phase shift (16-8-3)

34. With illustrations, describe how an inductor and resistor could be used for the following applications: (16-11)

    a. Filtering high and low frequencies **c.** Differentiation
    b. Integration

35. Explain briefly why an inductor acts as an open to an instantaneous change. Why does an inductor act like a short to dc? (16-7)

Practice Problems

36. Convert the following:

    a. 0.037 H to mH   **c.** 862 mH to H
    b. 1760 µH to mH **d.** 0.256 mH to µH

37. Calculate the impedance (Z) of the following series RL combinations:

    a. 22 MΩ, 25 µH, $f = 1$ MHz   **c.** 60 Ω, 0.05 H, $f = 1$ MHz
    b. 4 kΩ, 125 mH, $f = 100$ kHz

38. Calculate the voltage across a coil if: ($d = \Delta$ or delta)

    a. $d_i/d_t = 120$ mA/ms and $L = 2$ µH
    b. $d_i/d_t = 62$ µA/µs and $L = 463$ mH
    c. $d_i/d_t = 4$ A/s and $L = 25$ mH

39. Calculate the total inductance of the following series circuits:

    a. 75 µH, 61 µH, 50 mH **b.** 8 mH, 4 mH, 22 mH

40. Calculate the total inductance of the following parallel circuits:

    a. 12 mH, 8 mH **b.** 75 µH, 34 µH, 27 µH

41. Calculate the total inductance of the following series–parallel circuits:

    a. 12 mH in series with 4 mH, and both in parallel with 6 mH
    b. A two-branch parallel arrangement made up of 6 µH and 2 µH in series with one another, and 8 µH and 4 µH in series with one another

c. Two parallel arrangements in series with one another, made up of 1 μH and 2 μH in parallel and 4 μH and 15 μH in parallel

42. Determine the time constant of all the examples in Question 41, and state how long it will take in each example for current to build up to maximum.

43. In a series RL circuit, if $V_L = 12$ V and $V_R = 6$ V, calculate:

a. V_S **d.** Q
b. I if $Z = 14$ kΩ   **e.** Power factor
c. Phase angle

44. What value of inductance is needed to produce 3.3 kΩ of reactance at 15 kHz?

45. At what frequency will a 330 μH inductor have a reactance of 27 kΩ?

46. Calculate the impedance of the circuits seen in Figure 16-42.

47. Referring to Figure 16-43, calculate the voltage across the inductor for all five time constants after the switch has been closed.

48. Referring to Figure 16-44, calculate:

a. L_T
b. X_L
c. Z
d. I_{R1}, I_{L1}, and I_{L2}
e. θ
f. True power, reactive power, and apparent power
g. Power factor

FIGURE 16-42   **Calculating Impedance.**

FIGURE 16-43
Inductive Time
Constant Example.

FIGURE 16-44   Series *RL* Circuit.

49. Referring to Figure 16-45, calculate:

a. R_T f. I_{RT}, I_{LT}
b. L_T g. I_T
c. X_L h. θ
d. Z i. Apparent power
e. V_{RT}, V_{LT} j. PF

Troubleshooting Problem

50. What problems can occur with inductors, and how can the ohmmeter be used to determine these problems?

FIGURE 16-45   Parallel *RL* Circuit.

After completing this chapter, you will be able to:

1. Define mutual inductance and how it relates to transformers.
2. Describe the basic operation of a transformer.
3. Explain the differences between a loaded and unloaded transformer.
4. State the meaning of the coefficient of coupling.
5. List the three basic applications of transformers.

6. Describe how a transformer's turns ratio can be used to step up or step down voltage or current, or match impedances.
7. Describe the transformer dot convention as it relates to windings and phase.
8. List and explain the fixed and variable transformer types.
9. Explain how to test the windings of a transformer for opens, partial shorts, or complete shorts.
10. Describe the three basic transformer power losses.

Transformers

Woolen Mill Makes Minis

At the age of 24 in 1950, Kenneth Olsen went to work at MIT's Digital Computer Laboratory as a research assistant. For the next seven years, Olsen worked on the SAGE (Semi Automatic Ground Environment) project, which was a new computer system designed to store the constantly changing data obtained from radars tracking long-range enemy bombers capable of penetrating American air space. IBM won the contract to build the SAGE network and Olsen traveled to the plant in New York to supervise production. After completing the SAGE project, Olsen was ready to move on to bigger and better things; however, he didn't want to go back to MIT and was not happy with the rigid environment at a large corporation like IBM.

In September 1957, Olsen and fellow MIT colleague Harlan Anderson leased an old brick woolen mill in Maynard, Massachusetts, and after some cleaning and painting they began work in their new company, which they named Digital Equipment Corporation or DEC. After three years, the company came out with their first computer, the PDP 1 (programmed data processor), and its ease of use, low cost, and small size made it an almost instant success. In 1965, the desktop PDP 8 was launched, and since in that era miniskirts were popular, it was probably inevitable that the small machine was nicknamed a mini-computer. In 1969, the PDP 11 was unveiled, and like its predecessors, this number-crunching data processor went on to be accepted in many applications, such as tracking the millions of telephone calls received on the 911 emergency line, directing welding machines in automobile plants, recording experimental results in laboratories, and processing all types of data in offices, banks, and department stores. Just 20 years after introducing low-cost computing in the 1960s, DEC was second only to IBM as a manufacturer of computers in the United States.

Transformer
Device consisting of two or more coils that are used to couple electric energy from one circuit to another, yet maintain electrical isolation between the two.

When alternating current was introduced in Chapter 11, it was mentioned that an ac voltage could be stepped up to a larger voltage or stepped down to a smaller voltage by a device called a **transformer.** The transformer is an electrical device that makes use of electromagnetic induction to transfer alternating current from one circuit to another. The transformer consists of two inductors that are placed in very close proximity to one another. When an alternating current flows through the first coil or **primary winding** the inductor sets up a magnetic field. The expanding and contracting magnetic field produced by the primary cuts across the windings of the second inductor or **secondary winding** and induces a voltage in this coil.

By changing the ratio between the number of turns in the secondary winding to the number of turns in the primary winding, some characteristics of the ac signal can be changed or transformed as it passes from primary to secondary. For example, a low ac voltage could be stepped up to a higher ac voltage, or a high ac voltage could be stepped down to a lower ac voltage.

In this chapter we examine the operation, characteristics, types, testing, and applications of transformers.

Primary Winding
First winding of a transformer that is connected to the source.

Secondary Winding
Output winding of a transformer that is connected across the load.

17-1 MUTUAL INDUCTANCE

As was discussed in the previous chapter, self-inductance is the process by which a coil induces a voltage within itself. The principle on which a transformer is based is an inductive effect known as **mutual inductance,** which is the process by which an inductor induces a voltage in another inductor.

Figure 17-1 illustrates two inductors that are magnetically linked, yet electrically isolated from one another. As the alternating current continually rises, falls, and then rises in the opposite direction, a magnetic field will build up, collapse, and then build up in the opposite direction.

If a second inductor, or secondary coil (L_2), is in close proximity with the first inductor or primary coil (L_1), which is producing the alternating magnetic field, a voltage will be induced into the nearby inductor, which causes current to flow in the secondary circuit through the load resistor. This phenomenon is known as mutual inductance or transformer action.

As with self-inductance, mutual inductance is dependent on change. Direct current (dc) is a constant current and produces a constant or stationary magnetic field that does not change, as reviewed in Figure 17-2(a). Alternating current, however, is continually varying, and as the polarity of the magnetic field is dependent on the direction of current flow (left-hand rule), the magnetic field will also be alternating in polarity, as seen in Figure 17-2(b); it is this continual

Mutual Inductance
Ability of one inductor's magnetic lines of force to link with another inductor.

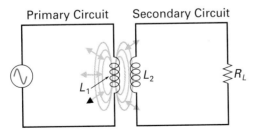

Magnetic Field Alternately Builds up and Collapses

FIGURE 17-1   Mutual Inductance.

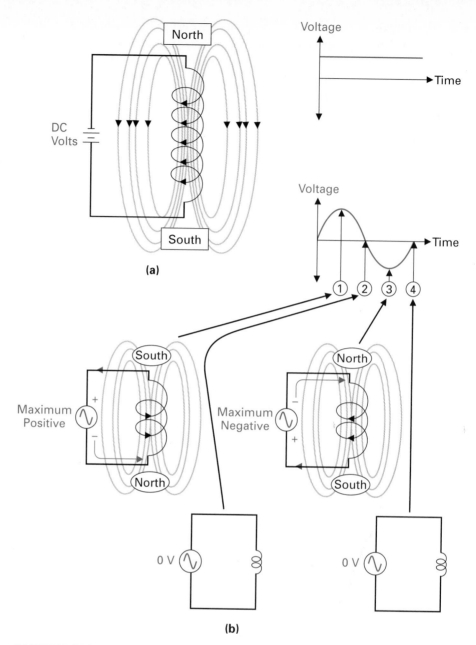

FIGURE 17-2 **Magnetic Field Produced by (a) DC across an Inductor and (b) AC across an Inductor.**

building up and collapsing of the magnetic field that cuts the adjacent inductor's conducting coils and induces a voltage in the secondary circuit. Mutual induction is possible only with ac and cannot be achieved with dc due to the lack of change.

Self-induction is a measure of how much voltage an inductor can induce within itself. Mutual inductance is a measure of how much voltage is induced in the secondary coil due to the change in current in the primary coil.

1. Define mutual inductance and how it differs from self-inductance.

2. True or false: Mutual induction is possible only with direct current flow.

17-2 BASIC TRANSFORMER

Figure 17-3(a) illustrates a basic transformer, which consists of two coils within close proximity to one another, to ensure that the second coil will be cut by the magnetic flux lines produced by the first coil, and thereby ensure mutual inductance. The ac voltage source is electrically connected (through wires) to the primary coil or winding, and the load (R_L) is electrically connected to the secondary coil or winding.

In Figure 17-3(b), the ac voltage source has produced current flow in the primary circuit, as illustrated. This current flow produces a north pole at the top of the primary winding and, as the ac voltage input swings more negative, the current increase causes the magnetic field being developed by the primary winding to increase. This expanding magnetic field cuts the coils of the secondary winding and induces a voltage, and a subsequent current flows in the secondary circuit, which travels up through the load resistor. The ac voltage follows a sinu-

(a)

(b)

(c)

FIGURE 17-3 Transformer Action.

soidal pattern and moves from a maximum negative to zero and then begins to build up toward a maximum positive.

In Figure 17-3(c), the current flow in the primary circuit is in the opposite direction due to the ac voltage increase in the positive direction. As voltage increases, current increases, and the magnetic field expands and cuts the secondary winding, inducing a voltage and causing current to flow in the reverse direction down through the load resistor.

You may have noticed a few interesting points about the basic transformer just discussed.

1. As primary current increases, secondary current increases, and as primary current decreases, secondary current also decreases. It can therefore be said that the frequency of the alternating current in the secondary is the same as the frequency of the alternating current in the primary.

2. Although the two coils are electrically isolated from one another, energy can be transferred from primary to secondary, because the primary converts electrical energy into magnetic energy, and the secondary converts magnetic energy back into electrical energy.

SELF-TEST REVIEW QUESTIONS FOR SECTION 17-2

1. True or false: A transformer achieves electrical isolation.
2. True or false: The transformer converts electrical energy to magnetic, and then from magnetic back to electrical.
3. What are the names given to the windings of a transformer?

17-3 TRANSFORMER LOADING

Let's now carry our discussion of the basic transformer a little further and see what occurs when the transformer is not connected to a load, as shown in Figure 17-4. Primary circuit current is determined by $I = V/Z$, where Z is the impedance of the primary coil (both its inductive reactance and resistance) and V is the applied voltage. Since no current can flow in the secondary, because an open in the circuit exists, the primary acts as a simple inductor, and the primary current

FIGURE 17-4   Unloaded Transformer.

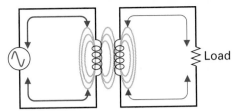

FIGURE 17-5   **Loaded Transformer.**

is small due to the inductance of the primary winding. This small primary current lags the applied voltage due to the counter emf by approximately 90°
because the coil is mainly inductive and has very little resistance.

When a load is connected across the secondary, as shown in Figure 17-5, a
change in conditions occurs and the transformer acts differently. The important
point that will be observed is that as we go from a no-load to a load condition
the primary current will increase due to mutual inductance. Let's follow the
steps one by one.

1. The ac applied voltage sets up an alternating magnetic field in the primary
   winding.

2. The continually changing flux of this primary field induces and produces a
   counter emf into the primary to oppose the applied voltage.

3. The primary's magnetic field also induces a voltage in the secondary winding, which causes current to flow in the secondary circuit through the load.

4. The current in the secondary winding produces another magnetic field that
   is opposite to the field being produced by the primary.

5. This secondary magnetic field feeds back to the primary and induces a voltage that tends to cancel or weaken the counter emf that was set up in the
   primary by the primary current.

6. The primary's counter emf is therefore reduced, so primary current can
   now increase.

7. This increase in primary current is caused by the secondary's magnetic
   field; consequently, the greater the secondary current, the stronger the secondary magnetic field, which causes a reduction in the primary's counter
   emf, and therefore a primary current increase.

In summary, an increase in secondary current ($I_s\uparrow$) causes an increase in
primary current ($I_p\uparrow$), and this effect in which the primary induces a voltage in
the secondary (V_s) and the secondary induces a voltage into the primary (V_p) is
known as *mutual inductance.*

SELF-TEST REVIEW QUESTIONS FOR SECTION 17-3

1. True or false: An increase in secondary current causes an increase in primary current.
2. True or false: The greater the secondary current, the greater the primary's
   counter emf.

17-4 COEFFICIENT OF COUPLING (*k*)

The voltage induced into the secondary winding is dependent on the mutual inductance between the primary and secondary, which is determined by how much of the magnetic flux produced by the primary actually cuts the secondary winding.

Coefficient of Coupling
The degree of coupling that exists between two circuits.

The **coefficient of coupling** (*k*) is a ratio of the number of magnetic lines of force that cut the secondary compared to the total number of magnetic flux lines being produced by the primary and is a figure between 0 and 1.

$$k = \frac{\text{flux linking secondary coil}}{\text{total flux produced by primary}}$$

If, for example, all the primary flux lines cut the secondary winding, the coefficient of coupling will equal 1. If only half of the total flux lines being produced cut the secondary winding, *k* will equal 0.5.

◾ EXAMPLE:

A primary coil is producing 65 μWb (microwebers) of magnetic flux. Calculate the coefficient of coupling if 52 μWb links with the secondary coil.

◾ *Solution:*

$$k = \frac{52 \ \mu\text{Wb}}{65 \ \mu\text{Wb}} = 0.8$$

This means that 80% of the magnetic flux lines being generated by the primary coil are linking with the secondary coil.

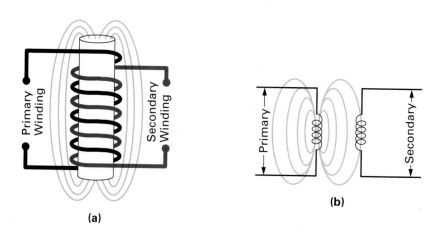

FIGURE 17-6   Coefficient of Coupling. (a) Wound on Same Core, Small Distance Between Primary and Secondary, High *k*. (b) Winding Far Apart, Small *k*.

The coefficient of coupling depends on:

1. How close together the primary and secondary are to one another
2. The type of core material used

Figure 17-6 illustrates why the primary and secondary have to be in close proximity to one another to achieve a high coefficient of coupling.

17-5 TRANSFORMER RATIOS AND APPLICATIONS

Basically, transformers are used for one of three applications:

1. To step up (increase) or step down (decrease) voltage
2. To step up (increase) or step down (decrease) current
3. To match impedances

In all three cases, any of the applications can be achieved by changing the ratio of the number of turns in the primary winding compared to the number of turns in the secondary winding. This ratio is appropriately called the turns ratio.

17-5-1 Turns Ratio

The **turns ratio** is the ratio between the number of turns in the secondary winding (N_s) and the number of turns in the primary winding (N_p).

Turns Ratio
Ratio of the number of turns in the secondary winding to the number of turns in the primary winding of a transformer.

$$\text{turns ratio} = \frac{N_s}{N_p}$$

Let us use a few examples to see how the turns ratio can be calculated.

EXAMPLE:

If the primary has 200 turns and the secondary has 600, what is the turns ratio (Figure 17-7)?

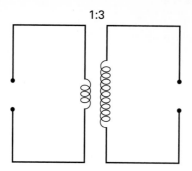

1:3

FIGURE 17-7 Step-Up Transformer Example.

■ *Solution:*

$$\text{turns ratio} = \frac{N_s}{N_p}$$

$$= \frac{600}{200}$$

$$= \frac{3\,(\text{secondary})}{1\,(\text{primary})}$$

$$= 3$$

This simply means that there are three windings in the secondary to every one winding in the primary. Moving from a small number (1) to a larger number (3) means that we *stepped up* in value. Stepping up always results in a turns ratio figure greater than 1, in this case, 3.

■ **EXAMPLE:**

If the primary has 120 turns and the secondary has 30 turns, what is the turns ratio (Figure 17-8)?

■ *Solution:*

$$\text{turns ratio} = \frac{N_s}{N_p}$$

$$= \frac{30}{120}$$

$$= \frac{1\,(\text{secondary})}{4\,(\text{primary})}$$

$$= 0.25$$

Said simply, there are four primary windings to every one secondary winding. Moving from a larger number (4) to a smaller number (1) means that we *stepped down* in value. Stepping down always results in a turns ratio figure of less than 1, in this case 0.25.

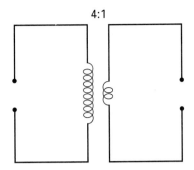

FIGURE 17-8   **Step-Down Transformer Example.**

17-5-2   *Voltage Ratio*

Transformers are used within the power supply unit of almost every piece of electronic equipment to step up or step down the 115 V ac from the outlet. Some electronic circuits require lower-power supply voltages, while other devices may require higher-power supply voltages. The transformer is used in both instances to convert the 115 V ac to the required value of voltage.

Step Up

If the secondary voltage (V_s) is greater than the primary voltage (V_p), the transformer is called a **step-up transformer** ($V_s > V_p$), as shown in Figure 17-9. The voltage is stepped up or increased in much the same way as a generator voltage can be increased by increasing the number of turns.

If the ac primary voltage is 100 V and the turns ratio is a 1:5 step up, the secondary voltage will be five times that of the primary voltage, or 500 V, because the magnetic flux established by the primary cuts more turns in the secondary and therefore induces a larger voltage.

In this example, you can see that the ratio of the secondary voltage to the primary voltage is equal to the turns ratio; in other words,

Step-Up Transformer
Transformer in which the ac voltage induced in the secondary is greater (due to more secondary windings) than the ac voltage applied to the primary.

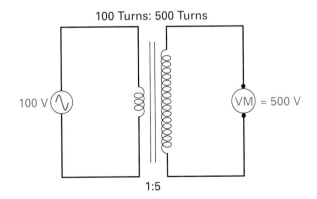

FIGURE 17-9   **Step-Up Transformer.**

$$\frac{V_s}{V_p} = \frac{N_s}{N_p}$$

or

$$\frac{500}{100} = \frac{500}{100}$$

To calculate V_s, therefore, we can rearrange the formula and arrive at

$$V_s = \frac{N_s}{N_p} \times V_p$$

In our example, this is

$$V_s = \frac{500}{100} \times 100 \text{ V}$$
$$= 500 \text{ V}$$

Step Down

Step-Down Transformer

Transformer in which the ac voltage induced in the secondary is less (due to fewer secondary windings) than the ac voltage applied to the primary.

If the secondary voltage (V_s) is smaller than the primary voltage (V_p), the transformer is called a **step-down transformer** ($V_s < V_p$), as shown in Figure 17-10. The secondary voltage will be equal to

$$V_s = \frac{N_s}{N_p} \times V_p$$

$$= \frac{10}{100} \times 1000 \text{ V} = 100 \text{ V}$$

▪ **EXAMPLE:**

Calculate the secondary voltage (V_s) if a 1:6 step-up transformer has 24 V ac applied to the primary.

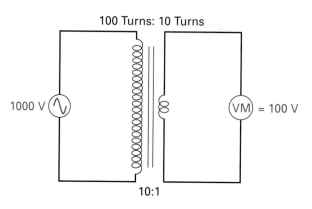

FIGURE 17-10 Step-Down Transformer.

$$V_s = \frac{N_s}{N_p} \times V_p$$

$$= \frac{6}{1} \times 24 \text{ V} = 144 \text{ V}$$

The coupling coefficient (k) in this formula is always assumed to be 1, which for most iron-core transformers is almost always the case. This means that all the primary magnetic flux is linking the secondary, and the secondary voltage is dependent on the number of secondary turns that are being cut by the primary magnetic flux.

The transformer can be used to transform the primary ac voltage into any other voltage, either up or down, merely by changing the transformer's turns ratio.

17-5-3 Power and Current Ratio

The power in the secondary of the transformer is equal to the power in the primary $(P_p = P_s)$. Power, as we know, is equal to $P = V \times I$, and if voltage is stepped up or down, the current automatically is stepped down or up, respectively, in the opposite direction to voltage to maintain the power constant.

For example, if the secondary voltage is stepped up $(V_s \uparrow)$, the secondary current is stepped down $(I_s \downarrow)$, so the output power is the same as the input power.

$$P_s = V_s \uparrow \times I_s \downarrow$$

This is an equal but opposite change. Therefore, $P_s = P_p$; and you cannot get more power out than you put in. The current ratio is therefore inversely proportional to the voltage ratio:

$$\frac{V_s}{V_p} = \frac{I_p}{I_s}$$

If the secondary voltage is stepped up, the secondary current goes down:

$$\frac{V_s \uparrow}{V_p} = \frac{I_p}{I_s \downarrow}$$

If the secondary voltage is stepped down, the secondary current goes up:

$$\frac{V_s \downarrow}{V_p} = \frac{I_p}{I_s \uparrow}$$

If the current ratio is inversely proportional to the voltage ratio, it is also inversely proportional to the turns ratio:

$$\frac{I_p}{I_s} = \frac{V_s}{V_p} = \frac{N_s}{N_p}$$

By rearranging the current and turns ratio, we can arrive at a formula for secondary current, which is

$$I_s = \frac{N_p}{N_s} \times I_p$$

■ **EXAMPLE:**

The step-up transformer in Figure 17-11 has a turns ratio of 5 to 1. Calculate:
a. Secondary voltage (V_s)
b. Secondary current (I_s)
c. Primary power (P_p)
d. Secondary power (P_s)

■ *Solution:*

The secondary has five times as many windings as the primary, and, consequently, the voltage will be stepped up by 5 between primary and secondary. If the secondary voltage is going to be five times that of the primary, the secondary current is going to decrease to one-fifth of the primary current.

a. $V_s = \dfrac{N_s}{N_p} \times V_p$

$= \dfrac{5}{1} \times 100 \text{ V}$

$= 500 \text{ V}$

b. $I_s = \dfrac{N_p}{N_s} \times I_p$

$= \dfrac{1}{5} \times 10 \text{ A}$

$= 2 \text{ A}$

c. $P_p = V_p \times I_p = 100 \text{ V} \times 10 \text{ A} = 1000 \text{ VA}$

d. $P_s = V_s \times I_s = 500 \text{ V} \times 2 \text{ A} = 1000 \text{ VA}$

Therefore, $P_p = P_s$.

Maximum Power Transfer
The maximum power will be absorbed by the load from the source, when the impedance of the load is equal to the impedance of the source.

17-5-4 Impedance Ratio

The **maximum power transfer** theorem, which has been discussed previously and is summarized in Figure 17-12, states that maximum power is transferred from source (ac generator) to load (equipment) when the impedance of the load is

FIGURE 17-11   Step-Up Transformer Example.

equal to the internal impedance of the source. If these impedances are different, a large amount of power could be wasted.

In most cases it is required to transfer maximum power from a source that has an internal impedance (Z_s) that is not equal to the load impedance (Z_L). In this situation, a transformer can be inserted between the source and the load to make the load impedance appear to equal the source's internal impedance.

For example, let's imagine that your car stereo system (source) has an internal impedance of 100 Ω and is driving a speaker (load) of 4 Ω impedance, as seen in Figure 17-13.

By choosing the correct turns ratio, the 4 Ω speaker can be made to appear as a 100 Ω load impedance, which will match the 100 Ω internal source impedance of the stereo system, resulting in maximum power transfer.

The turns ratio can be calculated by using the formula

$$\text{turns ratio} = \sqrt{\frac{Z_{Load}}{Z_{Source}}}$$

Z_{Load} = Load impedance in ohms.
Z_{Source} = Source impedance in ohms.

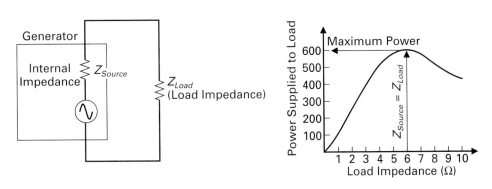

FIGURE 17-12   Maximum Power Transfer Theorem.

FIGURE 17-13   Impedance Matching.

In our example, this will calculate out to be

$$\text{turns ratio} = \sqrt{\frac{Z_L}{Z_s}}$$

$$= \sqrt{\frac{4}{100}} = \frac{\sqrt{4}}{\sqrt{100}}$$

$$= \frac{2}{10} = \frac{1}{5}$$

$$= 0.2$$

If the turns ratio is less than 1, a step-down transformer is required. A turns ratio of 0.2 means a step-down transformer is needed with a turns ratio of 5:1 ($\frac{1}{5} = 0.2$).

■ **EXAMPLE:**

Calculate the turns ratio needed to match the 22.2 Ω output impedance of an amplifier to two 16 Ω speakers connected in parallel.

■ *Solution:*

The total load impedance of two 16 Ω speakers in parallel will be

$$Z_{Load} = \frac{\text{product}}{\text{sum}} = \frac{16 \times 16}{16 + 16} = 8\,\Omega$$

The turns ratio will be

$$\text{turns ratio} = \sqrt{\frac{Z_{Load}}{Z_{Source}}}$$

$$= \sqrt{\frac{8 \ \Omega}{22.2 \ \Omega}}$$

$$= \sqrt{0.36} = 0.6$$

Therefore, a step-down transformer is needed with a turns ratio of 1.67 : 1.

SELF-TEST REVIEW QUESTIONS FOR SECTION 17-5

1. What is the turns ratio of a 402-turn primary and 1608-turn secondary, and is this transformer step up or step down?
2. State the formula for calculating the secondary voltage (V_s).
3. True or false: A transformer can, by adjusting the turns ratio, be made to step up both current and voltage between primary and secondary.
4. State the formula for calculating the secondary current (I_s).
5. What turns ratio is needed to match a 25 Ω source to a 75 Ω load?
6. Calculate V_s if $N_s = 200$, $N_p = 112$, and $V_p = 115$ V.

17-6 WINDINGS AND PHASE

The way in which the primary and secondary coils are wound around the core determines the polarity of the voltage induced into the secondary relative to the polarity of the primary. In Figure 17-14, you can see that if the primary and secondary windings are both wound in a clockwise direction around the core, the voltage induced in the primary will be in phase with the voltage induced in the secondary. Both the input and output will also be in phase with one another if the primary and secondary are both wound in a counterclockwise direction.

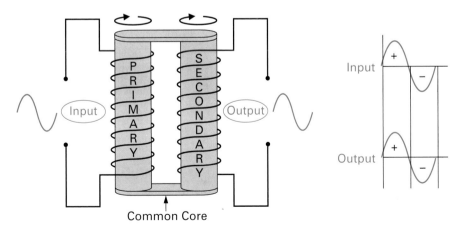

FIGURE 17-14 **Primary and Secondary in Phase.**

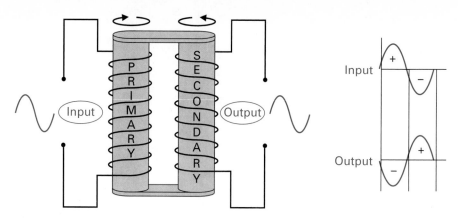

FIGURE 17-15   **Primary and Secondary out of Phase.**

In Figure 17-15, you will notice that in this case we have wound the primary and secondary in opposite directions, the primary being wound in a clockwise direction and the secondary in a counterclockwise direction. In this situation, the output ac sine-wave voltage is 180° out of phase with respect to the input ac voltage.

In a schematic diagram, there has to be a way of indicating to the technician that the secondary voltage will be in phase or 180° out of phase with the input. The **dot convention** is a standard used with transformers and is illustrated in Figure 17-16(a) and (b).

Dot Convention
A standard used with transformer symbols to indicate whether the secondary voltage will be in phase or out of phase with the primary voltage.

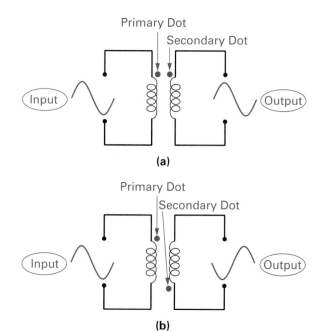

FIGURE 17-16   **Dot Convention. (a) In-Phase Dots. (b) Out-of-Phase Dots.**

A positive on the primary dot causes a positive on the secondary dot, and similarly, since ac is applied, a negative on a primary dot will cause a negative on the secondary dot. In Figure 17-16(a), you can see that when the top of the primary swings positive or negative, the top of the secondary will also follow suit and swing positive or negative, respectively, and you will now be able to determine that the transformer secondary voltage is in phase with the primary voltage.

In Figure 17-16(b), however, the dots are on top and bottom, which means that as the top of the primary winding swings positive, the bottom of the secondary will go positive, which means the top of the secondary will actually go negative, resulting in the secondary ac voltage being out of phase with the primary ac voltage.

SELF-TEST REVIEW QUESTIONS FOR SECTION 17-6

1. How does the winding of the primary and secondary affect the phase of the output with respect to the input?
2. Describe the transformer dot convention used on schematics.

17-7 · TRANSFORMER TYPES

We will treat transformers the same as every other electronic component and classify them as either fixed or variable turns ratio.

17-7-1 *Fixed Turns Ratio Transformers*

Fixed turns ratio transformers have a turns ratio that cannot be varied. They are generally wound on a common core to ensure a high k and can be classified by the type of core material used. The two types of fixed transformers are:

1. Air core [Figure 17-17(a)]
2. Iron or ferrite core [Figure 17-17(b)]

The air-core transformers typically have a nonmagnetic core, such as ceramic or a cardboard hollow shell, and are used in high-frequency applications. The electronic circuit symbol just shows the primary and secondary coils. The more common iron- or ferrite-core transformers concentrate the magnetic lines of force, resulting in improved transformer performance; they are symbolized by two lines running between the primary and secondary. The iron-core transformer's lines are solid, while the ferrite-core transformer's lines are dashed.

Figure 17-18 illustrates the physical appearance of a few types of fixed transformers.

FIGURE 17-17 **Fixed-Value Transformers. (a) Air-Core Transformers (Courtesy of Sencore). (b) Iron- or Ferrite-Core Transformers (Courtesy of Coilcraft).**

17-7-2 *Variable Turns Ratio Transformers*

Variable turns ratio transformers have a turns ratio that can be varied. They can be classified as follows:

1. Center-tapped secondary
2. Multiple-tapped secondary
3. Multiple winding
4. Single winding

Figure 17-19 illustrates the first two of these, the center-tapped and the multiple-tapped secondary types. Transformers can often have tapped secondaries. A tapped winding will have a lead connected to one of the loops other than the two end connections.

FIGURE 17-18 Physical Appearance of Some Transformer Types (Courtesy of Sencore).

(a)

(b)

FIGURE 17-19 Secondary Tapped Transformers. (a) Center Tapped (Courtesy of Sencore). (b) Multiple Tapped.

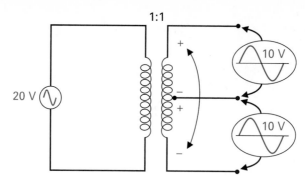

FIGURE 17-20 **Center-Tapped Secondary Transformer.**

Center-Tapped Secondary

Center-Tapped Transformer
A transformer that has a connection at the electrical center of the secondary winding.

If the tapped lead is in the exact center of the secondary, the transformer is said to have a center-tapped secondary, as shown in Figure 17-20. With the **center-tapped transformer,** the two secondary voltages are each half of the total secondary voltage. If we assume a 1:1 turns ratio (primary turns = secondary turns) and a 20 V ac primary voltage and therefore secondary voltage, each of the output voltages between either end of the secondary and the center tap will be 10-V waveforms, as shown in Figure 17-20. The two secondary outputs will be 180° out of phase with one another from the center tap. When the top of the secondary swings positive (+), the bottom of the secondary will be negative, and vice versa.

Figure 17-21 illustrates a typical power company utility pole, which as you can see makes use of a center-tapped secondary transformer. These transformers are designed to step down the high 4.8 kV from the power line to 120 V/240 V for commercial and residential customers. Within the United States, the actual secondary voltage may be anywhere between 225 and 245 V, depending on the demand. The demand is actually higher during the day and in the winter months (source voltage is pulled down by smaller customer load resistance). The multiple taps on the primary are used to change the turns ratio and compensate for power line voltage differences.

Neutral Wire
The conductor of a polyphase circuit or of a single-phase three-wire circuit that is intended to have a ground potential. The potential differences between the neutral and each of the other conductors are approximately equal in magnitude and are also equally spaced in phase.

The center tap of the secondary winding is connected to a copper earth grounding rod. This center tap wire is called the **neutral wire,** and within the building it is color coded with a white insulation. The two outside wires are color coded black and red. Residential or commercial appliances or loads that are designed to operate at 240 V are connected between the red wire and the black wire, while loads that are designed to operate at 120 V are connected between the red wire and the white wire (neutral) or the black wire and the white wire (neutral).

Multiple-Tapped Secondary

If several leads are attached to the secondary, the transformer is said to have a multiple-tapped secondary. This multitapped transformer will tap or pick off different values of ac voltage, as seen in Figure 17-22.

FIGURE 17-21 Center-Tapped Utility Pole Transformer.

Multiple Windings

The simple transformer has only one primary and one secondary. In some applications, transformers can have multiple primaries and even multiple secondaries.

Multiple primaries. The major application of this arrangement is to switch between two primary voltages and obtain the same secondary voltage. In Figure 17-23(a), the two primaries are connected in parallel, so the primary only has an

FIGURE 17-22   Multiple-Tapped Secondary Transformer.

(a) Primaries in Parallel

(b) Primaries in Series

(c)

FIGURE 17-23   Multiple Primary Windings Transformer.

equivalent 100 turns; and since the secondary has 10 turns, this 10:1 step-down ratio will result in a secondary voltage of

$$V_s = \frac{N_s}{N_p} \times V_p$$

$$= \frac{10}{100} \times 120 \text{ V}$$

$$= 12 \text{ V}$$

If the two primaries are connected in series, as shown in Figure 17-23(b), the now 200-turn primary will result in a secondary voltage of

$$V_s = \frac{N_s}{N_p} \times V_p$$

$$= \frac{10}{200} \times 240 \text{ V}$$

$$= 12 \text{ V}$$

Some portable electrical or electronic equipment, such as radios or shavers, have a switch that allows you to switch between 120 V ac (U.S. wall outlet voltage) and 240 V ac (European wall outlet voltage), as shown in Figure 17-23(c). By activating the switch, you can always obtain the correct voltage to operate the equipment, whether the wall socket is supplying 240 or 120 V.

FIGURE 17-24   Multiple Secondary Windings Transformer.

Multiple secondaries. In some applications, more than one secondary is wound onto a common primary and core. The advantage of this arrangement can be seen in Figure 17-24, where many larger and smaller voltages can be acquired from one primary voltage.

Single Winding (Autotransformer)

The single winding transformer or autotransformer is used in the automobile ignition system (switched dc pulse system) to raise the low 12 V battery voltage up to about 20 kV for the spark plugs, as shown in Figure 17-25(a). Autotransformers are constructed by winding one continuous coil onto a core, which acts as both the primary and secondary. This yields three advantages for the single winding transformer:

1. They are smaller.

2. They are cheaper.

3. They are lighter than the normal separated primary/secondary transformer types.

Its disadvantage, however, is that no electrical isolation exists between primary and secondary. It is normally used as a step-up transformer for the automobile, or in televisions to obtain 20 kV. This transformer type can also be arranged to step down voltage, as shown in Figure 17-25(b).

(a)

(b)

FIGURE 17-25 Single-Winding Transformer. (a) Step-Up. (b) Step-Down.

1. List the two types of fixed transformers.
2. List the four basic classifications for variable transformers.
3. True or false: The two outputs of a center-tapped secondary transformer will always be 90° out of phase with one another.
4. In what application would a multiple primary transformer be used?
5. In what application would a single-winding transformer be used?
6. What are the advantages of a single-winding transformer?

17-8   TRANSFORMER RATINGS

A typical transformer rating could read 1 kVA, 500/100, 60 Hz. The 500 normally specifies the maximum primary voltage, the 100 normally specifies the maximum secondary voltage, and the 1 kVA is the apparent power rating. In this example, the maximum load current will equal

$$I_s = \frac{\text{apparent power } (P_A)}{\text{secondary voltage } (V_s)} \qquad \left(\begin{array}{c} P = V \times I; \text{ therefore,} \\[4pt] I = \dfrac{P}{V} \end{array} \right)$$

$$= \frac{1\,\text{kVA}}{100\,\text{V}} = 10\,\text{A}$$

With this secondary voltage at 100 V and a maximum current of 10 A, the smallest load resistor that can be connected across the output of the secondary is

$$R_L = \frac{V_s}{I_s}$$

$$= \frac{100\,\text{V}}{10\,\text{A}}$$

$$= 10\,\Omega$$

Exceeding the rating of the transformer will cause overheating and even burning out of the windings.

EXAMPLE:

Calculate the smallest value of load resistance that can be connected across a 3 kVA, 600/200, 60 Hz step-down transformer.

Solution:

$$I = \frac{\text{apparent power } (P_A)}{\text{secondary voltage } (V_s)}$$

$$\frac{3 \text{ kVA}}{200 \text{ V}} = 15 \text{ A}$$

$$R_L = \frac{V_s}{I_s} = \frac{200 \text{ V}}{15 \text{ A}} = 13.3 \text{ }\Omega$$

SELF-TEST REVIEW QUESTIONS FOR SECTION 17-8

1. What do each of the values mean when a transformer is rated as a 10 kVA, 200/100, 60 Hz?
2. If a 100 Ω resistor is connected across the secondary of a transformer that is to supply 1 kV and is rated at a maximum current of 8 A, will the transformer overheat and possibly burn out?

17-9   TESTING TRANSFORMERS

Like inductors, transformers, which are basically two or more inductors, can develop one of three problems:

1. An open winding
2. A complete short in a winding
3. A short in a section of a winding

17-9-1   *Open Primary or Secondary Winding*

An open in the primary winding will prevent any primary current, and therefore there will be no induced voltage in the secondary and therefore no voltage will be present across the load. An open secondary winding will prevent the flow of secondary current, and once again, no voltage will be present across the load.

An open in the primary or secondary winding is easily detected by disconnecting the transformer from the circuit and testing the resistance of the windings with an ohmmeter. Like an inductor, a transformer winding should have a low resistance, in the tens to hundreds range. An open will easily be recognized because of its infinite (maximum) resistance.

17-9-2 *Complete Short or Section Short in Primary or Secondary Winding*

A partial or complete short in the primary winding of a transformer will result in an excessive source current that will probably blow the circuit's fuse or trip the circuit's breaker. A partial or complete short in the secondary winding will cause an excessive secondary current, which in turn will result in an excessive primary current.

In both instances, a short in the primary or secondary will generally burn out the primary winding unless the fuse or circuit breaker opens the excessive current path. An ohmmeter can be used to test for partial or complete shorts in transformer windings; however, all coils have a naturally low resistance which can be mistaken for a short. An inductive or reactive analyzer can accurately test transformer windings, checking the inductance value of each coil.

It is virtually impossible to repair an open, partially shorted, or completely shorted transformer winding, and therefore defective transformers are always replaced with a transformer that has an identical rating.

SELF-TEST REVIEW QUESTIONS FOR SECTION 17-9

1. Which test instrument could you use to test for a suspected open primary winding?
2. If a secondary winding had a resistance of 50 Ω, would this value indicate a problem?

17-10 TRANSFORMER LOSSES

Internal losses cause the power delivered from the secondary winding of a transformer in reality to be less than the power fed into the primary winding. $P_s < P_p$ in reality, due to transformer losses. There are basically three types of internal transformer loss: **copper loss,** core loss, and magnetic leakage.

Copper Loss
Also called I^2R loss, it is the power lost in transformers, generators, connecting wires, and other parts of a circuit because of the current flow (I) through the resistance (R) of the conductors.

17-10-1 *Copper Losses*

Due to ohmic resistance of the windings, heat ($I^2 \times R$) is dissipated in both the primary and secondary windings.

17-10-2 Core Losses

Hysteresis

Hysteresis
Amount that the magnetization of a material lags the magnetizing force due to molecular friction.

Eddy Currents
Small currents induced in a conducting core due to the variations in alternating magnetic flux.

During each ac cycle, the core is taken through a cycle of magnetization; hence energy is lost due to **hysteresis** and appears as heat in the core. This loss is proportional to frequency and is minimized by using a core of soft iron or an alloy such as stalloy or permalloy.

Eddy-Current Loss

The continuously changing flux induces voltages in the conducting core, which results in local **eddy currents** in the core that combine to produce a large circulating current, as shown in Figure 17-26(a), that opposes the main flux and also generates heat. This loss, which is proportional to frequency, is minimized by using a laminated core, as shown in Figure 17-26(b).

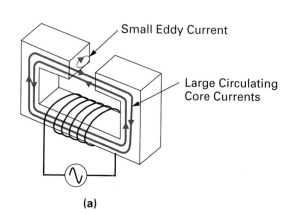

FIGURE 17-26 Eddy Currents. (a) Solid Core, Large Eddy Currents. (b) Laminated Core, Small Eddy Currents.

17-10-3 Magnetic Leakage

Not all the flux lines produced will cut the secondary winding. The result is that for a given turns ratio, the secondary terminal voltage will be slightly lower than expected due to **magnetic leakage.**

Magnetic Leakage
The passage of magnetic flux outside the path along which it can do useful work.

SELF-TEST REVIEW QUESTIONS FOR SECTION 17-10

1. List the four types of transformer losses.
2. Why will a laminated transformer core reduce eddy current loss?

SUMMARY

Transformers (Figure 17-27)

1. The transformer is an electrical device that makes use of electromagnetic induction to transfer alternating current from one circuit to another.

2. The transformer consists of two inductors that are placed in very close proximity to one another. When an alternating current flows through the first coil or primary winding the inductor sets up a magnetic field. The expanding and contracting magnetic field produced by the primary cuts across the windings of the second inductor or secondary winding and induces a voltage in this coil.

3. By changing the ratio between the number of turns in the secondary winding to the number of turns in the primary winding, some characteristics of the ac signal can be changed or transformed as it passes from primary to secondary.

Mutual Inductance

4. The principle on which a transformer is based is an inductive effect known as mutual inductance, which is the process by which an inductor induces a voltage in another inductor.

5. As with self-inductance, mutual inductance is dependent on change. Direct current (dc) is a constant current and produces a constant or stationary magnetic field that does not change. Alternating current, however, is continually varying, and as the polarity of the magnetic field is dependent on the direction of current flow (left-hand rule), the magnetic field will also be alternating in polarity.

6. Self-induction is a measure of how much voltage an inductor can induce within itself. Mutual inductance is a measure of how much voltage is induced in the secondary coil due to the change in current in the primary coil.

7. The basic transformer consists of two coils within close proximity to one another, to ensure that the second coil will be cut by the magnetic flux lines

FIGURE 17-27   Transformers.

$$I_p = \frac{V_p}{Z_p}$$

$$\text{Coupling coefficient } (k) = \frac{\text{flux linking secondary coil}}{\text{total flux produced by primary}}$$

$$\text{Turns ratio} = \frac{N_s}{N_p}$$

N_s = number of turns in the secondary
N_p = number of turns in the primary

$$V_s = \frac{N_s}{N_p} \times V_p$$

V_s = secondary voltage
V_p = primary voltage
Z_p = impedance of primary coil

$$I_s = \frac{N_p}{N_s} \times I_p$$

I_s = secondary current
I_p = primary current

$$\text{Turns ratio} = \sqrt{\frac{Z_L}{Z_s}}$$

Z_L = load impedance
Z_s = source impedance

$$I_s = \frac{P_A}{V_s}$$

P_A = apparent power, in volt-amperes (VA)
R_L = load resistance, in ohms (Ω)

$$R_L = \frac{V_s}{I_s}$$

Transformer Action

Step-Up Transformer

Step-Down Transformer

Maximum Power Transfer

Center-Tapped Secondary

Multiple-Tapped Secondary

FIGURE 17-27 (*continued*)

Multiple Primary

(a) Primaries in Parallel

(b) Primaries in Series

(c)

Multiple Secondary

$$V_S = \frac{N_S}{N_P} \times V_P = \frac{10}{50} \times 120 = 24 \text{ V}$$

$$V_S = \frac{N_S}{N_P} \times V_P = \frac{100}{50} \times 120 = 240 \text{ V}$$

$$V_S = \frac{N_S}{N_P} \times V_P = \frac{5}{50} \times 120 = 12 \text{ V}$$

Single Winding

Secondary (20 kV AC)

Tap

Primary (12 V AC)

(a)

Primary

Tap

Secondary

(b)

produced by the first coil, and thereby ensure mutual inductance. The ac voltage source is electrically connected (through wires) to the primary coil or winding, and the load (R_L) is electrically connected to the secondary coil or winding.

8. As primary current increases, secondary current increases, and as primary current decreases, secondary current also decreases. It can therefore be said that the frequency of the alternating current in the secondary is the same as the frequency of the alternating current in the primary.

9. Although the two coils are electrically isolated from one another, energy can be transferred from primary to secondary, because the primary converts electrical energy into magnetic energy, and the secondary converts magnetic energy back into electrical energy.

Transformer Loading

10. When the transformer is not connected to a load, primary circuit current is determined by $I = V/Z$, where Z is the impedance of the primary coil (both its inductive reactance and resistance) and V is the applied voltage. Since no current can flow in the secondary, because an open in the circuit exists, the primary acts as a simple inductor, and the primary current is small due to the inductance of the primary winding. This small primary current lags the applied voltage due to the counter emf by approximately 90° because the coil is mainly inductive and has very little resistance.

11. When a load is connected across the secondary, a change in conditions occurs and the transformer acts differently. The important point that will be observed is that as we go from a no-load to a load condition the primary current will increase due to mutual inductance. Let's follow the steps one by one.

    a. The ac applied voltage sets up an alternating magnetic field in the primary winding.
    b. The continually changing flux of this primary field induces and produces a counter emf into the primary to oppose the applied voltage.
    c. The primary's magnetic field also induces a voltage in the secondary winding, which causes current to flow in the secondary circuit through the load.
    d. The current in the secondary winding produces another magnetic field that is opposite to the field being produced by the primary.
    e. This secondary magnetic field feeds back to the primary and induces a voltage that tends to cancel or weaken the counter emf that was set up in the primary by the primary current.
    f. The primary's counter emf is therefore reduced, so primary current can now increase.
    g. This increase in primary current is caused by the secondary's magnetic field; consequently, the greater the secondary current, the stronger the secondary magnetic field, which causes a reduction in the primary's counter emf, and therefore a primary current increase.

12. In summary, an increase in secondary current ($I_s\uparrow$) causes an increase in primary current ($I_p\uparrow$), and this effect in which the primary induces a voltage in

the secondary (V_s) and the secondary induces a voltage into the primary (V_p) is known as *mutual inductance*.

Coefficient of Coupling (k)

13. The voltage induced into the secondary winding is dependent on the mutual inductance between the primary and secondary, which is determined by how much of the magnetic flux produced by the primary actually cuts the secondary winding.

14. The coefficient of coupling (k) is a ratio of the number of magnetic lines of force that cut the secondary compared to the total number of magnetic flux lines being produced by the primary and is a figure between 0 and 1.

15. The coefficient of coupling depends on:

    a. How close together the primary and secondary are to one another

    b. The type of core material used

Transformer Ratios and Applications

16. Basically, transformers are used for one of three applications:

    a. To step up (increase) or step down (decrease) voltage

    b. To step up (increase) or step down (decrease) current

    c. To match impedances

        In all three cases, any of the applications can be achieved by changing the ratio of the number of turns in the primary winding compared to the number of turns in the secondary winding. This ratio is appropriately called the turns ratio.

17. The turns ratio is the ratio between the number of turns in the secondary winding (N_s) and the number of turns in the primary winding (N_p).

18. Transformers are used within the power supply unit of almost every piece of electronic equipment to step up or step down the 115 V ac from the outlet.

19. Some electronic circuits require lower-power supply voltages, while other devices may require higher-power supply voltages. The transformer is used in both instances to convert the 115 V ac to the required value of voltage.

20. If the secondary voltage (V_s) is greater than the primary voltage (V_p), the transformer is called a step-up transformer ($V_s > V_p$).

21. If the secondary voltage (V_s) is smaller than the primary voltage (V_p), the transformer is called a step-down transformer ($V_s < V_p$).

22. The coupling coefficient (k) in this formula is always assumed to be 1, which for most iron-core transformers is almost always the case. This means that all the primary magnetic flux is linking the secondary, and the secondary voltage is dependent on the number of secondary turns that are being cut by the primary magnetic flux.

23. The power in the secondary of the transformer is equal to the power in the primary ($P_p = P_s$). Power, as we know, is equal to $P = V \times I$, and if voltage is stepped up or down, the current automatically is stepped down or up, respectively, in the opposite direction to voltage to maintain the power constant.

24. The maximum power transfer theorem states that maximum power is transferred from source (ac generator) to load (equipment) when the impedance of the load is equal to the internal impedance of the source. If these impedances are different, a large amount of power could be wasted.

25. In most cases it is required to transfer maximum power from a source that has an internal impedance (Z_s) that is not equal to the load impedance (Z_L). In this situation, a transformer can be inserted between the source and the load to make the load impedance appear to equal the source's internal impedance.

26. The way in which the primary and secondary coils are wound around the core determines the polarity of the voltage induced into the secondary relative to the polarity of the primary.

27. If the primary and secondary windings are both wound in a clockwise direction around the core, the voltage induced in the primary will be in phase with the voltage induced in the secondary. Both the input and output will also be in phase with one another if the primary and secondary are both wound in a counterclockwise direction.

28. If both the primary and secondary are wound in opposite directions, the primary being wound in a clockwise direction and the secondary in a counterclockwise direction, the output ac sine-wave voltage is 180° out of phase with respect to the input ac voltage.

29. The dot convention is a standard used with transformer symbols to indicate whether the secondary voltage will be in phase or out of phase with the primary voltage.

Transformer Types

30. Fixed turns ratio transformers have a turns ratio that cannot be varied. They are generally wound on a common core to ensure a high k and can be classified by the type of core material used. The two types of fixed transformers are:

    a. Air core   **b.** Iron or ferrite core

31. The air-core transformers typically have a nonmagnetic core, such as ceramic or a cardboard hollow shell, and are used in high-frequency applications.

32. The more common iron- or ferrite-core transformers concentrate the magnetic lines of force, resulting in improved transformer performance; they are symbolized by two lines running between the primary and secondary. The iron-core transformer's lines are solid, while the ferrite-core transformer's lines are dashed.

33. Variable turns ratio transformers have a turns ratio that can be varied. They can be classified as follows:

    a. Center-tapped secondary   **c.** Multiple winding
    b. Multiple-tapped secondary   **d.** Single winding

34. If the tapped lead is in the exact center of the secondary, the transformer is said to have a center-tapped secondary. With the center-tapped transformer, the two secondary voltages are each half of the total secondary voltage.

35. If several leads are attached to the secondary, the transformer is said to have a multiple-tapped secondary. This multitapped transformer will tap or pick off different values of ac voltage.

36. The simple transformer has only one primary and one secondary. In some applications, transformers can have multiple primaries and even multiple secondaries.

37. The single winding transformer or autotransformer is used in the automobile ignition system (switched dc pulse system) to raise the low 12 V battery voltage up to about 20 kV for the spark plugs.

Transformer Ratings

38. A typical transformer rating could read 1 kVA, 500/100, 60 Hz. The 500 normally specifies the maximum primary voltage, the 100 normally specifies the maximum secondary voltage, and the 1 kVA is the apparent power rating.

Testing Transformers

39. Like inductors, transformers, which are basically two or more inductors, can develop one of three problems.

a. An open winding
b. A complete short in a winding
c. A short in a section of a winding

40. An open in the primary winding will prevent any primary current, and therefore there will be no induced voltage in the secondary and therefore no voltage will be present across the load.

41. An open secondary winding will prevent the flow of secondary current, and once again, no voltage will be present across the load.

42. An open in the primary or secondary winding is easily detected by disconnecting the transformer from the circuit and testing the resistance of the windings with an ohmmeter. Like an inductor, a transformer winding should have a low resistance, in the tens to hundreds range. An open will easily be recognized because of its infinite (maximum) resistance.

43. A partial or complete short in the primary winding of a transformer will result in an excessive source current that will probably blow the circuit's fuse or trip the circuit's breaker.

44. A partial or complete short in the secondary winding will cause an excessive secondary current, which in turn will result in an excessive primary current.

45. In both instances, a short in the primary or secondary will generally burn out the primary winding unless the fuse or circuit breaker opens the excessive current path.

46. An ohmmeter can be used to test for partial or complete shorts in transformer windings; however, all coils have a naturally low resistance which can be mistaken for a short.

47. An inductive or reactive analyzer can accurately test transformer windings, checking the inductance value of each coil.

48. It is virtually impossible to repair an open, partially shorted, or completely shorted transformer winding, and therefore defective transformers are always replaced with a transformer that has an identical rating.

Transformer Losses

49. Internal losses cause the power delivered from the secondary winding of a transformer in reality to be less than the power fed into the primary winding.

50. $P_s < P_p$ in reality, due to transformer losses. There are basically three types of internal transformer loss: copper loss, core loss, and magnetic leakage.

51. Due to ohmic resistance of the windings, heat ($I^2 \times R$) is dissipated in both the primary and secondary windings.

52. During each ac cycle, the core is taken through a cycle of magnetization; hence energy is lost due to hysteresis and appears as heat in the core.

53. The continuously changing flux induces voltages in the conducting core, which results in local eddy currents in the core that combine to produce a large circulating current that opposes the main flux and also generates heat.

54. Not all the flux lines produced will cut the secondary winding. The result is that for a given turns ratio, the secondary terminal voltage will be slightly lower than expected due to magnetic leakage.

NEW TERMS		
	Autotransformer	Neutral wire
	Center-tapped secondary	Power ratio
	Coefficient of coupling	Primary winding
	Copper losses	Ratings
	Core losses	Secondary winding
	Current ratio	Single-winding transformer
	Dot convention	Step-down transformer
	Eddy-current losses	Step-up transformer
	Fixed transformer	Transformer
	Impedance ratio	Transformer action
	Magnetic leakage	Turns ratio
	Multiple-tapped secondary	Variable transformer
	Multiple winding	Voltage ratio
	Mutual inductance	

Multiple-Choice Questions

1. Transformer action is based on:
   a. Self-inductance
   b. Air between the coils
   c. Mutual capacitance
   d. Mutual inductance

2. An increase in transformer secondary current will cause a(an) _____ in primary current.
   a. Decrease   b. Increase

3. If 50% of the magnetic lines of force produced by the primary were to cut the secondary coil, the coefficient of coupling would be:
   a. 75   b. 50   c. 0.5   d. 0.005

4. A step-up transformer will always have a turns ratio _____, while a step-down transformer has a turns ratio _____.
   a. $< 1, > 1$ b. $> 1, > 1$ c. $> 1, < 1$ d. $< 1, < 1$

5. With an 80-V ac secondary voltage center-tapped transformer, what would be the voltage at each output, and what would be the phase relationship between the two secondary voltages?
   a. 20 V, in phase with one another   c. 40 V, in phase
   b. 30 V, 180° out of phase   d. 40 V, 180° out of phase

6. One application of the autotransformer would be:
   a. To obtain the final anode high-voltage supply for the cathode-ray tube in a television
   b. To obtain two outputs, 180° out of phase with one another
   c. To tap several different voltages from the secondary
   d. To obtain the same secondary voltage for different voltages

7. Alternating current can be used only with transformers because:
   a. It produces an alternating magnetic field.
   b. It produces a fixed magnetic field.
   c. Its magnetic field is greater than that of dc.
   d. Its rms is 0.707 of the peak.

8. Eddy-current losses are reduced with laminated iron cores because:
   a. The air gap is always kept to a minimum.
   b. The resistance of iron is always low.
   c. The laminations are all insulated from one another.
   d. Current cannot flow in iron.

9. Assuming 100% efficiency, the output power, P_s, is always equal to:
   a. P_p
   b. $V_s \times I_s$
   c. $0.5 \times P_p$
   d. Both (a) and (c)
   e. Both (a) and (b)

10. If the primary winding of a transformer were open, the result would be:
   a. No flux linkage between primary and secondary
   b. No primary or secondary current
   c. No induced voltage in the secondary
   d. All of the above

Essay Questions

11. What is a transformer? (Introduction)

12. Describe mutual inductance and how it relates to transformers. (17-1)

13. Why does loading the transformer's secondary circuit affect primary transformer current? (17-3)

14. Describe what a coefficient of coupling figure of 0.9 means. (17-4)

15. List the three basic applications of transformers, and then describe how each of these applications can be achieved by merely changing the transformer's turns ratio. (17-5)

16. Briefly describe how the dot convention is used in schematics to describe phase. (17-6)

17. Illustrate the schematic symbol and main points relating to the following transformers:
   a. Fixed air core (17-7-1)
   b. Fixed iron core (17-7-1)
   c. Fixed ferrite core (17-7-1)
   d. Center-tapped secondary (17-7-2)
   e. Four-output tapped secondary (17-7-2)
   f. Multiple primary and multiple secondary (17-7-2)
   g. Autotransformer (17-7-2)

18. Illustrate and explain the following transformer applications: (17-7)
   a. 120/240-V primary voltage to constant secondary voltage
   b. 20-kV secondary voltage using the autotransformer

19. Briefly describe what is meant by copper losses with transformers. (17-10)

20. Briefly describe the following iron and core losses and how they can be reduced: (17-10)
   a. Hysteresis   **b.** Eddy current

21. Briefly describe the transformer loss known as magnetic leakage. (17-10)

22. What meter could be used to check a suspected open primary coil? (17-9)

23. Could a transformer be considered a dc block? (17-1)

24. Would a step-up voltage transformer step up or step down current? What would happen to secondary power? (17-5)

25. Briefly describe how transformers are used to reduce I^2R losses in ac power distribution. (Chapter 11)

Practice Problems

26. Calculate the turns ratio of the following transformers and state whether they are step up or step down:
   a. P = 12T, S = 24T   **c.** P = 24T, S = 5T
   b. P = 3T, S = 250T   **d.** P = 240T, S = 120T

27. Calculate the secondary ac voltage for all the examples in Question 26 if the primary voltage equals 100 V.

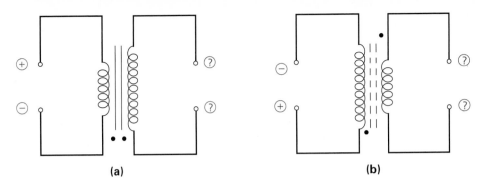

FIGURE 17-28   Dot Convention Examples.

28. Calculate the secondary ac current for all the examples in Question 26 if the primary current equals 100 mA.

29. What turns ratio would be needed to match a source impedance of 24 Ω to a load impedance of 8 Ω?

30. What turns ratio would be needed to step:

   a. 120 V to 240 V   b. 240 V to 720 V   c. 30 V to 14 V   d. 24 V to 6 V

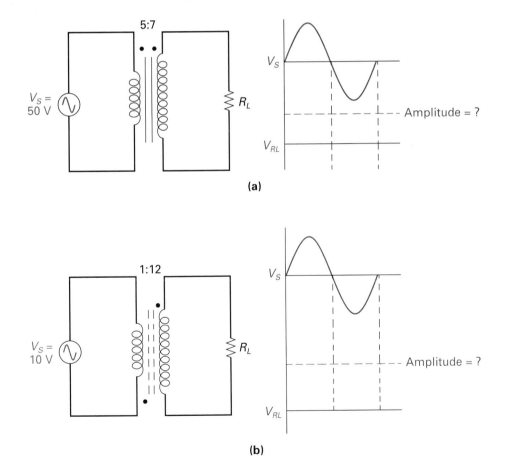

FIGURE 17-29   Input/Output Polarity Examples.

31. For a 24 V, 12-turn primary, and 16-, 2-, 1-, and 4-turn multiple secondary transformer, calculate each of the secondary voltages.

32. If a 2:1 step-down transformer has a primary input voltage of 120 V, 60 Hz, and a 2 kVA rating, calculate the maximum secondary current and smallest load resistor that can be connected across the output.

33. Indicate the polarity of the secondary voltages in Figure 17-28(a) and (b).

34. Referring to Figure 17-29(a) and (b), sketch the outputs, showing polarity and amplitude with respect to the inputs.

35. If a transformer is rated at 500 VA, 60 Hz, the primary voltage is 240 V ac, and the secondary voltage is 600 V ac, calculate:

a. Maximum load current **b.** Smallest value of R_L

O B J E C T I V E S

After completing this chapter, you will be able to:

1. Identify the difference between a series and parallel *RLC* circuit.

2. Explain the following as they relate to series *RLC* circuits:
   a. Impedance
   b. Current
   c. Voltage
   d. Phase angle
   e. Power

3. Explain the following as they relate to parallel *RLC* circuits:
   a. Voltage
   b. Current
   c. Impedance
   d. Phase angle
   e. Power

4. Define resonance, and explain the characteristics of:
   a. Series resonance
   b. Parallel resonance

5. Identify and explain the following *RLC* circuit applications:
   a. Low-pass filter
   b. High-pass filter
   c. Bandpass filter
   d. Band-stop filter

6. Describe complex numbers in both rectangular and polar form.

7. Perform complex number arithmetic.

8. Describe how complex numbers apply to ac circuits containing series–parallel *RLC* components.

Resistive, Inductive, and Capacitive (*RLC*) Circuits

The Fairchildren

On December 23, 1947, John Bardeen, Walter Brattain, and William Shockley first demonstrated how a semiconductor device, named the *transistor,* could be made to amplify. However, the device had mysterious problems and was very unpredictable. Shockley continued his investigations, and in 1951 he presented the world with the first reliable junction transistor. In 1956, the three shared the Nobel Prize in physics for their discovery, and much later, in 1972, Bardeen would win a rare second Nobel Prize for his research at the University of Illinois in the field of superconductivity.

Shockley left Bell Labs in 1955 to start his own semiconductor company near his home in Palo Alto and began recruiting personnel. He was, however, very selective, only hiring those who were bright, young, and talented. The company was a success, although many of the employees could not tolerate Shockley's eccentricities, such as posting everyone's salary and requiring that the employees rate one another. Two years later, eight of Shockley's most talented defected. The "traitorous eight," as Shockley called them, started their own company only a dozen blocks away, named Fairchild Semiconductor.

More than 50 companies would be founded by former Fairchild employees. One of the largest was started by Robert Noyce and two other colleagues from the group of eight Shockley defectors; they named their company Intel, which was short for "intelligence."

In this chapter we combine resistors (R), inductors (L), and capacitors (C) into series and parallel ac circuits. Resistors, as we have discovered, operate and react to voltage and current in a very straightforward way; the voltage across a resistor is in phase with the resistor current.

Inductors and capacitors operate in essentially the same way, in that they both store energy and then return it back to the circuit. However, they have completely opposite reactions to voltage and current. To help you remember the phase relationships between voltage and current for capacitors and inductors, you may wish to use the following memory phrase:

ELI the ICE man

This phrase states that voltage (symbolized E) leads current (I) in an inductive (L) circuit (abbreviated by the word "ELI"), while current (I) leads voltage (E) in a capacitive (C) circuit (abbreviated by the word "ICE").

In this chapter we study the relationships between voltage, current, impedance, and power in both series and parallel RLC circuits. We also examine the important RLC circuit characteristic called resonance, and see how RLC circuits can be made to operate as filters. In the final section we discuss how complex numbers can be used to analyze series and parallel ac circuits containing resistors, inductors, and capacitors.

18-1 SERIES RLC CIRCUIT

Figure 18-1 begins our analysis of series RLC circuits by illustrating the current and voltage relationships. The circuit current is always the same throughout a series circuit and can therefore be used as a reference. Studying the waveforms and vector diagrams shown alongside the components, you can see that the voltage across a resistor is always in phase with the current, while the voltage across the inductor leads the current by 90° and the voltage across the capacitor lags the current by 90°.

Now let's analyze the impedance, current, voltage, and power distribution of this circuit in a little more detail.

18-1-1 *Impedance*

Impedance is the total opposition to current flow and is a combination of both reactance (X_L, X_C) and resistance (R). An example circuit is illustrated in Figure 18-2(a).

Capacitive reactance can be calculated by using the formula

$$X_C = \frac{1}{2\pi f C}$$

In the example,

$$X_C = \frac{1}{2\pi \times 60 \times 10\ \mu\text{F}} = 265.3\ \Omega$$

Inductive reactance is calculated by using the formula

$$X_L = 2\pi f L$$

In the example,

$$X_L = 2\pi \times 60 \times 20\ \text{mH} = 7.5\ \Omega$$

(a)

(b)

FIGURE 18-1 Series *RLC* Circuit. (a) *RLC* Series Circuit Current: Current Flow Is Always the Same in All Parts of a Series Circuit. (b) *RLC* Series Circuit Voltages: *I* Is in Phase with V_R, *I* Lags V_L by 90°, and *I* Leads V_C by 90°.

Resistance in the example is equal to $R = 33\ \Omega$. Figure 18-2(b) illustrates these values of resistance and reactance in a vector diagram. In this vector diagram you can see that X_L is drawn 90° ahead of R, and X_C is drawn 90° behind R; the capacitive and inductive reactances are 180° out of phase with one another and counteract to produce a vector diagram, as shown in Figure 18-2(c). The difference between X_L and X_C is equal to 257.8, and since X_C is greater than X_L, the resultant reactive vector is capacitive. Reactance, however, is not in phase with resistance, and impedance is the vector sum of the reactive (X) and resistance (R) vectors. The formula, based on the Pythagorean theorem, illustrated in Figure 18-2(d), is

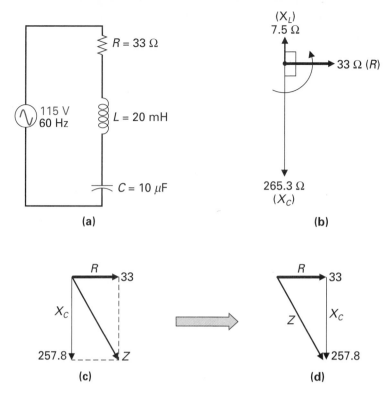

FIGURE 18-2 Series Circuit Impedance.

$$Z = \sqrt{R^2 + X^2}$$

In this example, therefore, the circuit impedance will be equal to

$$Z = \sqrt{R^2 + X^2} = \sqrt{33^2 + 257.8^2} = 260 \ \Omega$$

Since reactance (X) is equal to the difference between (symbolized $\sim$) X_L and X_C $(X_L \sim X_C)$, the impedance formula can be modified slightly to incorporate the calculation to determine the difference between X_L and X_C.

$$Z = \sqrt{R^2 + (X_L \sim X_C)^2}$$

Using our example with this new formula, we arrive at the same value of impedance, and since the difference between X_L and X_C resulted in a capacitive vector, the circuit is said to act capacitively.

$$
\begin{aligned}
Z &= \sqrt{R^2 + (X_L \sim X_C)^2} \\
&= \sqrt{33^2 + (7.5 \sim 265.3)^2} \\
&= \sqrt{33^2 + 257.8^2} \\
&= 260 \ \Omega
\end{aligned}
$$

If, on the other hand, the component values were such that the difference was an inductive vector, then the circuit would be said to act inductively.

18-1-2 *Current*

The current in a series circuit is the same at all points throughout the circuit, and therefore

$$I = I_R = I_L = I_C$$

Once the total impedance of the circuit is known, Ohm's law can be applied to calculate the circuit current:

$$I = \frac{V_S}{Z}$$

In the example, circuit current is equal to

$$I = \frac{V_S}{Z}$$
$$= \frac{115 \text{ V}}{260 \text{ }\Omega}$$
$$= 0.44 \text{ A} \quad \text{or} \quad 440 \text{ mA}$$

18-1-3 *Voltage*

Now that you know the value of current flowing in the series circuit, you can calculate the voltage drops across each component, as shown in Figure 18-3(a).

$$V_R = I \times R$$

$$V_L = I \times X_L$$

$$V_C = I \times X_C$$

Since none of these voltages are in phase with one another as shown in Figure 18-3(b), they must be added vectorially to obtain the applied voltage.

The formula, based on the Pythagorean theorem and illustrated in Figure 18-3(c) and (d), can be used to calculate V_S.

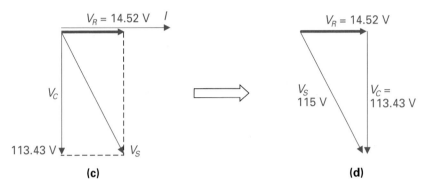

FIGURE 18-3 Series Voltage Drops.

$$V_S = \sqrt{V_R{}^2 + (V_L \sim V_C)^2}$$

In the example circuit, the applied voltage is, as we already know, 115 V.

$$V_S = \sqrt{15^2 + (3 \sim 117)^2}$$
$$= \sqrt{225 + 12{,}996}$$
$$= \sqrt{13{,}221}$$
$$= 115 \text{ V}$$

18-1-4 Phase Angle

As can be seen in Figure 18-3(c), there is a phase difference between V_S and I. This phase difference can be calculated with either of the following formulas.

$$\theta = \arctan \frac{V_L \sim V_C}{V_R}$$

$$\theta = \arctan \frac{X_L \sim X_C}{R}$$

In the example circuit, θ is

$$\theta = \arctan \frac{3.3\text{V} \sim 116.732\text{V}}{14.52\text{V}}$$

$$= \arctan 7.812$$

$$= 82.7°$$

Since the example circuit is capacitive (ICE), the phase angle will be $-82.7°$ since V_S lags I in a circuit that acts capacitively.

18-1-5 Power

The true power or resistive power (P_R) dissipated by a circuit can be calculated using the formula

$$P_R = I^2 \times R$$

which in our example will be

$$P_R = 0.44^2 \times 33 \ \Omega = 6.4 \text{ W}$$

The apparent power (P_A) consumed by the circuit is calculated by

$$P_A = V_S \times I$$

which in our example will be

$$P_A = 115 \text{ V} \times 0.44 = 50.6 \text{ volt-amperes (VA)}$$

The true or actual power dissipated by the resistor is, as expected, smaller than the apparent power that seems to be being used.

The power factor can be calculated, as usual, by

$$\text{PF} = \cos \theta = \frac{R}{Z} = \frac{P_R}{P_A}$$

PF of 0 = reactive circuit
PF of 1 = resistive circuit

In the example circuit, PF = 0.126, indicating that the circuit is mainly reactive.

EXAMPLE:

For a series circuit where $R = 10 \, \Omega$, $L = 5$ mH, $C = 0.05 \, \mu$F, and $V_S = 100$ V/2 kHz, calculate:

a. X_C f. Apparent power

b. X_L g. True power

c. Z h. Power factor

d. I i. Phase angle

e. V_R, V_C, and V_L

Solution:

a. $X_C = \dfrac{1}{2\pi f C} = 1.6 \text{ k}\Omega$

b. $X_L = 2\pi f L = 62.8 \, \Omega$

c. $Z = \sqrt{R^2 + (X_L \sim X_C)^2}$

$\quad = \sqrt{(10 \, \Omega)^2 + (1.6 \text{ k}\Omega \sim 62.8 \, \Omega)^2}$

$\quad = \sqrt{10 \, \Omega^2 + 1.54 \text{ k}\Omega^2}$

$\quad = 1.54 \text{ k}\Omega$ (capacitive circuit due to high X_C)

d. $I = \dfrac{V_S}{Z} = \dfrac{100 \text{ V}}{1.54 \text{ k}\Omega} = 64.9$ mA

e. $V_R = I \times R = 64.9 \text{ mA} \times 10 \, \Omega = 0.65$ V

$\quad V_C = I \times X_C = 64.9 \text{ mA} \times 1.6 \text{ k}\Omega = 103.9$ V

$\quad V_L = I \times X_L = 64.9 \text{ mA} \times 62.8 \, \Omega = 4.1$ V

f. Apparent power $= V_s \times I = 100 \text{ V} \times 64.9 \text{ mA} = 6.49$ VA

g. True power $= I^2 \times R = 64.9^2 \times 10 \, \Omega = 42.17$ mW

h. $\text{PF} = \dfrac{R}{Z} = \dfrac{10 \, \Omega}{1.5 \text{ k}\Omega} = 0.006$ (reactive circuit)

i. $\theta = \arctan \dfrac{V_L \sim V_C}{V_R}$

$\quad = \arctan \dfrac{4.1 \text{ V} \sim 103.9 \text{ V}}{0.65 \text{ V}}$

$\quad = \arctan 153.54$

$\quad = 89.63°$

Capacitive circuit (ICE); therefore, V_S lags I by $-89.63°$.

18-2   PARALLEL *RLC* CIRCUIT

Now that the characteristics of a series circuit are understood, let us connect a resistor, inductor, and capacitor in parallel with one another. Figure 18-4(a) and (b) show the current and voltage relationships of a parallel *RLC* circuit.

18-2-1   *Voltage*

As can be seen in Figure 18-4(a), the voltage across any parallel circuit will be equal and in phase. Therefore,

$$V_R = V_L = V_C = V_S$$

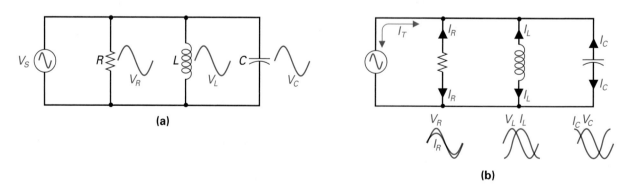

(a)

(b)

FIGURE 18-4   Parallel *RLC* Circuit. (a) *RLC* Parallel Circuit Voltage: Voltages across Each Component Are All Equal and in Phase with One Another in a Parallel Circuit. (b) *RLC* Parallel Circuit Currents: I_R Is in Phase with V_R, I_L Lags V_L by 90°, and I_C Leads V_C by 90°.

18-2-2 Current

With current, we must first calculate the individual branch currents (I_R, I_L, and I_C) and then calculate the total circuit current (I_T). An example circuit is illustrated in Figure 18-5(a), and the branch currents can be calculated by using the formulas

$$I_R = \frac{V}{R}$$

$$I_L = \frac{V}{X_L}$$

$$I_C = \frac{V}{X_C}$$

Figure 18-5(b) illustrates these branch currents vectorially, with I_R in phase with V_S, I_L lagging by 90°, and I_C leading I_R by 90°. The 180° phase difference between I_C and I_L results in a cancellation, as shown in Figure 18-5(c).

The total current (I_T) can be calculated by using the Pythagorean theorem on the right triangle, as illustrated in Figure 18-5(d).

$$I_T = \sqrt{I_R^2 + I_X^2}$$

$$(I_X = I_L \sim I_C)$$

$$= \sqrt{3.5^2 + 14.9^2}$$
$$= 15.3 \text{ A}$$

18-2-3 Phase Angle

As shown in Figure 18-5(b), there is a phase difference between the source voltage (V_S) and the circuit current (I_T). This phase difference can be calculated using the formula

$$\theta = \arctan \frac{I_L \sim I_C}{I_R}$$

$$I_R = \frac{V}{R} = \frac{115 \text{ V}}{33 \ \Omega} = 3.5 \text{ A}$$

$$I_L = \frac{V}{X_L} = \frac{115 \text{ V}}{7.5 \ \Omega} = 15.3 \text{ A}$$

$$X_L = 2\pi fL$$
$$= 2\pi \times 60 \times 20 \text{ mH} = 7.5 \ \Omega$$

$$I_C = \frac{V}{X_C} = \frac{115 \text{ V}}{265.3 \ \Omega} = 0.43 \text{ A}$$

$$X_C = \frac{1}{2\pi fL}$$
$$= \frac{1}{2\pi \times 60 \times 10 \ \mu\text{F}} = 265.3 \ \Omega$$

(a)

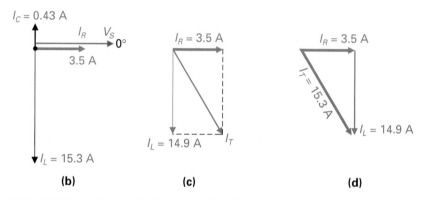

(b)   **(c)**   **(d)**

FIGURE 18-5   **Example *RLC* Parallel Circuit.**

$$= \arctan \frac{15.3 \text{ A} - 0.43 \text{ A}}{3.5 \text{ A}}$$
$$= \arctan 4.25$$
$$= 76.7°$$

Since this is an inductive circuit (ELI), the total current (I_T) will lag the source voltage (V_S) by $-76.7°$.

18-2-4   *Impedance*

Since the total current (I_T) is known, the impedance of all three components in parallel can be calculated by the formula

$$Z = \frac{V}{I_T}$$

$$= \frac{115 \text{ V}}{15.3 \text{ A}}$$

$$= 7.5 \ \Omega$$

18-2-5 Power

The true power dissipated can be calculated using

$$P_R = I_R^2 \times R$$

$$= 3.5^2 \times 33 \ \Omega$$

$$= 404.3 \text{ W}$$

the apparent power consumed by the circuit is calculated by

$$P_A = V_S \times I_T$$

$$= 115 \text{ V} \times 15.3 \text{ A}$$

$$= 1759.5 \text{ volt-amperes (VA)}$$

and the power factor can be calculated, as usual, with

$$\text{PF} = \cos\theta = \frac{P_R}{P_A}$$

In the example circuit, PF = 0.23.

SELF-TEST REVIEW QUESTIONS FOR SECTION 18-2

1. State the formulas for calculating the following in relation to a parallel *RLC* circuit:
   a. I_R d. I_L
   b. I_T e. θ
   c. I_C f. Z
2. State the formulas for:
   a. P_R c. P_A
   b. P_X d. PF

Resonance is a circuit condition that occurs when the inductive reactance (X_L) and the capacitive reactance (X_C) have been balanced. Figure 18-6 illustrates a parallel- and a series-connected LC circuit. If a dc voltage is applied to the input of either circuit, the capacitor will act as an open $(X_C = \text{infinite } \Omega)$ and the inductor will act as a short $(X_L = 0 \ \Omega)$.

If a low-frequency ac is now applied to the input, X_C will decrease from maximum, and X_L will increase from zero. As the ac frequency is increased further, the capacitive reactance will continue to fall $(X_C \downarrow \propto 1/f\uparrow)$ and the inductive reactance to rise $(X_L \uparrow \propto f\uparrow)$, as shown in Figure 18-7.

As the input ac frequency is increased further, a point will be reached where X_L will equal X_C, and this condition is known as *resonance*. The frequency at which $X_L = X_C$ in either a parallel or a series LC circuit is known as the *resonant frequency* (f_0) and can be calculated by the following formula, which has been derived from the capacitive and inductive reactance formulas:

$$f_0 = \frac{1}{2\pi \sqrt{LC}}$$

where f_0 = resonant frequency, in hertz (Hz)
$\quad L$ = inductance, in henrys (H)
$\quad C$ = capacitance, in farads (F)

▨ **EXAMPLE:**

Calculate the resonant frequency (f_0) of a series LC circuit if $L = 750$ mH and $C = 47$ μF.

▨ *Solution:*

$$f_0 = \frac{1}{2\pi \sqrt{L \times C}}$$

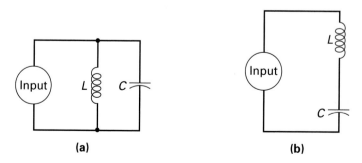

(a)   (b)

**FIGURE 18-6   Resonance. (a) Parallel *LC* Circuit.
(b) Series *LC* Circuit.**

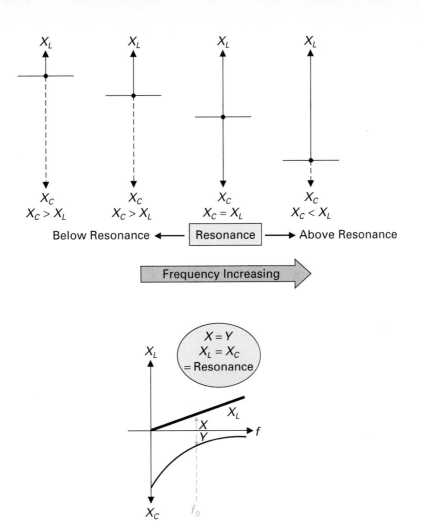

FIGURE 18-7 Frequency versus Reactance.

$$= \frac{1}{2\pi \sqrt{(750 \times 10^{-3}) \times (47 \times 10^{-6})}}$$

$$= 26.8 \text{ Hz}$$

18-3-1 *Series Resonance*

Figure 18-8(a) illustrates a series RLC circuit at resonance ($X_L = X_C$), or **series resonant circuit.** The ac input voltage causes current to flow around the circuit, and since all the components are connected in series, the same value of current (I_S) will flow through all the components. Since R, X_L, and X_C are all equal to 100 Ω and the current flow is the same throughout, the voltage dropped across each component will be equal, as illustrated vectorially in Figure 18-8(b).

The voltage across the resistor is in phase with the series circuit current (I_S); however, since the voltage across the inductor (V_L) is 180° out of phase with the

Series Resonant Circuit
A resonant circuit in which the capacitor and coil are in series with the applied ac voltage.

FIGURE 18-8   Series Resonant Circuit.

voltage across the capacitor (V_C), and both are equal to one another, V_L cancels V_C, when both are measured in series.

Three unusual characteristics occur when a circuit is at resonance, which do not occur at any other frequency.

(1). The first is that if V_L and V_C cancel, the voltage across L and C will measure 0 V on a voltmeter. Since there is effectively no voltage being dropped across these two components, all the voltage must be across the resistor ($V_R =$ 12 V). This is true; however, since the same current flows throughout the series circuit, a voltmeter will measure 12 V across C, 12 V across L, and 12 V across R, as shown in Figure 18-8(c). It now appears that the voltage drops around the series circuit (36 V) do not equal the voltage applied (12 V). This is not true, as V_L and V_C cancel, because they are out of phase with one another, so Kirchhoff's voltage law is still valid.

(2). The second unusual characteristic of resonance is that because the total opposition or impedance (Z) is equal to

$$Z = \sqrt{R^2 + (X_L \sim X_C)^2}$$

and the difference between X_L and X_C is 0 $(Z = \sqrt{R^2 + 0})$, the impedance of a series circuit at resonance is equal to the resistance value R $(Z = \sqrt{R^2} = R)$. Consequently, the applied ac voltage of 12 V is forcing current to flow through this series RLC circuit. Since current is equal to $I_s = V/Z$ and $Z = R$, the circuit current at resonance is dependent only on the value of resistance. The capacitor and inductor are invisible and are seen by the source as simply a piece of conducting wire with no resistance, as illustrated in Figure 18-8(d). Since only resistance exists in the circuit, current (I_s) and voltage (V_S) are in phase with one another, and as expected for a purely resistive circuit, the power factor will be equal to 1.

(3). To emphasize the third strange characteristic of series resonance, we will take another example, shown in Figure 18-9. The circuit current in this example is equal to $I = V/R = 12$ V/10 = 1.2 A, since $Z = R$ at resonance. Since the same current flows throughout a series circuit, the voltage across each component can be calculated.

$$V_R = I \times R = 1.2 \text{ A} \times 10 = 12 \text{ V}$$
$$V_L = I \times X_L = 1.2 \text{ A} \times 100 = 120 \text{ V}$$
$$V_C = I \times X_C = 1.2 \text{ A} \times 100 = 120 \text{ V}$$

FIGURE 18-9 Circuit Effects at Resonance.

As V_L is 180° out of phase with V_C, the 120 V across the capacitor cancels with the 120 V across the inductor, resulting in 0 V across L and C combined, as shown in Figure 18-9(b). Since L and C have the ability to store energy, the voltage across them individually will appear larger than the applied voltage.

If the resistance in the circuit is removed completely, as shown in Figure 18-9(c), the circuit current, which is determined by the resistance only, will increase to a maximum ($I\uparrow = V/R\downarrow$) and, consequently, cause an infinitely high voltage across the inductor and capacitor ($V\uparrow = I\uparrow \times R$). In reality, the ac source will have some value of internal resistance, and the inductor, which is a long length of thin wire ($R\uparrow$), will have some value of resistance, as shown in Figure 18-9(d), which limits the series resonant circuit current.

In summary, we can say that in a series resonant circuit:

1. The inductor and capacitor electrically disappear due to their equal but opposite effect, resulting in 0-V drops across the series combination, and the circuit consequently seems purely resistive.

2. The current flow is large because the impedance of the circuit is low and equal to the series resistance (R), which has the source voltage developed across it.

3. The individual voltage drops across the inductor or capacitor can be larger than the source voltage if R is smaller than X_L and X_C.

Quality Factor

As discussed previously regarding inductance, the Q factor is a ratio of inductive reactance to resistance and is used to express how efficiently an inductor will store rather than dissipate energy. In a series resonant circuit, the Q factor indicates the quality of the series resonant circuit, or is the ratio of the reactance to the resistance.

$$Q = \frac{X_L}{R}$$

or, since $X_L = X_C$,

$$Q = \frac{X_C}{R}$$

Another way to calculate the Q of a series resonant circuit is by using the formula

$$Q = \frac{V_L}{V_R} = \frac{V_C}{V_R}$$ (at resonance only)

FIGURE 18-10 Quality Factor at Resonance.

or, since $V_R = V_S$,

$$Q = \frac{V_L}{V_S} = \frac{V_C}{V_S}$$ (at resonance only)

If the Q and source voltage are known, the voltage across the inductor or capacitor can be found by transposition of the formula, as can be seen in the example in Figure 18-10.

The Q of a resonant circuit is almost entirely dependent on the inductor's coil resistance, because capacitors tend to have almost no resistance figure at all, only reactive, which makes them very efficient.

The inductor has a Q value of its own, and if only L and C are connected in series with one another, the Q of the series resonant circuit will be equal to the Q of the inductor, as shown in Figure 18-11. If the resistance is added in with L and C, the Q of the series resonant circuit will be less than that of the inductor's Q.

EXAMPLE:

Calculate the resistance of the series resonant circuit illustrated in Figure 18-12.

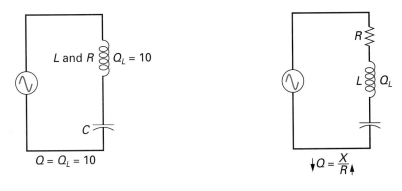

FIGURE 18-11 Resistance within Inductor.

FIGURE 18-12   Series Resonant Circuit Example.

Solution:

$$Q = \frac{V_L}{V_S} = \frac{100 \text{ V}}{10 \text{ V}} = 10$$

Since $Q = X_L/R$, $R = X_L/Q$; so if the inductive reactance can be found, the R can be determined.

$$\begin{aligned} X_L &= 2\pi \times f \times L \\ &= 2\pi \times 100 \times 8 \text{ mH} \\ &= 5 \ \Omega \end{aligned}$$

R will consequently equal

$$\begin{aligned} R &= \frac{X_L}{Q} \\ &= \frac{5}{10} \\ &= 0.5 \ \Omega \end{aligned}$$

Bandwidth

Bandwidth
Width of the group or band of frequencies between the half-power points.

Frequency Response Curve
A graph indicating a circuit's response to different frequencies.

Cutoff Frequency
Frequency at which the gain of the circuit falls below 0.707 of the maximum current or half-power (−3 dB).

A series resonant circuit is selective in that frequencies at resonance or slightly above or below will cause a larger current than frequencies well above or below the circuit's resonant frequency. The group or band of frequencies that causes the larger current is called the circuit's **bandwidth.**

Figure 18-13 illustrates a series resonant circuit and its bandwidth. The × marks on the curve illustrate where different frequencies were applied to the circuit and the resulting value of current measured in the circuit. The resulting curve produced is called a **frequency response curve,** as it illustrates the circuit's response to different frequencies. At resonance, $X_L = X_C$ and the two cancel, which is why maximum current was present in the circuit (100 mA) when the resonant frequency (100 Hz) was applied.

The bandwidth includes the group or band of frequencies that cause 70.7% or more of the maximum current to flow within the series resonant circuit; in this example, frequencies from 90 to 110 Hz cause 70.7 mA or more, which is 70.7% of maximum (100 mA), to flow. The bandwidth in this example is equal to

$$BW = 110 - 90 = 20 \text{ Hz}$$

(110 Hz and 90 Hz are known as **cutoff frequencies**).

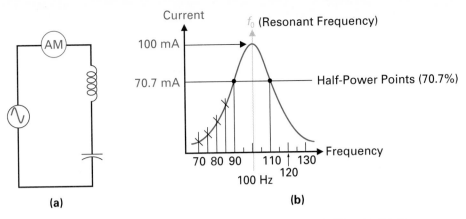

FIGURE 18-13 Series Resonant Circuit Bandwidth. (a) Circuit. (b) Frequency Response Curve.

Referring to the bandwidth curve in Figure 18-13(b), you may notice that 70.7% is also called the **half-power points,** although it does not exist halfway between 0 and maximum. This value of 70.7% is not the half-current point but the half-power point, as we can prove with a simple example.

Half-Power Point
A point at which power is 50%. This half-power point corresponds to 70.7% of the total current.

▨ **EXAMPLE:**

$R = 2$ kΩ and $I = 100$ mA; therefore, power $= I^2 \times R = 100$ mA$^2 \times 2$ k$\Omega = 20$ W. If the current is now reduced so that it is 70.7% of its original value, calculate the power dissipated.

▨ *Solution:*

$$P = I^2 \times R = 70.7 \text{ mA}^2 \times 2 \text{ k}\Omega = 10 \text{ W}$$

In summary, the 70.7% current points are equal to the 50% or half-power points. A circuit's bandwidth is the band of frequencies that exists between the 70.7% current points or half-power points.

The bandwidth of a series resonant circuit can also be calculated by use of the formula

$$\text{BW} = \frac{f_0}{Q_{f0}}$$

$f_0 =$ Resonant Frequency
$Q_{f0} =$ Quality factor at resonance

This formula states that the BW is proportional to the resonant frequency of the circuit and inversely proportional to the Q of the circuit.

Figure 18-14 illustrates three example response curves. In these three examples, the value of R is changed from 100 Ω to 200 to 400 Ω. This does not vary the resonant frequency, but simply alters the Q and therefore the BW. The resistance value will determine the Q of the circuit, and since Q is inversely pro-

FIGURE 18-14 **Bandwidth of a Series Resonant Circuit.**

portional to resistance, Q is proportional to current; consequently, a high value of Q will cause a high value of current.

In summary, the bandwidth of a series resonant circuit will increase as the Q of the circuit decreases ($BW\uparrow = f_0/Q\downarrow$), and vice versa.

SELF-TEST REVIEW QUESTIONS FOR SECTION 18-3-1

1. Define *resonance*.
2. What is series resonance?
3. In a series resonant circuit, what are the three rather unusual circuit phenomena that take place?
4. How does Q relate to series resonance?
5. Define *bandwidth*.
6. Calculate BW if $f_0 = 12 \text{ kHz}$ and $Q = 1000$.

18-3-2 Parallel Resonance

The **parallel resonant circuit** acts differently from the series resonant circuit, and these different characteristics need to be analyzed and discussed. Figure 18-15 illustrates a parallel resonant circuit. The inductive current could be calculated by using the formula

$$I_L = \frac{V_L}{X_L}$$
$$= \frac{10 \text{ V}}{1 \text{ k}\Omega}$$
$$= 10 \text{ mA}$$

Parallel Resonant Circuit
Circuit having an inductor and capacitor in parallel with one another, offering a high impedance at the frequency of resonance.

The capacitive current could be calculated by using the formula

$$I_C = \frac{V_C}{X_C}$$
$$= \frac{10 \text{ V}}{1 \text{ k}\Omega}$$
$$= 10 \text{ mA}$$

Looking at the vector diagram in Figure 18-15(b), you can see that I_C leads the source voltage by 90° (ICE) and I_L lags the source voltage by 90° (ELI), creating a 180° phase difference between I_C and I_L. This means that when 10 mA of current flows up through the inductor, 10 mA of current will flow in the opposite direction down through the capacitor, as shown in Figure 18-16(a). During the opposite alternation, 10 mA will flow down through the inductor and 10 mA will travel up through the capacitor, as shown in Figure 18-16(b).

If 10 mA arrives into point X and 10 mA of current leaves point X, no current can be flowing from the source (V_S) to the parallel LC circuit; the current is simply swinging or oscillating back and forth between the capacitor and inductor.

The source voltage (V_S) is needed initially to supply power to the LC circuit and start the oscillations; but once the oscillating process is in progress (assuming the ideal case), current is only flowing back and forth between induc-

(a)   **(b)**

FIGURE 18-15 Parallel Resonant Circuit. (a) Circuit. (b) Vector Diagram.

I_S *(Source Current)* = 0; Therefore Resistance = Maximum.

(Infinite) $R = \dfrac{V}{I}$ (=0)

(c)

FIGURE 18-16 **Current in a Parallel Resonant Circuit.**

tor and capacitor, and no current is flowing from the source. So the *LC* circuit appears as an infinite impedance and the source can be disconnected, as shown in Figure 18-16(c).

Flywheel Action

Flywheel Action
Sustaining effect of oscillation in an *LC* circuit due to the charging and discharging of the capacitor and the expansion and contraction of the magnetic field around the inductor.

Let's discuss this oscillating effect, called **flywheel action,** in a little more detail. The name is derived from the fact that it resembles a mechanical flywheel, which, once started, will keep going continually until friction reduces the magnitude of the rotations to zero.

The electronic equivalent of the mechanical flywheel is a resonant parallel-connected *LC* circuit. Figure 18-17(a) through (h) illustrate the continual energy transfer between capacitor and inductor, and vice versa. The direction of the circulating current reverses each half-cycle at the frequency of resonance. Energy is stored in the capacitor in the form of an electric field between the plates on one half-cycle, and then the capacitor discharges, supplying current to build up a magnetic field on the other half-cycle. The inductor stores its energy in the form of a magnetic field, which will collapse, supplying a current to charge the capacitor, which will then discharge, supplying a current back to the inductor, and so on. Due to the "storing action" of this circuit, it is sometimes related to the fluid analogy and referred to as a **tank circuit.**

Tank Circuit
Circuit made up of a coil and capacitor that is capable of storing electric energy.

The Reality of Tanks

Under ideal conditions a tank circuit should oscillate indefinitely if no losses occur within the circuit. In reality, the resistance of the coil reduces that 100% efficiency, as does friction with the mechanical flywheel. This coil resis-

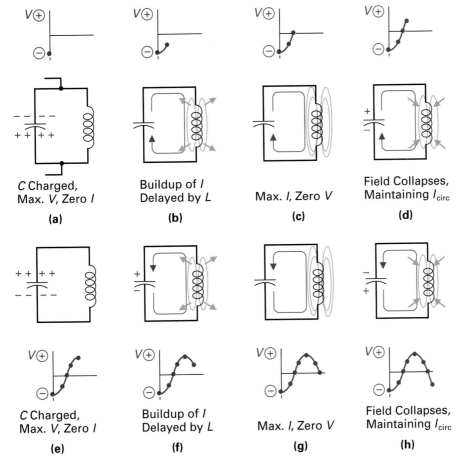

FIGURE 18-17 **Energy and Current in an *LC* Parallel Circuit at Resonance.**

tance is illustrated in Figure 18-18(a), and, unlike reactance, resistance is the opposition to current flow, with the dissipation of energy in the form of heat. As a small part of the energy is dissipated with each cycle, the oscillations will be reduced in size and eventually fall to zero, as shown in Figure 18-18(b).

If the ac source is reconnected to the tank, as shown in Figure 18-18(c), a small amount of current will flow from the source to the tank to top up the tank or replace the dissipated power. The higher the coil resistance is, the higher the loss and the larger the current flow from source to tank to replace the loss.

Quality Factor

In the series resonant circuit, we were concerned with voltage drops since current remains the same throughout a series circuit, so

$$Q = \frac{V_C \text{ or } V_L}{V_S} \quad \text{(at resonance only)}$$

(a)

(b)

(c)

FIGURE 18-18   Losses in Tanks.

In a parallel resonant circuit, we are concerned with circuit currents rather than voltage, so

$$Q = \frac{I_{\text{tank}}}{I_S}$$

(at resonance only)

The quality factor, Q, can also be expressed as the ratio between reactance and resistance:

$$Q = \frac{X_L}{R} \quad \text{(at any frequency)}$$

Another formula, which is the most frequently used when discussing and using parallel resonant circuits, is

$$Q = \frac{Z_{\text{tank}}}{X_L}$$

(at resonance only)

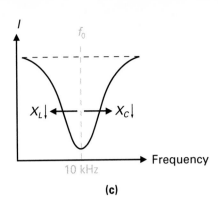

(a)	**(b)**	**(c)**

FIGURE 18-19 **Parallel Resonant Circuit Bandwidth.**

This formula states that the Q of the tank is proportional to the tank impedance. A higher tank impedance results in a smaller current flow from source to tank. This assures less power is dissipated, and that means a higher-quality tank.

Of all the three Q formulas for parallel resonant circuits, $Q = I_{tank}/I_S$, $Q = X_L/R$, or $Q = Z_{tank}/X_L$, the latter is the easiest to use as both X_L and the tank impedance can easily be determined in most cases where C, L, and R internal for the inductor are known.

Bandwidth

Figure 18-19 illustrates a parallel resonant circuit and two typical response curves. These response curves summarize what we have described previously, in that a parallel resonant circuit has maximum impedance [Figure 18-19(b)] and minimum current [Figure 18-19(c)] at resonance. The current versus frequency response curve shown in Figure 18-19(c) is the complete opposite to the series resonant response curve. At frequencies below resonance (< 10 kHz), X_L is low and X_C is high, and the inductor offers a low reactance, producing a high current path and low impedance. On the other hand, at frequencies above resonance (> 10 kHz), the capacitor displays a low reactance, producing a high current path and low impedance. The parallel resonant circuit is like the series resonant circuit in that it responds to a band of frequencies close to its resonant frequency.

The bandwidth (BW) can be calculated by use of the formula

$$BW = \frac{f_0}{Q_{f_0}}$$

f_0 = Resonant frequency

Q_{f_0} = Quality factor at resonance

■ EXAMPLE:

Calculate the bandwidth of the circuit illustrated in Figure 18-20.

$f_0 = 125 \text{ kHz}$

$Q_{f_0} = 75$

FIGURE 18-20   **Bandwidth Example.**

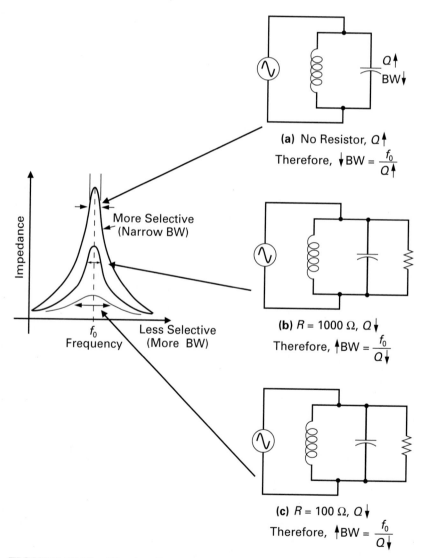

(a) No Resistor, $Q\uparrow$

Therefore, $\downarrow BW = \dfrac{f_0}{Q\uparrow}$

(b) $R = 1000\ \Omega$, $Q\downarrow$

Therefore, $\uparrow BW = \dfrac{f_0}{Q\downarrow}$

(c) $R = 100\ \Omega$, $Q\downarrow$

Therefore, $\uparrow BW = \dfrac{f_0}{Q\downarrow}$

More Selective
(Narrow BW)

Less Selective
(More BW)

Impedance

f_0
Frequency

FIGURE 18-21   **Varying Bandwidth by Loading the Tank Circuit.**

$$BW = \frac{f_0}{Q_{f_0}}$$

$$= \frac{125 \text{ kHz}}{75}$$

$$= 1.7 \text{ kHz}$$

$$\frac{1.7 \text{ kHz}}{2} = 0.85 \text{ kHz}$$

therefore the bandwidth extends from

$$f_0 + 0.85 \text{ kHz} = 125.85 \text{ kHz}$$

$$f_0 - 0.85 \text{ kHz} = 124.15 \text{ kHz}$$

$$BW = 124.15 \text{ kHz to } 125.85 \text{ kHz}$$

Selectivity

Circuits containing inductance and capacitance are often referred to as **tuned circuits** since they can be adjusted to make the circuit responsive to a particular frequency (the resonant frequency). **Selectivity,** by definition, is the ability of a tuned circuit to respond to a desired frequency and ignore all others. Parallel resonant *LC* circuits are sometimes too selective, as the *Q* is too large, producing too narrow a bandwidth, as shown in Figure 18-21(a) (BW↓ = f_0/Q↑).

In this situation, because of the very narrow response curve, a high resistance value can be placed in parallel with the *LC* circuit to provide an alternative path for line current. This process is known as *loading* or *damping* the tank and will cause an increase in line current and decrease in Q (Q↓ = I_{tank}/I_{line}↑). The decrease in Q will cause a corresponding increase in BW (BW↑ = f_0/Q↓), as shown by the examples in Figure 18-21, which illustrates a 1000-Ω loading resistor [Figure 18-21(b)] and a 100-Ω loading resistor [Figure 18-21(c)].

In summary, a parallel resonant circuit can be made less selective with a broader bandwidth if a resistor is added in parallel, providing an increase in current and a decrease in impedance, which widens the bandwidth.

Tuned Circuit
Circuit that can have its components' values varied so that the circuit responds to one selected frequency yet heavily attenuates all other frequencies.

Selectivity
Characteristic of a circuit to discriminate between the wanted signal and the unwanted signal.

SELF-TEST REVIEW QUESTIONS FOR SECTION 18-3-2

1. What are the differences between a series and a parallel resonant circuit?
2. Describe flywheel action.
3. Calculate the value of *Q* of a tank if $X_L = 50$ Ω and $R = 25$ Ω.
4. When calculating bandwidth for a parallel resonant circuit, can the series resonant bandwidth formula be used?
5. What is selectivity?

18-4 APPLICATIONS OF *RLC* CIRCUITS

In the previous chapters, you saw how *RC* and *RL* filter circuits are used as low- or high-pass filters to pass some frequencies and block others. There are basically four types of filters:

1. Low-pass filter, which passes frequencies below a cutoff frequency
2. High-pass filter, which passes frequencies above a cutoff frequency
3. Bandpass filter, which passes a band of frequencies
4. Band-stop filter, which stops a band of frequencies

18-4-1 Low-Pass Filter

Figure 18-22(a) illustrates how an inductor and capacitor can be connected to act as a low-pass filter. At low frequencies, X_L has a small value compared to the load resistor (R_L), so nearly all the low-frequency input is developed and appears at the output across R_L. Since X_C is high at low frequencies, nearly all the current passes through R_L rather than C.

At high frequencies, X_L increases and drops more of the applied input across the inductor rather than the load. The capacitive reactance, X_C, aids this low output at high-frequency effect by decreasing its reactance and providing an alternative path for current to flow.

Since the inductor basically blocks alternating current and the capacitor shunts alternating current, the net result is to prevent high-frequency signals from reaching the load. The way in which this low-pass filter responds to frequencies is graphically illustrated in Figure 18-22(b).

18-4-2 High-Pass Filter

Figure 18-23(a) illustrates how an inductor and capacitor can be connected to act as a high-pass filter. At high frequencies, the reactance of the capacitor (X_C) is low while the reactance of the inductor (X_L) is high, so all the high frequen-

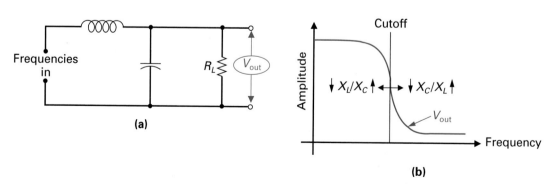

FIGURE 18-22   Low-Pass Filter. (a) Circuit. (b) Frequency Response.

(a)

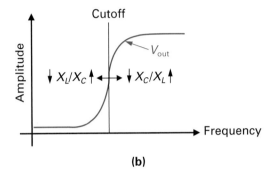

(b)

FIGURE 18-23 **High-Pass Filter. (a) Circuit.**
(b) Frequency Response.

cies are easily passed by the capacitor and blocked by the inductor, so they all are routed through to the output and load.

At low frequencies, the reverse condition exists, resulting in a low X_L and a high X_C. The capacitor drops nearly all the input, and the inductor shunts the signal current away from the output load.

18-4-3 *Bandpass Filter*

Figure 18-24(a) illustrates a series resonant **bandpass filter,** and Figure 18-24(b) shows a parallel resonant bandpass filter. Figure 18-24(c) shows the frequency response curve produced by the bandpass filter. At resonance, the series resonant LC circuit has a very low impedance and will consequently pass the resonant frequency to the load with very little drop across the L and C components.

Below resonance, X_C is high, and the capacitor drops a large amount of the input signal; above resonance, X_L is high and the inductor drops most of the input frequency voltage. This circuit will therefore pass a band of frequencies centered around the resonant frequency of the series LC circuit and block all other frequencies above and below this resonant frequency.

Figure 18-24(b) illustrates how a parallel resonant LC circuit can be used to provide a bandpass response. The series resonant circuit was placed in series with the output, whereas the parallel resonant circuit will have to be placed in parallel with the output to provide the same results. At resonance, the parallel resonant circuit or tank has a high impedance, so very little current will be shunted away from the output; it will be passed on to the output, and almost all the input will appear at the output across the load.

Bandpass Filter
Filter circuit that passes a group or band of frequencies between a lower and an upper cutoff frequency, while heavily attenuating any other frequency outside this band.

(a)

(b)

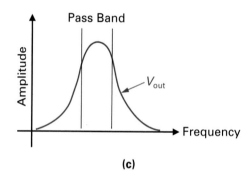

(c)

FIGURE 18-24 Bandpass Filter. (a) Series Resonant Bandpass Filter. (b) Parallel Resonant Bandpass Filter. (c) Frequency Response.

Above resonance, X_C is small, so most of the input is shunted away from the output by the capacitor; below resonance, X_L is small, and the shunting action occurs again, but this time through the inductor.

Figure 18-25 illustrates how a transformer can be used to replace the inductor to produce a bandpass filter. At resonance, maximum flywheel current flows within the parallel circuit made up of the capacitor and the primary of the transformer (L), which is known as a *tuned transformer*. With maximum flywheel current, there will be a maximum magnetic field, which means that there will be maximum power transfer between primary and secondary. So nearly all the input will be coupled to the output (coupling coefficient $k = 1$) and appear across the load at and around a small band of frequencies centered on resonance.

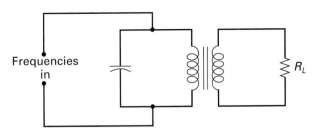

FIGURE 18-25   Parallel Resonant Bandpass Circuit Using a Transformer.

Above and below resonance, current within the parallel resonant circuit will be smaller. So the power transfer ability will be less, effectively keeping the frequencies outside the bandpass from appearing at the output.

18-4-4   Band-Stop Filter

Figure 18-26(a) illustrates a series resonant and Figure 18-26(b) a parallel resonant **band-stop filter.** Figure 18-26(c) shows the frequency response curve produced by a band-stop filter. The band-stop filter operates exactly the opposite to

Band-Stop Filter
A filter that attenuates alternating currents whose frequencies are between given upper and lower cutoff values while passing frequencies above and below this band.

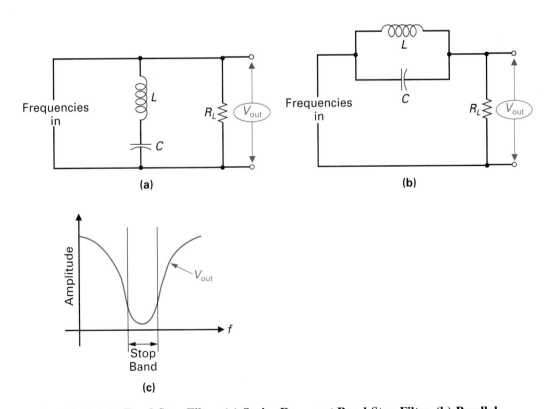

FIGURE 18-26   Band-Stop Filter. (a) Series Resonant Band-Stop Filter. (b) Parallel Resonant Band-Stop Filter. (c) Frequency Response.

FIGURE 18-27   Tuning in of Station by Use of a Bandpass Filter.

a bandpass filter in that it blocks or attenuates a band of frequencies centered on the resonant frequency of the *LC* circuit.

In the series resonant circuit in Figure 18-26(a), the *LC* impedance is very low at and around resonance, so these frequencies are rejected or shunted away from the output. Above and below resonance, the series circuit has a very high impedance, which results in almost no shunting of the signal away from the output.

In the parallel resonant circuit, in Figure 18-26(b), the *LC* circuit is in series with the load and output. At resonance, the impedance of a parallel resonant circuit will be very high, and the band of frequencies centered around resonance will be blocked. Above and below resonance, the impedance of the tank is very low, so nearly all the input is developed across the output.

Filters are necessary in applications such as television or radio, where we need to tune in (select or pass) one frequency that contains the information we desire, yet block all the millions of other frequencies that are also carrying information, as shown in Figure 18-27.

SELF-TEST REVIEW QUESTIONS FOR SECTION 18-4

1. Of the four types of filters, which:
   a. Would utilize the inductor as a shunt?
   b. Would utilize the capacitor as a shunt?
   c. Would use a series resonant circuit as a shunt?
   d. Would use a parallel resonant circuit as a shunt?
2. In what applications can filters be found?

18-5 COMPLEX NUMBERS

After reading this section you will realize that there is really nothing complex about **complex numbers.** The complex number system allows us to determine the *magnitude* and *phase angle* of electrical quantities by adding, subtracting, multiplying, and dividing phasor quantities, and is an invaluable tool in ac circuit analysis.

18-5-1 The Real Number Line

Real numbers can be represented on a horizontal line, known as the real number line, as in Figure 18-28. Referring to this line, you can see that positive numbers exist to the right of the center point corresponding to zero, while negative numbers exist to the left. This representation satisfied most mathematicians for a short time, as they could indicate numbers such as 2 or 5 as a point on the line. Numbers corresponding to the $\sqrt{9}$ could also be represented, as three points to the right of zero ($\sqrt{9} = +3$). However, a problem was reached if they wished to indicate a point corresponding to $\sqrt{-9}$. The -9 is not $+3$ [since $(+3) \times (+3) = +9$], and it is not -3 [since $(-3) \times (-3) = +9$]. So it was eventually realized that the square root of a negative number could not be indicated on the real number line, as it is not a real number.

18-5-2 The Imaginary Number Line

Mathematicians decided to call the square root of a negative number, such as $\sqrt{-4}$ or $\sqrt{-9}$, **imaginary numbers,** which are not fictitious or imaginary, but simply a particular type of number.

    Just as real numbers can be represented on a real number line, imaginary numbers can be represented on an imaginary number line, as shown in Figure 18-29. The imaginary number line is vertical, to distinguish it from the real number line, and when working with electrical quantities a $\pm j$ prefix, known as the ***j* operator**, is used for values that appear on the imaginary number line.

18-5-3 The Complex Plane

A complex number is the combination of a real and imaginary number and is represented on a two-dimensional plane called the **complex plane,** shown in Figure 18-30. Generally, the real number appears first, followed by the imaginary number. Here are some examples of complex numbers.

FIGURE 18-28   Real Number Line.

Positive Imaginary Numbers

Negative Imaginary Numbers   **FIGURE 18-29   Imaginary Number Line.**

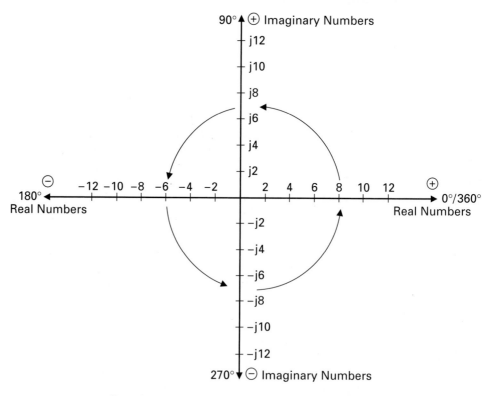

FIGURE 18-30   Complex Plane.

REAL NUMBERS	IMAGINARY NUMBERS
3	$+j4$
-2	$+j4$
-3	$-j2$

Complex numbers, therefore, are merely terms that need to be added as phasors, and all you have to do basically is draw a vector representing the real number and then draw another vector representing the imaginary number.

◼ **EXAMPLE:**

Find the points in the complex plane in Figure 18-31 that correspond to the following complex numbers.

$$W = 3 + j4$$
$$X = 5 - j7$$
$$Y = -4 + j6$$
$$Z = -3 - j5$$

◼ *Solution:*

By first locating the point corresponding to the real number on the horizontal line and then plotting it against the imaginary number on the vertical line, the points can be determined as shown in Figure 18-31.

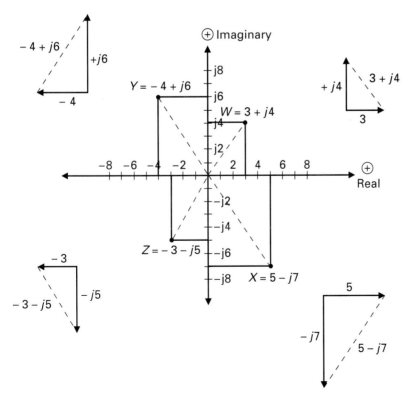

FIGURE 18-31 **Complex Numbers Examples.**

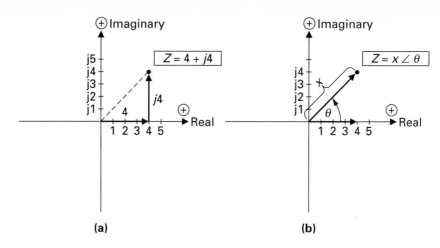

FIGURE 18-32   Representing Phasors. (a) Rectangular Notation. (b) Polar Notation.

Rectangular Coordinates
A Cartesian coordinate of a Cartesian coordinate system whose straight-line axes or coordinate planes are perpendicular.

Polar Coordinates
Either of two numbers that locate a point in a plane by its distance from a fixed point on a line and the angle this line makes with a fixed line.

A number like $3 + j4$ specifies two phasors in **rectangular coordinates,** so this system is the *rectangular representation of a complex number.* There are several other ways to describe a complex number, one of which is the polar representation of a complex number, using **polar coordinates,** which will be discussed next.

18-5-4   *Polar Complex Numbers*

Phasors can also be expressed in polar form, as shown in Figure 18-32, which compares rectangular and polar notation. With the rectangular notation in Figure 18-32(a), the horizontal coordinate is the real part and the vertical coordinate is the imaginary part of the complex number. With the polar notation shown in Figure 18-32(b), the magnitude of the phasor (x, or size) and the angle ($\angle \theta$, meaning "angle theta") relative to the positive real axis (measured in a counterclockwise direction) are stated.

▨   **EXAMPLE:**

Sketch the following polar numbers:

   a. $5 \angle 60°$

   b. $3 \angle 220°$

▨   *Solution:*

As you can see in Figure 18-33, an equivalent negative angle, which is calculated by subtracting the given positive angle from 360°, can also be used.

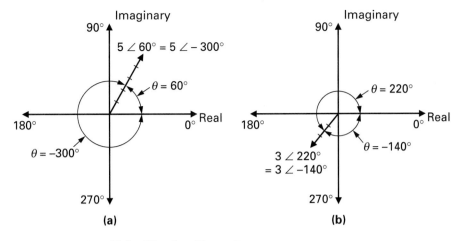

FIGURE 18-33   Polar Number Examples.

18-5-5   *Rectangular/Polar Conversions*

Many scientific calculators have a feature that allows you to convert rectangular numbers to polar numbers, and vice versa. These conversions are based on the Pythagorean theorem and trigonometric functions, discussed previously in a mini-math review.

Polar-to-Rectangular Conversion

The polar notation states the magnitude and angle, as shown in Figure 18-34(a). The following examples show how this conversion can be achieved.

EXAMPLE:

Convert the following polar numbers to rectangular form:

  a. $5 \angle 30°$ b. $18 \angle -35°$ c. $44 \angle 220°$

Solution:

  a. Real number $= 5 \cos 30° = 4.33$
     Imaginary number $= 5 \sin 30° = j2.5$
     Polar number, $5 \angle 30°$ = rectangular number, $4.33 + j2.5$

FIGURE 18-34   **Polar-to-Rectangular Conversion.**

b. Real number = 18 cos (−35°) = 14.74
   Imaginary number = 18 sin (−35°) = −j10.32
   Polar number, 18 ∠ −35° = rectangular number, 14.74 − j10.32

c. Real number = 44 cos 220° = −33.7
   Imaginary number = 44 sin 220° = −j28.3
   Polar number, 44 ∠ 220° = rectangular number, −33.7 − j28.3

Rectangular-to-Polar Conversion

The rectangular notation states the horizontal (real) and vertical (imaginary) sides of a triangle, as shown in Figure 18-35. The following examples show how the conversion can be achieved.

■ **EXAMPLE:**

Convert the following rectangular numbers to polar form:

   a. 4 + j3   b. 16 − j14

■ *Solution:*

a. Magnitude = $\sqrt{4^2 + 3^2}$ = 5
   Angle = arctan (3/4) = 36.9°
   Rectangular number, 4 + j3 = polar number, 5 ∠ 36.9°

b. Magnitude = $\sqrt{16^2 + (−14)^2}$ = 21.3
   Angle = arctan (−14/16) = − 41.2°
   Rectangular number, 16 − j14 = polar number, 21.3 ∠ −41.2°

18-5-6 *Complex Number Arithmetic*

Since a phase difference exists between real and imaginary (j) numbers, certain rules should be applied when adding, subtracting, multiplying, or dividing complex numbers.

Addition

The sum of two complex numbers is equal to the sum of their real and imaginary parts.

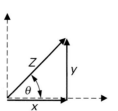

Polar Notation

Rectangular Notation

Magnitude, $Z = \sqrt{x^2 + y^2}$
Angle, $\theta = \arctan\left(\frac{y}{x}\right)$

FIGURE 18-35 **Rectangular-to-Polar Conversion.**

Add the following complex numbers:

   a. $(3 + j4) + (2 + j5)$
   b. $(4 + j5) + (2 - j3)$

Solution:

   a. $(3 + j4) + (2 + j5) = (3 + 2) + (j4 + j5) = 5 + j9$
   b. $(4 + j5) + (2 - j3) = (4 + 2) + (j5 - j3) = 6 + j2$

Subtraction

The difference of two complex numbers is equal to the difference between the separate real and imaginary parts.

EXAMPLE:

Subtract the following complex numbers:

   a. $(4 + j3) - (2 + j2)$
   b. $(12 + j6) - (6 - j3)$

Solution:

   a. $(4 + j3) - (2 + j2) = (4 - 2) + j(3 - 2) = 2 + j1$
   b. $(12 + j6) - (6 - j3) = (12 - 6) + j[6 - (-3)] = 6 + j9$

Multiplication

Multiplication of two complex numbers is achieved more easily if they are in polar form. The simple rule to remember is to multiply the magnitudes and then add the angles algebraically.

EXAMPLE:

Multiply the following complex numbers:

   a. $5 \angle 35° \times 7 \angle 70°$
   b. $4 \angle 53° \times 12 \angle -44°$

Solution:

   a. Multiply the magnitudes: $5 \times 7 = 35$.
      Algebraically add the angles: $\angle (35° + 70°) = \angle 105° = 35 \angle 105°$.
   b. Multiply the magnitudes: $4 \times 12 = 48$.
      Algebraically add the angles: $\angle [53° + (-44°)] = \angle 9° = 48 \angle 9°$.

Division

Division is also more easily carried out in polar form. The rule to remember is to divide the magnitudes, and then subtract the denominator angle from the numerator angle.

Divide the following complex numbers:

  a. $60 \angle 30°$ by $30 \angle 15°$ b. $100 \angle 20°$ by $5 \angle -7°$

▦ *Solution:*

  a. Divide the magnitudes: $60/30 = 2$.
     Subtract the denominator angle from the numerator angle: $\angle (30° - 15°) = \angle 15° = 2 \angle 15°$.

  b. Divide the magnitudes: $100/5 = 20$.
     Subtract the angles: $\angle [20° - (-7°)] = \angle 27° = 20 \angle 27°$.

18-5-7 *How Complex Numbers Apply to AC Circuits*

Complex numbers find an excellent application in ac circuits due to all the phase differences that occur between different electrical quantities, such as X_L, X_C, R, and Z as shown in Figure 18-36. The positive real number line, at an angle of $0°$, is used for resistance, which in this example is $3 \ \Omega$, as shown in Figure 18-36(a) and (b).

On the positive imaginary number line, at an angle of $90°$ $(+j)$, inductive reactance (X_L) is represented, which in this example is $j4 \ \Omega$ $(X_L = 4 \ \Omega)$. The voltage drop across an inductor (V_L) is proportional to its inductive reactance (X_L), and both are represented on the $+j$ imaginary number line, since the voltage drop across an inductor will always lead the current (which in a series circuit is always in phase with the resistance) by $90°$.

On the negative imaginary number line, at an angle of $-90°$ or $270°$ $(-j)$, capacitive reactance (X_C) is represented, which in this example is $-j2$ $(X_C = 2 \ \Omega)$. The voltage drop across a capacitor (V_C) is proportional to its capacitive reactance (X_C), and both are represented on the $-j$ imaginary number line since the voltage drop across a capacitor always lags current (charge and discharge) by $-90°$.

Series AC Circuits

Referring to Figure 18-36(a) and (b) once again, we can calculate total impedance simply by adding the phasors.

Z_T **(Rectangular).**   Total series impedance is equal to the sum of all the resistances and reactances:

$$
\begin{aligned}
Z &= R + (jX_L \sim jX_C) \\
&= 3 + (+j4 \sim -j2) \\
&= 3 + j2
\end{aligned}
$$

Z_T **(Polar).**   The total series impedance can be converted from rectangular to polar form:

FIGURE 18-36 Applying Complex Numbers to Series AC Circuits. (a) Impedance Phasors. (b) Series Circuit. (c) Voltage and Current Phasors.

$$\text{magnitude} = \sqrt{3^2 + 2^2} = 3.61 \ \Omega$$

$$\text{angle} = \arctan \frac{2}{3} = 33.7°$$

$$= 3.61 \ \angle \ 33.7°$$

Current. Once the magnitude of Z_T is known (3.61 Ω), I can be calculated. The source voltage of 12 V is a real positive number (0°) and is therefore represented as 12 $\angle$ 0°. Current is equal to

$$I = \frac{V_s}{Z_T} = \frac{12 \ \angle \ 0°}{3.61 \ \angle \ 33.7°}$$

Polar division   Divide the magnitudes: $\dfrac{12}{3.61} = 3.32$ A

Subtract the angles: $0° - 33.7° = -33.7°$

$I = 3.32 \angle -33.7°$

Phase angle.   The circuit current has an angle of −33.7°, which means that it lags V_T (inductive circuit, therefore ELI). This negative phase angle is expected since this series circuit is inductive ($X_L > X_C$), in which case current should lag voltage by some phase angle. This phase angle is less than 45° because the net reactance is less than the circuit resistance, and is shown in Figure 18-36(c).

Voltage drops.   The component voltage drops are calculated with the following formulas and shown in Figure 18-36(c).

$V_R = I \times R = (3.32 \angle -33.7°) \times (3 \angle 0°)$

Multiply the magnitudes: $3.32 \times 3 = 9.96$ V

Algebraically add the angles: $\angle (-33.7° + 0°) = -33.7°$

$V_R = 9.96 \angle -33.7°$

$V_L = I \times X_L = (3.32 \angle -33.7°) \times (4 \angle 90°)$

Multiply the magnitudes: $3.32 \times 4 = 13.28$ V

Algebraically add the angles: $\angle (-33.7 + 90°) = 56.3°$

$V_L = 13.28 \angle 56.3°$

$V_C = I \times X_C = (3.32 \angle -33.7°) \times (2 \angle -90°)$

Multiply the magnitudes: $3.32 \times 2 = 6.64$ V

Algebraically add the angles: $\angle (-33.7 + -90°) = -123.7°$

$V_C = 6.64 \angle -123.7°$

Phase relationships.   As shown in Figure 18-36(c), the voltage across an inductor leads the circuit current by +90°, while the voltage across a capacitor lags the circuit current by −90°. The source voltage acts as the zero reference phase and leads the circuit current and the voltage across the resistor (V_R) by 33.7°.

Source voltage.   Although the source voltage is known, it can be checked to verify all the previous calculations, since the sum of all the individual voltage drops should equal the source voltage.

POLAR $\longrightarrow$ RECTANGULAR

$V_L = 13.28 \angle 56.3 \quad = \quad 7.37 + j11.05$

$V_R = 9.96 \angle -33.7° \quad = \quad 8.29 - j5.53$

$V_C = 6.64 \angle -123.7° = \underline{-3.68 - j5.52}$

$ 11.98 + j0$

Series–Parallel AC Circuits

Figure 18-37(a) illustrates a series–parallel ac circuit containing an R_L branch, an R_C branch, and an RLC branch.

Impedance of each branch. The three branches will each have a value of impedance that will be equal to:

$$\text{RECTANGULAR} \longrightarrow \text{POLAR}$$

$$Z_1 = 10 + j5 \ = 11.2 \angle 26.6° \ \Omega$$

$$Z_2 = 25 - j15 = 29.2 \angle -31.0° \ \Omega$$

The third branch is capacitive since the difference between $-j30$ (X_{C2}) and $+j10$ (X_{L2}) is $-j20$.

$$Z_3 = 20 - j20 = 28.3 \angle -45° \ \Omega$$

Branch currents. The three branch currents, I_1, I_2, and I_3, are calculated by dividing the source voltage (V_s) by the individual branch impedances.

$$I_1 = \frac{V_S}{Z_1} = \frac{30 \angle 0°}{11.2 \angle 26.6°}$$

Divide magnitudes: $30 \div 11.2 = 2.68$

Subtract angles: $\angle (0° - 26.6°) = -26.6°$

$$I_1 = 2.68 \angle -26.6° = 2.4 - j1.2 \text{A}$$

$$I_2 = \frac{V_S}{Z_2} = \frac{30 \angle 0°}{29.2 \angle -31°} = 1.03 \angle +31° = 0.88 + j0.5 \text{ A}$$

$$I_3 = \frac{V_S}{Z_3} = \frac{30 \angle 0°}{28.3 \angle -45°} = 1.06 \angle +45° = 0.75 + j0.7 \text{ A}$$

Total current.

$$I_T = I_1 + I_2 + I_3$$
$$= (2.4 - j1.2) + (0.88 + j0.5) + (0.75 + j0.7)$$
$$= (2.4 + 0.88 + 0.75) + [-j1.2 + (+j0.5) + (+j0.7)]$$
$$= 4.03 \text{ A}$$

FIGURE 18-37 **Applying Complex Numbers to Series–Parallel AC.**

In polar form, this will equal $4.03 \angle 0°$ A.

Total impedance.

$$Z_T = \frac{V_S}{I_T} = \frac{30 \angle 0°}{4.03 \angle 0°} = 7.44 \angle 0° \ \Omega$$

$$\overline{\text{POLAR} \longrightarrow \text{RECTANGULAR}}$$

$$7.44 \angle 0° = 7.44 + j0$$

The complex ac circuit seen in Figure 18-37 is therefore equivalent to a 7.44-Ω resistor in series with no reactance.

SELF-TEST REVIEW QUESTIONS FOR SECTION 18-5

1. In complex numbers, resistance is a/an _____ term and reactance is a/an _____ term. (imaginary/real)
2. Convert the following rectangular number to polar form: $5 + j6$.
3. Convert the following polar number to rectangular form: $33 \angle 25°$.
4. What is a complex number?

SUMMARY

Resistive, Inductive, and Capacitive (*RLC*) Circuits (Figure 18-38)

1. Resistors, as we have discovered, operate and react to voltage and current in a very straightforward way; the voltage across a resistor is in phase with the resistor current.

2. Inductors and capacitors operate in essentially the same way, in that they both store energy and then return it back to the circuit. However, they have completely opposite reactions to voltage and current.

3. The phrase "ELI the ICE man" states that voltage (symbolized *E*) leads current (*I*) in an inductive (*L*) circuit (abbreviated by the word "ELI"), while current (*I*) leads voltage (*E*) in a capacitive (*C*) circuit (abbreviated by the word "ICE").

Series *RLC* Circuit

4. The circuit current is always the same throughout a series circuit and can therefore be used as a reference.

5. The voltage across a resistor is always in phase with the current, while the voltage across the inductor leads the current by 90° and the voltage across the capacitor lags the current by 90°.

6. Impedance is the total opposition to current flow and is a combination of both reactance (X_L, X_C) and resistance (R).

7. The capacitive and inductive reactances are 180° out of phase with one another and counteract.

FIGURE 18-38 *RLC* Circuits.

- **Series *RLC* Circuits**

$$X_C = \frac{1}{2\pi fC}$$

$$X_L = 2\pi fL$$

$$Z = \sqrt{R^2 + (X_L \sim X_C)^2}$$

$$I = \frac{V_s}{Z}$$

$$V_R = I \times R \quad V_L = I \times X_L \quad V_C = I \times X_C$$

$$V_s = \sqrt{V_R^2 + (V_L \sim V_C)^2}$$

- **Parallel *RLC* Circuits**

$$I_R = \frac{V}{R} \quad I_L = \frac{V}{X_L} \quad I_C = \frac{V}{X_C}$$

$$I_T = \sqrt{I_R^2 + I_X^2}$$

$$Z = \frac{V}{I_T}$$

- **Resonance ($X_L = X_C$)**

$$f_0 = \frac{1}{2\pi\sqrt{LC}}, \; f_0 = \text{resonant frequency}$$

$$Z = R$$

$$\text{BW} = \frac{f_0}{Q}, \text{BW} = \text{bandwidth}$$

- **Series Resonance**

$$Q = \frac{X_L}{R} = \frac{X_C}{R} = \frac{V_L}{V_R} = \frac{V_C}{V_R} = \frac{V_L}{V_s} = \frac{V_C}{V_s}$$

- **Parallel Resonance**

$$Q = \frac{I_{\text{tank}}}{I_s} = \frac{Z_{\text{tank}}}{X_L}$$

- **Power**

$$\text{Apparant power} = V_s \times I \text{ (volt-amperes)}$$

$$\text{True power} = I^2 \times R \text{ (watts)}$$

$$\text{PF} = \cos\theta = \frac{R}{Z} = \frac{P_R}{P_A}$$

- **Complex Numbers**

Polar-to-rectangular: Real number = magnitude × cos θ
Imaginary number = magnitude × sin θ
Rectangular-to-polar:
Magnitude = $\sqrt{(\text{real number})^2 + (\text{imaginary number})^2}$

Angle = $\arctan\left(\dfrac{\text{imaginary number}}{\text{real number}}\right)$

Addition = sum of the real and sum of the imaginary parts (rectangular form)

Subtraction = difference between the real and the difference between imaginary parts (rectangular form)

Multiplication = multiply the magnitudes and then add the angles algebraically (polar form)

Division = divide the magnitudes and then subtract the denominator angle from the numerator angle (polar form)

FIGURE 18-38 *(continued)*

Series *RLC* Circuits

FIGURE 18-38 (*continued*)

Parallel *RLC* Circuits

(a)

(b)

$$I_R = \frac{V}{R} = \frac{115\text{ V}}{33\ \Omega} = 3.5\text{ A}$$

$$I_L = \frac{V}{X_L} = \frac{115\text{ V}}{7.5\ \Omega} = 15.3\text{ A}$$

$$X_L = 2\pi fL = 2\pi \times 60 \times 20\text{ mH} = 7.5\ \Omega$$

$$I_C = \frac{V}{X_C} = \frac{115\text{ V}}{265.3\ \Omega} = 0.43\text{ A}$$

$$X_C = \frac{1}{2\pi fL} = \frac{1}{2\pi \times 60 \times 10\ \mu\text{F}} = 265.3\ \Omega$$

(a)

(b)   **(c)**   **(d)**

8. Impedance is the vector sum of the reactive (X) and resistance (R) vectors.

9. Since none of these voltages are in phase with one another in a series *RLC* circuit, they must be added vectorially to obtain the applied voltage.

Parallel *RLC* Circuit

10. The voltage across any parallel circuit will be equal and in phase.

11. With current, we must first calculate the individual branch currents (I_R, I_L, and I_C) and then calculate the total circuit current (I_T).

12. There is a phase difference between the source voltage (V_S) and the circuit current (I_T).

FIGURE 18-38 *(continued)*

Series Resonance

13. Once the total current (I_T) is known, the impedance of all three components in parallel can be calculated.

Resonance

14. Resonance is a circuit condition that occurs when the inductive reactance (X_L) and the capacitive reactance (X_C) have been balanced.

15. As the input ac frequency is increased, a point will be reached where X_L will equal X_C, and this condition is known as *resonance*. The frequency at which $X_L = X_C$ in either a parallel or a series LC circuit is known as the *resonant frequency* (f_0).

FIGURE 18-38 *(continued)*

Parallel Resonance

C Charged,
Max. *V*, Zero *I*

(a)

Buildup of *I*
Delayed by *L*

(b)

Max. *I*, Zero *V*

(c)

Field Collapses,
Maintaining I_{circ}

(d)

C Charged,
Max. *V*, Zero *I*

(e)

Buildup of *I*
Delayed by *L*

(f)

Max. *I*, Zero *V*

(g)

Field Collapses,
Maintaining I_{circ}

(h)

(a)

(b)

(c)

Series Resonance

16. A series resonant circuit is a resonant circuit in which the capacitor and coil are in series with the applied ac voltage.

17. In summary, we can say that in a series resonant circuit:

  a. The inductor and capacitor electrically disappear due to their equal but opposite effect, resulting in 0 V drops across the series combination, and the circuit consequently seems purely resistive.

  b. The current flow is large because the impedance of the circuit is low and equal to the series resistance (*R*), which has the source voltage developed across it.

c. The individual voltage drops across the inductor or capacitor can be larger than the source voltage if R is smaller than X_L and X_C.

18. The Q factor is a ratio of inductive reactance to resistance and is used to express how efficiently an inductor will store rather than dissipate energy. In a series resonant circuit, the Q factor indicates the quality of the series resonant circuit, or is the ratio of the reactance to the resistance.

19. The Q of a resonant circuit is almost entirely dependent on the inductor's coil resistance, because capacitors tend to have almost no resistance figure at all, only reactive, which makes them very efficient.

20. A series resonant circuit is selective in that frequencies at resonance or slightly above or below will cause a larger current than frequencies well above or below the circuit's resonant frequency.

21. The group or band of frequencies that causes the larger current is called the circuit's bandwidth. A frequency response curve is a graph indicating a circuit's response to different frequencies.

22. The bandwidth includes the group or band of frequencies that cause 70.7% or more of the maximum current to flow within the series resonant circuit.

23. The cutoff frequency is the frequency at which the gain of the circuit falls below 0.707 of the maximum current or half-power (− 3 dB).

24. The half-power point is a point at which power is 50%. This half-power point corresponds to 70.7% of the total current.

25. Bandwidth is proportional to the resonant frequency of the circuit and inversely proportional to the Q of the circuit. In summary, the bandwidth of a series resonant circuit will increase as the Q of the circuit decreases (BW $\uparrow = f_0 / Q \downarrow$), and vice versa.

Parallel Resonance

26. A parallel resonant circuit is a circuit having an inductor and capacitor in parallel with one another, offering a high impedance at the frequency of resonance.

27. The source voltage (V_S) is needed initially to supply power to the LC circuit and start the oscillations; but once the oscillating process is in progress (assuming the ideal case), current is only flowing back and forth between inductor and capacitor, and no current is flowing from the source.

28. Flywheel action is a sustaining effect of oscillation in an LC circuit due to the charging and discharging of the capacitor and the expansion and contraction of the magnetic field around the inductor. The name is derived from the fact that it resembles a mechanical flywheel, which, once started, will keep going continually until friction reduces the magnitude of the rotations to zero.

29. The electronic equivalent of the mechanical flywheel is a resonant parallel-connected LC circuit.

30. The direction of the circulating current reverses each half-cycle at the frequency of resonance. Energy is stored in the capacitor in the form of an

electric field between the plates on one half-cycle, and then the capacitor discharges, supplying current to build up a magnetic field on the other half-cycle. The inductor stores its energy in the form of a magnetic field, which will collapse, supplying a current to charge the capacitor, which will then discharge, supplying a current back to the inductor, and so on. Due to the "storing action" of this circuit, it is sometimes related to the fluid analogy and referred to as a tank circuit.

31. Under ideal conditions a tank circuit should oscillate indefinitely if no losses occur within the circuit. In reality, the resistance of the coil reduces that 100% efficiency, as does friction with the mechanical flywheel.

32. As a small part of the energy is dissipated with each cycle, the oscillations will be reduced in size and eventually fall to zero.

33. If the ac source is reconnected to the tank, a small amount of current will flow from the source to the tank to top up the tank or replace the dissipated power. The higher the coil resistance is, the higher the loss and the larger the current flow from source to tank to replace the loss.

34. In the series resonant circuit, we were concerned with voltage drops since current remains the same throughout a series circuit. In a parallel resonant circuit, we are concerned with circuit currents rather than voltage.

35. Of all the three Q formulas for parallel resonant circuits, $Q = I_{tank}/I_S$, $Q = X_L/R$, or $Q = Z_{tank}/X_L$, the latter is the easiest to use as both X_L and the tank impedance can easily be determined in most cases where C, L, and R internal for the inductor are known.

36. The current versus frequency response curve is the complete opposite to the series resonant response curve. At frequencies below resonance X_L is low and X_C is high, and the inductor offers a low reactance, producing a high current path and low impedance. On the other hand, at frequencies above resonance the capacitor displays a low reactance, producing a high current path and low impedance.

37. The parallel resonant circuit is like the series resonant circuit in that it responds to a band of frequencies close to its resonant frequency.

38. Circuits containing inductance and capacitance are often referred to as tuned circuits since they can be adjusted to make the circuit responsive to a particular frequency (the resonant frequency).

39. Selectivity, by definition, is the ability of a tuned circuit to respond to a desired frequency and ignore all others.

40. Parallel resonant LC circuits are sometimes too selective, as the Q is too large, producing too narrow a bandwidth. In this situation, because of the very narrow response curve, a high resistance value can be placed in parallel with the LC circuit to provide an alternative path for line current. This process is known as *loading* or *damping* the tank and will cause an increase in line current and decrease in Q ($Q\downarrow = I_{tank}/I_{line}\uparrow$). The decrease in Q will cause a corresponding increase in BW (BW $\uparrow = f_0/Q\downarrow$).

RLC Filter Circuits (Figure 18-39)

41. There are basically four types of filters:
  a. Low-pass filter, which passes frequencies below a cutoff frequency
  b. High-pass filter, which passes frequencies above a cutoff frequency
  c. Bandpass filter, which passes a band of frequencies
  d. Band-stop filter, which stops a band of frequencies

42. An inductor and capacitor can be connected to act as a low-pass filter. At low frequencies, X_L has a small value compared to the load resistor (R_L), so nearly all the low-frequency input is developed and appears at the output across R_L. Since X_C is high at low frequencies, nearly all the current passes through R_L rather than C.

At high frequencies, X_L increases and drops more of the applied input across the inductor rather than the load. The capacitive reactance, X_C, aids this low output at high-frequency effect by decreasing its reactance and providing an alternative path for current to flow.

Since the inductor basically blocks alternating current and the capacitor shunts alternating current, the net result is to prevent high-frequency signals from reaching the load.

43. An inductor and capacitor can be connected to act as a high-pass filter. At high frequencies, the reactance of the capacitor (X_C) is low while the reactance of the inductor (X_L) is high, so all the high frequencies are easily passed by the capacitor and blocked by the inductor, so they all are routed through to the output and load.

At low frequencies, the reverse condition exists, resulting in a low X_L and a high X_C. The capacitor drops nearly all the input, and the inductor shunts the signal current away from the output load.

44. A bandpass filter circuit passes a group or band of frequencies between a lower and an upper cutoff frequency, while heavily attenuating any other frequency outside this band. At resonance, the series resonant LC circuit has a very low impedance and will consequently pass the resonant frequency to the load with very little drop across the L and C components.

45. Below resonance, X_C is high, and the capacitor drops a large amount of the input signal; above resonance, X_L is high and the inductor drops most of the input frequency voltage. This circuit will therefore pass a band of frequencies centered around the resonant frequency of the series LC circuit and block all other frequencies above and below this resonant frequency.

46. A parallel resonant LC circuit can be used to provide a bandpass response. The series resonant circuit was placed in series with the output, whereas the parallel resonant circuit will have to be placed in parallel with the output to provide the same results. At resonance, the parallel resonant circuit or tank has a high impedance, so very little current will be shunted away from the output; it will be passed on to the output, and almost all the input will appear at the output across the load.

Above resonance, X_C is small, so most of the input is shunted away from the output by the capacitor; below resonance, X_L is small, and the shunting action occurs again, but this time through the inductor.

FIGURE 18-39 *RLC* Filter Circuits.

Low-Pass Filter

(a)

(b)

High-Pass Filter

(a)

(b)

Bandpass Filter

(a)

(b)

(c)

Bandstop Filter

(a)

(b)

(c)

47. A transformer can be used to replace the inductor to produce a bandpass filter. At resonance, maximum flywheel current flows within the parallel circuit made up of the capacitor and the primary of the transformer (L), which is known as a *tuned transformer.* With maximum flywheel current, there will be a maximum magnetic field, which means that there will be maximum power transfer between primary and secondary. So nearly all the input will be coupled to the output (coupling coefficient $k = 1$) and appear across the load at and around a small band of frequencies centered on resonance.

Above and below resonance, current within the parallel resonant circuit will be smaller. So the power transfer ability will be less, effectively keeping the frequencies outside the bandpass from appearing at the output.

48. A band-stop filter attenuates alternating currents whose frequencies are between given upper and lower cutoff values while passing frequencies above and below this band. The band-stop filter operates exactly the opposite to a bandpass filter in that it blocks or attenuates a band of frequencies centered on the resonant frequency of the LC circuit.

In the series resonant circuit the LC impedance is very low at and around resonance, so these frequencies are rejected or shunted away from the output. Above and below resonance, the series circuit has a very high impedance, which results in almost no shunting of the signal away from the output.

In the parallel resonant circuit, the LC circuit is in series with the load and output. At resonance, the impedance of a parallel resonant circuit will be very high, and the band of frequencies centered around resonance will be blocked. Above and below resonance, the impedance of the tank is very low, so nearly all the input is developed across the output.

49. Filters are necessary in applications such as television or radio, where we need to tune in (select or pass) one frequency that contains the information we desire, yet block all the millions of other frequencies that are also carrying information.

Complex Numbers

50. The complex number system allows us to determine the *magnitude* and *phase angle* of electrical quantities by adding, subtracting, multiplying, and dividing phasor quantities, and is an invaluable tool in ac circuit analysis.

51. Real numbers can be represented on a horizontal line, known as the real number line. Positive numbers exist to the right of the center point corresponding to zero, while negative numbers exist to the left. This representation satisfied most mathematicians for a short time, as they could indicate numbers such as 2 or 5 as a point on the line. However, a problem was reached if they wished to indicate a point corresponding to $\sqrt{-9}$. The $\sqrt{-9}$ is not +3 [since $(+3) \times (+3) = +9$], and it is not −3 [since $(-3) \times (-3) = +9$]. So it was eventually realized that the square root of a negative number could not be indicated on the real number line, as it is not a real number.

52. Mathematicians decided to call the square root of a negative number, such as $\sqrt{-4}$ or $\sqrt{-9}$, imaginary numbers, which are not fictitious or imaginary, but simply a particular type of number. Just as real numbers can be represented on a real number line, imaginary numbers can be represented on an imaginary number line. The imaginary number line is vertical, to distinguish it from the real number line, and when working with electrical quantities a $\pm j$ prefix, known as the *j operator*, is used for values that appear on the imaginary number line.

53. A complex number is the combination of a real and imaginary number and is represented on a two-dimensional plane called the complex plane. Generally, the real number appears first, followed by the imaginary number.

54. Complex numbers are merely terms that need to be added as phasors, and all you have to do basically is draw a vector representing the real number and then draw another vector representing the imaginary number.

55. A number like $3 + j4$ specifies two phasors in rectangular coordinates, so this system is the *rectangular representation of a complex number*.

56. There are several other ways to describe a complex number, one of which is the polar representation of a complex number, using polar coordinates.

57. With the rectangular notation the horizontal coordinate is the real part and the vertical coordinate is the imaginary part of the complex number. With the polar notation the magnitude of the phasor (x, or size) and the angle ($\angle \theta$, meaning "angle theta") relative to the positive real axis (measured in a counterclockwise direction) are stated.

58. Since a phase difference exists between real and imaginary (j) numbers, certain rules should be applied when adding, subtracting, multiplying, or dividing complex numbers.

59. The sum of two complex numbers is equal to the sum of their real and imaginary parts.

60. The difference of two complex numbers is equal to the difference between the separate real and imaginary parts.

61. Multiplication of two complex numbers is achieved more easily if they are in polar form. The simple rule to remember is to multiply the magnitudes and then add the angles algebraically.

62. Division is also more easily carried out in polar form. The rule to remember is to divide the magnitudes, and then subtract the denominator angle from the numerator angle.

63. Complex numbers find an excellent application in ac circuits due to all the phase differences that occur between different electrical quantities, such as X_L, X_C, R, and Z. The positive real number line, at an angle of $0°$, is used for resistance.

64. On the positive imaginary number line, at an angle of $90°$ $(+j)$, inductive reactance (X_L) is represented.

65. The voltage drop across an inductor (V_L) is proportional to its inductive reactance (X_L), and both are represented on the $+j$ imaginary number line, since the voltage drop across an inductor will always lead the current (which in a series circuit is always in phase with the resistance) by 90°.

66. On the negative imaginary number line, at an angle of −90° or 270° (−j), capacitive reactance (X_C) is represented.

67. The voltage drop across a capacitor (V_C) is proportional to its capacitive reactance (X_C), and both are represented on the −j imaginary number line since the voltage drop across a capacitor always lags current (charge and discharge) by −90°.

NEW TERMS

Bandpass filter	Imaginary number	Resonant frequency
Band-stop filter	j operator	Selectivity
Bandwidth	Parallel resonance	Series resonance
Complex numbers	Polar coordinate	Tank
Complex plane	Real number	Tank circuit
Flywheel action	Rectangular coordinate	Tuned transformer
Frequency response curve	Resonance	

REVIEW QUESTIONS

Multiple-Choice Questions

1. Capacitive reactance is _____ to frequency and capacitance, while inductive reactance is _____ to frequency and inductance.
   a. Proportional, inversely proportional
   b. Inversely proportional, proportional
   c. Proportional, proportional
   d. Inversely proportional, inversely proportional

2. Resonance is a circuit condition that occurs when:
   a. V_L equals V_C d. Both (a) and (c)
   b. X_L equals X_C e. Both (a) and (b)
   c. L equals C

3. As frequency is increased, X_L will _____, while X_C will _____.
   a. Decrease, increase   c. Remain the same, decrease
   b. Increase, decrease   d. Increase, remain the same

4. In an RLC series resonant circuit, with R = 500 Ω and X_L = 250 Ω, what would be the value of X_C?
   a. 2 Ω   c. 250 Ω
   b. 125 Ω   d. 500 Ω

5. At resonance, the voltage drop across both a series-connected inductor and a capacitor will equal:
   a. 70.7 V   c. 10 V
   b. 50% of the source   d. Zero

6. In a series resonant circuit the current flow is _____, as the impedance is _____ and equal to _____.

    a. Large, small, R **c.** Large, small, X
    b. Small, large, X **d.** Small, large, R

7. A circuit's bandwidth includes a group or band of frequencies that cause
_____ or more of the maximum current, or more than
_____ of the maximum power to appear at the output.

    a. 110, 90   **c.** 70.7%, 50%
    b. 50%, 70.7% **d.** Both (a) and (c)

8. The bandwidth of a circuit is proportional to the:

    a. Frequency of resonance **c.** Tank current
    b. Q of the tank   **d.** Two of the above

9. Series or parallel resonant circuits can be used to create:

    a. Low-pass filters   **c.** Bandpass and band-stop filters
    b. Low-pass and high-pass filters   **d.** All of the above

10. Flywheel action occurs in:

    a. A tank circuit   **d.** Two of the above
    b. A parallel LC circuit   **e.** None of the above
    c. A series LC circuit

11. $25 \angle 39°$ is an example of a complex number in:

    a. Polar form   **c.** Algebraic form
    b. Rectangular form **d.** None of the above

12. $3 + j10$ is an example of a complex number in:

    a. Polar form   **c.** Algebraic form
    b. Rectangular form **d.** None of the above

13. Which complex number form is usually more convenient for addition and
subtraction?

    a. Rectangular **b.** Polar

14. Which complex number form is usually more convenient for multiplication
and division?

    a. Rectangular **b.** Polar

15. In complex numbers, *resistance* is a real term, while *reactance* is a/an
_____.

    a. j term   **c.** Value appearing on the vertical axis
    b. Imaginary term **d.** All of the above

Essay Questions

16. Illustrate with phasors and describe the current and voltage relationships in
a series RLC circuit. (18-1)

17. Describe the procedure for the analysis of a series RLC circuit. (18-1)

18. Define resonance and give the formula for calculating the frequency of resonance. (18-3)

19. Describe the three unusual characteristics of a circuit that is at resonance. (18-3)

20. Define the following: (18-3)
    a. Flywheel action   c. Bandwidth
    b. Quality figure   d. Selectivity

21. Describe a frequency response curve. (18-4)

22. Illustrate with phasors and describe the current and voltage relationships in a parallel *RLC* circuit. (18-2)

23. Describe the differences between a series and a parallel resonant circuit. (18-3)

24. Explain how loading a tank affects bandwidth and selectivity. (18-3-2)

25. Illustrate the circuit and explain the operation of the following, with their corresponding response curves. (18-4)
    a. Low-pass filter   c. Bandpass filter
    b. High-pass filter   d. Band-stop filter

26. Describe why capacitive reactance is written as $-jX_C$ and inductive reactance is written as jX_L. (18-5)

27. How are capacitive and inductive reactances written in polar form? (18-5-4)

28. List the rules used to perform complex numbers: (18-5-6)
    a. Addition (rectangular)   c. Multiplication (polar)
    b. Subtraction (rectangular)   d. Division (polar)

29. Describe briefly how the real number and imaginary number lines are used for ac circuit analysis and what electrical phasors are represented at $0°$, $90°$, and $-90°$. (18-5-7)

30. Referring to Figure 18-36, describe why the series circuit current (I) is not in phase with the source voltage (V_S). (18-5-7)

Practice Problems

31. Calculate the values of capacitive and inductive reactance for the following when connected across a 60 Hz source:
    a. 0.02 µF   e. 4 mH
    b. 18 µF   f. 8.18 H
    c. 360 pF   g. 150 mH
    d. 2700 nF   h. 2H

32. If a 1.2 kΩ resistor, a 4 mH inductor, and an 8 µF capacitor are connected in series across a 120 V/60 Hz source, calculate:
    a. X_C g. V_C
    b. X_L h. Apparent power
    c. Z i. True power
    d. I j. Resonant frequency
    e. V_R k. Circuit quality factor
    f. V_L l. Bandwidth

33. If a 270 Ω resistor, a 150 mH inductor, and a 20 μF capacitor are all connected in parallel with one another across a 120 V/60 Hz source, calculate:

   a. X_L **f.** I_T
   b. X_C **g.** Z
   c. I_R **h.** Resonant frequency
   d. I_L **i.** Q factor
   e. I_C **j.** Bandwidth

34. Calculate the impedance of a series circuit if $R = 750$ Ω, $X_L = 25$ Ω, and $X_C = 160$ Ω.

35. Calculate the impedance of a parallel circuit with the same values as those of Question 34 when a 1 V source voltage is applied.

36. State the following series circuit impedances in rectangular and polar form:

   a. $R = 33$ Ω, $X_C = 24$ Ω **b.** $R = 47$ Ω, $X_L = 17$ Ω

37. Convert the following impedances to rectangular form:

   a. $25 \angle 37°$ **c.** $114 \angle -114°$
   b. $19 \angle -20°$ **d.** $59 \angle 99°$

38. Convert the following impedances to polar form:

   a. $-14 + j14$ **c.** $-33 - j18$
   b. $27 + j17$ **d.** $7 + j4$

39. Add the following complex numbers:

   a. $(4 + j3) + (3 + j2)$ **b.** $(100 - j50) + (12 + j9)$

40. Perform the following mathematical operations:

   a. $(35 \angle -24°) \times (13 \angle 50°)$ **c.** $(98 \angle 80°) \div (40 \angle 17°)$
   b. $(100 - j25) - (25 + j5)$

41. State the impedances of the circuits seen in Figure 18-40 in rectangular and polar form. What is Z_T in ohms and its phase angle?

42. Calculate in polar form the impedances of both circuits shown in Figure 18-41. Then combine the two impedances as if the circuits were parallel connected, using the product-over-sum method. Express the combined impedance in polar form.

(a)

(b)

(c)

FIGURE 18-40

(a)

$$Z_T = \frac{Z_1 \times Z_2}{Z_1 + Z_2}$$

(b)

FIGURE 18-41

43. Referring to Figure 18-42, calculate:
    a. Z_T (rectangular and polar)   c. Voltage drops
    b. Circuit current and phase angle   d. V_C, V_L, and V_R phase relationships

44. Sketch an impedance and voltage phasor diagram for the circuit in Figure 18-40.

45. Referring to Figure 18-43, calculate:
    a. Impedance of the two branches   c. Total current
    b. Branch currents   d. Total impedance

46. Sketch an impedance and current phasor diagram for the circuit shown in Figure 18-41.

47. Referring to Figure 18-40, verify that the sum of all the individual voltage drops is equal to the total voltage.

48. Referring to Figure 18-41, verify that the sum of all the branch currents is equal to the total current.

49. Determine whether Questions 43 and 45 would be easier to answer with or without the use of complex numbers.

50. Is the circuit in Figure 18-41 more inductive or capacitive?

FIGURE 18-42

FIGURE 18-43

A
Electronics Dictionary

Absorption Loss or dissipation of energy as it travels through a medium. For example, radio waves lose some of their electromagnetic energy as they travel through the atmosphere.

AC Abbreviation for "alternating current."

AC beta (β_{AC}) The ratio of a transistor's ac output current to input current.

Accelerate To go faster.

AC coupling Circuit or component that couples or passes the ac signal yet blocks any dc level.

AC/DC If indicated on a piece of equipment, it means that the equipment will operate from either an ac or dc supply.

AC generator Device that transforms or converts a mechanical input into an ac electrical power output.

Acoustic Relating to sound or the science of sound.

AC power supply Power supply that delivers one or more sources of ac voltage.

Activate To put ac voltage to work or make active. The application of an enabling signal.

Active band-pass filter A circuit that uses an amplifier with passive filter elements to pass only a band of input frequencies.

Active band-stop filter A circuit that uses an amplifier with passive filter elements to block a band of input frequencies.

Active component Component that amplifies a signal or achieves some sort of gain between input and output.

Active equipment Equipment that will transmit and receive, as opposed to passive equipment, which only receives.

Active filter A filter that uses an amplifier with passive filter elements to provide pass or rejection characteristics.

Active high-pass filter A circuit that uses an amplifier with passive filter elements to pass all frequencies above a cut-off frequency.

Active low-pass filter A circuit that uses an amplifier with passive filter elements to pass all frequencies below a cut-off frequency.

AC voltage Alternating voltage.

ADC Abbreviation for "analog-to-digital converter."

Adjustable resistor Resistor whose value can be changed.

Admittance (symbolized Y) Measure of how easily ac will flow through a circuit. It is equal to the reciprocal of impedance and is measured in siemens.

Aerial Another term used for antenna.

AF Abbreviation for "audio frequency."

AFC Abbreviation for "automatic frequency control."

AGC Abbreviation for "automatic gain control."

Air-core inductor An inductor that has no metal core.

Aligning tool Small, nonconductive, nonmagnetic screwdriver used to adjust receiver or transmitter tuned circuits.

Alkaline cell Also called an alkaline manganese cell, it is a primary cell that delivers more current than the carbon–zinc cell.

Alligator clip Spring clip normally found on the end of test leads and used to make temporary connections.

Alphanumeric Having numerals, symbols, and letters of the alphabet.

Alpha wave Also called alpha rhythm, it is a brain wave between 9 and 14 Hz.

Alternating current Electric current that rises from zero to a maximum in one direction, falls to zero, and then rises to a maximum in the opposite direction, and then repeats another cycle. The positive and negative alternations being equal.

Alternation One ac cycle consists of a positive and negative alternation.

Alternator Another term used to describe an ac generator.

AM Abbreviation for "amplitude modulation."

Ambient temperature Temperature of the air surrounding the components.

Ammeter Meter placed in the circuit path to measure the amount of current flow.

Ampere (A) Unit of electric current.

Ampere-hour If you multiply the amount of current, in amperes, that can be supplied for a given time frame in hours, a value of ampere-hours is obtained.

Ampere-hour meter Meter that measures the amount of current drawn in a unit of time (hours).

Ampere-turn Unit of magnetomotive force.

Ampere-turn per meter Base unit of magnetic field strength.

Amplification Process of making bigger or increasing the voltage, current, thus increasing the power of a signal.

Amplifier Circuit or device that achieves amplification.

Amplitude Magnitude or size an alternation varies from zero.

Amplitude distortion Changing of a wave shape so that it does not match its original form.

Analog Relating to devices or circuits in which the output varies in direct proportion to the input.

Analog data Information that is represented in a continuously varying form, as opposed to digital data, which have two distinct and discrete values.

Analog display Display in which a moving pointer or some other analog device is used to display the data.

Analog electrical circuits A power circuit designed to control power in a continuously varying form.

Analog electronic circuits Electronic circuits that represent information in a continuously varying form.

Anode Positive electrode or terminal.

Antenna Device that converts an electrical wave into an electromagnetic wave that radiates away from the antenna.

Antenna transmission line System of conductors connecting the transmitter or receiver to the antenna.

Apparent power The power value obtained in an ac circuit by multiplying together effective values of voltage and current, which reach their peaks at different times.

Arc Discharge of electricity through a gas. For example, lightning is a discharge of static electricity buildup through air.

Armature Rotating or moving component of a magnetic circuit.

Armstrong oscillator A tuned transformer oscillator developed by E. H. Armstrong.

Atom Smallest particle of an element.

Atomic number Number of positive charges or protons in the nucleus of an atom.

Atomic weight The relative weight of a neutral atom of an element, based on a neutral oxygen atom having an atomic weight of 16.

Attenuate To reduce in amplitude an action or signal.

Attenuation Loss or decrease in energy of a signal.

Audio Relating to all the frequencies that can be heard by the human ear; from the Latin word meaning "I hear."

Audio frequency Frequency that can be detected by a human ear. From approximately 20 Hz to 20 kHz.

Audio-frequency amplifier An amplifier that has one or more transistor stages designed to amplify audio-frequency signals.

Audio-frequency generator A signal generator that can be set to generate a sinusoidal AF signal voltage at any desired frequency in the audio spectrum.

Autotransformer Single-winding tapped transformer.

Average value Mean value found when the area of a wave above a line is equal to the area of the wave below the line.

Avionics Field of aviation electronics.

AWG Abbreviation for "American wire gauge."

Axial lead Component that has its connecting leads extending from either end of its body.

Balanced bridge Condition that occurs when a bridge is adjusted to produce a zero output.

Ballast resistor A resistor that increases in resistance when current increases. It can therefore maintain a constant current despite variations in line voltage.

Bandpass filter Filter circuit that passes a group or band of frequencies between a lower and an upper cutoff frequency, while heavily attenuating any other frequency outside this band.

Band-stop filter A filter that attenuates alternating currents whose frequencies are between given upper and lower cutoff values while passing frequencies above and below this band.

Bandwidth Width of the group or band of frequencies between the half-power points.

Bar graph display A left-to-right or low-to-high set of bars that are turned on in successive order based on the magnitude they are meant to represent.

Barometer Meter used to measure atmospheric pressure.

Barretter Temperature-sensitive device having a positive temperature coefficient; that is, as temperature increases, resistance increases.

Barrier potential or voltage The potential difference, or voltage, that exists across the junction.

Base The region that lies between an emitter and a collector of a transistor and into which minority carriers are injected.

Base biasing A transistor biasing method in which the dc supply voltage is applied to the base of the transistor via a base bias resistor.

Bass Low audio frequencies normally handled by a woofer in a sound system.

Battery DC voltage source containing two or more cells that convert chemical energy into electrical energy.

Battery charger Piece of electrical equipment that converts ac input power to a pulsating dc output, which is then used to charge a battery.

Baud Unit of signaling speed describing the number of elements per second.

_B-H_ curve Curve plotted on a graph to show successive states during magnetization of a ferromagnetic material.

Bias voltage The dc voltage applied to a semiconductor device to control its operation.

Bidirectional device A device that will conduct current in either direction.

Binary Number system having only two levels and used in digital electronics.

Binary word A numerical value expressed as a group of binary digits.

Bipolar family A group of digital logic circuits that make use of the bipolar junction transistor.

Bipolar transistor A bipolar junction transistor in which excess minority carriers are injected from an emitter region into a base region and from there pass into a collector region.

Bits An abbreviation for "binary digits."

Bleeder current Current drawn continuously from a voltage source. A bleeder resistor is generally added to lessen the effect of load changes or provide a voltage drop across a resistor.

Body resistance The resistance of the human body.

Bolometer Device whose resistance changes when heated.

Branch current A portion of the total current that is present in one path of a parallel circuit.

Breakdown voltage The voltage at which breakdown of a dielectric or insulator occurs.

Bridge rectifier circuit A full-wave rectifier circuit using four diodes that will convert an alternating voltage input into a direct voltage output.

BW Abbreviation for "bandwidth."

Bypass capacitor Capacitor that is connected to provide a low-impedance path for ac.

Byte Group of 8 binary digits or bits.

Cable Group of two or more insulated wires.

CAD Abbreviation for "computer-aided design."

Calculator Device that can come in a pocket (battery-operated) or desktop (ac power) size and is used to achieve arithmetic operations.

Calibration To determine by measurement or comparison with a standard the correct value of each scale reading.

Capacitance (C) Measured in farads, it is the ability of a capacitor to store an electrical charge.

Capacitance meter Instrument used to measure the capacitance of a capacitor or circuit.

Capacitive filter A capacitor used in a power supply filter system to supress ripple currents while not affecting direct currents.

Capacitive reactance (X_C) Measured in ohms, it is the ability of a capacitor to oppose current flow without the dissipation of energy.

Capacitor Device that stores electric energy in the form of an electric field that exists within a dielectric (insulator) between two conducting plates, each connected to a connecting lead. This device was originally called a condenser.

Capacitor-input filter Filter in which a capacitor is connected to shunt away low-frequency signals such as ripple from a power supply.

Capacitor microphone Microphone that contains a stationary and movable plate separated by air. The arriving sound waves cause movements in the movable plate, which affects the distance between the plates and therefore the capacitance, generating a varying AF electrical wave. Also called an electrostatic microphone.

Capacitor speaker Also called the electrostatic speaker, its operation depends on mechanical forces generated by an electrostatic field.

Capacity Output current capabilities of a device or circuit.

Capture The act of gaining control of a signal.

Carbon-film resistor Thin carbon film deposited on a ceramic form to create resistance.

Carbon microphone Microphone whose operation depends on the pressure variation in the carbon granules changing their resistance.

Carbon resistor Fixed resistor consisting of carbon particles mixed with a binder, which is molded and then baked. Also called a composition resistor.

Cardiac pacemaker Device used to control the frequency or rhythm of the heart by stimulating it through electrodes.

Carrier Wave that has one of its characteristics varied by an information signal. The carrier will then transport this information between two points.

Cascaded amplifier An amplifier that contains two or more stages arranged in a series manner.

Cassette Thin, flat, rectangular device containing a length of magnetic tape that can be recorded onto or played back.

Cathode Term used to describe a negative electrode or terminal.

Cathode ray tube A vacuum tube in which electrons emitted by a hot cathode are focussed into a narrow beam by an electron gun and then applied to a fluorescent screen. The beam can be varied in intensity and position to produce a pattern or picture on the screen.

Cell Single unit having only two plates that convert chemical energy into a dc electrical voltage.

Celsius-temperature scale Scale that defines the freezing point of water as 0°C and the boiling point as 100°C (named after Anders Celsius).

Center tap Midway connection between the two ends of a winding or resistor.

Center-tapped rectifier Rectifier that makes use of a center-tapped transformer and two diodes to achieve full-wave rectification.

Center-tapped transformer A transformer that has a connection at the electrical center of the secondary winding.

Ceramic capacitor Capacitor in which the dielectric material used between the plates is ceramic.

Cermet Ceramic-metal mixture used in film resistors.

Channel A path for a signal.

Charge Quantity of electrical energy stored in a battery or capacitor.

Charging current Current that flows to charge a capacitor or battery when a voltage is applied.

Chassis Metal box or frame in which components, boards, and units are mounted.

Chassis ground Connection to the metal box or frame that houses the components and associated circuitry.

Choke Inductor used to impede the flow of alternating or pulsating current.

Circuit Interconnection of many components to provide an electrical path between two or more points.

Circuit breaker Reusable fuse. This device will open a current-carrying path without damaging itself once the current value exceeds its maximum current rating.

Circuit failure A malfunction or breakdown of a component in the circuit.

Circular mil (cmil) A unit of area equal to the area of a circle whose diameter is one mil.

Clapp oscillator A series tuned colpitts oscillator circuit.

Class A amplifier An amplifier in which the transistor is in its active region for the entire signal cycle.

Class AB amplifier An amplifier in which the transistor is in its active region for slightly more than half the signal cycle.

Class B amplifier An amplifier in which the transistor is in its active region for approximately half the signal cycle.

Class C amplifier An amplifier in which the transistor is in its active region for less than half the signal cycle.

Clock Generally, a square waveform used for the synchronization and timing of several circuits.

Clock input (*C*) An accurate timing signal used for synchronizing the operation of digital circuits.

Closed circuit Circuit having a complete path for current to flow.

Closed-loop mode A control system containing one or more feedback control loops, in which functions of the controlled signals are combined with functions of the commands that tend to maintain prescribed relationships between the commands and the controlled signals.

Closed-looped voltage gain (*A_{CL}*) The voltage gain of an amplifier when it is operated in the closed loop mode.

Coaxial cable Transmission line in which a center signal-carrying conductor is completely covered by a solid dielectric, another conductor, and then an outer insulating sleeve.

Coefficient This is related to the ratio of change under certain conditions.

Coefficient of coupling The degree of coupling that exists between two circuits.

Coercive force (*H*) Magnetizing force needed to reduce the residual magnetism within a material to zero.

Coil Number of turns of wire wound around a core to produce magnetic flux (an electromagnet) or to react to a changing magnetic flux (an inductor).

Cold resistance The resistance of a device when cold.

Collector A semiconductor region through which a flow of charge carriers leaves the base of the transistor.

Collector characteristic curve A set of characteristic curves of collector voltage versus collector current for a fixed value of transistor base current.

Color code Set of colors used to indicate a component's value, tolerance, and rating.

Colpitts oscillator An *LC* tuned oscillator circuit in which two tank capacitors are used instead of a tapped coil.

Common Shared by two or more services, circuits, or devices. Although the term "common ground" is frequently used to describe two or more connections sharing a common ground, the term "common" alone does not indicate a ground connection, only a shared connection.

Common base A transistor configuration in which the base is common to both the input and output signal.

Common collector A transistor configuration in which the collector is common to both the input and output signal.

Common drain An FET configuration in which the drain is grounded and common to the input and output signal.

Common emitter A transistor configuration in which the emitter is common to both the input and output signal.

Common gate An FET configuration in which the gate is grounded and common to the input and output signal.

Common-mode input An input signal applied equally to both ungrounded inputs of a balanced amplifier. Also called in-phase input.

Common mode rejection The ability of a device to reject a voltage signal applied simultaneously to both input terminals.

Common-mode rejection ratio Abbreviated CMRR, it is the ratio of an operational amplifier's differential gain to common-mode gain.

Common source An FET configuration in which the source is grounded and common to the input and output signal.

Communication Transmission of information between two points.

Comparator An operational amplifier used without feedback to detect changes in voltage level.

Complementary DC amplifier An amplifier in which NPN and PNP transistors are used in an alternating sequence.

Complementary latch circuit A circuit containing a NPN and PNP transistor that once triggered ON will remain latched ON.

Complex numbers Numbers composed of a real number part and an imaginary number part.

Complex plane A plane whose points are identified by means of complex numbers.

Component (1) Device or part in a circuit or piece of equipment. (2) In vector diagrams, it can mean a part of a wave, voltage, or current.

Compound A material composed of united separate elements.

Computer Piece of equipment used to process information or data.

Conductance (*G*) Measure of how well a circuit or path conducts or passes current. It is equal to the reciprocal of resistance.

Conduction Use of a material to pass electricity or heat.

Conductivity Reciprocal of resistivity; it is the ability of a material to conduct current.

Conductors Materials that have a very low resistance and pass current easily.

Connection When two or more devices are attached together so that a path exists between them, there is said to be a connection between the two.

Connector Conductive device that makes a connection between two points.

Constant Fixed value.

Constant current Current that remains at a fixed, unvarying level despite variations in load resistance.

Contact Current-carrying part of a switch, relay, or connector.

Continuity Occurs when a complete path for electric current exists.

Continuity test Resistance test to determine whether a path for electric current exists.

Contrast The difference between the light and dark areas in a video picture. High contrast pictures have dark blacks and brilliant whites. Low contrast pictures have an overall gray appearance.

Control voltage A voltage signal that starts, stops, or adjusts the operation of a device, circuit, or system.

Conventional current flow A current produced by the movement of positive charges toward a negative terminal.

Copper loss Also I^2R loss, it is the power lost in transformers, generators, connecting wires, and other parts of a circuit because of the current flow (I) through the resistance (R) of the conductors.

Core Magnetic material within a coil used to intensify or concentrate the magnetic field.

Coulomb Unit of electric charge. One coulomb equals 6.24×10^{18} electrons.

Counter electromotive force Abbreviated "counter emf," or sometimes called back emf, it is the voltage generated in an inductor due to an alternating or pulsating current and is always of opposite polarity to that of the applied voltage.

Covalent bond A pair of electrons shared by two neighboring atoms.

Crossover distortion Distortion of a signal at the zero crossover point.

CRT Abbreviation for "cathode-ray tube."

Crystal A natural or synthetic piezoelectric or semiconductor material whose atoms are arranged with some degree of geometric regularity.

Current (I) Measured in amperes or amps, it is the flow of electrons through a conductor.

Current amplifier An amplifier designed to build up the current of a signal.

Current clamp A device used in conjunction with an ac ammeter, containing a magnetic core in the form of hinged jaws that can be snapped around the current-carrying wire.

Current divider A parallel network designed to proportionally divide the circuit's total current.

Current-limiting resistor Resistor that is inserted into the path of current flow to limit the amount of current to some desired level.

Current sink A circuit or device that absorbs current from a load circuit.

Current source A circuit or device that supplies current to a load circuit.

Current-to-voltage converter An amplifier circuit used for converting a low dc current into a proportional low dc voltage.

Cutoff The minimum value of bias voltage that stops output current in a transistor.

Cutoff frequency Frequency at which the gain of the circuit falls below 0.707 of the maximum current or half-power (−3 dB).

Cycle When a repeating wave rises from zero to some positive value, back to zero, and then to a maximum negative value before returning back to zero, it is said to have completed one cycle. The number of cycles occurring in one second is the frequency measured in hertz (cycles per second).

DAC Abbreviation for "digital-to-analog converter."

Damping Reduction in the magnitude of oscillation due to absorption.

Darlington pair A current amplifier consisting of two separate transistors with their collectors connected together and the emitter of one connected to the base of the other.

D'Arsonval movement When a direct current is passed through a small, light-weight, movable coil, the magnetic field produced interacts with a fixed permanent magnetic field that rotates while moving an attached pointer positioned over a scale.

DC Abbreviation for "direct current."

DC alpha Term for the dc current amplification factor of a transistor.

DC block Component used to prevent the passage of direct current, normally a capacitor.

DC generator Device used to convert a mechanical energy input into a dc electrical output.

DC offsets The change in input voltage that is required to produce an output voltage of zero when no input signal is present.

DC power supply A power line, generator, battery, power pack, or other dc source of power for electrical equipment.

Dead short Short circuit having almost no resistance.

Decibel (dB) One tenth of a bel (1×10^{-1} bel). The logarithmic unit for the difference in power at the output of an amplifier compared to the input.

Degenerative (negative) feedback Feedback in which a portion of the output signal is fed back 180° out of phase with the input signal. Also called degenerative feedback, inverse feedback, or stabilized feedback.

Depletion region A small layer on either side of the junction that becomes empty, or depleted, of free electrons or holes.

Depletion-type MOSFET A field-effect transistor with an insulated gate (MOS-FET) that can be operated in either the depletion or enhancement mode.

Design engineer Engineer responsible for the design of a product for a specific application.

Desoldering The process of removing solder from a connection.

Device Component or part.

DIAC A bidirectional diode that has a symmetrical switching mode.

Diagnostic Related to the detection and isolation of a problem or malfunction.

Dielectric Insulating material between two (*di*) plates in which the *electric* field exists.

Dielectric constant (*K*) The property of a material that determines how much electrostatic energy can be stored per unit volume when unit voltage is applied. Also called permittivity.

Dielectric strength The maximum potential a material can withstand without rupture.

Difference amplifier An amplifier whose output is proportional to the difference between the voltages applied to the two inputs.

Differential amplifier An amplifier whose output is proportional to the difference between the voltages applied to its two inputs.

Differential-mode input An input signal applied between the two ungrounded terminals of a balanced three-terminal system.

Differentiator A circuit whose output voltage is proportional to the rate of change of the input voltage. The output waveform is then the time derivative of the input waveform, and the phase of the output waveform leads that of the input by 90°.

Differentiator amplifier A circuit whose output voltage is proportional to the rate of change of the input voltage. The output waveform is the time derivative of the input waveform.

Diffusion A method of producing a junction by diffusing an impurity metal into a semiconductor at a high temperature.

Diffusion current The current that is present when the depletion layer is expanding.

Digital Relating to devices or circuits that have outputs of only two distinct levels or steps, for example, on–off, 0–1, open–closed, and so on.

Digital computer Piece of electronic equipment used to process digital data.

Digital data Data represented in digital form.

Digital display Display in which either light-emitting diodes (LEDs) or liquid crystal diodes (LCDs) are used and each of the segments can be either turned on or off.

Digital electrical (power) circuit An electrical circuit designed to turn power either ON or OFF (two-state).

Digital electronic circuits Electronic circuits that encode information into a group of pulses consisting of HIGH or LOW voltages.

Digital multimeter Multimeter used to measure amperes, volts, and ohms and indicate the results on a digital readout display.

DIP Abbreviation for "dual in-line package."

Direct coupling The coupling of two circuits by means of a wire.

Direct current (dc) Current flow in only one direction.

Direct-current amplifier An amplifier capable of amplifying dc voltages and slowly varying voltages.

Direct voltage (dc voltage) Voltage that causes electrons to flow in only one direction.

Discharge Release of energy from either a battery or capacitor.

Disconnect Breaking or opening of an electric circuit.

Discrete components Separate active and passive devices that were manufactured before being used in a circuit.

Display Visual presentation of information or data.

Display unit Unit designed to display digital data.

Dissipation Release of electrical energy in the form of heat.

Distortion An undesired change in the waveform of a signal.

Distributed capacitance Also known as self-capacitance, it is any capacitance other than that within a capacitor, for example, the capacitance between the coil of an inductor or between two conductors (lead capacitance).

Distributed inductance Any inductance other than that within an inductor, for example, the inductance of a length of wire (line inductance).

Domain Also known as a magnetic domain, it is a moveable magnetized area in a magnetic material.

Doping The process wherein impurities are added to the intrinsic semiconductor material either to increase the number of free electrons or to increase the number of holes.

Dot convention A standard used with transformer symbols to indicate whether the secondary voltage will be in phase or out of phase with the primary voltage.

Double-pole, double-throw switch (DPDT) Switch having two movable contacts (double pole) that can be thrown or positioned in one of two positions (double throw).

Double-pole, single-throw switch (DPST) Switch having two movable contacts (double pole) that can be thrown in only one position (single throw) to either make a dual connection, or opened to disconnect both poles.

Drain One of the field-effect transistor's electrodes.

Drain-feedback biased A configuration in which the gate receives a bias voltage fed back from the drain.

Dropping resistor Resistor whose value has been chosen to drop or develop a given voltage across it.

Dry Cell DC voltage-generating chemical cell using a nonliquid (paste) type of electrolyte.

Dual-gate d-MOSFET A metal oxide semiconductor FET having two separate gate electrodes.

Dual in-line package Package that has two (dual) sets or lines of connecting pins.

Dual-trace oscilloscope Oscilloscope that can simultaneously display two signals.

Dynamic Related to conditions or parameters that change, i.e. motion.

E-core Laminated form, in the shape of the letter E, onto which inductors and transformers are wound.

Eddy currents Small currents induced in a conducting core due to the variations in alternating magnetic flux.

Efficiency Ratio of output power to input power, normally expressed in percent.

Electrical components Components, circuits, and systems that manage the flow of *power*.

Electrical equipment Equipment designed to manage or control the flow of power.

Electrical wave Travelling wave prorogated in a conductive medium that is a variation in voltage or current and travels at slightly less than the speed of light.

Electric charge Electric energy stored on the surface of a material.

Electric current Electron movement or motion.

Electric field Also called a voltage field, it is a field or force that exists in the space between two different potentials or voltages.

Electrician Person involved with the design, repair, and assembly of electrical equipment.

Electricity Science that states that certain particles possess a force field, which with electrons is negative and with protons is positive. Electricity can be divided into two groups: static and dynamic. Static electricity deals with charges at rest, while dynamic electricity deals with charges in motion.

Electric motor Device that converts an electrical energy input into a mechanical energy output.

Electric polarization A displacement of bound charges in a dielectric when placed in an electric field.

Electroacoustic transducer Device that achieves an energy transfer from electric to acoustic (sound), and vice versa. Examples include a microphone and a loudspeaker.

Electroluminescence Transmission or conversion of electrical energy into light energy.

Electrolyte Electrically conducting liquid (wet) or paste (dry).

Electrolytic capacitor Capacitor having an electrolyte between the two plates; due to chemical action, a very thin layer of oxide is deposited on only the positive plate, which accounts for why this type of capacitor is polarized.

Electromagnet A magnet consisting of a coil wound on a soft iron or steel core. When current is passed through the coil a magnetic field is generated and the core is strongly magnetized to concentrate the magnetic field.

Electromagnetic communication Use of an electromagnetic wave to pass information between two points. Also called *wireless* communication.

Electromagnetic energy Radiant electromagnetic energy, such as radio and light waves.

Electromagnetic field Field having both an electric (voltage) and magnetic (current) field.

Electromagnetic induction The voltage produced in a coil due to relative motion between the coil and magnetic lines of force.

Electromagnetic spectrum List or diagram showing the entire range of electromagnetic radiation.

Electromagnetic wave Wave that consists of both an electric and magnetic variation.

Electromagnetism Relates to the magnetic field generated around a conductor when current is passed through it.

Electromechanical transducer Device that transforms electrical energy into mechanical energy (motor), and vice versa (generator).

Electromotive force Force that causes the motion of electrons due to a potential difference between two points.

Electron Smallest subatomic particle of negative charge that orbits the nucleus of the atom.

Electron flow A current produced by the movement of free electrons toward a positive terminal.

Electron-hole pair When an electron jumps from the valence shell or band to the conduction band, it leaves a gap in the covalent bond called a hole. This action creates an electron-hole pair.

Electronic components Components, circuits, and systems that manage the flow of *information*.

Electronic equipment Equipment designed to control the flow of information.

Electronics Science related to the behavior of electrons in devices.

Electron-pair bond or covalent bond A pair of electrons shared by two neighboring atoms.

Electrostatic Related to static electric charge.

Electrostatic field Force field produced by static electrical charges.

Element There are 107 different natural chemical substances, or elements, that exist on earth. These can be categorized as gas, solid, or liquid.

Emitter A transistor region from which charge carriers are injected into the base.

Emitter feedback The coupling from the emitter output to the base input in a transistor amplifier.

Emitter follower A grounded-collector transistor amplifier whose operation provides current gain and impedance matching.

Emitter modulator A modulating circuit in which the modulating signal is applied to the emitter of a bipolar transistor.

Encoder circuit A circuit that produces different output voltage codes, depending on the position of a rotary switch.

Energized Being electrically connected to a voltage source so that the device is activated.

Energy Capacity to do work.

Energy gap The space between two orbital shells.

Engineer Person who designs and develops materials to achieve desired results.

Engineering notation A floating-point system in which numbers are expressed as products consisting of a number that is greater than one multiplied by an appropriate power of ten that is some multiple of three.

Enhancement-mode MOSFET A field-effect transistor in which there are no charge carriers in the channel when the gate-source voltage is zero.

Equipment Term used to describe electrical or electronic units.

Equivalent resistance (R_{eq}) Total resistance of all the individual resistances in a circuit.

Error voltage A voltage that is proportional to the error that exists between input and output.

Excited state An energy level in which a nucleus may exist if given sufficient energy to reach this state from a lower state.

Extrinsic semiconductor A semiconductor whose electrical properties are dependent on impurities added to the semiconductor crystal.

Facsimile Electronic process whereby pictures or images are scanned and the graphical information is converted into electrical signals that can be reproduced locally or transmitted to a remote point, where a likeness or facsimile of the original can be produced.

Fahrenheit temperature scale Temperature scale that indicates the freezing point of water at 32°F and boiling point at 212°F.

Fall time Time it takes a negative edge of a pulse to fall from 90% to 10% of its peak value

Farad (F) Unit of capacitance.

Faraday's law 1. When a magnetic field cuts a conductor, or when a conductor cuts a magnetic field, an electric current will flow in the conductor if a closed path is provided over which the current can circulate. 2. Two other laws related to electrolytic cells.

FCC Abbreviation for "Federal Communications Commission."

Ferrite A powdered, compressed, and sintered magnetic material having high resistivity. The high resistance makes eddy-current losses low at high frequencies.

Ferrite bead Ferrite composition in the form of a bead.

Ferrite core Ferrite core normally shaped like a doughnut.

Ferrite-core inductor An inductor containing a ferrite core.

Ferrites Compound composed of iron oxide, a metallic oxide, and ceramic. The metal oxides include zinc, nickel, cobalt, or iron.

Ferrous Composed of and/or containing iron. A ferrous metal exhibits magnetic characteristics, as opposed to nonferrous metals.

Fiber optics Laser's light output carries information that is conveyed between two points by thin glass optical fibers.

Field-effect transistor Abbreviated FET, it is a transistor in which the resistance of the source to drain current path is changed by applying a transverse electric field to the gate.

Field strength The strength of an electric, magnetic, or electromagnetic field at a given point.

Filament Thin thread of, for example, carbon or tungsten, which when heated by the passage of electric current will emit light.

Filament resistor The resistor in a light bulb or electron tube.

Filter Network composed of resistor, capacitor, and inductors used to pass certain frequencies yet block others through heavy attenuation.

Fission Atomic or nuclear fission is the process of splitting the nucleus of heavy elements such as uranium and plutonium into two parts, which results in large releases of radioactivity and heat (fission or division of a nuclei).

Fixed biasing A constant value of bias for an FET in which the voltage is independent of the input signal strength.

Fixed component Component whose value or characteristics cannot be varied or changed.

Fixed-value capacitor A capacitor whose value is fixed and cannot be varied.

Fixed-value resistor A resistor whose value cannot be changed.

Floating ground Ground potential that is not tied or in reference to earth.

Flow-soldering Flow or wave soldering technique is used in large-scale electronic assembly to solder all the connections on a printed circuit board by moving the board over a wave or flowing bath of molten solder.

Fluorescent lamp Gas-filled glass tube that when bombarded by the flow of electric current causes the gas to ionize and then release light.

Flux A material used to remove oxide films from the surfaces of metals in preparation for soldering.

Flux density A measure of the strength of a wave.

Flywheel effect Sustaining effect of oscillation in an *LC* circuit due to the charging and discharging of the capacitor and the expansion and contraction of the magnetic field around the inductor.

Force Physical action capable of moving a body or modifying its movement.

Forward biased A small depletion region at the junction will offer a small resistance and permit a large current. Such a junction is forward biased.

Forward voltage drop (V_F) The forward voltage drop is equal to the junction's barrier voltage.

Free electrons Electrons that are not in any orbit around a nucleus.

Free-running Operating without any external control.

Frequency Rate or recurrences of a periodic wave normally within a unit of one second, measured in hertz (cycles/second).

Frequency drift A slow change in the frequency of a circuit due to temperature or frequency determining component value changes.

Frequency-division multiplex (FDM) Transmission of two or more signals over a common path by using a different frequency band for each signal.

Frequency-domain analysis A method of representing a waveform by plotting its amplitude versus frequency.

Frequency meter Meter used to measure the frequency or cycles per second of periodic waves.

Frequency multiplier A harmonic conversion circuit in which the frequency of the output signal is an exact multiple of the input signal.

Frequency response Indication of how well a device or circuit responds to the different frequencies applied to it.

Frequency response curve A graph indicating a circuit's response to different frequencies.

Friction The rubbing or resistance to relative motion between two bodies in contact.

Frictional electricity Generation of electric charges by rubbing one material against another.

Full-scale deflection (FSD) Deflection of a meter's pointer to the farthest position on the scale.

Full-wave center-tapped rectifier A rectifier circuit that make use of a center-tapped transformer to cause an output current to flow in the same direction during both half-cycles of the ac input.

Full-wave rectifier Rectifier that makes use of the full ac wave (in both the positive and negative cycle) when converting ac to dc.

Full-wave voltage-doubler circuit A rectifier circuit that doubles the output voltage by charging capacitors during both alternations of the ac input—making use of the full ac input wave.

Function generator Signal generator that can function as a sine, square, rectangular, triangular, or sawtooth waveform generator.

Fundamental frequency This sine wave is always the lowest frequency and largest amplitude component of any waveform shape and is used as a reference.

Fuse This circuit- or equipment-protecting device consists of a short, thin piece of wire that melts and breaks the current path if the current exceeds a rated damaging level.

Fuse holder Housing used to support a fuse with two connections.

Fusion Atomic or nuclear fusion is the process of melting the nuclei of two light atoms together to create a heavier nucleus, which results in a large release of energy (fusion or combining of nuclei).

Gain Increase in power from one point to another. Normally expressed in decibels.

Gamma rays High-frequency electromagnetic radiation from radioactive particles.

Ganged Mechanical coupling of two or more capacitors, switches, potentiometers, or any other components so that the activation of one control will operate all.

Gas Any aeriform or completely elastic fluid, which is not a solid or liquid. All gases are produced by the heating of a liquid beyond its boiling point.

Gate One of the field effect transistor's electrodes (also used for thyristor devices).

Geiger counter Device used to detect nuclear particles.

Generator Device used to convert a mechanical energy input into an electrical energy output.

Giga Prefix for 1 billion (10^9).

Graph origin Center of the graph where the horizontal axis and vertical axis cross.

Greenwich mean time (GMT) Also known as universal time, it is a standard based on the earth's rotation with respect to the sun's position. The solar time at the meridian of Greenwich, England, which is at zero longitude.

Ground An intentional or accidental conducting path between an electrical circuit or system and the earth, or some conducting body acting in the place of the earth.

Gunn diode A semiconductor diode that utilizes the Gunn effect to produce microwave frequency oscillation or to amplify a microwave-frequency signal.

Half-power point A point at which power is 50%. This corresponds to 70.7% of the total current.

Half-split method A troubleshooting technique used to isolate a faulty block in a circuit or system. In this method a mid-point is chosen and tested to determine which half is malfunctioning.

Half-wave rectifier circuit A circuit that converts ac to dc by only allowing current to flow during one-half of the ac input cycle.

Hall effect sensor A sensor that generates a voltage in response to a magnetic field.

Hardware Electrical, electronic, mechanical, or magnetic devices or components. The physical equipment.

Hartley oscillator An *LC* tuned oscillator circuit in which the tank coil has an intermediate tap.

Harmonic Sine wave that is smaller in amplitude and some multiple of the fundamental frequency.

Henry (H) Unit of inductance.

Hertz (Hz) Unit of frequency. One hertz is equal to one cycle per second.

Hexadecimal number system A base 16 number system, using the digits 0 through 9 and the six additional digits represented by A, B, C, D, E, and F.

High fidelity (hi-fi) Sound reproduction equipment that is as near to the original sound as possible.

High-pass filter Network or circuit designed to pass any frequencies above a critical or cutoff frequency and reject or heavily attenuate all frequencies below.

High *Q* Abbreviation for quality and generally related to inductors that have a high value of inductance and very little coil resistance.

High tension Lethal voltage in the kilovolt range and above.

High-voltage probe Accessory to the voltmeter that has added multiplier resistors within the probe to divide up the large potential being measured by the probe.

***H*-lines** Invisible lines of magnetic flux.

Hole A mobile vacancy in the valance structure of a semiconductor. This hole exists when an atom has less than its normal number of electrons, and is equivalent to a positive charge.

Hole flow Conduction in a semiconductor when electrons move into holes when a voltage is applied.

Hologram Three-dimensional picture created with a laser.

Holography Science dealing with three-dimensional optical recording.

Horizontal Parallel to the horizon or perpendicular to the force of gravity.

Horizontally polarized wave Electromagnetic wave that has the electric field lying in the horizontal plane.

Hot resistance The resistance of a device when hot due to the generation of heat by electric current.

Hybrid circuit Circuit that combines two technologies (passive and active or discrete and integrated components) onto one microelectronic circuit. Passive components are generally made by thin-film techniques, while active components are made utilizing semiconductor techniques. Integrated circuits can be mounted on the microelectronic circuit and connected to discrete components also on the small postage-stamp-size boards.

Hydroelectric Generation of electric power by the use of water in motion.

Hysteresis Amount that the magnetization of a material lags the magnetizing force due to molecular friction.

IC Abbreviation for "integrated circuit."

Imaginary number A complex number whose imaginary parts are not zero.

IMPATT diode A semiconductor diode that has a negative resistance characteristic produced by a combination of impact avalanche breakdown and charge carrier transmit time effects in a thin semiconductor chip.

Impedance (Z) Measured in ohms, it is the total opposition a circuit offers to current flow (reactive and resistive).

Impedance coupling The coupling of two signal amplifier circuits through the use of an impedance, such as a choke.

Impedance matching Matching of the source impedance to the load impedance causes maximum power to be transferred.

Incandescence State of a material when it is heated to such a high temperature that it emits light.

Incandescent lamp An electric lamp that generates light when an electric current is passed through its filament of resistance, causing it to heat to incandescence.

Induced current Current that flows due to an induced voltage.

Induced voltage Voltage generated in a conductor when it is moved through a magnetic field.

Inductance Property of a circuit or component to oppose any change in current as the magnetic field produced by the change in current causes an induced countercurrent to oppose the original change.

Inductive circuit Circuit that has a greater inductive reactance figure than capacitive reactance figure.

Inductive reactance (X_L) Measured in ohms, it is the opposition to alternating or pulsating current flow without the dissipation of energy.

Inductor Length of conductor used to introduce inductance into a circuit.

Inductor analyzer A test instrument designed to test inductors.

Infinite Having no limits.

Infinity Amount larger than any number can indicate.

Information Data or meaningful signals.

Infrared Electromagnetic heat radiation whose frequencies are above the microwave frequency band and below red in the visible band.

Ingot A mass of metal cast into a convenient shape.

Inhibit To stop an action or block data from passing.

Injection laser diode or ILD A semiconductor P-N junction diode that uses a lasing action to increase and concentrate the light output.

In phase Two or more waves of the same frequency whose maximum positive and negative peaks occur at the same time.

Input impedance Impedance seen when looking into the input terminals of a device.

Insulated When a nonconductive material is used to isolate conducting materials from one another.

Insulating material Material that will in nearly all cases prevent the flow of current due to its chemical composition.

Insulation resistance Resistance of the insulating material. The greater the insulation resistance, the better the insulator.

Insulator Material that has few electrons per atom and those electrons are close to the nucleus and cannot be easily removed.

Integrated When two or more components are combined (into a circuit) and then incorporated in one package or housing.

Integrator amplifier An amplifier whose output is the integral of its input with respect to time.

Integrator circuit A circuit with an output which is the integral of its input with respect to time.

Intermediate-frequency amplifier An amplifier that has one or more transistor stages designed to amplify intermediate frequency signals.

Intermittent Occurring at random intervals of time. An intermittent component, circuit, or equipment problem is undesirable and difficult to troubleshoot, as the problem needs to occur before isolation of the fault can begin.

Internal resistance No voltage source is 100% efficient in that not all the energy in is converted to electrical energy out; some is wasted in the form of heat dissipation. The internal series resistance of a voltage source represents this inefficiency.

Intrinsic semiconductor materials Pure semiconductor materials.

Inverting amplifier An operational amplifier in which the inverting signal or negative input is held virtually at ground by negative feedback and the output is inverted with respect to the input.

Inverting input The inverting or negative input of an operational amplifier.

Ion Atom that has an equal number of protons (+charge) and electrons (−charge) is considered a neutral atom. If more electrons exist in orbit around the atom it is considered a negative ion, whereas if less electrons are in orbit around the atom it is referred to as a positive ion.

Jack Socket or connector into which a plug may be inserted.

JFET A field-effect transistor made up of a gate region diffused into a channel region. When a control voltage is applied to the gate, the channel is depleted or enhanced, and the current between source and drain is thereby controlled.

***j* operator** A prefix used to indicate an imaginary number.

Joule The unit of work and energy.

Junction Contact or connection between two or more wires or cables.

Junction diode A semiconductor diode in which the rectifying characteristics occur at a junction between the *n*-type and *p*-type semiconductor materials.

Kilovolt-ampere 1000 volts at 1 ampere.

Kilowatt-hour 1000 watts for 1 hour.

Kilowatt-hour meter A meter used by electric companies to measure a customer's electric power use in kilowatt-hours.

Kinetic energy Energy associated with motion.

Kirchoff's current law The sum of the currents flowing into a point in a circuit is equal to the sum of the currents flowing out of that same point.

Kirchoff's voltage law The algebraic sum of the voltage drops in a closed-path circuit is equal to the algebraic sum of the source voltage applied.

Knee voltage The voltage at which a curve joins two relatively straight portions of a characteristic curve.

Lag Difference in time between two waves of the same frequency expressed in degrees, i.e. one waveform lags another by a certain number of degrees.

Laminated core Core made up of sheets of magnetic material insulated from one another by an oxide or varnish.

Lamp Device that produces light.

Laser Device that produces a very narrow, intense beam of light. The name is an acronym for "light amplification by stimulated emission of radiation."

Latched A bistable circuit action that causes the circuit to remain held or locked in its last activated state.

***LC* filter** A selective circuit which makes use of an inductance-capacitance network.

Lead The angle by which one alternating signal leads another in time, or a wire that connects two points in a circuit.

Lead-acid cell Cell made up of lead plates immersed in a sulfuric acid electrolyte.

Leakage current or reverse current (I_R) The extremely small current present at the junction.

Least significant bit (LSB) The right-most binary digit.

LED Abbreviation for "light-emitting diode."

Left-hand rule If the fingers of the left hand are placed around the wire so that the thumb points in the direction of the electron flow, the fingers will be pointing in the direction of the magnetic field being produced by the conductor.

Lenz's law The current induced in a circuit due to a change in the magnetic field is so directed as to oppose the change in flux, or to exert a mechanical force opposing the motion.

Lie detector Piece of electronic equipment, also called a polygraph, that determines whether a person is telling the truth by looking for dramatic changes in blood pressure, body temperature, breathing rate, heart rate, and skin moisture in response to certain questions.

Lifetime The time difference between an electron jumping into the conduction band and then falling back into a hole.

Light Electromagnetic radiation in a band of frequencies that can be received by the human eye.

Light-emitting diode A semiconductor diode that converts electric energy into electromagnetic radiation at visible and near infrared frequencies when its P-N junction is forward biased.

Light-emitting semiconductor devices Semiconductor devices that will emit light when an electrical signal is applied.

Limiter Circuit or device that prevents some portion of its input from reaching the output.

Linear Relationship between input and output in which the output varies in direct proportion to the input.

Linear scale A scale whose divisions are uniformly spaced.

Liquid crystal displays (LCDs) A digital display having two sheets of glass separated by a sealed quantity of liquid crystal material. When a voltage is applied across the front and back electrodes, the liquid crystals' molecules become disorganized, causing the liquid to darken.

Live Term used to describe a circuit or piece of equipment that is on and has current flow within it.

Load Source drives a load, and whatever component, circuit, or piece of equipment is connected to the source can be called a load and will have a certain load resistance, which will consequently determine the load current.

Load current The current that is present in the load.

Load impedance Total reactive and resistive opposition of a load.

Loading effect Large load resistance will cause a small load current to flow, and so the loading down of the source or loading effect will be small (light load), whereas a small load resistance will cause a large load current to flow from the source, which will load down the source (heavy load).

Load resistance The resistance of the load.

Locked To automatically follow a signal.

Lodestone A magnetite stone possessing magnetic polarity.

Logarithms The exponent that indicates the power to which a number is raised to produce a given number.

Logic Science dealing with the principles and applications of gates, relays, and switches.

Logic gate circuits Two-state (ON/OFF) circuits used for decision-making functions in digital logic circuits.

Loss Term used to describe a decrease in power.

Lower sideband A group of frequencies that are equal to the differences between the carrier and modulation frequencies.

Low-pass filter Network or circuit designed to pass any frequencies below a critical or cutoff frequency and reject or heavily attenuate all frequencies above.

Magnet Body that can be used to attract or repel magnetic materials.

Magnetic circuit breaker Circuit breaker that is tripped or activated by use of an electromagnet.

Magnetic coil Spiral of a conductor, which is called an electromagnet.

Magnetic core Material that exists in the center of the magnetic coil to either support the windings (nonmagnetic-material) or intensify the magnetic flux (magnetic-material).

Magnetic field Magnetic lines of force traveling from the north to the south pole of a magnet.

Magnetic flux The magnetic lines of force produced by a magnet.

Magnetic leakage The passage of magnetic flux outside the path along which it can do useful work.

Magnetic poles Points of a magnet from which magnetic lines of force leave (north pole) and arrive (south pole).

Magnetism Property of some materials to attract and repel others.

Magnetizing force Also called magnetic field strength, it is the magnetomotive force per unit length at any given point in a magnetic circuit.

Magnetomotive force Force that produces a magnetic field.

Mainframe Large computers that initially were only affordable to medium-sized and large businesses who had the space for them. Mini-computers came after mainframes and were affordable to any business, and now we have microcomputers which are easily affordable to anyone.

Majority carriers The type of carrier that constitutes more than half the total number of carriers in a semiconductor material. In *n*-type materials electrons are the majority carriers, whereas in *p*-type materials holes are the majority carriers.

Matched impedance Condition that occurs when the source impedance is equal to the load impedance, resulting in maximum power being transferred.

Matching Connection of two components or circuits so that maximum energy is transferred or coupled between the two.

Maximum power transfer A theorem that states maximum power will be transferred from source to load when the source resistance is equal to the load resistance.

Maxwell One magnetic line of force or flux is called a maxwell.

Measurement Determining the presence and magnitude of variables.

Medical electronics Branch of electronics involved with therapeutic or diagnostic practices in medicine.

Mercury cell Primary cell that has mercuric oxide cathode, a zinc anode, and a potassium hydroxide electrolyte.

Metal-film resistor A resistor in which a film of metal oxide or alloy is deposited on an insulating substrate.

Metal oxide resistor A metal film resistor in which an oxide of a metal (such as tin) is deposited as a film onto the substrate.

Meter (1) Any electrical or electronic measuring device. (2) In the metric system, it is a unit of length equal to 39.37 inches or 3.28 feet.

Meter FSD current The value of current needed to cause the meter movement to deflect the needle to its full-scale deflection (FSD) position.

Meter resistance The resistance of a meter's armature coil.

Mica capacitor Fixed capacitor that uses mica as the dielectric between its plates.

Microphone Electroacoustic transducer that responds and converts a sound wave input into an equivalent electrical wave out.

Microwave Term used to describe a band of very small wavelength radio waves within the UHF, SHF, and EHF bands.

Mil One thousandth of an inch (0.001 in.).

Miller-effect capacitance (C_M) An undesirable inherent capacitance that exists between the junctions of transistors.

Minority carriers The type of carrier that constitutes less than half the total number of carriers in a semiconductor material. In n-type materials holes are the minority carriers, whereas in p-type materials electrons are the minority carriers.

Mismatch Term used to describe a difference between the source impedance and load impedance, which will prevent maximum power transfer.

Modulation Process whereby an information signal is used to modify some characteristic of another higher-frequency wave known as a carrier.

Molecule Smallest particle of a compound that still retains its chemical characteristics.

MOS family A group of digital logic circuits that make use of metal oxide semiconductor field effect transistors (MOSFETs).

MOSFET A field-effect transistor in which the insulating layer between the gate electrode and the channel is a metal-oxide layer. Either a p or n substrate.

Most significant bit (MSB) The left-most binary digit.

Moving-coil microphone Microphone that makes use of a moving coil between a fixed magnetic field. Also called a dynamic microphone.

Moving-coil pickup Dynamic phonograph pickup that uses a coil between a fixed magnetic field, which is moved back and forth by the needle or stylus.

Moving-coil speaker Dynamic speaker that uses a coil placed between a fixed magnetic field and converts the electrical wave input into sound waves.

Multimeter Piece of electronic test equipment that can perform multiple tasks in that it can be used to measure voltage, current, or resistance.

Multiple-emitter input transistor A transistor specially constructed to have more than one emitter.

Multiplier resistor A resistor connected in series with the meter movement of a voltmeter.

Mutual inductance Ability of one inductor's magnetic lines of force to link with another inductor.

Navigation equipment Electronic equipment designed to aid in the direction of aircraft and ships to their destination.

n-channel d-type MOSFET A depletion type MOSFET having an *n*-type channel between its source and drain terminals.

n-channel JFET A junction field effect transistor having an *n*-type channel between source and drain.

Negative (neg.) (1) Some value less than zero. (2) Terminal that has an excess of electrons.

Negative charge An electric charge that has more electrons than protons.

Negative feedback Feedback in which a portion of the output signal is fed back 180° out of phase with the input signal.

Negative ground A system whereby the negative terminal of the voltage source is connected to the system's conducting chassis or body.

Negative ion Atom that has more than the normal neutral amount of electrons.

Negative resistance A resistance such that when the current through it increases, the voltage drop across the resistance decreases.

Negative temperature coefficient Effect that describes that if temperature increases, resistance or capacitance will decrease.

Neon bulb Glass envelope filled with neon gas, which when ionized by an applied voltage will glow red.

Network Combination and interconnection of components, circuits, or systems.

Neutral When an object is neither positive nor negative.

Neutral atom An atom in which the number of positive charges in the nucleus (protons) is equal to the number of negative charges (electrons) that surround the nucleus.

Neutral wire The conductor of a polyphase circuit or a single-phase three-wire circuit that is intended to have a ground potential. The potential differences between the neutral and each of the other conductors are approximately equal in magnitude and are also equally spaced in phase.

Neutron Subatomic particle residing within the nucleus and having no electrical charge.

Nickel-cadmium cell Most popular secondary cell; it uses a nickel oxide positive electrode and cadmium negative electrode.

Node Junction or branch point.

Noise Unwanted electromagnetic radiation within an electrical or mechanical system.

Noninverting amplifier An operational amplifier in which the input signal is applied to the ungrounded positive input terminal to give a gain that is greater than one, and make the output change in phase with the input voltage.

Noninverting input The noninverting or positive input of an operational amplifier.

Nonlinear scale A scale whose divisions are not uniformly spaced.

Normally closed (N.C.) Designation which states that the contacts of a switch or relay are connected normally; however, when activated, these contacts will open.

Normally open (N.O.) Designation which states that the contacts of a switch or relay are normally not connected; however, when activated, these contacts will close.

North pole Pole of a magnet out of which magnetic lines of force are assumed to originate.

Norton's theorem Any network of voltage sources and resistors can be replaced by a single equivalent current source (I_N) in parallel with a single equivalent resistance (R_N).

NPN transistor Negative-positive-negative transistor in which a layer of p-type conductive semiconductor is located between two n-type regions.

n-Type semiconductor A material that has more valence-band electrons than valence-band holes.

Nuclear energy Atomic energy or power released in a nuclear reaction when either a neutron is used to split an atom into smaller atoms (fission) or when two smaller nuclei are joined together (fusion).

Nuclear reactor Unit that maintains a continuous self-supporting nuclear reaction (fission).

Nucleus Core of an atom; it contains both positive (protons) and neutral (neutrons) subatomic particles.

Octave Interval between two sounds whose fundamental frequencies differ by a ratio of 2 to 1.

Ohm Unit of resistance, symbolized by the Greek capital letter omega (Ω).

Ohmmeter Measurement device used to measure electric resistance.

Ohm's law Relationship between the three electrical phenomena of voltage, current, and resistance, which states that the current flow within a circuit is directly proportional to the voltage applied across the circuit and inversely proportional to its resistance.

Ohms per volt Value that indicates the sensitivity of a voltmeter. The higher the ohms per volt rating, the more sensitive the meter.

Open circuit Break in the path of current flow.

Open collector output A type of output structure found in certain bipolar logic families. Resistive pull-ups are generally added to provide the high level output voltage.

Open-loop mode A control system that has no means of comparing the output with the input for control purposes.

Operational amplifier Special type of high-gain amplifier; also called an op amp.

Operator error An incorrect use of the controls of a circuit or system.

Optimum power transfer Since the ideal maximum power transfer conditions cannot always be achieved, most designers try to achieve maximum power transfer and have the source resistance and load resistance as close in value as possible.

Ordinary ground A connection in the circuit that is said to be at ground potential or zero volts and, because of its connection to earth, has the ability to conduct electrical current to and from the earth.

OR gate When either input A OR B is HIGH, the output will be HIGH.

Oscillate Continual repetition of or passing through a cycle.

Oscillator Electronic circuit that converts dc to a continuous alternating current out.

Oscilloscope Instrument used to view signal amplitude, period, and shape at different points throughout a circuit.

Out of phase When the maximum and minimum points of two or more waveforms do not occur at the same time.

Output Terminals at which a component, circuit, or piece of equipment delivers current, voltage, or power.

Output impedance Impedance measured across the output terminals of a device without the load connected.

Output power Amount of power a component, circuit, or system can deliver to its load.

Overload Situation that occurs when the load is greater than the component, circuit, or system was designated to handle (load resistance too small, load current too high), resulting in waveform distortion and/or overheating.

Overload protection Protective device such as a fuse or circuit breaker that automatically disconnects or opens a current path when it exceeds an excessive value.

Paper capacitor Fixed capacitor using oiled or waxed paper as a dielectric.

Parallax error The apparent displacement of an object's position caused by a shift in the point of observation of the object.

Parallel Also called shunt; circuit having two or more paths for current flow.

Parallel data transmission The transfer of information simultaneously over a set of parallel paths or channels.

Parallel resonant circuit Circuit having an inductor and capacitor in parallel with one another, offering a high impedance at the frequency of resonance.

Parallel-to-serial converter An IC that converts the parallel word applied to its inputs into a serial data stream.

Parasitic oscillations An unwanted self-sustaining oscillation or self generated random pulse.

Passband Band or range of frequencies that will be passed by a filter circuit.

Passive component Component that does not amplify a signal, such as a resistor or capacitor.

Passive filter A filter that contains only passive (nonamplifying) components and provides pass or rejection characteristics.

Passive system System that emits no energy; in other words it only receives; it does not transmit and consequently reveal its position.

p-channel (d-type) MOSFET A depletion type MOSFET having a p-type channel between its source and drain terminal.

p-channel JFET A junction field effect transistor having a p-type channel between source and drain.

Peak Maximum or highest-amplitude level.

Peak inverse voltage Abbreviated PIV, it is the maximum rated value of an ac voltage acting in the direction opposite to that in which a device is designed to pass current.

Peak to peak Difference between the maximum positive and maximum negative values.

Pentode A five-electrode electron tube that has an anode, cathode, control grid, and two additional electrodes.

Percent of regulation The change in output voltage that occurs between no-load and full-load in a dc voltage source. Dividing this change by the rated full-load value and multiplying the result by 100 gives percent regulation.

Percent of ripple The ratio of the effective or rms value of ripple voltage to the average value of the total voltage. Expressed as a percentage.

Period Time taken to complete one complete cycle of a periodic or repeating waveform.

Permanence Magnetic equivalent of electrical conductance and consequently equal to the reciprocal of reluctance, just as conductance is equal to the reciprocal of resistance.

Permanent magnet Magnet, normally made of hardened steel, that retains its magnetism indefinitely.

Permeability Measure of how much better a material is as a path for magnetic lines of force with respect to air, which has a permeability of 1 (symbolized by the Greek lowercase letter mu, μ).

Phase Angular relationship between two waves, normally between current and voltage in an ac circuit.

Phase angle Phase difference between two waves, normally expressed in degrees.

Phase-locked loop (PLL) circuit A circuit consisting of a phase comparator that compares the output frequency of a voltage controlled oscillator with an input frequency. The error voltage out of the phase comparator is then coupled via an amplifier and low-pass filter to the control input of the voltage controlled oscillator to keep it in phase, and therefore at exactly the same frequency as the input.

Phase shift Change in phase of a waveform between two points, given in degrees of lead or lag.

Phase-shift oscillator An *RC* oscillator circuit in which the 180° phase shift is achieved with several *RC* networks.

Phase splitter A circuit that takes a single input signal and produces two output signals that are 180° apart in phase.

Phonograph Piece of equipment used to reproduce sound.

Phosphor Luminescent material applied to the inner surface of a cathode ray tube that when bombarded with electrons will emit light.

Photoconduction A process by which the conductance of a material is changed by incident electromagnetic radiation in the light spectrum.

Photoconductive cell Material whose resistance decreases or conductance increases when light strikes it.

Photodetector Component used to detect or sense light.

Photodiode A semiconductor diode that changes its electrical characteristics in response to illumination.

Photometer Meter used to measure light intensity.

Photon Discrete portion of electromagnetic energy. A small packet of light.

Photoresistor Also known as a photoconductive cell or light-dependent resistor, it is a device whose resistance varies with the illumination of the cell.

Photovoltaic action A process by which a device generates a voltage as a result of exposure to radiation.

Photovoltaic cell Component, commonly called a solar cell, used to convert light energy into electric energy (voltage).

Pi Value representing the ratio between the circumference and diameter of a circle and equal to approximately 3.142 (symbolized by the lowercase Greek letter π).

Pierce crystal oscillator An oscillator circuit in which a piezoelectric crystal is connected in a tank between output and input.

Piezoelectric crystal Crystal material that will generate a voltage when mechanical pressure is applied and conversely will undergo mechanical stress when subjected to a voltage.

Piezoelectric effect The operation of a voltage between the opposite sides of a piezoelectric crystal as a result of pressure or twisting. Also, the reverse effect in which the application of a voltage to opposite sides causes deformation to occur at the frequency of the applied voltage.

Piezoresistive diaphragm pressure sensor A sensor that changes its resistance in response to pressure.

Pinch-off region A region on the characteristic curve of an FET in which the gate bias causes the depletion region to extend completely across the channel.

PIN diode A semiconductor diode that has a high-resistance intrinsic region between its low-resistance *p*-type and *n*-type regions.

Pitch Term used to describe the inflection or frequency scale of sounds. When the pitch is increased by one octave, twice the original frequency will be the result.

Plastic film capacitor A capacitor in which alternate layers of metal aluminum foil are separated by thin films of plastic dielectric.

Plate Conductive electrode in either a capacitor or battery.

Plug Movable connector that is normally inserted into a socket.

P-N Junction The point at which two opposite doped materials come in contact with one another.

PNP transistor Positive-negative-positive transistor in which a layer of n-type conductive semiconductor is located between two p-type regions.

Polar coordinates Either of two numbers that locate a point in a plane by its distance from a fixed point on a line and the angle this line makes with a fixed line.

Polarity Term used to describe positive and negative charges.

Polarized electrolytic capacitor An electrolytic capacitor in which the dielectric is formed adjacent to one of the metal plates, creating a greater opposition to current in one direction only.

Positive Point that attracts electrons, as opposed to negative, which supplies electrons.

Positive charge The charge that exists in a body which has fewer electrons than normal.

Positive ground A system whereby the positive terminal of the voltage source is connected to the system's conducting chassis or body.

Positive ion Atom that has lost one or more of its electrons and therefore has more protons than electrons, resulting in a net positive charge.

Positive shunt clipper A circuit that has a diode connected in shunt with the lead or output, and its orientation is such that it will clip off the positive alternation of the ac input.

Potential difference (pd) Voltage difference between two points, which will cause current to flow in a closed circuit.

Potential energy Energy that has the potential to do work because of its position relative to others.

Potentiometer Three-lead variable resistor that through mechanical turning of a shaft can be used to produce a variable voltage or potential.

Power Amount of energy converted by a component or circuit in a unit of time, normally seconds. It is measured in units of watts (joules/second).

Power amplifier An amplifier designed to deliver maximum output power to a load.

Power dissipation Amount of heat energy generated by a device in one second when current flows through it.

Power factor Ratio of actual power to apparent power. A pure resistor has a power factor of 1 or 100%, while a capacitor has a power factor of 0 or 0%.

Power loss Ratio of power absorbed to power delivered.

Power supply Piece of electrical equipment used to deliver either ac or dc voltage.

Prefix Name used to designate a factor or multiplier.

Pressure The application of force to something by something else in direct contact with it.

Primary First winding of a transformer that is connected to the source, as opposed to the secondary that is connected to the load.

Primary cell Cell that produces electrical energy through an internal electro-chemical action; once discharged, it cannot be reused.

Printed circuit board (PCB) Insulating board that has conductive tracks printed onto the board to make the circuit.

Propagation Traveling of electromagnetic, electrical, or sound waves through a medium.

Propagation time Time it takes for a wave to travel between two points.

Proportional Having the same or constant ratio.

Protoboard An experimental arrangement of a circuit on a board. Also called a breadboard.

Proton Subatomic particle within the nucleus that has a positive charge.

***p*-Type Semiconductor** A material that has more valence-band holes than valence-band electrons.

Pull-up resistor A resistor connected between a signal output and positive V_{CC} to pull the signal line HIGH when it is not being driven LOW.

Pulse Rise and fall of some quantity for a period of time.

Pulse fall time Time it takes for a pulse to decrease from 90% to 10% of its maximum value.

Pulse repetition frequency The number of times per second that a pulse is transmitted.

Pulse repetition time The time interval between the start of two consecutive pulses.

Pulse rise time Time it takes for a pulse to increase from 10% to 90% of its maximum value.

Pulse width, pulse length, or pulse duration The time interval between the leading edge and trailing edge of a pulse at which the amplitude reaches 50% of the peak pulse amplitude.

Push-pull amplifier A balanced amplifier that uses two similar equivalent amplifying transistors working in phase opposition.

Pythagorean theorem A theorem in geometry: The square of the length of the hypotenuse of a right triangle equals the sum of the squares of the lengths of the other two sides.

Q Quality factor of an inductor or capacitor; it is the ratio of a component's reactance (energy stored) to its effective series resistance (energy dissipated).

Quiescent point The voltage or current value that sets up the no-signal input or operating point bias voltage.

Race condition An unpredictable bistable circuit condition.

Radar Acronym for "radio detection and ranging"; it is a system that measures the distance and direction of objects.

Radioastronomy Branch of astronomy that studies the radio waves generated by celestial bodies and uses these emissions to obtain more information about them.

Radio broadcast Transmission of music, voice, and other information on radio carrier waves that can be received by the general public.

Radio communication Term used to describe the transfer of information between two or more points by use of radio or electromagnetic waves.

Radio-frequency (RF) amplifier An amplifier that has one or more transistor stages designed to amplify radio-frequency signals.

Radio-frequency generator A generator capable of supplying RF energy at any desired frequency in the radio spectrum.

Radio-frequency probe A probe used in conjunction with an ac meter to measure high-frequency RF signals.

RC Abbreviation for "resistance–capacitance."

RC **Circuit** Circuit containing both a resistor and capacitor.

RC **coupling** A coupling method in which resistors are used as the input and output impedances of the two stages. A coupling capacitor is generally used between the stages to couple the ac signal and block the dc supply bias voltages.

RC **filter** A selective circuit which makes use of a resistance-capacitance network.

RC **time constant** In one time constant, which is equal to the product of resistance and capacitance in seconds, a capacitor will have charged or discharged 63.2% of the maximum applied voltage.

Reactance (*X*) Opposition to current flow without the dissipation of energy.

Reactive power Also called imaginary power or wattless power, it is the power value obtained by multiplying the effective value of current by the effective value of voltage and the sine of the angular phase difference between current and voltage.

Real number A number that has no imaginary part.

Receiver Unit or piece of equipment used for the reception of information.

Recombination The combination and resultant neutralization of particles or objects having unlike charges. For example, a hole and an electron, or a positive ion and negative ion.

Rectangular coordinates A Cartesian coordinate of a Cartesian coordinate system whose straight-line axes or coordinate planes are perpendicular.

Rectangular wave Also known as a pulse wave; it is a repeating wave that only alternates between two levels or values and remains at one of these values for a small amount of time relative to the other.

Rectification Process that converts alternating current (ac) into direct current (dc).

Rectifier A device that converts alternating current into a unidirectional or dc current.

Rectifier circuit Achieves rectification.

Rectifier diodes or rectifiers Junction diodes that achieve rectification.

Reed relay Relay that consists of two thin magnetic strips within a glass envelope with a coil wrapped around the envelope so that when it is energized the relay's contacts or strips will snap together, making a connection between the leads attached to each of the reed strips.

Regulator A device that maintains a desired quantity at a predetermined voltage.

Relative Not independent; compared with or with respect to some other value of a measured quantity.

Relaxation oscillator An oscillator circuit whose frequency is determined by an *RL* or *RC* network, producing a rectangular or sawtooth output waveform.

Relay Electromechanical device that opens or closes contacts when a current is passed through a coil.

Reluctance Resistance to the flow of magnetic lines of force.

Remanence Amount a material remains magnetized after the magnetizing force has been removed.

Reset and carry An action that occurs in any column that reaches its maximum count.

Reset state A circuit condition in which the output is reset to binary 0.

Residual magnetism Magnetism remaining in the core of an electromagnet after the coil current has been removed.

Resistance Symbolized R and measured in ohms (Ω), it is the opposition to current flow with the dissipation of energy in the form of heat.

Resistive power (True power) The average power consumed by a circuit during one complete cycle of alternating current.

Resistive temperature detector (RTD) A temperature detector consisting of a fine coil of conducting wire (such as platinum) that will produce a relatively linear increase in resistance as temperature increases.

Resistivity Measure of a material's resistance to current flow.

Resistor Component made of a material that opposes the flow of current and therefore has some value of resistance.

Resistor color code Coding system of colored stripes on a resistor that indicates the resistor's value and tolerance.

Resonance Circuit condition that occurs when the inductive reactance (X_L) is equal to the capacitive reactance (X_C).

Resonant circuit Circuit containing an inductor and capacitor tuned to resonate at a certain frequency.

Resonant frequency Frequency at which a circuit or object will produce a maximum amplitude output.

Reverse biased A large depletion region at the junction will offer a large resistance and permit only a small current. Such a junction is reverse biased.

Reverse leakage current (I_R) The undesirable flow of current through a device in the reverse direction.

Reverse voltage drop (V_R) The reverse voltage drop is equal to the source voltage (applied voltage).

RF Abbreviation for "radio frequency."

RF amplifier/frequency tripler circuit A radio frequency amplifier circuit in which the frequency of the output is three times that of the input frequency.

RF amplifier/multiplier circuit A radio frequency amplifier circuit in which the frequency of the output is an exact multiple of the input frequency.

RF local oscillator circuit The radio frequency oscillator in a superheterodyne receiver.

Rheostat Two-terminal variable resistor used to control current.

Right angle triangle A triangle having a 90° or square corner.

Ripple frequency The frequency of the ripple present in the output of a dc source.

Rise time Time it takes a positive edge of a pulse to rise from 10% to 90% of its high value.

RL differentiator An RL circuit whose output voltage is proportional to the rate of change of the input voltage.

RL filter A selective circuit of resistors and inductors that offers little or no opposition to certain frequencies while blocking or attenuating other frequencies.

RL integrator An RL circuit with an output proportionate to the integral of the input signal.

RMS Abbreviation for "root mean square."

RMS value Rms value of an ac voltage, current, or power waveform is equal to 0.707 times the peak value. The rms value is the effective or dc value equivalent of the ac wave.

Rotary switch Electromechanical device that has a rotating shaft connected to one terminal that is capable of making or breaking a connection.

R-2R ladder circuit A network or circuit composed of a sequence of L networks connected in tandem. This R-$2R$ circuit is used in digital-to-analog converters.

Saturation The condition in which a further increase in one variable produces no further increase in the resultant effect.

Saturation point The point beyond which an increase in one of two quantities produces no increase in the other.

Sawtooth wave Repeating waveform that rises from zero to a maximum value linearly and then falls to zero and repeats.

Scale Set of markings used for measurement.

Schematic diagram Illustration of the electrical or electronic scheme of a circuit, with all the components represented by their respective symbols.

Schottky diode A semiconductor diode formed by contact between a semiconductor layer and a metal coating. Hot carriers (electrons for n-type material and holes for p-type material) are emitted from the Schottky barrier of the semiconductor and move to the metal coating. Since majority carriers predominate, there is essentially no injection or storage of minority carriers to limit switching speed.

Scientific notation Numbers are entered and displayed in terms of a power of 10. For example:

Number	Scientific Notation
7642	7.642×10^3
64,000,000	64×10^6
0.0012	1.2×10^{-3}
0.000096	96×10^{-6}

Secondary Output winding of a transformer that is connected across the load.

Secondary cells Electrolytic cells for generating electricity. Once discharged the cell may be restored or recharged by sending an electric current through the cell in the opposite direction to that of the discharge current.

Selectivity Characteristic of a circuit to discriminate between the wanted signal and the unwanted signal.

Self biasing Gate bias for an FET in which a resistor is used to drop the supply voltage and provide gate bias.

Self-inductance The property that causes a counter electromotive force to be produced in a conductor when the magnetic field expands or collapses with a change in current.

Semiconductors Materials that have properties that lie between insulators and conductors.

Serial data transmission The transfer of information sequentially through a single path or channel.

Serial-to-parallel converter An IC that converts a serial data stream applied to its input into a parallel output word.

Series circuit Circuit in which the components are connected end to end so that current has only one path to follow throughout the circuit.

Series clipper A circuit that will clip off part of the input signal. Also known as a limitor since the circuit will limit the ac input. A series clipper circuit has a clipping or limiting device in series with the load.

Series-parallel network Network or circuit that contains components that are connected in both series and parallel.

Series resonance Condition that occurs when the inductive and capacitive reactances are equal and both components are connected in series with one another, and therefore the impedance is minimum.

Series resonant circuit A resonant circuit in which the capacitor and coil are in series with the applied ac voltage.

Series switching regulators A regulator circuit containing a power transistor in series with the load, that is switched ON and OFF to regulate the dc output voltage delivered to the load.

Set-reset flip-flop A bistable multivibrator circuit that has two inputs that are used to either SET or RESET the output.

Set state A circuit condition in which the output is set to binary 1.

Seven-segment display Component that normally has eight LEDs, seven of which are mounted into segments or bars that make up the number 8, and the eighth LED is used as a decimal point.

Shells or bands An orbital path containing a group of electrons that have a common energy level.

Shield Metal grounded cover that is used to protect a wire, component, or piece of equipment from stray magnetic and/or electric fields.

Shock A sudden pain, convulsion, unconsciousness, or death produced by the passage of an electrical current through the body.

Short circuit Also called a short; it is a low-resistance connection between two points in a circuit, typically causing a large amount of current flow.

Shorted out Term used to describe a component that has either internally malfunctioned, resulting in a low-resistance path through the component, or a component that has been bypassed by a low-resistance path.

Shunt clipper A circuit that will clip off part of the input signal. Also known as a limitor since the circuit will limit the ac input. A shunt clipper circuit has a clipping or limiting device in shunt with the load.

Shunt resistor A resistor connected in parallel or shunt with the meter movement of an ammeter.

Side bands A band of frequencies on both sides of the carrier frequency of a modulated signal produced by modulation.

Signal Conveyor of information.

Signal-to-noise ratio Ratio of the magnitude of the signal to the magnitude of the noise, normally expressed in decibels.

Signal voltage Rms or effective voltage value of a signal.

Silicon (Si) Nonmetallic element (atomic number 14) used in pure form as a semiconductor.

Silicon controlled rectifier Abbreviated SCR, it is a three-junction, three-terminal unidirectional P-N-P-N thyristor that is normally an open circuit. When triggered with the proper gate signal it switches to a conducting state and allows current to flow in one direction.

Silicon transistor Transistor using silicon as the semiconducting material.

Silver (Ag) Precious metal that does not easily corrode and is more conductive than copper.

Silvered mica capacitor Mica capacitor with silver deposited directly onto the mica sheets, instead of using conducting metal foil.

Silver solder Solder composed of silver, copper, and zinc with a melting point lower than silver but higher than the standard lead-tin solder.

Simplex Communication in only one direction at a time, for example, facsimile and television.

Simulcast Broadcasting of a program simultaneously in two different forms, for example, a program on both AM and FM.

Sine Sine of an angle of a right-angle triangle is equal to the opposite side divided by the hypotenuse.

Sine wave Wave whose amplitude is the sine of a linear function of time. It is drawn on a graph that plots amplitude against time or radial degrees relative to the angular rotation of an alternator.

Single in-line package (SIP) Package containing several electronic components (generally resistors) with a single row of external connecting pins.

Single-pole, double-throw (SPDT) Three-terminal switch or relay in which one terminal can be thrown in one of two positions.

Single-pole, single-throw (SPST) Two-terminal switch or relay that can either open or close one circuit.

Single sideband (SSB) AM radio communication technique in which the transmitter suppresses one sideband and therefore only transmits a single sideband.

Single-throw switch Switch containing only one set of contacts, which can be either opened or closed.

Sink Device, such as a load, that consumes power, or conducts away heat.

Sintering Process of bonding either a metal or powder by cold-pressing it into a desired shape and then heating to form a strong, cohesive body.

Sinusoidal Varying in proportion to the sine of an angle or time function; for example, alternating current (ac) is sinusoidal.

SIP Abbreviation for "single in-line package."

Skin effect Tendency of high-frequency (rf) currents to flow near the surface layer of a conductor.

Slide switch Switch having a sliding bar, button, or knob.

Slow-acting relay Slow-operating relay that when energized may not pull up the armature for several seconds.

Slow-blow fuse Fuse that can withstand a heavy current (up to ten times its rated value) for a small period of time without opening.

Snap switch Switch containing a spring under tension or compression that causes the contacts to come together suddenly when activated.

SNR Abbreviation for "signal-to-noise ratio."

Soft magnetic material Ferromagnetic material that is easily demagnetized.

Software Program of instructions that directs the operation of a computer.

Solar cell Photovoltaic cell that converts light into electric energy. They are especially useful as a power source for space vehicles.

Solder Metallic alloy that is used to join two metal surfaces.

Soldering Process of joining two metallic surfaces to make an electrical contact by melting solder (usually tin and lead) across them.

Soldering gun Soldering tool having a trigger switch and pistol shape that at its tip has a fast-heating resistive element for soldering.

Soldering iron Soldering tool having an internal heating element that is used for soldering.

Solenoid Coil and movable iron core that when energized by an alternating or direct current will pull the core into a central position.

Solid conductor Conductor having a single solid wire, as opposed to strands.

Solid state Pertaining to circuits and devices that use solid semiconductors such as silicon. Solid-state electronics devices have a solid material between their input and output pins (transistors, diodes), whereas vacuum-tube electronics uses tubes, which have a vacuum between input and output.

Sonar Acronym for "sound navigation and ranging." A system using sound waves to determine a target's direction and distance.

Sonic Pertaining to the speed of sound waves.

Sound wave Traveling wave propagated in an elastic medium that travels at a speed of approximately 1133 ft/s.

Source One of the field effect transistor's electrodes.

Source follower An FET amplifier in which the signal is applied between the gate and drain and the output is taken between the source and drain. Used to handle large-input signals and applications requiring a low-input capacitance.

Source impedance Impedance a source presents to a load.

South pole Pole of a magnet into which magnetic lines of force are assumed to enter.

Spark Momentary discharge of electric energy due to the breakdown of air or some other dielectric material separating two terminals.

SPDT Abbreviation for "single-pole, double-throw."

Speaker Also called a loudspeaker; it is an electroacoustic transducer that converts an electrical wave input into a mechanical sound wave (acoustic) output into the air.

Specification sheet or data sheet Details the characteristics and maximum and minimum values of operation of a device.

Spectrum The frequency spectrum displays all the frequencies and their applications.

Spectrum analyzer Instrument that can display the frequency domain of a waveform, plotting amplitude against frequency of the signals present.

Speed of light Physical constant—the speed at which light travels through a vacuum—equal to 186,282.397 miles/s, 2.997925×10^8 m/s, 161,870 nautical miles/s, or 328 yards/μs.

Speed of sound Speed at which a sound wave travels through a medium. In air it is equal to about 1133 ft/s or 334 m/s, while in water it is equal to approximately 4800 ft/s or 1463 m/s. Also known as sonic speed.

Speedup capacitor Capacitor connected in a circuit to speed up an action due to its inherent behavior.

SPST Abbreviation for "single-pole, single-throw."

Square wave Wave that alternates between two fixed values for an equal amount of time.

Square-wave generator A circuit that generates a continuously repeating square wave.

S-R latch Another name for *S-R* flip-flop, so called because the output remains latched in the set or reset state even though the input is removed.

Staircase-wave generator A signal generator circuit that produces an output signal voltage that increases in steps.

Standard Exact value used as a basis for comparison or calibration.

Static Crackling noise heard on radio receivers caused by electric storms or electric devices in the vicinity.

Static electricity Electricity at rest or stationary.

Stator Stationary part of some rotating device.

Statute mile Distance unit equal to 5280 ft or 1.61 km.

Step-down transformer Transformer in which the ac voltage induced in the secondary is less (due to fewer secondary windings) than the ac voltage applied to the primary.

Stepper motor A motor that rotates in small angular steps.

Step-up transformer Transformer in which the ac voltage induced in the secondary is greater (due to more secondary windings) than the ac voltage applied to the primary.

Stereo sound Sound system in which the sound is delivered through at least two channels and loudspeakers arranged to give the listener a replica of the original performance.

Stranded conductor Conductor composed by a group of twisted wires.

Stray capacitance Undesirable capacitance that exists between two conductors such as two leads or a lead and a metal chassis.

Subassembly Components contained in a unit for convenience in assembling or servicing the equipment.

Subatomic Particles such as electrons, protons, and neutrons that are smaller than atoms.

Substrate The mechanical insulating support on which a device is fabricated.

Superconductor Metal such as lead or niobium that, when cooled to within a few degrees of absolute zero, can conduct current with no electrical resistance.

Superheterodyne receiver Radio-frequency receiver that converts all RF inputs to a common intermediate frequency (IF) before demodulation.

Superhigh frequency (SHF) Frequency band between 3 and 30 GHz, so designated by the Federal Communications Commission (FCC).

Superposition theorem Theorem designed to simplify networks containing two or more sources. It states: In a network containing more than one source, the current at any point is equal to the algebraic sum of the currents produced by each source acting separately.

Supersonic Faster than the speed or velocity of sound (Mach 1).

Supply voltage Voltage produced by a power source or supply.

SW Abbreviation for "shortwave."

Sweep generator Test instrument designed to generate a radio-frequency voltage that continually and automatically varies in frequency within a selected frequency range.

Swing Amount a frequency or amplitude varies.

Switch Manual, mechanical, or electrical device used for making or breaking an electric circuit.

Switching power supply A dc power supply that makes use of a series switching regulator controlled by a pulse-width-modulator to regulate the output voltage.

Switching transistor Transistor designed to switch either on or off.

Synchronization Also called sync; it is the precise matching or keeping in step of two waves or functions.

Synchronous Two or more circuits or devices in step or in phase.

Sync pulse or signal Pulse waveform generated to synchronize two processes.

System Combination or linking of several parts or pieces of equipment to perform a particular function.

Tachometer Instrument that produces an output voltage that indicates the angular speed of the input in revolutions per minute.

Tank circuit Circuit made up of a coil and capacitor that is capable of storing electric energy.

Tantalum capacitor Electrolytic capacitor having a tantalum foil anode.

Tap Electrical connection to some point, other than the ends, on the element of a resistor or coil.

Tapered Nonuniform distribution of resistance per unit length throughout the element.

Technician Expert in troubleshooting circuit and system malfunctions. Along with a thorough knowledge of all test equipment and how to use it to diagnose problems, the technician is also familiar with how to repair or replace faulty components. Technicians basically translate theory into action.

Telegraphy Communication between two points by sending a series of coded current pulses either through wires or by radio.

Telemetry Transmission of instrument reading to a remote location either through wires or by radio waves.

Telephone Apparatus designed to convert sound waves into electrical waves, which are then sent to and reproduced at a distant point.

Telephone line Wires existing between subscribers and central stations.

Telephony Telecommunications system involving the transmission of speech information, therefore allowing two or more persons to converse verbally.

Teletypewriter Electric typewriter that like a teleprinter can produce coded signals corresponding to the keys pressed or print characters corresponding to the coded signals received.

Television (TV) System that converts both audio and visual information into corresponding electric signals that are then transmitted through wires or by radio to a receiver, which reproduces the original information.

Telex Teletypewriter exchange service.

Temperature coefficient of frequency Rate frequency changes with temperature.

Temperature coefficient of resistance Rate resistance changes with temperature.

Tera (T) Prefix that represents 10^{12}.

Terminal Connecting point for making electric connections.

Tesla (T) SI unit of magnetic flux density (1 tesla = 1 Wb/m^2).

Test Sequence of operations designed to verify the correct operation or malfunction of a system.

Tetrode A four-electrode electron tube that has an anode, cathode, control grid, and an additional electrode.

Thermal relay Relay activated by a heating element.

Thermal stability The ability of a circuit to maintain stable characteristics despite changes in the ambient temperature.

Thermistor Temperature-sensitive semiconductor that has a negative temperature coefficient of resistance (as temperature increases, resistance decreases).

Thermocouple Temperature transducer consisting of two dissimilar metals welded together at one end to form a junction that when heated will generate a voltage.

Thermometry Relating to the measurement of temperature.

Thermostat Temperature-sensitive device that opens or closes a circuit.

Thévenin's theorem Theorem that replaces any complex network with a single voltage source in series with a single resistance. It states: Any network of resistors can be replaced with an equivalent voltage source (V_{Th}) and an equivalent series resistance (R_{Th}).

Thick-film capacitor Capacitor consisting of two thick-film layers of conductive film separated by a deposited thick-layer dielectric film.

Thick-film resistor Fixed-value resistor consisting of a thick-film resistive element made from metal particles and glass powder.

Thin-film capacitor Capacitor in which both the electrodes and the dielectric are deposited in layers on a substrate.

Thin-film detector (TFD) A temperature detector containing a thin layer of platinum, and used for very precise temperature readings.

Three-phase supply AC supply that consists of three ac voltages that are 120° out of phase with one another.

Threshold Minimum point at which an effect is produced or indicated.

Thyristor A semiconductor switching device in which bistable action depends on P-N-P-N regenerative feedback. A thyristor can be bidirectional or uni-directional and have from two to four terminals.

Time constant Time needed for either a voltage or current to rise to 63.2% of the maximum or fall to 36.8% of the initial value. The time constant of an *RC* circuit is equal to the product of *R* and *C*, while the time constant of an *RL* circuit is equal to the inductance divided by the resistance.

Time-division multiplex (TDM) Transmission of two or more signals on the same path but at different times.

Time-domain analysis A method of representing a waveform by plotting its amplitude versus time.

Tinned Coated with a layer of tin or solder to prevent corrosion and simplify the soldering of connections.

Toggle switch Spring-loaded switch that is put in one of two positions, either on or off.

Tolerance Permissible deviation from a specified value, normally expressed as a percentage.

Tone A term describing both the bass and treble of a sound signal.

TO package Cylindrical, metal can type of package for some semiconductor components.

Toroidal coil Coil wound on a doughnut-shaped core.

Torque Moving force.

Totem pole circuit A transistor circuit containing two transistors connected one on top of the other with two inputs and one output.

Transconductance Also called mutual conductance, it is the ratio of a change in output current to the initiating change in input voltage.

Transducer Any device that converts energy from one form to another.

Transformer Device consisting of two or more coils that are used to couple electric energy from one circuit to another, yet maintain electrical isolation between the two.

Transformer coupling Also called inductive coupling, it is the coupling of two circuits by means of mutual inductance provided by a transformer.

Transient suppressor diode A device used to protect voltage sensitive electronic devices in danger of destruction by high energy voltage transients.

Transistance The characteristic achieved by a transistor that makes possible the control of voltages and currents so as to achieve gain or switching action.

Transistor (TRANSfer resISTOR) Semiconductor device having three main electrodes called the emitter, base, and collector that can be made to either amplify or rectify.

Transmission Sending of information.

Transmission line Conducting line used to couple signal energy between two points.

Transmitter Equipment used to achieve transmission.

Transorb Absorb transients. Another name for transient suppressor diode.

Treble High audio frequencies normally handled by a tweeter in a sound system.

TRIAC A bidirectional gate-controlled thyristor that provides full-wave control of ac power.

Triangular wave A repeating wave that has equal positive and negative ramps that have linear rates of change with time.

Triangular-wave generator A signal generator circuit that produces a continuously repeating triangular wave output.

Trigger Pulse used to initiate a circuit action.

Triggering Initiation of an action in a circuit which then functions for a predetermined time, for example, the duration of one sweep in a cathode-ray tube.

Trigonometry The study of the properties of triangles and trigonometric functions and of their applications.

Trimmer Small-value variable resistor, capacitor, or inductor.

Triode A three-electrode vacuum tube that has an anode, cathode, and control grid.

Troubleshooting The process of locating and diagnosing malfunctions or breakdowns in equipment by means of systematic checking or analysis.

Tune To adjust the resonance of a circuit so that it will select the desired frequency.

Tuned circuit Circuit that can have its components' values varied so that the circuit responds to one selected frequency yet heavily attenuates all other frequencies.

Tunnel diode A heavily doped junction diode that has negative resistance in the forward direction of its operating range due to quantum mechanical tunnelling.

Turns ratio Ratio of the number of turns in the secondary winding to the number of turns in the primary winding of a transformer.

Twin-T sine-wave oscillator An oscillator circuit that makes use of two T-shaped feedback networks.

Two phase Two repeating waveforms having a phase difference of 90°.

UHF Abbreviation for "ultrahigh frequency."

Ultrasonic Signals that are just above the range of human hearing of approximately 20 kHz.

Uncharged Having a normal number of electrons and therefore no electrical charge.

Unidirectional device A device that will conduct current in only one direction.

Unijunction transistor Abbreviated UJT, it is a P-N device that has an emitter connected to the P-N junction on one side of the bar, and two bases at either end of the bar. Used primarily as a switching device.

Upper sideband A group of frequencies that are equal to the sums of the carrier and modulation frequencies.

VA Abbreviation for "volt-ampere."

Vacuum tube Electron tube evacuated to such a degree that its electrical characteristics are essentially unaffected by the presence of residual gas or vapor. Eventually replaced by the transistor for amplification and rectification.

Valence shell Outermost shell formed by electrons.

Varactor diode A P-N semiconductor diode that is reverse biased to increase or decrease its depletion region width and vary device capacitance.

Variable Quantity that can be altered or controlled to assume a number of distinct values.

Variable capacitor Capacitor whose capacitance can be changed by varying the effective area of the plates or the distance between the plates.

Variable resistor *See* Rheostat *and* Potentiometer.

Variable-tuned RF amplifier A tuned radio frequency amplifier in which the tuned circuit(s) can be adjusted to select the desired station carrier frequency.

Variable-value capacitor A capacitor whose value can be varied.

Variable-value inductor An inductor whose value can be varied.

VCR Abbreviation for "videocassette recorder."

Vector Quantity that has both magnitude and direction. They are normally represented as a line, the length of which indicates magnitude and the orientation of which, due to the arrowhead on one end, indicates direction.

Vector diagram Arrangement of vectors showing the phase relationships between two or more ac quantities of the same frequency.

Vertical-channel E-MOSFET An enhancement type MOSFET that, when turned ON, form a vertical channel between source and drain.

Very high frequency (VHF) Electromagnetic frequency band from 30 to 300 MHz as set by the FCC.

Very low frequency (VLF) Frequency band from 3 to 30 kHz as set by the FCC.

Video Relating to any picture or visual information, from the Latin word meaning "I see."

Video amplifier An amplifier that has one or more transistor stages designed to amplify video signals.

Video signals A signal that contains visual information for television or radar systems.

Virtual ground A ground for voltage but not for current.

Voice coil Coil attached to the diaphragm of a moving coil speaker, which is moved through an air gap between the pole pieces of a permanent magnet.

Voice synthesizer Synthesizer that can simulate speech in any language by stringing together phonemes.

Volt (V) Unit of voltage, potential difference, or electromotive force. One volt is the force needed to produce 1 ampere of current in a circuit containing 1 ohm of resistance.

Voltage (*V or E*) Term used to designate electrical pressure or the force that causes current to flow.

Voltage amplifier An amplifier designed to build up the voltage of a signal.

Voltage divider Fixed or variable series resistor network that is connected across a voltage to obtain a desired fraction of the total voltage.

Voltage divider biasing A biasing method used with amplifiers in which a series arrangement of two fixed-value resistors are connected across the voltage source. The result is that a desired fraction of the total voltage is obtained at the center of the two resistors and is used to bias the amplifier.

Voltage drop Voltage or difference in potential developed across a component or conductor due to the loss of electric pressure as a result of current flow.

Voltage follower An operational amplifier that has a direct feedback to give unity gain so that the output voltage follows the input voltage. Used in applications where a very high input impedance and very low output impedance are desired.

Voltage gain Also called voltage amplification, it is equal to the difference between the output voltage level and the input signal voltage level. This value is normally expressed in decibels, which are equal to 20 times the logarithm of the ratio of the output voltage to the input voltage.

Voltage rating Maximum voltage a component can safely withstand without breaking down.

Voltage regulator A device that maintains the output voltage of a voltage source within required limits despite variations in input voltage and load resistance.

Voltage source A circuit or device that supplies voltage to a load circuit.

Voltaic cell Primary cell having two unlike metal electrodes immersed in a solution that chemically interacts with the plates to produce an emf.

Volt-ampere (VA) Unit of apparent power in an ac circuit containing reactance. Apparent power is equal to the product of voltage and current.

Voltmeter Instrument designed to measure the voltage or potential difference. Its scale can be graduated in kilovolts, volts, or millivolts.

Volume Magnitude or power level of a complex audio frequency (AF) wave, expressed in volume units (VU).

VOM meter Abbreviation for volt-ohm-milliamp meter.

VRMS Abbreviation for "volts root-mean-square."

Wall outlet Spring-contact outlet mounted on the wall to which a portable appliance is connected to obtain electric power.

Watt (W) Unit of electric power required to do work at a rate of 1 joule/second. One watt of power is expended when 1 ampere direct current flows through a resistance of 1 ohm. In an ac circuit, the true power is effective volts multiplied by effective amperes, multiplied by the power factor.

Wattage rating Maximum power a device can safely handle continuously.

Watt-hour (Wh) Unit of electrical work, equal to a power of 1 watt being absorbed continuously for 1 hour.

Wattmeter A meter used to measure electric power in watts.

Wave Electric, electromagnetic, acoustic, mechanical, or other form whose physical activity rises and falls or advances and retreats periodically as it travels through some medium.

Waveform Shape of a wave.

Waveguide Rectangular or circular metal pipe used to guide electromagnetic waves at microwave frequencies.

Wavelength (λ) Distance between two points of corresponding phase and is equal to waveform velocity or speed divided by frequency.

Weber (Wb) Unit of magnetic flux. One weber is the amount of flux that when linked with a single turn of wire for an interval of 1 second, will induce an electromotive force of 1 V.

Wet cell Cell using a liquid electrolyte.

Wetting The coating of a contact surface.

Wheatstone bridge A four-arm, generally resistive bridge that is used to measure resistance.

Wideband amplifier Also called broadband amplifier, it is an amplifier that has a flat response over a wide range of frequencies.

Wien-Bridge oscillator An *RC* oscillator circuit in which a Wien Bridge determines frequency.

Winding One or more turns of a conductor wound to form a coil.

Wire Single solid or stranded group of conductors having a low resistance to current flow.

Wire gauge American wire gauge (AWG) is a system of numerical designations of wire sizes, with the first being 0000 (the largest size) and then going to 000, 00, 0, 1, 2, 3, and so on up to the smallest sizes of 40 and above.

Wireless Term describing radio communication that requires no wires between the two communicating points.

Wirewound resistor Resistor in which the resistive element is a length of high-resistance wire or ribbon, usually nichrome, wound onto an insulating form.

Wire-wrapping Method of prototyping in which solderless connections are made by wrapping wire around a rectangular terminal.

Woofer Large loudspeaker designed primarily to reproduce low audio-frequency signals at large power levels.

Work Work is done anytime energy is transformed from one type to another, and the amount of work done is dependent on the amount of energy transformed.

X Symbol for reactance.

X axis Horizontal axis.

Y Symbol for admittance.

Y axis Vertical axis.

Z axis Axis perpendicular to both the X and Y axes.

Zener diode A semiconductor diode in which a reverse breakdown voltage current causes the diode to develop a constant voltage.

Zeroing Calibrating a meter so that it shows a value of zero when zero is being measured.

A (amp) Ampere
AC Alternating current
AC/DC Alternating current or direct current
A/D Analog to digital
ADC Analog-to-digital converter
AF Audio frequency
AFC Automatic frequency control
aft Automatic fine tuning
AGC Automatic gain control
Ah Ampere-hour
A_i Current gain
AM Amplitude modulation
AM/FM Amplitude modulation or frequency modulation
AMM Analog multimeter
antilog Antilogarithm
A_P Power gain
apc Automatic phase control
A_V Voltage gain
AVC Automatic volume control
AWG American wire gauge
B Flux density
BCD Binary coded decimal
bfo Beat frequency oscillator
BJT Bipolar junction transistor
BW Bandwidth
c Centi (10^{-2})
C Capacitance; capacitor
CAD Computer-aided design
CAM Computer-aided manufacturing
CATV Cable TV
CB Common-base configuration
CB Citizen's band
CC Common-collector configuration
CE Common-emitter configuration
cm Centimeter
cmil (cir mil) Circular mil
CPU Central processing unit

C (Q) Coulomb
CR, cr Junction diode
CRO Cathode ray oscilloscope
CRT Cathode ray tube
C_T Total capacitance
cw Continuous wave transmission
d Deci (10^{-1})
D/A or D-A Digital to analog
DC Direct current
di or Δi Change in current
DIP Dual in-line package
DMM Digital multimeter
DPDT Double pole, double throw
DPST Double pole, single throw
dt or Δt Change in time
DTL Diode-transistor logic
dv or Δr Change in voltage
DVM Digital voltmeter
E DC or rms difference in potential
e Instantaneous difference in potential
ECG Electrocardiogram
ECL Emitter-coupled logic
EHF Extremely high frequency
EHV Extra high voltage
ELF Extremely low frequency
EMF Electromotive force
EMI Electromagnetic interference
EW Electronic warfare
f Frequency
FET Field-effect transistor
FF Flip-flop
fil Filament
FM Frequency modulation
4PDT Four pole, double throw
4PST Four pole, single throw
f_r Frequency at resonance
FSD Full-scale deflection
fsk Frequency-shift keying

G Gravitational force
G Conductance
G Giga (10^9)
H Henry
H Magnetic field intensity
H Magnetizing flux
h Hecto (10^2)
h Hybrid
HF High frequency
hp Horsepower
Hz Hertz
I Current
i Instantaneous current
I_B DC base current
I_C DC collector current
IC Integrated circuit
I_E Total emitter current
I_{eff} Effective current
IF Intermediate frequency
I_{max} Maximum current
I_{min} Minimum current
I/O Input/output
IR Infrared
I_R Current through resistance
I_S Secondary current
I_T Total current
JFET Junction FET
K Coefficient of coupling
k Kilo (10^3)
kHz Kilohertz
kV Kilovolt
kVA Kilovoltampere
kW Kilowatt
kWh Kilowatt-hour
L Coil; inductance
LC Inductance–capacitance
LCD Liquid crystal display
L-C-R Inductance–capacitance–resistance
LDR Light-dependent resistor
LED Light-emitting diode
LF Low frequency
L_M Mutual inductance
LNA Low-noise amplifier
LO Local oscillator
LSI Large-scale integration
L_T Total inductance

M Mega (10^6)
M Mutual conductance
M Mutual inductance
m Milli (10^{-3})
mA Milliampere
mag Magnetron
max Maximum
MF Medium frequency
mH Millihenry
MHz Megahertz
min Minimum
mm Millimeter
mmf Magnetomotive force
MOS Metal oxide semiconductor
MOSFET Metal-oxide semiconductor field effect transistor
MPU Microprocessor unit (µP)
MSI Medium scale integrated circuit
mV Millivolt
mW Milliwatt
N Number of turns in an inductor
N Revolutions per minute
n Nano (10^{-9})
N Negative
nA Nanoampere
NC No connection
NC Normally closed
NEG, neg Negative
nF Nanofarad
nH Nanohenry
nm Nanometer
NO Normally open
NPN Negative–positive–negative
ns Nanosecond
nW Nanowatt
OP AMP Operational amplifier
P Pico (1×10^{-12})
P Power
p Instantaneous power
P Positive; peak
PA Public address or power amplifier
pA Picoampere
PAM, pam Pulse-amplitude modulation
P_{ap} Apparent power

P_{av} Average power
PCB Printed circuit board
PCM, pcm Pulse-code modulation
PDM Pulse-duration modulation
pF Picofarad
PLL Phase-locked loop
PM Phase modulation; permanent magnet
PNP Positive–negative–positive
POT, pot Potentiometer
P-P Peak to peak
PPM Pulse-position modulation
PRF Pulse repetition frequency
PRT Pulse repetition time
PW Pulse width
PWM, pwm Pulse-width modulation
Q Charge or quality
q Instantaneous charge
R Resistance
R Reluctance
RAM Random-access memory
RC Resistance-capacitance
rcvr Receiver
rect Rectifier
ref Reference
rf Radio frequencies
RF Radio frequency
RFI Radio frequency interference
R_L Load resistor
RLC Resistance–inductance–capacitance
RMS Root mean square
ROM Read-only memory
rpm Revolutions per minute
SCR Silicon controlled rectifier
SHF Superhigh frequency
SIP Single in-line package
SNR Signal-to-noise ratio
SPDT Single pole, double throw
SPST Single pole, single throw
sq cm Square centimeter
SSB Single sideband
SW Shortwave
SWR Standing-wave ratio
SYNC Synchronous
T Tera (10^{12})
T Torque

T Transformer
t Time (seconds)
TC Time constant; temperature coefficient
TE Transverse electric
temp Temperature
THz Teraherz
TM Transverse magnetic
TR Transmit–receive
TTL Transistor–transistor logic
TV Television
TWT Travelling wave tube
UHF Ultrahigh frequency
UHV Ultrahigh voltage
UJT Unijunction transistor
UV Ultraviolet
V Vacuum tube
V, v Volt
v Instantaneous voltage
VA Voltampere
V_{av} Voltage (average value)
V_{BE} Fixed base-emitter voltage
V_c Capacitive voltage
V_{CE} Fixed collector-emitter voltage
VCO Voltage-controlled oscillator
VHF Very high frequency
V_{in} Input voltage
V_L Inductive voltage
VLF Very low frequency
V_m, V_{max} Maximum voltage
VOM Volt-ohm-milliammeter
V_{out} Output voltage
V_p Primary voltage
V_s Source voltage
VSWR Voltage standing wave ratio
V_T Total voltage
W Watt
X_C Capacitive reactance
X_L Inductive reactance
Y Admittance
Z Impedance
Z_{in} Input impedance
Z_o Output impedance
Z_p Primary impedance
Z_s Secondary impedance
Z_T Total impedance
kΩ Kilohm

$M\Omega$ Megohm

μA Microampere

μF **(mfd)** Microfarad

μH Microhenry

μV Microvolt

μW Microwatt

Ω Ohm

λ Wavelength

° Degrees

°C Degrees Centigrade

°F Degrees Fahrenheit

APPENDIX

C
The Scientific Calculator

Problem solving is an integral part of your electronics study, for it is only through the solutions of problems that decisions are tested and verified.

C.1 CALCULATOR SELECTION

It is very important that your "full-function" scientific or engineering calculator have the capabilities necessary for your particular application. For this reason, Table C-1 lists all the functions that your calculator should have, with some blank columns so that you can check off these functions against several different types.

Calculators use a keypad to enter the data to be processed and a display to show the result. There are two systems of data entry:

1. *Algebraic entry system* (AES): Here the basic arithmetic operations ($+, -, \div, \times$, and so on) are placed between the two numbers; for example, $8 + 4$.

2. *Reverse Polish notation* (RPN): This system is based on the work of Jan Lukasiewicz, a Polish mathematician. With this system, the previous example would be entered as 8 enter, 4 enter, and then +.

The programmable calculator is another feature you may want to consider, and once this tool is mastered you will find it invaluable in applications where you have to repeatedly solve a function for many different values.

Before you purchase a calculator, talk to your instructor and other students, and shop around to find out what works best for you.

TABLE C-1

KEYPAD SYMBOLS	FUNCTION	DESCRIPTION							
$+$	Add								
$-$	Subtract	Simple arithmetic							
$\times$	Multiply	operations							
$\div$	Divide								
CHS or $+/-$	Change sign								
y^x or x^y	Exponential	Raise number to power							
$1/x$	Reciprocal	Calculate the reciprocal							
$\sqrt{x}$	Square root	Calculate square root							
x^2	Square x	Square a number							
FIX	Fix point notation								
SCI	Scientific notation	Display and rounding							
ENG	Engineering notation								
STO	Store	Memory							
RCL	Recall								
$x\blacktriangle y$ or EXC	Exchange x and y								
EE or EEX	Enter exponent								
log	Common logarithm	Logarithmic							
10^x or INV LOG	Common antilog	and							
ln x	Natural logarithm	exponential							
e^x or INV LN	Natural antilogarithm	functions							
SIN	Sine	Trigonometric							
COS	Cosine	functions							
TAN	Tangent								
SIN⁻¹ or INV SIN	Arc sine	Inverse							
COS⁻¹ or INV COS	Arc cosine	trigonometric							
TAN⁻¹ or INV TAN	Arc tangent	functions							
DEG or DRG	Degree	Angular mode							
RAD	Radian	selection							
$\rightarrow$P	Rectangular to polar	Polar/rectangular							
$\rightarrow$R	Polar to rectangular	coordinate conversion							

Electronic Schematic Symbols

RESISTORS

fixed-value resistor

variable resistor

voltage-sensitive resistor
(varistor)

SOURCES

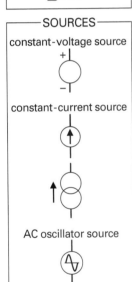

constant-voltage source

constant-current source

AC oscillator source

BATTERIES

single-cell battery

multiple-cell battery

CIRCUIT PROTECTORS

fuse

circuit breaker

CAPACITORS

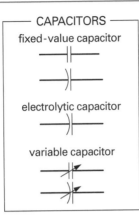

fixed-value capacitor

electrolytic capacitor

variable capacitor

CRYSTALS

piezoelectric crystal

LAMPS

incandescent lamp

signal lamp

flashing signal lamp

neon lamp, DC type

neon lamp, AC type

fluorescent lamp,
two-terminal

GROUND

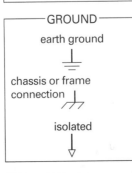

earth ground

chassis or frame
connection

isolated

INDUCTORS

fixed-value inductor

fixed-value inductor
with magnetic core

variable inductor

AUDIO DEVICES

loudspeaker

Microphone

AMPLIFIERS

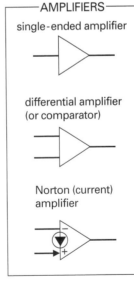

single-ended amplifier

differential amplifier
(or comparator)

Norton (current)
amplifier

WAVEGUIDE

circular waveguide

rectangular waveguide

flexible waveguide

twisted waveguide

TRANSFORMERS

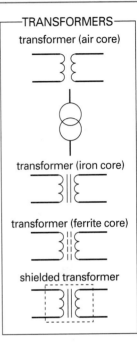

transformer (air core)

transformer (iron core)

transformer (ferrite core)

shielded transformer

CABLE

Two-conductor cable
with grounded shield

coaxial cable with
grounded shield

twisted pair

STRIPLINE

unbalanced stripline

balanced stripline

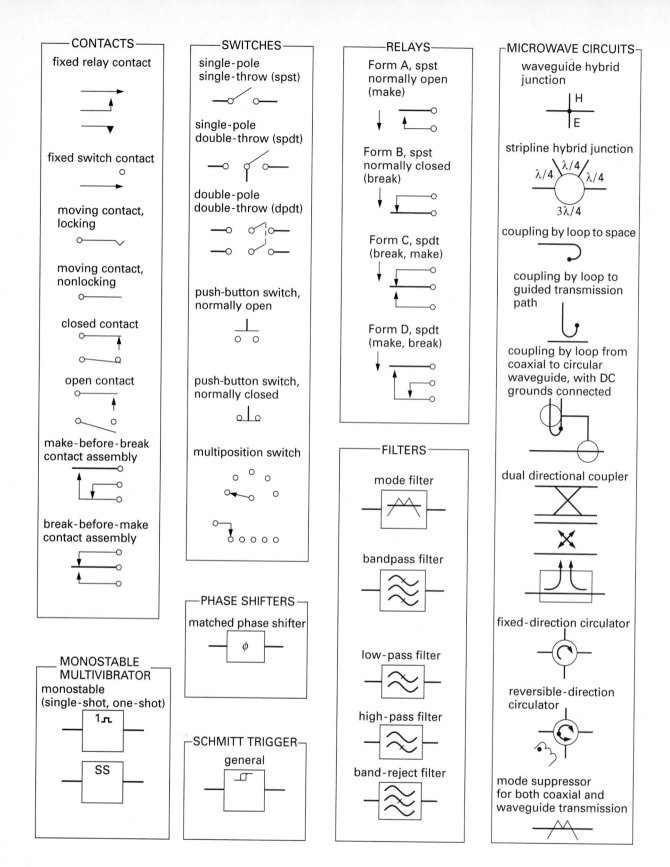

CONTACTS

fixed relay contact

fixed switch contact

moving contact, locking

moving contact, nonlocking

closed contact

open contact

make-before-break contact assembly

break-before-make contact assembly

SWITCHES

single-pole single-throw (spst)

single-pole double-throw (spdt)

double-pole double-throw (dpdt)

push-button switch, normally open

push-button switch, normally closed

multiposition switch

PHASE SHIFTERS

matched phase shifter

MONOSTABLE MULTIVIBRATOR

monostable (single-shot, one-shot)

1 ⊓

SS

SCHMITT TRIGGER

general

RELAYS

Form A, spst normally open (make)

Form B, spst normally closed (break)

Form C, spdt (break, make)

Form D, spdt (make, break)

FILTERS

mode filter

bandpass filter

low-pass filter

high-pass filter

band-reject filter

MICROWAVE CIRCUITS

waveguide hybrid junction

H

E

stripline hybrid junction

$\lambda/4$ $\lambda/4$ $\lambda/4$

$3\lambda/4$

coupling by loop to space

coupling by loop to guided transmission path

coupling by loop from coaxial to circular waveguide, with DC grounds connected

dual directional coupler

fixed-direction circulator

reversible-direction circulator

mode suppressor for both coaxial and waveguide transmission

DIODES

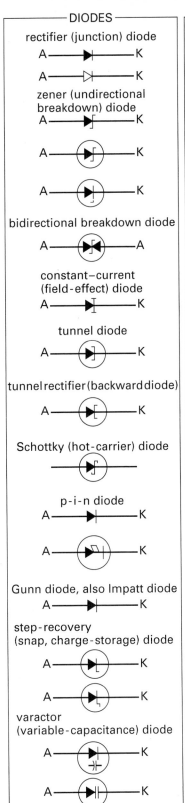

rectifier (junction) diode

A ——▶|—— K

A ——▷|—— K

zener (undirectional breakdown) diode

A ——▶|—— K

A ——(▶|)—— K

A ——(▶|)—— K

bidirectional breakdown diode

A ——(▶◀)—— A

constant−current (field-effect) diode

A ——▶|—— K

tunnel diode

A ——(▶|)—— K

tunnel rectifier (backward diode)

A ——(▶|)—— K

Schottky (hot-carrier) diode

——(▶|)——

p-i-n diode

A ——▶|—— K

A ——(▶|)—— K

Gunn diode, also Impatt diode

A ——▶|—— K

step-recovery (snap, charge-storage) diode

A ——(▶|)—— K

A ——(▶|)—— K

varactor (variable-capacitance) diode

A ——(▶|)—— K

A ——(▶||)—— K

TRANSISTORS

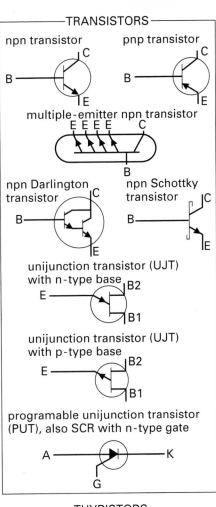

npn transistor

pnp transistor

multiple-emitter npn transistor

npn Darlington transistor

npn Schottky transistor

unijunction transistor (UJT) with n-type base

unijunction transistor (UJT) with p-type base

programable unijunction transistor (PUT), also SCR with n-type gate

A ———(▶|)——— K

G

THYRISTORS

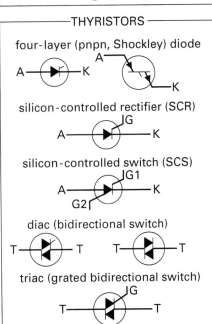

four-layer (pnpn, Shockley) diode

silicon-controlled rectifier (SCR)

silicon-controlled switch (SCS)

diac (bidirectional switch)

triac (grated bidirectional switch)

FIELD-EFFECT TRANSISTORS (FETs)

n-channel   **p-channel**

junction-gate (JFET)

three-terminal depletion-type insulated-gate (IGFET)

three-terminal depletion-type IGFET, substrate tied to source

four-terminal depletion-type IGFET

four-terminal enhancement-type IGFET

five-terminal dual-gate depletion-type IGFET

five-terminal dual-gate enhancement-type IGFET

OPTOELECTRONIC DIODES

light-emitting diode (LED)

A ———— K

photodiode

A ———— K

npn bidirectional photodiode
(photo-duo-diode)

T ———— T

pnp bidirectional photodiode
(photo-duo-diode)

T ———— T

pnp two-segment
photodiode,
with common cathode

A ———— A

pnp four-quadrant
photodiode,
with common cathode

A
A ———— A
K
A

PHOTOTRANSISTORS

npn phototransistor,
no base connection

C
E

npn phototransistor,
with base connection

C
B
E

OPTICALLY COUPLED
ISOLATORS

with photodiode output

with phototransistor output,
no base connection

with phototransistor output,
and base connection

with photo-Darlington output,
no base

with photo-Darlington
output, and base

with photodiode and amplifier-
transistor output

with NAND-gate-photo-
detector output

GATES

dual-input   multiple-input

AND gate

&   &

NAND gate, negated output

&   &

NAND gate, negated inputs

&   &

OR gate

≥1   ≥1

NOR gate, negated output

≥1   ≥1

NOR gate, negated inputs

≥1   ≥1

exclusive-OR gate

= 1

inverter gate

1

FLIP-FLOPS

R-S (set-reset)
flip-flop

R
FF
S

R   Q
FF
S   Q̄

toggle-flip-flop

FF
T

T FF   Q
Q̄

J-K flip-flop

J
C FF
K

J   Q
C FF
K   Q̄

D-type flip-flop

D
FF
C

D   Q
FF
CLK   Q̄

E
Conversions, Constants, Units, and Prefixes

Greek Alphabet

NAME	CAPITAL	LOWERCASE	DESIGNATES
Alpha	A	α	Angles, coefficients, attenuation constant, absorption factor, area.
Beta	B	β	Angles, coefficients, phase constant.
Gamma	Γ	γ	Specific quantity, angles, electrical conductivity, propagation constant, complex propagation constant (cap).
Delta	Δ	δ	Density, angles, increment or decrement (cap or small), determinant (cap), permittivity (cap).
Epsilon	E	ε	Dielectric constant, permittivity, base of natural (Napierian) logarithms, electric intensity.
Zeta	Z	ζ	Coordinate, coefficients.
Eta	H	η	Intrinsic impedance, efficiency, surface charge density, hysteresis, coordinates.
Theta	Θ	Θ	Reluctance, phase angle.
Iota	I	ι	Unit vector.
Kappa	K	κ	Susceptibility, coupling coefficient.
Lambda	Λ	λ	Wavelength, attenuation constant, permeance (cap).
Mu	M	μ	Prefix *micro-*, permeability, amplification factor.
Nu	N	ν	Reluctivity, frequency.
Xi	Ξ	ξ	Coordinates.
Omicron	O	o	—
Pi	Π	π	3.1416 (circumference divided by diameter).
Rho	P	ρ	Resistivity, volume charge density, coordinates.
Sigma	Σ	σ	Surface charge density, complex propagation constant, electrical conductivity, leakage coefficient, sign of summation (cap).
Tau	T	τ	Time constant, volume resistivity, time-phase displacement, transmission factor, density.
Upsilon	γ	υ	—
Phi	Φ	ϕ	Magnetic flux, angles, scalar potential (cap).
Chi	X	χ	Electric susceptibility, angles.
Psi	Ψ	ψ	Dielectric flux, phase difference, coordinates, angles.
Omega	Ω	ω	Angular velocity ($2\pi f$), resistance in ohms (cap), solid angles (cap).

Symbols and Units

QUANTITY	SYMBOL	UNIT	SYMBOL
Charge	Q	coulomb	C
Current	I	ampere	A
Voltage, potential difference	V	volt	V
Electromotive force	δ	volt	V
Resistance	R	ohm	Ω
Conductance	G	mho (siemens)	A/V, or mho (S)
Reactance	X	ohm	Ω
Susceptance	B	mho	A/V, or mho
Impedance	Z	ohm	Ω
Admittance	Y	mho	A/V, or mho
Capacitance	C	farad	F
Inductance	L	henry	H
Energy, work	W	joule	J
Power	P	watt	W
Resistivity	ρ	ohm-meter	Ωm
Conductivity	σ	mho per meter	mho/m
Electric displacement	D	coulomb per square meter	C/m^2
Electric field strength	E	volt per meter	V/m
Permittivity (absolute)	ε	farad per meter	F/m
Relative permittivity	ε_r	(numeric)	
Magnetic flux	Φ	weber	Wb
Magnetomotive force	$\mathfrak{F}$	ampere (ampere-turn)	A
Reluctance	$\mathscr{R}$	ampere per weber	A/Wb
Permeance	$\mathscr{P}$	weber per ampere	Wb/A
Magnetic flux density	B	tesla	T
Magnetic field strength	H	ampere per meter	A/m
Permeability (absolute)	μ	henry per meter	H/m
Relative permeability	μ_r	(numeric)	
Length	l	meter	m
Mass	m	kilogram	kg
Time	t	second	s
Frequency	f	hertz	Hz
Angular frequency	ω	radian per second	rad/s
Force	F	newton	N
Pressure	p	newton per square meter	N/m^2
Temperature (absolute)	T	Kelvin	K
Temperature (International)	t	degree Celsius	°C

SI Symbols and Units

UNIT	SYMBOL	UNIT	SYMBOL	UNIT	SYMBOL
ampere	A	kelvin	K	milliliter	mL
ampere-hour	Ah	kiloampere	kA	millimeter	mm
ampere per meter	A/m	kilobaud	kBd	milliohm	mΩ
angstrom	Å	kilobit	kb	millirad	mrd
attoampere	aA	kiloelectronvolt	keV	milliradian	mrad
bar	bar	kilogauss	kG	milliroentgen	mR
barn	b	kilogram	kg	millisecond	ms
baud	Bd	kilohertz	kHz	millisiemens	mS
bel	B	kilohm	kΩ	millivolt	mV
bit	b	kilometer	km	milliwatt	mW
bit per second	b/s	kilometer per hour	km/h	minute (time)	min
calorie	cal	kilovolt	kV	nanoampere	nA
candela	cd	kilovoltampere	kVA	nanofarad	nF
candela per square meter	cd/m^2	kilowatt	kW	nanohenry	nH
		kilowatthour	kWh	nanometer	nm
centimeter	cm	lambert	L	nanosecond	ns
circular mil	cmil	liter	L	nanovolt	nV
coulomb	C	lumen	lm	nanowatt	nW
cubic centimeter	cm^3	lumen per square foot	lm/ft^2	nautical mile	nmi
cubic foot per minute	ft^3/min	lumen per square meter	lm/m^2	neper	Np
cubic meter	m^3			oersted	Oe
cubic meter per second	m^3/s	lumen per watt	lm/W	ohm	Ω
curie	Ci	lumen second	lm·s	ohm per volt	Ω/V
decibel	dB	lux	lx	picoampere	pA
degree (plane angle)	°	maxwell	Mx	picofarad	pF
degree Celsius	°C	megabar	Mbar	picosecond	ps
degree Fahrenheit	°F	megabit	Mb	picowatt	pW
degree Rankine	°R	megaelectronvolt	MeV	pound	lb
dyne	dyn	megahertz	MHz	rad	rd
electronvolt	eV	megampere	MA	radian	rad
erg	erg	megavolt	MV	revolution per minute	r/min
farad	F	megawatt	MW	revolution per second	r/s
femtoampere	fA	megawatt-hour	MWh	roentgen	R
femtovolt	fV	megohm	MΩ	second (time)	s
femtowatt	fW	meter	m	siemens	S
foot	ft	microampere	μA	square foot	ft^2
foot per minute	ft/min	microbar	μbar	square inch	in.^{2}
foot per second	ft/s	microfarad	μF	square meter	m^2
gauss	G	microhenry	μH	square mile	mi^2
gigabit	Gb	microhm	$\mu\Omega$	square yard	yd^2
gigaelectronvolt	GeV	micrometer	μm	steradian	sr
gigahertz	GHz	microsecond	μs	teraelectronvolt	TeV
gigaohm	GΩ	microsiemens	μS	terahertz	THz
gigawatt	GW	microvolt	μV	teraohm	TΩ
gilbert	Gb	microwatt	μW	tesla	T
gram	g	mil	mil	var	var
henry	H	mile	mi	volt	V
hertz	Hz	mile per hour	mi/h	volt-ampere	VA
horsepower	hp	milliampere	mA	watt	W
hour	h	millibar	mbar	watt-hour	Wh
inch	in.	milligauss	mG	watt per steradian	W/sr
inch per second	in./s	milligram	mg	weber	Wb
joule	J	millihenry	mH	yard	yd

Metric Conversion Chart (Approximations)

WHEN YOU KNOW	MULTIPLY BY	TO FIND
Length		
millimeters (mm)	0.04	inches (in.)
centimeters (cm)	0.4	inches (in.)
meters (m)	3.3	feet (ft)
meters (m)	1.1	yards (yd)
kilometers (km)	0.6	miles (mi)
Area		
sq. centimeters (cm^2)	0.16	sq. inches ($in.^2$)
sq. meters (m^2)	1.2	sq. yards (yd^2)
sq. kilometers (km^2)	0.4	sq. miles (mi^2)
hectares (ha) ($10\,000\ m^2$)	2.5	acres
Mass (weight)		
grams (g)	0.035	ounces (oz)
kilograms (kg)	2.2	pounds (lb)
tonnes ($1\,000$ kg) (t)	1.1	short tons
Volume		
milliliters (mL)	0.03	fluid ounces (fl oz)
liters (L)	2.1	pints (pt)
liters (L)	1.06	quarts (qt)
liters (L)	0.26	gallons (gal)
cubic meters (m^3)	35	cubic feet (ft^3)
cubic meters (m^3)	1.3	cubic yards (yd^3)
Temperature (exact)		
Celsius (°C)	$9/5(°C) + 32$	Fahrenheit (°F)
Temperature (exact) to Metric		
Fahrenheit (°F)	$5/9(°F - 32)$	Celsius (°C)
Length		
inches (in.)	*2.5	centimeters (cm)
feet (ft)	30	centimeters (cm)
yards (yd)	0.9	meters (m)
miles (mi)	1.6	kilometers (km)
Area		
sq. inches ($in.^2$)	6.5	sq. centimeters (cm^2)
sq. feet (ft^2)	0.09	sq. meters (m^2)
sq. yards (yd^2)	0.8	sq. meters (m^2)
sq. miles (mi^2)	2.6	sq. kilometers (km^2)
acres	0.4	hectares (ha)

*in. = 2.54 cm exactly

WHEN YOU KNOW	MULTIPLY BY	TO FIND
	Mass (weight)	
ounces (oz)	28	grams (g)
pounds (lb)	0.45	kilograms (kg)
short tons (2 000 lb)	0.9	tonnes (t)
	Volume	
teaspoons (tsp)	5	milliliters (mL)
tablespoons (tbsp)	15	milliliters (mL)
fluid ounces (fl oz)	30	milliliters (mL)
cups (c)	0.24	liters (L)
pints (pt)	0.47	liters (L)
quarts (qt)	0.95	liters (L)
gallons (gal)	3.8	liters (L)
cubic feet (ft^3)	0.03	cubic meters (m^3)
cubic yards (yd^3)	0.76	cubic meters (m^3)

MULTIPLICATION FACTOR	PREFIX	SYMBOL
$1\ 000\ 000\ 000\ 000\ 000\ 000 = 10^{18}$	exa	E
$1\ 000\ 000\ 000\ 000\ 000 = 10^{15}$	peta	P
$1\ 000\ 000\ 000\ 000 = 10^{12}$	tera	T
$1\ 000\ 000\ 000 = 10^{9}$	giga	G
$1\ 000\ 000 = 10^{6}$	mega	M
$1\ 000 = 10^{3}$	kilo	k
$100 = 10^{2}$	hecto	h
$10 = 10$	deka	da
$0.1 = 10^{-1}$	deci	d
$0.01 = 10^{-2}$	centi	c
$0.001 = 10^{-3}$	milli	m
$0.000\ 001 = 10^{-6}$	micro	μ
$0.000\ 000\ 001 = 10^{-9}$	nano	n
$0.000\ 000\ 000\ 001 = 10^{-12}$	pico	p
$0.000\ 000\ 000\ 000\ 001 = 10^{-15}$	femto	f
$0.000\ 000\ 000\ 000\ 000\ 001 = 10^{-18}$	atto	a

Standard Resistor Values

RESISTANCE TOLERANCE

0.1% 0.25% 0.5%	1%	2% 5%	10%	0.1% 0.25% 0.5%	1%	2% 5%	10%	0.1% 0.25% 0.5%	1%	2% 5%	10%
10.0	10.0	10	10	17.6	—	—	—	30.9	30.9	—	—
10.1	—	—	—	17.8	17.8	—	—	31.2	—	—	—
10.2	10.2	—	—	18.0	—	18	18	31.6	31.6	—	—
10.4	—	—	—	18.2	18.2	—	—	32.0	—	—	—
10.5	10.5	—	—	18.4	—	—	—	32.4	32.4	—	—
10.6	—	—	—	18.7	18.7	—	—	32.8	—	—	—
10.7	10.7	—	—	18.9	—	—	—	33.2	33.2	33	33
10.9	—	—	—	19.1	19.1	—	—	33.6	—	—	—
11.0	11.0	11	—	19.3	—	—	—	34.0	34.0	—	—
11.1	—	—	—	19.6	19.6	—	—	34.4	—	—	—
11.3	11.3	—	—	19.8	—	—	—	34.8	34.8	—	—
11.4	—	—	—	20.0	20.0	20	—	35.2	—	—	—
11.5	11.5	—	—	20.3	—	—	—	35.7	35.7	—	—
11.7	—	—	—	20.5	20.5	—	—	36.1	—	36	—
11.8	11.8	—	—	20.8	—	—	—	36.5	36.5	—	—
12	—	12	12	21.0	21.0	—	—	37.0	—	—	—
12.1	12.1	—	—	21.3	—	—	—	37.4	37.4	—	—
12.3	—	—	—	21.5	21.5	—	—	37.9	—	—	—
12.4	12.4	—	—	21.8	—	—	—	38.3	38.3	—	—
12.6	—	—	—	22.1	22.1	22	22	38.8	—	—	—
12.7	12.7	—	—	22.3	—	—	—	39.2	39.2	39	39
12.9	—	—	—	22.6	22.6	—	—	39.7	—	—	—
13.0	13.0	13	—	22.9	—	—	—	40.2	40.2	—	—
13.2	—	—	—	23.2	23.2	—	—	40.7	—	—	—
13.3	13.3	—	—	23.4	—	—	—	41.2	41.2	—	—
13.5	—	—	—	23.7	23.7	—	—	41.7	—	—	—
13.7	13.7	—	—	24.0	—	24	—	42.2	42.2	—	—
13.8	—	—	—	24.3	24.3	—	—	42.7	—	—	—
14.0	14.0	—	—	24.6	—	—	—	43.2	43.2	43	—
14.2	—	—	—	24.9	24.9	—	—	43.7	—	—	—
14.3	14.3	—	—	25.2	—	—	—	44.2	44.2	—	—
14.5	—	—	—	25.5	25.5	—	—	44.8	—	—	—
14.7	14.7	—	—	25.8	—	—	—	45.3	45.3	—	—
14.9	—	—	—	26.1	26.1	—	—	45.9	—	—	—
15.0	15.0	15	15	26.4	—	—	—	46.4	46.4	—	—
15.2	—	—	—	26.7	26.7	—	—	47.0	—	47	47
15.4	15.4	—	—	27.1	—	27	27	47.5	47.5	—	—
15.6	—	—	—	27.4	27.4	—	—	48.1	—	—	—
15.8	15.8	—	—	27.7	—	—	—	48.7	48.7	—	—
16.0	—	16	—	28.0	28.0	—	—	49.3	—	—	—
16.2	16.2	—	—	28.4	—	—	—	49.9	49.9	—	—
16.4	—	—	—	28.7	28.7	—	—	50.5	—	—	—
16.5	16.5	—	—	29.1	—	—	—	51.1	51.1	51	—
16.7	—	—	—	29.4	29.4	—	—	51.7	—	—	—
16.9	16.9	—	—	29.8	—	—	—	52.3	52.3	—	—
17.2	—	—	—	30.1	30.1	30	—	53.0	—	—	—
17.4	17.4	—	—	30.5	—	—	—	53.6	53.6	—	—

0.1% 0.25% 0.5%	1%	2% 5%	10%	0.1% 0.25% 0.5%	1%	2% 5%	10%	0.1% 0.25% 0.5%	1%	2% 5%	10%
54.2	—	—	—	66.5	66.5	—	—	81.6	—	—	—
54.9	54.9	—	—	67.3	—	—	—	82.5	82.5	82	82
56.6	—	—	—	68.1	68.1	68	68	83.5	—	—	—
56.2	56.2	56	56	69.0	—	—	—	84.5	84.5	—	—
56.9	—	—	—	69.8	69.8	—	—	85.6	—	—	—
57.6	57.6	—	—	70.6	—	—	—	86.6	86.6	—	—
58.3	—	—	—	71.5	71.5	—	—	87.6	—	—	—
59.0	59.0	—	—	72.3	—	—	—	88.7	88.7	—	—
59.7	—	—	—	73.2	73.2	—	—	89.8	—	—	—
60.4	60.4	—	—	74.1	—	—	—	90.9	90.9	91	—
61.2	—	—	—	75.0	75.0	75	—	92.0	—	—	—
61.9	61.9	62	—	75.9	—	—	—	93.1	93.1	—	—
62.6	—	—	—	76.8	76.8	—	—	94.2	—	—	—
63.4	63.4	—	—	77.7	—	—	—	95.3	95.3	—	—
64.2	—	—	—	78.7	78.7	—	—	96.5	—	—	—
64.9	64.9	—	—	79.6	—	—	—	97.6	97.6	—	—
65.7	—	—	—	80.6	80.6	—	—	98.8	—	—	—

(Available in Multiples of 0-1, 10, 100, 1 k, and 1 m

F
Codes

Numeration

ARABIC	ROMAN	GREEK	BINARY NUMBERS	ARABIC	ROMAN	GREEK	BINARY NUMBERS
1	I	α'	1	50	L	ν'	110010
2	II	β'	10	60	LX	ξ	111100
3	III	γ'	11	70	LXX	ο'	1000110
4	IV	δ'	100	80	LXXX	π'	1010000
5	V	ε'	101	90	XC	ϙ'	1011010
6	VI	ς'	110	99	XCIX	ϙθ'	1100011
7	VII	ζ'	111	100	C	ρ'	1100100
8	VIII	η'	1000	111	CXI	ρια'	1101111
9	IX	θ'	1001	150	CL	ρν'	10010110
10	X	ι'	1010	200	CC	σ'	11001000
11	XI	ια'	1011	300	CCC	τ'	100101100
12	XII	ιβ'	1100	400	CD	ν'	110010000
13	XIII	ιγ'	1101	500	D	φ'	111110100
14	XIV	ιδ'	1110	600	DC	χ'	1001011000
15	XV	ιε'	1111	700	DCC	ψ'	1010111100
16	XVI	ις'	10000	800	DCCC	ω'	1100100000
17	XVII	ιζ'	10001	900	CM	ϡ'	1110000100
18	XVIII	ιη'	10010	1,000	M	,α	1111101000
19	XIX	ιθ'	10011	1,111	MCXI	,αρια'	10001010111
20	XX	κ'	10100	1,500	MD	,αφ'	10111011100
21	XXI	κα'	10101	2,000	MM	,β	11111010000
22	XXII	κβ'	10110	3,000	MMM	,γ	101110111000
23	XXIII	κγ'	10111	4,000	MV̄	,δ	111110100000
24	XXIV	κδ'	11000	5,000	V̄	,ε	1001110001000
25	XXV	κε'	11001	6,000	V̄M	,ς	1011101110000
29	XXIX	κθ'	11101	10,000	X̄	,ι	10011100010000
30	XXX	λ'	11110	20,000	X̄X	,κ	100111000100000
40	XL	μ'	101000	100,000	C̄	,ρ	11000011010100000

Decimal Number System / Binary Number System

100 Place	10 Place	1 Place	64 Place	32 Place	16 Place	8 Place	4 Place	2 Place	1 Place
		0	0	0	0	0	0	0	0
		1	0	0	0	0	0	0	1
		2	0	0	0	0	0	1	0
		3	0	0	0	0	0	1	1
		4	0	0	0	0	1	0	0
		5	0	0	0	0	1	0	1
		6	0	0	0	0	1	1	0
		7	0	0	0	0	1	1	1
		8	0	0	0	1	0	0	0
		9	0	0	0	1	0	0	1
	1	0	0	0	0	1	0	1	0
	1	1	0	0	0	1	0	1	1
	1	2	0	0	0	1	1	0	0
	1	3	0	0	0	1	1	0	1
	1	4	0	0	0	1	1	1	0
	1	5	0	0	0	1	1	1	1
	1	6	0	0	1	0	0	0	0
	1	7	0	0	1	0	0	0	1
	1	8	0	0	1	0	0	1	0
	1	9	0	0	1	0	0	1	1
	2	0	0	0	1	0	1	0	0
	2	1	0	0	1	0	1	0	1
	2	2	0	0	1	0	1	1	0
	2	3	0	0	1	0	1	1	1
	2	4	0	0	1	1	0	0	0
	2	5	0	0	1	1	0	0	1
	2	6	0	0	1	1	0	1	0
	2	7	0	0	1	1	0	1	1
	2	8	0	0	1	1	1	0	0
	2	9	0	0	1	1	1	0	1
	3	0	0	0	1	1	1	1	0
	3	1	0	0	1	1	1	1	1
	3	2	0	1	0	0	0	0	0
	3	3	0	1	0	0	0	0	1
	3	4	0	1	0	0	0	1	0
	3	5	0	1	0	0	0	1	1
	3	6	0	1	0	0	1	0	0
	3	7	0	1	0	0	1	0	1
	3	8	0	1	0	0	1	1	0
	3	9	0	1	0	0	1	1	1
	4	0	0	1	0	1	0	0	0
	4	1	0	1	0	1	0	0	1
	4	2	0	1	0	1	0	1	0
	4	3	0	1	0	1	0	1	1
	4	4	0	1	0	1	1	0	0
	4	5	0	1	0	1	1	0	1
	4	6	0	1	0	1	1	1	0
	4	7	0	1	0	1	1	1	1
	4	8	0	1	1	0	0	0	0
	4	9	0	1	1	0	0	0	1
	5	0	0	1	1	0	0	1	0
	5	1	0	1	1	0	0	1	1
	5	2	0	1	1	0	1	0	0
	5	3	0	1	1	0	1	0	1
	5	4	0	1	1	0	1	1	0
	5	5	0	1	1	0	1	1	1
	5	6	0	1	1	1	0	0	0
	5	7	0	1	1	1	0	0	1
	5	8	0	1	1	1	0	1	0
	5	9	0	1	1	1	0	1	1
	6	0	0	1	1	1	1	0	0
	6	1	0	1	1	1	1	0	1
	6	2	0	1	1	1	1	1	0
	6	3	0	1	1	1	1	1	1

100 Place	10 Place	1 Place	64 Place	32 Place	16 Place	8 Place	4 Place	2 Place	1 Place
	6	4	1	0	0	0	0	0	0
	6	5	1	0	0	0	0	0	1
	6	6	1	0	0	0	0	1	0
	6	7	1	0	0	0	0	1	1
	6	8	1	0	0	0	1	0	0
	6	9	1	0	0	0	1	0	1
	7	0	1	0	0	0	1	1	0
	7	1	1	0	0	0	1	1	1
	7	2	1	0	0	1	0	0	0
	7	3	1	0	0	1	0	0	1
	7	4	1	0	0	1	0	1	0
	7	5	1	0	0	1	0	1	1
	7	6	1	0	0	1	1	0	0
	7	7	1	0	0	1	1	0	1
	7	8	1	0	0	1	1	1	0
	7	9	1	0	0	1	1	1	1
	8	0	1	0	1	0	0	0	0
	8	1	1	0	1	0	0	0	1
	8	2	1	0	1	0	0	1	0
	8	3	1	0	1	0	0	1	1
	8	4	1	0	1	0	1	0	0
	8	5	1	0	1	0	1	0	1
	8	6	1	0	1	0	1	1	0
	8	7	1	0	1	0	1	1	1
	8	8	1	0	1	1	0	0	0
	8	9	1	0	1	1	0	0	1
	9	0	1	0	1	1	0	1	0
	9	1	1	0	1	1	0	1	1
	9	2	1	0	1	1	1	0	0
	9	3	1	0	1	1	1	0	1
	9	4	1	0	1	1	1	1	0
	9	5	1	0	1	1	1	1	1
	9	6	1	1	0	0	0	0	0
	9	7	1	1	0	0	0	0	1
	9	8	1	1	0	0	0	1	0
	9	9	1	1	0	0	0	1	1
1	0	0	1	1	0	0	1	0	0
1	0	1	1	1	0	0	1	0	1
1	0	2	1	1	0	0	1	1	0
1	0	3	1	1	0	0	1	1	1
1	0	4	1	1	0	1	0	0	0
1	0	5	1	1	0	1	0	0	1
1	0	6	1	1	0	1	0	1	0
1	0	7	1	1	0	1	0	1	1
1	0	8	1	1	0	1	1	0	0
1	0	9	1	1	0	1	1	0	1
1	1	0	1	1	0	1	1	1	0
1	1	1	1	1	0	1	1	1	1
1	1	2	1	1	1	0	0	0	0
1	1	3	1	1	1	0	0	0	1
1	1	4	1	1	1	0	0	1	0
1	1	5	1	1	1	0	0	1	1
1	1	6	1	1	1	0	1	0	0
1	1	7	1	1	1	0	1	0	1
1	1	8	1	1	1	0	1	1	0
1	1	9	1	1	1	0	1	1	1
1	2	0	1	1	1	1	0	0	0
1	2	1	1	1	1	1	0	0	1
1	2	2	1	1	1	1	0	1	0
1	2	3	1	1	1	1	0	1	1
1	2	4	1	1	1	1	1	0	0
1	2	5	1	1	1	1	1	0	1
1	2	6	1	1	1	1	1	1	0
1	2	7	1	1	1	1	1	1	1

A-67

Char								Char							
•	0	1	0	0	0	0	0	P	1	0	1	0	0	0	0
!	0	1	0	0	0	0	1	Q	1	0	1	0	0	0	1
"	0	1	0	0	0	1	0	R	1	0	1	0	0	1	0
#	0	1	0	0	0	1	1	S	1	0	1	0	0	1	1
$	0	1	0	0	1	0	0	T	1	0	1	0	1	0	0
%	0	1	0	0	1	0	1	U	1	0	1	0	1	0	1
&	0	1	0	0	1	1	0	V	1	0	1	0	1	1	0
'	0	1	0	0	1	1	1	W	1	0	1	0	1	1	1
(	0	1	0	1	0	0	0	X	1	0	1	1	0	0	0
)	0	1	0	1	0	0	1	Y	1	0	1	1	0	0	1
*	0	1	0	1	0	1	0	Z	1	0	1	1	0	1	0
+	0	1	0	1	0	1	1	[	1	0	1	1	0	1	1
,	0	1	0	1	1	0	0	/	1	0	1	1	1	0	0
–	0	1	0	1	1	0	1	]	1	0	1	1	1	0	1
.	0	1	0	1	1	1	0	^	1	0	1	1	1	1	0
/	0	1	0	1	1	1	1	_	1	0	1	1	1	1	1
0	0	1	1	0	0	0	0	'	1	1	0	0	0	0	0
1	0	1	1	0	0	0	1	a	1	1	0	0	0	0	1
2	0	1	1	0	0	1	0	b	1	1	0	0	0	1	0
3	0	1	1	0	0	1	1	c	1	1	0	0	0	1	1
4	0	1	1	0	1	0	0	d	1	1	0	0	1	0	0
5	0	1	1	0	1	0	1	e	1	1	0	0	1	0	1
6	0	1	1	0	1	1	0	f	1	1	0	0	1	1	0
7	0	1	1	0	1	1	1	g	1	1	0	0	1	1	1
8	0	1	1	1	0	0	0	h	1	1	0	1	0	0	0
9	0	1	1	1	0	0	1	i	1	1	0	1	0	0	1
:	0	1	1	1	0	1	0	j	1	1	0	1	0	1	0
;	0	1	1	1	0	1	1	k	1	1	0	1	0	1	1
<	0	1	1	1	1	0	0	l	1	1	0	1	1	0	0
=	0	1	1	1	1	0	1	m	1	1	0	1	1	0	1
>	0	1	1	1	1	1	0	n	1	1	0	1	1	1	0
?	0	1	1	1	1	1	1	o	1	1	0	1	1	1	1
@	1	0	0	0	0	0	0	p	1	1	1	0	0	0	0
A	1	0	0	0	0	0	1	q	1	1	1	0	0	0	1
B	1	0	0	0	0	1	0	r	1	1	1	0	0	1	0
C	1	0	0	0	0	1	1	s	1	1	1	0	0	1	1
D	1	0	0	0	1	0	0	t	1	1	1	0	1	0	0
E	1	0	0	0	1	0	1	u	1	1	1	0	1	0	1
F	1	0	0	0	1	1	0	v	1	1	1	0	1	1	0
G	1	0	0	0	1	1	1	w	1	1	1	0	1	1	1
H	1	0	0	1	0	0	0	x	1	1	1	1	0	0	0
I	1	0	0	1	0	0	1	y	1	1	1	1	0	0	1
J	1	0	0	1	0	1	0	z	1	1	1	1	0	1	0
K	1	0	0	1	0	1	1	(	1	1	1	1	0	1	1
L	1	0	0	1	1	0	0	¦	1	1	1	1	1	0	0
M	1	0	0	1	1	0	1	)	1	1	1	1	1	0	1
N	1	0	0	1	1	1	0	~	1	1	1	1	1	1	0
O	1	0	0	1	1	1	1	DEL	1	1	1	1	1	1	1

International Morse Code

A	· —	N	— ·	Á	· — — · —	8	— — — · ·
B	— · · ·	O	— — —	Ä	· — · —	9	— — — — ·
C	— · — ·	P	· — — ·	É	· · — · ·	0	— — — — —
D	— · ·	Q	— — · —	N̄	— — · — —	, (comma)	— — · · — —
E	·	R	· — ·	Ö	— — — ·	.	· — · — · —
F	· · — ·	S	· · ·	Ü	· · — —	?	· · — — · ·
G	— — ·	T	—	1	· — — — —	;	— · — · — ·
H	· · · ·	U	· · —	2	· · — — —	:	— — — · · ·
I	· ·	V	· · · —	3	· · · — —	' (apostrophe)	· — — — — ·
J	· — — —	W	· — —	4	· · · · —	- (hyphen)	— · · · · —
K	— · —	X	— · · —	5	· · · · ·	/	— · · — ·
L	· — · ·	Y	— · — —	6	— · · · ·	parenthesis	— · — — · —
M	— —	Z	— — · ·	7	— — · · ·	underline	· · — — · —

G
Decibels

Decibels (dB) were originally based on the response of the human ear to sound waves. The decibel is a measure of the smallest "volume change" an ear can detect; this is illustrated in Figure G-1. Referring to the graph, you can see that a sound level of 1 mW can hardly be heard (a whisper) but is used as a reference of 0 db. The power has to be increased to 1.3 mW before the ear can detect that a change has occurred in the sound level. This point is labeled 1 dB since it is the smallest detectable change.

If the sound level is further increased, the next change in power level is detected at 1.6 mW, and this is labeled the 2 dB point. As the power level is further increased and the points plotted, you can see that, to notice a change in sound level, 1 dB, requires more and more power. For example, a power

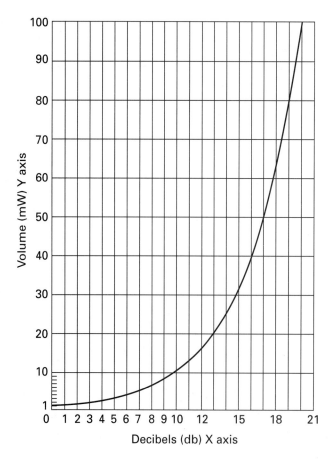

Whisper = 0 dB
Conversation = 45 dB
Industrial Equipment = 90 dB
Very Loud Music = 110 dB
Threshold of Pain = 130 dB

Decibels X	Y Volume
0	1.0
1	1.3
2	1.6
3	2.0
4	2.5
5	3.2
6	4.0
7	5.0
8	6.3
9	7.9
10	10.0
11	12.5
12	15.8
13	20.0
14	25.1
15	31.6
16	39.8
17	50.1
18	63.1
19	79.4
20	100.0

FIGURE G-1

increase from 10 to 20 mW is a 3-dB change (10 to 13 dB), while a change from 50 to 100 mW is also detected as a 3-dB change.

G.1 WATTS TO DECIBELS

Decibels are therefore a measure of change. To calculate the amount of change, two power levels need to be known: a starting power, which will be called P_{in}, and an ending power, which will be called P_{out}. The following formula can be used to calculate the amount of change.

$$dB = 10 \times \log \frac{P_{out}}{P_{in}}$$

The final decibel value will be an indication of the amount of change between input and output, not a measure of the actual output power. The value could come out as a +dB, indicating a gain in power, or a –dB figure, signifying a loss in power between input and output.

Table G-1 lists the common power ratios.

EXAMPLE:

Calculate the power gain of an amplifier that has an input of 10 mW and an output of 1 watt.

Solution:

$$dB = 10 \times \log \frac{P_{out}}{P_{in}}$$

$$= 10 \times \log \frac{1000 \text{ mW}}{10 \text{ mW}} = +20 \text{ dB}$$

TABLE G-1   Common Power Ratios

GAIN (+dB)	LOSS (–dB)
+3 dB = ×2	−3 dB = ÷2 or ½
+6 dB = ×4	−6 dB = ÷4 or ¼
+9 dB = ×8	−9 dB = ÷8
+12 dB = ×16	−12 dB = ÷16
+15 dB = ×32	−15 dB = ÷32
+10 dB = ×10	−10 dB = ÷10
+20 dB = ×100	−20 dB = ÷100
+30 dB = ×1000	−30 dB = ÷1,000
+40 dB = ×10,000	−40 dB = ÷10,000
+50 dB = ×100,000	−50 dB = ÷100,000

An amplifier gain of 20 dB does not mean that the output is 20 times greater than the input; we already know that the output is 100 times greater than the input (10 mW to 1000 mW). It simply means that a 20-dB gain has occurred between input and output, and 20 dB equates to × 100.

◼ EXAMPLE:

A cable has an input of 1.5 kW and an output of 1.2 kW. Calculate the cable loss.

◼ *Solution:*

$$dB = 10 \times \log \frac{P_{out}}{P_{in}}$$

$$= 10 \times \log \frac{1.2 \text{ kW}}{1.5 \text{ kW}}$$

$$= -1 \text{ dB}$$

In this example, the cable has a negative gain or loss of −1 dB between input and output.

◼ EXAMPLE:

Calculate the gain of the following radio receiver.

$$\text{Antenna gain} = +7 \text{ dB}$$
$$\text{Coaxial cable} = -2 \text{ dB}$$
$$\text{Preamplifier} = +45 \text{ dB}$$
$$\text{Amplifier} = +15 \text{ dB}$$
$$\text{Speaker} = -5 \text{ dB}$$

◼ *Solution:*

Total gain in decibels is +60 dB or a gain of 1,000,000 between the antenna input and the speaker output.

Up to now, two power levels have had to be known in order to calculate the amount of gain or loss. The power at a point, however, can be expressed in decibels with respect to a reference. Telephone companies have established a standard of 1 mW as a reference power, and all other power levels are expressed with respect to the reference. Power levels below 1 mW are indicated as −dBm, and positive power levels above 1 mW are written as +dBm.

G.2 DECIBELS TO WATTS

By transposition of the original power formula, we can obtain the following:

$$\boxed{dB = 10 \log \frac{P_{out}}{P_{in}}}$$

$$P_{out} = P_{in} \text{ antilog } \frac{dB}{10}$$

$$P_{in} = \frac{P_{out}}{\text{antilog } (dB/10)}$$

▧ **EXAMPLE:**

Calculate the output power of a 35-dB amplifier if the input is 4 mW.

▨ *Solution:*

$$P_{out} = P_{in} \text{ antilog } \frac{dB}{10}$$

$$= 4 \text{ mW} \times \text{antilog } \frac{35}{10} = 12.65 \text{ watts}$$

▧ **EXAMPLE:**

Calculate the input power needed for a 30-dB amplifier to produce a 2-watt output.

▨ *Solution:*

$$P_{in} = \frac{P_{out}}{\text{antilog } (dB/10)}$$

$$= \frac{2W}{\text{antilog } (30/10)} = 2 \text{ mW}$$

G.3 VOLTAGE OR CURRENT GAIN IN DECIBELS

A voltage or current gain or loss can also be expressed in decibels. The equation for voltage can be derived by substituting P_{out} and P_{in} in the power formula with V^2/R to obtain:

$$dB = 20 \log \frac{V_{out}}{V_{in}}$$

Similarly, a current formula can be derived:

$$dB = 20 \log \frac{I_{out}}{I_{in}}$$

Table G-2 makes a comparison between the common decibel values for power and voltage ratios. Looking at this table you will notice that

$$dB = 10 \log \left(\frac{P_{out}}{P_{in}} \right)$$

$$dB = 20 \log \frac{I_{out}}{I_{in}}$$

$$dB = 20 \log \frac{V_{out}}{V_{in}}$$

The corresponding voltage or current decibel values are twice those of the same power ratio. For example, double the power is 3 dB, whereas double the voltage or current is 6 dB. Also, 100 times the power is 20 dB, whereas 100 times the voltage or current is 40 dB. This is because the voltage and current decibel formulas have a multiplying factor of 20 rather than 10.

In summary, the decibel is a measure of change between input and output, and knowing the decibel figure for an amplifier or radio receiver, for example, helps give the user a better idea of the performance of these circuits or devices.

TABLE G-2   Common Power and Voltage Ratios

DECIBELS		POWER RATIO	VOLTAGE/CURRENT RATIO
Gain:	40	10,000	100
	20	100	10
	6	4	2
	3	2	1.4
	1	1.2	1.1
Loss:	−1	1/1.2	1/1.1
	−3	1/2	1/1.4
	−6	1/4	1/2
	−20	1/100	1/10
	−40	1/10,000	1/100

H
Frequency Spectrum

The term "frequency" describes the number of alternations occurring in one second. Direct current (dc) is a steady or constant current that does not alternate and is therefore listed as zero cycles per second or 0 Hz. This appendix illustrates in detail the entire range of frequencies from the lowest subaudible frequency to the highest cosmic rays, along with their applications.

The following page is an overall summary of the complete range of frequencies or spectrum, as it is normally called. The subsequent pages cover each band or section of frequencies in a lot more detail and list the different frequency applications.

This and all of the other appendixes will be useful as references throughout your course of electronic study.

OVERVIEW OF FREQUENCY SPECTRUM

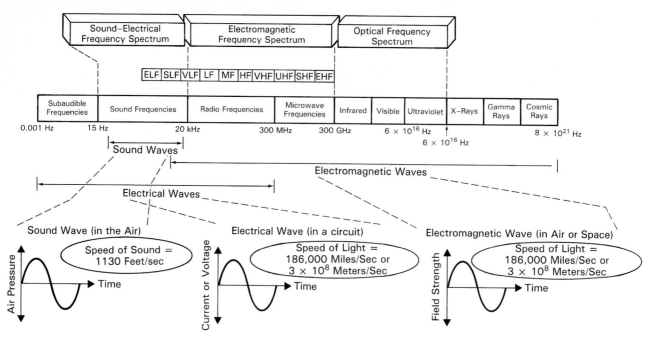

0 Hz	16 →16 kHz	16→ 30 kHz	10 kHz→ 300 GHz
Direct Current (DC) Motors, Relays, Supply Voltages	Audio Frequencies (AC) Motors, Amplifiers, Music Equipment, Speakers, Microphones, Oscillators	Sound (Ultrasonic) Frequencies Sonar, Music, Speech	Electromagnetic (Radio) Frequencies Voice Communications Television Navigation Medical, Scientific, and Military

300 GHz→ 4×10^{14}	4×10^{14}→ 7.69×10^{14}	4×10^{14}→6×10^{16}	9.375×10^{15}→3×10^{19}
Infrared (R) Heating, Photography, Sensing, Military	Visible Color, Photography, Movies, TV	Ultraviolet (UV) Sterilizing, Medical	X–Rays Medical, Gauge Thickness, Inspection

3×10^{19} → 5×10^{20}	5×10^{20} →
Gamma Rays Deeper Penetrating Than X–Rays, Detection of Radiation	Cosmic Rays Present in Outer Space

THE SOUND SPECTRUM

VLF (Very Low–Frequency Band)

3 kHz

30 kHz

LF (Low–Frequency Band)

30

International Fixed Public (IFP)

Maritime Mobile

30 kHz

---

59 | **61** | **70** | **90**

International Fixed Public (IFP)	Standard Frequency	(IFP)	(IFP)	Loran C. Radionavigation
			Decca Maritime Mobile Navigation (British)	
Maritime Mobile		MARITIME MOBILE	Radiolocation (Radar)	

---

110 | **130** | **160**

Loran C. Radionavigation	(IFP)	(IFP)	(IFP)
	Maritime Mobile	Maritime Mobile (MM)	Maritime Mobile (MM)
	Radar (Radio Detection and Ranging)		

---

200 | **285** | **300**

| (IFP) | Aeronautical Mobile Communications | Maritime Radio–navigation |
| Maritime Mobile (MM) | Maritime Radionavigation | Aeronautical Radio–navigation |

300 kHz

MF (Medium–Frequency Band)

300 kHz

3 kHz

HF (High–Frequency Band)

VHF (Very High–Frequency Band)

UHF (Ultra High–Frequency Band)

SHF (Super High–Frequency Band)

EHF (Extremely High–Frequency Band)

Optical Spectrum

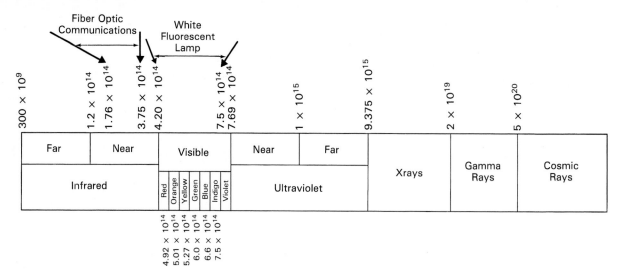

DEFINITIONS AND CODES

USAGE

- Exclusively used by government, state, and federal agencies
- Shared by government and nongovernment agencies
- Nongovernment; publicly used frequencies

TYPES OF EMISSION

Amplitude (AM;Amplitude Modulation)
A0–Steady Unmodulated Pure Carrier
A1–Telegraphy on Pure Continuous Waves
A2–Amplitude Tone–modulated Telegraphy
A3–AM Telephony Including Single and Double Sideband
   with full, Reduced, or Suppressed Carrier
A4–Facsimile (Pictures)
A5–Television

Frequency FM;Frequency Modulation)
F0–Steady Unmodulated Pure Carrier
F1–Carrier–Shift Telegraphy
F2–Audio–frequency–shift Telegraphy
F3–Frequency or Phase–modulated Telephony
F4–Facsimile
F5–Television

Pulse (PM;Pulse Modulation)
P0–Pulsed Radar
P1–Telegraphy: On/Off Keying of Pulsed Carrier
P2–Telegraphy: Pulse Modulation of Pulse Carrier,
   Pulse Width, Phase or Position Tone Modulated
P3–Telephony: Amplitude (PAM), Width (PWM),
   Phase or Position (PPM) Modulated Pulses

ABBREVIATED TERMS

(R), International major air route (air traffic control)
(OR), Off route military (air traffic control)
  S   Assigned satellite frequency
  ↑   Earth to space communication (uplink)
  ↓   Space to earth communication (downlink)

ISM: industrial, scientific, medical
Passive communication equipment:
   generates no electromagnetic radiation (receive only)
Active communication equipment:
   generates electromagnetic radiation (transmits)

APPLICATIONS

- International Fixed: public radiocommunication service. Transmitters designed for one frequency.

- Aeronautical Fixed: a service intended for the transmission of air navigation and preparation for safety of flight.

- Fixed Satellite: satellites that maintain a geostationary orbit (same position above the earth)

- Radiolocation: radio waves used to detect an object's direction, position, or motion.

- Radio Astronomy: radio waves emitted by celestial bodies that are used to obtain data about them.

- Space Operations

- Space Research

APPLICATIONS (continued)

- Mobile: Radio service between a fixed location and one or more mobile stations or between mobile stations.

- Aeronautical Mobile: As above, except mobile stations are aircraft.

- Maritime Mobile: As above, except mobile stations are marine.

- Land Mobile: As above, except mobile stations are automobiles.

- Mobile Satellite

- Aeronautical Mobile Satellite

- Maritime Mobile Satellite

- Radionavigation (Aeronautical and Land): Navigational use of radiolocation equipment such as direction finders, radio compass, radio homing beacons, etc.

- Aeronautical Radionavigation

- Maritime Radionavigation

- Satellite Radionavigation

- Aeronautical Radionavigation Satellite

- Maritime Radionavigation Satellite

- Inter Satellite

APPLICATIONS (continued)

- Broadcasting: Transmission of speech, music, or visual programs for commercial or public service purposes.

- Broadcasting Satellite

- Amateur: a frequency used by persons licensed to operate radio transmitters as a hobby. Person is also called a radio ham.

- Amateur Satellite: Communication by radio hams via satellite.

- Citizen's: a radiocommunications service of fixed, land, and mobile stations intended for short-distance personal or business purposes.

- Standard Frequency: Highly accurate signal broadcasted by the national bureau of standards (NBS) radio station (WWV) to provide frequency, time, solar flare, and other standards.

- Standard Frequency Satellite: NBS broadcast via satellite.

- Meteorological Aids: a radio service in which the emission consists of signals used solely for meteorological use.

- Meteorological Satellite: Meteorological broadcast via satellite.

- Earth Exploration Satellite: Radio frequencies used for earth exploration.

I

Answers to Self-Test Review Questions

STR 1-1 1. Current, voltage, resistance, power
2. Resistors, capacitors, inductors
3. Amplifier, oscillator   4. Communications
STR 1-2 1. Ohm (Georg S. Ohm)   2. Ampere (André M. Ampère)
3. Volt (Alessandro Volta)   4. Watt (James Watt)
STR 1-3   Lasers and robotics

STR 2-1 1. Elements are made up of similar atoms; compounds are made up of similar molecules.
2. Protons, neutrons, electrons   3. Copper
4. Like charges repel; unlike charges attract.
STR 2-2 1. Amp
2. Current = Q/t (number of coulombs divided by time in seconds).
3. The direction of flow is different: negative to positive is termed electron current flow; positive to negative is known as conventional current flow.
4. a. Set to highest scale and work down.
   b. Connect positive lead to positive circuit point and negative lead to negative circuit point.
   c. Place meter in the path of current flow.
STR 2-3 1. Volts   2. 3000 kV
3. a. Set to highest scale and work down.
   b. Connect positive lead to positive circuit point and negative lead to negative circuit point.
   c. Place meter across component.
4. They are directly proportional.
STR 2-4 1. False
2. a. Electrons are far away from nucleus.
   b. Many electrons per atom.
   c. Incomplete valence shell.
3. Copper   4. 28.6 millisiemens
STR 2-5 1. True   2. Mica
3. The voltage needed to cause current to flow through a material
4. Small
STR 2-6 and 2-7 1. A material whose characteristics fall between a conductor and an insulator
2. Silicon   3. No

4. A short circuit provides an unintentional path for current to flow. A closed circuit has a complete path for current.

STR 3-1 1. It is the opposition to current flow.
2. The difference is in how much or how little current will flow.
3. They are inversely proportional.   4. Small

STR 3-2 1. A circuit is said to have a resistance of 1 ohm when 1 volt produces a current of 1 ampere.
2. $I = V/R = 24\text{ V}/6\ \Omega = 4\text{ A}$.
3. A memory aid to help remember Ohm's law:

4. Current is proportional to voltage and inversely proportional to resistance.
5. $V = I \times R = 25\text{ mA} \times 1\text{ k}\Omega = (25 \times 10^{-3}) \times (1 \times 10^{3}) = 25\text{ V}$.
6. $R = V/I = 12\text{ V}/100\ \mu\text{A} = 12\text{ V}/100 \times 10^{-6} = 120\text{ k}\Omega$.

STR 3-3 1. Type of conducting material used, cross-sectional area, length, temperature
2. The resistance of a conductor is proportional to its length and resistivity, and inversely proportional to its cross-sectional area.
3. True   4. True

STR 3-4 1. Light, heat, magnetic, chemical, electrical, mechanical
2. When energy is transformed, work is done. Power is the rate at which work is done or energy is transformed.
3. $W = Q \times V, P = V \times I$
4. When 1 kW of power is used in 1 hour

STR 4-1 1. Carbon composition, carbon film, metal film, wirewound, metal oxide, thick film
2. SIPs have one row of connecting pins, while DIPs have two rows of connecting pins.
3. Rheostat has two terminals; potentiometer has three terminals.
4. Linear means that the resistance changes in direct proportion to the amount of change of the input, while a tapered potentiometer varies nonuniformly.
5. True   6. Photoresistor

STR 4-2 1. Ohmmeter   2. As $22 \times 1000 = 22,000 = 22\text{ k}\Omega$
3. As $470 \times 1 = 470\ \Omega$
4. Short together meter leads and adjust zero ohms control; turn off power and remove part; connect meter leads across component and adjust scale for midscale reading and multiply if necessary.

STR 4-3 1. General purpose are $\pm5\%$ or greater; precision are $\pm2\%$ or less.
2. $222 \times 0.1, \pm0.25\% = 22.2\ \Omega, \pm0.25\%$

3. Yellow, violet, red, silver

4. True (General purpose, ±20% tolerance)

STR 4-4
and 4-5 1. $P = I^2 \times R$ 2. Small, large

STR 4-6 1. Opens   2. Ohmmeter

STR 5-1 1. Between 10 and 50 kΩ   2. See Section 5.1.2.

STR 5-2 1. To supply an accurate dc voltage that can be varied

2. Current, voltage, resistance

3. Sine, square, rectangular, triangular and sawtooth

4. Oscilloscope   5. A cathode ray tube or CRT

6. Period = 20 μs/div × 4 div
   = 80 μs
   Frequency = 1/80 μs = 12.5 kHz

7. Peak = 2 V/cm × 2 cm = 4 V peak
   Peak-to-peak = 2 V/cm × 4 cm
   = 8 V pp

8. Allows comparisons of phase, amplitude, shape, and timing

STR 5-3 1. It is designed to accommodate and interconnect components to form experimental circuits.

2. A power supply voltage is generally connected across these two points so that voltages can be tapped off to supply the circuit.

STR 5-4 1. Heat and solder   2. Tin, lead   3. See Section 5.4.5.

4. Desoldering braid, desoldering bulb, spring-loaded plunger

---

STR 6-1 1. Load resistance is the resistance of a device, circuit, or system. Load current is the amount of current drawn from the source by the device, circuit, or system. A dc voltage source is one that will supply a constant polarity output voltage, while a dc current is a unidirectional (one direction) current.

2. False   3. They are inversely proportional.

4. A load that will draw a large amount of current.

STR 6-2-1 1. Friction forces electrons out of their orbits to create a difference in potential.

2. Due to the piezoelectric effect, compounds such as quartz will produce a voltage when pressure is applied.

STR 6-2-2 1. Thermocouple

2. Used to control a gas shutoff valve, as in a water heater

STR 6-2-3 1. Photocell

2. Used as a light meter on a camera to help control exposure

STR 6-2-4 1. Lodestone   2. Like poles repel; unlike poles attract.

3. Permanent, temporary, electromagnets   4. Dc electric generator

STR 6-2-5 1. Negative plate, positive plate, and electrolyte

2. Carbon–zinc, alkaline–manganese, mercury, silver oxide, lithium

3. Lead–acid, nickel–cadmium   4. False

5. Maximum power transfer is possible when the internal resistance of the battery is equal to the load resistance.

6. No battery is a perfect voltage source and is therefore not 100% efficient. This inefficiency is represented by an internal resistance.

STR 6-2-6 1. Ac voltage 2. False 3. False
   4. More accurate voltage source, easy to vary output, never runs down

STR 6-3 1. Indicate the maximum current that can flow without blowing the fuse and the maximum voltage value that will not cause an arc.
   2. Fast-blow fuses open almost instantly after an excessive current occurs, whereas a slow-blow fuse will have a certain time delay until it opens.
   3. Thermal, magnetic, thermomagnetic
   4. SPST, SPDT, DPST, DPDT, NOPB, NCPB, Rotary, and DIP

STR 7-1 and 7-2 1. A circuit in which current has only one path 2. 8 A
STR 7-3 1. $R_T = R_1 + R_2 + R_3 + \cdots$
   2. $R_T = R_1 + R_2 + R_3 = 2\ \text{k}\Omega + 3\ \text{k}\Omega + 4700 = 9.7\ \text{k}\Omega$
STR 7-4 1. True 2. True
   3. $R_T = R_1 + R_2 = 6 + 12 = 18\ \Omega; I_T = V_S/R_T = 18/18 = 1\ \text{A}$
   $V_{R1} = 1\ \text{A} \times 6\ \Omega = 6\ \text{V}$
   $V_{R2} = 1\ \text{A} \times 12\ \Omega = 12\ \text{V}$
   4. $V_X = (R_X/R_T) \times V_S$ 5. Potentiometer 6. No
STR 7-5 1. $P = I \times V$ or $P = V^2/R$ or $P = I^2 \times R$
   2. $P = V^2/R = 12^2/12 = 144/12 = 12\ \text{W}$
   3. Wirewound, at least 12 W, ideally a 15 W
   4. $P_T = P_1 + P_2 = 25\ \text{W} + 3800\ \text{mW} = 28.8\ \text{W}$
STR 7-6 1. Component will open, component's value will change, and component will short.
   2. No current will flow; source voltage dropped across it.
   3. False
   4. No voltage drop across component, yet current still flows in circuit; resistance equals zero for component.

STR 8-1 and 8-2 1. When two or more components are connected across the same voltage source so that current can branch out over two or more paths
   2. False 3. $V_{R1} = V_S = 12\ \text{V}$ 4. No
STR 8-3 1. The sum of all currents entering a junction is equal to sum of all currents leaving that junction.
   2. $I_2 = I_T - I_1 = 4\ \text{A} - 2.7\ \text{A} = 1.3\ \text{A}$
   3. $I_X = (R_T/R_X) \times I_T$
   4. $I_T = V_T/R_T = 12/1\ \text{k}\Omega = 12\ \text{mA}; I_1 = 1\ \text{k}\Omega/2\ \text{k}\Omega \times 12\ \text{mA} = 6\ \text{mA}$

STR 8-4 1. $R_T = \dfrac{R_1 \times R_2}{R_1 + R_2}$ 2. $R_T = \dfrac{1}{(1/R_1) + (1/R_2) + (1/R_3)} + \cdots$

3. $R_T = \dfrac{\text{common value of resistors } (R)}{\text{number of parallel resistors } (n)}$

4. $R_T = \dfrac{1}{(1/2.7 \text{ k}\Omega) + (1/24 \text{ k}\Omega) + (1/1 \text{ M}\Omega)} = 2.421 \text{ k}\Omega$

STR 8-5 1. True   2. $P_1 = I_1 \times V = 2 \text{ mA} \times 24 \text{ V} = 48 \text{ mW}$
3. $P_T = P_1 + P_2 = 22 \text{ mW} + 6400 \text{ }\mu\text{W} = 28.4 \text{ mW}$ 4. Yes

STR 8-6 1. No current will flow in the open branch; total current will decrease.
2. Maximum current is through shorted branch; total current will increase.
3. Will cause a corresponding opposite change in branch current and total current
4. False

STR 9-1
and 9-2 1. By tracing current to see if it has one path (series connection) or more than one path (parallel connection)

2. $R_{1,2} = R_1 + R_2 = 12 \text{ k}\Omega + 12 \text{ k}\Omega = 24 \text{ k}\Omega; R_{1,2,3} = \dfrac{R_{1,2} \times R_3}{R_{1,2} + R_3} =$

$\dfrac{24 \text{ k}\Omega \times 6 \text{ k}\Omega}{24 \text{ k}\Omega + 6 \text{ k}\Omega} = \dfrac{144 \text{ k}\Omega}{30 \text{ k}\Omega} = 4.8 \text{ k}\Omega$

3. Step A: Find equivalent resistances of series-connected resistors. Step B: Find equivalent resistances of parallel-connected combinations. Step C: Find equivalent resistances of remaining series-connected resistances.
4. $R_{1,2} = R_1 + R_2 = 470 + 330 = 800 \text{ }\Omega$

$R_{1,2,3} = \dfrac{R_{1,2} \times R_3}{R_{1,2} + R_3} = \dfrac{800 \times 270}{800 + 270} = \dfrac{216 \text{ k}\Omega}{1.07 \text{ k}\Omega} = 201.9 \text{ }\Omega$

470 Ω R_1

R_3 270 Ω

330 Ω R_2

FIGURE 9-1, 4

STR 9-3 1. Step 1: Find total resistance (Steps A, B and C). Step 2: Find total current. Step 3: Find voltage drop with $I_T \times R_X$.
2. The voltage drops previously calculated would not change since the ratio of the series resistor to the series equivalent resistors remains the same, and therefore the voltage division will remain the same.

STR 9-4,
9-5, and 1. Find total resistance; find total current; find the voltage across
9-6   each series and parallel combination resistors; find the current
        through each branch of parallel resistors; find the total and indi-
        vidual power dissipated.
     2. This is a do-it-yourself question; each answer will vary.
STR 9-7 1. When a load resistance changes the circuit and lowers output
        voltage

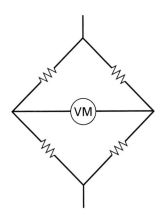

FIGURE 9-7, 2

     2. Used to check for an unknown resistor's resistance
     3. R
     4. Current divider for a digital-to-analog converter
STR 9-8 1. If component is in series with circuit, no current will flow and
        source voltage will be across bad component. If component is par-
        allel, no current will flow in that branch.
     2. Total resistance will decrease and bad component will have 0 V
        dropped across it.
     3. Will cause the circuit's behavior to vary
STR 9-9 1. Small, large   2. Voltage source
     3. Voltage, resistor   4. Current

STR 10-1 1. Ammeter, voltmeter, ohmmeter
     2. a. R_m is the resistance of the moving coil within a meter.
        b. I_m is the value of movement current needed to cause FSD.
STR 10-2 1. Shunt parallel

     2. $R_{sh} = \dfrac{V_{Rsh}}{I_{Rsh}} = \dfrac{50\,\text{mV}}{29\,\text{mA}} = 1.724\,\Omega$

STR 10-3 1. Multiplier, series
     2. Ohms-per-volt rating is the reciprocal of the FSD current of the
        coil. The larger this value is, the less current it takes to cause FSD,
        so the more sensitive it is.

STR 10-4 1. Ammeters and voltmeters get their power from the circuit under test, whereas an ohmmeter has its own battery. They are all used to measure different electrical quantities.

2. Short meter leads and adjust zero ohm control; ensure no power to component; connect meter leads across component and read resistance on scale, multiplying by range selected; disconnect one lead of component if resistance is being measured while component is in circuit.

STR 10-5 1. a. Ohmmeter   c. Voltmeter   e. Ammeter and Voltmeter
         b. Ammeter   d. Ohmmeter

2. Volt-ohm-milliammeter.

STR 11-1 1. a. Alternating current   b. Direct current
2. Dc; current only flows in one direction
3. Ac   4. Dc
5. Power transfer, information transfer
6. Dc flows in one direction, whereas ac first flows in one direction and then in the opposite direction.

STR 11-2-1 1. Ac generators can be larger, less complex, and cheaper to run; transformers can be used with ac to step up/down, so low-current power lines can be used; easy to change ac to dc, but hard the other way around.

2. False   3. $P = I^2 \times R$
4. A device that can step up or down ac voltages
5. 120 V ac   6. Ac, dc

STR 11-2-2 1. The property of a signal or message that conveys something meaningful to the recipient; the transfer of information between two points

2. Sound, electromagnetic, electrical
3. a. Sound wave   c. Electrical wave
     b. Electromagnetic wave
4. 1133 feet per second; 186,000 miles per second
5. Sound, electromagnetic, electrical
6. a. Microphone   c. Antenna
     b. Speaker   d. Human ear

STR 11-3 1.

2. Time, frequency

3. (sound wave); $\lambda(mm) = \dfrac{344.4 \text{ m/s}}{f(\text{Hz})}$;

(Electromagnetic wave): $\lambda(m) = \dfrac{3 \times 10^8 \text{ m/s}}{f(\text{Hz})}$;

different because sound waves travel at a different speed than do electromagnetic waves.

4. Odd

STR 11-4 1. Electronic   2. Electrical

STR 12-1 1. True: no
2. RF probe is used to measure high-frequency electrical waves and uses a special high-frequency rectifier. HV probe is used to measure high voltages and uses multiplier resistors.
3. To analyze the frequency of a periodic wave at its input jack and display the frequency
4. The frequencies generated: AF = 20 Hz to 20 kHz; RF = 3 kHz to 3 GHZ

STR 12-2 1. Electronics, test equipment, troubleshooting techniques, and system repair
2. (1) Symptom recognition
(2) List probable faulty units
(3) Localizing faulty section
(4) Localizing faulty circuit
(5) Localizing faulty component
(6) Repair, final testing, and analysis of failure

STR 13-1 1. Two plates, dielectric   2. Dielectric
STR 13-2 1. True   2. None
STR 13-3 1. B
2. Field strength (V/cm) = charge difference (V), volts/distance between plates (d), meters
3. The displacement of an atom's electrons within the dielectric toward the positive plate
4. "Di" because it is between two plates; "electric" because an electric field exists within it
STR 13-4 1. The farad
2. Capacitance, C (farads) = charge, Q (coulombs)/voltage, V (volts)
3. $30{,}000 \times 10^{-6} = 0.03 \times 10^{0} = 0.03$ F
4. $C = Q/V = 17.5/9 = 1.94$ F
STR 13-5 1. Plate area, distance between plates, type of dielectric used
2. $C = \dfrac{(8.85 \times 10^{-12}) \times K \times A}{d}$
3. Double   4. Double
STR 13-6 1. Glass   2. False   3. Large, small   4. Nanoamperes (b)
STR 13-7 1. $C_T = \dfrac{1}{(1/C_1) + (1/C_2) + (1/C_3)} = \dfrac{1}{(1/2\ \mu F) + (1/3\ \mu F) + (1/5\ \mu F)}$
$= 0.968\ \mu F$ or 1 μF
2. $C_T = C_1 + C_2 + C_3 = 7$ pF $+ 2$ pF $+ 14$ pF $= 23$ pF
3. $V_{CX} = (C_T/C_X) \times V_T$ 4. True
STR 13-8 1. Mica, ceramic, paper, plastic, electrolytic
2. a. Electrolytic; b. ceramic   3. Air, mica, ceramic, plastic
4. a. Air; b. mica, ceramic, plastic   5. Electrolytic

STR 13-9 1. 470 pF, 2% tolerance   2. 0.47 μF or 470 nF, 5% tolerance

STR 13-10 1. The time it takes a capacitor to charge to 63.2%
2. 63.2%   3. 36.8%   4. False

STR 13-11 1. Capacitive opposition is low for ac (short) and high for dc (open).
2. True

STR 13-12 1. It means that the voltage and current maximums and minimums occur at different times.
2. False   3. True   4. False

STR 14-1 1. Opposition to current flow without the dissipation of energy
2. $X_C = 1/2\pi fC$
3. When frequency or capacitance goes up, there is more charge and discharge current; so X_C is lower.
4. $X_C = 1/2\pi fC = 1/2\pi \times 4\ \text{kHz} \times 4\ \mu\text{F} = 9.95\ \Omega$

STR 14-2 1. Current leads voltage by some phase angle less than 90%.
2. An arrangement of vectors to illustrate the magnitude and phase relationships between two or more quantities of the same frequency
3. Z = total opposition to current flow; $Z = \sqrt{R^2 + X_C^2}$
4. a. 0°   b. 90°   c. Between 0 and 90°

STR 14-3 1. Resistor current is in phase with voltage; capacitor current is 90° out of phase (leading) with voltage.
2. No
3. a. $I_T = \sqrt{I_R^2 + I_C^2}$; b. $Z = V_S/I_T$; also $Z = (R \times X_C)/\sqrt{R^2 + X_C^2}$
4. Lead

STR 14-4 1. 0.5 μF   2. Capacitance meter or analyzer

STR 14-5 1. a. Filter   b. Voltage divider   c. Differentiator
2. Capacitor, resistor   3. Long   4. Differentiator

STR 15-1-1
and 15-1-2 1. False   2. Electron
3. To determine the direction of the magnetic force
4. North

STR 15-1-3
and 15-1-4 1. Coil   2. True   3. Dc   4. False
5. Used to determine the magnetic polarity; placing your left-hand fingers in the direction of the current, your thumb points in the direction of the north magnetic pole.
6. Ac

STR 15-1-5
and 15-1-6 1. Number of lines of force (or maxwells) in webers
2. Number of magnetic lines of flux per square meter
3. The magnetic pressure that produces the magnetic field
4. Magnetomotive force divided by length of coil
5. Opposition or resistance to the establishment of a magnetic field
6. A measure of how easily a material will allow a magnetic field to be set up within it

7. True
8. The state beyond which an electromagnet is incapable of further magnetic strength

STR 15-1-7 1. Magnetic-type circuit breaker, relays
2. An NO relay is open when off; NC is closed when off.
3. A reed relay has an electromagnet around it, while a reed switch needs an external magnet to operate.
4. Home security circuits

STR 15-2 1. The voltage induced or produced in a coil as the magnetic lines of force link with the turns of a coil
2. Faraday's: When the magnetic flux linking a coil is changing, an emf is induced.
Lenz's: The current induced in a coil due to the change in the magnetic flux is such as to oppose the cause producing it.
3. Sine wave
4. Because it converts sound waves of air pressure (acoustical) to electrical waves (electro)

STR 16-1 1. When the current-carrying coil of a conductor induces a voltage within itself
2. The induced voltage, which opposes the applied emf; $V_{\text{ind}} = L \times (\Delta i/\Delta t)$
3. $V_{\text{ind}} = L \times (\Delta i/\Delta t) = 2\text{ mH} \times 4\text{ kA/s} = 8\text{ V}$

STR 16-2 1. Different applications: An electromagnet is used to generate a magnetic field; an inductor is used to oppose any changes of circuit current.
2. True

STR 16-3 1. Number of turns, area of coil, length of coil, core material used

2. $L = \dfrac{N^2 \times A \times \mu}{l}$

STR 16-4 1. False
2. a. $L_T = L_1 + L_2 + L_3 + \cdots$
   b. $L_T = 1/(1/L_1) + (1/L_2) + (1/L_3) + \cdots$
3. a. $L_T = L_1 + L_2 = 4\text{ mH} + 2\text{ mH} = 6\text{ mH}$

   b. $L_T = \dfrac{L_1 \times L_2}{L_1 + L_2}$ (using product over sum) $= \dfrac{4\text{ mH} \times 2\text{ mH}}{4\text{ mH} + 2\text{ mH}}$

   $= \dfrac{8\text{ mH}}{6\text{ mH}} = 1.33\text{ mH}$

STR 16-5 1. Air core, iron core, ferrite core   2. Air core
3. Chemical compound, basically powdered iron oxide and ceramic
4. Toroidal-type inductors have a greater inductance.
5. Ferrite-core variable inductor
6. Core permeability

STR 16-6 1. Current in an inductive circuit builds up in the same way that voltage does in a capacitive circuit, but the capacitive time constant is proportional to resistance, where the inductive time constant is inversely proportional to resistance.

2. True   3. False

4. It will continuously oppose the alternating current.

STR 16-7 1. The opposition to current flow offered by an inductor without the dissipation of energy: $X_L = 2 \times \pi \times f \times L$

2. The higher the frequency (and therefore the faster the change in current) or the larger the inductance of the inductor, the larger the magnetic field created, the larger the counter emf will be to oppose applied emf; so the inductive reactance (opposition) will be large also.

3. Inductive reactance is an opposition, so it can be used in Ohm's law in place of resistance.

4. False

STR 16-8 1. False

2. $V_S = \sqrt{V_R^2 + V_L^2} = \sqrt{4^2 + 2^2} = \sqrt{16 + 4} = \sqrt{20} = 4.47$ V

3. The total opposition to current flow offered by a circuit with both resistance and reactance: $Z = \sqrt{R^2 + X_L^2}$

4. +45°

5. Power consumption above the zero line, caused by positive current and voltage or negative current and voltage

6. Quality factor of an inductor that is the ratio of the energy stored in the coil by its inductance to the energy dissipated in the coil by the resistance: $Q = X_L/R$

7. True power is energy dissipated and lost by resistance; reactive power is energy consumed and then returned by a reactive device.

8. PF = true power (P_R)/apparent power (P_A) or PF = R/Z or PF = cos θ

STR 16-9 1. False   2. $I_T = \sqrt{I_R^2 + I_L^2}$

STR 16-10 1. a. With an ohmmeter; resistance will be infinite instead of low.

b. With an inductor analyzer to check its inductance value

2. An open

STR 16-11 1. *RL* integrator, *RL* differentiator, *RL* filter

2. Can be made to function as either a high- or a low-pass filter

3. The integrator will approximate a dc level; the differentiator will output spikes of current.

STR 17-1 1. Mutual inductance is the process by which an inductor induces a voltage in another inductor, whereas self-inductance is the process by which a coil induces a voltage within itself.

2. False

STR 17-2 1. True   2. True   3. Primary and secondary

STR 17-3 1. True   2. False

STR 17-4 1. It is the ratio of the number of magnetic lines of force that cut the secondary compared to the total number of magnetic flux lines being produced by the primary: *k* = flux linking secondary coil/total flux produced by primary.

2. True
3. How close together the primary and secondary are to one another and the type of core material used

STR 17-5 1. Turns ratio = $N_s/N_p = 1608/402 = 4$; step up
2. $V_s = N_s/N_p \times V_p$ 3. False   4. $I_s = N_p/N_s \times I_p$
5. Turns ratio = $\sqrt{Z_L/Z_S} = \sqrt{75\ \Omega/25\ \Omega} = \sqrt{3} = 1.732$
6. $V_s = (N_s/N_p) \times V_p = (200/112) \times 115 = 205.4$ V

STR 17-6 1. If both primary and secondary are wound in the same direction, output is in phase with input; but if primary and secondary are wound in different directions, the output will be 180° out of phase with the input.
2. When there is a positive on the primary dot, there will be a positive on the secondary dot. If there is a negative on the primary dot, there will be a negative on the secondary dot.

STR 17-7 1. Air core, iron or ferrite core
2. Center-tapped secondary, multiple-tapped secondary, multiple winding, single winding
3. False
4. To be able to switch between two primary voltages and obtain the same secondary voltage
5. Automobile ignition system
6. Smaller, cheaper, lighter than normal separated primary/secondary transformer types

STR 17-8 1. 10 kVA is the apparent power rating, 200 V the maximum primary voltage, 100 V the maximum secondary voltage, at 60 cycles per second (Hz)
2. $R_L = V_s/I_s = 1$ kV/8 A = 125 Ω; 125 > 100, so the transformer will overheat and possibly burn out.

STR 17-9 1. An ohmmeter
2. Probably not since 50 Ω is a normal coil resistance.

STR 17-10 1. Copper losses, hysteresis, eddy-current loss, magnetic leakage
2. Eddy currents cannot cross the laminated parts of the core, so reduced eddy currents mean reduced opposition to main flux and reduced heat, resulting is a reduced loss.

STR 18-1 1. Calculate the inductive and capacitive reactance (X_L and X_C), the circuit impedance (Z), the circuit current (I), the component voltage drops (V_R, V_L, and V_C), and the power distribution and power factor (PF).
2. a. $Z = \sqrt{R^2 + (X_L \sim X_C)^2}$
b. $I = V_s/Z$
c. Apparent power = $V_s \times I$ (volt-amperes)
d. $V_S = \sqrt{V_R^2 + (V_L \sim V_C)^2}$
e. True power = $I^2 \times R$ (watts)   f. $V_R = I \times R$
g. $V_L = I \times X_L$ h. $V_C = I \times X_C$
i. $\theta = \arctan \dfrac{V_L \sim V_C}{V_R}$ j. PF = $\cos \theta$

STR 18-2 1. a. $I_R = V/R$ c. $I_C = V/X_C$
      b. $I_T = \sqrt{I_R{}^2 + I_X{}^2}$ d. $I_L = V/X_L$
    2. a. $P_R = I^2 \times R$ c. $P_A = V_S \times I_T$
      b. $P_X = I^2 \times X_L$ d. $PF = \cos \theta$

STR 18-3-1 1. A circuit condition that occurs when the inductive reactance (X_L) and the capacitive reactance (X_C) have been balanced
    2. A series RLC circuit that at resonance X_L equals X_C, so V_L and V_C will cancel, and $Z = R$
    3. Voltage across L and C will measure 0; impedance only equals R; voltage drops across inductor or capacitor can be higher than source voltage
    4. Q factor indicates the quality of the series resonant circuit, or is the ratio of the reactance to the resistance
    5. Group or band of frequencies that causes the larger current flow
    6. $BW = f_0/Q = 12\ kHz/1000 = 12\ Hz$

STR 18-3-2 1. In a series resonant RLC circuit, source current is maximum; in a parallel resonant circuit, source current is minimum at resonance.
    2. Oscillating effect with continual energy transfer between capacitor and inductor
    3. $Q = X_L/R = 50\ \Omega/25\ \Omega = 2$ 4. Yes
    5. Ability of a tuned circuit to respond to a desired frequency and ignore all others

STR 18-4 1. a. High pass   c. Band stop
      b. Low pass   d. Band pass
    2. Television, radio, and other communications equipment

STR 18-5 1. Real, imaginary
    2. Magnitude $= \sqrt{5^2 + 6^2} = 7.81$
      Angle $= \arctan\ (6/5) = 50.2°$
      Rectangular number $5 + j6 = $ polar number, $7.81\ \angle\ 50.2°$
    3. Real number $= 33 \cos 25° = 29.9$
      Imaginary number $= 33 \sin 25° = 13.9$
      Polar number, $33\ \angle\ 25° = $ rectangular number, $29.9 + j13.9$
    4. Combination of a real and imaginary number

Answers to Odd-Numbered Problems

Chapter 1

1. d   **3.** d   **5.** a   **7.** c   **9.** b   **11.** c   **13.** b   **15.** c

(The answers to Essay Questions 16 through 30 can be found in the sections indicated that follow the questions.)

Chapter 2

1. a   **5.** b   **9.** a   **13.** b   **17.** d

3. d   **7.** a   **11.** a   **15.** d   **19.** d

(The answers to Essay Questions 21 through 30 can be found in the sections indicated that follow the questions.)

31. $G = 1/100\ \Omega = 0.01$ or 10 millisiemens

33. $I = Q/t = 3/4 = 0.75$ amps or 750 milliamps

35. **a.** 1.473 V   **b.** 7.143 kV   **c.** 139 V   **d.** 390 kV

37. 24,000 V/70 kV = 0.343 cm

39. **a.** 200,000 kV/meter   **b.** 25 kV/mm

Chapter 3

1. a   **3.** a   **5.** d   **7.** a   **9.** a   **11.** b   **13.** c   **15.** d

(The answers to Essay Questions 16 through 40 can be found in the sections indicated that follow the questions.)

41. **a.** $I = V/R = 120\ \text{V}/6\ \Omega = 20$ amps

43. $R = V/I = 120\ \text{V}/10\ \text{A} = 12\ \Omega$

45. $P = I \times V, I = P/V$
  a. $300/120 = 2.5$ A   **c.** $60/120 = 0.5$ A
  b. $100/120 = 833$ mA   **d.** $25/120 = 208.3$ mA

47. **a.** Volts   **b.** Milliamps   **c.** Kilowatts   **d.** Megaohm

49. 13

51. $I = V/R = 12\ \text{V}/12\ \text{mA} = 1\ \text{k}\Omega$

53. $R = V/I = 120\ \text{V}/500\ \text{mA} = 240\ \Omega$
  $P = I \times V = 500\ \text{mA} \times 120\ \text{V} = 60\ \text{W}$

55. $P = V \times I$
  a. $120\ \text{V} \times 20\ \text{mA} = 2.4\ \text{W}$ **c.** $9\ \text{V} \times 100\ \mu\text{A} = 900\ \mu\text{W}$
  b. $12\ \text{V} \times 2\ \text{A} = 24\ \text{W}$ **d.** $1.5\ \text{V} \times 4\ \text{mA} = 6\ \text{mW}$

57. **a.** 1 kW   **c.** 1.25 MW
  b. 345 mW   **d.** 1250 μW

120 V $\lessgtr$ 6 Ω

FIGURE Ch. 3, 41(b)

59. a. $7500 = 7.5$ kW

$$\text{Energy Consumed} = 7.5 \text{ kW} \times 1 \text{ hr}$$
$$= 7.5 \text{ kWh}$$

b. $25 \text{ W} = 0.025 \text{ kW}$

$$\text{Energy Consumed} = 0.025 \text{ kW} \times 6 \text{ hrs}$$
$$= 0.15 \text{ kWh}$$

c. $127{,}000 \text{ W} = 127 \text{ kW}$

$$\text{Energy Consumed} = 127 \text{ kW} \times 0.5 \text{ hr}$$
$$= 63.5 \text{ kWh}$$

Chapter 4

1. b   **5.** a   **9.** c   **13.** c   **17.** b

3. c   **7.** c   **11.** b   **15.** c   **19.** d

(The answers to Essay Questions 21 through 35 can be found in the sections indicated that follow the questions.)

37. a. $P = I^2 \times R = 50 \text{ mA}^2 \times 10 \text{ k}\Omega = 25 \text{ W};$ **b.** no

39. Question 31

    a. $1.2 \text{ M}\Omega \times 0.1 = \pm120 \text{ k}\Omega = 1.08 \text{ M}\Omega$ to $1.32 \text{ M}\Omega$

    b. $10 \times 0.05 = \pm0.5 = 9.5 \text{ }\Omega$ to $10.5 \text{ }\Omega$

    c. $27 \text{ k}\Omega \times 0.2 = \pm5.4 \text{ k}\Omega = 21.6 \text{ k}\Omega$ to $32.4 \text{ k}\Omega$

    d. $273 \text{ k}\Omega \times 0.005 = \pm1.365 \text{ k}\Omega = 271.635 \text{ k}\Omega$ to $274.365 \text{ k}\Omega$

    Question 32

    a. $33 \times 0.2 = \pm6.6 = 26.4 \text{ }\Omega$ to $39.6 \text{ }\Omega$

    b. $22.5 \text{ k}\Omega \times 0.02 = \pm450 = 22.05 \text{ k}\Omega$ to $22.95 \text{ k}\Omega$

    c. $10 \text{ k}\Omega \times 0.05 = \pm500 = 9.5 \text{ k}\Omega$ to $10.5 \text{ k}\Omega$

    d. $910 \times 0.1 = \pm91 = 819 \text{ }\Omega$ to $1001 \text{ }\Omega$

41. See Section 4-6, #3.   **43.** In tolerance

45. Calibrate the ohmmeter first, turn off power and remove part, multiply reading by the scale used, remember resistor tolerance, avoid touching conductive parts of probes or component leads.

Chapter 5

1. b. $I = V/R = 24 \text{ V}/20 \text{ k}\Omega = 1.2 \text{ mA}$

3. b   **5.** c   **7.** a   **9.** d

(The answers to Essay Questions 16 through 30 can be found in the sections indicated that follow the questions.)

21. 15 is half of scale, which is 300 mA maximum. Therefore $I = 150 \text{ mA}$

23. 200 mV

Chapter 6

1. b   **5.** c   **9.** a   **13.** a   **17.** b

3. d   **7.** b   **11.** d   **15.** c   **19.** a

(The answers to Essay Questions 21 through 30 can be found in the sections indicated that follow the questions.)

31. $I = P/V = 50 \text{ W}/120 \text{ V} = 0.4167 \text{ A}$ (either **a** or **d** could be used).

33. a. $150 \text{ Ah} \times 3600 = 54 \times 10^4$ coulombs

    b. $150 \text{ Ah} \div 5 \text{ A} = 30$ hours

FIGURE Ch. 6, 33

35. a. $3.5 \text{ A} \times 4 \text{ hours} = 14 \text{ Ah}, 14 \text{ Ah} \times 3600 = 50.4 \times 10^3$ Ampere-seconds (As).

    b. $14 \text{ Ah} \div 2 \text{ A} = 7$ hours

Chapter 7

1. d   **3.** c   **5.** a   **7.** d   **9.** a   **11.** c   **13.** d   **15.** d

(The answers to Essay Questions 16 through 25 can be found in the sections indicated that follow the questions.)

27. $I = \dfrac{V_S}{R_T}, R_T = R_1 + R_2 = 40 + 35 = 75 \ \Omega$

    $I = 24/75 = 320 \text{ mA}; 150 \ \Omega$ (double 75 Ω) needed to half current

29. $40 \ \Omega, 20 \ \Omega, 60 \ \Omega$. (Any values can be used, as long as the ratio remains the same.)

31. $I_{R1} = I_T = 6.5 \text{ mA}$

33. $P_T = P_1 + P_2 + P_3 = 120 + 60 + 200 = 380 \text{ W}; I_T = P_T/V_S = 380/120 = 3.17 \text{ A}$

    $V_1 = P_1/I_T = 120 \text{ W}/3.17 \text{ A} = 38 \text{ V}; V_2 = P_2/I_T = 60 \text{ W}/3.17 \text{ A} = 18.9 \text{ V}$

    $V_3 = P_3/I_T = 200 \text{ W}/3.17 \text{ A} = 63.1 \text{ V}$

35. a. $R_T = R_1 + R_2 + R_3 = 22 \text{ k}\Omega + 3.7 \text{ k}\Omega + 18 \text{ k}\Omega = 43.7 \text{ k}\Omega$

       $I = V/R = 12 \text{ V}/43.7 \text{ k}\Omega = 274.6 \ \mu\text{A}$

    b. $R_T = V/I = 12 \text{ V}/10 \text{ mA} = 1.2 \text{ k}\Omega$

       $P_T = V \times I = 12 \text{ V} \times 10 \text{ mA} = 120 \text{ mW}$

    c. $R_T = R_1 + R_2 + R_3 + R_4 = 5 + 10 + 6 + 4 = 25 \ \Omega; V_S = I \times$

       $R_T = 100 \text{ mA} \times 25 \ \Omega = 2.5 \text{ V}$

       $V_{R_1} = I \times R_1 = 100 \text{ mA} \times 5 = 500 \text{ mV}, V_{R_2} = I \times R_2 =$

       $100 \text{ mA} \times 10 = 1 \text{ V}, V_{R_3} = I \times R_3 = 100 \text{ mA} \times 6 = 600 \text{ mV}$

       $V_{R_4} = I \times R_4 = 100 \text{ mA} \times 4 = 400 \text{ mV}, P_1 = I \times V_1 = 100 \text{ mA} \times 500 \text{ mV} =$

       50 mW

       $P_2 = I \times V_2 = 100 \text{ mA} \times 1 \text{ V} = 100 \text{ mW}, P_3 = I \times V_3 = 100 \text{ mA} \times 600 \text{ mV} =$

       60 mW

       $P_4 = I \times V_4 = 100 \text{ mA} \times 400 \text{ mV} = 40 \text{ mW}$

    d. $P_T = P_1 + P_2 + P_3 + P_4 = 12 \text{ mW} + 7 \text{ mW} + 16 \text{ mW} + 3 \text{ mW} = 38 \text{ mW}$

       $I = P_T/V_S = 38 \text{ mW}/12.5 \text{ V} = 3.04 \text{ mA}$

       $R_1 = P_1/I^2 = 12 \text{ mW}/(3.04 \text{ mA})^2 = 1.3 \text{ k}\Omega$

       $R_2 = P_2/I^2 = 7 \text{ mW}/(3.04 \text{ mA})^2 = 757.4 \ \Omega$

$$R_3 = P_3/I^2 = 16 \text{ mW}/(3.04 \text{ mA})^2 = 1.73 \text{ k}\Omega$$
$$R_4 = P_4/I^2 = 3 \text{ mW}/(3.04 \text{ mA})^2 = 324.6 \text{ }\Omega$$

37. Zero voltage drop across the shorted component, while there is also an increase in voltage across the others.

39. a. No current at all (zero)   **c.** Measure source voltage
   b. Go to infinity (∞)   **d.** No voltage across any other component

Chapter 8

1. b   **3.** d   **5.** c   **7.** b   **9.** a

(The answers to Essay Questions 11 through 20 can be found in the sections indicated that follow the questions.)

21. $R_T = R/\text{no. of } R\text{'s} = 30 \text{ k}\Omega/4 = 7.5 \text{ k}\Omega$

23. $R_T = R/\text{no. of } R\text{'s} = 25/3 = 8.33 \text{ }\Omega, I_T = V_S R_T = 10/8.33 = 1.2 \text{ A}$
   $I_1 = I_2 = I_3 = R_T/R_x \times I_T = 8.33/25 \times 1.2 = I_T/\text{no. of } R\text{'s} = 1.2 \text{ A}/3 = 400 \text{ mA}$

25. $I_T = V_S/R_T = 14/700 = 20 \text{ mA}; I_X = I_T/\text{no. of } R = 20 \text{ mA}/3 = 6.67 \text{ mA}$

27. a. $R_T = \dfrac{R_1 \times R_2}{R_1 + R_2} = \dfrac{33 \times 22}{33 + 22} = \dfrac{726 \text{ k}\Omega}{55 \text{ k}\Omega} = 13.2 \text{ k}\Omega$

   b. $I_T = V_S/R_T = 20/13.2 \text{ k}\Omega = 1.5 \text{ mA}$
   c. $I_1 = R_T/R_1 \times I_T = (13.2 \text{ k}\Omega/33 \text{ k}\Omega) \times 1.5 \text{ mA} = 600 \text{ }\mu\text{A}$
   $I_2 = R_T/R_2 \times I_T = 13.2 \text{ k}\Omega/22 \text{ k}\Omega \times 1.5 \text{ mA} = 900 \text{ }\mu\text{A}$
   d. $P_T = I_T \times V_S = 1.5 \text{ mA} \times 20 = 30 \text{ mW}$
   e. $P_1 = I_1 \times V_1, (V_1 = V_S = 20 \text{ V}), 600 \text{ }\mu\text{A} \times 20 = 12 \text{ mW}$
   $P_2 = I_2 \times V_2, (V_2 = V_S = 20 \text{ V}), 900 \text{ }\mu\text{A} \times 20 = 18 \text{ mW}$

29. a. $R_T = \dfrac{R_1 \times R_2}{R_1 + R_2} = \dfrac{22 \text{ k}\Omega \times 33 \text{ k}\Omega}{22 \text{ k}\Omega + 33 \text{ k}\Omega} = \dfrac{726 \text{ k}\Omega}{55 \text{ k}\Omega} = 13.2 \text{ k}\Omega,$

   $I_T = \dfrac{V_S}{R_T} = \dfrac{10}{13.2 \text{ k}\Omega} = 757.6 \text{ }\mu\text{A}$

   $I_1 = \dfrac{R_T}{R_1} \times I_T = \dfrac{13.2 \text{ k}\Omega}{22 \text{ k}\Omega} \times 757.6 \text{ }\mu\text{A} = 454.56 \text{ }\mu\text{A}$

   $I_2 = \dfrac{R_T}{R_1} \times I_T = \dfrac{13.2 \text{ k}\Omega}{33 \text{ k}\Omega} \times 757.6 \text{ }\mu\text{A} = 303.04 \text{ }\mu\text{A}$

   b. $R_T = \dfrac{1}{(1/R_1) + (1/R_2) + (1/R_3)} =$

   $\dfrac{1}{(1/220 \text{ }\Omega) + (1/330 \text{ }\Omega) + (1/470 \text{ }\Omega)} = 103 \text{ }\Omega$
   $I_T = V_S/R_T = 10/103 = 97 \text{ mA}, I_1 = R_T/R_1 \times I_T = 103/220 \times 97 \text{ mA} = 45.4 \text{ mA}$
   $I_2 = (R_T/R_2) \times I_T = (103/330) \times 97 \text{ mA} = 30.3 \text{ mA}$
   $I_3 = (R_T/R_2) \times I_T = (103/470) \times 97 \text{ mA} = 21.3 \text{ mA}$

31. a. $G_T = \dfrac{1}{R_1} + \dfrac{1}{R_2} + \dfrac{1}{R_3} = \dfrac{1}{5} + \dfrac{1}{5} + \dfrac{1}{5} = 0.6 \text{ S}, R_T = \dfrac{1}{G} = \dfrac{1}{0.6} = 1.67 \text{ }\Omega$

   b. $G_T = \dfrac{1}{R_1} + \dfrac{1}{R_2} = \dfrac{1}{200} + \dfrac{1}{200} = 10 \text{ mS}, R_T = \dfrac{1}{G} = \dfrac{1}{10 \text{ mS}} = 100 \text{ }\Omega$

c. $G_T = \dfrac{1}{R_1} + \dfrac{1}{R_2} + \dfrac{1}{R_3} = \dfrac{1}{1\ M\Omega} + \dfrac{1}{500\ M\Omega} + \dfrac{1}{3.3\ M\Omega} =$

1.305 µS, $R_T = \dfrac{1}{G} = \dfrac{1}{1.305\ \mu S} = 766.3\ k\Omega$

d. $G_T = \dfrac{1}{R_1} + \dfrac{1}{R_2} + \dfrac{1}{R_3} = \dfrac{1}{5} + \dfrac{1}{3} + \dfrac{1}{2} = 1.033\ S,$

$R_T = \dfrac{1}{G} = \dfrac{1}{1.033} = 967.7\ m\Omega$

33. a. $R_T = \dfrac{R_1 \times R_2}{R_1 + R_2} = \dfrac{15 \times 7}{15 + 7} = \dfrac{105}{22} = 4.77\ \Omega$

b. $R_T = \dfrac{1}{(1/R_1) + (1/R_2) + (1/R_3)} = \dfrac{1}{(1/26\ \Omega) + (1/15\ \Omega) + (1/30\ \Omega)} = 7.22\ \Omega$

c. $R_T = \dfrac{R_1 \times R_2}{R_1 + R_2} = \dfrac{5.6\ k\Omega \times 2.2\ k\Omega}{5.6\ k\Omega + 2.2\ k\Omega} = \dfrac{12.32\ M\Omega}{7.8\ k\Omega} = 1.58\ k\Omega$

d. $R_T = \dfrac{1}{(1/R_1) + (1/R_2) + (1/R_3) + (1/R_4) + (1/R_5)} =$

$\dfrac{1}{(1/1\ M\Omega) + (1/3\ M\Omega) + (1/4.7\ M\Omega) + (1/10\ M\Omega) + (1/33\ M\Omega)} = 596.5\ k\Omega$

35. a. $I_2 = I_T - I_1 - I_3 = 6\ mA - 2\ mA - 3.7\ mA = 300\ \mu A$

b. $I_T = I_1 + I_2 + I_3 = 6\ A + 4\ A + 3\ A = 13\ A$

c. $R_T = \dfrac{R_1 \times R_2}{R_1 + R_2} = \dfrac{5.6\ M \times 3.3\ M}{5.6\ M + 3.3\ M} = \dfrac{18.48\ M\Omega}{8.9\ M\Omega} = 2.08\ M\Omega$

$V_S = I_T \times R_T = 100\ mA \times 2.08\ M\Omega = 208\ kV,$

$I_1 = \dfrac{R_T}{R_1} \times I_T = \dfrac{2.08\ M\Omega}{5.6\ M\Omega} \times 100\ mA = 37\ mA,$

$I_2 = \dfrac{R_T}{R_2} \times I_T = \dfrac{2.08\ M\Omega}{3.3\ M\Omega} \times 100\ mA = 63\ mA.$

d. $I_1 = V_{R_1}/R_1 = 2/200\ k\Omega = 10\ \mu A,\ I_2 = I_T - I_1 = 100\ mA - 10\ \mu A = 99.99\ mA$

$R_2 = \dfrac{V_{R_2}}{I_2} = \dfrac{2}{99.99\ mA} = 20.002\ \Omega,\ P_T = I_T \times V_S = 100\ mA \times 2\ V =$

200 mW

37. a

39. Total current would increase, and the branch current with the shorted resistor would increase to 20 V/1 Ω = 20 A.

Chapter 9

1. c **5.** c **9.** c **13.** a **17.** c

3. b **7.** a **11.** b **15.** a **19.** c

(The answers to Essay Questions 21 through 35 can be found in the sections indicated that follow the questions.)

37. $V_A = V_S = 100 \text{ V}, V_B = V_S - V_R = 100 - 11.125 = 88.875 \text{ V}$
$V_C = V_S - V_{R_1} - V_{R_2} = 44.3, V_D = I_{R_4} \times R_4 = 4.43 \text{ mA} \times 2.5 \text{ k}\Omega = 11.075 \text{ V},$
$V_e = 0 \text{ V}$

39. a. $V_{RL} = \dfrac{R_L}{R_T} \times V_S = \dfrac{R_L}{R_L + R_{\text{int}}} \times V_S = \dfrac{25}{25 + 15} \times 15 = 9.375$

b. $V_{RL} = \dfrac{R_L}{R_T} \times V_S = \dfrac{R_L}{R_L + R_{\text{int}}} \times V_S = \dfrac{2.5 \text{ k}\Omega}{2.5 \text{ k}\Omega + 15} \times 15 = 14.91 \text{ V}$

c. $V_{RL} = \dfrac{R_L}{R_T} \times V_S = \dfrac{R_L}{R_L + R_{\text{int}}} \times V_S = \dfrac{2.5 \text{ M}\Omega}{2.5 \text{ M}\Omega + 15} \times 15 = 14.99991 \text{ V}$

41. a. For $V_1 : R_T = R_1 + R_{2,3}, R_{2,3} = \dfrac{2 \text{ k}\Omega \times 6 \text{ k}\Omega}{2 \text{ k}\Omega + 6 \text{ k}\Omega} = \dfrac{12 \text{ M}\Omega}{8 \text{ k}\Omega} = $

$1.5 \text{ k}\Omega, R_T = 8 \text{ k}\Omega + 1.5 \text{ k}\Omega = 9.5 \text{ k}\Omega, I_T = V_S/R_T = $

$\dfrac{28}{9.5 \text{ k}\Omega} = 2.95 \text{ mA}, I_{R_2} = R_{2,3}/R_2 \times I_T = \dfrac{1.5 \text{ k}\Omega}{2 \text{ k}\Omega} \times 2.95 \text{ mA} = 2.21 \text{ mA}$

For $V_2 : R_T = R_3 + R_{1,2}, R_{1,2} = \dfrac{8 \text{ k}\Omega \times 2 \text{ k}\Omega}{8 \text{ k}\Omega + 2 \text{ k}\Omega} = \dfrac{16 \text{ M}\Omega}{10 \text{ k}\Omega} = $

$1.6 \text{ k}\Omega, R_T = 6 \text{ k}\Omega + 1.6 \text{ k}\Omega = 7.6 \text{ k}\Omega, I_T = V_S/R_T = $
$20/7.6 \text{ k}\Omega = 2.63 \text{ mA}, I_{R_2} = R_{1,2}/R_2 \times I_T = 1.6 \text{ k}\Omega/2 \text{ k}\Omega \times$
$2.63 \text{ mA} = 2.1 \text{ mA}, I_{R_2} \text{ total} = 2.21 \text{ mA} + 2.1 \text{ mA} = 4.31 \text{ mA}$

b. For $V_1 : R_T = R_1 + R_{2,3} + R_4, R_{2,3} = \dfrac{R_2 \times R_3}{R_2 + R_3} = \dfrac{15 \times 75}{15 + 75} = \dfrac{1125}{90} = 12.5 \text{ }\Omega$

$R_T = 10 + 12.5 + 5 = 27.5, I_T = V_S/R_T = 3.5/27.5 = 127 \text{ mA},$
$I_{R_3} = (R_{2,3}/R_3) \times I_T = (12.5/75) \times 127 \text{ mA} = 21 \text{ mA}$

For $V_2 : R_T = R_2 + \dfrac{1}{(1/R_{1,4}) + (1/R_3)}, R_{1,4} = R_1 + R_4 = 10 \text{ }\Omega + 5 \text{ }\Omega = $

$15 \text{ }\Omega, R_T = 15 \text{ }\Omega + \dfrac{1}{(1/15 \text{ }\Omega) + (1/75 \text{ }\Omega)} = 27.5 \text{ }\Omega, I_T = V_S/R_T = $

$1.5 \text{ V}/27.5 \text{ }\Omega = 54.5 \text{ mA}, I_{R_3} = (R_{1,3,4}/R_3) \times I_T = (12.5 \text{ }\Omega/75 \text{ }\Omega) \times 54.5 \text{ mA} = $
$9 \text{ mA}, I_{R_3}$
$\text{total} = 21 \text{ mA} - 9 \text{ mA} = 12 \text{ mA}.$

43. $V_{\text{TH}} = 5 \text{ V}, R_{\text{TH}} = 3 \text{ k}\Omega$
$I_{RL} = V_{\text{TH}}/R_T = 5/(3 \text{ k}\Omega + 1 \text{ k}\Omega) = 5/4 \text{ k}\Omega = 1.25 \text{ mA}$
$I_N = I_{R_2} = V_S/R_2 = 5/3 \text{ k}\Omega = 1.67 \text{ mA} \qquad R_N = 3 \text{ k}\Omega$

45. a. $V = I \times R = 5 \text{ mA} \times 5 \text{ M}\Omega = 25 \text{ kV}$
b. $V = I \times R = 10 \text{ A} \times 10 \text{ k}\Omega = 100 \text{ kV}$
c. $V = I \times R = 0.0001 \text{ A} \times 2.5 \text{ k}\Omega = 250 \text{ mV}$

47. This answer will vary with each person.

49. $V_{\text{TH}} = \dfrac{R_4}{R_T} \times V_S, R_T = R_{1,2} + R_4, R_{1,2} = \dfrac{R_1 \times R_2}{R_1 + R_2} = \dfrac{2 \times 3}{2 + 3} = \dfrac{6}{5} = 1.2 \text{ }\Omega$

$R_T = 1.2 + 7 = 8.2 \text{ }\Omega, V_{\text{TH}} = \dfrac{7}{8.2} \times 20 = 17.1 \text{ V}$

$R_{\text{TH}} = \dfrac{1}{(1/R_1) + (1/R_2) + (1/R_4)} = \dfrac{1}{(1/2) + (1/3) + (1/7)} = 1.024 \text{ }\Omega$

FIGURE Ch. 9, 43

a. If R_2 shorts, $V_{TH} = 20$ V, $R_{TH} = 0 \, \Omega$

b. If R_2 opens, $V_{TH} = \dfrac{R_4}{R_T} \times V_S$, $R_T = R_1 + R_4 = 2 + 7 = 9 \, \Omega$,

$$V_{TH} = \frac{7}{9} \times 20 = 15.56 \text{ V}, \quad R_{TH} = \frac{R_1 \times R_4}{R_1 + R_4} = \frac{2 \times 7}{2 + 7} = \frac{14}{9} = 1.56 \, \Omega$$

Chapter 10

1. a **3.** c **5.** a **7.** a **9.** b **11.** d **13.** d **15.** d

(The answers to Essay Questions 16 through 25 can be found in the sections indicated that follow the questions.)

27. $60,000 \, \Omega/\text{V} \times 1000 \text{ V} = 60 \text{ M}\Omega$

29. $I_M = 1/50,000 \, \Omega/\text{V} = 20 \, \mu\text{A}$
$R = V/I = 10 \text{ V}/20 \, \mu\text{A} = 500 \text{ k}\Omega$
$R = V/I = 100 \text{ V}/20 \, \mu\text{A} = 5 \text{ M}\Omega$

31. a. $25 \text{ V} \times 20,000 \, \Omega/\text{V} = 500 \text{ k}\Omega$ **b.** $100 \text{ V} \times 20,000 \, \Omega/\text{V} = 2 \text{ M}\Omega$
  c. $1.5 \text{ V} \times 20,000 \, \Omega/\text{V} = 30 \text{ k}\Omega$

33. Follow same formulas as Figure 10-14.

35. a. $V_m = I_m \times R_m = 50 \, \mu\text{A} \times 1500 \, \Omega = 75 \text{ mV}$

FIGURE Ch. 10, 35(a)

$I_{RSH_1} = 200 \, \mu\text{A} - 50 \, \mu\text{A} = 150 \, \mu\text{A}$
$R_{SH_1} = 75 \text{ mV}/150 \, \mu\text{A} = 500 \, \Omega$
$I_{RSH_2} = 200 \text{ mA} - 50 \, \mu\text{A} = 199.95 \text{ mA}$
$R_{SH_2} = 75 \text{ mV}/199.95 \text{ mA} = 375 \, \Omega$
$I_{RSH_3} = 2 \text{ A} - 50 \, \mu\text{A} = 1.99995 \text{ A}$
$R_{SH_3} = 75 \text{ mV}/1.99995 \text{ A} = 37.5 \, \Omega$
$I_{RSH_4} = 20 \text{ A} - 50 \, \mu\text{A} = 19.99995 \text{ A}$
$R_{SH_4} = 75 \text{ mV}/19.99995 \text{ A} = 3.75 \, \Omega$
R_{SH_1} is now $500 \, \Omega$, R_{SH_2} is now $375 \, \Omega$
R_{SH_3} is now $37.5 \, \Omega$, R_{SH_4} is now $3.75 \, \Omega$

b. Decrease the shunt resistor values, and increase the multiplier resistor values appropriately.

Chapter 11

1. c **5.** b **9.** a **13.** b **17.** c
3. d **7.** b **11.** d **15.** b **19.** d

(The answers to Essay Questions 21 through 30 can be found in the sections indicated that follow the questions.)

31. Time $= \dfrac{1}{\text{frequency}}$: **a.** $1/27 \text{ kHz} = 37 \, \mu\text{s}$; **b.** $1/3.4 \text{ MHz} = 294 \text{ ns}$;

  c. $1/25 \text{ Hz} = 40 \text{ ms}$; **d.** $1/365 \text{ Hz} = 2.7 \text{ ms}$; **e.** $1/60 \text{ Hz} = 16.67 \text{ ms}$;

  f. $\dfrac{1}{200 \text{ kHz}} = 5 \, \mu\text{s}$

33. a. RMS $= 0.707 \times \text{peak} = 0.707 \times 22 \text{ V} = 15.554 \text{ V}$
  b. Average $= 0.637 \times \text{peak} = 0.637 \times 22 \text{ V} = 14.014 \text{ V}$
  c. Peak to peak $= 2 \times \text{peak} = 2 \times 22 \text{ V} = 44 \text{ V}$

35. Time = distance/velocity = 60 miles/186,000 miles/s = 322.58 μs

37. PRT = 1/PRF = 1/400 kHz = 2.5 μs **39.** 50%

Chapter 12

1. b **3.** d **5.** e **7.** c **9.** a

(The answers to Essay Questions 11 through 25 can be found in the sections indicated that follow the questions.)

27. $V_{p-p} = 2 \times V_p$, $V_p = V_{rms} \times 1.414 = 6 \times 1.414 = 8.484$; $V_{p-p} = 2 \times 8.484 = 16.968$, 2 V/cm × 8 would only show 16 V, so 5 V/cm is lowest setting. $t = 1/f = 1/350$ kHz = 2.857 μs; 0.2 μs/cm × 10 would only show 2 μs, so 0.5 μs/cm is lowest setting.

29. $t = 5.5$ cm × 1 μs/cm = 5.5 μs; frequency = $1/t$ = 1/5.5 μs = 181.818 kHz

39. OL will be displayed on all ranges.

Chapter 13

1. b **5.** c **9.** b **13.** d **17.** b
3. a **7.** d **11.** a **15.** d **19.** c

(The answers to Essay Questions 21 through 39 can be found in the sections indicated that follow the questions.)

41. Field strength = V/d = 6 V/32 μm = 187.5 kV/m

43. $C = \dfrac{(8.85 \times 10^{-12}) \times k \times A}{d} = \dfrac{(8.85 \times 10^{-12}) \times 2.5 \times 0.008 \text{ m}^2}{0.00095 \text{ m}} = 186.3 \text{ pF}$

45. $C_T = \dfrac{1}{(1/C_1) + (1/C_2) + (1/C_3)} =$

$\dfrac{1}{(1/0.025 \text{ μF}) + (1/0.04 \text{ μF}) + (1/0.037 \text{ μF})} = 0.0109 \text{ μF}$

$V_{C1} = \dfrac{C_T}{C_1} \times V_T = \dfrac{0.0109 \text{ μF}}{0.025 \text{ μF}} \times 12 \text{ V} = 5.2 \text{ V}$

$V_{C2} = \dfrac{C_T}{C_2} \times V_T = \dfrac{0.0109 \text{ μF}}{0.04 \text{ μF}} \times 12 \text{ V} = 3.27 \text{ V}$

$V_{C3} = \dfrac{C_T}{C_3} \times V_T = \dfrac{0.0109 \text{ μF}}{0.037 \text{ μF}} \times 12 \text{ V} = 3.53 \text{ V}$

47. a. 47 F **b.** 34×10^{-7} F or 3.4 μF

49. a. $V_{1TC} = 63.2\%$ of $V_S = 0.632 \times 10$ V = 6.32 V; 5TC = 5 × 84 ms = 420 ms
b. $V_{1TC} = 63.2\%$ of $V_S = 0.632 \times 10$ V = 6.32 V; 5TC = 5 × 16.8 ms = 84 ms
c. $V_{1TC} = 63.2\%$ of $V_S = 0.632 \times 10$ V = 6.32 V; 5TC = 5 × 4.08 ms = 20.4 ms
d. $V_{1TC} = 63.2\%$ of $V_S = 0.632 \times 10$ V = 6.32 V; 5TC = 5 × 980 μs = 4.9 ms

Chapter 14

1. b **3.** b **5.** d **7.** b **9.** d **11.** a **13.** e **15.** b

(The answers to Essay Questions 16 through 25 can be found in the sections indicated that follow the questions.)

27. $V_S = \sqrt{V_R^2 + V_C^2} = \sqrt{6^2 + 12^2} = \sqrt{36 + 144} = \sqrt{180} = 13.4$ V

29. a. $I_R = V/R = 12$ V/4 M$\Omega = 3$ μA

    b. $I_C = V/X_C = 12$ V/1.3 k$\Omega = 9.23$ mA

    c. $I_T = \sqrt{I_R^2 + I_C^2} = \sqrt{(3 \ \mu A)^2 + (9.23 \ mA)^2} = 9.23$ mA

    d. $Z = V/I_T = 12/9.23$ mA $= 1.3$ kΩ

    e. $\theta = \arctan(R/X_C) = \arctan(4 \ M\Omega/1.3 \ k\Omega) = 89.98°$

31. $C = \dfrac{1}{2\pi f X_C} = \dfrac{1}{2\pi(20 \ kHz) \ 10 \ k\Omega} = 795.8$ pF

33.

FIGURE Ch. 14, 33(a)

    b. $Z = \sqrt{R^2 + X_C^2} = \sqrt{40^2 + 33^2} = 51.9$ Ω

        $I = V/Z = 24$V/51.9 $\Omega = 462.4$ mA

        $V_R = I \times R = 462.4$ mA $\times 40$ $\Omega = 18.496$ V

        $V_C = I \times X_C = 462.4$ mA $\times 33$ $\Omega = 15.2592$ V

        $I_R = I_C = I_T = 462.4$ mA

        $\theta = \arctan(X_C/R) = \arctan(33 \ \Omega/40 \ \Omega) \ 39.5°$

35. a. $Z = \sqrt{R^2 + X_C^2} = \sqrt{1 \ M^2 + 2.5 \ M^2} = 2.69$ MΩ

        $I = V/Z = 12/2.69$ M$\Omega = 4.5$ μA

        $V_R = I \times R = 4.5$ μA $\times 1$ M$\Omega = 4.5$ V

        $V_C = I \times X_C = 4.5$ μA $\times 2.5$ M$\Omega = 11.25$ V

    b. $Z = \sqrt{R^2 + X_C^2} = \sqrt{300^2 + 200^2} = 360.6$ Ω

        $I = V/Z = 50$ V/360.6 $\Omega = 138.7$ mA

        $V_R = I \times R = 138.7$ mA $\times 300$ $\Omega = 41.61$ V

        $V_C = I \times R = 138.7$ mA $\times 200$ $\Omega = 27.74$ V

37. Lag, 90

39. $X_C = \dfrac{1}{2\pi f c} = \dfrac{1}{2\pi(35 \ kHz) \ 10 \ \mu F} = 455$ mΩ.

    $Z = \sqrt{R^2 + X_C^2} = \sqrt{100 \ k\Omega^2 + 455 \ m\Omega^2} = 100$ kΩ.

    $I = V/Z = 24$ V/100 k$\Omega = 240$ μA

    True power $= I^2 \times R = 240$ $\mu A^2 \times 100$ k$\Omega = 5.76$ mW

    Reactive power $= I^2 \times X_C = 240$ $\mu A^2 \times 455$ m$\Omega = 26$ nVAR

    Apparent power $= \sqrt{P_R^2 + P_X^2} = 5.76$ mVA

    Power factor $= R/Z = 100$ k$\Omega/100$ k$\Omega = 1$

41. Short: contact from plate to plate

    Open: lead disconnected from its plate

    Breakdown: deterioration of dielectric

43. Capacitor value change, capacitor leakage: leakage current will flow between plates.

Dielectric absorption: will not fully discharge, leaving residual charge.
Equivalent series resistance: resistance of leads, lead to plate connection, electrolyte causing effective circuit capacitance value to change.

45. Shorted, Open

Chapter 15

1. a	**5.** c	**9.** c	**13.** d	**17.** a	**21.** c	**25.** d	**29.** b	**33.** d
3. b	**7.** b	**11.** b	**15.** c	**19.** a	**23.** a	**27.** a	**31.** d	

(The answers to Essay Questions 35 through 55 can be found in the sections indicated that follow the questions.)

57. $\text{mmf} = I \times N$; $760 \text{ mA} \times 25 = 19 \text{ A} \cdot \text{t}$

59. $\mu = \text{relative } \mu \times \mu^0$
  a. $90 \times 4\pi \times 10^{-7} = 1.13 \times 10^{-4} \text{ H/m}$
  b. $50 \times 4\pi \times 10^{-7} = 6.28 \times 10^{-5} \text{ H/m}$
  c. $450 \times 4\pi \times 10^{-7} = 5.65 \times 10^{-4} \text{ H/m}$

61. $\text{mmf} = I \times N = 4 \times 50 = 200$ ampere-turns

63. $\text{mmf} = I \times N$; $I = V/R = 9/23 = 391 \text{ mA}$; $391 \text{ mA} \times 50 = 19.55$ ampere-turns

65. $H = \text{mmf}/l = 19.55/0.7 = 27.9$ ampere-turns/meter

Chapter 16

1. c	**5.** a	**9.** a	**13.** e	**17.** a	**21.** b	**25.** a
3. b	**7.** d	**11.** d	**15.** d	**19.** b	**23.** a	

(The answers to Essay Questions 26 through 35 can be found in the sections indicated that follow the questions.)

37. a. $22 \text{ M}\Omega$ **b.** $78.6 \text{ k}\Omega$ **c.** $314.2 \text{ k}\Omega$

39. a. $L_T = L_1 + L_2 + L_3 = 75 \ \mu\text{H} + 61 \ \mu\text{H} + 50 \text{ mH} = 50.136 \text{ mH}$
  b. $L_T = L_1 + L_2 + L_3 = 8 \text{ mH} + 4 \text{ mH} + 22 \text{ mH} = 34 \text{ mH}$

41. a. 4.36 mH **b.** $4.7 \ \mu\text{H}$ **c.** $3.82 \ \mu\text{H}$

43. a. $V_S = \sqrt{V_R^2 + V_L^2} = \sqrt{12^2 + 6^2} = \sqrt{144 + 36} = \sqrt{180} = 13.4 \text{ V}$
  b. $I = V_S/Z = 13.4 \text{ V}/14 \text{ k}\Omega = 957.1 \ \mu\text{A}$
  c. $\angle = \arctan V_L/V_R = \arctan 2 = 63.4°$.
  d. $Q = V_L/V_R = 12/6 = 2$. **e.** $\text{PF} = \cos \theta = 0.448$.

45. $f = X_L/2\pi L = 27 \text{ k}\Omega/2\pi 330 \ \mu\text{H} = 13.02 \text{ MHz}$

47. $\tau = \dfrac{L}{R} = \dfrac{400 \text{ mH}}{2 \text{ k}\Omega} = 200 \ \mu\text{s}$

V_L will start at 12 V and then exponentially drop to 0 V

Time	Factor	V_S	V_L
0	1.0	12	12
1 T_C	0.365	12	4.416
2 T_C	0.135	12	1.62
3 T_C	0.05	12	0.6
4 T_C	0.018	12	0.216
5 T_C	0.007	12	0.084

49. a. $R_T = R_1 + R_2 = 250 + 700 = 950 \ \Omega$

  b. $L_T = L_1 + L_2 = 800 \ \mu H + 1200 \ \mu H = 2 \ mH$

  c. $X_L = 2\pi f L = 2 \ \pi (350 \ Hz) \ 2 \ mH = 4.4 \ \Omega$

  d. $Z = \dfrac{R \times X_L}{\sqrt{R^2 + X_L^2}} = \dfrac{950 \times 4.4}{\sqrt{950^2 + 4.42^2}} = 4.39$

  e. $V_{RT} = V_{LT} = V_S = 20 \ V$

  f. $I_{RT} = \dfrac{V_S}{R_T} = \dfrac{20 \ V}{950 \ \Omega} = 21 \ mA, \ I_{LT} = \dfrac{V_{S1}}{X_{LT}} = \dfrac{20 \ V}{4.4 \ \Omega} = 4.5 \ A$

  g. $I_T = \sqrt{I_R^2 + I_L^2} = \sqrt{21 \ mA^2 + 4.5 \ A^2} = 4.5 \ A$

  h. $\theta = \arctan (R/X_L) = \arctan 950/4.4 = 89.7°$

  i. $P_R = I^2 \times R = 21 \ mA^2 \times 950 = 418.95 \ mW$
  $P_X = I^2 \times X_L = 4.5 \ A^2 \times 4.4 = 89.1 \ VAR$
  $P_A = \sqrt{P_R^2 + P_X^2} = \sqrt{418.95 \ mW^2 + 89.1 \ A^2} = 89.1 \ VA$

  j. $PF = P_R/P_A = 418.95 \ mW/89.1 \ W = 0.0047$

Chapter 17

1. d **3.** c **5.** d **7.** a **9.** a

(The answers to Essay Questions 11 through 25 can be found in the sections indicated that follow the questions.)

27. a. $V_S = \dfrac{N_S}{N_P} \times V_P = 24/12 \times 100 \ V = 200 \ V$

  b. $V_S = \dfrac{N_S}{N_P} \times V_P = 250/3 \times 100 \ V = 8.33 \ kV$

  c. $V_S = \dfrac{N_S}{N_P} \times V_P = 5/24 \times 100 \ V = 20.83 \ V$

  d. $V_S = \dfrac{N_S}{N_P} \times V_P = 120/240 \times 100 \ V = 50 \ V$

29. Turns ratio $= \sqrt{Z_L/Z_S} = \sqrt{8 \ \Omega/24 \ \Omega} = \sqrt{1/3} = 0.58$

31. For 16 turns: $V_S = (N_S/N_P) \times V_P = (16/12) \times 24 \ V = 32 \ V$
  For 2 turns: $V_S = (N_S/N_P) \times V_P = (2/12) \times 24 \ V = 4 \ V$
  For 1 turn: $V_S = (N_S/N_P) \times V_P = (1/12) \times 24 \ V = 2 \ V$
  For 4 turns: $V_S = (N_S/N_P) \times V_P = (4/12) \times 24 \ V = 8 \ V$

33. Follow polarity dots.

35. a. $I_S = $ apparent power$/V_S = 500 \ VA/600 \ V = 833.3 \ mA$

  b. $R_L = V_S/I_S = 600 \ V/833.3 \ mA = 720 \ \Omega$

Chapter 18

1. b **3.** b **5.** d **7.** c **9.** c **11.** a **13.** a **15.** d

(The answers to Essay Questions 16 through 30 can be found in the sections indicated that follow the questions.)

31. a. $X_C = \dfrac{1}{2\pi f C} = \dfrac{1}{2\pi(60)0.02\mu F} = 132.6 \ k\Omega$

b. $X_C = \dfrac{1}{2\pi f C} = \dfrac{1}{2\pi(60)18\mu F} = 147.4\ \Omega$

c. $X_C = \dfrac{1}{2\pi f C} = \dfrac{1}{2\pi(60)360\ pF} = 7.37\ M\Omega$

d. $X_C = \dfrac{1}{2\pi f C} = \dfrac{1}{2\pi(60)2700\ nF} = 982.4\ \Omega$

e. $X_L = 2\pi f L = 2\pi(60)4\ mH = 1.5\ \Omega$
f. $X_L = 2\pi f L = 2\pi(60)8.18\ H = 3.08\ k\Omega$
g. $X_L = 2\pi f L = 2\pi(60)150\ mH = 56.5\ \Omega$
h. $X_L = 2\pi f L = 2\pi(60)2\ H = 753.98\ \Omega$

33. a. $X_L = 2\pi f L = 2\pi(60\ Hz)150\ mH = 56.5\ \Omega$

b. $X_C = \dfrac{1}{2\pi f c} = \dfrac{1}{2\pi(60\ Hz)20\ \mu F} = 132.6\ \Omega$

c. $I_R = V/R = 120\ V/270\ \Omega = 444.4\ mA$
d. $I_L = V/X_L = 120\ V/56.5\ \Omega = 2.12\ A$
e. $I_C = V/X_C = 120\ V/132.6\ \Omega = 905\ mA$
f. $I_T = \sqrt{I_R^2 + I_X^2} = \sqrt{(444.4\ mA)^2 + (1.215)^2} = 1.29\ A$
g. $Z = V/I_T = 120\ V/1.29\ A = 93.02\ \Omega$

h. Resonant frequency $= \dfrac{1}{2\pi\sqrt{LC}} = \dfrac{1}{2\pi\sqrt{150\ mH \times 20\ \mu F}} = 91.89\ Hz$

i. $X_L = 2\pi f L$
$\quad = 6.28 \times 91.89\ Hz \times 150\ mH$
$\quad = 86.54\ \Omega$

$\quad Q = \dfrac{X_L}{R} = \dfrac{86.54\ \Omega}{270\ \Omega} = 0.5769$

j. $BW = \dfrac{f_0}{Q} = \dfrac{91.89\ Hz}{0.5769} = 159.28\ Hz$

35. Using a source voltage of 1 volt:
$Z = V/I_T, I_T = \sqrt{I_R^2 + I_X^2}, I_R = V/R = 1/750 = 1.33\ mA$
$I_L = V/X_L = 1/25 = 40\ mA, I_C = V/X_C = 1/160 = 6.25\ mA$
$I_X = I_L - I_C = 40\ mA - 6.25\ mA = 33.75\ mA$
$I_T = \sqrt{(1.33\ mA)^2 + (33.75\ mA)^2} = 33.78\ mA$
$Z = 1/33.78\ mA = 29.6\ \Omega$

37. a. Real number $= 25 \cos 37° = 19.97$; imaginary number $= 25 \sin 37° = 15, 19.97 + j15$
   b. Real number $= 19 \cos - 20° = 17.9$; imaginary number $= 19 \sin -20° = -6.5, 17.9 - j6.5$
   c. Real number $= 114 \cos - 114° = -46.4$; imaginary number $= 114 \sin - 114° = -104.1, -46.4 - j104.1*$
   d. Real number $= 59 \cos 99° = +9.2$; imaginary number $= 59 \sin 99° = 58.3, +9.2 + j58.3$

*These examples were for practice purposes only; real numbers are always positive when they represent impedances.

39. a. $(4 + j3) + (3 + j2) = (4 + 3) + (j3 + j2) = 7 + j5$

    b. $(100 - j50) + (12 + j9) = (100 + 12) + (-j50 + j9) = 112 - j41$

41. a. $Z_T = 73 - j23$, $\sqrt{73^2 + 23^2} = 76.5$, $\angle = \arctan(23/73) =$
       $-17.5°$, $76.5 \angle -17.5°$; $Z_T = 76.5\ \Omega$ at $-17.5°$ phase angle

    b. $Z_T = 40 + j15$, $\sqrt{40^2 + 15^2} = 42.7$, $\angle = \arctan(15/40) = 20.6°$,
       $42.7 \angle 20.6°$; $Z_T = 42.7\ \Omega$ at $20.6°$ phase angle

    c. $Z_T = 8\ k\Omega - j3\ k\Omega + j20\ k\Omega = 8\ k\Omega + 17\ k\Omega$,
       $\sqrt{8\ k\Omega^2 + 17\ k\Omega^2} = 18.8\ k\Omega$, $\angle = \arctan(17\ k\Omega/8\ k\Omega) = 64.8°$,
       $18.8\ k\Omega \angle 64.8°$; $Z_T = 18.8\ k\Omega$ at $64.8°$ phase angle

43. a. $Z_T = 47 - j40 + j30 = 47 - j10$, $\sqrt{47^2 + 10^2} = 48.05$, $\angle = \arctan(-10/47) = -12°$,
       $Z_T = 48.05 \angle -12°$

    b. $I = \dfrac{V_S}{Z_T} = \dfrac{20 \angle 0°}{48.05 \angle -12} = \dfrac{20}{48.05 \angle 0 - (-12)°} = 416.2\ \text{mA} \angle 12°$

    c. $V_R = I \times R = 416.2\ \text{mA} \angle 12° \times 47 \angle 0 = 416.2\ \text{mA} \times 47 \angle$
       $(12 + 0) = 19.56\ \text{V} \angle 12°$
       $V_C = I \times X_C = 416.2\ \text{mA} \angle 12° \times 40 \angle -90° = 416.2\ \text{mA} \times$
       $40 \angle (12° - 90°) = 16.65 \angle -78°$
       $V_L = I \times X_L = 416.2\ \text{mA} \angle 12° \times 30 \angle 90° = 416.2\ \text{mA} \times$
       $30 \angle (12° + 90°) = 12.49\ \text{V} \angle 102°$

    d. V_C lags I_T by $90°$, V_L leads I by $90°$, V_R is in phase with I.

45. a. $Z_1 = 0 + j27 - j17 = j10$, $\sqrt{0^2 + 10^2} = 10 \angle = 90°$, $Z_1 = 10 \angle 90°$; $Z_2 = 37 - j20$
       $= 42.06 \angle -28.39°$; Z combined $=$ Product/Sum; Product $= 420.59 \angle$
       $61.61°$;
       Sum $= 37 - j10 = 38.33 \angle -15.12°$; Z combined $= 10.97 \angle 76.73°$

    b. $I_1 = 3.72 \angle -34.48\ \text{A}$; $I_2 = 884 \angle 83.91\ \text{mA}$;

    c. $I_T = 3.39 \angle -21.21°\ \text{A}$;

    d. $Z_T = 29.52 \angle 21.21°\ \Omega$

47. $V_R = 19.56 \angle 12° = 19.56 \cos 12° + j19.56 \sin 12° = 19.13 + j4.07$
    $V_C = 16.65 \angle -78° = 16.65 \cos - 78° + j16.65 \sin -78° = 3.46 - j16.29$
    $V_L = 12.49 \angle 102° = 12.49 \cos 102° + j12.49 \sin 102° = -2.6 + j12.22$

49. Easier with use of complex numbers

Index